ANNUAL REVIEW OF GENETICS

ANNUAL REVIEW OF GENETICS

VOLUME 24, 1990

ALLAN CAMPBELL, *Editor*
Stanford University, Stanford

BRUCE S. BAKER, *Associate Editor*
Stanford University, Stanford

ELIZABETH W. JONES, *Associate Editor*
Carnegie Mellon University, Pittsburgh

ANNUAL REVIEWS INC.   4139 EL CAMINO WAY   PO BOX 10139   PALO ALTO, CALIFORNIA 94303-0897

ANNUAL REVIEWS INC.
Palo Alto, California, USA

International Standard Serial Number: 0066–4197
International Standard Book Number: 0–8243–1224-4
Library of Congress Catalog Card Number: 63-8847

Annual Review and publication titles are registered trademarks of Annual Reviews
Inc.

∞ The paper used in this publication meets the minimum requirements of American National Standard for Information Sciences—Permanence of Paper for Printed
Library Materials, ANSI Z39.48-1984.

Annual Reviews Inc. and the Editors of its publications assume no responsibility
for the statements expressed by the contributors to this *Review*.

Typesetting by Kachina Typesetting Inc., Tempe, Arizona; John Olson, President
Typesetting Coordinator, Janis Hoffman

PRINTED AND BOUND IN THE UNITED STATES OF AMERICA

Annual Review of Genetics
Volume 24 (1990)

CONTENTS

v (continued)

SOME RELATED ARTICLES IN OTHER *ANNUAL REVIEWS*

From the *Annual Review of Biochemistry*, Volume 59, 1990:

Chemical Nucleases: New Reagents in Molecular Biology, *D. S. Sigman and C.-h. B. Chen*

Structure, Function, and Diversity of Class I Major Histocompatibility Complex Molecules, *P. J. Bjorkman and P. Parham*

DNA Helicases, *S. W. Matson and K. A. Kaiser-Rogers*

T Cell Receptor Gene Diversity and Selection, *M. M. Davis*

Self-Splicing of Group I Introns, *T. R. Cech*

Regulation of Vaccinia Virus Transcription, *B. Moss*

Sequence-Directed Curvature of DNA, *P. J. Hagerman*

The Family of Collagen Genes, *E. Vuorio and B. de Crombrugghe*

DNA Recognition by Proteins with the Helix-Turn-Helix Motif, *S. C. Harrison and A. K. Aggarwal*

From the *Annual Review of Biophysics and Biophysical Chemistry*, Volume 19, 1990:

Zinc Finger Domains: Hypotheses and Current Knowledge, *J. M. Berg*

Spectroscopic Analysis of Genetically Modified Photosynthetic Reaction Centers, *W. J. Coleman and D. C. Youvan*

From the *Annual Review of Cell Biology*, Volume 6, 1990

Genetic Analysis of Mammalian Cell Differentiation, *H. Gourdeau and R. E. K. Fournier*

Control of Globin Gene Transcription, *T. Evans, G. Felsenfeld, and M. Reitman*

Regulation of Transposition in Bacteria, *N. Kleckner*

Transgenic Mouse Models of Self-Tolerance and Autoreactivity by the Immune System, *D. Hanahan*

Nuclear Proto-Oncogenes *FOS* and *JUN*, *L. J. Ransone and I. M. Verma*

Histone Function in Transcription, *M. Grunstein*

Positive Effects on Eukaryotic Gene Expression, *C. Wilson, H. J. Bellen, and W. J. Gehring*

From the *Annual Review of Entomology*, Volume 35, 1990

Baculovirus Diversity and Molecular Biology, *G. W. Blissard and G. F. Rohrmann*

Ecological Genetics and Host Adaptation in Herbivorous Insects: The Experimental Study of Evolution in Natural and Agricultural Systems, *S. Via*

(*continued*) vii

From the *Annual Review of Immunology*, Volume 8, 1990

Evolution of Class-I MHC Genes and Proteins: From Natural Selection to Thymic Selection, *D. A. Lawlor, J. Zemmour, P. D. Ennis, P. Parham*

Cellular and Genetic Aspects of Antigenic Variation in Trypanosomes, *G. A. M. Cross*

Genetic and Mutational Analysis of the T Cell Antigen Receptor, *J. D. Ashwell and R. D. Klausner*

Transmission of Signals from T Lymphocyte Antigen Receptor to the Genes Responsible for Cell Proliferation and Immune Function, *K. Ullman, J. P. Northrop, C. L. Verweij and G. R. Crabtree*

Regulation of HIV-1 Gene Expression, *W. C. Greene*

Host Genetic Control of Spontaneous and Induced Immunity to Friend Murine Retrovirus Infection, *B. Chesebro, M. Miyazawa and W. J. Britt*

Developmental Biology of T Cells in T Cell Receptor Transgenic Mice, *H. von Boehmer*

Regulation of Major Histocompatibility Complex Class-II Genes: X, Y, and Other Letters of the Alphabet, *C. Benoist and D. Mathis*

From the *Annual Review of Microbiology*, Volume 44, 1990:

Methylotrophs: Genetics and Commercial Applications, *M. E. Lidstrom and D. I. Stirling*

Mycoplasmal Genetics, *K. Dybvig*

Mechanism and Regulation of Bacterial Ribosomal RNA Processing, *A. K. Srivastava and D. Schlessinger*

Antigenic Variation of a Relapsing Fever *Borrelia* Species, *A. G. Barbour*

Reciprocity in the Interactions between the Poxviruses and Their Host Cells, *S. Dales*

Molecular Evolution and Genetic Engineering of Protein Domains Involving Aspartate Transcarbamoylase, *J. R. Wild and M. E. Wales*

Coronavirus: Organization, Replication and Expression of Genome, *M. M. C. Lai*

General Microbiology of *recA:* Environmental and Evolutionary Significance, *R. V. Miller and T. A. Kokjohn*

Sexual Differentiation in Malaria Parasites, *P. Alano and R. Carter*

The Generation of Genetic Diversity in Malaria Parasites, *G. A. McConkey, A. P. Waters, and T. F. McCutchan*

The Genetics of Bacterial Spore Germination, *A. Moir and D. A. Smith*

Environmental Application of Nucleic Acid Hybridization, *G. S. Sayler and A. C. Layton*

Flavivirus Genome Organization, Expression, and Replication, *T. J. Chambers, C. S. Hahn, R. Galler, and C. M. Rice*

Regulation of the Cell Division Cycle and Differentiation in Bacteria, *A. Newton and N. Ohta*

ERRATUM

Volume 23 (1989)

In "*Rhizobium* Genetics," by Sharon R. Long, page 486, second paragraph, 1st and 2nd sentences, should read:

Most strains in the *Rhizobiaceae* have large plasmids, and in *Rhizobium* these appear to be the primary sites of genes for symbiosis. In *Bradyrhizobium* and *Azorhizobium, sym* genes have not been detected on plasmids . . .

For the convenience of readers, a detachable order form/envelope is bound into the back of this volume.

GEORGE W. BEADLE

Annu. Rev. Genet. 1990. 24:1–4

G. W. BEADLE

Adrian M. Srb

Section of Genetics and Development, Cornell University, Ithaca, New York 14853

George Wells Beadle was born in 1903 and died in 1989. During his lifetime, the science of genetics changed dramatically. Beadle contributed definitively to that change. Perhaps more than that of any other investigator, his work prepared the way for the transition from neoclassical to molecular genetics. His contributions were recognized with a Nobel Prize and many other high honors.

Beadle did his work with three among the relatively few organisms that offer well-developed facilities for genetic research, and learned these facilities in excellent environments for scientific accomplishment. His Ph.D. research was with maize. The studies were carried out under the direction of Professor R. A. Emerson, and in the company of some remarkable fellow students at Cornell, including Barbara McClintock, Marcus Rhoades, and Charles Burnham. In postdoctoral work, Beadle turned to Drosophila—his major studies with that organism being carried out in collaboration with Boris Ephrussi. The evolution of Beadle's interests, which in part parallels the changing course of genetic science at that time, can be seen in the circumstance that Emerson was a magnificent formal geneticist, primarily interested in inheritance patterns, mapping and variations on the classical Mendelian ratios. Ephrussi, in contrast, while a highly competent geneticist, might better be described as a developmentalist who was adept at using genetic materials and techniques to address the problems that interested him.

The third experimental organism with which Beadle worked, and the one with which he made his definitive contributions, was Neurospora. Indeed, largely due to Beadle and his coworkers, that organism has become one of the better established and most useful objects of genetic research. Until Beadle, Neurospora was relatively obscure in the scientific world. The mycologist B. O. Dodge had found Neurospora to have intriguing possibilities for fungal

1

0066-4197/90:1215-0001$02.00

genetic studies but his vision of the possibilities was different than emerged in Beadle's work.

When I first came to Stanford in 1941 to do graduate work with Beadle, who had come there from Harvard as a professor in 1937, the transition of his work with Drosophila to that with Neurospora was nearly complete. A refrigerator or two held bottles containing Drosophila pupae in storage, but in the laboratory Neurospora mutants were starting to emerge for study. In an informal seminar, or discussion with his students and coworkers, I remember Beadle saying, in words as nearly as I can recall, that he was "tired of hearing other biologists say that geneticists only work with trivial attributes of organisms, like kernel color in maize or eye color in Drosophila." Clearly Beadle felt challenged to do something about such a criticism, justified or not.

Beadle had in mind to study genetic control of the steps in metabolic pathways. What was truly revolutionary in his plan was a practicable, although by modern standards not highly efficient, scheme for obtaining a large number of appropriate *conditional mutants* for study. The mutants that were sought, and in fact found, were nutritional mutants, each representing a genetically determined block at a particular step in a metabolic pathway. The standard, or wild-type, Neurospora has simple nutritional requirements, little more than a few salts and sugar being needed. From this *minimal medium* the organism makes all the biochemicals necessary for its existence—amino acids, nucleic acids, vitamins, complex structural components, and so on. By manipulating nutritional conditions, Beadle's scheme permitted detection, preservation, and study of mutant Neurosporas representing alteration of genes that control steps in essential biosyntheses.

In essence, after treatment of experimental material with mutagens, appropriate mutants can be preserved on media supplemented with essential metabolites, and detected by transfer to the minimal medium for the unmutated organism. Systematic studies with particular nutritional substances, or the absence of them, refine the relationship of the mutant gene and its unmutated allele to particular biochemical reactions. One can view the Neurospora nutritional mutants as special instances of *conditional lethals* now so useful and widely employed in molecular genetics and other research on many organisms.

Insight into Beadle's innovative research with Neurospora emerges from his immediately precedent work with Drosophila. Shortly after Beadle arrived at Stanford, E. L. Tatum, had come as a Research Associate to work on the biochemical genetics of eye pigmentation in Drosophila. The work was an extension of investigations carried out by Beadle and Ephrussi in Paris.

Those studies had utilized different eye color mutants. Reciprocal transplants of imaginal discs were made between larvae of different genotype. In particular instances, the occurrence of wild-type eye pigmentation in a derivative adult Drosophila revealed the presence of diffusible precursor substances that had accumulated behind genetic blocks in the mutants. In spite of the elegance of these sophisticated complementation experiments, the experimental system had clear limitations. The chemistry of the substances involved was extremely difficult to resolve. More importantly, the system as such could not be adapted to a broad attack on problems of gene action.

At this point the association with Tatum, the biochemist in the team, was particularly fortunate. A microbiologist as well as biochemist, Tatum was well aware of a growing body of literature that showed a remarkable diversity of growth requirements among microorganisms, some them quite closely related. A number of findings showed that in nature many of these organisms satisfied their growth factor requirements by some form of symbiotic relationship between strains or species. The similarity between such a situation and that shown in the transplant experiments is fairly obvious—at least it was so to Beadle and Tatum. And out of such a background, the Nobel prize-winning work with Neurospora was generated.

The various details of the Neurospora studies, and their extensions, have been often and well reviewed, and have become part of standard text book fare. They need not be summarized here. What needs to be said is that the volume and excellence of the work are in part due to Beadle's ability to attract outstanding associates. In the group that assembled early on at Stanford were, for example, Norman Horowitz, David Bonner, Herschel Mitchell. and Mary Houlahan Mitchell. Outstanding post doctorals like Francis Ryan were also drawn to Stanford, as was a full complement of graduate students.

Beadle was well aware that in the long run his work with Neurospora would benefit from information about the organism in addition to that derived from experiments on the biochemical mutants per se. At one point he enticed Barbara McClintock to come to Stanford to have a look at the chromosomes of Neurospora. He fostered some detailed studies of heterocaryosis. And he persuaded his graduate students to do at least minimal mapping of the mutants with which they worked. Indeed, the fact that Neurospora became a *major experimental object* in the broad sense of the words is largely due to him and his generosity with strains of Neurospora that he had developed.

The year 1946 was the beginning of the end of Beadle's experimental career. At that time he moved to Cal Tech as successor to T. H. Morgan as head of the Department of Biology. Highly successful in that position, he accepted the presidency of the University of Chicago in 1961, where he served with distinction until retirement in 1968.

When Beadle retired from the presidency of the University of Chicago, he came full circle and began again to do experiments with maize. These had to do with the origin of corn, a problem that had interested his first scientific mentor, R. A. Emerson. Energetic and incisive as ever, Beadle attacked this problem. He became an enthusiastic proponent of the theory that the important crop plant maize (corn) derives from teosinte. At the last seminar he ever gave at Cornell, on a visit made after his retirement, he described how he had disposed of arguments that teosinte was too useless a food source to tempt primitive native peoples to take it through its first stages as a crop plant. Beadle told how he had shown that the very tough kernels of teosinte could be popped like popcorn, thus making them edible, and had shown in addition the possibility of gathering enough of the kernels to make them a feasible source of food. Being Beadle, of course he had done genetic and cytogenetic studies, some in his graduate student days, some after retirement, that bore on the problem. He inevitably knew all the relevant studies by others, as reported in the literature, even to archaeological and cultural anthropological information.

Beadle was not only a great scientist and highly successful academic administrator. He was also an attractive and interesting human being. Born and raised on a farm near Wahoo. Nebraska, he was the kind of person whose fancy was tickled by his derivation from a community with an amusing name. He was also pleased to have come from a small place that in addition had nurtured such notables as Wahoo Sam Crawford, a famous old-time professional baseball hero; Howard Hanson, American composer and Director of the Eastman School of Music; and Darryl Zanuck, a historic director of films. He often expressed gratitude to the local schoolteacher who had inspired and encouraged him to go to college at the University of Nebraska. He was similarly grateful to F. D. Keim, a professor of agronomy at the university. Keim, an unusually perceptive talent scout for young men of scientific promise, persuaded Beadle to go to Cornell for graduate study.

Beadle's tastes were rather simple. He once told me that for pleasure-reading his favorite author was H. H. Munro. Beadle liked activities that involved physical exertion. In all these activities he showed his basic competiveness. He was a ferocious if not highly skilled tennis player. He invited his graduate students to compete with him in a now, and even then, outmoded athletic event called the standing broad jump. An enthusiastic gardener, he frequently wanted to place a small bet as to whether some other gardening friend could produce a larger pumpkin or grow earlier edible sweet corn. The very few bets he lost were cheerfully paid.

Annu. Rev. Genet. 1990. 24:5–36

MOLECULAR GENETICS OF *ASPERGILLUS* DEVELOPMENT

William E. Timberlake

Departments of Genetics and Plant Pathology, University of Georgia, Athens, Georgia 30602

KEY WORDS: regulatory cascades, gene regulation, morphogenesis, fungi, sporulation

CONTENTS

INTRODUCTION

The fungi are a diverse, interesting, and important group of organisms whose biological activities affect the daily lives of most people (113, 114). In addition to their practical importance, a few fungal species represent highly manipulable model systems for the study of cellular processes. The pioneering work of Beadle & Tatum (17) with the Euascomycete *Neurospora crassa* provided a basis for much of modern molecular genetics. The Hemiascomycete *Saccharomyces cerevisiae* ("yeast") is one of the most intensively investigated eukaryotes, and studies of yeast have helped shape many of the fundamental concepts of modern molecular, cellular, and developmental biology. Specialized biological features of other fungi, such as *Aspergillus nidulans, Coprinus cinereus, Podospora anserina, Schizophyllum commune, Sordaria fimicola,* and *Ustilago maydis,* have contributed, for example, to our understanding of genetic recombination; DNA repair; gene structure, function, and regulation; cell-cell communication; self recognition; and signal transduction.

Aspergillus nidulans and *N. crassa* are closely related fungi that are well suited for studying mechanisms controlling development and cell differentiation in multicellular eukaryotes. These filamentous organisms have sophisticated genetic systems, including DNA-mediated transformation systems, and have been studied intensively for many years (32, 86, 84). Extensive background information exists on their biology and genetics. The asexual reproductive pathway of *A. nidulans,* leading to the formation of spores called conidia, is particularly amenable to experimental analysis, and a substantial body of information has accumulated concerning the genetic mechanisms controlling *Aspergillus* conidiation. This review summarizes what is known about the genes that regulate entry into the conidial pathway and control the steps leading to formation of the complex, multicellular reproductive apparatus, the conidiophore, and the mature conidia. Emphasis is placed on how these regulatory genes interact with one another and with the many genes whose products are directly responsible for the assembly and specialized physiology of the conidiophore and conidia. Springer & Yanofsky (101) have recently presented a comprehensive analysis of the morphological events and genetic controls of conidiophore development in *Neurospora.*

GROWTH AND DEVELOPMENT IN *ASPERGILLUS*

Life Cycle

Figure 1 depicts the life cycle of *A. nidulans.* Like most fungi, the vegetative cells of *Aspergillus* are haploid, multinucleate filaments, called hyphae, that grow by apical extension and multiply by branching. These cells are specialized for the acquisition of nutrients from the environment and often produce extracellular hydrolases that mobilize growth substrates. When genetically distinct, but compatible, homokaryotic strains are mixed, they form heterokaryons by hyphal anastamosis and nuclear division and movement. Diploid nuclei form spontaneously during growth, and heterozygous diploid strains can be obtained by plating dilutions of uninucleate conidia from heterokaryons in selective medium. Diploids are quite stable and grow and conidiate well. However, vegetative diploids cannot be induced to undergo meiosis, in contrast to *S. cerevisiae* and related yeasts. Instead, in *Aspergillus,* premeiotic diploid nuclei are formed naturally during sexual reproduction in ascus mother cells and exist only transiently. Asexual reproduction involves formation of uninucleate, haploid conidia by multicellular conidiophores. New genes can be assigned to linkage groups and mapped by a variety of classical and molecular techniques that exploit the sexual and asexual reproductive cycles, the ability to form diploids and to recover haploids from them, and the existence of strains containing reciprocal translocations (9, 22, 32, 71, 85, 87, 92, 113, 115, 116).

Control of Growth and Development

Aspergillus can be maintained in the vegetative state or induced to enter either the sexual or asexual reproductive pathway. When conidia or ascospores are inoculated into liquid growth medium and incubated with aeration and agita-

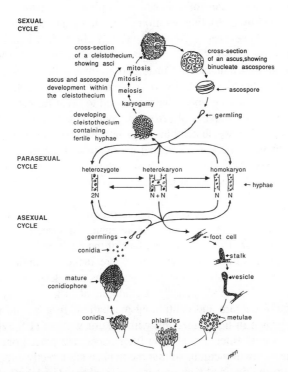

Figure 1 Life cycle of *Aspergillus nidulans*. The center portion of the figure illustrates vegetative hyphae that may grow as homokaryons, heterokaryons, or diploids. Diploids can be maintained by selection for complementing markers or induced to "haploidize" by treatment with benomyl. The upper portion of the figure illustrates the sexual reproductive cycle that is employed for meiotic mapping and other genetic procedures. The lower portion of the figure illustrates the asexual reproductive cycle that is used for routine propagation of the organism. Ascospores and conidia can be stored in the dry state for long periods of time. (Reproduced from 113, with permission.)

tion they germinate and grow vegetatively. Under conditions of nitrogen and carbon source sufficiency, specialized reproductive structures are not formed unless the hyphae are exposed directly to an air interface (27, 61). However, some conidiophores may be produced in liquid medium under certain conditions (66, 91, 98, 99) and this may be affected by genetic background (27). Colonies grown on the surface of agar-solidified media are initially vegetative and subsequently form conidiophores a few mm behind the growing front. A radial section from a colony contains conidiophores at all stages of development, and this property has been exploited in the preparation of samples for microscopy (21, 76). Cleistothecia (fruiting bodies) form within colonies upon extended incubation, especially in the older areas near the centers (27).

Individual *Aspergillus* colonies are too heterogeneous to be of much use for biochemical analyses so methods have been developed that favor more synchronous formation of cleistothecia or conidiophores. Cleistothecium formation is stimulated by restricting aeration (32). Media that promote cleistothecium formation have been formulated (1, 62). In addition, acleistothecial strains exist (e.g. 30, 48, 49, 58). Many auxotrophic mutants, for example, tryptophan-requiring mutants, are unable to produce selfed cleistothecia (14, 53), presumably due to an inability to supplement the auxotrophy in developing fruiting bodies. This property has been used to favor formation of crossed cleistothecia (14). Champe et al (28) and Champe & El-Zayat (26) isolated an *A. nidulans* hormone that induces sexual development and could be of value for inducing synchronous sexual sporulation. Procedures for producing large numbers of cleistothecia at similar stages of development have been described (28) and used to isolate useful quantities of RNA from ascospores (51).

Methods for synchronizing conidiophore production have been described. In one procedure, conidia are inoculated onto agar-solidified medium in a thin layer of soft top agar (10, 121). The surface of the medium is then covered by a layer of liquid medium and the cultures are incubated until the colonies reach the desired size. Submersion of the colonies completely and indefinitely inhibits initiation of conidiogenesis. Development is induced by decanting away the liquid medium and continuing incubation. In a second procedure, cultures are grown in liquid medium with (61) or without (10) agitation and then harvested onto filter papers. The filter papers are placed onto liquid or agar-solidified growth medium so that the hyphae are directly exposed to air, thus inducing conidiation. Kinetics of conidiophore development and conidium production are highly reproducible (10, 61, 121).

Conidiation in some *A. nidulans* strains is light-dependent (L. N. Yager & J. Mooney, personal communication). This property depends on the presence of the wild-type allele of the *velvet* gene (*veA*; 52). Most *A. nidulans* laboratory strains are *veA1*, because this mutation leads to prolific conidiation and facilitates genetic manipulations, and are therefore light-independent.

Conidiophore Development

Ultrastructural aspects of spore differentiation in conidial fungi have been reviewed by Cole (36). Conidiophore development in *A. nidulans* has been examined at the ultrastructural level by Oliver (80) and by Mims et al (76). Conidiophore development can be divided into the five morphologically distinct stages shown in Figure 2. Development begins when a conidiophore stalk initial arises from a specialized hyphal compartment, called the foot cell, that anchors the mature conidiophore in the growth substratum and provides cytoplasmic connection with the remainder of the thallus. During stage 1, the stalk grows as a thick-walled aerial hypha (Figure 2A) that, like vegetative hyphae, elongates by apical growth. Elongation continues until the stalk reaches a height of about 100 μm. Apical vesicles, which are thought to contain cell wall precursors and perhaps cell wall polymerizing enzymes (41), are abundant near the tip of the growing stalk where they fuse with the plasma membrane. When growth stops, signaling entrance into stage 2 of development, vesicles decrease rapidly in number and become dispersed around the periphery of the tip. The tip subsequently swells to form the conidiophore vesicle (Figure 2B), which is about 10 μm in diameter. Clutterbuck (30) determined that the height of the conidiophore from the base of the stalk to the tip of the vesicle was 98.9 ± 4.0 μm (Figure 3). This value is species-specific and varies little in response to different growth conditions, implying that it is genetically determined.

Conidiophores arising from heterokaryons are usually homokaryotic, consistent with either the foot cells being initially uninucleate or being multinucleate but rarely heterokaryotic. Clutterbuck & Spathas (35) showed that most condiophores contain nuclei of two types when heterokaryons are formed between strains carrying complementing defects in nuclear movement

Figure 2 Scanning electron microscopic analysis of conidiophore development in *A. nidulans* showing the various cell types (reproduced from 76, with permission). (A) Young conidiophore stalk just prior to vesicle formation. (B) Developing vesicle at tip of conidiophore. (C) Developing metulae. (D) Developing phialides (arrowheads). (E) Tip of a mature conidiophore bearing numerous chains of conidia.

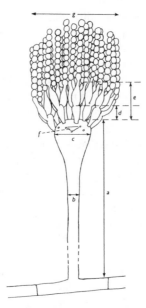

Figure 3 Diagram of an *A. nidulans* condiophore illustrating the sizes of the various cell types (reproduced from 29, with permission). The measurements (in μm; n=20–50) of the parameters indicated by the lettering are (a; conidiophore stalk height) 98.4 ±4.0; (b; stalk diameter) 4.4 ±0.1; (c; conidiophore vesicle diameter) 10.2 ±0.2; (d; metula length) 5.18 ±0.08; (e; metula + phialide length) 11.52 ±0.10; (f; metular triangle area) 2.54 ±0.12 (μm²); (g; conidial column diameter) 46.7 ±0.9.

(*apsA⁻* and *apsB⁻* strains), thereby encouraging more uniform distribution of nuclear types in the thallus. Thus, it is probable that foot cells are frequently multinucleate, but that nuclei are not freely intermingled in most heterokaryons. Repeated mitotic divisions occur during stalk elongation and the nuclei move into the vesicle as it forms.

In stage 3, numerous buds are produced synchronously on the surface of the vesicle (Figure 2C), giving rise to cells called primary sterigmata or metulae. Clutterbuck (30) and Oliver (80) reported that repeated mitotic divisions occur in the conidiophore vesicle and, following these divisions, a single nucleus migrates into each fully developed metula. Mims et al (76), on the other hand, observed that a final, more or less synchronous mitotic division occurs within the conidiophore vesicle. Nuclei appeared to subtend the metular bud initials and to undergo an oriented division such that one daughter nucleus is retained in the vesicle and one enters the differentiation metula, analogous to the oriented nuclear divisions of budding yeasts. Nuclei further down the vesicle and in the stalk also divide at this time, even though the products of these divisions are not utilized in formation of metulae. Other organelles also enter the differentiating metulae before these cells become delimited by formation of septa produced by centripetal deposition of cell wall material near their bases. These septa, like those in hyphae, possess central pores providing cytoplasmic continuity between the cells. The pores are guarded by Woronin bodies on both sides. Woronin bodies have been proposed to serve as emergency safety plugs that can seal off cells following breakage to prevent protoplasmic loss (37, 89).

Clutterbuck (30) used the area of the "sterigmata triangle" (2.54 ± 0.12 μm^2), a term used to describe the triangle formed by three adjacent metulae (Figure 3), the estimated surface area of the conidiophore vesicle, and the assumptions that the sterigmata are in close spacing and cover one hemisphere of the approximately spherical vesicle to calculate that an average conidial head bears 31 sterigmata. However, examination of conidial heads by scanning electron microscopy has shown that the actual number of cells is about twice this value, probably because the conidiophore vesicle is not spherical and metulae arise from more than just the uppermost surface.

During stage 4, the distal tips of metulae produce buds that develop into uninucleate, sporogenous phialides (Figure 2D). Phialides become delimited by incomplete septa that provide cytoplasmic connection to the metulae. Additional phialides often develop from the sides of metulae.

Conidia are formed during stage 5 by repeated divisions of the phialide nucleus (Figure 2E). A conidium initial develops as a cytoplasmic protrusion from the tip of a phialide (96; Figure 4). The phialide nucleus enters the neck region and divides mitotically. One daughter nucleus enters the differentiating conidium and becomes arrested in the G1 phase of the cell cycle, whereas the other nucleus moves back down the neck of the phialide. Once the conidium initial is fully expanded, a septum develops in the phialide neck by radial invagination of the plasma membrane and deposition of wall material. This septum initially possesses a central pore that closes as the septum thickens. As the conidium becomes delimited, the phialide nucleus again undergoes division to produce another spore. As this process continues, long, clonal chains of spores are formed. Thus, phialides, like stem cells, repeatedly produce differentiated cells while retaining their own cellular identity. In addition, it has been discovered that two genes whose products are required for pigment accumulation in the spore walls, *wA* and *yA,* are apparently transcribed in phialide cells, not in differentiating spores (69, 79). It has been shown that the product of *yA,* conidial laccase, accumulates in the spore walls. These observations imply that the phialide secretes enzymes needed for spore wall synthesis. In this respect the *Aspergillus* phialide resembles the *Bacillus* spore mother cell (64), which is also in part responsible for spore wall synthesis.

Conidia undergo maturation, which is defined as differentiation occurring after septum formation and closure have delimited the conidium from the

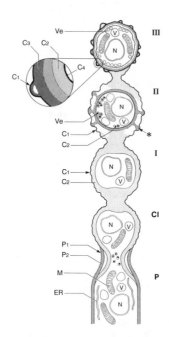

Figure 4 Diagram of the stages of conidium differentiation (reproduced from 96, with permission). Each conidium as well as the phialide (P) contains a nucleus (N), mitochondria (M), endoplasmic reticulum (ER), vacuoles (V), and vesicles (Ve). Successive stages of differentiation (I, II, and III) are shown as a conidium initial (CI) displaces the chain of conidia from the phialide. The wall of the conidium initial and Stage I conidium is composed of an inner (C2) layer and an outer (C1) rodlet layer. The C2 layer condenses during Stage II leaving projections (*) in contact with C1. The mature (Stage III) conidium wall is thinner than Stage I or II walls and has two additional layers (C3 and C4). The outer layer is crenulated in regions subtended by projections from the C2 and C3 layers. Cytoplasmic vesicles (Ve) are prevalent in Stage II conidia whereas Stage III conidia contain peripheral vesicles with contents that stain for carbohydrates.

phialide (96). Conidium maturation occurs in three stages (Figure 4). Stage I conidia are fully delimited and possess walls consisting of an inner (C2) and outer layer (C1). During Stage II, C2 condenses and C1 becomes crenulated. In the transition from Stage II to Stage III, regions of C1 not in contact with the projections from C2 collapse into the condensed regions of C2. As a result, C1 becomes appressed to C2, except where it is subtended by C2 projections. During stage III, a third wall layer, designated C3, forms between C1 and C2. Finally, the innermost wall, C4, forms, probably by deposition of wall material from plasma membrane-associated vesicles present in Stage III conidia. C4 is the only layer that appears to be provided by the spore itself. Formation of the final wall layers makes the spores difficult to fix for electron microscopy, indicating that these final stages of differentiation make the conidia impermeable. Sewall et al (96) speculated that formation of the C3 and C4 wall layers seals the spore protoplast off from the environment and that their deposition is required for the establishment of spore dormancy.

Molecular Genetic Manipulations

The DNA-mediated transformation system of *A. nidulans* is well developed and facilitates genomic manipulations similar to those used with *S. cerevisiae* (16, 20, 103, 106, 122). Methods and approaches for genetic engineering of these and related species have been reviewed recently (88, 114). Several features of the *Aspergillus* molecular genetic system are relevant to studies of development. First, genes can be cloned directly by complementation of mutations by using either cosmid or plasmid vectors (50, 123). It may often be possible to complement mutations in one ascomycetous species with DNA from a different species (120). Due to the small sizes of most fungal genomes (e.g. 55, 107), the number of transformants that must be screened to have a high probability of identifying a particular sequence is not prohibitive.

Second, following recovery in *Escherichia coli* of recombinant DNA molecules from complemented strains, the complementing activities can be rapidly mapped to small restriction fragments (111) that can then be used to inactivate genes by homologous recombination events (20, 21, 32, 69, 74, 90, 93). It appears that Repeat-Induced Point (RIP) mutation-mediated gene inactivation, which has been used to good advantage in *N. crassa* (94, 95), does not occur at high frequency in *A. nidulans*. It has been possible to make very large deletions (~38 kb) in *Aspergillus* by transformation with linear DNA fragments containing flanking sequences attached to a selective marker (8). Site-directed, null mutations can be subjected to allelism tests to ensure that the gene of interest, as opposed to a suppressor, has been cloned (21, 69, 123).

Third, cloned genes can be overexpressed in cells either by increasing their copy number or by placing them under the control of a strong, regulatable promoter, for example, the promoter from the *A. nidulans* catabolic alcohol dehydrogenase *(alcA)* gene (43, 63). Gene amplification has been used to titrate away molecules involved in gene regulation and has permitted determination of how loss of such molecules alters patterns of gene expression (54). Developmental regulatory genes have been placed under the control of the *alcA* promoter to study the effects of misscheduled expression on growth, development, and cell differentiation (2, 65, 77). The extremely low level of activity of this promoter under repressive conditions makes it possible to use it to control even those genes whose products are highly toxic when produced in an inappropriate cell type.

Finally, *cis*-acting regulatory elements can be attached to the *E. coli lacZ* structural gene as a convenient means of investigating gene regulation (118). β-galactosidase levels can then be measured in cell extracts or visualized by in situ staining, thus allowing determination of the cellular specificity of gene expression during development (3, 47, 117).

REGULATION OF CONIDIATION

Developmental Competence

Hyphae of *Aspergillus* grown in submerged culture do not normally form conidiophores or cleistothecia, but instead grow vegetatively until reaching stationary phase. Repression of conidiation in submerged culture, however, is not absolute. For example, it can be partially relieved by nutrient limitation or heat shock of germinating spores (25, 66, 91, 100). Nevertheless, almost all experiments concerned with *Aspergillus* conidiation have exploited the ability to suppress conidiation completely by growing the cells in submerged culture and the ability to initiate development synchronously by exposing the cells directly to air (10, 61, 121).

Axelrod et al (11) observed that colonies derived from individual conidia were able to respond to developmental induction only after a defined period of growth had transpired. In their experiments, colonies initiated from single conidia were grown in submerged culture for various times and then induced to conidiate by harvesting them onto growth medium-saturated membrane filters and exposing them directly to air. The interval between induction and conidiophore development was then measured. Conidiophore development in these types of experiments is usually taken to begin at the time of vesicle formation because it is difficult to distinguish between conidiophore initials and nonsporogenous aerial hyphae prior to that stage. As shown in Figure 5, the time following induction required for initiation of development decreases until 20 hr, with conidiation always occurring at about 25 hr. However, after 20 hr of vegetative growth, the time required for development following induction remains constant at approximately 5 hr. These results indicate that prior to 20 hr the cells are incapable of responding to induction and that after 20 hr the response to induction, that is formation of conidiophore vesicles, takes 5 hr. Cells that have the capacity to respond immediately to induction are said to be developmentally competent (11). The time between induction of competent cells and formation of morphologically distinguishable conidiophores is referred to as the maturation period. The lengths of the competence acquisition and maturation periods are temperature-dependent (121).

Competence acquisition appears to be endogenously determined, as opposed to being the result of external changes in growth medium, because growth of cells at very low densities or continuous replacement of the growth medium has no effect on the time to conidiation (82). In addition, competence

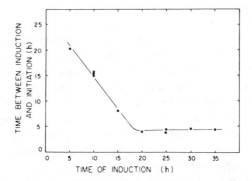

Figure 5   Acquisition of developmental competence (reproduced from ref. 11, with permission). Conidia were inoculated onto growth medium at zero time and covered with liquid medium to inhibit conidiation. At various times thereafter individual plates were induced to conidiate by decanting the liquid medium and exposing the hyphae to air. The colonies were subsequently examined microscopically to establish the time of initiation of conidiation as detected by the presence of conidiophore vesicles in the cultures.

acquisition is under genetic control. Axelrod et al (11) obtained precocious mutants that sporulated prior to 25 hr by using a filtration enrichment procedure. One of these mutants, BB142, became developmentally competent 2 hr before the wild-type strain and was unaltered in germination, growth rate, or nutritional requirements. The mutation behaved as a simple Mendelian factor. Thus, the BB142 mutation identifies a gene that is involved in setting the time required for acquisition of developmental competence. Additional mutagenesis of this strain produced a derivative exhibiting a more extreme phenotype, becoming competent 3.5 hr before the wild-type strain (57).

Another class of mutants was isolated based on the supposition that genes whose products are required for acquisition of competence might be unessential for conidiophore development once a strain has become competent. Temperature-sensitive mutations in such loci are expected to produce conditionally aconidial strains that become thermoinsensitive prior to induction. Such mutants can be experimentally distinguished from other classes of aconidial mutants by growing them at permissive temperature under repressive conditions and then subjecting them to a simultaneous shift to inductive conditions and restrictive temperature. Preinduction mutants conidiate because they have passed their thermosensitive period, whereas postinduction mutants, with defective genes needed for conidiophore development, do not. Three mutants meeting this criterion have been described and partially characterized (23, 24, 121). They define three unlinked loci, designated *acoA, B,* and *C*. Butnick et al (23, 24) showed that this class of *aco* mutants secretes the antibiotic diorcinol and that mutants are also blocked in sexual development. Second-site suppressors were found for *acoB202* and *acoC193* that restored the ability to conidiate. Some of these suppressors also restored the ability to form cleistothecia and eliminated diorcinol production. Butnick et al (24) interpreted these results to indicate that preinduction mutations and their suppressors affect steps in a single metabolic pathway whose intermediates include one effector specific for asexual sporulation and a second effector specific for sexual sporulation, implying that the controls of the two sporulation pathways are intimately interrelated. Champe et al (28) identified and partially purified a factor called psi that is overproduced and secreted in aconidial, acleistothecial strains such as the *acoC193* mutant, and inhibits conidiation while stimulating the sexual reproductive pathway. Psi has been resolved into multiple, related components with the formula $C_{18}H_{30-32}O_{2-3}$ (26). PsiA1 appears to be a lactone whose ring is opened by alcohol addition and has been proposed to be the metabolic precursor of other more active psi components. Whether the inhibitory effect of psi on conidiation is direct or is an indirect consequence of its stimulation of the sexual reproductive pathway is unknown.

Given that the timing of competence acquisition appears to be an endogenous, genetically determined property of the spores, then a mechanism must exist to keep track of the time that has transpired following germination. A number of physiological changes occur during precompetent growth, including changes in DNA extractability related to cellular iron content (45, 46); inducibility of nitrate reductase and extracellular protease (40); the ability to take up glucose (57) and fructose (56); and cellular cAMP levels (33). None of these alterations, however, has been shown to be causally related to acquisition of competence, so their relevance to this phenomenon is unclear.

A simple explanation for the control of developmental competence is that a developmental repressor becomes diluted during early growth. In one model, a fixed amount of repressor would be produced during the final stages of conidium differentiation and stored in the spore. The proposed repressor is neither synthesized nor degraded during vegetative growth and must be diluted to a critical concentration before development can ensue. The repressor could be almost anything, but one possibility is that it is a negatively acting transcription factor that prevents expression of genes required for conidiation. Mutations that affect the conidial level of such a repressor would be expected to alter the time required for acquisition of competence, whereas loss-of-function repressor mutations would be expected to lead to precocious development and might be lethal. Loss-of-function mutations in genes that are maintained in the inactive state by the repressor might be expected to interfere with or prevent competence acquisition. Cloning and characterization of precompetence genes such as *acoA*, *B*, and *C*, and genes that affect the timing of competence acquisition should help to clarify the mechanisms underlying this phenomenon.

Tamame et al (104, 105) reported that treatment of *A. nidulans* hyphae with low levels of 5-azacytidine (5-AC) leads to high-frequency production of strains showing the "fluffy" phenotype. This phenotype is characterized by unrestricted proliferation of undifferentiated aerial hyphae that give rise to large, cottonlike colonies. These colonies fail to show normal growth inhibition upon encountering adjacent colonies, which they overgrow (39), and form a few conidiophores only after an extended growth period. Fluffy mutations induced by 5-AC map to a single locus, designated *fluF*. *fluF* mutants are recessive in diploids, codominant in heterokaryons, and show simple segregation patterns. The locus has been mapped to chromosome VIII and is tightly linked to the centromere (105). Even though the *A. nidulans* genome does not contain detectable levels of methylcytosine (7), a simple interpretation of these results is that treatment of germlings with 5-AC leads to demethylation of *fluF*. In one model, the 5-AC-induced alterations in *fluF* methylation would lead to a change in developmental potential similar to what might be expected for a mutant unable to acquire developmental competence, i.e. an ability to respond to developmental induction signals. It is thus possible that *fluF* exists in a hemimethylated state in conidia and becomes fully methylated during the first few nuclear divisions following germination. Full methylation of *fluF* is required for cells to respond to developmental induction, and treatment with 5-AC prevents full methylation. Once *fluF* becomes fully demethylated, remethylation is a rare event; reversion occurs at a frequency of $<10^{-4}$ (105). This model makes formal predictions that can be readily tested once the *fluF* gene has been cloned.

Developmental Induction

The environmental signals inducing *A. nidulans* conidiation are poorly understood. Submersion of cells in growth medium strongly inhibits conidiation, whereas exposure of hyphae to an air interface rapidly induces development (78). Thus, in colonies growing on agar-solidified medium only those hyphae at the surface produce conidiophores; submerged cells exhibit only vegetative growth. When colonies are covered with liquid growth medium, no development occurs until the liquid medium is removed. It is possible that exposure to air induces osmotic changes that serve as inductive signals. Arguing against this possibility are the observations that conidiation occurs in surface cultures even when the atmosphere is saturated with water (112), and that conidiation does not occur in submerged culture when hyphae are subjected to osmotic shock by addition of sorbitol or NaCl to the medium (K. Y. Miller & W. E. Timberlake, unpublished results). Some conidiation in submerged culture may occur following heat shock or nutrient limitation (25, 66, 91, 98; J. Aguirre, personal communication). However, both elevated temperature and nutrient limitation inhibit conidiation in surface colonies. Indeed, poor supplementation of auxotrophic strains can make them nearly aconidial. Surface colonies grown on membranes can be transferred frequently to fresh growth medium without inhibiting conidiation, indicating that nutrient limitation of the colonies is not required to induce the asexual reproductive pathway. However, it is difficult to eliminate the possibility that some cells within the mycelium experience nutrient limitation due to competition and that these are the cells that are induced to undergo development.

Recently it has been shown that conidiation requires exposure of hyphae to red light and that induction is at least partially reversible by exposure to far red light, reminiscent of phytochrome-mediated responses in higher plants (L. N. Yager & J. Mooney, personal communication). This phenomenon is observed only in strains that carry the wild-type allele of the *velvet (veA)* locus. This locus was identified by Käfer (52) who found that a *veA* mutant (designated *veA⁻* or *veA1*) did not produce the abundant nonsporogenous aerial hyphae normally formed by wild-type colonies but instead conidiated more profusely, making it easier to replicate and otherwise manipulate *veA1* strains. Subsequently, the *veA1* mutation was incorporated into most commonly used *A. nidulans* strains and came to be largely ignored, to the point where the *veA* allele present in strains is often not specified in the genotype. Champe et al (27) described a number of differences between *veA⁺* and *veA1* strains, including differing sensitivities of sexual development to high temperatures, quantitative differences in cleistothecium production at normal temperatures, and differences in ability to conidiate in submerged culture. Interestingly, an abundant, sporulation-specific protein, designated p36, is produced by *veA⁺*, but not by *veA1*, strains and it is possible that this protein is encoded by the *veA* locus. The *veA1* mutation is recessive in diploids and the *veA* gene has been cloned (L. N. Yager & J. Mooney, personal communication).

Given these observations, it seems likely that the *veA1* mutation results in no or reduced *veA* activity and that *veA* represses development. This repression is relieved by exposure to red light. The observation that repression is partially reestablished by exposure to far red light suggests that the *veA* product is retained in the cells in an inactive state and is reconverted to the active state. *veA*-mediated light effects are exerted at multiple steps in the pathway regulating conidiation (L. N. Yager & J. Mooney, personal communication; 77) (Figure 6). The *brlA* regulatory gene (see below) is activated in a *veA⁺* strain only after exposure to light, showing that *veA* repression affects the regulatory pathway prior to *brlA* activation. It is also possible that *veA* repression is exerted downstream of *brlA*.

Interpretation of *veA* effects is complicated by the pleiotropic activity of this locus. For example, *veA* is temperature-sensitive in that conidiation by *veA⁺* strains becomes light-independent at 42°C. By contrast, *veA1* strains are

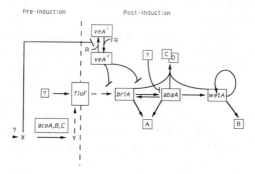

Figure 6   A model for genetic control of *A. nidulans* conidiation. Solid arrows and bars indicate positive and negative interactions, respectively, inferred from genetic analyses. Dashed arrows ·are hypothetical. The *velA* product is proposed to exist in active (a) and inactive (i) states that are interconvertable by exposure to red (R) or far red (FR) light. However, there is no evidence that VeA is itself a chromophore. *fluF* is placed over the line separating pre- and post-induction controls because temperature shift analyses to determine the times during which it is required have not been done due to the lack of a thermosensitive allele.

acleistothecial at 42°C, whereas *veA*$^+$ strains are self-fertile. In addition, many of the temperature-sensitive *aco* mutations that were isolated in a *veA*$^+$ background are partially suppressed by *veAl* (24), indicating that these loci interact with *veA* (Figure 6). These complications may be resolved with the molecular cloning and subsequent characterization of the *veA* gene and *aco* genes.

Conidiophore Development

MUTATIONAL STUDIES   Mutations that prevent initiation of conidiation, for example by blocking competence acquisition, are referred to as class I mutations (109). Class II mutations, by contrast, directly affect the formation of conidiophore or conidia, or both, after development has been initiated. Thus, the postinduction *aco* mutants of Axelrod et al (12) analyzed by Yager et al (121) fit into class II. Clutterbuck (29, 30) divided class II mutations into four categories based on an extensive analysis of conidiation-defective mutants. This analysis was restricted to mutations that had a qualitative effect on conidiophore development without altering growth rate or nutritional requirements and that were thus likely to occur in genes having pathway-specific functions. The four categories of mutations were (*a*) those that modify conidia; (*b*) those that alter conidiophore structure, but do not completely block spore formation; (*c*) those that completely block spore formation; and (*d*) those that affect conidiophore pigmentation.

The first group (class IIa) (conidial modification mutants) includes mutants with altered conidial colors (32, 38, 86). These define loci such as *yellow (yA), white (wA),* and *fawn (fwA)* that are known or presumed to encode or control enzymes catalyzing synthesis of the conidial wall pigment (31, 34, 58, 59, 61). It also includes mutations that affect conidial wall structure. For example, *wet-white (wetA⁻)* mutants produce unpigmented conidia with defective walls that break down during the conidial maturation process (96). *Dark (drkA⁻)* conidia are also structurally abnormal, having an outer layer that encloses the entire conidial chain rather than the individual conidia (29, 30). Another class of mutants having defective conidia is characterized by an inability to replicate older colonies. These mutants produce superficially normal spores that rapidly lose viability, perhaps due to a failure to shut down respiration (68).

Class IIb (conidiophore structure mutants) comprises oligosporogenous mutants of which two, *stunted (stuA⁻)* and *medusa (medA⁻),* have been studied. *Stunted* mutants produce very small conidiophores with unthickened walls. These conidiophores produce some viable spores, perhaps by vesicular budding, although minute metulae and phialides may be produced (A. J. Clutterbuck, personal communication). *Medusa* mutants produce supernumerary layers of sterigmata. Both of these mutants are also self-sterile, although cross-fertile, again suggesting an intimate relationship between the control of the asexual and sexual reproductive pathways.

Class IIc (aconidial mutants) comprises two types of mutants, *bristle (brlA⁻)* and *abacus (abaA⁻).* *Bristle* mutants produce conidiophore stalks that grow indeterminately, failing to produce conidiophore vesicles or any subsequent cell types. *Abacus* mutants produce initially normal conidiophores, but the phialides branch and proliferate rather than forming conidia, producing rodlike cells with swellings at intervals. Neither of these mutants produces conidia, nor do they express conidium-specific biochemical markers (21, 79).

Class IId (conidiophore pigmentation mutants) includes only one type of mutant, *ivory (ivoA,B).* *Ivory* mutants were identified in an *abaA⁻* background by their failure to accumulate the grey-brown pigment normally found in the conidiophore stalk and vesicle. Subsequently, the *ivoA* and *ivoB* genes have been shown to mediate two steps in synthesis of a melaninlike pigment from N-acetyl-6-hydroxytryptophan (70).

A striking feature of the results of Clutterbuck (29, 30) is the low number of genes identified, even though many independent representatives of each type of mutant were obtained. These results indicate that there are fairly few loci having required functions that are specifically directed toward conidiophore development or spore differentiation. However, several types of mutants might have been missed in these screens. For example, strains that lacked a diffusible substance supplied by adjacent colonies or strains that produced a toxic metabolite that interfered with vegetative growth would probably have been missed (29, 30). Similarly, mutations in genes having redundant or incremental functions would result in no or minor changes in phenotype that would be difficult to detect by simple visual examination. Finally, loss-of-function mutations in negatively acting regulatory genes might be expected to be lethal due to activation of sporulation functions in vegetative cells (2, 3, 77) (see below). Noting these types of limitations, Martinelli & Clutterbuck (68) estimated the number of loci specifically required for conidiophore development and spore formation by comparing the frequency of occurrence of developmentally abnormal strains with the average frequency of mutation of numerous loci with readily detectable mutant phenotypes. Elimination of all slow-growing conidiation mutants from the sample led to an estimate of 45 to 150 loci, whereas inclusion of slow-growing mutants led to an estimate of 1000 loci.

Clutterbuck (34) proposed a functional classification system based on the behavior of conidiation mutants. In this system, loci that regulate the conversion from vegetative growth to conidiation are designated "strategic." Strategic loci include *acoA, B,* and *C, veA,* and perhaps *fluF.* Loci that regulate the orderly assembly of the conidiophore are designated "tactical," and include *brlA, abaA,* and probably *wetA, medA,* and *stuA.* Loci that directly determine the structure and secondary physiological characteristics of the conidiophore or spores, such as *wA, yA,* and *ivoA* and *B,* are designated "auxiliary." Finally, the many genes required for normal vegetative growth and for conidiophore development are termed "support" loci. Molecular analysis of conidiation-specific genes has provided support for this classification system and led to formulation of explicit models for genetic regulation of conidiophore development (Figure 6).

MOLECULAR ESTIMATES OF REGULATED GENES   Timberlake (108) used RNA-DNA hybridization analysis to estimate the number of diverse transcripts present in hyphae, conidiating cultures (containing hyphae, conidiophores, and spores), and mature spores. Hybridization of an excess of poly(A)$^+$RNA from hyphae with purified single-copy DNA showed that about 13% of the genome is transcribed as mRNA. This equates to ~6000 diverse mRNAs based on the complexity of *A. nidulans* single-copy DNA (2.6×10^4 kbp; 13, 107), the assumption of asymmetric transcription, and an estimated average mRNA size of 1200 nt. Hybridization of excess poly(A)$^+$RNA with cDNA led to a similar estimate and further showed that the RNAs occur in three modally distributed abundance classes comprised of ~1, ~20, and ~200 copies of each transcript per nucleus. A recurrent subtractive hybridization procedure, termed "cascade hybridization," was used to purify cDNAs complementary to poly(A)$^+$RNA sequences that were present only in conidiating cultures or in purified spores. The kinetics of hybridization of these cDNAs with the RNAs from which they were derived and their fractionation histories were used to estimate the abundance and diversity of mRNAs that accumulate specifically during development. The results showed that 7% of the mass of mRNA in conidiating cultures consists of ~1000 diverse sequences that are not present in hyphae or mature spores. Similarly, 11% of the mass of mRNA in conidiating cultures consisting of ~200 diverse sequences (81) that are not present in hyphae but are present in mature spores. These development-specific mRNAs are distributed into abundance classes similar to those for vegetative mRNAs. Thus, approximately 1200 diverse mRNAs with varying cellular concentrations accumulate specifically during development. By using a similar approach, it was shown that few mRNAs present in vegetative cells are lost during conidiation.

There are several possible explanations for the differences between the genetic (45–150) and molecular (~1000) estimates of the number of genes specifically involved in development. First, estimates from both approaches are based on a number of untested assumptions and are subject to substantial experimental variation. Thus, either or both could simply be in error. Second, some of the genes detected by molecular analysis could have no, redundant, or incremental functions and would thus not be readily detected by mutational analysis. In favor of this argument is the observation of Aramayo et al (8) that deletion of a 38-kbp region of the A. nidulans genome containing multiple spore-specific genes had no readily detectable effect on phenotype. Numerous yeast genes that are induced during sporulation are dispensable (42). Some conidiation-defective mutants having significantly, but subtly, altered phenotypes might have gone undetected by simple direct or microscopic examination. Third, mutations in genes encoding spore-specific mRNAs that are not utilized until germination or in negatively acting regulatory genes would probably have been overlooked in previous screens. Fourth, the estimate from mutational analyses is based on exclusion of mutants with reduced growth rates. It is possible that mRNAs that accumulate preferentially during development are present at low levels in hyphae and contribute to vegetative growth. Finally, it is possible that developmentally important genes are not regulated at the level of mRNA accumulation and would therefore not have been detected by molecular analysis. Arguing against this is the observation that several conidiation-specific genes with known functions, including *ivoB*, *wA*, *yA*, *brlA*, *abaA*, and *wetA*, code for mRNAs that accumulate specifically during development (18, 19, 21, 69, 79). Regardless of the exact number, it is clear that numerous genes are activated during conidiation. A few genes have been identified by molecular cloning that are inactivated during development (44). Two major unanswered questions are (*a*) the functions of these genes and (*b*) how their expression is controlled in time and space.

MOLECULAR CLONING OF DEVELOPMENTALLY IMPORTANT GENES
Developmentally important *Aspergillus* genes have been cloned by two

techniques. In the first, genomic clones were selected by hybridization with a probe that had been enriched in developmentally induced mRNA sequences by subtractive hybridization (124). Boylan et al (21) used differential hybridization (102) to isolate cDNA clones corresponding to developmentally regulated mRNAs. In the second, *Aspergillus* cosmid- and plasmid-cloning vectors were developed (50, 123) and used to isolate *yA, brlA, abaA, wetA, wA, stuA, veA,* and *ivoB,* among other genes, by genetic complementation of mutations following transformation of *Aspergillus* mutants with libraries containing wild-type DNA inserts (18, 19, 69, 79, 111; B. L. Miller, personal communication; J. Mooney & L. N. Yager, personal communication). Proof that the complementing plasmids or cosmids contained the desired wild-type gene, as opposed to a suppressor of the mutation, has typically come from mapping the complementing activity within the cloned insert (111), making a gene disruption by homologous recombination between plasmid and genomic DNA sequences (74), and subsequently subjecting the induced mutation to complementation tests with a known mutant allele of the gene under study (21, 69, 70).

CLUSTERING OF SPORE-SPECIFIC GENES   In the course of selecting chromosomal recombinant DNA clones containing genes whose expression is regulated during conidiation, Zimmermann et al (124) obtained evidence that conidium-specific genes are nonrandomly associated with one another in the *Aspergillus* genome. First, many fewer recombinant phage plaques gave positive autoradiographic signals when hybridized with largely spore-specific cDNA probes than was expected on the basis of the complexity of the cDNA and the average length of the cloned DNA inserts. Second, several randomly selected clones that contained at least one spore-specific gene hybridized with several spore-specific poly(A)$^+$RNAs, a highly improbable event if the RNAs are not closely related in sequence and the spore-specific RNA coding regions are randomly dispersed in the genome. Clustering of spore-specific genes was investigated further by Orr & Timberlake (81) who used two experimental approaches to estimate the degree of nonrandom association of conidium-specific transcription units. In one approach, 30 recombinant DNA clones containing at least one spore-specific gene were examined for the presence of additional, similarly regulated genes. The distribution of spore-specific genes deviated significantly from that predicted from the assumption that the genes are randomly dispersed in the genome. An average clone with an insert size of 15.1 kbp contained 2.3 spore-specific genes, instead of the 1.1 predicted from random dispersal. More than 50% of the clones contained two or more spore-specific genes, in contrast to the statistical prediction of only 7%. Twelve of the clones examined had three or more spore-specific RNA coding region. The probability of this occurring by chance association is negligible.

In the second experimental approach, samples from differently sized, quasi-random recombinant DNA libraries were hybridized with spore-specific cDNA to determine the fraction of clones at each fragment length that did not contain any spore-specific coding regions. Many fewer clones yielded positive hybridization signals with spore-specific cDNA than predicted, consistent with the proposed nonrandom arrangement of the genes.

Timberlake & Barnard (110) and Gwynne et al (44) elucidated the structure of a large cluster of spore-specific genes, designated SpoC1. The SpoC1 region is 38 kbp in length and codes for 19–20 transcripts, many of which arise from nested transcription of three regions. If minor, overlapping transcripts are ignored, the region contains nine independent transcription units that show strong developmental regulation. Eight code for RNAs that accumulate specifically in conidia. The remaining transcription unit, designated L8B, codes for an RNA that accumulates in phialides and not in spores. The SpoC1 region was found to be delimited by 1.1-kbp direct repeats partially complementary to a transcript that is present at somewhat higher levels in conidiating cultures and in conidia than in hyphae. Two additional transcription units adjacent to the repeated elements and within the cluster are less strongly regulated than those near the center of the cluster. One SpoC1 mRNA codes for a conidial polypeptide and the polyadenylated state and polysomal association of the remaining transcripts indicates that they probably code for polypeptides as well (8).

Timberlake & Barnard (110) proposed a model for regulation of SpoC1 genes in which expression is repressed in hyphae by localized features of the chromatin. Alterations in chromatin structure, perhaps occurring in the pre-spore nucleus during or immediately following mitosis in the phialide, were proposed to allow all of the genes to become available for transcription simultaneously. Each gene would then be transcribed at a rate appropriate to its biological function as determined by the intrinsic efficiency of its promoter. Gwynne et al (44) pointed out that the existence of the exceptional L8B gene, which is activated earlier during development and in a different cell type from its neighbors, indicated that the SpoC1 region is in a transcriptionally competent conformation before expression of the spore-specific genes and concluded that this model must have an additional level of control superimposed on it to allow for independent control of the genes in the cluster. The transition from strong regulation in the center of the cluster to moderate and then weak regulation near the borders of the cluster was taken to be suggestive of a regulatory mechanism with a chromatin-level component, although completely independent regulatory models could not be ruled out.

Miller et al (73) tested the effect of chromosomal position on expression of SpoC1 genes by moving a gene from near the center of the cluster to several different chromosomal locations. With only one exception, strains containing the gene at ectopic chromosomal sites produced elevated levels of mRNA in hyphae. In addition, insertion of the constitutively transcribed *argB* gene to any of three different positions within SpoC1, in either orientation, caused a fivefold or more reduction in *argB* transcript levels in hyphae. These results were taken to support the hypothesis that repression of SpoC1 gene activity in hyphae is mediated by a property of the SpoC1 chromosomal region. However, the SpoC1 gene investigated showed some developmental regulation regardless of its chromosomal location. This regulation is mediated by sequences 5' of the transcription initiation site (B. L. Miller, personal communication). Based on these observations, Miller et al (73) proposed a two-component model in which transcription of genes within the SpoC1 cluster is repressed in hyphae by regional regulatory elements and induced (or derepressed) during development by *trans*-acting factors that interact with DNA sequences adjacent to individual genes. Regional repression is necessary to achieve the negligible levels of transcription of SpoC1 genes observed in hyphae. Genes near the borders of the cluster could be less subject to this control than genes near the center because of distance-dependent attenuation of the repressive effect, resulting in low levels of hyphal expression. According to this model, activation of the cluster occurs in two steps. In the first, the regional repression system is inactivated early during development, leading to a low level of transcription of the conidium-specific genes and permitting higher levels of expression of the exceptional L8B gene. In the second, the gene-specific systems are activated in prespore nuclei, leading to fully induced transcription of the conidium-specific genes. The transformation system available for *A. nidulans* should permit experimental testing of this model.

Even though the SpoC1 region comprises 0.15% of the *Aspergillus* genome, encodes a significant proportion of the mRNA stored in spores, and is highly regulated during conidiation, the functions of SpoC1 genes are unknown. Aramayo et al (8) attempted to determine the functions of these genes by deleting the entire 38-kbp SpoC1 region through a gene-transplacement technique (73, 90). The resultant mutant was indistinguishable from the wild-type strain. It was concluded that the genes are dispensable for growth and development, at least in the laboratory. It is of course possible that genes with redundant functions exist, masking the phenotype of the deletion strain. Indeed, Timberlake & Barnard (110) detected genomic sequences that cross-hybridized weakly with SpoC1 DNA. It is also possible that SpoC1 genes have subtle functions whose loss leads to only minor phenotypic changes. Regardless of the biological functions of SpoC1 genes, the mechanisms controlling their expression remain of considerable interest.

CLONING AND CHARACTERIZATION OF REGULATORY GENES   The functional classification scheme of Clutterbuck (34) implies that genes such as *acoA,B,C, veA, brlA, abaA, wetA, medA,* and *stuA* regulate either the conversion from vegetative growth to asexual reproduction or the actual steps of conidiophore development and spore maturation. Of these genes, *brlA*, *abaA*, *wetA*, and *stuA* have been studied in some depth, whereas the other genes have either not been cloned or have been cloned but not yet subjected to extensive molecular analysis. Our current understanding of the activities of these genes is summarized in the following sections.

brlA   *brlA* codes for a 2.1-knt polyadenylated mRNA that is absent from vegetative cells and accumulates during conidiophore development beginning at about the time conidiophore vesicles form (21). *brlA* mRNA contains rather long 5' and 3' untranslated regions (340 and 380 nt, respectively) and possesses a single long open reading frame beginning with AUG. That this open reading frame corresponds to the BrlA polypeptide is supported by the observation that an antibody against a predicted polypeptide from the BrlA carboxyl terminus precipitates BrlA polypeptide made by in vitro transcription of the gene and in vitro translation of the resultant RNA (T. H. Adams & W. E. Timberlake, unpublished results). BrlA consists of 432 amino acid residues and is rich in proline (10%) and serine (13%) (2). It contains near its carboxyl terminus a repeated sequence that is similar to the Cys_2-His_2 Zn(II) coordination sites first recognized in *Xenopus laevis* transcription factor IIIA (TFIIIA) (75), suggesting that BrlA is a nucleic acid binding protein. Consistent with this possibility, the polypeptide binds tightly to heparin-agarose, phosphocellulose, and DNA-agarose (H. Deising, T. H. Adams, & W. E. Timberlake, unpublished results). In addition, disruption of either Zn finger by site-directed mutations that convert finger cysteines to serines results in complete loss of brlA activity (3). Neither the direct targets nor the precise biochemical activities of BrlA have been elucidated.

Aguirre et al (6) fused the 5' flanking region and untranslated leader of *brlA* to the *E. coli lacZ* gene and introduced the fusion gene into *Aspergillus* by transformation. Hyphae and conidiophores were then stained cytochemically for β-galactosidase to determine the cellular specificity of *brlA* expression. Staining was observed only in the conidiophore vesicles, metulae, phialides, and immature spores. A possible problem of interpretation of these results is that *E. coli* β-galactosidase might not be localized in the same way as the BrlA protein itself. Nevertheless, the results indicate that *brlA* transcription is restricted to conidiophore cells. Accumulation of *E. coli* β-galactosidase was not examined in cleistothecial cells due to the presence of high endogenous levels of β-galactosidase. However, Jurgenson & Champe (51) obtained results indicating that *brlA* is involved in regulating gene expression during sexual reproduction. Although most genes that are selectively activated during conidiation are not activated during sexual reproduction, one gene, designated Spo28, codes for an RNA that accumulates both in conidia and ascospores. This transcript fails to accumulate during conidiation in *brlA1* strains, as expected, but also fails to accumulate in ascospores. The requirement for *brlA+* activity for Spo28 transcript accumulation in ascospores implies, although does not prove, that *brlA* must be activated during ascosporogenesis, even though *brlA* mutations have no obvious effects on ascospore formation or germination.

Adams et al (2) constructed an *A. nidulans* strain that permitted controlled transcription of *brlA* in hyphae by fusion of the gene with the promoter (p) from the *A. nidulans alcA* gene, encoding catabolic alcohol dehydrogenase (43, 83). The *alcA*(p)::*brlA* fusion gene was then induced in hyphae by transferring cells grown in medium containing glucose as carbon source, which represses the *alcA*(p), to medium containing threonine as carbon source, which induces the *alcA*(p). In these experiments, the normal conidiation pathway was repressed by maintaining the cells in submerged culture. Induction of *brlA* caused the hyphal tips to differentiate into phialidelike structures that gave rise to conidia, showing that *brlA* expression is sufficient to direct the essential developmental pathway leading to conidium formation. In addition, *brlA* expression led to activation of the *abaA* and *wetA* regulatory genes, the *yA* and *ivoB* structural genes, and numerous developmentally regulated genes with unknown functions. It was concluded that the product of *brlA* is a positively acting molecule that controls a major component of the *A. nidulans* asexual reproductive pathway. As *brlA* expression alone fails to lead to the formation of normal conidiophores, parallel, independent regulatory pathways must exist that augment the *brlA*-directed pathway inferred from the results of the ectopic expression studies. It is possible that genes such as *stuA* and *medA* regulate pathways that add levels of complexity to the central conidiation induced by *brlA*.

abaA   *abaA* codes for 2.8–3.0 knt polyadenylated mRNAs that differ at their 5' and 3' ends. These RNAs are absent from vegetative cells and accumulate during conidiophore development, beginning at about the time phialides form. *brlA* mutations are epistatic to *abaA* mutations (67) and *abaA* mRNA accumulation is dependent on *brlA+* activity (21), indicating that *brlA* directly or indirectly activates *abaA*. As with *brlA*, *abaA* mRNAs have unusually long 5' and 3' untranslated regions (140–160 and 130–350 nt, respectively) and contain a single long open reading frame beginning with AUG (77). The gene contains two short introns with typical fungal splice donor and acceptor sequences (15). The conceptual AbaA polypeptide is 796 amino acid residues long (~90 kd) and is somewhat rich in histidine (6%) and proline (8%). Although AbaA contains a potential "leucine zipper" (60), it appears to lack the adjacent "basic region" consensus sequence found in the leucine zipper family of DNA-binding proteins (5, 119). It is possible that the leucine repeat mediates AbaA dimerization, but that either the protein does not bind to DNA or binds to DNA via a different mechanism from that proposed for proteins in the Jun-Fos-C/EBP class (119).

Mirabito et al (77) constructed a strain containing the nutritionally inducible *alcA*(p)::*abaA* fusion gene on chromosome III. Forced expression of *abaA* in spores or growing hyphae had several effects, including inhibition of spore germination and hyphal elongation, formation of abnormal vacuoles, production of thickened septations, and induction of *brlA* and *wetA* transcription. The latter result showed that *brlA* and *abaA* are mutually inductive. *wetA* induction required *abaA*$^{+}$ and *wetA*$^{+}$, but not *brlA*$^{+}$, activities. It was concluded that *abaA* is a positive feedback regulator of *brlA* and that *wetA* is in part autogenously regulated. Sporulation-specific genes were divided into four classes based on the effects of *brlA*, *abaA*, or *wetA* mutations on their expression during normal (21) or artificially induced (2, 77) development (Figure 6). Class A genes are activated by either *brlA* or *abaA* independent of the other genes. Class B genes are also induced by either *brlA* or *abaA* activation, but their expression is dependent on *abaA*$^{+}$ and *wetA*$^{+}$ activities. Class C genes are induced by forced expression of either *brlA* or *abaA*, whereas class D genes are detectably induced only by forced expression of *abaA*. Although expression of class C and D genes during artificially induced development depends on *brlA*$^{+}$, *abaA*$^{+}$, and *wetA*$^{+}$ activities, they differ in their patterns of expression in *brlA*, *abaA*, and *wetA* mutants during normal development in that class D, but not class C, genes are expressed independent of *brlA*$^{+}$ activity (21). These results imply that a *brlA*-, *abaA*-, and *wetA*-independent regulatory pathway is activated during normal development, but not during development induced artificially by forced expression of *brlA* or *abaA*.

brlA induction in vegetative cells leads to formulation of functional phialidelike cells, and conidium production by these cells is *abaA*-dependent (2). By contrast, artificial activation of *abaA* fails to induce spore formation, even though it does activate *brlA* and leads to accumulation of *brlA*, *abaA*, and *wetA* transcripts to the same levels as those occurring after artificial activation of *brlA* (77). These results imply that the proper order of *brlA* and *abaA* activation is essential for the formation of sporogenous cells. It is thus possible that *brlA* expression "conditions" cells so that they can respond appropriately to *abaA*.

Adams & Timberlake (4) investigated the *cis*-acting elements that regulate *abaA* expression by fusing an 822-bp DNA fragment from the *abaA* 5'-flanking region to the *E. coli lacZ* gene. Deletion analysis showed that this region contains elements that repress transcription in vegetative cells and immature conidiophores and activate transcription later during development. A 45-bp region encompassing the *abaA* transcription initiation sites contains directly repeated sequences related to the mammalian initiator (Inr) element

(97). This small promoter element is sufficient for correct transcription initiation and for a significant level of developmental induction, but is active at a relatively high level in vegetative cells; upstream elements are required to repress the Inr. BrlA may be involved in displacing a repressor from the *abaA* promoter region during development. Negative regulation of *abaA* has significant implications concerning the identification of developmental regulatory genes in *Aspergillus* in that previous screens for developmental mutants were not designed to detect negatively acting regulatory genes. The existence of a tractable DNA-mediated transformation system for *A. nidulans* will make the design of such screens relatively straightforward.

wetA   *wetA* codes for a 2.1-knt polyadenylated mRNA that, like *brlA* and *abaA* mRNAs, is absent from hyphae, but, unlike these mRNAs, accumulates in mature conidia (21). It is not known if *wetA* mRNA also accumulates in conidiophore cells, but the requirement for *brlA*$^+$, *abaA*$^+$, and *wetA*$^+$ activity for expression of class C and D genes implies that *wetA* is expressed in phialides. *brlA* and *abaA* mutations are epistatic to *wetA* mutations (67) and *wetA* mRNA accumulation is dependent on *brlA*$^+$ and *abaA*$^+$ activities (21), indicating that these genes directly or indirectly activate *wetA*. However, forced expression of *abaA* in hyphae of a *brlA*$^-$ strain also activates *wetA*, showing that *brlA* is not required. As with *brlA* and *abaA*, *wetA* has long 5' and 3' untranslated regions (284 and 140 nt, respectively) (65). The *wetA* gene consists of a single exon with the capacity to encode a basic polypeptide of 525 amino acid residues (~59 kd). The biochemical function(s) of WetA is not known.

Marshall & Timberlake (65) constructed a strain containing an *alcA*(p):: *wetA* fusion gene. Hyphal activation of *wetA* under conditions that suppress conidiation inhibits, but does not stop, vegetative growth. *wetA* activation does not induce either *brlA* or *abaA* transcription, but does cause the accumulation of class B transcripts. Thus, the *abaA* requirement for B gene expression reported by Mirabito et al (77) was due to the dependency of *wetA* expression on *abaA* activation. Class B transcripts accumulate specifically in conidia, suggesting a role for their products in the spore maturation process or possibly in spore germination.

stuA   The *stuA* gene has been cloned and partially characterized (72). This gene, unlike *brlA*, *abaA*, and *wetA*, is transcribed during vegetative growth beginning at about the time at which the cells acquire developmental competence. Following induction of conidiation, *stuA* transcription shifts from an upstream to a downstream start site, leading to the accumulation of a smaller mRNA. However, the larger and smaller *stuA* mRNAs are predicted to encode the same polypeptide, so the significance of the transcriptional shift is unclear. The *stuA* mutant was isolated due to the formation of highly abnormal conidiophores. However, *stuA* mutants are also self-sterile and show a vegetative defect that leads to senescence in older parts of colonies (T. H. Adams, personal communication). The *stuA* requirement for normal maintenance of vegetative cells, sexual reproduction, and conidiation supports the idea that these developmental pathways have overlapping controls. How *stuA* is involved in these diverse processes should become clearer as it is analyzed further.

A MODEL FOR DEVELOPMENTAL REGULATION

Figure 6 presents a model for genetic regulation of *Apsergillus* conidiophore development. *AcoA, B,* and *C* are the only genes known to have functions in the preinduction period. One possibility is that these genes encode enzymes catalyzing conversion of a metabolite "X", which inhibits conidiation, to "Y", which is not inhibitory. Loss-of-function mutations in these genes would lead to accumulation of X or other metabolites that inhibit conidiation. One such metabolite could be psi factor, which inhibits conidiation while stimulating sexual development (28). *veA1* is a weak suppressor of *aco* preinduction mutations, suggesting that X exerts its effect through the product of *veA*. *veA*$^+$ is required for light control of conidiation; *veA1* mutants conidiate in complete darkness, whereas *veA*$^+$ strains are aconidial. A simple interpretation of the *veA* effect is that the gene encodes a repressor of conidiation that acts as early steps. This repressor is inactivated directly or indirectly by red light and can be partially reactivated by far red light. Accumulation of metabolite X in preinduction *aco* mutants could prevent inactivation of the *veA* repressor. It is possible that *fluF* has preinduction and postinduction functions and is an inducer of *brlA,* but given the paucity of information about its regulation and biochemical activities its position in the model is quite tentative.

brlA is activated early during conidiophore development and is required for the conversion from polarized apical growth by the conidiophore stalk to the swelling required for formation of the conidiophore vesicle. *brlA* activates class A genes, whose products are presumably utilized in formation of conidiophore structures, and the *abaA* regulatory gene. *abaA* transcription is repressed in vegetative cells and *brlA* activation appears to involve displacement of one or more *abaA* repressors (4). *abaA* further activates class A genes and, in addition, feeds back to activate *brlA.* The positive feedback loop formed by *brlA* and *abaA* could serve to amplify inductive signals or to make the pathway independent of inductive signals. Arguing against the latter possibility is the observation that artificially induced development in an *alcA*(p)::*brlA* strain requires continuous induction; when developing cells are returned to medium that inactivates the *alcA* promoter the wild-type copy of *brlA* becomes inactive and the cells return to vegetative activities (T. H. Adams & W. E. Timberlake, unpublished results). *abaA* also activates the *wetA* regulatory gene which in turn activates class B genes (65), whose products are presumably involved in spore differentiation. *wetA*⁺ is necessary for its own induction. Positive autoregulation of *wetA* could be required to maintain the gene in the active state in the absence of *abaA* product in differentiating spores that have become separated from the phialide. Class C and D genes require wild-type copies of the *brlA, abaA,* and *wetA* regulatory genes during artificially induced development, but D genes can be activated during normal development in *brlA*⁻ mutants, implying the existence of additional or more complicated regulatory pathways. The requirement for all three genes implies that their products must accumulate in the same cell type, probably the phialide. Thus, *wetA* could control those aspects of spore differentiation that are mediated by the phialide and those that are intrinsic to the spore itself. The proposed pathway regulating conidiophore development is attractive in that the timing and extent of expression of the regulatory genes and their targets are determined as development proceeds by intrinsically controlled changes in the relative concentrations of regulatory gene products in the various conidiophore cell types. This model will certainly undergo many modifications as additional results are obtained. Nevertheless, it provides a useful framework for the design of informative experiments.

CONCLUSIONS AND PROSPECTS

It should be apparent from the above discussion that although much has been learned about the genetic controls of development in *Aspergillus,* numerous questions remain unanswered. For example, the biochemical activities of the products of major regulatory genes have yet to be determined. Molecular studies of these genes and their *cis*-acting regulatory elements may reveal additional control elements that will need to be characterized genetically and biochemically. Genes involved in acquisition of developmental competence need to be investigated further. We need to understand the environmental cues that induce conidiation. Almost nothing is known about the genes and products that contribute directly to conidiophore form and function. Similarly, the mechanisms controlling spore dormancy and germination remain to be elucidated. The molecular genetic system of *A. nidulans* makes it possible to adopt experimental approaches that are available in only a few experimental systems. It can be expected that continued molecular genetic studies of *Aspergillus* development will provide important insights into the processes controlling development in fungi and in eukaryotes in general.

ACKNOWLEDGMENTS

I thank Drs. Tom Adams, John Clutterbuck, and Meg Marshall for their critical comments on the manuscript and Drs. Charles Mims and Tommy Sewall for help with the figures. Ms. Kathy Vinson provided invaluable assistance in preparation of the manuscript. I also thank my many colleagues in the laboratory for their numerous contributions to the work summarized here, Dr. Ron Morris for introducing me to *Aspergillus* as an experimental system, and Dr. Bob Goldberg for many stimulating discussions on development over the past 15 years. Work in my laboratory on *Aspergillus* development has been funded by grants from the National Institutes of Health, National Science Foundation, and United States Department of Agriculture.

Literature Cited

1. Acha, I. G., Villanueva, J. R. 1961. A selective medium for the formation of ascospores by *Aspergillus nidulans*. *Nature* 189:328

2. Adams, T. H., Boylan, M. T., Timberlake, W. E. 1988. *brlA* is necessary and sufficient to direct conidiophore development in *Aspergillus nidulans*. *Cell* 54:353–62

3. Adams, T. H., Deising, H., Timberlake, W. E. 1990. *brlA* requires both zinc fingers to induce development. *Mol. Cell. Biol.* 10:1815–17

4. Adams, T. H., Timberlake, W. E. 1990. Developmentally regulated transcription from *Aspergillus* Inr elements. *Mol. Cell. Biol.* In press

5. Agre, P., Johnson, P. F., McKnight, S. L. 1989. Cognate DNA binding specificity retained after leucine zipper exchange between GCN4 and C/EBP. *Science* 246:922–26

6. Aguirre, J., Adams, T. H., Timberlake, W. E. 1990. Spatial control of developmental regulatory genes in *Aspergillus nidulans*. *Exp. Mycol.* In press

7. Antequera, F., Tamame, M., Villanueva, J. R., Santos, T. 1984. DNA methylation in the fungi. *J. Biol. Chem.* 259:8033–36

8. Aramayo, R., Adams, T. H., Timberlake, W. E. 1989. A large cluster of highly expressed genes is dispensable for growth and development in *Aspergillus nidulans*. *Genetics* 122:65–71

9. Arst, H. N. Jr., Jones, S. A., Bailey, C. R. 1981. A method for the selection of deletion mutations in the L-proline catabolism gene cluster of *Aspergillus nidulans*. *Genet. Res.* 38:171–95

10. Axelrod, D. E. 1972. Kinetics of differentiation of conidiophores and conidia by colonies of *Aspergillus nidulans*. *J. Gen. Microbiol.* 73:181–84

11. Axelrod, D. E., Gealt, M., Pastushok, M. 1973. Gene control of developmental competence in *Aspergillus nidulans*. *Dev. Biol.* 34:9–15

12. Axelrod, D. E., Peterson, D., Hasner, M. 1976. Temperature-sensitive sporulation mutants of the fungus *Aspergillus nidulans*. *Am. Soc. Microbiol. Abstr.* 76:119

13. Bainbridge, B. W. 1971. Macromolecular composition and nuclear division during spore germination in *Aspergillus nidulans*. *J. Gen. Microbiol.* 66:319–25

14. Bainbridge, B. W. 1974. A simple and rapid technique for obtaining a high proportion of hybrid cleistothecia in *Aspergillus nidulans*. *Genet. Res.* 23:115–17

15. Ballance, D. J. 1986. Sequences important for gene expression in filamentous fungi. *Yeast* 2:229–36

16. Ballance, D. J., Buxton, F. P., Turner, G. 1983. Transformation of *Aspergillus nidulans* by the orotidine-5'-phosphate decarboxylase gene of *Neurospora crassa*. *Biochem. Biophys. Res. Commun.* 112:284–89

17. Beadle, G. W., Tatum, E. L. 1941. Genetic control of biochemical reactions in *Neurospora*. *Proc. Natl. Acad. Sci. USA* 27:499–506

18. Birse, C. E., Clutterbuck, J. 1990. Cloning and expression of the *Aspergillus nidulans ivoB* gene coding for a developmental phenol oxidase. Submitted for publication

19. Birse, C. E., Clutterbuck, J. 1990. N-acetyl-6-hydroxytryptophan oxidase, a developmentally controlled phenol oxidase from *Aspergillus nidulans*. Submitted for publication

20. Botstein, D., Davis, R. W. 1981. Principles and practice of recombinant DNA research with yeast. In *The Molecular Biology of the Yeast Saccharomyces. Metabolism and Gene Expression*, ed. J. N. Strathern, E. W. Jones, J. R. Broach, 2:607–37. Cold Spring Harbor: Cold Spring Harbor Lab.

21. Boylan, M. T., Mirabito, P. M., Willett, C. E., Zimmerman, C. R., Timberlake, W. E. 1987. Isolation and physical characterization of three essential conidiation genes from *Aspergillus nidulans. Mol. Cell. Biol.* 7:3113–18

22. Brody, H., Carbon, J. 1989. Electrophoretic karyotype of *Aspergillus nidulans. Proc. Natl. Acad. Sci. USA* 86:6260–63

23. Butnick, N. Z., Yager, L. N., Hermann, T. E., Kurtz, M. B., Champe, S. P. 1984. Mutants of *Aspergillus nidulans* blocked at an early stage of sporulation secrete an unusual metabolite. *J. Bacteriol.* 160:533–40

24. Butnick, N. Z., Yager, L. N., Kurtz, M. B., Champe, S. P. 1984. Genetic analysis of mutants of *Aspergillus nidulans* blocked at an early stage of sporulation. *J. Bacteriol.* 160:541–45

25. Carter, B. L., Bull, A. T. 1969. Studies of fungal growth and intermediary carbon metabolism under steady and nonsteady state conditions. *Biotechnol. Bioeng.* 11:785–804

26. Champe, S. P., El-Zayat, A. A. E. 1989. Isolation of a sexual sporulation hormone from *Aspergillus nidulans. J. Bacteriol.* 171:3982–88

27. Champe, S. P., Kurtz, M. B., Yager, L. N., Butnick, N. J., Axelrod, D. E. 1981. Spore formation in *Aspergillus nidulans:* competence and other developmental processes. In *The Fungal Spore: Morphogenetic Controls*, ed. H. R. Hohl, G. Turian, pp. 255–76. New York: Academic

28. Champe, S. P., Rao, P., Chang, A. 1987. An endogenous inducer of sexual development in *Aspergillus nidulans. J. Gen. Microbiol.* 133:1383–87

29. Clutterbuck, A. J. 1969. Cell volume per nucleus in haploid and diploid strain of *Aspergillus nidulans. J. Gen. Microbiol.* 55:291–99

30. Clutterbuck, A. J. 1969. A mutational analysis of conidial development in *Aspergillus nidulans. Genetics* 63:317–27

31. Clutterbuck, A. J. 1972. Absence of laccase from yellow-spored mutants of *Aspergillus nidulans. J. Gen. Microbiol.* 70:423–35

32. Clutterbuck, A. J. 1974. *Aspergillus nidulans.* In *Handbook of Genetics*, ed. R. C. King, pp. 447–510. New York: Plenum

33. Clutterbuck, A. J. 1975. Cyclic AMP levels during growth and conidiation. *Aspergillus News Lett.* 12:13–15

34. Clutterbuck, A. J. 1977. The genetics of conidiation in *Aspergillus nidulans.* In *Genetics and Physiology of Aspergillus*, ed. J. A. Pateman, J. E. Smith, pp. 305–17. New York: Academic

35. Clutterbuck, A. J, Spathas, D. H. 1984. Genetic and environmental modification of gene expression in the brlA12 variegated position effect mutants of *Aspergillus nidulans. Genet. Res.* 43:123–38

36. Cole, G. T. 1986. Models of cell differentiation in conidial fungi. *Microbiol. Rev.* 50:95–132

37. Collinge, A. J., Markham, P. 1985. Woronin bodies rapidly plug septal pores of servered *Penicillium chrysogenum* hyphae. *Exp. Mycol.* 9:80–85

38. Dorn, G. L. 1967. A revised map of the eight linkage groups of *Aspergillus nidulans. Genetics* 56:619–31

39. Dorn, G. L. 1970. Genetic and morphological properties of undifferentiated and invasive variants of *Aspergillus nidulans. Genetics* 66:267–79

40. Gealt, M. A., Axelrod, D. E. 1974. Coordinate regulation of enzyme inducibility and developmental competence in *Aspergillus nidulans. Dev. Biol.* 41:224–32

41. Gooday, G. W. 1983. The hyphal tip. In *Fungal Differentiation. A Contemporary Synthesis*, ed. J. E. Smith, pp. 315–56. New York: Dekker

42. Gottlin-Ninfa, E., Kaback, D. B. 1986. Isolation and function analysis of sporulation-induced transcribed sequences from *Saccharomyces cerevisiae. Mol. Cell. Biol.* 6:2165–97

43. Gwynne, D. I., Buxton, F. P., Sibley, S., Davies, R. W., Locking, R. A., et al. 1987. Comparison of the *cis*-acting control regions of two coordinately controlled genes involved in ethanol utilization in *Aspergillus nidulans. Gene* 51:205–16

44. Gwynne, D. I., Miller, B. L., Miller, K. Y., Timberlake, W. E. 1984. Structure and regulated expression of the SpoC1 gene cluster from *Aspergillus nidulans. J. Mol. Biol.* 180:91–109

45. Hall, N. E. L., Axelrod, D. E. 1977. Interference of cellular ferric ions with DNA extraction and the application to methods of DNA determination. *Anal. Biochem.* 79:425–30

46. Hall, N. E. L., Axelrod, D. E. 1978. Sporulation competence in *Aspergillus*

nidulans: a role for iron in development. *Cell Differ.* 7:73–82
47. Hamer, J. E., Timberlake, W. E. 1987. Functional organization of the *Aspergillus nidulans trpC* promoter. *Mol. Cell. Biol.* 7:2352–59
48. Houghton, J. A. 1970. A new class of slow growing non-perithecial mutants of *Aspergillus nidulans. Genet. Res.* 16:285–92
49. Houghton, J. A. 1971. Biochemical investigations of the slow growing non-perithecial *(sgp)* mutants of *Aspergillus nidulans. Genet. Res.* 17:237–44
50. Johnstone, I. L., Hughes, S. G., Clutterbuck, A. J. 1985. Cloning an *Aspergillus nidulans* developmental gene by transformation. *EMBO J.* 4:1307–11
51. Jurgenson, J. E., Champe, S. P. 1990. The sexual and asexual spores of *Aspergillus nidulans* contain partially overlapping sets of mRNAs. *Exp. Mycol.* 14:89–93
52. Käfer, E. 1965. Origins of translocation in *Aspergillus nidulans. Genetics* 52:217–32
53. Käfer, E. 1977. Meiotic and mitotic recombination in *Aspergillus* and its chromosomal aberrations. *Adv. Genet.* 19:33–131
54. Kelly, J. M., Hynes, M. J. 1987. Multiple copies of the amdS gene of *Aspergillus nidulans* cause titration of trans-acting regulatory proteins. *Curr. Genet.* 12:21–31
55. Krumlauf, R., Marzluf, G. A. 1979. Characterization of the sequence complexity and organization of the *Neurospora crassa* genome. *Biochemistry* 18:3705–13
56. Kurtz, M. B. 1980. Regulation of fructose transport during growth of *Aspergillus nidulans. J. Gen. Microbiol.* 118:389–96
57. Kurtz, M. B., Champe, S. P. 1979. Genetic control of transport loss during development of *Aspergillus nidulans. Dev. Biol.* 70:82–88
58. Kurtz, M. B., Champe, S. P. 1981. Dominant spore color mutants of *Aspergillus nidulans* defective in germination and sexual development. *J. Bacteriol.* 148:629–38
59. Kurtz, M. B., Champe, S. P. 1982. Purification and characterization of the conidial laccase of *Aspergillus nidulans. J. Bacteriol.* 151:1338–45
60. Landschulz, W. H., Johnson, P. F., McKnight, S. L. 1988. The leucine zipper: a hypothetical structure common to a new class of DNA binding proteins. *Science* 240:1759–64
61. Law, D. J., Timberlake, W. E. 1980.

Developmental regulation of laccase levels in *Aspergillus nidulans. J. Bacteriol.* 144:509–17
62. Leal, J. A., Villanueva, J. R. 1962. An improved selective medium for the formation of ascospores by *Aspergillus nidulans. Nature* 193:1106
63. Lockington, R. A., Sealy-Lewis, H. M., Scazzocchio, C., Davies, R. W. 1985. Cloning and characterization of the ethanol utilization regulon in *Aspergillus nidulans. Gene* 33:137–49
64. Losick, R., Youngman, P., Piggot, P. J. 1986. Genetics of endospore formation in *Bacillus. Annu. Rev. Genet.* 20:625–69
65. Marshall, M. A., Timberlake, W. E. 1990. The *Aspergillus nidulans wetA* gene regulates spore-specific gene expression. *Mol. Cell. Biol.* Submitted
66. Martinelli, S. D. 1976. Conidiation of *Aspergillus nidulans* in submerged culture. *Trans. Br. Mycol. Soc.* 67:121–28
67. Martinelli, S. D. 1979. Phenotypes of double conidiation mutants of *Aspergillus nidulans. J. Gen. Microbiol.* 114:277–87
68. Martinelli, S. D., Clutterbuck, A. J. 1971. A quantitative survey of conidiation mutants in *Aspergillus nidulans. J. Gen. Mirobiol.* 69:261–68
69. Mayorga, M. E., Timberlake, W. E. 1990. Isolation and molecular characterization of the *Aspergillus nidulans wA* gene. *Genetics.* In press
70. McCorkindale, N. J., Hayes, D., Johnston, G. A., Clutterbuck, A. J. 1983. N-acetyl-6-hydroxytryptophan a natural substrate of a monophenal oxidase from *Aspergillus nidulans. Phytochemistry* 22:1026–28
71. McCully, K. S., Forbes, E. 1965. The use of p-fluorophenylalanine with 'master strains' of *Aspergillus nidulans* for assigning genes to linkage groups. *Genet. Res.* 6:352–59
72. Miller, B. L. 1990. The developmental genetics of asexual reproduction in *Aspergillus nidulans. Semin. Dev. Biol.* In press
73. Miller, B. L., Miller, K. Y., Roberti, K. A., Timberlake, W. E. 1987. Position-dependent and -independent mechanisms regulate cell-specific expression of the SpoC1 gene cluster of *Aspergillus nidulans. Mol. Cell. Biol.* 7:427–34
74. Miller, B. L., Miller, K. Y., Timberlake, W. E. 1985. Direct and indirect gene replacements in *Aspergillus nidulans. Mol. Cell. Biol.* 5:1714–21
75. Miller, J., McLachlan, A. D., Klug, A. 1985. Repetitive zinc-binding domains in the protein transcription factor IIIA

from *Xenopus oocytes*. *EMBO J.* 41: 1609–14

76. Mims, C. W., Richardson, E. A., Timberlake, W. E. 1988. Ultrastructural analysis of conidiophore development in the fungus *Aspergillus nidulans* using freeze-substitution. *Protoplasma* 44:132–41

77. Mirabito, P. M., Adams, T. H., Timberlake, W. E. 1989. Interactions of three sequentially expressed genes control temporal and spatial specificity in *Aspergillus* development. *Cell* 57:859–68

78. Morton, A. G. 1961. The induction of sporulation in mould fungi. *Proc. R. Soc. London Ser. B* 153:548–69

79. O'Hara, E. B., Timberlake, W. E. 1989. Molecular characterization of the *Aspergillus nidulans* yA locus. *Genetics* 121:249–54

80. Oliver, P. T. P. 1972. Conidiophore and spore development in *Aspergillus nidulans*. *J. Gen. Microbiol.* 73:45–54

81. Orr, W. C., Timberlake, W.E. 1982. Clustering of spore-specific genes in *Aspergillus nidulans*. *Proc. Natl. Acad. Sci. USA* 79:5976–80

82. Pastushok, M., Axelrod, D. E. 1976. Effect of glucose, ammonium and media maintenance on the time of conidiophore initiation by surface colonies of *Aspergillus nidulans*. *J. Gen. Microbiol.* 94: 221–24

83. Pateman, J. A., Doy, C. H., Olsen, J. E., Norris, U., Creaser, E. H., Hynes, M. 1983. Regulation of alcohol dehydrogenase (ADH) and aldehyde dehydrogenase (AldDH) in *Aspergillus nidulans*. *Proc. R. Soc. London Ser. B* 217:243–64

84. Perkins, D. D., Radford, A., Newmeyer, D., Bjorkman, M. 1982. Chromosomal loci of *Neurospora crassa*. *Microbiol. Rev.* 46:426–570

85. Pontecorvo, G. 1958. *Trends in Genetic Analysis*. New York: Columbia Univ. Press

86. Pontecorvo, G., Roper, J. A., Hemmons, L. M., MacDonald, K. D., Bufton, A. W. J. 1953. The genetics of *Aspergillus nidulans*. *Adv. Genet.* 5: 141–238

87. Pritchard, R. H. 1955. The linear arrangement of a series of alleles of *Aspergillus nidulans*. *Heredity* 9:343–71

88. Rambosek, J., Leach, J. 1987. Recombinant DNA in filamentous fungi: progress and prospects. *CRC Crit. Rev. Biotechnol.* 6:357–93

89. Reichle, R. E., Alexander, J. V. 1965. Multiperforate septations, Woronin bodies and septal plugs in *Fusarium*. *J. Cell Biol.* 24:489–96

90. Rothstein, R. J. 1983. One-step gene disruption in yeast. *Methods Enzymol.* 101:202–11

91. Saxena, R. K., Sinha, U. 1973. Conidiation of *Aspergillus nidulans* in submerged liquid culture. *J. Gen. Appl. Microbiol.* 19:141–46

92. Scazzocchio, C., Sdrin, N., Ong, G. 1982. Positive regulation in a eukaryote, a study of the uaY gene of *Aspergillus nidulans:* I. Characterization of alleles, dominance and complementation studies, and a fine structure map of the *uaY-oxpA* cluster. *Genetics* 100:185–208

93. Scherer, S., Davis, R. W. 1979. Replacement of chromosome segments with altered DNA sequences constructed in vitro. *Proc. Natl. Acad. Sci. USA* 76:4951–55

94. Selker, E. U., Cambareri, E. B., Jensen, B. C., Haack, K. R. 1987. Rearrangement of duplicated DNA in specialized cells of *Neurospora*. *Cell* 51:741–52

95. Selker, E. U., Garrett, P. W. 1988. DNA sequence duplications trigger gene inactivation in *Neurospora crassa*. *Proc. Natl. Acad. Sci. USA* 85:6870–74

96. Sewall, T., Mims, C. W., Timberlake, W. E. 1990. Conidial differentiation in wild type and wetA⁻ strains of *Aspergillus nidulans*. *Dev. Biol.* 138:499–508

97. Smale, S. T., Baltimore, D. 1989. The "initiator" as a transcription control element. *Cell* 57:103–13

98. Smith, J. E. 1978. Asexual sporulation in filamentous fungi. In *The Filamentous Fungi*, ed. J. E. Smith, D. R. Berry, pp. 214–39. London: Edward Arnold

99. Smith, J. E., Anderson, J. G. 1973. Differentiation in the *Aspergilli*. *Symp. Soc. Gen. Microbiol.* 23:295–337

100. Smith, J. E., Anderson, J. G., Deans, S. O., Davis, B. 1977. A sexual development in *Aspergillus*. In *Genetics and Physiology of Aspergillus*, ed. J. E. Smith, J. A. Pateman, pp. 23–58. London: Academic

101. Springer, M. L., Yanofsky, C. 1989. A morphological and genetic analysis of conidiophore development in *Neurospora crassa*. *Genes Dev.* 3:559–71

102. St. John, T. P., Davis, R. W. 1979. Isolation of galactose-inducible DNA sequences from *Saccharomyces cerevisiae* by differential plaque filter hybridization. *Cell* 16:443–52

103. Struhl, K. 1983. The new yeast genetis. *Nature* 305:391–97

104. Tamame, M., Antequera, F., Santos, E.

1988. Developmental characterization and chromosomal mapping of the 5-azacytidine-sensitive fluF locus of *Aspergillus nidulans*. *Mol. Cell. Biol.* 8:3043–50

105. Tamame, M., Antequera, F., Villanueva, J. R., Santos, T. 1983. High-frequency conversion to a "fluffy" developmental phenotype in *Aspergillus* spp. by 5-azacytidine treatment: Evidence for involvement of a single nuclear gene. *Mol. Cell. Biol.* 3:2287–97

106. Tilburn, J., Scazzocchio, C., Taylor, G. T., Zabicky-Zissman, J. H., Lockington, R. A., Davies, R. W. 1983. Transformation by integration in *Aspergillus nidulans*. *Gene* 26:205–21

107. Timberlake, W. E. 1978. Low repetitive DNA content in *Aspergillus nidulans*. *Science* 202:973–75

108. Timberlake, W. E. 1980. Developmental gene regulation in *Aspergillus nidulans*. *Dev. Biol.* 78:497–510

109. Timberlake, W. E. 1987. Molecular genetic analysis of development in *Aspergillus nidulans*. In *Genetic Regulation of Development*, ed. W. F. Loomis, pp. 63–82. New York: Liss

110. Timberlake, W. E., Barnard, E. C. 1981. Organization of a gene cluster expressed specifically in the asexual spores of *A. nidulans*. *Cell* 26:29–37

111. Timberlake, W. E., Boylan, M. T., Cooley, M. B., Mirabito, P. M., O'Hara, E. B., Willett, C. E. 1985. Rapid identification of mutation-complementing restriction fragments from *Aspergillus nidulans* cosmids. *Exp. Mycol.* 9:351–55

112. Timberlake, W. E., Hamer, J. E. 1986. Regulation of gene activity during conidiophore development in *Aspergillus nidulans*. In *Genetic Engineering*, ed. J. K. Setlow, A. Hollaender, pp. 1–29. New York: Plenum

113. Timberlake, W. E., Marshall, M. A. 1988. Genetic regulation of development in *Aspergillus nidulans*. *Trends Genet.* 4:162–69

114. Timberlake, W. E., Marshall, M. A. 1989. Genetic engineering in filamentous fungi. *Science* 244:1313–17

115. Tomsett, A. B., Cove, D. J. J. 1979. Deletion mapping of the niiA niaD gene region of *Aspergillus nidulans*. *Genet. Res.* 34:19–32

116. Upshall, A., Giddings, B., Mortimore, I. D. 1977. The use of benlate for distinguishing between haploid and diploid strains of *Aspergillus nidulans* and *Aspergillus terreus*. *J. Gen. Microbiol.* 100:413–18

117. van den Hondel, C. A. M. J. J., van Gorcom, R. F. M., Goosen, T., van den Broek, H. W. J., Timberlake, W. E., Pouwels, P. H. 1985. Development of a system for analysis of regulation signals in *Aspergillus*. In *Molecular Genetics of Filamentous Fungi*, ed. W. E. Timberlake, pp. 29–38. New York: Liss

118. van Gorcom, R. F. M., Pouwels, P. H., Goosen, T., Visser, J., van den Broek, H. W. J., et al. 1985. Expression of the *Escherichia coli* β-galactosidase gene in *Aspergillus nidulans*. *Gene* 40:99–106

119. Vinson, C. R., Sigler, P. B., McKnight, S. L. 1989. Scissors-grip model for DNA recognition by a family of leucine zipper proteins. *Science* 246:911–19

120. Weltring, K.-M., Turgeon, B. G., Yoder, O. C., VanEtten, H. D. 1988. Isolation of a phytoalexin-detoxification gene from the plant pathogenic fungus *Nectria haematococca* by detecting its expression in *Aspergillus nidulans*. *Gene* 68:335–44

121. Yager, L. N., Kurtz, M. B., Champe, S. P. 1982. Temperature-shift analysis of conidial development in *Aspergillus nidulans*. *Dev. Biol.* 93:92–103

122. Yelton, M. M., Hamer, J. E., Timberlake, W. E. 1984. Transformation of *Aspergillus nidulans* by using a trpC plasmid. *Proc. Natl. Acad. Sci. USA* 81:1470–74

123. Yelton, M. M., Timberlake, W. E., van den Hondel, C. A. M. J. J. 1985. A cosmid for selecting genes by complementation in *Aspergillus nidulans*: selection of the developmentally regulated yA locus. *Proc. Natl. Acad. Sci. USA* 82:834–38

124. Zimmermann, C. R., Orr, W. C., Leclerc, R. F., Barnard, E. C., Timberlake, W. E. 1980. Molecular cloning and selection of genes regulated in *Aspergillus* development. *Cell* 21:709–15

Annu. Rev. Genet. 1990. 24:37–66

MOLECULAR GENETICS OF POLYKETIDES AND ITS COMPARISON TO FATTY ACID BIOSYNTHESIS

David A. Hopwood and David H. Sherman[1]

John Innes Institute, John Innes Centre for Plant Science Research, Norwich NR4 7UH, England

KEY WORDS: antibiotics, multienzyme complex, multifunctional enzyme, chalcone synthase, *Streptomyces*

CONTENTS

[1]Present address: Institute for Advanced Studies in Biological Process Technology, and Department of Microbiology, University of Minnesota, 1479 Gortner Avenue, St Paul, Minnesota 55108

0066-4197/90/1215-0037$02.00

INTRODUCTION: WHAT ARE POLYKETIDES?

Figure 1 shows the structures of six metabolites of bacteria, fungi, and plants. Their structures are strikingly diverse, yet they share a common pattern of biosynthesis. This begins by the successive addition of simple carboxylic acid units to build a carbon chain. The building units are of various kinds: commonly, they are residues of acetate, propionate, and butyrate, but sometimes are more complex. Each unit contributes two carbon atoms to the assembly of the linear chain, of which the β-carbon always carries a keto group. Typically, some of these keto groups are reduced to hydroxyls, or removed, after addition of a unit to the chain, but most are not. The occurrence of the remaining keto groups at many of the alternate carbon atoms of the chain gives rise to the name "polyketide" for this family of compounds.

The second carbon atom donated by each unit to the growing polyketide chain carries different substituents depending on its origin: only hydrogen in the case of acetate residues; a methyl or ethyl group for propionate or butyrate; and others for the rarer building units. This side-chain variation, together with the varied fate of each of the keto groups, the possibility of chirality at one or more carbon atoms, and the total length of the chain (from 6 carbons for triacetic acid lactone to 50 for brevetoxin: Figure 1), accounts for the huge potential diversity built into the primary structures of the mole-

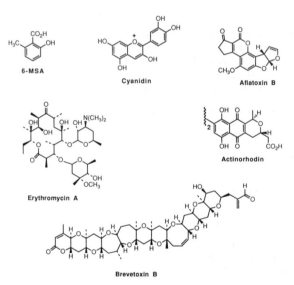

Figure 1   Structures of 6-methylsalicylic acid (6-MSA); a flavanone (cyanidin); aflatoxin B; erythromycin A; actinorhodin; and brevetoxin B. Note that all carbon atoms except that of the O-methyl of aflatoxin and the sugar carbons of erythromycin are of polyketide origin.

Table 1   Selected examples of the occurrence and biological properties of polyketides

Organisms	Examples of polyketides produced	Review or reference
EUKARYOTES		
Angiosperms	Flavonoids (usually as glycosides) responsible for much of the flavor of food and drink of plant origin and for the colors of many flowers	62, Table 2
	Urishiols as irritants in poison ivy	62
	Stilbenes and isoflavonoids as phytoalexins (antifungal agents) produced in response to fungal infection	101
Fungi	Mycotoxins, including aflatoxins and ochratoxins from *Aspergillus* spp., ergochromes from *Claviceps*, patulin from *Penicillium*	100
	Antibiotics, such as griseofulvin from *Penicillium*	100
	Polyketide-derived melanins in infection by phytopathogenic fungi	4
Marine organisms	Brevetoxins from *Gymnodium* (dinoflagellate); polyunsaturated hydrocarbons from brown algae; macrolides from sponges; polypropionate from molluscs	26
Actinomycetes	Hundreds of diverse antibiotics, including: tetracyclines; anthracyclines (e.g. daunorubicin, an anticancer agent); macrolides (e.g. erythromycin); polyethers (e.g. monensin); ansamycins (e.g. rifamycin); avermectin (an anti-helminthic agent)	Table 2
	Pharmacologically active compounds, e.g. the immunosuppressant macrolide FK506	106
	Possible signal molecules (e.g. pamamycin)	55
Myxobacteria	Several diverse antibiotics	79
Rhizobia	Putative metabolites synthesized by products of *nod E,F,G* genes of *R. leguminosarum* and *R. meliloti*	Table 2
Mycobacterium phlei	6-Methylsalicylic acid for iron uptake	77
Other bacteria	A few antibiotics, e.g. pseudomonic acid (mupirocin) from *Pseudomonas;* aurantinin from *Bacillus*	19, 69

cules by the genetic "programming" of the polyketide synthases (PKS) that catalyze chain building. The challenge to understand this programming is a major motivation of our current research and the stimulus to write this review.

THE OCCURRENCE AND BIOLOGICAL ROLES OF POLYKETIDES

Polyketide-derived compounds abound in both prokaryotes and eukaryotes where they play an amazing variety of roles (Table1). They owe their chemical diversity both to the programming of the PKS and to events occurring after chain assembly, such as formation of aromatic, ether, or macrolide ring systems, addition of moieties such as methyl groups, terpene chains, or sugar residues, and many others. Polyketides are examples of secondary metabolites, so-called because they occur sporadically, and are probably adaptive only under particular conditions, in contrast to the primary metabolites that provide the structure and energy requirements of all living cells (62). Our knowledge of their occurrence is inevitably very incomplete, and reflects man's interests and needs. For example, the bases of the striking polyketide-derived colors of the flowers and the flavors of other parts of higher plants are obvious subjects for study, as are the phytoalexins that help to defend crop plants against fungal disease; among fungi, the mycotoxins are the most important polyketides from an anthropocentric point of view; and in prokaryotes, the polyketide-derived antibiotics of the mycelial actinomycetes are overwhelmingly significant.

While the evolutionary advantage of many actinomycete antibiotics may well be as agents of competition, or of the defense of the previously elaborated food source when the vegetative colony is undergoing programmed degeneration to give rise to the reproductive stage (14), some actinomycete polyketides with an (incidental, perhaps) antimicrobial activity may well function as signals for colonial development; e.g. pamamycin that stimulates aerial mycelium formation in *Streptomyces alboniger* (55). An interesting parallel is perhaps the interaction of *Rhizobium* spp. with the roots of their plant hosts. Plant-derived polyketides—the flavonoids—secreted from the roots are needed for the activation of the nodulation genes of *Rhizobium* (50); and some of the *nod* genes resemble the PKS genes of *Streptomyces* (6), suggesting a role for a further series of polyketides, this time *Rhizobium*-derived, in host recognition.

BIOCHEMISTRY OF POLYKETIDE SYNTHESIS: STRUCTURE OF THE SYNTHASES

In a penetrating and lucid review (8), Birch described how he was led in the early 1950s to propose and elaborate the polyketide hypothesis, first adumbrated by Collie (16), and its relationship to the biosynthesis of long-chain fatty acids. Figure 2 illustrates the concepts. For the most abundant class of fatty acids—those with unbranched, saturated carbon chains—chain building starts by condensation of an acetyl unit (the "starter", with two carbon atoms) with a malonyl unit (the "extender", with three carbons) to yield a C_4 intermediate (an acetoacetyl residue), one carbon having been eliminated as CO_2 during condensation; hence, the malonyl unit donates an acetyl residue to the product. The distal keto group of the growing chain is now removed by a cycle of three reactions: ketoreduction to a hydroxyl, followed by dehydration to an enoyl system, and finally enoyl reduction to give alkyl functionality. There follow further rounds of condensation, reduction, dehydration, and reduction, until the chain reaches its final length, typically C_{16} (for palmitic acid) or C_{18} (for stearic acid).

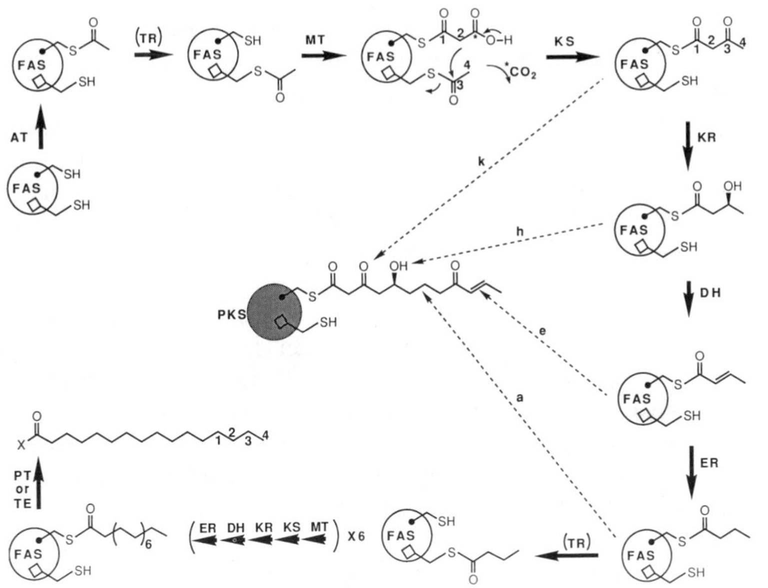

Figure 2 Schematic representation of fatty acid and polyketide biosynthesis. The circle labelled FAS or PKS represents the fatty acid or polyketide synthase, carrying two thiol groups, one on the β-ketoacyl synthase (condensing enzyme) (◊) and the other on the acyl carrier protein (●). The reaction steps are labelled: AT, acetyl transferase; (TR), acyl transfer reaction, not unambiguously assigned to a specific enzyme component; MT, malonyl transferase; KS, β-ketoacyl synthase; KR, ketoreductase; DH, dehydrase; ER, enoyl reductase; PT, or TE, palmityl transferase or thioesterase, involved in chain termination to produce palmityl CoA (X = CoA) or free palmitic acid (X = OH) respectively. 1,2,3,4 designate carbon atoms of malonate and acetate that contribute to chain building, while the asterisk labels the carbon of malonate that is eliminated as CO_2. k,h,e,a, represent the various possibilities that follow each condensation step to give keto, hydroxyl, enoyl or alkyl functionality at specific points in the product of a polyketide synthase.

The fatty acid synthase (FAS) that catalyzes this series of reactions varies in its structural organization in different life forms (64). In *Escherichia coli* (and probably most other bacteria), as well as in higher plants, it is a multienzyme complex (a so-called type II FAS). It consists of separate polypeptides for condensation (the β-ketoacyl synthase), ketoreduction, dehydration, and enoyl reduction, together with a small polypeptide, the acyl carrier protein (ACP) to which the growing carbon chain is intermittently attached via a phosphopantetheine "arm" (see below). In addition there are acetyl and malonyl transferases to catalyze attachment of the building units to the FAS and a palmityl transferase to detach the completed fatty acid chain from it. In vertebrates, in contrast, corresponding biochemical functions are carried out by distinct domains on a large multifunctional polypeptide (type I FAS); in the yeast *Saccharomyces cerevisiae* the type I FAS consists of two polypeptides carrying five and three functions, respectively. *Mycobacterium smegmatis* is a so-far unique example where type I and type II FASs coexist (9, 10), both capable of producing unusually long fatty acids.

In the simplest form outlined above fatty acid synthesis involves few programmed choices on the part of the FAS; carbon chain length—determined by the number of successive additions of acetyl residues—is the primary one. However, the situation is in reality somewhat more complex. Thus, organisms contain some branched fatty acids that result from the selection by the FAS of alternative starter units, such as an isopropyl or isobutyryl instead of an acetyl unit, or methylmalonate instead of malonate as extender unit. Moreover, the synthesis of unsaturated fatty acids, which make up a proportion of total membrane lipids, depends, in bacteria, on a lack of enoyl reduction after the dehydration step at one, or sometimes two, points in chain assembly so that a double bond remains (in mammals, such double bonds are introduced by specific desaturases after chain building is complete, rather than being programmed into chain assembly).

Polyketide synthesis may be considered to involve an extension—sometimes to an extreme degree—of the limited set of choices made by a FAS, so that the PKS needs to be more highly programmed. Figure 3 shows an example, that of monensin A in which the PKS makes 37 choices during chain assembly: a choice of 13 building units in a specific order from acetyl, propionyl, and butyryl residues; a choice of four possibilities for the chemistry after each of 12 steps in chain building, giving keto, hydroxyl, enoyl, or alkyl functionality at particular positions in the chain; and the correct choice of stereochemistry at 12 positions where R or S configurations of methyl or hydroxyl groups, or E or Z double bonds, are specifically introduced. Thus,

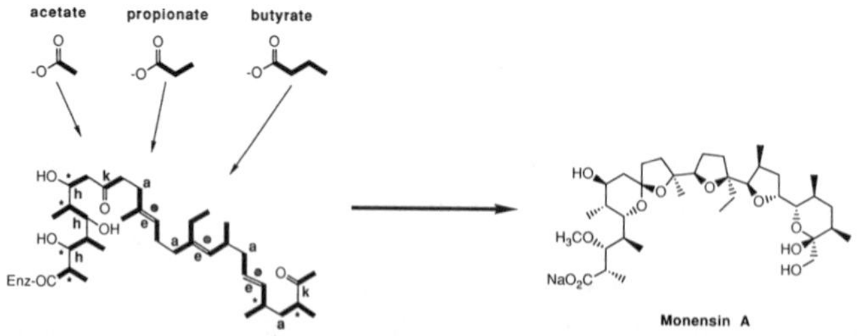

Figure 3   Biosynthesis of monensin A. It arises from a hypothetical polyketide precursor (left) derived by the condensation of five acetyl, seven propionyl and one butyryl residue. For meaning of k,h,e,a, see Figure 2. The asterisks indicate six chiral carbon atoms with a specific R or S stereochemistry of the methyl or hydroxyl; and ● indicates the three double bonds with E (*trans*) rather than Z (*cis*) configuration.

the monensin PKS constructs just one out of an enormous number of theoretically possible carbon chains. Understanding the basis of this programming is the challenge that we believe can now be realistically addressed by the combination of biochemistry and chemistry with the currently developing molecular genetics of polyketide producers.

A few further biochemical points deserve mention before we discuss the genetics of polyketide biosynthesis. One concerns the fidelity of PKS programming. It is impressive, but need not be absolute. For example, while the major polyether produced by *Streptomyces cinnamonensis* is monensin A, a minority of monensin B occurs, explicable by incorporation of a propionate residue instead of butyrate as the fifth extender unit (30). The programming of starter unit choice can also be imprecise; thus the *a* and *b* avermectins, coproduced by *Streptomyces avermitilis,* result from the choice of a 2-methylbutyrate and an isobutyrate residue, respectively, as starter. On the other hand, the difference between the series *1* and *2* avermectins may result from variation in chemistry after addition of the third building unit, to give an enoyl or hydroxyl function, repectively.

The above account took it for granted that the PKS determines the functionality at each point in the chain by programmed choices made at the corresponding stage during chain assembly, rather than introducing a specific pattern of reduction and dehydration after the chain is complete. The former is known to occur in fatty acid biosynthesis and there is indirect evidence for it in 6-methylsalicylic acid synthesis (118). The question was tested experimentally in the complex polyketides by the groups of Cane (12) and Hutchinson (117), who showed incorporation of chain-elongation intermediates, already modified by reduction and dehydration as appropriate, into erythromycin and the polyketide-derived precursor of tylosin, respectively. Further recent evidence came through isolation of a series of incomplete polyketide precursors of mycinamicin (53) and tylosin (45) from blocked mutants of the antibiotic producers. In both cases these early intermediates showed the proper pattern of reduction, dehydration, and stereochemistry found in the final macrolide product. These data are suggestive, but not conclusive. Moreover, work is required in other polyketide systems to test the generality of the findings.

We need also to be aware of variations in the handling of the starter and extender units by different FAS/PKS enzymes. The general scheme for fatty acid synthesis has the starter acetyl transferred from its CoA ester by the acetyl transferase to the thiol group of the 4'-phosphopantetheine arm of the ACP and thence (by an unknown mechanism) to the thiol of the active-site cysteine of the β-ketoacyl synthase. Condensation now occurs with the malonyl extender unit, which has meanwhile been attached to the vacated ACP. Thus the chain "grows" on the ACP but is later transferred back to the

β-ketoacyl synthase to become the substrate for the next round of chain growth by condensation with a further malonyl-ACP. However, only recently (49) the first step, to the C_4 stage, in fatty acid synthesis in *E. coli* was shown to involve a dedicated β-ketoacyl synthase III that catalyzes condensation of malonyl-ACP with acetyl-CoA without the formation of acetyl-ACP. (Subsequent extension from the C_4 to the C_{16} or C_{18} stage is carried out by either of two other condensing enzymes, β-ketoacyl synthases I and II, which use ACP-linked substrates in the usual way.) For polyketides, the first relevant biochemical evidence was for the condensing enzyme involved in anthocyanin synthesis in parsley (chalcone synthase) in which, surprisingly, no ACP nor specific acyl transferase appeared to be involved; all of the substrates for chain building are the CoA esters of the building units rather than ACP-linked residues (87). Hence the importance of the recent deduction, from molecular genetic analysis of the *Streptomyces* polyketides granaticin and tetracenomycin, that an ACP is encoded within the group of genes for the PKS (6, 94).

The number and roles of the FAS acyl transferases also vary (64). Thus, FAS of *E. coli* has separate acetyl and malonyl transferases for chain building and a palmityl transferase for termination, whereas in *S. cerevisiae* one domain of the type I FAS handles acetate, but another catalyzes both malonyl and palmityl transfers. In contrast, the vertebrate FAS has a single catalytic site that handles both acetyl and malonyl residues for chain building, while chain termination takes place via hydrolysis by a thioesterase domain, rather than palmityl transfer to CoA, as in the other systems.

THE GENES FOR FATTY ACID AND POLYKETIDE BIOSYNTHESIS

A mixture of in vivo genetics and gene cloning has yielded information about the number and arrangement of the genes involved in biosynthesis of polyketides and fatty acids from a variety of prokaryotes and eukaryotes.

Fatty Acid Biosynthesis in Escherichia coli

Isolation and mapping of fatty acid auxotrophs of *E. coli* (17,76) served to identify five genes (*fabA,B,D,E,F*) on the chromosomal linkage map, of which only two, *fabD* and *fabF*, are closely linked. Of the three genes that mapped alone, two may not code for components of the FAS itself: *fabE* encodes a component of acetyl CoA carboxylase, that synthesizes malonyl CoA, and *fabA* encodes a dehydrase involved uniquely in the anerobic biosynthesis of unsaturated fatty acids. The remaining "lone" gene, *fabB*, encodes one of the three identified condensing enzymes, β-ketoacyl synthase I. This left the closely linked *fabD* and *fabF* genes, for the malonyl transferase and β-ketoacyl synthase II, respectively. Recently, two other FAS genes have been identified between *fabD* and *fabF: acpP* for an ACP and *fabG* for a ketoreductase (M. Rawlings, T. Larson & J. E. Cronan Jr., personal communication). It will be very interesting to know if genes for enoyl reductase, dehydrase, and acetyl and palmityl transferases also lie in the same cluster as the other four genes.

Polyketide Antibiotic Biosynthesis by Actinomycetes

Table 2 summarizes much information about the genes for polyketide biosynthesis in members of the closely related genera *Streptomyces, Saccharopolyspora*, and *Amycolatopsis*. The most striking generalization is that genes for the various steps show a marked tendency to clustering (42,46,63,91). (This is also true of genes for the biosynthesis of antibiotics of other chemical classes.) For at least four polyketides—actinorhodin in *S. coelicolor* (59), tetracenomycin in *S. glaucescens* (67), erythromycin in *Saccharopolyspora erythrea* (99) and oxytetracycline in *S. rimosus* (7)—all the essential genes form a single cluster, proved by linkage mapping to be chromosomal for actinorhodin (83), erythromycin (113), and oxytetracycline (81). Groups of adjacent genes have been identified in several other examples and it remains possible that the complete set of biosynthetic genes are clustered, but this has yet to be proved: examples include the macrolides tylosin (3), spiramycin (82), carbomycin (25), and avermectin (103), together with some other polyketides such as daunorubicin (104) and a presumptive spore pigment of *S. coelicolor* (recognized by sequencing the *whiE* genes that showed them to code for a PKS; 20). For two of these compounds—avermectin (47) and the *whiE* product (13)—linkage mapping has again demonstrated a chromosomal location for the genes. It is worth emphasizing the chromosomal linkage of the biosynthetic genes in these examples given that the genes for an antibiotic of another chemical class, methylenomycin, are certainly borne on the large linear plasmid SCP1 (52) and that claims have been made for plasmid linkage of other antibiotic biosynthetic genes, including those for polyketides, on evidence that appears to us unconvincing (41). For tylosin production by *S. fradiae* the genetic evidence is inconclusive. A group of mutations in different steps in tylosin biosynthesis were transferred at higher frequency in matings than would have been likely for chromosomal genes, suggesting an extrachromosomal location (102), but critical evidence is lacking. The genes do not appear to reside on the large linear plasmid of *S. fradiae* (36). Perhaps they lie on some other kind of mobile genetic element.

Putative Polyketides Involved in Nodulation by Rhizobium

The ability of rhizobia to form intimate symbiotic relationships with leguminous plants, involving root nodules containing nitrogen-fixing "bacte-

Table 2 Genetics of polyketide biosynthesis

Organism	Compound(s)	Genetics	References
Fungi			
Aspergillus flavus and *A. parasiticus*	Aflatoxins	Various nonproducing mutants, including some (for *A. flavus*) presumptively blocked in polyketide chain assembly	5
Penicillium patulum	Patulin	6-MSA synthase (PKS) gene cloned and sequenced; reveals domains for condensing enzyme, acetyl/malonyltransferase, ketoreductase, ACP (See Figure 4)	2
Angiosperms			
Antirrhinum majus, Petunia hybrida, Zea mais	Anthocyanins	Mutations in the chalcone synthase gene (except for *Petunia*) and several later steps in anthocyanin synthesis isolated and mapped: no significant close linkages between the different genes. Chalcone synthase genes from *A. majus* and *Z. mais*, as well as many other plants, sequenced; multiple genes with different roles in some plants	35, 54, 70, 80, 84
Arachis hypogaea	Stilbene phytoalexins	Resveratrol synthase gene sequenced	86
Actinomycetes			
Streptomyces coelicolor	Actinorhodin	Entire set of *act* genes lies on a ca 22-kb segment of chromosomal DNA; five contiguous genes for PKS components (condensing enzyme, ACP, ketoreductase, cyclase/dehydrase) identified by sequence analysis, gene disruption and mutant complementation[a]	1a, 59, 60, 83, 119, M. Fernandez-Moreno & F. Malpartida (personal communication)
Streptomyces violaceoruber	Granaticin	Six contiguous genes for components of the PKS (condensing enzyme, ACP, ketoreductase, cyclase/dehydrase) identified by homology with *act* genes, gene disruption and sequence analysis[a]	61, 94
Streptomyces glaucescens	Tetracenomycin	Entire set of *tcm* genes and a resistance gene lie on a 24-kb segment. Four contiguous genes for PKS components (condensing enzyme, ACP, cyclase/O-methyltransferase) identified by sequence analysis and mutant complementation[a]	6, 61, 67; C. R. Hutchinson, (personal communication)

Organism	Product	Description	References
Streptomyces coelicolor	Unknown spore pigment	Six open reading frames with homology to corresponding genes in *tcm* PKS cluster identified by sequence analysis in a segment of chromosomal DNA that complements *whiE* (spore pigment deficient) mutants and hybridizes with *act* PKS region[a]	20, 61
Streptomyces peucetius	Daunorubicin	Five nonoverlapping segments of DNA homologous to *act* and *tcm* PKS genes identified by hybridization. At least one segment ("group IV") carries genes for daunorubicin resistance and biosynthesis, including PKS genes, as demonstrated by production of E-rhodomycinone in *S. lividans*	104; C. R. Hutchinson (personal communication)
Streptomyces rimosus	Oxytetracycline	Entire set of *otc* genes lies on a 47-kb segment of chromosomal DNA bounded by two resistance genes (*otrA*, *otrB*). Genes for PKS components, identified by homology with *act* genes and mutant complementation, followed by sequence analysis[a]	7, 11, 61, D. H. Sherman (unpublished data)
Saccharopolyspora erythrea	Erythromycin	Cluster of *ery* genes surrounds *ermE* resistance gene on the chromosome, including genes for the sugar moieties and PKS genes; the latter occur as a series of partially repeated sequences containing multiple genes for condensing enzymes, ACPs, acyl transferases, ketoreductases and perhaps a dehydrase	23, 24, 99, 111, 112, 113; S. Donadio & L. Katz (personal communication)
Streptomyces fradiae	Tylosin	Genes for many biosynthetic steps, bounded by two resistance genes (*tlrB*, *tlrC*), identified in a 90-kb segment by sequence analysis, gene disruption and mutant complementation. Probably multiple genes for components of the PKS, as evidenced by repeated DNA sequences in region that complements a *tylG* (PKS) mutant. Some suggestion of location of the *tyl* cluster on a mobile genetic element, but not on the large linear plasmid of the strain	3, 36, 102
Streptomyces thermotolerans	Carbomycin	Some PKS genes (*carG*) and genes for late steps in carbomycin biosynthesis, together with two resistance genes (*carA*, *carB*) identified in a region of at least 70 kb by complementation of *car* and *tyl* mutants. Indications of multiple genes for PKS components as evidenced by repeated DNA sequences in the *carG* region	25; B. E. Schoner (personal communication)

Organism	Product	Description	References
Streptomyces ambofaciens	Spiramycin	Some PKS genes and genes for late steps in spiramycin biosynthesis together with a resistance gene (*srmB*) identified in a 40-kb segment by mutant complementation and gene disruption. Indication of multiple genes for PKS components	82, N. Rao (personal communication)
Streptomyces avermitilis	Avermectin	Mapping experiments located PKS genes and genes for later steps close together on the chromosome. Genes for several biosynthetic steps, including PKS genes, identified in ca 70-kb segment through mutant complementation by cloned DNA	47, 103
Streptomyces cinnamonensis; Streptomyces griseus; Streptomyces curacoi; Streptomyces roseofulvus; Streptomyces griseus	Monensin; nonactin; curamycin; frenolicin; griseusin	DNA that hybridizes with *act* PKS genes cloned and partially sequenced, revealing genes resembling the condensing enzyme, ACP and (except for *S. curacoi*) ketoreductase genes for other PKSs; not yet certain that these represent biosynthetic genes for the relevant antibiotic	93; T. Arrowsmith; R. Plater; and S. Bergh (personal communications); D. H. Sherman & M. J. Bibb (unpublished data)
Amycolatopsis mediterranei	Rifamycin	Three genes for late biosynthetic steps mapped to the same arc of the chromosome. No information on PKS genes	28
Streptomyces griseus	Candicidin	A phosphate-regulated gene for PABA synthase (producing the starter for polyketide synthesis) cloned. No information on PKS genes	29, 78
Rhizobia			
Rhizobium leguminosarum b.v. *viciae* and b.v. *trifolii*; *R. meliloti*	Specificity signals for legume recognition	Within clusters of plasmid-borne nodulation genes, *nodE* would encode a condensing enzyme, *nodF* an acyl carrier protein and *nodG*, where present (*R. meliloti*), a keto-reductase; perhaps involved in biosynthesis of a polyketide moiety added to root hair branching factor made by *nodA,B,C* products	6, 21, 43, 85, 92, 98, 105

---

[a] See Figure 6

roids," is determined by the *nod* genes, which are borne on large plasmids. They include genes needed to generate signals that lead the host to become infected and to develop the nodule. A subset of the *nod* genes are particularly interesting because they determine the host-specificity of the process (98, 105), ensuring, for example, that peas are nodulated only by *Rhizobium leguminosarum* b.v. *viciae*, clovers by *R. l.* b.v. *trifolii*, and alfalfa by *R. meliloti*. A striking recent conclusion from sequencing studies is that two genes in each of these three examples appear to encode components of a PKS: *nodE* a condensing enzyme and *nodF* an ACP; in addition, *R. meliloti* has a *nodG* gene whose product would be a ketoreductase (6, 21, 43, 85, 92, 98). The first visible plant reaction in response to exposure to *Rhizobium* is root-hair deformation, followed by invasion of the root hairs by the bacteria. This deformation appears to be stimulated by a compound synthesized by the products of other *nod* genes *(nodA,B,C)* that are common to all species. A working hypothesis (J. A. Downie, personal communication) is that the *nodA,B,C* compound is modified by the *nodE,F,(G)* products, by addition of a unique polyketide side chain, to give it the specificity to react only with root hairs of the appropriate host.

Anthocyanin and Phytoalexin Biosynthesis in Flowering Plants

In several plants, notably maize *(Zea mais)*, snapdragon *(Antirrhinum majus)*, and *Petunia*, genetic mapping of flower-color mutants has identified unlinked or loosely linked genes controlling various steps in the anthocyanin biosynthetic pathway, only one of which encodes the information for polyketide assembly itself. This is because the successive condensation of three malonyl-CoAs onto a coumaroyl-CoA starter unit to yield a "chalcone" (which is further modified by lactone ring formation and a series of hydroxylations and glycosylations to produce the pigmented products) is catalyzed by a single polypeptide (chalcone synthase: CHS) that is essentially equivalent to a condensing enzyme alone of a typical FAS. As we saw above, no ACP nor acyl transferases are involved in carbon chain building; and no reduction or dehydration steps are needed. Following the original isolation of a cDNA clone from a parsley tissue culture (80), CHS genes have been studied in numerous plants (70). Perhaps the most interesting finding is the recognition of multiple CHS genes that play specialized roles in the biosynthesis of different flavonoids, such as anthocyanin pigments and phytoalexins, sometimes in different organs and tissues. Thus, *Petunia* and *Phaseolus* have seven or more CHS genes (54, 84), some closely linked. Different members of the gene family are active in particular tissues, or in response to specific signals such as exposure to UV or visible light, or fungal infection. In pea, specific regulatory genes control expression of different CHS isoforms in petal and root tissue (35). CHS is becoming an excellent subject for analyzing the relationship between plant development and the biosynthesis of secondary metabolites.

MOLECULAR GENETICS OF FATTY ACID AND POLYKETIDE BIOSYNTHESIS

Penetrating biochemical studies of type I and type II FAS systems, which revealed the structure and complex properties of the enzymes, set the scene for recent analysis of the cloned genes. Meanwhile, developments in gene cloning and sequencing in microbial polyketide producers is yielding information about the number and roles of the genes that code for PKSs of various types.

Vertebrate Type I FAS Systems

Isolation of cDNA clones by screening a λgt11 library with antibodies to the thioesterase domain of rat mammary gland FAS led to sequencing of the gene (1). The deduced amino acid sequence revealed six of the expected seven catalytic domains on the 2505 amino acid multifunctional protein (Figure 4). The active site Cys of the β-ketoacyl synthase is at position 161 in a region identical to eight amino acid residues around the chicken β-ketoacyl synthase active site defined by the biochemical activity of an isolated domain (40, 75). The active site Ser of the acetyl/malonyl transferase is at amino acid 581 in a 13-residue motif identical to that of the goat enzyme in a biochemically characterized domain (65). About 1000 amino acid residues separate the acyl transferase from the next series of defined functional regions; within this gap the dehydrase activity is postulated to occur, even though positive in vitro studies or meaningful comparisons with other bacterial, yeast, or animal dehydrases are so far lacking. The first of two NADPH-binding sites (characteristic of reductases) is centered at Gly-1670, near the Lys-1698 that can react with pyridoxal phosphate to inhibit enoyl reductase activity (1, 74), while the second is centered at Gly-1888 and is assigned ketoreductase activity because it is insensitive to pyridoxal phosphate. The consensus sequence for nucleotide binding sites, Gly-X-Gly-X-X-Gly (114), is upheld in both cases. The 4'-phosphopantetheine-binding Ser-2151 of the ACP lies within a motif previously identified as the rat ACP domain (57), while Ser-2302 coincides with the active site of rat FAS thioesterase (68). Similar results on rat FAS have been described from another laboratory (89a).

The complete chicken liver FAS cDNA sequence was also recently reported (40, 116); it shows high colinear similarity with the rat FAS (70–80% over the various domains: Figure 4). In this study, the proposed dehydrase domain showed some similarity to previously sequenced bacterial, yeast, and

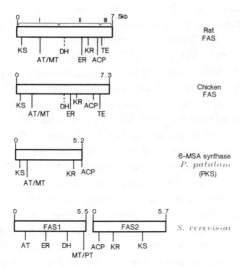

Figure 4 Organization of catalytic sites and domains (I, II, III) on type I FAS and PKS polypeptides. For explanation of enzymatic functions, see caption to Figure 2. Sources of information: rat FAS (1); chicken FAS (40); 6-MSA synthase (2); *S. cerevisiae* FAS1 and FAS2 (88,89).

animal (non-FAS) dehydrases and fell in the region identified through mapping by proteolytic digestion of the isolated FAS protein as containing the dehydrase (109). Further investigation (perhaps using site-directed mutagenesis) is needed to establish the precise substrate binding site of the dehydrase.

Studies of these type I systems have provided a working model (109) for fatty acid chain construction and termination that involves a head-to-tail orientation of two identical subunits, each carrying three major functional domains (Figure 4). Domain I contains the acetyl/malonyl transferase and β-ketoacyl synthase activities and thus catalyzes fatty acid chain initiation and extension; Domain II contains the ACP, dehydrase, enoyl, and ketoreductase activities and is the reaction center for the iterative reduction, dehydration, reduction that occurs with the acyl chain bound to the ACP 4'-phosphopantetheine arm; and Domain III contains the thioesterase activity that terminates chain construction under normal circumstances at the palmitic acid stage by hydrolysis of the link between the mature fatty acid chain and the ACP.

The Fungal Type I FAS

The type I FAS of the yeast, *S. cerevisiae*, is a multifunctional protein encoded by two unlinked, intron-free, genes, FAS1 and FAS2, for the pentafunctional β-subunit, and trifunctional α-subunit, respectively, which aggregate to form the $\alpha_6\beta_6$ FAS (58). The order of the catalytic domains in FAS1 was determined by complementation of defined *fas1* mutants with overlapping FAS1 subclones. Two groups have published nucleotide sequences of FAS1 (15, 89) and FAS2 (66, 88). The sequence of the gene revealed an open reading frame (ORF) corresponding to a protein of 1845 amino acids (89). The order of the catalytic domains of FAS1 (and FAS2) differs from those on the rat and chicken multifunctional proteins (Figure 4). Thus, the N-terminal region of the FAS1 of *S. cerevisiae* contains acetyltransferase activity with an active site at Ser-274 (counting the Met start as 1). Mutant complementation assigned enoyl reductase activity to the next domain, with a putative active site motif centered on Ser-773. About 300 amino acids separate it from the dehydrase domain, correlating with mutant complementation. However, again, the likely substrate binding site has not been identified for the dehydrase. The most significant difference between the animal type I FAS and FAS1 of *S. cerevisiae* is the bifunctional malonyl/palmityl-transferase domain at the C-terminus of the yeast protein, with active site Ser-1808 in a region showing significant similarity to the animal acetyl/malonyltransferase active sites. This is particularly interesting in light of the different carbon-chain termination mechanisms of animal and yeast FASs.

The ACP, ketoreductase, and β-ketoacyl synthase functional regions are on the 1887 amino acid FAS2 polypeptide (88). The phosphopantetheine-binding Ser of the ACP was assigned to position 180, but DNA that complemented mutants in which the ACP domain lacked a pantetheinyl side chain also mapped to a position corresponding to the C-terminal region of FAS2. Thus, two different segments of the protein must interact, as the native enzyme is folded, to give a functional ACP. The ketoreductase was mapped to the N-terminal half of FAS2, but without identification of the active site residue. The active site of the β-ketoacyl synthase is at Cys-1305 in a region resembling those of animal and bacterial FAS and PKS condensing enzymes.

The complete DNA sequence of the FAS2 gene from another fungus, *Penicillium patulum*, reveals a protein sequence with 50–70% end-to-end similarity to the *S. cerevisiae* FAS2 over the various domains (115). A notable difference is the presence of two short introns in the *P. patulum* FAS2 gene. A homolog of the FAS1-encoded β-subunit of *S. cerevisiae* has also been identified in *P. patulum* (88).

The Type I PKS of Penicillium patulum

Penicillium patulum represents the first case where FAS and PKS have been characterized from the same organism. This fungus produces 6-methylsalicylic acid (Figure 1), one of the simplest polyketides, derived from one acetate starter and three malonate extender units as a precursor of patulin. Its PKS would require a condensing enzyme, acetyl and malonyl-transferases, a ketoreductase and dehydrase, but no enoyl reductase, and perhaps a cyclase, together with an additional function to mediate chain termination (e.g. thioesterase or terminal transferase). Comparison of the 6-methylsalicylic acid synthase (6-MSAS) deduced protein sequence (2) with type I and type II FAS and PKS systems has allowed a preliminary assignment of functional domains and putative active site residues. The overall structure of the 6-MSAS Type I system more closely resembles that of the animal FAS multifunctional proteins than the fungal α/β subunit type I system (Figure 4). The β-ketoacyl synthase active site Cys is identified at position 204 from the predicted ATG start, based on the high amino acid similarity with the rat (1) and chicken (40,116) FAS condensing enzyme domains, and the FabB of $E.$ $coli$ β-ketoacyl synthase I (51, 96), as well as the deduced protein sequences of the gra (94) and tcm (6) PKS condensing enzymes (see below). The next functional domain can be deduced by its similarity with the animal acetyl/ malonyl transferase domains; the active site would be Ser-653, within a highly characteristic motif. The data suggest, therefore, that chain initiation and extension in the 6-MSAS PKS are similar to the animal FAS systems. The central region of the 6-MSAS shows no significant similarity with the central regions of the chicken and rat FASs, which appear to contain the dehydrase domain (see above). However, a convincing candidate for ketoreductase activity is near the C-terminus, where a nucleotide binding site motif (Gly-Leu-Gly-Val-Leu-Gly) at positions 1419–1424 fits the Gly-X-Gly-X-X-Gly consensus for the ketoreductase domains in the animal FAS systems and is included in a stretch of 40–50 amino acid residues with high similarity to the act (33) and gra (94) PKS ketoreductases. At the extreme C-terminus of the 6-MSAS protein is a predicted ACP region, with a likely phosphopantetheine-binding Ser-1732, based on similarity to other PKS and FAS ACPs (Figure 5). An important unanswered question in polyketide systems in general is the mechanism of chain termination. Sequence comparisons yield no evidence for a thioesterase domain in the 6-MSAS, but presumably an enzymatic step would be required for chain release from the ACP. Whether or not a specific cyclase activity is required to catalyze ring closure of the nascent tetraketide is unknown, but this would not itself explain chain termination because cyclization would involve an internal ring formation, leaving the nascent product still attached to the PKS.

The Type II FAS of E. coli

Apart from $fabA$, for the β-hydroxydecanoyl thioester dehydrase (18) involved in the anaerobic biosynthesis of unsaturated fatty acids, the first typical FAS gene to be sequenced was $fabB$ (51), which encodes β-ketoacyl synthase I, one of three condensing enzymes that operate in the biosynthesis of both saturated and unsaturated fatty acids (17). The $fabB$ gene was cloned

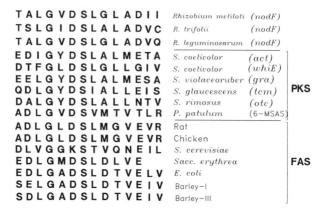

Figure 5   Amino acid sequences of ACPs of type II FAS and PKS enzymes, and ACP domains of type 1 FASs, around the 4'-phosphopantetheine-binding Ser(S). Sources of information: *nodF* (21,43,85,92); *act* (M. Fernandez-Moreno & F. Malpartida, personal communication); *whiE* (20); *gra* (94); *tcm* (6); *otc* (D. H. Sherman, unpublished data); 6-MSAS (2); rat (1); chicken (40); *S. cerevisiae* (88); *S. erythrea* and *E. coli* (32); barley (34).

by mutant complementation (21a, 51); its product, ketoacyl synthase I, was characterized biochemically using ^{3}H-cerulenin, which irreversibly binds to most FAS condensing enzymes (72). (Interestingly, β-ketoacyl synthase III of *E. coli* is not affected by cerulenin: 49.) The same inhibitor was subsequently used to pinpoint the active site Cys of the FabB protein (96). Recently, the clustered genes for an ACP *(acpP)*, a ketoreductase *(fabG)* and a malonyl transferase *(fabD)* have been sequenced (M. Rawlings, T. Larson & J. E. Cronan Jr., personal communication), but not yet the gene *(fabF)* in the same cluster for β-ketoacyl synthase II. The FabG protein shows high sequence similarity with ketoreductase domains of animal type I FASs and with type II PKS ketoreductases from the *act* and *gra* systems of *Streptomyces* (see below). The *fabD* product resembles the acetyl/malonyl transferase domain of the rat FAS, particularly around the active site Ser.

Type II Plant FAS Systems

In plants, multiple β-ketoacyl synthases and ACPs appear to occur widely. In barley chloroplasts two distinct cerulenin-binding proteins are found (M. Siggaard-Andersen & P. von Wettstein-Knowles, personal communication). Whether or not these correspond to the β-ketoacyl synthases I and II of spinach leaf (95) is unknown. For ACPs, several recent molecular genetic studies (34, 38, 71) have shown that isoforms are differentially expressed in various plant tissues, with as many as three identified in barley. The precise roles of these ACP isoforms in fatty acid biosynthesis or other metabolic processes (note that an ACP is involved in oligosaccharide synthesis in *E. coli:* 107) remains to be established.

Type II PKS Systems in Streptomyces

Actinorhodin of *S. coelicolor* A3(2) (Figure 1) is an acetate-derived antibiotic that has been subjected to considerable genetic analysis. The genes for its entire synthesis from primary metabolites lie on a segment of ca 22kb (59, 60), within which mutations blocking biosynthesis at different stages were mapped. This led to recognition of the DNA encoding the PKS because it complemented mutants of classes *act*I and *act*III, deduced to be blocked in carbon chain building because they failed to secrete any biosynthetic intermediates, but could convert to actinorhodin intermediates secreted by later blocked mutants (83). A 2.2-kb *Bam*HI fragment was shown to contain at least part of the *act*I region, and an adjacent 1.1-kb *Bam*HI fragment corresponded to *act*III (60). A nearby region complemented *act*VII mutants, deduced to be defective in carbon chain cyclization because they accumulated an incorrectly cyclized shunt product, mutactin (119). The *act*III (33) and *act*I/VII regions (M. Fernandez-Moreno & F. Malpartida, personal communication) were sequenced. Used as probes in Southern blots, *act*I and *act*III DNA revealed bands in genomic DNA of several other polyketide-producing streptomycetes (61), identifying, among others, the PKS genes (*gra*) of the granaticin producer, *S. violaceoruber,* which were cloned and sequenced (94). Meanwhile, mutant complementation (67) and hybridization with *act*I (61) identified the tetracenomycin PKS genes *(tcm)* of *S. glaucescens* for sequencing (6). Corresponding genes *(otc)* from *S. rimosus,* the oxytetracycline producer, were identified by similar evidence (11, 61) and partially sequenced (D. H. Sherman, unpublished data). In addition, sequencing of DNA that complements spore–pigment deficient *S. coelicolor* mutants *(whiE)* led to the recognition of a set of ORFs whose sequences clearly correspond to those of known PKS genes (20).

A common feature of the organization of the sequenced *act, gra, tcm, otc,* and *whiE* PKS gene clusters is a group of three characteristic ORFs, labelled ORF1-2-3 in Figure 6 (*whiE* ORFIII-IV-V). The deduced protein sequences of the first two ORFs show particularly high similarity with the FabB condensing enzyme of *E. coli* (51, 96) (Figure 7). A potential active site Cys occurs only in the first ORF in each case. This fact, along with the possibility of translational coupling between the first two ORFs (because of overlapping 3' stop/5' start codons), suggests that the gene products are stoichiometrically produced to form a heterodimer, rather than being two distinct condensing enzymes. The deduced protein sequence of the third ORF in each cluster resembles type II FAS ACPs from bacteria and plants and ACP domains of animal type I FASs, with a characteristic motif centered on a potential

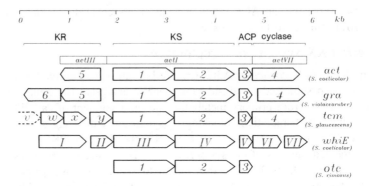

Figure 6   Arrangement of ORFs in the PKS gene clusters for actinorhodin *(act)* in *S. coelicolor* (M. Fernandez-Moreno & F. Malpartida, personal communication); granaticin *(gra)* in *S. violaceoruber* (94); tetracenomycin *(tcm)* in *S. glaucescens* (6; C. R. Hutchinson, personal communication); the presumptive spore pigment *(whiE* product) in *S. coelicolor* (20); and oxytetracycline *(otc)* in *S. rimosus* (D. H. Sherman, unpublished data). Above the ORFs are shown the locations of the corresponding *act* mutant classes.

4'-phosphopantetheine-binding Ser (Figure 5). This result is highly significant in revealing that PKSs involved in biosynthesis of complex metabolites in *Streptomyces* utilize an ACP-mediated pathway like that of fatty acids, in contrast to plant chalcone synthases that lack an ACP requirement.

Immediately downstream of these three ORFs is a fourth ORF that shows relatedness between the clusters (it is also present in the *otc* cluster but has not been sequenced). Here, however, the resemblances are only in specific domains of the deduced protein sequences. The *act* and *gra* ORF4 sequences are similar throughout the length of the proteins, but each resembles *tcm* ORF4 only in the N-terminal half of the molecules. The deduced protein product of the corresponding *whiE* ORFVI is only about half the length of the *act, gra,* and *tcm* ORF4 proteins (Figure 6) and resembles their N-terminal domains over its entire length. Since *act* and *gra* ORF4 complement *act*VII mutants, deduced to be defective in cyclization and an accompanying dehydration because they accumulate mutactin (119; see above), we postulate that *act* and *gra* ORF4 encode a bifunctional cyclase/dehydrase (D. H. Sherman, F. Malpartida, D. A. Hopwood, et al, in preparation). In contrast, the *tcm* ORF4 protein seems to be a cyclase/O-methyltransferase because its C-terminal half resembles bovine hydroxyindole O-methyltransferase (48), and *tcm* ORF4 complements *tcm* mutants lacking a specific O-methylation step (H. Motamedi & C. R. Hutchinson, personal communication). The *whiE* ORFVI perhaps is a monofunctional polyketide cyclase.

The divergently transcribed *act*III region (*act* ORF5) would encode a ketoreductase, based on similarity to known ketoreductases (33) and chemical evidence (1a). *gra* ORF5 and ORF6 encode deduced protein sequences that

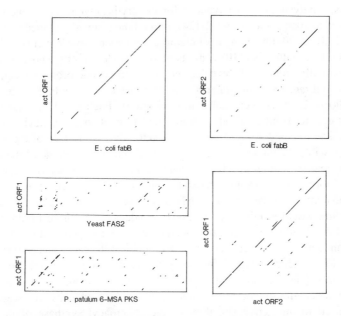

Figure 7   COMPARE/DOTPLOT analysis (22) of deduced protein sequences of various PKS and FAS genes using a window of 40 residues and stringency of 19. Sources of information; *act* ORFs (M. Fernandez-Moreno & F. Malpartida, personal communication); *E. coli* FabB (51); *S. cerevisiae* FAS2 (88); *P. patulum* 6-MSAS (2).

are very similar to the *act*III gene product (94). Interestingly, *act* ORF5- and *gra* ORF5- but not *gra* ORF6-deduced protein sequences contain the consensus nucleotide binding Gly-X-Gly-X-X-Ala motif (90). This motif is also present in the yeast FAS2 ketoreductase domain, which shows no obvious sequence relationship to the animal type I FAS, type I PKS 6-MSAS of *P. patulum* or type II PKS ketoreductases. In the animal and yeast systems the cofactor is NADPH, and we therefore expect that this will also be true for the *act*- and *gra*-encoded enzymes. The *tcm* PKS cluster reveals no sequence homologous with the *act* ketoreductase (61), consistent with a lack of a reductive step in polyketide chain building. (The lack of a putative dehydrase domain in *tcm* ORF4 would also follow from this—see above.)

Macrolide Antibiotic Biosynthesis and the Module Hypothesis

Numerous molecular genetic studies are in progress on macrolide antibiotics, including erythromycin, tylosin, spiramycin, carbomycin, and avermectin, by the generation and characterization of blocked mutants, complementation and gene disruption, and most recently, sequence analysis of the PKS regions (Table 2). The most remarkable feature emerging from the various systems is the existence of multiple sets of PKS genes within the biosynthetic gene clusters. Analysis is most advanced for the erythromycin *(ery)* gene cluster predicted to span some 50–60 kb (24), around the originally cloned resistance gene (108). A 30–kb segment shown to complement *eryA* mutants defective in polyketide synthesis (110) consists of six repeated motifs whose sequence (only partially completed) revealed the ability to encode proteins resembling the *gra* and *tcm* PKS components, and enzymes of various FAS systems (S. Donadio & L. Katz, personal communication). Each motif, similar to but clearly distinct from each other, was found to contain β-ketoacyl synthase, acyltransferase, and ACP sequences. A ketoreductase in five of the repeats and a possible dehydrase in one have also been detected. Some of the repeat units appear to contain different organizations of components and occasional gene fusions that would generate bifunctional proteins (e.g. ACP/ acyltransferase). Bearing in mind that assembly of the erythromycin polyketide would require six successive condensations, these findings have led to the module hypothesis (S. Donadio & L. Katz, personal communication) on which a specific repeat unit could handle each successive chain elongation residue required for biosynthesis of complex polyketides like macrolides (and perhaps polyethers?). After each chain extension, with reduction/dehydration or not as appropriate, the acyl substrate would be transferred to the active site thiol of the β-ketoacyl synthase of the next module, where condensation would occur with the specific chain elongation unit required for that particular step of the pathway. Completion of the *eryA* DNA sequence will allow this hypothesis to be tested by specific gene– disruption experiments that could lead to the accumulation of isolable chain-building intermediates, such as those already found in conventional blocked mutants for mycinomicin and tylosin (45, 53).

Plant Chalcone and Resveratrol Synthases

Substrate-specificity studies on purified chalcone synthase (CHS) from parsley (44, 87) suggested that the mechanism of flavonoid biosynthesis differed from that of fatty acids and polyketides: CHS utilizes 4-coumaroyl-CoA instead of acetyl-ACP as starter and malonyl-CoA instead of malonyl-ACP for chain elongation. CHS is a dimer of a rather small polypeptide (42 kd) without a 4'-phosphopantetheine-binding residue (56). This striking finding suggested that flavonoid metabolism in plants involves an independent system for the condensation process. CHS cDNA and deduced amino acid sequences from a variety of plants are greater than 66% identical at the nucleotide level and more than 80% similar at the amino acid level (70), but the protein sequences do not resemble those of FAS and PKS β-ketoacyl synthases. Perhaps these two types of condensing enzymes are evolutionarily unrelated.

Resveratrol synthase, involved in stilbene phytoalexin biosynthesis (86), catalyzes the formation of different secondary metabolites by alternative cyclization of the same tetraketide precursor as that produced by CHS. An evolutionary relationship with CHS is supported by the existence of conserved introns at identical positions in the genes and an overall similarity of 70–74% at the (deduced) protein level. Defined regions of amino acid changes between the two types of synthase suggest that differences in protein folding determine their different catalytic properties (86).

SPECIFICITY OF FAS AND PKS SYSTEMS: THE PROBLEM OF "PROGRAMMING"

The β-ketoacyl synthase appears to play a role in determining chain length of saturated and unsaturated fatty acids in *E. coli*. After synthase III has produced the C_4 chain, either synthase I or II can extend it to C_{16} or C_{18}, but they have vastly reduced activity on C_{18} substrates (27, 31), thus providing a mechanism for limiting the production of unusually long fatty acids. The substrate specificity of synthase II determines chain elongation of palmitoleate (C_{16}, one double bond) to *cis*-vaccenate, the major unsaturated fatty acid (C_{18}) of *E. coli*. Synthases I and II of spinach leaf also seem to play a role in determining saturated fatty acid chain length (95). Another interesting case is found in the host specificity-determining NodE proteins of *Rhizobium* (98), which closely resemble FAS and PKS condensing enzymes (6). Perhaps the NodE proteins from different rhizobia specify the construction of carbon chains of different length, and it is this factor that, in part, determines host specificity. Overall, these results suggest that the β-ketoacyl synthases of at least some FAS/PKS systems may play an essential role in determining the length of the polyketide chain.

Termination is another possible factor in the programming of carbon chain length. Thus, inactivation of the chicken FAS thioesterase domain resulted in elongation from the normal C_{16} chain to C_{18}, C_{20}, and C_{22}, at a progressively slower rate (57). The mechanism of chain termination in polyketide biosynthesis is poorly understood: it is unclear whether the fully elongated chain is released by a thioesterase or an acyltransferase. It is curious that candidate PKS components for acyl transfer and chain termination have not been identified in the *gra* (94) and *tcm* (6) PKS systems. Conceivably, in these cases, the corresponding FAS components of *Streptomyces* carry out acyl transfer and chain termination also in the polyketide pathway. Analysis of the FAS and PKS systems in the same *Streptomyces* should illuminate this question. It is significant that an acyltransferase domain has been found in each of the *eryA* PKS modules and on the 6-MSAS (see above), suggesting that these steps are indeed essential for PKS function. The chalcone synthases and their relatives seem to be exceptions in this regard.

The ACP also appears to be involved in programming. Although this small component might seem to play a passive part in the overall biosynthetic mechanism, some intriguing evidence suggests a crucial role in determining product structure. A novel fatty acid distribution was found in vitro using FAS of *E. coli* with the native ACP replaced by that of avocado and spinach (73, 97), with accumulation of C_{12} to C_{18} saturated and unsaturated fatty acids containing a hydroxyl group, presumably resulting from a failure to dehydrate. These results may imply that the ACP plays a role in determining the extent of the reduction, dehydration, reduction cycle after each condensation step. The recent elucidation of the three-dimensional structure of the FAS ACP of *E. coli* (39) has identified a hydrophobic cleft that is the likely site for acyl chain binding. It will be interesting to see if specific contacts between the growing chain and amino acid residues in the cleft could play a role in programming.

Another potentially critical aspect of the FAS/PKS biosynthetic cycle is the transacylation reaction that transfers the acyl chain from the ACP phosphopantetheine arm to the β-ketoacyl synthase active site (TR in Figure 2). It is not known whether this step is catalyzed by a unique transacylase activity or results from an activity of the β-ketoacyl synthase itself. Transacylation may perhaps play a role in programming by ensuring that only correctly modified acyl chain intermediates are transferred between the two active sites during chain construction. This could be especially significant for polyketides where the variations in the postcondensation reductive cycle result in vast differences in the final product of the PKS.

It is clear from progress in studying fatty acid and polyketide biosynthesis that a full complement of experimental approaches, including classical and molecular genetics, biochemistry, and chemistry, will continue to elucidate the complex processes mediated by FAS and PKS systems. The structure and organization of PKS gene clusters from representative examples of many antibiotic classes are currently being characterized. It is now possible to clone individual PKS genes of *Streptomyces* (e.g. for ACP or β-ketoacyl synthase) and obtain overexpression in *E. coli* (D. H. Sherman & D. A. Hopwood, unpublished data; H. Gramajo & M. J. Bibb, and P. F. Leadlay, personal communications), thus providing these gene products in quantities sufficient for in vitro analyses as well as for the production of antibodies to study the spatial and temporal expression of FAS and PKS components during development of *Streptomyces*. These studies should shed further light on the mechanisms that allow specificity between highly related components involved in different biosynthetic pathways even within the same organism, e.g. for fatty acids, actinorhodin, and the *whiE*-encoded spore pigment in *S. coelicolor*.

Molecular genetic analyses will also provide the information to engineer new biosynthetic pathways, using a variety of techniques. It should be possible to alter specific regions of a PKS gene, or to exchange complete ORFs of different PKS clusters, or even whole modules in the modular systems, in order to investigate the possibility of producing modified compounds, or perhaps new structural classes altogether. These experiments will increase our understanding of the programming of particular PKS systems and provide an experimental and theoretical base to design biosynthetic pathways using rational approaches.

ACKNOWLEDGMENTS

We thank J. E. Cronan Jr., N. K. Davis, M. Fernandez-Moreno, C. R. Hutchinson, L. Katz, F. Malpartida, and E. Schweizer for DNA sequence data prior to publication, and Mervyn Bibb, David Cane, Keith Chater, Stephen Gould, Chaitan Khosla, Tobias Kieser, and Tom Simpson for useful comments on the manuscript. Research in the authors' laboratory was supported by NIH grant GM 39784-03 and by the Agricultural and Food Research Council and the John Innes Foundation.

Literature Cited

1. Amy, C. M., Witkowski, A., Naggert, J., Williams, B., Randhawa, Z., Smith, S. 1989. Molecular cloning and sequencing of cDNAs encoding the entire rat fatty acid synthase. *Proc. Natl. Acad. Sci. USA* 86:3114–18

1a. Bartel, P. L., Zhu, C.-B., Lampel, J. S., Dosch, D. C., Connors, N. C., et al. 1990. Biosynthesis of anthraquinones by interspecies cloning of actinorhodin biosynthesis genes in streptomycetes: clarification of actinorhodin gene functions. *J. Bacteriol.* In press

2. Beck, J., Ripka, S., Siegner, A., Schiltz, E., Schweizer, E. 1990. The multifunctional 6–methylsalicylic acid synthase gene of *Penicillium patulum:* its gene structure relative to that of other polyketide synthases. Submitted

3. Beckmann, R. J., Cox, K., Seno, E. T. 1989. A cluster of tylosin biosynthetic genes is interrupted by a structurally unstable segment containing four repeated sequences. See Ref. 37, pp. 176–86

4. Bell, A. A., Wheeler, M. H. 1986. Biosynthesis and functions of fungal melanins. *Annu. Rev. Phytopathol.* 24:411-51

5. Bennett, J. W., Papa, K. E. 1988. The aflatoxigenic *Aspergillus* spp. *Adv. Plant Pathol.* 6:263–80

6. Bibb, M. J., Biró, S., Motamedi H., Collins, J. F., Hutchinson, C. R. 1989. Analysis of the nucleotide sequence of the *Streptomyces glaucescens tcmI* genes provides key information about the enzymology of polyketide antibiotic biosynthesis. *EMBO J.* 8:2727–36

7. Binnie, D., Warren, M., Butler, M. J. 1989. Cloning and heterologous expression in *Streptomyces lividans* of *Streptomyces rimosus* genes involved in oxytetracycline biosynthesis. *J. Bacteriol.* 171:887–95

8. Birch, A. J. 1967. Biosynthesis of polyketides and related compounds. *Science* 156:202–6

9. Bloch, K. 1977. Control mechanisms for fatty acid synthesis in *Mycobacterium smegmatis*. In *Advances in Enzymology*, ed. A. Meisler, 45:2-84. New York: Wiley

10. Bloch, K., Vance, D. 1977. Control mechanisms in the synthesis of saturated fatty acids. *Annu. Rev. Biochem.* 46:263–98

11. Butler, M. J., Friend, E. J., Hunter, I. S., Kaczmarek, F. S., Sugden, D. A., Warren, M. 1989. Molecular cloning of resistance genes and architecture of a linked gene cluster involved in biosynthesis of oxytetracycline by *Streptomyces rimosus*. *Mol. Gen. Genet.* 215:231–38

12. Cane, D. E., Yang, C. 1987. Macrolide biosynthesis. 4. Intact incorporation of a chain-elongation intermediate into erythromycin. *J. Am. Chem. Soc.* 109:1255–57

13. Chater, K. F. 1972. A morphological and genetic mapping study of white colony mutants of *Streptomyces coelicolor*. *J. Gen. Microbiol.* 72:9–28

14. Chater, K. F., Merrick, M. J. 1976. Streptomycetes. In *Developmental Biology of Prokaryotes, Studies in Microbiology*, ed. N. G. Carr, J. L. Ingraham, S. C. Rittenberg, 1:93–114. Oxford: Blackwell

15. Chirala, S. S., Kuziora, M. A., Spector, D. M., Wakil, S. J. 1987. Com-

plementation of mutations and nucleotide sequence of *FAS1* gene encoding β subunit of yeast fatty acid synthase. *J. Biol. Chem.* 262:4231–40

16. Collie, J. N. 1907. Derivatives of the multiple keten group. *J. Chem. Soc.* 91:1806–13

17. Cronan, J. E., Rock, C. A. 1987. Biosynthesis of membrane lipids. In *Escherichia coli and Salmonella typhimurium. Cellular and Molecular Biology,* ed. J. L., Ingraham, K. B. Low, B., Magasanik, M., Schaechter, 1:474–97. Washington DC: Am. Soc. Microbiol.

18. Cronan, J. E., Li, W-B., Coleman, R., Narasimhan, M., de Mendoza, D., Schwab, J. M. 1988. Derived amino acid sequence and identification of active site residues of *Escherichia coli* β-hydroxydecanoyl thioester dehydrase. *J. Biol. Chem.* 263:4641–46

19. Crimmin, M. J., O'Hanlon, P. J., Rogers, N. H. 1985. The chemistry of pseudomonic acid. Part 7. Stereochemical control in the preparation of C-2-substituted menic acid esters via the Peterson olefination. *J. Chem. Soc. Perkin Trans. 1* 3:541–47

20. Davis, N. K., Chater, K. F. 1990. Spore colour in *Streptomyces coelicolor* A3(2) involves the developmentally regulated synthesis of a compound biosynthetically related to polyketide antibiotics. *Molec. Microbiol.* In press

21. Debellé, F., Sharma, S. B. 1986. Nucleotide sequence of *Rhizobium meliloti* RCR2011 genes involved in host specificity of nodulation. *Nucleic Acids Res.* 14:7453–72

21a. de Mendoza, D., Ulrich, A. K., Cronan, J. E. 1983. Thermal regulation of membrane fluidity in *Escherichia coli. J. Biol. Chem.* 258:2098–101

22. Devereux, J., Haeberli, P., Smithies, O. 1984. A comprehensive set of sequence analysis programs for the VAX. *Nucleic Acids Res.* 12:387–95

23. Dhillon, N., Hale, R. S., Cortes, J., Leadlay, P. F. 1989. Molecular characterization of a gene from *Saccharopolyspora erythraea (Streptomyces erythraeus)* which is involved in erythromycin biosynthesis. *Molec. Microbiol.* 3:1405–14

24. Donadio, S., Tuan, J. S., Staver, M. J., Weber, M. J., Paulus, T. J., et al. 1989. Genetic studies on erythromycin biosynthesis in *Saccharopolyspora erythraea.* See Ref. 37, pp. 53–59

25. Epp, J. K., Huber, M. L., Turner, J. R., Schoner, B. E. 1989. Molecular cloning and expression of carbomycin bio-

synthetic and resistance genes from *Streptomyces thermotolerans.* See Ref. 37, pp.35–39

26. Garson, M. J. 1989. Biosynthetic studies on marine natural products. *Natl. Products Rep.* 6:143–70

27. Garwin, J. L., Klages, A. L., Cronan, J. E. 1980. Structural, enzymatic, and genetic studies of β-ketoacyl-acyl carrier protein synthases I and II of *Escherichia coli. J. Biol. Chem.* 255:11949–56

28. Ghisalba, O., Auden, J. A. L., Schupp, T., Nüesch, J. 1984. The rifamycins: properties, biosynthesis and fermentation. In *Biotechnology of Industrial Antibiotics,* ed. E. J. Vandamme, pp. 281–327. New York: Marcel Dekker

29. Gil, J. A., Hopwood, D. A. 1983. Cloning and expression of a *p*-aminobenzoic acid synthetase gene of the candicidin-producing *Streptomyces griseus. Gene* 25:119–32

30. Gorman, M., Chamberlin, J. W., Hamill, R. L. 1968. Monensin, a new biologically active compound. V. Compounds related to Monensin. *Antimicrob. Agents Chemother.* 1967:363–68

31. Greenspan, M. D., Birge, C. H., Powell, G., Hancock, W. S., Vagelos, P. R. 1970. Enzyme specificity as a factor in regulation of fatty acid chain length in *Escherichia coli. Science* 170:1203–4

32. Hale, R. S., Jordan, K. N., Leadlay, P. F. 1987. A small, discrete acyl carrier protein is involved in de novo fatty acid biosynthesis in *Streptomyces erythraeus. FEBS Lett.* 224:133–36

33. Hallam, S. E., Malpartida, F., Hopwood, D. A. 1988. Nucleotide sequence, transcription and deduced function of a gene involved in polyketide antibiotic synthesis in *Streptomyces coelicolor. Gene* 74:305–20

34. Hansen, L. 1987. Three cDNA clones for barley leaf acyl carrier proteins I and III. *Carlsberg Res. Commun.* 52:381–92

35. Harker, C. L., Ellis, T. H. N., Coen, E. S. 1990. Identification and genetic regulation of the chalcone synthase multigene family in pea. *Plant Cell* 2:185–94

36. Hershberger, C. L., Arnold, B., Larson, J., Skatrud, P., Reynolds, P., et al. 1989. Role of giant linear plasmids in the biosynthesis of macrolide and polyketide antibiotics. See Ref. 37, pp. 147–55

37. Hershberger, C. L., Queener, S. W., Hegeman, G. 1989. *Genetics and Molecular Biology of Industrial Microorganisms.* Washington DC: Am. Soc. Microbiol.

38. Høj, P. B., Svendsen, I. B. 1984. Barley chloroplasts contain two acyl carrier proteins coded for by different genes. *Carlsberg Res. Commun.* 49:483–92

39. Holak, T. A., Kearsley, S. K., Kim, Y., Prestegard, J. H. 1988. Three-dimensional structure of acyl carrier protein determined by NMR pseudoenergy and distance geometry calculations. *Biochemistry* 27:6135–42

40. Holzer, K. P., Liu, W., Hammes, G. G. 1989. Molecular cloning and sequencing of chicken liver fatty acid synthase cDNA. *Proc. Natl. Acad. Sci. USA* 86:4387–91

41. Hopwood, D. A. 1978. Extrachromosomally determined antibiotic production. *Annu. Rev. Microbiol.* 32:373–92

42. Hopwood, D. A. 1986. Cloning and analysis of antibiotic biosynthetic genes in *Streptomyces*. In *Biological, Biochemical and Biomedical Aspects of Actinomycetes*, ed. G. Szabo, S. Biro, M. Goodfellow, pp. 3–14. Budapest: Akademiai Kiado

43. Horvath, B., Kondorosi, E., John, M., Schmidt, J., Török, I., et al. 1986. Organization, structure and symbiotic function of *Rhizobium meliloti* nodulation genes determining host specificity for alfalfa. *Cell* 46:335–43

44. Hrazdina, G., Kreuzaler, F., Hahlbrock, K., Grisebach, H. 1976. Substrate specificity of flavanone synthase from cell suspension cultures of parsley and structure of release products *in vitro*. *Arch Biochem. Biophys.* 175:392–99

45. Huber, M. L. B., Paschal, J. W., Leeds, J. P., Kirst, H. A., Wind, J. A., et al. 1990. Branched chain fatty acids produced by mutants of *Streptomyces fradiae*, putative precursors of the lactone ring of tylosin. *Antimicrob. Agents Chemother.* 34:1535–41

46. Hunter, I. S., Baumberg, S. 1989. Molecular genetics of antibiotic formation. In *Microbial Products: New Approaches*, ed. S. Baumberg, I. Hunter, M. Rhodes, pp. 121–62. Cambridge: CUP

47. Ikeda, H., Kotaki, H., Ōmura, S. 1987. Genetic studies of avermectin biosynthesis in *Streptomyces avermitilis*. *J. Bacteriol.* 169:5615–21

48. Ishida, I., Obinata, M., Deguchi, T. 1987. Molecular cloning and nucleotide sequence of cDNA encoding hydroxyindole *O*-methyltransferase of bovine pineal glands. *J. Biol. Chem.* 262:2895–99

49. Jackowski, S., Murphy, C. M., Cronan, J. E., Rock, C. O. 1989. Acetoacetyl-acyl carrier protein synthase: a target for the antibiotic thiolactomycin. *J. Biol. Chem.* 264:7624–29

50. Johnston, A. W. B. 1989. The symbiosis between *Rhizobium* and legumes. In *Genetics of Bacterial Diversity*, ed. D. A. Hopwood, K. F. Chater, pp. 393–414. London: Academic

51. Kauppinen, S., Siggaard-Andersen, M., von Wettstein-Knowles, P. 1988. β-ketoacyl-ACP synthase I of *Escherichia coli*: nucleotide sequence of the *fabB* gene and identification of the cerulenin binding residue. *Carlsberg Res. Commun.* 53:357–70

52. Kinashi, H., Shimaji, M., Sakai, A. 1987. Giant linear plasmids in *Streptomyces* which code for antibiotic biosynthesis genes. *Nature* 328:454–56

53. Kinoshita, K., Takenaka, S., Hayashi, M. 1988. Isolation of proposed intermediates in the biosynthesis of mycinamicins. *J. Chem. Soc., Chem. Commun.* 943–49

54. Koes, R. E., Spelt, C. E., Mol, J. N. M. 1989. The chalcone synthase multigene family of *Petunia hybrida* (V30): differential, light-regulated expression during flower development and UV-light induction. *Plant Mol. Biol.* 12:213–25

55. Kondo, S., Yasui, K., Natsume, M., Katayama, M., Marumo, S. 1988. Isolation, physico-chemical properties and biological activity of pamamycin-607, an aerial mycelium-inducing substance from *Streptomyces alboniger*. *J. Antibiot.* 41:1196–204

56. Kreuzaler, F., Ragg, H., Heller, W., Tesch, R., Witt, I., et al. 1979. Flavanone synthase from *Petroselinum hortense*: molecular weight, subunit composition, size of messenger RNA and absence of pantetheinyl residue. *Eur. J. Biochem.* 99:89–96

57. Libertini, L. J., Smith, S. 1979. Synthesis of long chain acyl-enzyme thioesters by modified fatty acid synthetases and their hydrolysis by a mammary gland thioesterase. *Arch. Biochem. Biophys.* 192:47–60

58. Lynen, F. 1980. On the structure of fatty acid synthetase of yeast. *Eur. J. Biochem.* 112:431–42

59. Malpartida, F., Hopwood, D. A. 1984. Molecular cloning of the whole biosynthetic pathway of a *Streptomyces* antibiotic and its expression in a heterologous host. *Nature* 309:462–64

60. Malpartida, F., Hopwood, D. A. 1986. Physical and genetic characterisation of the gene cluster for the antibiotic acti-

norhodin in *Streptomyces coelicolor* A3(2). *Mol. Gen. Genet.* 205:66–73

61. Malpartida, F., Hallam, S. E., Kieser, H. M., Motamedi, H., Hopwood, D. A., et al. 1987. Homology between *Streptomyces* genes coding for synthesis of different polyketides used to clone antibiotic biosynthetic genes. *Nature* 325:818–21

62. Mann, J. 1987. *Secondary Metabolism.* Oxford: Clarendon

63. Martín, J. F., Liras, P. 1989. Organization and expression of genes involved in the biosynthesis of antibiotics and other secondary metabolites. *Annu. Rev. Microbiol.* 43:173–206

64. McCarthy, A. D., Hardie, D. G. 1984. Fatty acid synthase—an example of protein evolution by gene fusion. *Trends Biochem. Sci.* 9:60–63

65. Mikkelsen, J., Højrup, P., Rasmussen, M. M., Roepstorff, P., Knudsen, J. 1985. Amino acid sequence around the active-site serine residue in the acyltransferase domain of goat mammary fatty acid synthetase. *Biochem. J.* 227:21–27

66. Mohamed, A. H., Chirala, S. S., Mody, N. H., Huang, W-Y., Wakil, S. J. 1988. Primary structure of the multifunctional α subunit protein of yeast fatty acid synthase derived from *FAS2* gene sequence. *J. Biol. Chem.* 263:12315–25

67. Motamedi, H., Hutchinson, C. R. 1987. Cloning and heterologous expression of a gene cluster for the biosynthesis of tetracenomycin C, the anthracycline antitumor antibiotic of *Streptomyces glaucescens. Proc. Natl. Acad. Sci. USA* 84:4445–49

68. Naggert, J., Witkowski, A., Mikkelsen, J., Smith, S. 1988. Molecular cloning and sequencing of a cDNA encoding the thioesterase domain of the rat fatty acid synthetase. *J. Biol. Chem.* 263:1146–50

69. Nakagawa, A., Konda, Y., Hatano, A., Harigaya, Y., Onda, M., Ōmura, S. 1988. Structure and biosynthesis of novel antibiotics, aurantinins A and B produced by *Bacillus aurantinus. J. Org. Chem.* 53:2660–61

70. Niesbach-Klösgen, U., Barzen, E., Bernhardt, J., Rohde, W., Schwarz-Sommer, Z., et al. 1987. Chalcone synthase genes in plants: a tool to study evolutionary relationships. *Mol. Evol.* 26:213–25

71. Ohlrogge, J. B., Kuo, T. M. 1985. Plants have isoforms for acyl carrier protein that are expressed differently in different tissues. *J. Biol. Chem.* 260:8032–37

72. Ōmura, S. 1981. Cerulenin. *Methods Enzymol.* 72:520–32

73. Overath, P., Stumpf, P. K. 1964. Fat metabolism in higher plants. XXIII. Properties of a soluble fatty acid synthetase from avocado mesocarp. *J. Biol. Chem.* 239:4103–10

74. Poulose, A. J., Kolattukudy, P. E. 1983. Sequence of a tryptic peptide from the NADPH binding site of the enoyl reductase domain of fatty acid synthase. *Arch. Biochem. Biophys.* 220:652–56

75. Poulose, A. J., Bonsall, R. F., Kolattukudy, P. E. 1984. Specific modification of the condensation domain of fatty acid synthase and the determination of the primary structure of the modified active site peptides. *Arch. Biochem. Biophys.* 230:117–28

76. Raetz, C. R. H. 1986. Molecular genetics of membrane phospholipid synthesis. *Annu. Rev. Genet.* 20:253–95

77. Ratledge, C. 1984. Metabolism of iron and other metals by mycobacteria. In *The Mycobacteria. A Sourcebook, Part A,* ed. G. P. Kubica, L. G. Wayne, pp. 603–27. New York: Marcel Dekker

78. Rebollo, A., Gil, J. A., Liras, P., Asturias, J. A., Martin, J. F. 1989. Cloning and characterization of a phosphate-regulated promoter involved in phosphate control of candicidin biosynthesis. *Gene* 79:47–58

79. Reichenbach, H., Gerth, K., Irschik, H., Kunze, B., Höfle, G. 1988. Myxobacteria: a source of new antibiotics. *Trends Biotechnol.* 6:115–21

80. Reimold, U., Kröger, M., Kreuzaler, F., Hahlbrock, K. 1983. Coding and 3' non-coding nucleotide sequence of chalcone synthase mRNA and assignment of amino acid sequence of the enzyme. *EMBO J.* 2:1801–5

81. Rhodes, P. M., Winskill, N., Friend, E. J., Warren, M. 1981. Biochemical and genetic characterization of *Streptomyces rimosus* mutants impaired in oxytetracycline biosynthesis. *J. Gen. Microbiol.* 124:329–38

82. Richardson, M. A., Kuhstoss, S., Huber, M., Ford, L., Godfrey, O., et al. 1989. Cloning of genes involved in spiramycin biosynthesis from *Streptomyces ambofaciens.* See Ref. 37, pp. 40–43

83. Rudd, B. A. M., Hopwood, D. A. 1979. Genetics of actinorhodin biosynthesis by *Streptomyces coelicolor* A3(2). *J. Gen Microbiol.* 114:35–43

84. Ryder, T. B., Hedrick, S. A., Bell, J. N., Liang, X., Clouse, S. D., Lamb, C. J. 1987. Organization and differential activation of a gene family encoding the

plant defense enzyme chalcone synthase in *Phaseolus vulgaris. Mol. Gen. Genet.* 210:219–33

85. Schofield, P. R., Watson, J. M. 1986. DNA sequence of *Rhizobium trifolii* nodulation genes reveals a reiterated and potentially regulatory sequence preceeding *nodABC* and *nodFE. Nucleic Acids Res.* 14:2891–03

86. Schröder, G., Brown, J. W. S., Schröder, J. 1988. Molecular analysis of resveratrol synthase cDNA genomic clones and relationship with chalcone synthase. *Eur. J. Biochem.* 172:161–69

87. Schüz, R., Heller, W., Hahlbrock, K. 1983. Substrate specificity of chalcone synthase from *Petroselinum hortense. J. Biol. Chem.* 258:6730–34

88. Schweizer, E., Müller, G., Roberts, L. M., Schweizer, M., Rösch, J., et al. 1987. Genetic control of fatty acid synthetase biosynthesis and structure in lower fungi. *Fat Sci. Technol.* 89:570–77

89. Schweizer, M., Roberts, L. M., Höltke, H-J., Takabayashi, K., Höllerer, E., et al. 1986. The pentafunctional FAS1 gene of yeast: its nucleotide sequence and order of catalytic domains. *Mol. Gen. Genet.* 203:479–86

89a. Schweizer, M., Takabayashi, K., Laux, T., Beck K- F., Schreglmann, R. 1989. Rat mammary gland fatty acid synthase: localization of the constituent domains and two functional polyadenylation/termination signals in the cDNA. *Nucleic Acids Res.* 17:567–586

90. Scrutton, N. S., Berry, A., Perham, R. N. 1990. Redesign of the coenzyme specificity of a dehydrogenase by protein engineering. *Nature* 343:38–43

91. Seno, E. T., Baltz, R. H. 1989. Structural organization and regulation of antibiotic biosynthesis and resistance genes in actinomycetes. In *Regulation of Secondary Metabolism in Actinomycetes,* ed. S. Shapiro, pp. 1-48. Boca Raton, Florida: CRC

92. Shearman, C. A., Rossen, L., Johnston, A. W. B., Downie, J. A. 1986. The *Rhizobium leguminosarum* nodulation gene *nodF* encodes a polypeptide similar to acyl-carrier protein and is regulated by *nodD* plus a factor in pea root exudate. *EMBO J.* 5:647–52

93. Sherman, D. H., Malpartida, F., Bibb, M. J., Bibb, M. J., Hopwood, D. A., et al. 1989. Cloning and analysis of genes for the biosynthesis of polyketide antibiotics in *Streptomyces* species. In *Proc. 8th Int. Biotechnol. Symp.* ed. G.

Durand, L. Bobichon, J. Florent, 1:123–37. Paris: Soc. Fr. Microbiol.

94. Sherman, D. H., Malpartida, F., Bibb, M. J., Kieser, H. M., Bibb, M. J., Hopwood, D. A. 1989. Structure and deduced function of the granaticin-producing polyketide synthase gene cluster of *Streptomyces violaceoruber* TÜ22. *EMBO J.* 8:2717–25

95. Shimakata, T., Stumpf, P. K. 1982. Isolation and function of spinach leaf β-ketoacyl-[acyl-carrier-protein] synthases. *Proc. Natl. Acad. Sci. USA* 79:5808–12

96. Siggaard-Andersen, M. 1988. Role of *Escherichia coli* β-ketoacyl-ACP synthase I in unsaturated fatty acid synthesis *Carlsberg Res. Commun.* 53:371–79

97. Simoni, R. D., Criddle, R. S., Stumpf, P. K. 1967. Fat metabolism in higher plants. XXXI Purification and properties of plant and bacterial acyl carrier proteins. *J. Biol. Chem.* 242:573–81

98. Spaink, H. P., Weinman, J., Djordjevic, M. A., Wijffelman, C. A., Okker, R. J. H., Lugtenberg, J. J. 1989. Genetic analysis and cellular localization of the *Rhizobium* host specificity-determining NodE protein. *EMBO J.* 8:2811–18

99. Stanzak, R., Matsushima, P., Baltz, R. H., Rao, R. N. 1986. Cloning and expression in *Streptomyces lividans* of clustered erythromycin biosynthesis genes from *Streptomyces erythreus. Biol technology* 4:229–32

100. Steyn, P. S. 1980. *The Biosynthesis of Mycotoxins: A Study in Secondary Metabolism.* New York: Academic

101. Stoessl, A. 1982. Biosynthesis of phytoalexins. In *Stilbenes and Isoflavonoids as Phytoalexins,* ed. J. A. Bailey, J. W. Mansfield, pp. 133–80. Glasgow: Blackie

102. Stonesifer, J., Matsushima, P., Baltz, R. H. 1986. High frequency conjugal transfer of tylosin genes and amplifiable DNA in *Streptomyces fradiae. Mol. Gen. Genet.* 202:348–55

103. Streicher, S. L., Ruby, C. L., Paress, P. S., Sweasy, J. B., Danis, S. J., et al. 1989. Cloning genes for avermectin biosynthesis in *Streptomyces avermitilis.* See Ref. 37, pp. 44–52

104. Stutzman-Engwall, K. J., Hutchinson, C. R. 1989. Multigene families for anthracycline antibiotic production in *Streptomyces peucetius. Proc. Natl. Acad. Sci. USA* 86:3135–39

105. Surin, B. P., Downie, J. A. 1989. *Rhizobium leguminosarum* genes required for expression and transfer of host

specific nodulation. *Plant Mol. Biol.* 12:19–29

106. Tanaka, H., Kuroda, A., Marusawa, H., Hatanaka, H., Kino, T., et al. 1987. Structure of FK-506: A novel immunosuppressant isolated from *Streptomyces*. *J. Am. Chem. Soc.* 109:5031–33

107. Therisod, H., Weissborn, A. C., Kennedy, E. P. 1986. An essential function for acyl carrier protein in the biosynthesis of membrane-derived oligosaccharides of *Escherichia coli. Proc. Natl. Acad. Sci. USA* 83:7236–40

108. Thompson, C. J., Kieser, T., Ward, J. M., Hopwood, D. A. 1982. Physical analysis of antibiotic-resistance genes from *Streptomyces* and their use in vector construction. *Gene* 20:51–62

109. Tsukamoto, Y., Wakil, S. J. 1988. Isolation and mapping of the β-hydroxyacyl dehydratase activity of chicken liver fatty acid synthase. *J. Biol. Chem.* 263:16225–29

110. Tuan, J. S., Weber, J. M., Staver, M. J., Leung, J. O., Donadio, S., Katz, L. 1990. Cloning of genes involved in erythromycin biosynthesis from *Saccharopolyspora erythraea* using a novel actinomycete-*Escherichia coli* cosmid. Gene. 90:21–29

111. Vara, J., Lewandowska-Skarbek, M., Wang, Y-G., Donadio, S., Hutchinson, C. R. 1989. Cloning of genes governing the deoxysugar portion of the erythromycin biosynthesis pathway in *Saccharopolyspora erythraea (Streptomyces erythreus). J. Bacteriol.* 171:5872–81

112. Weber, J. M., Losick, R. 1988. The use of a chromosome integration vector to map erythromycin resistance and production genes in *Saccharopolyspora erythraea (Streptomyces erythraeus). Gene* 68:173–80

113. Weber, J. M., Wierman, C. K., Hutchinson, C. R. 1985. Genetic analysis of erythromycin production in *Streptomyces erythreus. J. Bacteriol.* 164:425–33

114. Wierenga, R. K., Hol. W. G. J. 1983. Predicted nucleotide-binding properties of p21 protein and its cancer-associated variant. *Nature* 302:842–44

115. Wiesner, P., Beck, J., Beck, K-F., Ripka, S., Müller, G., et al. 1988. Isolation and sequence analysis of the fatty acid synthetase *FAS2* gene from *Penicillium patulum. Eur. J. Biochem.* 177:69–79

116. Yuan, Z., Liu, W., Hammes, G. G. 1988. Molecular cloning and sequencing of DNA complementary to chicken liver fatty acid synthase mRNA. *Proc. Natl. Acad. Sci. USA* 85:6328–31

117. Yue, S., Duncan, J. S., Yamamoto, Y., Hutchinson, C. R. 1987. Macrolide biosynthesis. Tylactone formation involves the processive addition of three carbon units. *J. Am. Chem. Soc.* 109:1253–55

118. Zamir, L. O. 1980. The biosynthesis of patulin and penicillic acid. In *The Biosynthesis of Mycotoxins: A Study in Secondary Metabolism*, ed. P. S. Steyn, pp. 223–68. New York: Academic

119. Zhang, H-I., He, X-g., Adefarati, A., Gallucci, J., Cole, S. P., et al. 1990. Mutactin, a novel polyketide from *Streptomyces coelicolor*. Structure and biosynthetic relationship to actinorhodin. *J. Organ. Chem.* 55:1682–84

Annu. Rev. Genet. 1990. 24:67–90

GENETICS OF EXTRACELLULAR PROTEIN SECRETION BY GRAM-NEGATIVE BACTERIA

A. P. Pugsley, C. d'Enfert, I. Reyss, and M. G. Kornacker

Unité de Génétique Moléculaire, UA CNRS 1149, Institut Pasteur, 25 rue du Dr. Roux, 75724 Paris Cedex 15, France

KEY WORDS: general export pathway, bacterial membranes, protein targeting, secretion genes, signal peptide

CONTENTS

67

0066-4197/90/1215-0067$02.00

PULLULANASE AS A MODEL EXTRACELLULAR PROTEIN

Gram-negative bacteria export proteins to the cell envelope (the inner and outer membranes and the periplasm between them) or secrete them into the medium. The latter phenomenon is restricted to a small number of proteins; outer-membrane and periplasmic proteins are not found outside the cell in quantities that cannot be accounted for by spontaneous partial lysis (59). Extracellular proteins produced by Gram-negative bacteria have to cross two lipid bilayer membranes, a property they share with nuclear-encoded proteins that are imported into mitochondria or plastids (60).

Until recently, extracellular protein secretion by bacteria received little attention compared to the considerable effort devoted to understanding how *Escherichia coli* exports proteins to the cell envelope. It is now clear that *E. coli* exports most of its cell-envelope proteins by the general export pathway composed of certain cytoplasmic and inner membrane proteins (export factors) that function together to ensure efficient recognition of and translocation across the inner membrane. Most of these export factors were initially identified by studying mutations that block or severely impede protein export (67). Exported proteins have at least one uninterrupted sequence of 12 or more hydrophobic or neutral residues, the signal sequence, that directs them to and facilitates their insertion into the inner membrane (60). Proteins with a single, N-terminal signal sequence (signal peptide) that is proteolytically removed (by signal peptidase) during translocation are either released into the periplasm or inserted into the outer membrane. Unprocessed signal sequences remain embedded in the inner membrane, with the final protein topology being determined by the orientation and number of alternating signal sequences (membrane anchors) and stop transfer signals in the protein (see 60 for review).

This review is primarily concerned with how the general export pathway has been modified or extended to enable proteins to be selectively and efficiently secreted beyond the outer membrane. Certain pathogenic strains of *E. coli* (but not the laboratory K-12 strain) secrete extracellular α-hemolysin or heat-stable enterotoxin (HST). α-hemolysin is one of a growing number of characterized extracellular proteins that do not have signal peptides and that are secreted by a general export pathway-independent mechanism (48) described at the end of this review. HST, on the other hand, has a signal peptide, but its secretion pathway has not been characterized (59). Studies on extracellular protein secretion have concentrated instead on hydrolases and toxins produced by bacteria other than *E. coli*. The genes for over 30 such proteins have now been cloned and characterized; their sequences indicate that most encode precursors with signal peptides similar to those found in precursors of periplasmic or outer-membrane proteins. Most extracellular proteins do not have any other region of significant hydrophobicity that might function as a stop transfer-membrane anchor signal. Furthermore, with a few exceptions discussed below, the expression of these genes in *E. coli* leads to the accumulation of the soluble, mature polypeptides in the periplasm, rather than to their extracellular secretion (reviewed in 59).

Signal peptides on extracellular protein precursors should be recognized by general export factors and thereby initiate translocation across the inner membrane and into the periplasm (see below). *E. coli* may thus be unable to secrete these proteins extracellularly because it lacks secretion factors specifically required for their translocation across the outer membrane. The secretion process might thus be represented as a two-step process (Figure 1a). The model is supported by the demonstration of bona fide periplasmic intermediates in the extracellular secretion of heat-labile enterotoxin by *Vibrio cholerae* (30) and of aerolysin by *Aeromonas salmonicida* (77). The model predicts that factors necessary for the second step in the secretion pathway should be located in the outer membrane (or in the periplasm) and that *E. coli* producing these factors might be able to secrete proteins extracellularly.

Much of this review is devoted to studies on the secretion of the enzyme pullulanase produced by Gram-negative bacteria belonging to the genus Klebsiella. Pullulanases cleave $\alpha(1-6)$ linkages in branched starch to release linear dextrins composed of $\alpha(1-4)$-linked glucose residues. Pullulan, an $\alpha(1-6)$-linked maltotriose polymer that can be used by Klebsiella as a carbon source, is too large to diffuse across the outer membrane, and so pullulanase must be located either at the cell surface (with its catalytic site exposed in the medium) or free in the medium. In fact, the ca 117kd pullulanase

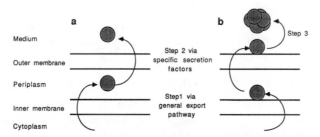

Figure 1   **a:** Simple model for two-step extracellular protein secretion by an extended, signal peptide-dependent, general export pathway. The model is based on evidence that periplasmic intermediates may occur naturally in certain systems and when extracellular protein release is blocked by mutations (see text). The model is not meant to imply that true periplasmic intermediates are in a folded conformation before they enter step 2, although this may be the case. **b:** Modified version of the two-step model that takes into account specific features unique to pullulanase (see text).

polypeptide is first anchored to the cell surface (= exposition) and is then slowly released into the medium (17, 75). Synthesis of the enzyme increases when maltose, pullulan, or starch are present, and is repressed by glucose (56, 75).

We chose pullulanase as a model extracellular protein for two reasons: the producing strains are closely related to *E. coli* and its synthesis seems to be regulated in the same way as a number of genes involved in maltodextrin metabolism common to *E. coli* and Klebsiella (reviewed in 69). We reasoned that these features would facilitate the reconstitution of the pullulanase-secretion pathway in *E. coli* K-12, the bacterium of choice for molecular studies on complex processes such as protein targeting and the only bacterium in which the general protein-export pathway has been extensively characterized (reviewed in 67). In the following sections, we review the current status of the genetic analysis of pullulanase secretion and then discuss the similarities and differences between it and the secretion of other proteins by other Gram-negative bacteria.

SIGNAL PEPTIDES IN EXTRACELLULAR SECRETION

Prepullulanase has a Signal Peptide

The pullulanase structural gene *(pulA)* of strain ATCC15050 of *Klebsiella plantolytica* (formerly *K. pneumoniae*) was found to reside in a 6.1 kb-*Eco*RI fragment of chromosomal DNA whose introduction into *E. coli* led to maltose-dependent pullulanase synthesis, as in Klebsiella (52). Essentially identical results were reported independently by Takizawa & Murooka using a slightly larger DNA fragment from strain W70 of *K. aerogenes* (70). Further studies of the cloned ATCC15050 DNA revealed several features important to the characterization of the pullulanase secretion pathway. One of these, deduced first from the DNA sequence of the 5' end of *pulA* and then confirmed experimentally, was that the predicted first 19 amino acids of pullulanase form a typical signal peptide (8, 61).

Signal Peptides as Extracellular Secretion Signals?

Although all signal peptides have the same basic features (a positively charged or neutral N-terminus and a highly hydrophobic, uncharged central region followed by a processing site for either of two different signal peptidases), almost no two signal peptides have the same primary sequence (60). Nevertheless, probably all are recognized by at least one general export factor, a signal peptide receptor. Do signal peptides of extracellular proteins initiate translocation across the inner membrane via the general export pathway, as predicted by the two-step secretion model, or do their signal peptides contain specific information that deflects them into an alternative, extracellular protein-specific pathway? Several lines of mainly circumstantial evidence argue that the general export pathway is indeed used by extracellular proteins. For example, signal peptides of extracellular proteins do not appear to have special features distinguishing them from "classical" signal peptides, and they are efficiently recognized by the general export pathway in *E. coli* (see above). Furthermore, signal peptides are usually processed early during translocation or even cotranslationally (60), and are thus unlikely to influence subsequent events such as translocation across the outer membrane.

Thus, it is unlikely that signal peptides of extracellular protein precursors could contain specific information for extracellular secretion. How can this be tested directly? One approach replaces the signal peptide of an extracellular protein by one from an outer-membrane or periplasmic protein. We predict that such a change would not affect extracellular secretion. Another approach tests whether conditions that block the general export pathway also block extracellular secretion. We have shown that processing of prepullulanase can be inhibited by expressing a *malE-lacZ* gene fusion whose product jams the general export machinery (61), possibly at a site involving the *secY(prlA)* gene product (5), and conversely that the products of certain *pulA-lacZ* gene fusions inhibit processing of other exported precursors in an apparently identical fashion (13). However, these experiments were performed in strains of *E. coli* devoid of most of the pullulanase-specific secretion factors (see below) and therefore cannot be interpreted as showing that pullulanase is normally exported by the general export pathway. Likewise, prepullulanase processing requires the *secA* gene product (A. P. Pugsley et al, unpublished data), but this result is again difficult to interpret because the *secA* mutation almost certainly affects the export of pullulanase secretion factors located in the cell envelope (see below).

Pullulanase is a Lipoprotein

The pentapeptide comprising the last four residues of the pullulanase signal peptide (Leu-Leu-Ser-Gly) and the first residue of mature pullulanase (Cys) is typically found in the same position in a subclass of exported bacterial proteins, the lipoproteins (reviewed in 64). These proteins are exported via the general export pathway (74) and are processed by lipoprotein signal peptidase specific for precursors containing the conserved pentapeptide in which Cys^{+1} is modified to glycerylcysteine and then eventually fatty acylated (64). Pullulanase is anchored to the cell surface solely by these fatty acids prior to its release into the medium. Furthermore, free extracellular pullulanase exists as large protein micelles that are held together by the fatty acids (38, 54, 62; A. P. Pugsley et al, unpublished data). These features are incorporated into a modified model for extracellular secretion (Figure 1b). The fatty acylation of pullulanase should prevent it from being released as a soluble periplasmic intermediate at the end of step 1, but steps 1 and 2 in pullulanase secretion may nevertheless be fundamentally similar to those depicted in Figure 1a as applying to other proteins secreted by an extended general export pathway.

Klebsiella species produce approximately 20 lipoproteins of which only pullulanase appears to be anchored to the cell surface or released extra-cellularly. Studies to determine the role of the fatty acylated cysteine are discussed below, but it is interesting to note that pullulanase is not unique because at least four other types of lipoproteins produced by Gram-negative bacteria are similarly located. The best characterized of these are the TraT proteins encoded by conjugative plasmids of *E. coli* and related bacteria (57). TraT proteins are at least partially exposed on the cell surface (49) but it is not known whether the entire polypeptide is thus exposed (in which case the fatty acids would be the sole membrane anchor) or whether the polypeptide chain also spans the outer membrane. Other possible cell-surface lipoproteins include the H8 antigen of uropathogenic Neisseria (3), a chitobiase produced by *Vibrio harveyii* (34), and an endoglucanase produced by *Fibrobacter succinogenes* (51). The endoglucanase is of particular interest because it, like pullulanase, is found in both a cell-associated form and as large extracellular complexes that might be pullulanaselike protein micelles (51).

GENETIC ANALYSIS OF PULLULANASE SECRETION FACTORS

Although expression of the cloned *pulA* gene from ATCC15050 in *E. coli* K-12 resulted in pullulanase synthesis, the enzyme was not exposed at the cell surface or released into the medium (52). Instead, correctly processed, fatty acylated enzyme accumulated on the periplasmic side of the inner membrane (62), indicating that *E. coli* probably lacks specific factors required to lead pullulanase from this intermediate location across the outer membrane and ultimately into the medium.

The cloning of genes required for the second step in pullulanase secretion (exposition) was facilitated by another important feature revealed by analysis of the DNA cloned from ATCC15050, namely that like other genes in the maltose regulon of *E. coli* and *K. oxytoca* UNF5023 (formerly *K. pneumoniae*) (69), *pulA* expression in vivo depends on the presence of maltose and on a functional *malT* allele coding for the maltose regulon-activator protein (8). As expected, replacement of the promoter of the chromosomal *malPQ* operon by a DNA fragment carrying *pulAp* led to MalT-dependent synthesis of amylomaltase, the *malQ* gene product. Surprisingly, however, amylomaltase synthesis was also MalT-dependent when the same fragment was inserted in reverse orientation, indicating the presence of another MalT-dependent pro-

moter (originally called *malXp* and subsequently renamed *pulCp*) adjacent to
and divergent from *pulAp* (8). We reasoned that *pulCp* might control the
expression of pullulanase-secretion genes that would be required only when
pulA was expressed, and that these genes were missing from the cloned
fragment of ATCC15050 DNA. Indeed, subsequent experiments in strain
ATCC15050 showed that the expression of *pulA* alone (under *lac* promoter
control) was sufficient for pullulanase synthesis but not for cell-surface
exposition; expression of at least one maltose-inducible gene was required for
the latter to occur (12). We therefore introduced a plasmid gene bank from
strain UNF5023 of *K. oxytoca* into *E. coli* and selected for growth on pullulan
as the sole carbon source. The plasmid derived from one such clone contained
22 kb of DNA derived from *K. oxytoca* (16). This clone exhibited both
cell-surface pullulanase exposition and subsequent extracellular release (16).

The cloned DNA fragment was reduced to 19.2 kb without affecting either
of these features, but transposons inserted almost anywhere throughout this
fragment abolished either pullulanase synthesis or exposition (Figure 2). This
result was independently confirmed by a search for nonsecreting mutants in a
transposon Tn*10*-mutagenized bank of strain K21 of *K. plantolytica* (39). All
but 2 of the 17 independent mutations characterized were located upstream or
downstream of *pulA* within a 22.5 kb-fragment of the chromosome. None of
these mutations appeared to have any effect other than preventing pullulanase
from reaching the cell surface or the medium.

In subsequent studies, we concentrated on the second step in the secretion

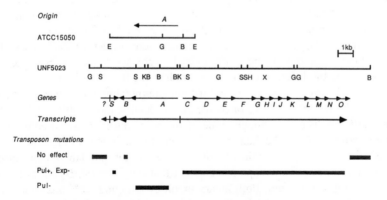

Figure 2 Genetic map of the cloned chromosomal DNA fragments from Klebsiella strains
ATCC15050 and UNF5023 showing the location and orientation of the *pul* genes (italics) and the
probable organization of the transcription units; putative MalT-dependent transcripts are shown in
thick arrows. Restriction endonuclease sites are indicated as follows: G, *Bgl*II; S, *Sma*I; K, *Kpn*I;
B, *Bam*HI; H, *Hind*III; X, *Xho*I. Also shown is a summary of the effects of transposon Tn5 and
miniMu insertions in the cloned DNA from UNF5023. (The figure is based on sequence and other
data presented in references 8, 12, 14–16, 52, 63 and 65 and on unpublished results.)

pathway leading to the cell surface-anchored form of the enzyme, rather than extracellular release. The reasons were that exposition is easy to detect and quantitate by a number of enzymatic and cytochemical techniques, the extracellular enzyme is unstable (62) and *E. coli* naturally releases outer-membrane blebs, especially after the end of exponential growth when pullulanase release is maximal.

An Operon of Secretion Genes in the Maltose Regulon

Alignment of the DNA fragment from UNF5023 with that from ATCC15050 revealed that most of the additional DNA in the fragment derived from UNF5023 was located downstream from *pulCp* (Figure 2). MalT-dependent transcription initiated at *pulCp* was shown to extend over at least 11kb by analyzing *malQ* and *lacZ* operon fusions (16, 63). Subsequent identification of secretion genes in the operon was based on sequencing, subcloning, and deletion analysis and on the effects of the insertion of linkers or Tn*tac-1*, a transposon with an inducible *tac* promoter at one extremity that abolishes its polar effect on the transcription of downstream genes (10). The products of most of the 13 genes in the operon (Figures 2 and 3) have been visualized by at least one of four different techniques and their location determined by subcellular fractionation (Figure 3). As an alternative approach, fusions between almost every gene in the operon and the reporter protein gene *phoA* (coding for alkaline phosphatase; see 67) were constructed by Tn*phoA* mutagenesis. The detection of hybrid proteins whose sizes corresponded to those predicted from the positions of the Tn*phoA* confirmed that the genes in the *pulC* operon were capable of being expressed and that the predicted translation initiation sites and reading frames were correct. In some cases, the subcellular locations of the PhoA hybrids were determined as possible indications of the location of the gene product themselves (Figure 3).

Our analysis of the operon is still incomplete (notably in the absence of any data on the function of *pulE* or on the direct identification of the *pulF*, *pulH*, and *pulK* products). Although some of the techniques employed could give rise to artefacts (65), the data are sufficiently unambiguous to indicate that at least 12 of the 13 genes in the operon are essential for pullulanase exposition, that all 12 code for envelope proteins but that only one *(pulD)* codes for an outer-membrane protein. Preliminary topology mapping with *phoA* and *lacZ* gene fusions suggest that 8 of the remaining 11 inner-membrane proteins adopt a type II configuration (60), with the C-terminal bulk of their sequences exposed in the periplasm. PulL protein appears to span the membrane once via a centrally positioned hydrophobic domain, whereas PulF and PulO proteins span the membrane several times (15, 63, 65; A. P. Pugsley et al, unpublished data).

Transcripts:

Genes:

Stem-loops and direct repeats:

	?	pulS	pulB	pulA	pulC	pulD	pulE	pulF	pulG	pulH	pulI	pulJ	pulK	pulL	pulM	pulN	pulO
Predicted protein (kDa):	?	14	19	118	31	71	55	44	15	18	13	22	36	44	18	28	25
Protein identified:																	
under lacZp or other promoter control	−	+	−	+	−	−	−	−	+	−	−	−	−	−	−	−	−
in minicells		+	−	+	+	+	+	−	−	−	−	−	−	−	+	+	−
under T7 gene 10 promoter control				+	+	+	+	−	+	−	−	−	−	−	+	+	+
by immunoblotting				+	+	+								+	+		
as alkaline phosphatase hybrid				+	+	+			+	+		+	+	+	+	+	+
Protein location as determined:																	
under lacZp or other promoter control		OM		OM IM	OM	OM											
by immunoblotting				OM IM	OM												
under exclusive T7 gene 10 promoter control				IM	IM	OM		IM	IM			IM	IM	IM	IM	IM	IM
as alkaline phosphatase hybrid					IM	P		IM	IM		IM	IM	IM				
Requirement for exposition shown by:																	
gene deletion or interruption	−	+	−	+	−	+		+	+	+						+	
nonpolar linker insertion					+	+		+	+								
nonpolar transposon insertion					−	+							+			+	−
Structural features of protein:																	
signal peptide	−	+	−	+	−	+	−										
membrane anchors		0	0	0	1	0	0	≥5	1	1	1	1	3	1	1	1	≥3
fatty acids		+		+													
predicted protein location		OM	C	IM	IM OM/P	OM/P	C	IM	IM	IM	IM	IM	IM	IM	IM	IM	IM

Figure 3 Characterization of *pul* gene products. The figure represents a map of the *pul* DNA region (not to scale) and the characteristics of the *pul* gene products, as determined by the listed techniques. Protein localization was determined by subcellular fractionation of membrane vesicles by centrifugation through isopycnic sucrose gradients. Note that the location of pullulanase varies according to the expression of other *pul* genes (see text) and cannot be determined by subcellular fractionation (62). In contrast to PulD protein, PulD-PhoA hybrids are soluble in the periplasm, probably because removal of the C-terminus of PulD blocks its ability to insert into the outer membrane (15). The figure also shows the probable organization of the *pul* transcription units, the location of palindromes, and the direct repeats located immediately 3′ of *pulO*. (Data are from references 14–16, 63, 65 and unpublished observations.)

Secretion Gene pulS is not Part of the Maltose Regulon

Transposon analysis of the cloned *pul* gene complex (Figure 2) showed that a region located downstream from *pulA* is also required for pullulanase secretion. Sequencing, subcloning, and deletion analysis of the cloned DNA stretching beyond the 3' end of *pulA* showed that a single secretion gene *(pulS)* is in fact located in this region. Surprisingly, *pulS* is transcribed in the opposite direction to *pulA* and its expression is MalT- and maltose-independent (14). *pulA* is separated from *pulS* by a region of DNA containing an open reading frame (*pulB*; Figure 2) whose gene product has not been identified and whose function, other than its noninvolvement in pullulanase exposition in *E. coli* K12, is undetermined.

pulS is the exception among the pullulanase secretion genes in that it is not part of the maltose regulon and is expressed even when pullulanase is not being produced. Indeed, pulS protein may have another function in addition to that in pullulanase secretion, and there is no obvious reason why *pulS* should be close to *pulA*. No major gene rearrangements occurred during the cloning of the secretion genes because strain ATCC15050 also carries *pulS* in the same location (14) and because Tn*10* insertions downstream from *pulA* in the chromosome of strain K21 abolish pullulanase secretion, presumably by inactivating *pulS* (39). It would be interesting to determine whether *pulS* is in the same orientation in these strains as in UNF5023 or in the opposite direction and under *pulAp* control, or whether constitutive *pulS* expression is a prerequisite for pullulanase secretion.

Are Other Genes Required for Pullulanase Secretion?

Although we have reconstituted the complete pullulanase secretion pathway in *E. coli* by introducing the cloned *pul* gene complex, both *E. coli* and Klebsiella may carry other, as yet unidentified genes (besides those coding for known components of the general export pathway) that are also required for pullulanase secretion. Indeed, two of the Tn*10* insertions in the K21 chromosome of *K. plantolytica* that prevented pullulanase secretion were apparently located outside the *pul* gene complex (39). A search for similar mutations in a bank of Tn*5*-mutagenized *E. coli* carrying a plasmid bearing the entire *pul* gene complex was unsuccessful, however, but the number of clones screened (8600) was too low to be statistically significant.

A Revised Model for Pullulanase Secretion

We can now incorporate a number of specific features into the model presented earlier for pullulanase secretion (Figure 4). As discussed, steps 1 and 2 are probably independent, with the products of the *pulA*-linked genes being required solely for step 2. Step 3 appears to be purely spontaneous because none of the characterized *pul* mutations specifically affect pullulanase release.

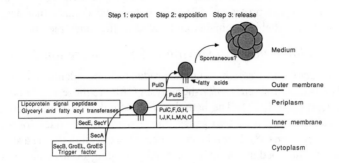

Figure 4   A refined model for pullulanase secretion based on the three-step secretion model in Figure 1. The direct involvement of SecA, SecB, SecE, SecY, GroEL, GroES and Trigger Factor proteins and of other components of the general protein export pathway remains to be tested.

The *pul* genes are as numerous as the *sec* and other genes whose products are components of the general export pathway in *E. coli* (67). In the following sections, we consider whether this is a general feature or unique to pullulanase.

GENETIC STUDIES ON THE SECRETION OF OTHER EXTRACELLULAR PROTEINS

Our approaches to characterize the pullulanase-secretion pathway have been applied to the study of many other extracellular proteins secreted by Gram-negative bacteria. The following discussion is divided into three sections dealing with proteins whose synthesis is alone sufficient for extracellular release, those that require a single, specific secretion-factor gene, and finally those that require several additional factors. The reader is reminded that we are considering only extracellular proteins synthesized as precursors with signal peptides.

Unassisted Extracellular Release

The literature contains several reports of proteins that are extracellular both when synthesized by the natural producing strain and when produced by *E. coli* expressing the cloned structural gene. In many cases one cannot yet exclude the possibility that a second gene in the cloned DNA fragment codes for a protein required for extracellular release (21) or that extracellular release is nonselective (59). The remaining examples, the IgA-specific proteases of Neisseria and Haemophilus and the serine protease of *Serratia marcescens,* are synthesized as large precursors (25, 53, 58) that insert into the outer membrane with their long, N-terminal catalytic domains initially exposed on the cell surface and then released by autocatalytic processing (53, 58) or via the action of other proteases (53). The C-terminal domain that acts as the secretion factor (helper) remains embedded in the outer membrane.

Extracellular Release Involving a Single Secretion Factor

S. marcescens and Proteus vulgaris produce almost identical extracellular hemolysins that, unlike E. coli α-hemolysin (18), are made as precursors with signal peptides (68, 71). The hemolysin structural gene is the second in an iron-regulated operon of two genes, the first of which codes for an outer-membrane protein (68). Hemolysin is released extracellularly via a short-lived cell-surface intermediate, as in the original producing strains when both genes are expressed in E. coli (68, 71), but the hemolysins remain inactive and intracellular (probably loosely associated with the periplasmic faces of the two membranes (68)) when the second gene is inactivated. Thus, this second gene codes for a secretion factor that may also be involved in hemolysin activation. The organization of these hemolysin determinants is reminiscent of that of the pul gene complex in that it ensures that the secretion factor is only made when needed. The data do not exclude the involvement of other envelope components common to both E. coli and Serratia/Proteus in hemolysin secretion, but do exclude the possible involvement of hemolysin-specific secretion factors in step 1 of the secretion pathway (see Figure 1a).

Extracellular Release Involving Several Secretion Factors

Many species of Gram-negative bacteria that secrete several extracellular proteins (e.g. Pseudomonas aeruginosa, S. marcescens, A. hydrophila, Vibrio and Erwinia sp) have been screened for mutants defective in the release of one or more proteins. One of the most striking observations repeatedly made with these mutants is that different extracellular proteins are affected to different extents, with some being totally unaffected (2, 29, 32, 33, 78). One explanation is that these bacteria may possess several fully or partially independent pathways for extracellular secretion, including some that do not involve the general export pathway (26, 44, 55) and others that fall into the two categories described above. Among the mutations that reduce the extracellular release of proteins secreted by an extended general export pathway, some have a greater effect on certain extracellular proteins than on others. A plausible and intriguing explanation is that while several proteins may share components of a common machinery for translocation across the outer membrane, other components of this machinery may be specific for one or a small subgroup of extracellular proteins.

Only a small number of genes required for extracellular secretion have been cloned, and only one ($hlyB^{V.c}$ of V. cholerae) has been sequenced. This gene is close to the hemolysin structural gene of V. cholerae, but appears to have its own promoter and may thus be expressed even when hemolysin is not being produced (1). Expression of $hlyA$-$B^{V.c}$ alone in $E. coli$ does not lead to extracellular hemolysin secretion but inactivation of $hlyB^{V.c}$ in the $V. cholerae$ chromosome causes hemolysin to accumulate in the periplasm (1). The $hlyB$ gene product contains two highly hydrophobic regions, one at the N-terminus that may be a signal peptide (1) and another near the center of the polypeptide that could anchor the protein in a type I conformation (60) in the cytoplasmic membrane. Preliminary fractionation data suggest, however, that HlyB protein is located in the outer membrane (1). Among the other characterized secretion genes, xcp-1 of $P. aeruginosa$ encodes a reportedly inner-membrane protein (4) whereas $outJ$ of $Erwinia chrysanthemi$ encodes an 83kd-protein from which the signal peptide is apparently removed when the gene is expressed in $E. coli$ (35). $outJ$ could thus encode a periplasmic or outer-membrane protein. A 68kd-periplasmic protein is absent from $outJ$ mutants, but it is not clear whether this protein is a processed form of the $outJ$ gene product or a protein whose synthesis or export is also affected by the $outJ$ mutation.

Apart from $hlyB^{Vc}$, the genes required for extracellular protein release in the bacteria discussed here do not appear to be linked to any genes coding for extracellular proteins, and may be located at several different chromosomal sites (4, 20, 33, 78). This may reflect the fact that the various extracellular proteins secreted by these bacteria are produced in response to different stimuli. Secretion genes required for more than one extracellular protein may therefore need to be expressed constitutively and, hence, no selective pressure exists for the establishment or maintenance of genetically linked systems.

POSSIBLE ROLES OF SECRETION FACTORS

Secretion Factors in the Outer Membrane and Periplasm

According to the model presented in Figure 1, proteins might be secreted extracellularly by a two-step pathway. We predicted that the second step would minimally comprise a secretion factor located in the outer membrane. Such a component has been found in all extensively characterized secretion pathways. Another feature of the model is that the secretion machinery must be able to distinguish extracellular from normally periplasmic proteins. IgA proteases and the serine protease of Serratia are specifically secreted because they and their specific secretion factors (helpers) are synthesized as a single polypeptide (see above). Relatively little is known about how proteins are inserted into the outer membrane, but it seems to be a general feature that the extreme N-terminus is on the periplasmic side of the membrane followed immediately by the first transmembrane domain. One would therefore predict that the longer N-terminal catalytic domain of the IgA and serine proteases would also be located on the periplasmic side of the outer membrane, but this is not the case. The C-terminal helper domain, which is integrated into the outer membrane, must therefore facilitate translocation of this N-terminal domain through this membrane to its penultimate location on the cell surface.

In the case of the hemolysins of Serratia and Proteus, the integral outer-membrane secretion factor must somehow distinguish between hemolysin and all other periplasmic proteins (implying that the hemolysin carries a specific sorting signal; see below) and translocate it to the cell surface. The PulD protein may perform the same function(s) in the more complex pullulanase-secretion pathway, but its sequence is unrelated to that of the hemolysin secretion factor.

The term translocation is used here in its loosest sense, and does not imply that anything is known about how extracellular proteins actually cross the outer membrane. Several different mechanisms may exist, depending on the organism and on the composition of its cell envelope. One potential advantage of the simple systems such as that for the hemolysin of Serratia is the possibility of reconstituting them in vitro and thereby learning more about how proteins cross lipid bilayer membranes. In contrast to protein transloca-tion across the inner membrane by the general export pathway (60), transloca-tion across the outer membrane does not necessarily depend on the absence of higher ordered structure in the secreted protein, since the subunits of entero-toxin of *V. cholerae* assemble in the periplasm prior to translocation across the outer membrane (31). This implies that this protein at least is translocated across the outer membrane by a fundamentally different mechanism from that of most other protein-translocation pathways (60). It will be important to determine whether other proteins secreted by an extended general export pathway adopt significant secondary or higher-ordered structure during transit through the periplasm. If they remain unfolded, one might suspect that molecular chaperones are involved, (66), similarly to fimbrial subunit secre-tion (45). So far, however, none of the characterized secretion genes has been shown to encode periplasmic proteins.

Secretion Factors in the Inner Membrane and Cytoplasm

Given that only a single outer-membrane protein is apparently required for the specific extracellular release of certain hemolysins, it is puzzling that other extracellular proteins should require many more secretion factors. Why, for example, is the pullulanase-secretion pathway so complex? Certain *pul* gene products may be specifically required to translocate the fatty acids of pullulan-ase to the cell surface, an idea we are currently testing by studying the secretion of a pullulanase derivative that is not fatty acylated. Since these fatty acids may temporarily anchor to the outer face of the inner membrane (62), one would expect pullulanase secretion factors to be similarly located, which they are. Transport to the cell surface may occur at sites where the inner and outer membranes are in contact. Since the existence of natural contact sites is questionable (37), one function of the *pul* gene products may be to form an envelope-spanning complex (and locally deform the peptidoglycan) specifically involved in pullulanase secretion. The inner-membrane secretion factors might even interact with the secreted protein as it is extruded across the inner membrane. Such close association between steps 1 and 2 in the secretion pathway, and the implied molecular chaperoning of the secreted protein by membrane-associated proteins, would, however, be incompatible with the model shown in Figure 1a, which predicts the existence of free periplasmic intermediates for other proteins secreted by an extended general export pathway.

As already discussed, it is unlikely that secretion factors such as those encoded by genes in the *pul* complex could replace all or even some of the inner membrane components of the general export pathway. However, the fact that the *pulF, pulL* and *pulO* gene products may be inner-membrane proteins with relatively large cytoplasmic domains means that their noninvolvement in secretion step 1 must be tested experimentally. It should soon be possible to develop an in vitro system based on mixed, purified components of the general export pathway and *pul* gene products to determine the requirements for pullulanase translocation across the inner membrane. It should also be possible to determine whether pullulanase exported to the end of step 1 in the secretion pathway when all other *pul* genes are not expressed can be subsequently translocated to the cell surface when these genes are turned on (15).

Are there other explanations for the presence of secretion factors in the inner membrane? One possibility is that certain pleiotropic secretion mutations affect inner-membrane proteins that are not directly involved in extracellular release but are important for membrane integrity or function and are needed, among other things, for efficient secretion. Changes in envelope composition and phage sensitivity (32, 33) could be other manifestations of such changes. This explanation probably does not apply to the *pulC-O* and *hlyB*[V,c] gene products since they are only produced when required.

Another related possibility is that inner-membrane proteins provide energy for outer-membrane translocation. This novel idea comes from recent studies by Wong & Buckley (76) showing that a proton gradient is required for aerolysin to cross the outer membrane of *A. salmonicida*. As these authors noted, there is no evidence that a proton gradient can be maintained across the outer membranes of Gram-negative bacteria, which have permanently open pores, even though some of their data would be consistent with the existence of such a gradient. More work is required to elucidate this fascinating new aspect of extracellular protein secretion.

Finally, none of the identified factors required for extracellular protein secretion have been shown to reside in the cytoplasm. However, the *pulE* gene product is predicted to be cytoplasmic because it is devoid of hydrophobic sequences that might cause it to insert into lipid bilayers. One intriguing possibility is that this protein acts as a molecular chaperone to ensure that pullulanase does not fold prior to translocation across the inner membrane; i.e. that it supplements the activities of the SecB, GroEL/GroES, and Trigger Factor proteins that perform this function for other proteins exported by the general export pathway (60, 66, 67).

EXTRACELLULAR SORTING SIGNALS

The specificity of extracellular protein secretion presumably depends on the recognition of extracellular protein(s) by components of the secretion machinery, implying that extracellular proteins have sorting signals. These signals will be easy to identify and manipulate only if they are composed of a sequence of amino acids, rather than of juxtapositioned residues from different regions in the folded polypeptide (patch signals) such as those that comprise sorting signals of some secreted proteins in higher organisms (60). Patch signals are more likely to be found in proteins that enter step 2 of the secretion pathway in folded rather than unfolded conformations (see above).

Sorting Signals in Pullulanase

Analysis of the predicted sequences of pullulanases from two strains of Klebsiella (36, 40) revealed the following potentially important features (Figure 5):

1. Eight short sequences that are found in other amylases. These are distributed throughout the central region of the pullulanase polypeptide, which might constitute a large catalytic domain (40).
2. Seven cysteine residues, of which six could be involved in the observed disulfide bridge formation (40) that might occur prior to step 2 in the secretion pathway, and the other is fatty acylated (36, 38, 54, 61, 62; A. P. Pugsley et al, unpublished data). Extracellular secretion is reduced but not abolished when Cys^{+1} is replaced by another amino acid (54; A. P. Pugsley et al, unpublished data). This indicates that pullulanase can be translocated across the outer membrane, albeit with reduced efficiency, without remaining membrane-anchored via the fatty acids. Sequence analysis showed that the signal peptide of a Cys^{+1} to Ser mutant pullulanase was processed within the hydrophobic core of the signal peptide, which may have caused the reduced efficiency of extracellular secretion (54). It should be recalled, however, that K21 of *K. plantolytica* efficiently

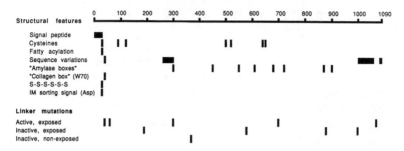

Figure 5   Some features of the pullulanase polypeptide sequence and the effects of 6 bp-linker insertions in *pulA*. Sequence variations were detected by comparing the predicted sequences of pullulanases from strains UNF5023 (40) and W70 (36). (Data are from reference 40 and unpublished observations; see text for further details.)

secretes unacylated pullulanase that is correctly processed by lipoprotein signal peptidase (38).

3. A short collagenlike sequence close to the N-terminus of pullulanase from strain W70/FG9 (9), which is missing from UNF5023 pullulanase and thus unimportant for extracellular sorting or activity (40).

4. A series of six Serine residues located close to the N-terminus, whose removal (by site-directed mutagenesis) does not affect pullulanase activity or secretion (M. G. Kornacker, unpublished data).

5. An acidic aspartate residue at position +2. Asp is also present in this position in lipoproteins of *E. coli* that are sorted to the inner membrane. Asp is proposed to be an inner-membrane sorting signal since outer-membrane lipoproteins do not have acidic residues at positions +1 through +5 (79). This would be consistent with the suggestion that pullulanase is temporarily anchored to the inner membrane at the end of step 1 in the secretion pathway. Replacement of Asp^{+2} by an uncharged Asparagine reduced but did not totally abolish cell-surface exposition (M. G. Kornacker, unpublished result), implying that the secretion machinery may override the absence of the inner-membrane sorting signal in pullulanase. Studies on the role of Asp^{+2} are continuing in our laboratory.

These analyses are complemented by linker mutagenesis and gene-fusion studies. In the former, we have inserted 6 bp-linkers at known sites in *pulA* (40) to obtain mutations that prevent exposition without affecting enzymatic activity. Their obtention should enable us to select suppressor mutations in *pul* genes coding for proteins that recognize the extracellular sorting signal(s). No such mutations have yet been obtained, however, but only one of the insertions obtained so far blocks exposition (Figure 5), suggesting that sorting is less sensitive to sequence changes than enzymatic activity.

All 3' *pulA* deletions tested so far code for highly unstable truncated pullulanase polypeptides that can be stabilized by fusing *phoA* to the end of the truncated *pulA* gene (40). These hybrids could be detected at the cell surface by immunofluorescence tests with antipullulanase serum, provided they contained at least 656 N-terminal residues of pullulanase. This suggests that secretion signals are located in the N-terminal half of the pullulanase polypeptide. However, the PhoA parts of these hybrids either did not reach the cell surface or did so only very inefficiently (13, 40, 41). This may be due to physical properties of alkaline phosphatase (probably its dimeric state). Indeed, one particular hybrid protein in which pullulanase-derived sequences were fused to β-lactamase was efficiently sorted into the pullulanase-secretion pathway and became anchored to the cell surface (41). Although this hybrid contains 834 N-terminal residues of prepullulanase, and therefore does not significantly advance our attempts to locate the sorting signal, the result shows that nothing intrinsic to normally intracellular proteins prevents them from being secreted extracellularly. It may now be possible to further define the pullulanase-sorting signal(s) by creating internal deletions in this *pulA-bla* gene fusion.

Sorting Signals in Other Extracellular Proteins

The search for sorting signals in two other extracellular proteins has been more successful than with pullulanase. Thus, results from gene-deletion studies suggest that extracellular sorting signals are located near the N-termini of exotoxin A of *P. aeruginosa* (residues 1–30; 27) and of hemolysin of Serratia (residues 1–269; 68).

SIGNAL PEPTIDE-INDEPENDENT EXTRACELLULAR SECRETION

So far, we have dealt only with extracellular proteins synthesized with signal peptides that are presumed to channel them into the general export pathway. As we have seen, this pathway is extended by supplementary factors that are probably required to translocate the proteins across the outer membrane. In cases like pullulanase, many supplementary secretion factors are required. In recent years, it has become clear that many Gram-negative bacteria have also developed or acquired additional, simpler, signal peptide-independent pathways for protein secretion. All of the novel pathways characterized so far are basically similar to the archetypical α-hemolysin-secretion pathway discovered in hemolytic *E. coli*. Recent rapid progress in studies of this and related secretion pathways has provided a clearer picture of their general characteristics.

Secretion Factors

An early observation made in studies with α-hemolysin of *E. coli* was that genes necessary for extracellular secretion are adjacent to the hemolysin structural gene ($hlyA^{E \cdot c}$)(72). This has subsequently been proven with other related systems such as the metalloprotease-secretion pathway of *E. chrysanthemi* (44) and the adenylate cyclase/hemolysin-secretion pathway of *Bordetella pertussis* (23). In the hemolysin of *E. coli*, two secretion genes ($hlyB^{E \cdot c}$ and $hlyD$) are located downstream from, and coexpressed with $hlyA$ (19). The metalloprotease system is somewhat different because there are three secretion genes (*prtD, prtE,* and *prtF*) in an operon located upstream from the protease structural genes (*prtB* and *prtC*) that have their own promoters (44). Thus in this case at least, extracellular protease could be synthesized when secretion genes are not expressed, although how these genes are regulated is not known precisely. *prtD* and *prtE* code for inner-membrane proteins that are ca 20% homologous to the *hlyB* and *hlyD* gene products, respectively (44), while *prtF* codes for an outer-membrane protein (73; P. Delepelaire, personal communication). The absence of a gene coding for an outer-membrane protein in the α-hemolysin system has until recently been difficult to explain because preliminary fractionation data indicated that HlyB and HlyD were located in the inner membrane (47). How then is hemolysin translocated across the outer membrane? This question now seems to have been answered by the observation that a gene normally present in nonhemolytic *E. coli, tolC,* is 20% homologous to *prtF* and codes for an outer-membrane protein that is absolutely required for hemolysin secretion (73).

Although the genetic data on the identification of the secretion factors in the metalloprotease and hemolysin systems are unambiguous, they do not exclude the involvement of other, as yet unidentified factors. As discussed below, the secretion signals are located at the extreme C-termini of α-hemolysin and *prtB*-encoded metalloprotease, which means that secretion is only initiated once the nascent chains are released from the ribosome. By analogy with the general export pathway, one might expect these proteins to be translocated as unfolded polypeptides, implying that they might bind to cytoplasmic chaperones in order to remain unfolded or loosely folded prior to translocation. Components of these secretion pathways are almost certainly assembled in the envelope via the general export pathway, making it difficult to distinguish between direct and indirect effects of mutations affecting general protein export on hemolysin or metalloprotease secretion.

Nevertheless, the simplest and most attractive hypothesis is that the three envelope proteins form a complex spanning both membranes and the periplasm (with local peptidoglycan deformation) that somehow translocates the extracellular proteins from the cytosol into the medium. Periplasmic intermediates have never been convincingly demonstrated and probably do not exist. HlyB$^{E.c}$ protein has two closely spaced sequences typically found in proteins that have ATP-binding and kinase activities, respectively (28, 42), which presumably explains why ATP is required for hemolysin secretion (24). This region is probably exposed on the cytoplasmic side of the inner membrane (7), and is required for HlyB activity (42), indicating that HlyB may be functionally similar to the HisP ATPase that forms part of the histidine permease in *Salmonella typhimurium* (6; reviewed in 7). ATP hydrolysis may be used to drive α-hemolysin secretion or to unfold the polypeptide prior to translocation. PrtD protein has a similar ATP-binding site and presumably performs the same function as HlyB (44).

Secretion Signals

As already mentioned, extracellular proteins secreted by the α-hemolysin-type pathways do not have N-terminal signal peptides (or any other region of high overall hydrophobicity) (11, 18, 22), but have a secretion signal located at the extreme C-terminus instead. This signal was discovered first in α-hemolysin (47) and then in PrtB protease (P. Delepelaire & C. Wandersman, in preparation) by deletion and gene-fusion techniques that removed N-terminal segments of these polypeptides. The exact sequence of amino acids constituting these secretion signals remains to be determined. That of α-hemolysin was initially reported to comprise at most the last 27 amino acids (46), but this view was later challenged by the finding that sequences located further upstream in the polypeptide increased secretion efficiency (43). Attempts to define a conserved structure in this region of α-hemolysin and related toxins that could be secreted by *E. coli* expressing *hlyB* and *hlyD*, led to the proposal that it was composed of three consecutive structural motifs (an amphiphilic helix, a region containing several charged residues, and a region rich in hydroxylated amino acids) (43). However, this organization is not present at the C-termini of the PrtB protease or the adenylate cyclase (11, 22). This is not surprising, however, because the signals are presumably recognized by specific receptors on the respective secretion factors. One would thus predict that these proteins would not be efficiently secreted by the heterologous secretion factor complexes, as appears to be the case (50; P. Delepelaire & C. Wandersman, in preparation). One would also predict that these secretion signals would be nonfunctional if they were located other than at the C-terminus (or possibly the N-terminus) of a protein because they might not remain accessible to the receptor.

CONCLUDING REMARKS

In this review, we have discussed the results of genetic studies on extracellular protein secretion by Gram-negative bacteria. We have shown that systems with varying degrees of complexity may be involved in the secretion of different extracellular proteins. Some of these systems are built onto the general protein-export pathway. The extent to which these systems are related to each other remains to be determined. Other proteins are secreted extracellularly without entering into the general export pathway. Components of the characterized systems in this group are clearly related but are not highly homologous. Still other systems remain to be characterized genetically while others may yet be discovered. Clearly, the phenomena of protein export and secretion in bacteria are far more complex than expected.

ACKNOWLEDGMENTS

We are indebted to colleagues who contributed or continue to contribute to the study of the pullulanase system, including Maxime Schwartz, who started it all, Christine Chapon, Pierre Cornelis, Susan Michaelis, Isabelle Poquet, Evelyne Richet, Olivier Raibaud, Antoinette Ryter, Dominique Vidal-Ingigliardi and Cécile Wandersman. We also thank Philippe Delepelaire, Sylvie Létoffé, and Cécile Wandersman for communicating ideas and unpublished results on α-hemolysin and metalloprotease secretion. Our work is financed by the CNRS, the Institut Pasteur and the Fondation pour la Recherche Médicale.

Literature Cited

1. Alm, R. A., Manning, P. A. 1990. Characterization of the *hlyB* gene and its role in the production of the El Tor haemolysin of *Vibrio cholerae* 01. *Mol. Microbiol.* 4:413–25
2. Andro, T., Chambost, J. P., Kotoujansky, A., Cattaneo, J., Bertheau, Y., et al. 1984. Mutants of *Erwinia chrysanthemi* defective in secretion of pectinase and cellulase. *J. Bacteriol.* 160:1199–203
3. Baehr, W., Gotschlich, E. C., Hitchcock, P. J. 1989. The virulence-associated gonococcal H-8 gene encodes 14 tandemly repeated pentapeptides. *Mol. Microbiol.* 3:49–55
4. Bally, M., Wretlind B., Lazdunski, A. 1989. Protein secretion in *Pseudomonas aeruginosa:* molecular cloning and characterization of the *xcp-1* gene. *J. Bacteriol.* 171:4342–48
5. Bieker, K. L., Silhavy, T. J. 1989. PrlA is important for the translocation of exported proteins across the cytoplasmic membrane of *Escherichia coli. Proc. Natl. Acad. Sci. USA* 86:968–72
6. Bishop, L., Agbayani, R. Jr., Ambudkar, S. V., Maloney, P. C., Ferro-Luzzi Ames, G. 1989. Reconstitution of a bacterial periplasmic permease in proteoliposomes and demonstration of ATP hydrolysis concomitant with transport. *Proc. Natl. Acad. Sci. USA* 86:6953–57
7. Blight, M. A., Holland, I. B. 1990. Structure and function of haemolysin B, P-glycoprotein and other members of a novel family of membrane translocators. *Mol. Microbiol.* 4:873–80
8. Chapon, C., Raibaud, O. 1985. Structure of two divergent promoters, located in front of the gene encoding pullulanase in *Klebsiella pneumoniae* and positively regulated by the *malT* product. *J. Bacteriol.* 164:639–45
9. Charalambous, B. M., Keen, J. N., McPherson, M. J. 1988. Collagen-like sequences stabilize homotrimers of a bacterial hydrolase. *EMBO J.* 7:2903–9
10. Chow, W-Y., Berg, D. E. 1988. Tn5tac1, a derivative of transposon Tn5 that generates conditional mutations. *Proc. Natl. Acad. Sci. USA* 85:6468–72
10a. Das, R., Robbins, P. W., eds. 1987. *Protein Transfer and Organelle Biogenesis.* San Diego: Academic
11. Delepelaire, P., Wandersman, C. 1989. Protease secretion by *Erwinia chrysanthemi.* Proteases B and C are synthesized as zymogens without a signal peptide. *J. Biol. Chem.* 284:9083–89
12. d'Enfert, C., Chapon, C., Pugsley, A.

P. 1987. Export and secretion of the lipoprotein pullulanase by *Klebsiella pneumoniae*. *Mol. Microbiol.* 1:107–16

13. d'Enfert, C., Pugsley, A. P. 1987. A gene fusion approach to the study of pullulanase export and secretion in *Escherichia coli*. *Mol. Microbiol.* 1:159–68

14. d'Enfert, C., Pugsley, A. P. 1989. *Klebsiella pneumoniae pulS* gene encodes an outer membrane lipoprotein required for pullulanase secretion. *J. Bacteriol.* 171:3673–79

15. d'Enfert, C., Reyss, I., Wandersman, C., Pugsley, A. P. 1989. Protein secretion by Gram-negative bacteria. Characterization of two membrane proteins required for pullulanase secretion by *Escherichia coli* K-12. *J. Biol. Chem.* 264:17462–68

16. d'Enfert, C., Ryter, A., Pugsley, A. P. 1987. Cloning and expression in *Escherichia coli* of the *Klebsiella pneumoniae* genes for production, surface localization and secretion of the lipoprotein pullulanase. *EMBO J.* 6:3531–38

17. Eisele, B., Rasched, I. R., Wallenfels, K. 1962. Molecular characterization of pullulanase from *Aerobacter aerogenes*. *Eur. J. Biochem.* 26:62–67

18. Felmlee, T., Pellett, S., Lee, E. Y., Welch, R. A. 1985. *Escherichia coli* hemolysin is released extracellularly without cleavage of a signal peptide. *J. Bacteriol.* 163:88–93

19. Felmlee, T., Pellett, S., Welch, R. A. 1985. Nucleotide sequence of an *Escherichia coli* chromosomal hemolysin. *J. Bacteriol.* 163:94–105

20. Filloux, A., Bally, M., Murgier, M., Wretlind, B., Lazdunski, A. 1989. Cloning of *xcp* genes located at the 55 min region of the chromosome and involved in protein secretion in *Pseudomonas aeruginosa*. *Mol. Microbiol.* 3:261–65

21. Givshov, M., Olsen, L., Molin, S. 1988. Cloning and expression in *Escherichia coli* for the gene for extracellular phospholipase A1 from *Serratia liquefaciens*. *J. Bacteriol.* 170:5855–62

22. Glaser, P., Ladant, D., Sezer, O., Pichot, F., Ullmann, A., Danchin, A. 1988. The calmodulin-sensitive adenylate cyclase of *Bordetella pertussis:* cloning and expression in *Escherichia coli*. *Mol. Microbiol.* 2:19–30

23. Glaser, P., Sakamoto, H., Bellalou, J., Ullmann, A., Danchin, A. 1988. Secretion of cyclolysin, the calmodulin-sensitive adenylate cyclase-haemolysin bifunctional protein of *Bordetella pertussis*. *EMBO J.* 7:3997–4004

24. Goebel, W., Hacker, J., Knapp, S., Then, J., Wagner, W., et al. 1984. Structure, function and regulation of the plasmid-encoded hemolysin determinant of *E. coli*. In *Plasmids in Bacteria*, ed. D. R. Helinski, S. N. Cohen, D. B. Clewell, D. A. Jackson, A. Hollaender, pp 791–805. New York: Plenum

25. Grundy, F. J., Plaut, A., Wright, A. 1987. *Haemophilus influenzae* immunoglobulin A1 protease genes: cloning by plasmid integration-excision, comparative analysis, and localization of secretion determinants. *J. Bacteriol.* 169:4442–50

26. Guzzo, J., Murgier, M., Filloux, A., Lazdunski, A. 1990. Cloning of the *Pseudomonas aeruginosa* alkaline protease gene and secretion of the protease into the medium by *Escherichia coli*. *J. Bacteriol.* 172:942–48

27. Hamood, A. N., Olson, J. C., Vincent, T. S., Iglewski, B. H. 1989. Regions of toxin A involved in toxin A excretion in *Pseudomonas aeruginosa*. *J. Bacteriol* 171:1817–24

28. Higgins, C. F., Hiles, I. D., Salmond, G. C., Gill, D. R., Downie, J. A., et al. 1986. A family of related ATP-binding subunits coupled to many distinct biological processes in bacteria. *Nature* 323:448–50

29. Hines, D. W., Saurugger, P. N., Ihler, G. M., Benedik, M. J. 1988. Genetic analysis of extracellular proteins of *Serratia marcescens*. *J. Bacteriol.* 170:4141–46

30. Hirst, T. R., Holmgren, J. 1987. Transient entry of enterotoxin subunits into the periplasm occurs during their secretion from *Vibrio cholerae*. *J. Bacteriol.* 169:1037–45

31. Hirst, T. R., Holmgren, J. 1987. Conformation of protein secreted across bacterial outer membranes: a study of enterotoxin translocation from *Vibrio cholerae*. *Proc. Natl. Acad. Sci. USA* 84:7418–22

32. Howard, S. P., Buckley, J. T. 1983. Intracellular accumulation of extracellular proteins by pleiotropic export mutants of *Aeromonas hydrophila*. *J. Bacteriol.* 154:413–18

33. Ichige, A., Oishi, K., Mizushima, S. 1988. Isolation and characterization of mutants of a marine *Vibrio* strain that are defective in production of extracellular proteins. *J. Bacteriol.* 170:3537–42

34. Jannatipour, M., Soto-Gil, R. W., Childers, L. C., Zyskind, J. W. 1987.

Translocation of *Vibrio harveyi* N, N'-diacetylchitobiase to the outer membrane of *Escherichia coli. J. Bacteriol.* 169:3785–91

35. Ji, J., Hugouvieux-Cotte-Pattat, N., Robert-Baudouy, J. 1989. Molecular cloning of the *outJ* gene involved in pectate lyase secretion by *Erwinia chrysanthemi. Mol. Microbiol.* 3:285–93

36. Katsugari, N., Takizawa, N., Murooka, Y. 1987. Entire nucleotide sequence of the pullulanase gene of *Klebsiella aerogenes* W70. *J. Bacteriol.* 169: 2301–6

37. Kellenberger, E. 1990. The "Bayer bridges" confronted with results from improved electron microscopy methods. *Mol. Microbiol.* 4:697–705

38. Kornacker, M. G., Boyd, A., Pugsley, A. P., Plastow, G. S. 1989. *Klebsiella pneumoniae* strain K21: evidence for the rapid secretion of an unacylated form of pullulanase. *Mol. Microbiol.* 3:497–503

39. Kornacker, M. G., Boyd, A., Pugsley, A. P., Plastow, G. S. 1989. A new regulatory locus of the maltose regulon in *Klebsiella pneumoniae* strain K21 identified by the study of pullulanase secretion mutants. *J. Gen. Microbiol.* 135: 397–408

40. Kornacker, M. G., Pugsley, A. P. 1990. Molecular characterization of *pulA* and its product, pullulanase, a secreted enzyme of *Klebsiella pneumoniae* UNF5023. *Mol. Microbiol.* 4:73–85

41. Kornacker, M. G., Pugsley, A. P. 1990. The normally periplasmic enzyme β-lactamase is specifically and efficiently translocated through the *Escherichia coli* outer membrane when it is fused to the cell surface enzyme pullulanase. *Mol. Microbiol.* In press

42. Koronakis, V., Koronakis, E., Hughes, C. 1988. Comparison of the haemolysin secretion protein HlyB from *Proteus vulgaris* and *Escherichia coli;* site-directed mutagenesis causing impairment of export function. *Mol. Gen. Genet.* 213:551–55

43. Koronakis, V., Koronakis, E., Hughes, C. 1989. Isolation and analysis of the C-terminal signal directing export of *Escherichia coli* hemolysin across both bacterial membranes. *EMBO J.* 8:545–605

44. Létoffé, S., Delepelaire, P., Wandersman, C. 1990. Protease secretion by *Erwinia chrysanthemi:* the specific secretion functions are analogous to those of *E. coli* α-hemolysin. *EMBO J.* 9:1375–82

45. Lindberg, F., Tennent, J. M., Hultgren, S. J., Lund, B., Normark, S. 1989. PapD, a periplasmic transport protein in P-pilius biogenesis. *J. Bacteriol.* 171: 6052–58

46. Mackman, N., Baker, K., Gray, L., Haigh, R., Nicaud, J-M., Holland, I. B. 1987. Release of a chimeric protein into the medium from *Escherichia coli* using the C-terminal secretion signal of haemolysin. *EMBO J.* 6:2835–41

47. Mackman, N., Nicaud, J-M., Gray, L., Holland, I. B. 1985. Identification of polypeptides required for the export of haemolysin 2001 from *E. coli. Mol. Gen. Genet.* 201:529–36

48. Mackman, N., Nicaud, J-M., Gray, L., Holland, I. B. 1986. Secretion of haemolysin by *Escherichia coli. Curr. Topi. Microbiol. Immunol.* 125:159–82

49. Manning, P. A., Beutin, L., Achtman, M. 1980. Outer membrane of *Escherichia coli:* properties of the F sex factor *traT* protein which is involved in surface exclusion. *J. Bacteriol.* 142: 285–94

50. Masure, R. H., Au, D. C., Gross, M. K., Donovan, M. G., Storm, D. R. 1990. Secretion of the *Bordella pertussis* adenylate cyclase from *Escherichia coli* containing the hemolysin operon. *Biochemistry* 29:140–45

51. McGavin, M. J., Forsberg, C. W., Crosby, B., Bell, A. W., Dignard, D., Thomas, D. Y. 1989. Structure of the *cel-3* gene from *Fibrobacter succinogenes* S85 and characteristics of the encoded gene product, endoglucanase 3. *J. Bacteriol.* 171:5587–95

52. Michaelis, S., Chapon, C., d'Enfert, C., Pugsley, A. P., Schwartz, M. 1985. Characterization and expression of the structural gene for pullulanase, a maltose-inducible secreted protein of *Klebsiella pneumoniae. J. Bacteriol.* 164:633–38

53. Miyazaki, H., Yanagida, H., Hirinouchi, S., Beppu, T. 1989. Characterization of the precursor of *Serratia marcescens* serine protease and COOH-terminal processing during excretion through the outer membrane of *Escherichia coli. J. Bacteriol.* 17:6566–72

54. Murooka, Y., Ikeda, R. 1989. Biosynthesis and secretion of pullulanase, a lipoprotein from *Klebsiella aerogenes. J. Biol. Chem.* 264:17524–31

55. Nakahama, K., Yoshimura, K., Marumoto, R., Kikuchi, M., Lee, I. S., et al. 1986. Cloning and sequencing of *Serratia* protease gene. *Nucleic Acids Res.* 14:5843–55

56. Ohba, R., Ueda, S. 1982. An inductive effector in production of extracellular pullulanase by *Aerobacter aerogenes*. *Agric. Biol. Chem.* 150:53–61
57. Perumal, N. B., Minkley, E. G. Jr. 1984. The product of F sex factor *traT* surface exclusion gene is a lipoprotein. *J. Biol. Chem.* 259:5357–60
58. Pohlner, J., Halter, R., Beyreuther, K., Meyer, T. F. 1987. Gene structure and extracellular secretion of *Neisseria gonorrhoeae* IgA protease. *Nature* 325: 458–62
59. Pugsley, A. P. 1987. Protein secretion across the outer membrane of Gram-negative bacteria. See Ref. 10a, pp. 607–52
60. Pugsley, A. P. 1989. *Protein Targeting.* San Diego: Academic
61. Pugsley, A. P., Chapon, C., Schwartz, M. 1986. Extracellular pullulanase of *Klebsiella pneumoniae* is a lipoprotein. *J. Bacteriol.* 166:1083–88
62. Pugsley, A. P., Kornacker, M. G., Ryter, A. 1990. Analysis of the subcellular location of pullulanase produced by *Escherichia coli* carrying the *pulA* gene from *Klebsiella pneumoniae* strain UNF5023. *Mol. Microbiol.* 4:59–72
63. Pugsley, A. P., Reyss, I. 1990. Five genes at the 3' end of the *Klebsiella pneumoniae pulC* operon are required for pullulanase secretion. *Mol. Microbiol.* 4:365–78
64. Regue, M., Wu, H. C. 1987. Synthesis and export of lipoproteins in bacteria. See Ref. 10a, pp. 587–606
65. Reyss, I., Pugsley, A. P. 1990. Five additional genes in the *pulC-O* operon of the Gram-negative bacterium *Klebsiella oxytoca* UNF5023 that are required for pullulanase secretion. *Mol. Gen. Genet.* In press
66. Rothman, J. E. 1989. Polypeptide chain binding proteins: catalysis of protein folding and related processes in cells. *Cell* 59:591–601
67. Schatz, P., Beckwith, J. 1990. Genetics of protein secretion by bacteria. *Annu. Rev. Genet.* 24:
68. Schiebel, E., Schwartz, H., Braun, V. 1990. Subcellular location and unique secretion of the hemolysin of *Serratia marcescens*. *J. Biol. Chem.* 264:16311–20
69. Schwartz, M. 1987. The maltose reg-

ulon. In *Escherichia coli and Salmonella typhimurium. Cellular and Molecular Biology.* ed. F. C. Neidhardt, pp. 1482–1502. Washington, DC: Am. Soc. Microbiol.
70. Takizawa, N., Murooka, Y. 1985. Cloning of the pullulanase gene and overproduction of pullulanase in *Escherichia coli* and *Klebsiella aerogenes*. *Appl. Environ. Microbiol.* 49:294–95
71. Uphoff, T. S., Welch, R. A. 1990. Nucleotide sequence of the *Proteus mirabilis* calcium independent hemolysin genes (*hmpA* and *B*) reveals sequence similarly with the *Serratia marcescens* hemolysin genes (*shcA* and *B*). *J. Bacteriol.* In press
72. Wagner, W., Vogel, M., Goebel, W. 1983. Transport of hemolysin across the outer membrane of *Escherichia coli* requires two functions. *J. Bacteriol.* 154:200–10
73. Wandersman, C., Delepelaire, P. 1990. TolC, an *Escherichia coli* outer membrane protein required for hemolysin secretion. *Proc. Natl. Acad. Sci. USA.* 87:4776–80
74. Watanabe, T., Hayashi, S., Wu, H. C. 1988. Synthesis and export of the outer membrane lipoprotein in *Escherichia coli* mutants defective in generalized protein export. *J. Bacteriol.* 170:4001–7
75. Wöhner, G., Wöber, G. 1978. Pullulanase, an enzyme of starch catabolism, is associated with the outer membrane of Klebsiella. *Arch. Microbiol.* 116:303–10
76. Wong, K. R., Buckley, J. T. 1989. Proton motive force involved in protein transport across the outer membrane of *Aeromonas salmonicida*. *Science* 246: 654–56
77. Wong, K. R., Green, M. J., Buckley, J. T. 1989. Extracellular secretion of cloned aerolysin and phospholipase by *Aeromonas salmonicida*. *J. Bacteriol.* 171:2523–27
78. Wretlind, B., Pavlovskis, O. R. 1984. Genetic mapping and characterization of *Pseudomonas aeruginosa* mutants defective in the formation of extracellular proteins. *J. Bacteriol.* 158:801–8
79. Yamaguchi, K., Yu, F., Inouye, M. 1988. A single amino acid determinant of membrane localization of lipoproteins in *E. coli. Cell* 53:423–32

Annu. Rev. Genet. 1990. 24:91–113

CONTROL OF MITOCHONDRIAL GENE EXPRESSION IN *SACCHAROMYCES CEREVISIAE*

Maria C. Costanzo and Thomas D. Fox

Section of Genetics and Development, Cornell University, Ithaca, New York 14853–2703

KEY WORDS: transcription, RNA processing, translation, yeast

CONTENTS

INTRODUCTION

The principal function of the mitochondrial genetic system of yeast is to produce seven protein subunits of respiratory chain enzyme complexes. These

0066-4197/90/1215-0091$02.00

complexes also contain the products of numerous nuclear genes, leading to the notion that mitochondrial and nuclear gene expression must be coordinated at some level, or levels, to achieve balanced accumulation of subunits. While this is a very attractive idea, it is at present unknown how, or even whether, mitochondrial gene expression is tightly coupled to the expression of nuclear genes. This is a particularly interesting issue in yeast, since the levels of respiratory chain complexes vary with environmental conditions (e.g. they are reduced when glucose is abundant or when oxygen is absent).

Virtually all protein components of the yeast mitochondrial genetic system are themselves the products of nuclear genes. These nuclearly coded proteins are synthesized in the cytoplasm and imported into the organelles. This leads to a complicated and peculiar situation in which possibly a hundred or so nuclear genes are required to allow the expression of seven mitochondrial genes. The advantages, if any, of this form of organization are not clear. However, the mitochondrial genetic systems of all eukaryotic organisms studied appear to share these features.

In this review we adopt the simplifying assumption that the control of mitochondrial gene expression can be studied by concentrating on the required nuclear genes and their targets of action coded in the mitochondrial genome. (This is likely to be an oversimplification, since mitochondrial genetic activity can also influence the expression of nuclear genes (117). Thus, this review is somewhat limited in scope. However, other (and overlapping) aspects of mitochondrial biogenesis (2, 50, 60, 145) and evolution (58) have been recently reviewed elsewhere.

MITOCHONDRIAL GENES

A large body of thoroughly reviewed genetic and biochemical evidence (2, 33, 40, 60) has demonstrated that yeast mtDNA has a circular map of about 75 to 80 kilobases, depending on the strain, and encodes seven major protein components of respiratory chain enzymes: cytochrome c oxidase subunits I, II and III,[1] apo-cytochrome b, and subunits 6, 8 and 9 of the F_0 component of the mitochondrial ATPase. The only other major protein encoded in mtDNA is a component of the mitochondrial ribosomal small subunit termed VAR1 that is required for mitochondrial protein synthesis. In addition, mtDNA contains single genes encoding large (21S) and small (15S) rRNAs, a full set of 24 tRNAs, and an RNA component of an RNase P-like enzyme involved in tRNA processing.

[1]The genes coding cytochrome oxidase subunits I, II and III have generally been termed *oxi3*, *oxi1*, and *oxi2*, respectively (see 40)! This is very confusing. Here we adopt the names *COX1*, *COX2*, and *COX3* for the respective subunit genes, using uppercase for wild-type and lowercase for mutants.

Several low-abundance proteins are also encoded by open reading frames present in introns of the genes coding cytochrome oxidase subunit I*(COX1)*, cytochrome *b(COB)*, and the 21S rRNA. Some of these intron-encoded proteins, termed maturases, are required for intron splicing (79), while others play a role in the transposition of the introns encoding them (41, 121). However, these intron-encoded functions are completely dispensable in strains that lack the introns (138). There are also three open reading frames, identified by sequencing, whose function remains unknown (60).

Mutations in both mitochondrial and nuclear genes that interfere with the function or expression of mitochondrial genes lead to a respiratory-deficient phenotype, termed Pet$^-$, most directly manifested as an inability to grow on nonfermentable carbon sources such as glycerol and ethanol. Mitochondrial mutations can be limited lesions affecting single enzymatic functions (frequently termed *mit*$^-$) or large deletions that are associated with the complete loss of mitochondrial protein synthesis (termed *rho*$^-$ or cytoplasmic petite). Curiously, the sequences retained in the mtDNA of *rho*$^-$ strains are present as mixtures of high mol wt concatemers and variable amounts of oligomeric circles (40, 88). In the limit case of mtDNA deletions, termed *rho*^{o}, the entire chromosome is lost. The viability of such strains on fermentable carbon sources demonstrates that the mitochondrial genome is not required for important cellular functions other than respiration.

Mutational analysis has probably identified all mitochondrial genes whose products are directly involved in respiration. However, it is quite possible that mitochondrial genes required for the maintenance of the complete *(rho*$^+$*)* genome remain to be discovered, since mutations in such genes would lead to *rho*$^-$ deletions that would be indistinguishable from the very high background of spontaneous *rho*$^-$ mutants (up to several percent of cells in a culture). In this connection, it is important to note that at least low levels of mitochondrial protein synthesis are required to maintain *rho*$^+$ mtDNA (but not *rho*$^-$) (47, 106, 107), suggesting that an unknown mitochondrially coded protein might play a direct role in DNA replication or general recombination.

MITOCHONDRIAL TRANSFORMATION AND GENE REPLACEMENT

Our understanding of mitochondrial gene expression and other organellar phenomena will be greatly aided by the recent discovery that high-velocity microprojectile bombardment can deliver DNA to yeast mitochondria in vivo, resulting in genetic transformation (53, 72). A particularly useful feature of this system is that synthetic *rho*$^-$ strains, containing only defined plasmid sequences, can be obtained by transformation of *rho*^{o} recipients (53). By combining the ability to produce synthetic *rho*$^-$ strains with known features of mitochondrial recombination and segregation (40, 120), it is now possible to replace wild-type genes with in vitro-generated mutant alleles (52; L. S. Folley, M. C. Costanzo & T. D. Fox, unpublished results).

Following the mating of two haploid yeast cells, homologous recombination between their mtDNAs occurs at high frequency in the resulting zygote. Gene replacements can be achieved by a two-step procedure taking advantage of this high-frequency recombination. First, a synthetic rho^- transformant carrying an in vitro-generated mutant gene can be identified by its ability to produce respiring diploids when crossed to an appropriate tester point mutant. Then, the mutation carried on the synthetic rho^- mtDNA can be transferred to rho^+ mtDNA by mating the haploid rho^- to a wild-type strain: double recombination results in replacement of wild-type genetic information by the in vitro-generated mutation in a significant fraction of the resulting diploid cells.

In the studies discussed below, mitochondrial transformation and gene replacement have not played a major role. However, in future work these methods will allow the same strategies that have been so productive in the study of yeast nuclear genes to be applied in the mitochondrion.

TRANSCRIPTION AND TRANSCRIPTIONAL CONTROL

Yeast mitochondria appear to contain a single RNA polymerase of two or possibly three subunits. This enzyme recognizes a simple promoter sequence both in vitro and in vivo. No nuclear genes controlling the transcription of specific mitochondrial genes have been identified.

Promoter Structure

There are 19 or 20 sites of transcription initiation on the mitochondrial genome, which were initially identified by in vitro capping and sequencing of 5'-triphosphate RNA ends (23). A highly conserved nine-nucleotide sequence (the consensus is 5'-ATATAAGTA-3'), whose 3' nucleotide corresponds to the mRNA 5' end, is found at all transcriptional startpoints (23, 43, 115). The critical region required for promoter function, as defined by in vitro transcription (43) of deletion mutant promoters, is between nucleotides -10 and $+2$ relative to the transcriptional startpoint (6, 132).

The importance of each nucleotide within this consensus has been assessed by point mutagenesis studies combined with in vitro transcription and promoter binding assays (7, 8, 10, 130, 132), as well as by comparison of nuclear DNA sequences that function in vitro as mitochondrial promoters (89). In general, the consensus sequence seems to be the optimal promoter, and certain nucleotides within it, particularly those at positions -2, -4, -6, and -7, are more important than others. The only nonrespiratory mitochondrial mutation known to affect a promoter is an A to T substitution at position −4 of the *COX2* promoter, which abolishes transcription of the gene in vivo (17). This mutation provides strong support for the in vitro analysis.

RNA Polymerase

The mitochondrial RNA polymerase is composed of a 145-kd core subunit and at least one specificity factor (reviewed in 133). The core alone has only nonspecific transcriptional activity and does not detectably bind to promoters (73, 129, 143, 151). Nucleotide sequence analysis of the gene encoding the core, *RPO41* (59), revealed the surprising fact that it is homologous to the RNA polymerases of bacteriophages T3 and T7 (91). The promoters recognized by these phage enzymes are also single blocks of highly conserved sequence adjacent to the transcription start point (101).

Unlike the T3 and T7 RNA polymerases, the mitochondrial enzyme needs a specificity factor to recognize its promoters in vitro (73, 129, 131, 143, 151). A 43-kd protein, purified by SDS-gel electrophoresis and renatured, conferred upon the purified core subunit the ability to initiate transcription specifically in at least eight mitochondrial promoters (129, 131). This factor had no transcriptional activity of its own and did not bind to DNA (129, 131). Purification of another specificity factor, a 70-kd DNA-binding protein, has also been reported (143).

The *RPO41* gene encoding RNA polymerase core was isolated using antiserum directed against a purified RNA polymerase activity to screen a λ expression library (59). As expected, disruption of the gene leads to an inability to respire (59). *rpo41* disruption mutants also lose their wild-type (*rho⁺*) mitochondrial genome (59), although *rho⁻* deletion derivatives can be maintained (46, 65). This is not surprising since mitochondrial translation is required to maintain *rho⁺* mtDNA (107). Whether *RPO41* activity is required directly for replication of the *rho⁺* genome is unknown (see 46, 133).

Mutations that specifically cause a decrease in mitochondrial mRNA levels have been sought in collections of temperature-sensitive *pet* mutants (86, 102). One such mutation, *pet-ts798*, fails to complement an *rpo41* disruption (86, 87). A gene, termed *MTF1*, that suppressed the *pet-ts798* mutation when present on a high-copy plasmid was thought to potentially encode a specificity factor (86). However, this now seems unlikely since disruption of this gene did not affect the amount of RNA polymerase holoenzyme obtained from mutant cells (S. H. Jang & J. A. Jaehning, personal communication). Other mutations that cause a decrease in mitochondrial mRNA levels may identify genes for the specificity factor(s) or other proteins required for transcription (85, 86, 102). In an alternative genetic approach, at least one second-site suppressor of the *cox2* promoter mutation (17) has been isolated. This mutation, located in a nuclear gene (V. L. Cameron, personal communication), could potentially identify the subunit of RNA polymerase that interacts with promoter DNA.

Modulation of Transcription

The transcription of yeast mitochondrial genes is far from uniform, in part due to differences between their promoters. Measurements of transcription in vivo (104) and promoter competition assays in vitro (9, 150) indicate as much as a 20-fold difference in strength between the strongest and weakest promoters. Promoter strength is affected not only by nucleotide sequence, but also by relative location: *cis* effects from a strong promoter (at distances of up to about 400 bases) can inhibit the activity of a weaker promoter by several fold (9). Localized DNA bending may also influence promoter strength (130). Another important factor in the transcriptional expression of many mitochondrial genes is their location in polygenic transcription units (reviewed in 60). Careful measurement of pulse-labeled mRNA levels showed less mRNA corresponding to promoter-distal genes than to promoter-proximal genes within the same transcription unit; in some cases this difference is as much as 17-fold (104). This phenomenon has been ascribed to attenuation of RNA polymerase elongation (104), although selective degradation of downstream transcripts within the pulse-labeling period (10 min) cannot be excluded.

Mitochondrial transcription is apparently modulated in response to environmental conditions. Steady-state levels of most mitochondrial mRNAs are three- to sixfold higher in derepressed than in glucose-repressed cells (103). The proportion of some individual mRNAs (for example, the *COX3* mRNA) to total RNA remains constant under glucose-repressed and derepressed conditions; other mRNAs increase proportionally in the absence of glucose (4, 154). Modulation of mRNA levels by oxygen availability has been less well studied. The *COX3* mRNA is reduced by only two- to threefold during anaerobic growth (D. L. Marykwas & T. D. Fox, unpublished results), while levels of *COB* and *COX1* mRNAs apparently remain constant (T. Pillar, personal communication).

A direct measurement of the rate of 21S rRNA synthesis in vivo revealed that it was 7.2-fold faster in derepressed cells than in glucose-repressed cells (103). Thus, increased transcription rate is probably at least one factor in the increased steady-state abundance of mitochondrial RNAs in derepressed cells. Changes in cellular level of mtDNA under various conditions would presumably result in apparent changes in transcription rates. However, at least for glucose repression, the magnitude of this variation (0- to 3-fold) (57, 74) cannot wholly account for the differences in steady-state RNA levels (103). Changes in the level of RNA polymerase could also alter transcription rates, especially since it has been estimated that yeast mitochondria contain roughly one RNA polymerase per promoter site (S. H. Jang & J. A. Jaehning, personal communication). Indeed, the level of *RPO41* mRNA is reduced in glucose-grown cells (91). However, *RPO41* transcription is apparently not the limiting factor in glucose-repressed cells since 10- to 15-fold overproduction of the core subunit does not lead to increased mitochondrial mRNA levels (S. H. Jang & J. A. Jaehning, personal communication).

RNA PROCESSING

Posttranscriptional processing of primary transcripts plays a prominent role in the expression of yeast mitochondrial genes. As noted above, most genes are initially transcribed onto polycistronic RNAs, and evidence has been obtained for numerous subsequent processing events (reviewed in 60). While the known processing events for mRNAs, rRNAs, and tRNAs appear to involve the action of nuclear genes, no nuclear mutants have been reported that are generally defective in the processing of any of these classes of molecules.

Specific Effects at mRNA Termini

The *COB* gene is initially transcribed onto a precursor that also contains the tRNA^{Glu} 1,100 bases upstream of *COB* coding sequences (22). This precursor is processed to yield mature tRNA^{Glu} and the *COB* mRNA, which has a long (954 bases) 5' leader starting 145 bases downstream of the 3' end of the tRNA^{Glu} (34, 35). Mutations in the nuclear gene *CBP1* drastically reduce the steady-state level of mature *COB* mRNA, although they appear not to block transcription of *COB* sequences and have no effect on the level of tRNA^{Glu} (34). This phenotype suggests that the *CBP1* gene product, which is a mitochondrially located protein (148), has no role in the 5' or 3' processing of tRNA^{Glu}, but is required either to stabilize the *COB* mRNA against rapid degradation or to generate a stable mRNA from an unstable precursor.

Whatever the CBP1 protein does, it appears to act through a site in the precursor RNA near the 5' end of the mature *COB* mRNA. This has been shown genetically by examining the properties of novel mRNAs encoded by rearranged mitochondrial genes. Chimeric mRNAs containing the 5' leader from the ATPase subunit 9 mRNA attached to the *COB* structural gene were stable and expressed in *cbp1* mutants (34). However, chimeric mRNAs with the 5' third of the *COB* leader (250 to 300 bases) attached to either the *COB* (35) or ATPase 9 (T. M. Mittelmeier & C. L. Dieckmann, personal communication) structural gene failed to accumulate in *cbp1* mutants, although they were functional in wild-type nuclear backgrounds. Taken together, these data indicate that CBP1 functions either to antagonize degradation of the mature *COB* mRNA by acting at a site in the 5' 250 bases of the *COB* leader,

or to generate a stable mRNA from an unstable precursor by processing the 5'
end. Consistent with these possibilities, an independent study of *cob* deletion
mutations indicated that the presence of the normal mature 5' end was
necessary for accumulation of stable *cob* transcripts in wild-type *(CBP1)*
nuclear backgrounds (38).

The 3' ends of mitochondrial mRNAs lie within a conserved consensus
sequence (142) identified as the dodecamer 5'-AAUAAUAUUCUU-3',
which serves as a processing site (114). Two mitochondrial mutations strongly confirm the importance of this sequence for gene expression. One is a
double point mutation in the dodecamer at the end of the open reading frame
(termed *FIT1*) in the *omega* intron of the 21S rRNA (156). The product of this
open reading frame is required for efficient gene conversion of the intron in
crosses (41, 121), and the dodecamer mutation abolishes both 3' processing
of this mRNA and gene conversion (156). The second mutation is a 195-
basepair deletion downstream of the ribosomal protein-coding gene *VAR1* that
removes its dodecamer and dramatically reduces *VAR1* expression. Transcripts of this mutant gene are reduced in abundance and have extended 3'
ends (155).

A dominant mutation, termed *SUV3-1*, was isolated as a suppressor of the
var1 dodecamer deletion mutation and found to have a variety of intriguing
effects (155). *SUV3-1* did not detectably alter the structure of the mutant *var1*
mRNA or greatly change its level (H. M. Conrad-Webb, R. A. Butow & P.
S. Perlman, personal communication), but apparently increased its translation. This result suggests that a proper 3' end is required for efficient VAR1
translation in wild-type, and that the *SUV3-1* mutation relieves this requirement. Surprisingly, however, the *SUV3-1* mutation blocked normal processing at the wild-type dodecamer site of the intron-coded *FIT1* mRNA (and
omega gene conversion) although it did not affect 3' processing of other
mRNAs (155). While it is difficult to interpret these results in terms of a
specific role for the wild-type *suv3* gene product, they confirm the importance
of proper 3' processing for mRNA function and suggest a link between this
processing and translation.

Control of Intron Splicing by Nuclear Genes

Yeast mitochondrial introns are probably examples of molecular parasites (3,
82, 100). Consistent with this notion, strains whose mitochondrial genes lack
all known introns respire normally (138). However, mutations that block
splicing lead to nonrespiratory phenotypes, a fact that has allowed extensive
genetic analysis of yeast mitochondrial splicing.

The removal of introns from primary transcripts involves RNA catalysis,
since self-splicing occurs in vitro (reviewed in 18). However, splicing in vivo
is dependent on proteins encoded either by the introns themselves (maturases;

reviewed in 79) or by nuclear genes (138), or both (66). Thus, expression of the intron-containing genes *COB* and *COX1* is subject to complex genetic controls at the level of splicing.

A number of nuclear genes appear to be required solely for the splicing of single mitochondrial introns. The clearest example of such a gene is *CBP2*. Analysis of transcripts from *cbp2* mutants indicated that they specifically failed to splice the terminal *COB* intron (93). Moreover, mitochondrially inherited mutations that suppressed the *cbp2* respiratory defect were precise deletions of the terminal intron sequences from mtDNA, arguing strongly that *CBP2* has no other important cellular function (68). In vitro, the terminal *COB* intron catalyzed its own excision, but only under nonphysiological conditions (55, 119). However, in the presence of CBP2 protein, partially purified from yeast mitochondria, the intron was spliced in vitro under conditions more closely resembling those of the mitochondrial matrix, demonstrating that CBP2 acts directly to assist RNA catalysis (54). Other nuclear genes required specifically for splicing introns in *COB* (80, 81) and *COX1* (137) have been described but have not been shown to act directly.

In *Neurospora crassa,* mitochondrial tyrosyl-tRNA synthetase has been shown genetically and biochemically to act directly to promote intron splicing (1). In yeast, a nuclear gene, termed *NAM2*, which interacts functionally with intron-encoded maturases, codes for the mitochondrial leucyl-tRNA synthetase (66). While *nam2* null mutations block protein synthesis and lead to *rho⁻* deletions, reduced *NAM2* expression allows some mitochondrial translation but appears to block *COB* and *COX1* splicing, suggesting that it may act directly (66). (The fact that splicing of precursors by maturases depends upon mitochondrial translation of the precursors leads to uncertainties in interpreting such genetic data; see 5, 61).

An intriguing nuclear gene, *MSS116*, required for *COB* and *COX1* splicing has recently been shown to be homologous to mammalian eIF-4A and other proteins thought to have helicase activity (136). *MSS116* appears to have at least one other function in addition to splicing, since the respiratory defect caused by a null *mss116* mutation was not suppressed in a strain lacking all known introns. However, the null mutation did not block mitochondrial translation, suggesting it may have a direct effect on splicing.

tRNA Processing and Base Modification

Mitochondrial tRNA precursors are processed in vivo by precise endonucleolytic cuts at the 5' and 3' ends (14) prior to the addition of CCA at the 3' end. An RNase P that correctly processes 5' ends in vitro has been partially purified from mitochondria and shown to consist of an RNA encoded in mtDNA and at least one protein that must be encoded in the nucleus (70,

99). An endonuclease that correctly processes 3' ends in vitro has also been partially purified (20). It is unknown whether this enzyme contains RNA.

At least two kinds of base modifications in mitochondrial tRNAs, those generating N^2,N^2-dimethylguanosine and N^6-(Δ^2-isopentenyl)adenosine, are carried out by the same nuclearly coded enzymes that correspondingly modify cytoplasmic tRNAs (37, 44, 110). The function of these modifications is unclear since mutations in the respective genes, *TRM1* and *MOD5*, have no apparent effect on mitochondrial gene expression.

Modulation of RNA Processing

Expression of mitochondrial genes could be modulated in response to physiological conditions at the level of RNA processing, although this issue has not been extensively examined. Although processing of most mitochondrial transcripts is relatively unaffected by environmental conditions, processing of *COX1* mRNA precursors appears to be reduced by both glucose repression (154) and anaerobic conditions (T. Pillar, personal communication).

No Evidence To Date for RNA Editing in Yeast Mitochondria

Posttranscriptional editing of RNA sequences has been shown to occur in the mitochondria of trypanosomes (11, 140), higher plants (29, 63, 67) and the slime mold *Physarum polycephalum* (R. Mahendran, M. Spottswood & D. L. Miller, personal communication). Although no evidence suggests that RNA editing occurs in yeast mitochondria (a disappointment or a relief, depending on your point of view), relatively few direct comparisons between genomic DNA and cDNA sequences have been made.

TRANSLATION AND TRANSLATIONAL CONTROL

The majority of nuclear genes identified to date as playing a direct role in mitochondrial gene expression are required for translation. Most of these genes encode general components of the mitochondrial translation system, such as ribosomal proteins and tRNA synthetases, that function exclusively in mitochondria. However, at least two tRNA synthetases that function in mitochondria are coded by the same genes that encode the cytoplasmic enzymes. Such shared components, which include the tRNA-modification enzymes mentioned above, represent a clear link between the nucleo-cytoplasmic and mitochondrial genetic systems. Surprisingly, in addition to the expected general components of the mitochondrial system, translation of at least three (and probably all) mitochondrial mRNAs requires mRNA-specific positive activators encoded by nuclear genes. Since no one has developed an mRNA-dependent in vitro translation system from mitochondria, functional analysis has had to rely primarily on genetics.

General Features of the Mitochondrial Translation System

Yeast mitochondrial ribosomes contain approximately 77 distinct polypeptides (98), all of which, with the exception of VAR1, are nuclear gene products. Twelve of the nuclear genes have been identified and examined. Four encode proteins whose sequences are detectably similar to ribosomal proteins of *E. coli:* the MRP2 protein is homologous to S14 (106), MRP7 has a region of strong similarity to L27 (47), while MRP3 and MRP4 may correspond to S19 and S2, respectively (145). However, the other eight known nuclearly coded mitochondrial ribosomal proteins, MRP1 (32, 106), MRP13 (118), PET123 (65, 94), YmL31 (62), YMR-31 (92), YMR-44 (92), MRP20, and MRP49 (K. Fearon & T. L. Mason, personal communication), exhibit no detectable homology to characterized ribosomal proteins from other organisms.

Elimination of single ribosomal proteins by disruption or deletion of the genes *MRP1* (107), *MRP2* (106), *MRP7* (47), *PET123* (65), and *MRP20* (K. Fearon & T. L. Mason, personal communication) completely blocks mitochondrial translation, as expected. However, disruption of *MRP13,* which encodes a small subunit protein, does not block mitochondrial translation per se, but rather greatly slows the recovery of cells from catabolite repression (J. A. Partaledis & T. L. Mason, personal communication). Disruption of *MRP49* produces a cold-sensitive nonrespiratory phenotype (K. Fearon & T. L. Mason, personal communication).

Nuclear genes encoding ten mitochondrial tRNA synthetases have been identified to date. Eight encode enzymes that function only in mitochondria, since mutations in the genes produce a viable but nonrespiratory phenotype (56, 66, 77, 108, 116, 144–146). In contrast, the genes coding histidinyl- (111) and valyl-tRNA synthetases (19) each specify both mitochondrial and cytoplasmic forms of the enzymes. In both cases, longer transcripts including sequences that code for N-terminal mitochondrial targeting signals (2) direct synthesis of enzymes that function in the organelle. Shorter transcripts initiating downstream of the targeting sequence initiation codon specify the cytoplasmic enzymes. Mutations affecting the targeting-sequence initiation codon prevent synthesis of only the longer form and produce a viable Pet$^-$ phenotype, while downstream mutations that disrupt enzymatic function are lethal (19, 111).

Genes encoding homologs to *E. coli*-elongation factors Tu (109) and G (145) have also been identified in yeast nuclear DNA. Mutations in these genes lead to a Pet$^-$ phenotype, indicating that these factors function only in mitochondria.

Analysis of informational suppressors has identified components required for translational fidelity in yeast mitochondria. Recessive suppressors of mitochondrial ochre mutations lie in at least two nuclear genes (some alleles of which are unlinked noncomplementing mutations) that may encode small subunit ribosomal proteins (12, 13). Mitochondrially inherited ochre suppressors include mutations affecting the structure of the 15S rRNA (139). Mitochondrial frameshift suppressors have been identified as mutations in the 15S rRNA (149) and a mitochondrially encoded serine-tRNA (71). Temperature-sensitive and drug-resistance mutations have also been used to study the function of the 21S rRNA (30, 31). In a puzzling case of dosage-dependent suppression, multiple copies of the yeast nuclear genes encoding eIF-4A have been found to suppress a *cox3* missense mutation (84).

The general features of translation initiation in yeast mitochondria are unclear. Most major mRNAs have 5' leaders of at least several hundred bases and all of these long leaders have at least one AUG upstream of the initiator AUG codon (see 60 and citations therein). The only exception is the *COX2* mRNA whose leader is short (54 bases) and lacks upstream AUGs (15). There does not appear to be a mitochondrial ribosome binding site with a fixed position relative to the start of translation, like those of prokaryotes. While sites in 5' leaders complementary to the 3' end of the 15S rRNA have been noted (83), their positions vary between -8 and -107 relative to initiator AUG codons. Furthermore, a chimeric mRNA lacking any of these sites was translated efficiently in vivo (26). A nuclear gene encoding a protein with homology to *E. coli* IF2 that functions in mitochondria has been identified (145). More recently, a site-directed mutation that changed the initiator ATG of the *COX3* gene to ATA has been inserted into otherwise wild-type mtDNA, as described above (L. S. Folley & T. D. Fox, unpublished results). Nuclear suppressors of this mutation have been obtained and may, by analogy with results obtained in the cytoplasmic translation system (24, 39), identify components required for initiation codon selection.

mRNA-Specific Positive Control of Translation

An unusual feature of the yeast mitochondrial translation system is the existence of nuclearly encoded mRNA-specific translational activators. The most intensively studied of these activators control translation of the *COX3* mRNA. They were identified by recessive mutations in the three nuclear genes *PET494, PET54,* and *PET122* that specifically block accumulation of the COXIII protein, leading to a Pet⁻ phenotype (16, 27, 42, 51, 75, 76, 105). (There may be additional genes involved in *COX3* specific activation, yet to be identified.) The defect in *COX3* gene expression is posttranscriptional, since the *COX3* mRNA is present at wild-type levels in the mutant strains (27, 51, 76, 105). That the defect in *COX3* expression was at the level of translation was demonstrated by isolating mitochondrial suppressors of null mutations in all three nuclear genes: these suppressors were *COX3* gene rearrangements that generated novel chimeric *COX3* mRNAs bearing 5'-untranslated leaders derived from other mitochondrial mRNAs (25, 27, 51).

The existence of such bypass suppressors also demonstrated that *PET494*, *PET54*, and *PET122* have no other important cellular functions (although splicing of an optional *COX1* intron is diminished in *pet54* mutants (147)).

The site(s) at which *PET494*, *PET54*, and *PET122* act was mapped genetically to the 5' two-thirds (about 400 bases) of the *COX3* mRNA leader by selecting (as a suppressor of a *cbs1* mutation (128); see below) a mtDNA rearrangement that generated a chimeric mRNA bearing this portion of the *COX3* leader fused to the *COB* structural gene. Translation of the "reporter" protein, cytochrome *b*, from this chimeric mRNA was dependent on *PET494*, *PET54*, and *PET122* (26). These results, taken together with the fact that the PET494, PET54, and PET122 proteins are all located in mitochondria (25, 28, 112; T. W. McMullin & T. D. Fox, unpublished results), suggest that the three proteins could mediate an interaction between the *COX3* mRNA leader and a component of the mitochondrial translation system.

The PET122 protein was shown to interact functionally with the small mitochondrial ribosomal subunit by analysis of unlinked allele-specific suppressors that partially compensated for the loss of PET122 carboxy terminal amino acids (65). The suppressors were in genes coding the small subunit ribosomal proteins PET123 (94) and MRP1 (32, 64, 106), both of which are required for the translation of all mitochondrial mRNAs. Furthermore, PET123 and MRP1 appear to interact closely with each other, since certain combinations of suppressor mutations generate synthetic defects in mitochondrial translation (64).

The three translational activators PET494, PET54, and PET122 appear to be largely associated with the mitochondrial inner membrane, based on immunological studies of submitochondrial fractions (T. W. McMullin & T. D. Fox, unpublished results). Thus, one could envisage these proteins as bringing the *COX3* mRNA together with the ribosomal small subunit to promote synthesis of COXIII near its site of assembly into cytochrome oxidase.

Translation of the *COB* mRNA also depends on specific activators coded by nuclear genes. Mutations in two such genes, *CBS1* and *CBS2*, posttranscriptionally block accumulation of apo-cytochrome *b* and can be suppressed by mitochondrial mutations that replace the *COB* mRNA 5' leader with a leader from another mRNA (126, 128). Furthermore, *CBS1* can act through the *COB* 5' leader to promote COXIII translation from a chimeric mRNA (127) and both the CBS1 and CBS2 proteins are imported into mitochondria in vitro (78; G. Rödel, personal communication). Thus, the function of *CBS1* and *CBS2* appears to be analogous to that of the *COX3* specific activators discussed above. A third nuclear gene, *CBP6*, also encodes a mitochondrially located protein required posttranscriptionally for normal cytochrome *b* accumulation (36). This gene may be required for translation, although *cbp6* mutations have not been shown to be suppressible by *COB* leader-substitution mutations.

Accumulation of the COXII protein is blocked posttranscriptionally by mutations in at least three nuclear genes. One of these, *PET111* (formerly *PETE11;*16, 42), encodes a mitochondrially located protein (141) required for *COX2* mRNA translation (122) that may act analogously to the *COX3-* and *COB*-specific activators described above. However, mitochondrial suppressors of *pet111* mutations replaced not only the *COX2* mRNA 5' leader but also the COXII amino terminus with sequences from other genes (122). Thus the data do not yet distinguish whether *PET111* acts on the *COX2* mRNA leader to promote translation, or on the nascent pre-COXII polypeptide chain (124) to relieve a hypothetical translational block (122). The action of two other nuclear genes required posttranscriptionally for COXII accumulation, *PET112* (51) and *SCO1* (134), is even less clear. Neither *pet112* nor *sco1* point mutations were suppressed by the mitochondrial suppressors of *pet111*, suggesting that the blocks are posttranslational. Furthermore, deletion of *SCO1* eliminated both COXI and COXII (135).

The precise locations of sites for translational activation in mitochondrial mRNA leaders have yet to be determined. However, a mitochondrial point mutation that results in insertion of a U in the 5' leader of the ATPase 9 mRNA, 88 bases upstream of the initiation codon, causes temperature-sensitive translation (113). Among other possibilities, this mutation might affect a site for an ATPase 9 translational activator. In this connection, it has been noted that in four relatively unrelated species of budding yeasts, the sequence 5'-TCTAA-3' occurs in the *COX2*-leader coding regions between 18 and 37 bases upstream of the initiation codon (C. M. Hardy & G. D. Clark-Walker, personal communication).

Modulation of Translation

Steady-state levels of mitochondrial mRNAs vary relatively little in response to glucose repression or oxygen deprivation (see above). In contrast, the apparent rates of synthesis (measured by in vivo labeling in the presence of cycloheximide) of most major mitochondrial gene products vary over a wider range. For example, COXI, COXII, COXIII, and cytochrome *b* were labeled at least 19 times more slowly in anaerobically than aerobically grown cells (152). Taken together, these findings suggest that modulation of mitochondrial gene expression in response to oxygen occurs at least partially at the translational level. Similarly, the release of glucose repression greatly stimulates translation of COXIII (45) but has little effect on its mRNA levels (154).

This apparent translational modulation is generally paralleled by the expression of nuclear genes coding components of the mitochondrial translation system, where the latter has been studied. However, these genes are not under tight coordinate regulation and are probably controlled by several different mechanisms. Transcription of the ribosomal protein genes *MRP1*, *MRP2*, and *MRP13* is glucose-repressed by several fold (106, 118). Expression of the ribosomal protein MRP7 is decreased ninefold by glucose repression although its mRNA levels are unchanged (47). Expression of the *COX3* translational activator *PET494* is glucose-repressed by several fold at the transcriptional level and repressed by oxygen deprivation at the translational level (90). Expression of *PET54* parallels that of *PET494* (Z. Shen & T. D. Fox, unpublished results) although the level of its control is unknown. The *COB*-specific translational activator *CBS1* is subject to glucose repression, and its transcription is reduced ninefold in the absence of oxygen (49); however, transcription of *CBS2* is only reduced twofold in the absence of oxygen (96).

If any theme emerges from studies of the expression of nuclear-encoded specific translational activators, it is the possibility that in several cases they may themselves be controlled at the translational level. While direct evidence for translational control has only been reported for *PET494* (90), the anatomy of the mRNAs encoding the other activators is suggestive (69). Both *PET54* and *PET122* are transcribed from two promoters, one giving rise to an mRNA with a long 5' untranslated leader containing short open reading frames and the other giving rise to transcripts with extremely short (0–11 bases) 5' leaders (28; K. A. Burke & J. E. McEwen, personal communication; J. D. Ohmen, K. A. Burke & J. E. McEwen, personal communication). In addition, the mRNA leaders of *PET111* (141), *CBS1* (49), and *SCO1* (G. Rödel, personal communication) each contain 1 to 4 short open reading frames.

POSTTRANSLATIONAL EVENTS

Functional expression of mitochondrial genes is not complete until the gene products are assembled into enzymatically active complexes. While posttranslational processes are poorly understood and largely beyond the scope of this review, they certainly bear mention. Two mitochondrial gene products, COX-II and ATPase subunit 6, undergo amino-terminal processing (97, 124). Two nuclear genes required for COXII processing have been identified, of which one is also required for processing of the nuclearly encoded mitochondrial protein cytochrome b_2 (123, 124).

The correct folding of mitochondrial proteins imported from the cytoplasm (2) has recently been shown to depend on at least one heat-shock protein, homologous to the *E. coli groEL* product (21, 95). Mitochondrially synthesized proteins may also require assistance for assembly, possibly by these same factors. In addition to general folding factors, it appears that certain nuclear gene products act in the assembly of specific enzyme complexes. For example, mutations in the gene termed *COX10* (M. P. Nobrega, F. G. Nobrega, & A. Tzagoloff, personal communication), that is identical to

PET1030 (42, 48), specifically block assembly of cytochrome c oxidase. Intriguingly, the *COX10/PET1030* product, which is not a cytochrome oxidase subunit, is homologous to a predicted protein coded by *ORF1* of a cytochrome oxidase operon of *Paracoccus denitrificans* (125; M. P. Nobrega, F. G. Nobrega, & A. Tzagoloff, personal communication; B. Fernandes, personal communication). A second nuclear gene required for oxidase assembly, termed *COX11*, is homologous to *ORF3* of the same bacterial operon (A. Tzagoloff, personal communication). Two other nuclear genes, *CBP3* and *CBP4*, are specifically required for the assembly of coenzyme QH_2-cytochrome c reductase (153; M. D. Crivellone & A. Tzagoloff, personal communication).

CONCLUSIONS

The mechanics of mitochondrial gene expression in yeast are surprisingly complex and difficult to rationalize. Regulation occurs at both the transcriptional and translational levels, and possibly at the level of mRNA processing as well. It remains to be determined what level, or levels, of control are actually rate limiting for the expression of mitochondrial genes under different environmental conditions. The overall rates of transcription and translation may be affected by modulation of expression of the nuclear genes encoding components of the mitochondrial genetic system. However, there is no evidence as yet for a common regulatory system coordinating expression of these genes.

Transcriptional control in mitochondria appears to be relatively global. However, translation is subject to mRNA-specific positive controls. This is puzzling since it does not appear to simplify the task of achieving balanced synthesis of respiratory chain components. It is unclear, for example, why the translation of cytochrome oxidase subunits II and III should be independently controlled since they are needed in equimolar amounts in the enzyme. Perhaps the state of our ignorance about the regulation of genes for respiratory enzyme complexes can best be characterized by a statement of the obvious: the level of a complex under a given set of conditions depends on the level of whichever subunit is present in limiting quantity, whether mitochondrially or nuclearly encoded. If expression of genes for the various subunits is not tightly coordinated, then different subunits may be limiting in different situations, and no single regulatory system is in control under all conditions.

ACKNOWLEDGMENTS

We thank our many colleagues who communicated results prior to publication, and J. A. Jaehning and A. Tzagoloff for helpful discussions. Research in the authors' laboratory is supported by a grant (GM29362) from the US National Institutes of Health.

Literature Cited

1. Akins, R. A., Lambowitz, A. M. 1987. A protein required for splicing group I introns in Neurospora mitochondria is mitochondrial tyrosyl-tRNA synthetase or a derivative thereof. *Cell* 50:331–45
2. Attardi, G., Schatz, G. 1988. Biogenesis of mitochondria. *Annu. Rev. Cell Biol.* 4:289–333
3. Augustin, S., Müller, M. W., Schweyen, R. J. 1990. Reverse self-splicing of group II intron RNAs in vitro. *Nature* 343:383–86
4. Baldacci, G., Zennaro, E. 1982. Mitochondrial transcripts in glucose-repressed cells of *Saccharomyces cerevisiae*. *Eur. J. Biochem.* 127:411–16
5. Ben Asher, E., Groudinsky, O., Dujardin, G., Altamura, N., Kermorgant, M., et al. 1989. Novel class of nuclear genes involved in both mRNA splicing and protein synthesis in *Saccharomyces cerevisiae* mitochondria. *Mol. Gen. Genet.* 215:517–28
6. Biswas, T. K., Edwards, J. C., Rabinowitz, M., Getz, G. S. 1985. Characterization of a yeast mitochondrial promoter by deletion mutagenesis. *Proc. Natl. Acad. Sci. USA* 82:1954–58
7. Biswas, T. K., Getz, G. S. 1986. A critical base in the yeast mitochondrial nonanucleotide promoter. *J. Biol. Chem.* 261:3927–30
8. Biswas, T. K., Getz, G. S. 1986. Nucleotides flanking the promoter sequence influence the transcription of the yeast mitochondrial gene coding for ATPase subunit 9. *Proc. Natl. Acad. Sci. USA* 83:270–74
9. Biswas, T. K., Getz, G. S. 1988. Promoter-promoter interactions influencing transcription of the yeast mitochondrial gene, Olil, coding for ATPase subunit 9: *cis* and *trans* effects. *J. Biol. Chem.* 263:4844–51
10. Biswas, T. K., Ticho, B., Getz, G. S. 1987. In vitro characterization of the yeast mitochondrial promoter using single-base substitution mutants. *J. Biol. Chem.* 262:13690–96
11. Blum, B., Bakalara, N., Simpson, L. 1990. A model for RNA editing in kinetoplastid mitochondria: "guide" RNA molecules transcribed from maxicircle DNA provide the edited information. *Cell* 60:189–98
12. Boguta, M., Mieszczak, M., Zagórski, W. 1988. Nuclear omnipotent suppressors of premature termination codons in mitochondrial genes affect the 37S mitoribosomal subunit. *Curr. Genet.* 13:129–35
13. Boguta, M., Zoladek, T., Putrament, A.

1986. Nuclear suppressors of the mitochondrial mutation *oxi1-V25* in *Saccharomyces cerevisiae*. Genetic analysis of the suppressors: absence of complementation between non-allelic mutants. *J. Gen. Microbiol.* 132:2087–97
14. Bordonne, R., Dirheimer, G., Martin, R. P. 1987. Transcription initiation and RNA processing of a yeast mitochondrial tRNA gene cluster. *Nucleic Acids Res.* 15:7381–94
15. Bordonne, R., Dirheimer, G., Martin, R. P. 1988. Expression of the *oxi1* and the maturase-related *RF1* genes in yeast mitochondria. *Curr. Genet.* 13:227–33
16. Cabral, F., Schatz, G. 1978. Identification of cytochrome *c* oxidase subunits in nuclear yeast mutants lacking the functional enzyme. *J. Biol. Chem.* 253:4396–401
17. Cameron, V. L., Fox, T. D., Poyton, R. O. 1989. Isolation and characterization of a yeast strain carrying a mutation in the mitochondrial promoter for *COX2*. *J. Biol. Chem.* 264:13391–94
18. Cech, T. R., Bass, B. L. 1986. Biological catalysis by RNA. *Annu. Rev. Biochem.* 55:599–629
19. Chatton, B., Walter, P., Ebel, J.-P., Lacroute, F., Fasiolo, F. 1988. The yeast *VAS1* gene encodes both mitochondrial and cytoplasmic valyl-tRNA synthetases. *J. Biol. Chem.* 263:52–57
20. Chen, J.-Y., Martin, N. C. 1988. Biosynthesis of tRNA in yeast mitochondria: an endonuclease is responsible for the 3'-processing of tRNA precursors. *J. Biol. Chem.* 263:13677–82
21. Cheng, M. Y., Hartl, F.-U., Martin, J., Pollock, R. A., Kalousek, F., et al. 1989. Mitochondrial heat-shock protein hsp60 is essential for assembly of proteins imported into yeast mitochondria. *Nature* 337:620–25
22. Christianson, T., Edwards, J. C., Mueller, D. M., Rabinowitz, M. 1983. Identification of a single transcriptional initiation site for the glutamic tRNA and *COB* genes in yeast mitochondria. *Proc. Natl. Acad. Sci. USA* 80:5564–68
23. Christianson, T., Rabinowitz, M. 1983. Identification of multiple transcriptional initiation sites on the yeast mitochondrial genome by in vitro capping with guanylyltransferase. *J. Biol. Chem.* 258:14025–33
24. Cigan, A. M., Pabich, E. K., Feng, L., Donahue, T. F. 1989. Yeast translation initiation suppressor *sui2* encodes the α

subunit of eukaryotic initiation factor 2 and shares sequence identity with the human α subunit. *Proc. Natl. Acad. Sci. USA* 86:2784–88

25. Costanzo, M. C., Fox, T. D. 1986. Product of *Saccharomyces cerevisiae* nuclear gene *PET494* activates translation of a specific mitochondrial mRNA. *Mol. Cell. Biol.* 6:3694–703

26. Costanzo, M. C., Fox, T. D. 1988. Specific translational activation by nuclear gene products occurs in the 5′ untranslated leader of a yeast mitochondrial mRNA. *Proc. Natl. Acad. Sci. USA* 85:2677–81

27. Costanzo, M. C., Seaver, E. C., Fox, T. D. 1986. At least two nuclear gene products are specifically required for translation of a single yeast mitochondrial mRNA. *EMBO J.* 5:3637–41

28. Costanzo, M. C., Seaver, E. C., Fox, T. D. 1989. The *PET54* gene of *Saccharomyces cerevisiae:* Characterization of a nuclear gene encoding a mitochondrial translational activator and subcellular localization of its product. *Genetics* 122:297–305

29. Covello, P. S., Gray, M. W. 1989. RNA editing in plant mitochondria. *Nature* 341:662–66

30. Cui, Z., Mason, T. L. 1989. A single nucleotide substitution at the *rib2* locus of the yeast mitochondrial gene for 21S rRNA confers resistance to erythromycin and cold-sensitive ribosome assembly. *Curr. Genet.* 16:273–79

31. Daignan-Fornier, B., Bolotin-Fukuhara, M. 1988. Mutational study of the rRNA in yeast mitochondria: functional importance of T$_{1696}$ in the large rRNA gene. *Nucleic Acids Res.* 16:9299–306

32. Dang, H., Franklin, G., Darlak, K., Spatola, A., Ellis, S. R. 1990. Discoordinate expression of the yeast mitochondrial ribosomal protein MRP1. *J. Biol. Chem.* 265:7449–59

33. de Zamaroczy, M., Bernardi, G. 1986. The primary structure of the mitochondrial genome of *Saccharomyces cerevisiae*—a review. *Gene* 47:155–77

34. Dieckmann, C. L., Koerner, T. J., Tzagoloff, A. 1984. Assembly of the mitochondrial membrane system: *CBP1*, a yeast nuclear gene involved in 5′ end processing of cytochrome *b* pre-mRNA. *J. Biol. Chem.* 259:4722–31

35. Dieckmann, C. L., Mittelmeier, T. M. 1987. Nuclearly encoded CBP1 interacts with the 5′ end of mitochondrial cytochrome *b* pre-mRNA. *Curr. Genet.* 12:391–97

36. Dieckmann, C. L., Tzagoloff, A. 1985. Assembly of the mitochondrial membrane system: *CBP6*, a yeast nuclear gene necessary for the synthesis of cytochrome *b*. *J. Biol. Chem.* 260:1513–20

37. Dihanich, M. E., Najarian, D., Clark, R., Gillman, E. C., Martin, N. C., et al. 1987. Isolation and characterization of *MOD5*, a gene required for isopentenylation of cytoplasmic and mitochondrial tRNAs of *Saccharomyces cerevisiae*. *Mol. Cell. Biol.* 7:177–84

38. Dobres, M., Gerbl-Rieger, S., Schmelzer, C., Mueller, M. W., Schweyen, R. J. 1985. Deletions in the *cob* gene of yeast mtDNA and their phenotypic effect. *Curr. Genet.* 10:283–90

39. Donahue, T. F., Cigan, A. M., Pabich, E. K., Valavicius, B. C. 1988. Mutations at a Zn(II) finger motif in the yeast eIF-2β gene alter ribosomal start-site selection during the scanning process. *Cell* 54:621–32

40. Dujon, B. 1981. Mitochondrial genetics and function. In *The Molecular Biology of the Yeast Saccharomyces, Life Cycle and Inheritance*, ed. J. N. Strathern, E. W. Jones, J. R. Broach, pp. 505–635. Cold Spring Harbor, NY: Cold Spring Harbor Lab. Press

41. Dujon, B. 1989. Group I introns as mobile genetic elements: facts and mechanistic speculations—a review. *Gene* 82:91–114

42. Ebner, E., Mason, T. L., Schatz, G. 1973. Mitochondrial assembly in respiration-deficient mutants of *Saccharomyces cerevisiae*. II. Effect of nuclear and extrachromosomal mutations on the formation of cytochrome *c* oxidase. *J. Biol. Chem.* 248:5369–78

43. Edwards, J. C., Levens, D., Rabinowitz, M. 1982. Analysis of transcriptional initiation of yeast mitochondrial DNA in a homologous in vitro transcription system. *Cell* 31:337–46

44. Ellis, S. R., Morales, M. J., Li, J.-M., Hopper, A. K., Martin, N. C. 1986. Isolation and characterization of the *TRM1* locus, a gene essential for the m$^2_2$G modification of both mitochondrial and cytoplasmic tRNA in *Saccharomyces cerevisiae*. *J. Biol. Chem.* 261:9703–9

45. Falcone, C., Agostinelli, M., Frontali, L. 1983. Mitochondrial translation products during release from glucose repression in *Saccharomyces cerevisiae*. *J. Bacteriol.* 153:1125–32

46. Fangman, W. L., Henly, J. W., Brewer, B. J. 1990. *RPO41*-independent maintenance of [*rho*⁻] mitochondrial DNA in *Saccharomyces cerevisiae*. *Mol. Cell. Biol.* 10:10–15

47. Fearon, K., Mason, T. L. 1988. Structure and regulation of a nuclear gene in *Saccharomyces cerevisiae* that specifies MRP7, a protein of the large subunit of the mitochondrial ribosome. *Mol. Cell. Biol.* 8:3636–46

48. Fernandes, B. L. 1988. *Characterization of PET1030, a nuclear yeast gene required for the assembly of the mitochondrial enzyme cytochrome c oxidase.* PhD thesis. Cornell Univ., Ithaca, NY. 115 pp.

49. Forsbach, V., Pillar, T., Gottenöf, T., Rödel, G. 1989. Chromosomal localization and expression of *CBS1*, a translational activator of cytochrome *b* in yeast. *Mol. Gen. Genet.* 218:57–63

50. Forsburg, S. L., Guarente, L. 1989. Communication between mitochondria and the nucleus in regulation of cytochrome genes in the yeast *Saccharomyces cerevisiae. Annu. Rev. Cell Biol.* 5:153–80

51. Fox, T. D., Costanzo, M. C., Strick, C. A., Marykwas, D. L., Seaver, E. C., et al. 1988. Translational regulation of mitochondrial gene expression by nuclear genes of *Saccharomyces cerevisiae. Philos. Trans. R. Soc. London Ser. B* 319:97–105

52. Fox, T. D., Folley, L. S., Mulero, J. J., McMullin, T. W., Thorsness, P. E., et al. 1990. Analysis and manipulation of yeast mitochondrial genes. In *Guide to Yeast Genetics and Molecular Biology. Methods Enzymol.* vol. 194, ed. C. Guthrie, G. R. Fink. Orlando, FL: Academic. pp. 149–65

53. Fox, T. D., Sanford, J. C., McMullin, T. W. 1988. Plasmids can stably transform yeast mitochondria lacking endogenous mtDNA. *Proc. Natl. Acad. Sci. USA* 85:7288–92

54. Gampel, A., Nishikimi, M., Tzagoloff, A. 1989. CBP2 protein promotes in vitro excision of a yeast mitochondrial group I intron. *Mol. Cell. Biol.* 9:5424–33

55. Gampel, A., Tzagoloff, A. 1987. In vitro splicing of the terminal intervening sequence of *Saccharomyces cerevisiae* cytochrome *b* pre-mRNA. *Mol. Cell. Biol.* 7:2545–51

56. Gampel, A., Tzagoloff, A. 1989. Homology of aspartyl- and lysyl-tRNA synthetases. *Proc. Natl. Acad. Sci. USA* 86:6023–27

57. Goldthwaite, C. D., Cryer, D. R., Marmur, J. 1974. Effect of carbon source on the replication and transmission of yeast mitochondrial genomes. *Mol. Gen. Genet.* 133:87–104

58. Gray, M. W. 1989. Origin and evolution of mitochondrial DNA. *Annu. Rev. Cell Biol.* 5:25–50

59. Greenleaf, A. L., Kelly, J. L., Lehman, I. R. 1986. Yeast *RPO41* gene product is required for transcription and maintenance of the mitochondrial genome. *Proc. Natl. Acad. Sci. USA* 83:3391–94

60. Grivell, L. A. 1989. Nucleo-mitochondrial interactions in yeast mitochondrial biogenesis. *Eur. J. Biochem.* 182:477–93

61. Grivell, L. A., Schweyen, R. J. 1989. RNA splicing in yeast mitochondria: taking out the twists. *Trends Genet.* 5:39–41

62. Grohmann, L., Graack, H.-R., Kitakawa, M. 1989. Molecular cloning of the nuclear gene for mitochondrial ribosomal protein YmL31 from *Saccharomyces cerevisiae. Eur. J. Biochem.* 183:155–60

63. Gualberto, J. M., Lamattina, L., Bonnard, G., Weil, J.-H., Grienenberger, J.-M. 1989. RNA editing in wheat mitochondria results in the conservation of protein sequences. *Nature* 341:660–62

64. Haffter, P., McMullin, T. W., Fox, T. D. 1990. Functional interactions among two yeast mitochondrial ribosomal proteins and an mRNA-specific translational activator. *Genetics*. In press

65. Haffter, P., McMullin, T. W., Fox, T. D. 1990. A genetic link between an mRNA-specific translational activator and the translation system in yeast mitochondria. *Genetics* 125:495–503

66. Herbert, C. J., Labouesse, M., Dujardin, G., Slonimski, P. P. 1988. The NAM2 proteins from *S. cerevisiae* and *S. douglasii* are mitochondrial leucyl-tRNA synthetases, and are involved in mRNA splicing. *EMBO J.* 7:473–83

67. Hiesel, R., Wissinger, B., Schuster, W., Brennicke, A. 1989. RNA editing in plant mitochondria. *Science* 246:1632–34

68. Hill, J., McGraw, P., Tzagoloff, A. 1985. A mutation in yeast mitochondrial DNA results in a precise excision of the terminal intron of the cytochrome *b* gene. *J. Biol. Chem.* 260:3235–38

69. Hinnebusch, A. G. 1988. Novel mechanisms of translational control in *Saccharomyces cerevisiae. Trends Genet.* 4:169–74

70. Hollingsworth, M. J., Martin, N. C. 1986. RNase P activity in the mitochondria of *Saccharomyces cerevisiae* depends on both mitochondrion and nucleus-encoded components. *Mol. Cell. Biol.* 6:1058–64

71. Hüttenhofer, A., Weiss-Brummer, B.,

Dirheimer, G., Martin, R. P. 1990. A novel type of +1 frameshift suppressor: a base substitution in the anticodon stem of a yeast mitochondrial serine-tRNA causes frameshift suppression. *EMBO J.* 9:551–58

72. Johnston, S. A., Anziano, P. Q., Shark, K., Sanford, J. C., Butow, R. A. 1988. Mitochondrial transformation in yeast by bombardment with microprojectiles. *Science* 240:1538–41

73. Kelly, J. L., Lehman, I. R. 1986. Yeast mitochondrial RNA polymerase. Purification and properties of the catalytic subunit. *J. Biol. Chem.* 261:10340–47

74. Kelly, R., Phillips, S. L. 1983. Comparison of the levels of the 21S mitochondrial rRNA in derepressed and glucose-repressed *Saccharomyces cerevisiae*. *Mol. Cell. Biol.* 3:1949–57

75. Kloeckener-Gruissem, B., McEwen, J. E., Poyton, R. O. 1987. Nuclear functions required for cytochrome *c* oxidase biogenesis in *Saccharomyces cerevisiae:* multiple trans-acting nuclear genes exert specific effects on the expression of each of the cytochrome *c* oxidase subunits encoded on mitochondrial DNA. *Curr. Genet.* 12:311–22

76. Kloeckener-Gruissem, B., McEwen, J. E., Poyton, R. O. 1988. Identification of a third nuclear protein-coding gene required specifically for posttranscriptional expression of the mitochondrial *COX3* gene in *Saccharomyces cerevisiae*. *J. Bacteriol.* 170:1399–402

77. Koerner, T. J., Myers, A. M., Lee, S., Tzagoloff, A. 1987. Isolation and characterization of the yeast gene coding for the α subunit of mitochondrial phenylalanyl-tRNA synthetase. *J. Biol. Chem.* 262:3690–96

78. Körte, A., Forsbach, V., Gottenöf, T., Rödel, G. 1989. In vitro and in vivo studies on the mitochondrial import of CBS1, a translational activator of cytochrome *b* in yeast. *Mol. Gen. Genet.* 217:162–67

79. Kotylak, Z., Lazowska, J., Hawthorne, D. C., Slonimski, P. P. 1985. Intron encoded proteins of mitochondria: key elements of gene expression and genomic evolution. In *Achievements and Perspectives in Mitochondrial Research,* ed. E. Quagliariello, E. C. Slater, F. Palmieri, C. Saccone, A. M. Kroon, pp. 1–20. Amsterdam: Elsevier

80. Kreike, J., Schulze, M., Ahne, F., Lang, B. F. 1987. A yeast nuclear gene, *MRS1,* involved in mitochondrial RNA splicing: nucleotide sequence and mutational analysis of two overlapping open reading frames on opposite strands. *EMBO J.* 6:2123–29

81. Kreike, J., Schulze, M., Pillar, T., Körte, A., Rödel, G. 1986. Cloning of a nuclear gene *MRS1* involved in the excision of a single group I intron (bI3) from the mitochondrial *COB* transcript in *S. cerevisiae. Curr. Genet.* 11:185–91

82. Lambowitz, A. M. 1989. Infectious introns. *Cell* 56:323–26

83. Li, M., Tzagoloff, A., Underbrink-Lyon, K., Martin, N. C. 1982. Identification of the paromomycin-resistance mutation in the 15S rRNA gene of yeast mitochondria. *J. Biol. Chem.* 257:5921–28

84. Linder, P. L., Slonimski, P. P. 1989. An essential yeast protein, encoded by duplicated genes *TIF1* and *TIF2* and homologous to the mammalian translation initiation factor eIF-4A, can suppress a mitochondrial missense mutation. *Proc. Natl. Acad. Sci. USA* 86:2286–90

85. Lisowsky, T. 1990. Molecular analysis of the mitochondrial transcription factor mtf2 of *Saccharomyces cerevisiae. Mol. Gen. Genet.* 220:186–90

86. Lisowsky, T., Michaelis, G. 1989. Mutations in the genes for mitochondrial RNA polymerase and a second mitochondrial transcription factor of *Saccharomyces cerevisiae. Mol. Gen. Genet.* 219:125–28

87. Lisowsky, T., Schweizer, E., Michaelis, G. 1987. A nuclear mutation affecting mitochondrial transcription in *Saccharomyces cerevisiae. Eur. J. Biochem.* 164:559–63

88. Locker, J., Lewin, A., Rabinowitz, M. 1979. The structure and organization of mitochondrial DNA from petite yeast. *Plasmid* 2:155–81

89. Marczynski, G. T., Schultz, P. W., Jaehning, J. A. 1989. Use of yeast nuclear DNA sequences to define the mitochondrial RNA polymerase promoter in vitro. *Mol. Cell. Biol.* 9:3193–202

90. Marykwas, D. L., Fox, T. D. 1989. Control of the *Saccharomyces cerevisiae* regulatory gene *PET494:* transcriptional repression by glucose and translational induction by oxygen. *Mol. Cell. Biol.* 9:484–91

91. Masters, B. S., Stohl, L. L., Clayton, D. A. 1987. Yeast mitochondrial RNA polymerase is homologous to those encoded by bacteriophages T3 and T7. *Cell* 51:89–99

92. Matsushita, Y., Kitakawa, M., Isono, K. 1989. Cloning and analysis of the nuclear genes for two mitochondrial

ribosomal proteins in yeast. *Mol. Gen. Genet.* 219:119–24

93. McGraw, P., Tzagoloff, A. 1983. Assembly of the mitochondrial membrane system: characterization of a yeast nuclear gene involved in processing of a cytochrome *b* pre-mRNA. *J. Biol. Chem.* 258:9459–68

94. McMullin, T. W., Haffter, P., Fox, T. D. 1990. A novel small subunit ribosomal protein of yeast mitochondria that interacts functionally with an mRNA-specific translational activator. *Mol. Cell. Biol.* 10:4590–95

95. McMullin, T. W., Hallberg, R. L. 1988. A highly evolutionarily conserved mitochondrial protein is structurally related to the protein encoded by the *Escherichia coli groEL* gene. *Mol. Cell. Biol.* 8:371–80

96. Michaelis, U., Schlapp, T., Rödel, G. 1988. Yeast nuclear gene *CBS2*, required for translational activation of cytochrome *b*, encodes a basic protein of 45 kDa. *Mol. Gen. Genet.* 214:263–70

97. Michon, T., Galante, M., Velours, J. 1988. NH2-terminal sequence of the isolated yeast ATP synthase subunit 6 reveals posttranslational cleavage. *Eur. J. Biochem.* 172:621–25

98. Mieszczak, M., Kozlowski, M., Zagorski, W. 1988. Protein composition of *Saccharomyces cerevisiae* mitochondrial ribosomes. *Acta Biochim. Pol.* 35:105–18

99. Morales, M. J., Wise, C. A., Hollingsworth, M. J., Martin, N. C. 1989. Characterization of yeast mitochondrial RNase P: an intact RNA subunit is not essential for activity in vitro. *Nucleic Acids Res.* 17:6865–81

100. Mörl, M., Schmelzer, C. 1990. Integration of group II intron bI1 into a foreign RNA by reversal of the self-splicing reaction in vitro. *Cell* 60:629–36

101. Morris, C. E., McGraw, N. J., Joho, K., Brown, J. E., Klement, J. F., et al. 1987. Mechanisms of promoter recognition by the bacteriophage T3 and T7 RNA polymerases. In *RNA Polymerase and the Regulation of Transcription*, ed. W. S. Reznikoff, R. R. Burgess, J. E. Dahlberg, C. A. Gross, M. P. Record, Jr., M. P. Wickens, pp. 47–58. New York: Elsevier

102. Mueller, D. M., Biswas, T. K., Backer, J., Edwards, J. C., Rabinowitz, M., et al. 1987. Temperature sensitive *pet* mutants in yeast *Saccharomyces cerevisiae* that lose mitochondrial RNA. *Curr. Genet.* 11:359–67

103. Mueller, D. M., Getz, G. S. 1986. Steady state analysis of mitochondrial RNA after growth of yeast *Saccharomyces cerevisiae* under catabolite repression and derepression. *J. Biol. Chem.* 261:11816–22

104. Mueller, D. M., Getz, G. S. 1986. Transcriptional regulation of the mitochondrial genome of yeast *Saccharomyces cerevisiae*. *J. Biol. Chem.* 261:11756–64

105. Müller, P. P., Reif, M. K., Zonghou, S., Sengstag, C., Mason, T. L., et al. 1984. A nuclear mutation that posttranscriptionally blocks accumulation of a yeast mitochondrial gene product can be suppressed by a mitochondrial gene rearrangement. *J. Mol. Biol.* 175:431–52

106. Myers, A. M., Crivellone, M. D., Tzagoloff, A. 1987. Assembly of the mitochondrial membrane system: *MRP1* and *MRP2*, two yeast nuclear genes coding for mitochondrial ribosomal proteins. *J. Biol. Chem.* 262:3388–97

107. Myers, A. M., Pape, L. K., Tzagoloff, A. 1985. Mitochondrial protein synthesis is required for maintenance of intact mitochondrial genomes in *Saccharomyces cerevisiae*. *EMBO J.* 4:2087–92

108. Myers, A. M., Tzagoloff, A. 1985. *MSW*, a yeast gene coding for mitochondrial tryptophanyl-tRNA synthetase. *J. Biol. Chem.* 260:15371–77

109. Nagata, S., Tsunetsugu-Yokota, Y., Naito, A., Kaziro, Y. 1983. Molecular cloning and sequence determination of the nuclear gene coding for mitochondrial elongation factor Tu of *S. cerevisiae*. *Proc. Natl. Acad. Sci. USA* 80:6192–96

110. Najarian, D., Dihanich, M. E., Martin, N. C., Hopper, A. K. 1987. DNA sequence and transcript mapping of *MOD5*: features of the 5' region which suggest two translational starts. *Mol. Cell. Biol.* 7:185–91

111. Natsoulis, G., Hilger, F., Fink, G. R. 1986. The *HTS1* gene encodes both the cytoplasmic and mitochondrial histidine tRNA synthetases of *S. cerevisiae*. *Cell* 46:235–43

112. Ohmen, J. D., Kloeckener-Gruissem, B., McEwen, J. E. 1988. Molecular cloning and nucleotide sequence of the nuclear *PET122* gene required for expression of the mitochondrial *COX3* gene in *S. cerevisiae*. *Nucleic Acids Res.* 16:10783–802

113. Ooi, B. G., Lukins, H. B., Linnane, A. W., Nagley, P. 1987. Biogenesis of mitochondria: a mutation in the 5'-untranslated region of yeast mitochondrial *oli1* mRNA leading to impairment in translation of subunit 9 of the

mitochondrial ATPase complex. *Nucleic Acids Res.* 15:1965–77

114. Osinga, K. A., De Vries, E., Van der Horst, G., Tabak, H. F. 1984. Processing of yeast mitochondrial messenger RNAs at a conserved dodecamer sequence. *EMBO J.* 3:829–34

115. Osinga, K. A., Tabak, H. F. 1982. Initiation of transcription of genes for mitochondrial ribosomal RNA in yeast: comparison of the nucleotide sequence around the 5'-ends of both genes reveals a homologous stretch of 17 nucleotides. *Nucleic Acids Res.* 10:3617–23

116. Pape, L. K., Koerner, T. J., Tzagoloff, A. 1985. Characterization of a yeast nuclear gene *(MST1)* coding for the mitochondrial threonyl-tRNA synthetase. *J. Biol. Chem.* 260:15362–70

117. Parikh, V. S., Morgan, M. M., Scott, R., Clements, L. S., Butow, R. A. 1987. The mitochondrial genotype can influence nuclear gene expression in yeast. *Science* 235:576–80

118. Partaledis, J. A., Mason, T. L. 1988. Structure and regulation of a nuclear gene in *Saccharomyces cerevisiae* that specifies MRP13, a protein of the small subunit of the mitochondrial ribosome. *Mol. Cell. Biol.* 8:3647–60

119. Partono, S., Lewin, A. S. 1988. Autocatalytic activities of intron 5 of the *cob* gene of yeast mitochondria. *Mol. Cell. Biol.* 8:2562–71

120. Perlman, P. S., Birky, C. W. J., Strausberg, R. L. 1979. Segregation of mitochondrial markers in yeast. *Methods Enzymol.* 56:139–54

121. Perlman, P. S., Butow, R. A. 1989. Mobile introns and intron-encoded proteins. *Science* 246:1106–9

122. Poutre, C. G., Fox, T. D. 1987. *PET111*, a *Saccharomyces cerevisiae* nuclear gene required for translation of the mitochondrial mRNA encoding cytochrome *c* oxidase subunit II. *Genetics* 115:637–47

123. Pratje, E., Guiard, B. 1986. One nuclear gene controls the removal of transient pre-sequences from two yeast proteins: one encoded by the nuclear the other by the mitochondrial genome. *EMBO J.* 5:1313–17

124. Pratje, E., Mannhaupt, G., Michaelis, G., Beyreuther, K. 1983. A nuclear mutation prevents processing of a mitochondrially encoded membrane protein in *Saccharomyces cerevisiae*. *EMBO J.* 2:1049–54

125. Raitio, M., Jalli, T., Saraste, M. 1987. Isolation and analysis of the genes for cytochrome *c* oxidase in *Paracoccus denitrificans*. *EMBO J.* 6:2825–33

126. Rödel, G. 1986. Two yeast nuclear genes, *CBS1* and *CBS2*, are required for translation of mitochondrial transcripts bearing the 5'-untranslated *COB* leader. *Curr. Genet.* 11:41–45

127. Rödel, G., Fox, T. D. 1987. The yeast nuclear gene *CBS1* is required for translation of mitochondrial mRNAs bearing the *cob* 5'-untranslated leader. *Mol. Gen. Genet.* 206:45–50

128. Rödel, G., Körte, A., Kaudewitz, F. 1985. Mitochondrial suppression of a yeat nuclear mutation which affects the translation of the mitochondrial apocytochrome *b* transcript. *Curr. Genet.* 9:641–48

129. Schinkel, A. H., Groot Koerkamp, M. J. A., Tabak, H. F. 1988. Mitochondrial RNA polymerase of *Saccharomyces cerevisiae:* composition and mechanism of promoter recognition. *EMBO J.* 7:3255–62

130. Schinkel, A. H., Groot Koerkamp, M. J. A., Teunissen, A. W. R. H., Tabak, H. F. 1988. RNA polymerase induces DNA bending at yeast mitochondrial promoters. *Nucleic Acids Res.* 16:9147–63

131. Schinkel, A. H., Groot Koerkamp, M. J. A., Touw, E. P. W., Tabak, H. F. 1987. Specificity factor of yeast mitochondrial RNA polymerase. Purification and interaction with core RNA polymerase. *J. Biol. Chem.* 262:12785–91

132. Schinkel, A. H., Groot Koerkamp, M. J. A., Van der Horst, G. T. J., Touw, E. P. W., Osinga, K. A., et al. 1986. Characterization of the promoter of the large ribosomal RNA gene in yeast mitochondria and separation of mitochondrial RNA polymerase into two different functional components. *EMBO J.* 5:1041–47

133. Schinkel, A. H., Tabak, H. F. 1989. Mitochondrial RNA polymerase: dual role in transcription and replication. *Trends Genet.* 5:149–54

134. Schulze, M., Rödel, G. 1988. *SCO1*, a yeast nuclear gene essential for accumulation of mitochondrial cytochrome coxidase subunit II. *Mol. Gen. Genet.* 211:492–98

135. Schulze, M., Rödel, G. 1989. Accumulation of the cytochrome *c* oxidase subunits I and II in yeast requires a mitochondrial membrane-associated protein, encoded by the nuclear *SCO1* gene. *Mol. Gen. Genet.* 216:37–43

136. Séraphin, B., Simon, M., Boulet, A., Faye, G. 1989. Mitochondrial splicing requires a protein from a novel helicase family. *Nature* 337:84–87

137. Séraphin, B., Simon, M., Faye, G. 1988. *MSS18,* a yeast nuclear gene involved in the splicing of intron al5β of the mitochondrial *coxI* transcript. *EMBO J.* 7:1455–64

138. Séraphin, G., Boulet, A., Simon, M., Faye, G. 1987. Construction of a yeast strain devoid of mitochondrial introns and its use to screen nuclear genes involved in mitochondrial splicing. *Proc. Natl. Acad. Sci. USA* 84:6810–14

139. Shen, Z., Fox, T. D. 1989. Substitution of an invariant nucleotide at the base of the highly conserved "530-loop" of 15S rRNA causes suppression of mitochondrial ochre mutations. *Nucleic Acids Res.* 17:4535–39

140. Simpson, L., Shaw, J. 1989. RNA editing and the mitochondrial cryptogenes of kinetoplastid protozoa. *Cell* 57:355–66

141. Strick, C. A., Fox, T. D. 1987. *Saccharomyces cerevisiae* positive regulatory gene *PET111* encodes a mitochondrial protein that is translated from an mRNA with a long 5' leader. *Mol. Cell. Biol.* 7:2728–34

142. Thalenfeld, B. E., Bonitz, S. G., Nobrega, F. G., Macino, G., Tzagoloff, A. 1983. *oli1* transcripts in wild type and in cytoplasmic "petite" mutant of yeast. *J. Biol. Chem.* 258:14065–68

143. Ticho, B. S., Getz, G. S. 1988. The characterization of yeast mitochondrial RNA polymerase: a monomer of 150,000 daltons with a transcription factor of 70,000 daltons. *J. Biol. Chem.* 263:10096–103

144. Tzagoloff, A., Akai, A., Kurkulos, M., Repetto, B. 1988. Homology of yeast mitochondrial leucyl-tRNA synthetase and isoleucyl- and methionyl-tRNA synthetases of *Escherichia coli. J. Biol. Chem.* 263:850–56

145. Tzagoloff, A., Dieckmann, C. L. 1990. *PET* genes of *Saccharomyces cerevisiae. Microbiol. Rev.* In press

146. Tzagoloff, A., Vambutas, A., Akai, A. 1989. Characterization of *MSM1,* the structural gene for yeast mitochondrial methionyl-tRNA synthetase. *Eur. J. Biochem.* 179:365–71

147. Valencik, M. L., Kloeckener-Gruissem, B., Poyton, R. O., McEwen, J. E. 1989. Disruption of the yeast nuclear *PET54* gene blocks excision of mitochondrial intron a15β from pre-

mRNA for cytochrome *c* oxidase subunit I. *EMBO J.* 8:3899–904

148. Weber, E. R., Dieckmann, C. L. 1990. Identification of the CBP1 polypeptide in mitochondrial extracts from *Saccharomyces cerevisiae. J. Biol. Chem.* 265:1594–600

149. Weiss-Brummer, B., Hüttenhofer, A. 1989. The paromomycin resistance mutation (*par*I-454) in the 15S rRNA gene of the yeast *Saccharomyces cerevisiae* is involved in ribosomal frameshifting. *Mol. Gen. Genet.* 217:362–69

150. Wettstein-Edwards, J., Ticho, B. S., Martin, N. C., Najarian, D., Getz, G. S. 1986. In vitro transcription and promoter strength analysis of five mitochondrial tRNA promoters in yeast. *J. Biol. Chem.* 261:2905–11

151. Wilcoxen, S. E., Peterson, C. R., Winkley, C. S., Keller, M. J., Jaehning, J. A. 1988. Two forms of *RPO41*-dependent RNA polymerase: regulation of the RNA polymerase by glucose repression may control yeast mitochondrial gene expression. *J. Biol. Chem.* 263:12346–51

152. Woodrow, G., Schatz, G. 1979. The role of oxygen in the biosynthesis of cytochrome *c* oxidase of yeast mitochondria. *J. Biol. Chem.* 254:6088–93

153. Wu, M., Tzagoloff, A. 1989. Identification and characterization of a new gene *(CBP3)* required for the expression of a yeast coenzyme H₂-cytochrome *c* reductase. *J. Biol. Chem.* 264:11122–30

154. Zennaro, E., Grimaldi, L., Baldacci, G., Frontali, L. 1985. Mitochondrial transcription and processing of transcripts during release from glucose repression in "resting cells" of *Saccharomyces cerevisiae. Eur. J. Biochem.* 147:191–96

155. Zhu, H., Conrad-Webb, H., Liao, X. S., Perlman, P. S., Butow, R. A. 1989. Functional expression of a yeast mitochondrial intron-encoded protein requires RNA processing at a conserved dodecamer sequence at the 3' end of the gene. *Mol. Cell. Biol.* 9:1507–12

156. Zhu, H., Macreadie, I. G., Butow, R. A. 1987. RNA processing and expression of an intron-encoded protein in yeast mitochondria: role of a conserved dodecamer sequence. *Mol. Cell. Biol.* 7:2530–37

Annu. Rev. Genet. 1990. 24:115–32

GENETICS OF RESPONSE TO SLOW VIRUS (PRION) INFECTION

David T. Kingsbury

Department of Microbiology, The George Washington University Medical Center, 2300 I Street, N.W., Washington, D.C. 20037

KEY WORDS: scrapie, Creutzfeldt-Jakob disease, neurological disease, prion protein, murine H-2 complex

CONTENTS

INTRODUCTION

For several decades there has been speculation about the fundamental nature of the etiologic agent of the neurodegenerative diseases of scrapie, kuru, Creutzfeldt-Jakob disease (CJD), Gerstmann-Straussler syndrome (GSS), transmissible mink encephalopathy, bovine spongiform encephalopathy, and chronic wasting disease of deer and elk (28, 51). Scrapie of sheep and goats has been the most widely studied of the diseases, especially in experimental systems. The mystery of the etiology of these diseases derives from the

115

enigmatic properties of the transmissible element itself. Because the agent was historically thought to be a virus, the techniques and paradigms of virology have been generally applied to its characterization. However, the unusual properties of the transmissible element make it clear that these are not simple viral diseases.

Work in several laboratories, especially in the past ten years, has revealed that the agent of these disorders is a novel proteinaceous pathogen, encoded in the host genome (35, 54, 56). Prusiner (52) coined the term *prion* and defined the agents as "small *pro*teinaceous *in*fectious particles that resist inactivation by procedures which modify nucleic acids." The current evidence indicates that prions are composed largely, and perhaps entirely, of an abnormal isoform of a normal brain protein termed the prion protein, or PrP (the cellular isoform is referred to as PrP^C, the scrapie isoform as PrP^{Sc}, the CJD isoform as PrP^{CJD}, etc). Characterization of PrP^C has been limited by the low concentrations in the brain and as a consequence, its normal function remains undetermined. In contrast, the most significant difficulty in obtaining a purified PrP^{Sc} or PrP^{CJD} preparation for full characterization has been the tendency of the protein to form rod-shaped insoluble amyloids (55). The dispersal of the scrapie form of PrP into detergent lipid protein complexes (DLPC) has greatly contributed to clarifying the role of PrP^{Sc} in the development of the disease (25–27). The most significant direct evidence for the role of PrP^{Sc} in the transmission of the disease comes from the ability of polyclonal antibodies raised against purified PrP^{Sc} to partially neutralize scrapie infectivity partitioned into DLPC, and from the copurification of PrP^{Sc} and scrapie infectivity through the use of affinity chromatography on antibody columns containing anti-PrP monoclonal antibodies (26). These results, together with other lines of evidence, have established that PrP^{Sc} is a major and necessary component of the transmissible agent (reviewed in 54).

Understanding the details of the genetic control of prion diseases is central to unraveling the biochemistry and epidemiology of the causative element. The prion diseases are of particular interest because at least two, GSS and familial CJD, seem to lie within the domains of both infectious (transmissible) and genetic diseases (53). Although most CJD cases appear to be sporadic, more like the result of environmental insult than an infectious agent, approximately 5% to 10% of CJD cases are ostensibly genetic in origin (8). GSS appears to be a purely genetic disease, being inherited in an autosomal dominant fashion (2, 32, 46), yet transmissible material may be obtained from the brains of diseased patients and the abnormal form of PrP is present (6, 46). While a few cases of CJD have resulted from iatrogenic transmission (i.e. corneal transplants and implanted brain electrodes), the only known "natural" mechanism of disease transmission in humans is in cases of kuru, where ritualistic cannibalism involving infected brains has been implicated (28). Likewise, during the past two years, throughout much of Britain, there has been a dramatic spread of scrapie to cattle and other animals, resulting from the ingestion of scrapie-contaminated feed (1, 47). Despite intense speculation, no other mechanism of natural transmission has been documented.

There has been considerable controversy about the genetic control of scrapie and CJD. This stems from the lack of agreement about the source of the genetic information that encodes the transmissible agent. Despite widespread agreement about the involvement of PrP, and the chromosomal location of the gene encoding the protein (24), several investigators continue to insist on an exogenous agent (45, 48). Therefore, the literature still contains disagreements regarding the actual molecules even that constitute the agent itself (45, 48). Many of the studies reviewed below have used the classical viral paradigm in the design and interpretation of experiments. Consequently, the unusual nature of the unconventional agents has led to a variety of alternative speculations regarding their mode of replication and mode of action. The prion model represents a radical departure from conventional wisdom and predicts an agent with little or no nucleic acid of its own. "Replication" occurs through the conversation, via posttranslational modification, of the normal endogenous protein PrP^C, which is encoded by the host. Therefore, the genetic make-up of the host is critical to the properties of the agent recovered from a given "infection" (6, 59). While there are well-defined host-range restrictions on the transmission of prion diseases to some species (29), CJD, GSS, and scrapie appear to be virtually identical in the same host. For example, experimental CJD in goats is indistinguishable from naturally occurring scrapie, both clinically and neuropathologically (31).

Beginning with the earliest studies of the scrapie agent, apparent differences were observed in isolates obtained from different sheep. These differences were most apparent in the host range of the individual isolate and the change in that host range as a result of serial passage. This apparent strain variation between scrapie isolates has been documented by several investigators (see for example, 28, 29, 50). These data have been widely interpreted to support the existence of strains of scrapie and to argue that the agent carries an informational molecule of some type. The alternative argument, that these phenotypic changes are a form of "host-induced modification," would also adequately explain most, but not all, of the published observations.

HOST GENETICS AND THE INFLUENCE OF HOST GENES

Genetic Control of Incubation Time

The influence of the host's genetic background on the susceptibility and incubation time of scrapie was first studied over 20 years ago in different flocks of naturally infected sheep (30). While it was recognized that the disease was transmissible from sheep to sheep and from sheep to goats, the original genetic analysis suggested that the disease was inherited and due to "a single autosomal dominant gene" (50). The detailed genetic analysis of scrapie in sheep has been complicated by what the investigators identified as maternal inheritance and the apparent coinfection of flocks with multiple "strains" of scrapie. Unfortunately, the original data are not all available for reexamination. A more thorough and controlled analysis of the genetics of host control of scrapie infection has been carried out in the experimental system in mice. Dickinson and collaborators (18, 21, 22), Kingsbury and collaborators (12, 37, 38), and Carlson et al (11, 13) have examined this system in some detail. The susceptibility of different strains of mice to scrapie and CJD infection has been measured by observing the time from intracerebral inoculation to the onset of clinical disease, a period referred to as the incubation time. It is universally agreed that all strains of mice examined to date are susceptible, but the incubation periods range from 15 weeks to well over 40 weeks (18, 37). The experimental results from this work, and their interpretation, vary between different groups. Dickinson and his collaborators have reported on extensive genetic analysis of a limited number of mouse strains, screened with multiple independent isolates of the scrapie agent. They concluded that incubation time was controlled by a single gene, which they designated *sinc* (for *s*crapie *inc*ubation), with only minor influences from other host genes. *Sinc* was interpreted to be a single genetic locus with two alleles, designated *s7* and *p7*. F_1 hybrid mice between long- and short-incubation period parents showed incubation periods intermediate between the two in most cases. In F_1 mice resulting from the cross between C57BL and VM mice, the incubation period exceeded that of either parent, a phenomenon termed "overdominance" (22, 23). Subsequent work from the same laboratory reported the genetic control of the incubation time to be more complex. In their initial experiments, the C57BL mice, which have a short incubation period when challenged with the ME7 strain of scrapie, were said to be homozygous for the *sinc* gene short-incubation-time allele (i.e. *s7/s7*) and another strain, the VM strain, the prolonged-incubation (when challenged with ME7) allele (*p7/p7*) of the same gene. When these mice were challenged with another "strain" of scrapie (strain 22A), the situation became reversed. These latter results have not been reproducible by other investigators using the same isolates (4).

In contrast to the results just described, Kingsbury et al (37, 38) failed to observe either intermediate incubation times or widespread overdominance. In all cases, the incubation time of the longest-incubation-time parent was dominant. The incubation period only rarely significantly exceeded that of the longest-incubation-time parent; however, in some crosses overdominance did occur. This work was not limited to scrapie, but also included the examination

Table 1   Influence of H-2 haplotype on CJD incubation time

Mouse Strain	No	Onset of disease (days ± SE)	H	I-A	I-E	S	D	Qa-2	Tla	Qa-1
B10.AKM	7	98 ± 3.0	k	k	k	k	q	a	a	a
B10.Q	7	107 ± 0.6	q	q	q	q	q	a	b	b
C57/BL10J	7	138 ± 1.0	b	b	b	b	b	a	b	b
B10.BR	8	146 ± 5.0	k	k	k	k	k	b	b	b
B10.A(3R)	7	142 ± 5.7	b	b	k	d	d	a	a	a
B10.A(4R)	7	138 ± 0.8	k	k	b	b	b	a	b	b
B10.A(5R)	5	146 ± 6.0	b	b	k	d	d	a	a	a
B10.MSn	6	130 ± 5.9	f	f	f	f	f	b	d	b
B10.S	5	138 ± 9.0	s	s	s	s	s	a	b	b
B10.P	8	144 ± 7.2	p	p	p	p	p	b	c	a
B10.A	7	153 ± 9.9	k	k	d	d	d	a	a	a
B10.PL	10	169 ± 1.2	u	u	u	u	d	a	a	a
B10.D2/nSn	7	150 ± 0.8	d	d	d	d	d	a	c	b

Weanling mice were inoculated intracerebrally with 10^4 ID50 units of the CDJ isolate Fukuoka-1 (adapted from ref. 37).

of the genetic control of the incubation time of the Fukuoka-1 isolate of CJD. This work showed that the mouse incubation time appeared to be controlled by at least two loci (37). One, designated *Pid-1* (*p*rion *i*ncubation *d*etermi-nant), maps within the murine major histocompatibility complex (H-2) on mouse chromosome 17. The second, originally designated *Pid-2* and now as *Prn-i*, maps to mouse chromosome 2, and exerts a profound effect on the incubation time. Speculation about the possible identity of this gene is discussed below.

The *Pid-1* locus (37) was identified in a group of congenic mice, differing only in their H-2 haplotypes. This work mapped the *Pid-1* locus near the D region of the H-2 complex on mouse chromosome 17, between the S and L regions. The q allele at the D locus resulted in shorter incubation times whereas the d allele resulted in long incubation times. The p, s, b, and k alleles gave intermediate incubation times. These results are summarized in Table 1.

Prn-i was identified through the extended incubation period in the I/LnJ mouse (12, 37). I/LnJ mice have incubation periods of greater than 300 days for scrapie and well over 400 days for CJD. When these mice are crossed with B10.Q (CJD incubation period approximately 110 days) the resulting F_1 offspring have incubation periods virtually identical to the I/LnJ parents. Scrapie incubation times in offspring from a cross between NZW and I/LnJ mice also exhibit the I/LnJ long incubation time with no intermediate times noted (37). Further genetic analysis of the *Prn-i* locus clearly showed a single genetic locus that is autosomal dominant (36). These experiments also included an examination of both scrapie and CJD in a controlled group of mice. Subsequent molecular analysis of DNA from short- and long-incubation-time mice clearly established a linkage between the *Prn-i* locus and *Prn-p* the gene for PrP, the prion protein (11, 36). DNA restriction analysis revealed the presence of an *Xba I* restriction fragment length polymorphism (RFLP) that correlated with the incubation-time phenotype. In short-incubation-time animals, the *Prn-p* DNA probe (the hamster or mouse cDNA clone for the prion protein gene, PrP) hybridized only with a 3.8-kb DNA fragment, whereas in the I/LnJ mice the probe reacted with a 5.5-kb fragment. F1 offspring of the two displayed one copy of each gene (11, 36). Extensive screening of additional mouse strains and the development of additional RFLPs with other restriction enzymes identified at least six *Prn-p* haplotypes (11), designated *Prn-p^{a-f}*. The most common haplotype, associated with the short-incubation-time C57BL/6, BALB/c, NZW etc (see Table 2 in 11) mice has been designated the *a* haplotype, the long-incubation-time allele found in I/LnJ mice the *b* haplotype and a series of intermediate-incubation-time mice represent the *c–f* haplotypes. This work shows that *Prn-i* (the incubation time gene) and *Prn-p* (the prion protein gene) are very closely linked, and well may be the same gene. These genes have been mapped to mouse chromosome 2 and fall between beta-2-microglobulin and agouti (11). Additional mapping is in progress, as are attempts to determine the gene order or identity of *Prn-i* and *Prn-p*.

Work by Carp et al (14) and by Hunter et al (33) have provided good evidence that *Prn-p* and *sinc* are the same gene. However, the analysis of the genotypes within this complex has become more confusing. As mentioned earlier, flanking polymorphic restriction sites define six *Prn-p* haplotypes, but there are only two observed phenotypes for incubation time. This means that while there are two phenotypic alleles of *sinc* as described by Dickinson and colleagues, there are at least six genotypic alleles.

Recently, work by Scott et al (59) has added much additional information. These investigators produced transgenic mice that carried multiple copies of the hamster PrP gene. Under these conditions the incubation time and host susceptibility were controlled by the transgene and not the parental murine gene. While definitive proof is still needed that no other material (e.g. the gene encoding the hypothetical *Prn-i* region) was incorporated into the DNA clone used to produce the transgenic animals, *Prn-i* and *Prn-p*, if not identical, clearly lie adjacent on the chromosome. The current evidence strongly suggests that they are identical. The details of the transgene studies are described below in the section on the genetic control of PrP.

Additional evidence linking the gene for PrP with incubation-time control comes from work comparing three species of hamsters. Several different scrapie isolates have been passed in Syrian hamsters and most of the biochemical properties of the prion protein have been obtained from the study of material purified from Syrian hamster brain. Recent work (43) compared various parameters of scrapie in Syrian, Armenian, and Chinese hamsters. The results demonstrated that each species had a unique incubation time when inoculated with scrapie prions. Passage studies confirmed earlier work in other species, and showed that the host species, and not the source of the scrapie prions, determined the incubation time. Detailed study of the prion proteins from the three hamster species demonstrated that each was unique, but that the differences were located within a hydrophilic segment of 11 amino acids that coincides with the location of the amino acid difference in the I/LnJ mouse and in the human GSS PrP gene (32, 66) (see below).

Genetic Control of PrP

Several laboratories have reported obtaining recombinant DNA clones that encode some or all of the information for the scrapie protein PrP (15, 49, 57). The most widely studied clone was selected from a scrapie-infected hamster-brain cDNA library through identification with an oligonucleotide probe corresponding to the N-terminus of the protein (3, 49). Southern blotting of hamster genomic DNA with this cDNA identified a single copy gene with an identical restriction map in both infected and uninfected brain DNA. A comparable single copy gene was also identified in both mouse and human DNA. Messenger RNA is present in both normal and infected tissue and the mRNA levels do not seem to vary during the course of disease. Expression of the PrP gene is clearly tissue specific. These data and the coincidence of the computer-translated amino acid sequence of the long open reading frame of the clone with the known amino acid sequence of hamster-scrapie PrP show that PrP is a host-encoded product. Figure 1 diagrams the known organization of the hamster *Prn-p* gene. To date no specific prion-associated nucleic acids have been identified in purified preparations of PrP from scrapie or CJD.

The observed differences in the CJD proteins from different strains of mice suggested that there might be some polymorphism within the PrP gene (36). Mouse DNAs were analyzed by Southern blotting for restriction fragment length polymorphisms within the gene and the PrP genes from different mouse strains were sequenced (42, 65). As described earlier, I/LnJ and several other strains showed unique restriction patterns. These results indicate that I/LnJ DNA has an *XbaI* restriction site that is not present in the B10.Q. The F1 mice clearly have a single copy of each marker. The presence of the *XbaI* site does not correlate with the H-2 haplotype in B10 or other strains and no work has been done to relate the H-2 linked marker with a specific RFLP.

Carlson et al (11) screened 55 inbred mouse strains for *Prn-p* related RFLPs. As described above, six constellations of RFLPs were detected,

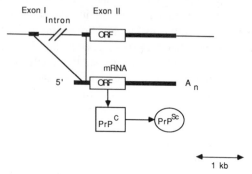

Figure 1   Organization and expression of the hamster prion protein (PrP) gene *Prn-p*. Figure adapted from (54) and including the newly recognized fact that PrP^{Sc} is not synthesized directly from the mRNA but is a posttranslational product of PrP^C. The open reading frame is designated ORF and the untranslated regions of the mRNA are represented by heavy solid dark lines.

defining six haplotypes of the prion protein gene. Mice with $b,c,d,e,$ and f haplotypes all share the 5.2 kb-*Xbal Prn-p* fragment that was initially used to define the *Prn-p^b* haplotype and then the *sinc^{p7}* allele. In all of the haplotypes examined the overall structure of the *Prn-p* gene was essentially similar and most of the RFLP markers lie in regions flanking the open reading frame. Physical mapping has placed the first exon containing most of the 5' untranslated (UT) sequence between 11.5 and 23.5 kb upstream from the exon 2 splice acceptor site (Figure 1) (64). A *BamH1* restriction site polymorphism, detected with a probe derived from the 5' UT sequences, differentiates the *a, d,* and *e* haplotypes of *Prn-p* (10.9-kb fragment) from the *b, c,* and *f* alleles (12.2-kb fragment). All of the other polymorphisms demonstrated were randomly distributed around exon 2. A single restriction enzyme polymorphism was identified within the exon 2 open reading frame resulting from *BstEII* digestion. It corresponds to the Thr to Val transition identified at codon 189 in the translated sequence (65). The overall organization of the *Prn-p* gene appears to be similar in all mice, but evidence shows that the *Prn-p^b* haplotype is unique in two respects. First, the PrP derived from *Prn-p^b* mice has two amino acid changes, a phenylalanine for leucine at position 108 and the valine for threonine at position 189, mentioned above. The second difference associated with every *Prn-p^b* mouse strain tested is their prolonged incubation time following inoculation with scrapie or CJD. These findings are compatible with the prion hypothesis and the argument that *Prn-p* and *Prn-i* are two aspects of the same gene.

Scott et al (59) have collected considerable additional information about this system. Their transfer of hamster *Prn-p* to the mouse and the subsequent production of hamster PrP have greatly modified perceptions of this system.

This work has clearly and definitively demonstrated that the PrP gene controls not only the production of PrP but also the susceptibility to disease caused by scrapie derived from other species and the incubation time. These investigators produced three transgenic mouse lines containing the PrP gene derived from Syrian hamsters. The genetic construction was built around a *NotI* fragment containing both exons 1 and 2 as well as 5' and 3' flanking sequences of 24 and 6 kb, respectively. This purified cloned insert was microinjected to produce transgenic mice and three stable lines were derived. A single transgenic mouse line was also derived using the untranslated leader sequence and the open reading frame, but these animals failed to show any effect of the hamster gene. Two of the three lines derived from the large *NotI* fragment showed 4–8 hamster-gene copies per mouse genome and one line, Tg 81, harbored 30–50 copies of the transgene, most likely integrated as a tandem array as observed for other transgene introductions (7, 41).

Northern and Western blot analysis of brains from these transgenic animals revealed that they produced hamster PrP mRNA of identical size to that seen in hamsters, and hamster-specific PrP itself. The Westerns were developed with species-specific antihamster antibodies to eliminate any background from mouse PrP. Surprisingly, the results appeared to indicate that, despite the high copy number in the cells of these animals, the amount of PrP produced was not significantly higher than in control hamsters. These animals, when inoculated with hamster scrapie, showed a clear reduction in incubation period. In the most extreme case of line Tg 81, the high copy number animals, the incubation period for hamster scrapie fell to 70–75 days while control animals had incubation periods of more than 600 days. The converse experiment, inoculation with mouse scrapie, demonstrated that the presence of the hamster PrP gene interfered with the infection, especially in the high copy number line Tg 81. Examination of the scrapie agent resulting from the infection of these mice revealed that they had not only become sensitive to hamster scrapie, but also produced hamster scrapie prions following infection. Furthermore, the titers of scrapie infectivity obtained were much higher than those obtained in the usual mouse experimental system and approximated those seen in hamsters.

An examination of the pathology of the scrapie brains from the transgenic animals was consistent with the other findings. There were several neuropathological features characteristic of hamster scrapie and not normally found in the mouse disease. Likewise, amyloid plaques immunopositive for hamster-specific monoclonal antibodies, were seen in several parts of the brain. This is not characteristic for murine scrapie in which amyloid plaque formation is unusual.

The results just described indicate that the genetic control of most aspects of scrapie infection is associated with the gene for the prion protein. Clearly, the characteristics of host susceptibility, incubation time, amyloid plaque production, and distribution of pathological lesions, as well as the species-specificity of the infectious material produced were all under the influence of this single transgenic insert. It will be interesting to extend this work to include the inoculation of human brain material or CJD lines derived from human brains. The Fukuoka-1 isolate has not been successfully transferred to hamsters despite repeated attempts. In contrast, Manuelidis et al (44) have reported the transmission of several human cases to hamsters. One inference of the prion hypothesis is that scrapie and CJD should be essentially identical diseases in the same host. It is possible that these transgenic mice will provide an effective vehicle for testing that aspect of the hypothesis.

Familial Transmission of GSS

Gerstmann-Straussler syndrome is a rare, vertically transmitted disease thought to be a purely genetic variant of CJD. The brains of GSS patients contain protease-resistant prion protein indistinguishable from that seen in brains of patients dying from CJD (6, 39, 58). Likewise, GSS has been successfully transmitted to experimental animals (46).

Hsiao et al (32) examined the prion protein genes of different patients with GSS and compared those to the genes in their unaffected family members. The results of this work were striking. Following sequence determination of each of the alleles of the PrP open reading frame in a GSS patient, one of the two was plainly a previously unrecognized form that showed an altered codon (a silent adenine to guanine shift in position 3 at position 117 of the protein, which altered a PvuII site present in 90% of the population (67). In that same allele there had also been a cytosine to thymine substitution in position 2 of codon 102, resulting in the substitution of a leucine for proline. This single base change also created a specific marker in the form of a DdeI restriction site. No other changes were noted in this patient and the other allele was identical to a previously published sequence thought to be the wild-type form (40).

Extensive Southern blotting of DdeI-digested DNA from normal controls as well as from family members from two pedigrees with a high incidence of GSS revealed that the codon 102 change (the presence of the DdeI site) was present in every case, and was never found in unaffected individuals. The codon 107 change was present in one pedigree but absent in the other, suggesting that it was an unlinked silent mutation.

These results establish that GSS is truly a genetic disorder. Furthermore, the changes seen in this disease resemble changes seen in the mouse, where the PrP gene not only encodes the prion protein but is also associated with susceptibility to disease and incubation time.

PROPERTIES ATTRIBUTED TO AGENT GENES

Host Range

Considerable attention has been paid to the host range of the agents of scrapie, CJD, and kuru. Gibbs et al (29) reported that the most obvious difference among the 112 CJD and 14 kuru isolates examined in their work was the range of susceptibility of different hosts upon primary isolation. Indeed, the most widely studied strain of human agent, the Fukuoka-1 CJD isolate, has a dramatically different host range from any other commonly examined isolate (62). Fukuoka-1, which most likely was isolated from a GSS patient rather than a spontaneous case of CJD (61), was readily adapted to mice and rats directly from human brain tissue. The only other isolates that show this property have also been derived from the same general geographic region, whereas isolates from the United States and Western Europe have not been reproducibly adapted to mice, regardless of their passage history. In general, almost all cases of CJD adapt readily to squirrel monkeys and chimpanzees, whereas the susceptibility of other primate species is highly variable. Likewise, transmission of CJD to domestic cats has only been successful in a very limited number of cases, despite attempts with a variety of isolates (29). There is some evidence for strain differences between kuru isolates with regard to host range. For example, one strain of kuru, the Eiru isolate, caused disease in mink, whereas several other isolates caused no such disease following prolonged incubation. Based on this limited work, the evidence for strain variation is almost identical to that of CJD.

From the earliest studies of the scrapie agent apparent differences were observed in isolates obtained from different sheep and these were most pronounced in the host range of the individual isolate and the changes in that host range as a result of passage (reviewed in 36). This apparent strain variation between scrapie isolates has been documented by several authors through several passages and multiple hosts (28, 29, 50). When two strains of scrapie were compared in a nearly identical sequence of passages through monkeys, striking differences appeared. When the mouse-adapted Compton strain was passed through a spider monkey there was no alteration in its ability to return to mice, the only change being an apparent prolongation of the incubation time, from 4 to 14 months. In contrast, the passage of this strain through squirrel, capuchin, cynomologous, or rhesus monkeys resulted in a dramatic change in the pathogenicity for mice. Contrasted to the normal 4-month incubation period following mouse-to-mouse transmission, there was no clinical disease in the mice 24 months after inoculation with the monkey brain tissue. There was, however, histopathological evidence for infection and, eventually, blind serially passed mouse brain tissue showed infectivity in mice after prolonged incubation. Furthermore, when the squirrel- and cynomologous monkey-passed material was inoculated into sheep and goats, no disease developed even after prolonged incubation.

When similar passage studies were carried out on the American isolate of scrapie, C-506, quite different results were obtained. Passage through squirrel and spider monkeys had no apparent effect on its subsequent transmissibility to sheep, goats, mice, or rats. However, some evidence suggested that passage of C-506 through a cynomolgous monkey did alter the subsequent transmission to mice (29). These investigators also observed that passage of scrapie through mink and ferrets eliminated its infectivity for mice.

In addition to the changes in transmissibility associated with interspecies transfer, the incubation time also changes dramatically as subsequent passages are done in the same species. The work on primate-passage of scrapie isolates cited above contrasts somewhat with the earlier experience obtained from the study of the disease in sheep, goats, mice, and rats (50). In the earlier studies the transmission patterns were more consistent and led the author to summarize the work with a set of standards to be expected when scrapie is transmitted across the species barrier. Those standards included, among others: (a) "The incubation period will shorten and become fixed;" and, (b) "the clinical signs and histopathological lesions characteristic for that species will be fixed." While not universally reproducible, these observations are consistent with the notion that the host plays the dominant or complete role in scrapie infection and in controlling the properties of the infectious agent. The work described above shows that host range and incubation time between different hosts depends upon passage history and the genetic make-up of the host. Variations between experimental animals, most of which were not inbred, could easily explain the results.

Thermal Stability of Prions

The use of heat stability as a genetic marker encoded by the CJD agent itself was reported by Walker et al (63). In this work a 10% suspension of CJD-infected mouse-brain homogenate was heated to 80°C for 30 min and inoculated intracerebrally into mice. Following death of the infected mice, the stability of the infecting agent was reexamined and shown to be significantly greater than that of the parental material (63). An expanded examination of this property and attempts to isolate additional heat-resistant strains have been limited by the technical problems of dealing with such crude brain preparations. Although additional mutants may have been isolated, the sensitivity and reproducibility of this marker limit its practical usefulness in crude preparations (D. T. Kingsbury, unpublished observations).

Variation in the Pathological Lesions

The brain lesions arising during infection can include vacuolation of either or both the grey and white matter, astrocytic hypertrophy, and, on occasion, easily observed amyloid deposits. The extent and distribution of lesions vary widely in different infected animals. However, some investigators (17, 18) have described experimental systems in which these factors may be controlled well enough to be used as genetic markers. By controlling the parameters of dose, scrapie "strain," method of preparation of the inoculum, the route of infection, and the genotype and sex of the host, Dickinson and his associates used the distribution of lesions in different brain regions as genotypic markers (9, 18). In their studies, the density of vacuolation varied as much as threefold throughout various portions of the brain. Using this parameter as a genetic marker, these investigators noted a high degree of genetic instability in certain "strains" of the scrapie agent. In these cases, the highest "mutation" rate, termed class III stability, depended on the murine host strain as the selective force (9). In all of this work, the authors interpret their results in the context of strain mixtures and high mutability, which is totally consistent with the data. Also consistent with these data is the concept that the phenotype of the scrapie agent is a function of the host and that as host strains or species change, so does the resulting agent. The dependence of the distribution of brain lesions on the route of inoculation significantly complicates the reproducibility and interpretation of the results. Despite the above arguments for an independent scrapie genome, there has been no reported work using mutagens to induce these mutations. Without exception, every report of mutant strains of scrapie or CJD has involved the results of passing material without mutagenesis. Clearly, the demonstration of an increased mutation rate as a result of treatment with known mutagens would add strength to the arguments for an independent scrapie or CJD genome.

In addition to the distribution of vacuolation in the brain, other markers have been used to differentiate strains of the scrapie agent. The most commonly used has been the production of amyloid plaques. Bruce & Dickinson (10) demonstrated that the appearance and density of amyloid lesions depended primarily on the murine host strain used with a lesser, but still significant, influence of the strain of the agent. The production of amyloid plaques was directly correlated with the incubation time and the genetic control of incubation time appeared to be identical with the genetic control of amyloid plaque production, with some mouse strains consistently producing plaques with a higher frequency than others. These findings are consistent with the finding by DeArmond et al (16) that the amyloid deposits are immunologically identifiable as accumulations of the scrapie-prion protein, and the work of Scott et al (59) showing that pathology depended upon the host genotype and not upon the scrapie agent causing the infection.

As with scrapie, amyloid deposition and lesion distribution differ measurably between different CJD cases. Although the distribution of pathological changes has not been studied as carefully for CJD as in scrapie, for some time wide variations have been noted in the distribution of lesions in different cases, including the variable occurrence of white- and grey-matter vacuolation. Likewise, the pattern of lesions and amyloid deposition in experimentally infected mice resembles that described earlier for scrapie (60). This work also described the apparent relationship between incubation time and the extent of amyloid deposition. Despite considerable variation between the incubation times of several strains when first adapted to mice, the incubation time became very quickly fixed after the first passage, following very closely the observations of Pattison (51) on scrapie.

Competition between Scrapie Isolates

One observation attributed to scrapie-controlled genetic information has been the ability of one scrapie isolate to interfere with the replication of another during coinfection. Kimberlin & Walker (34) examined the conditions controlling the interference between a short- and a long-incubation-time isolate in Compton White mice. It had been shown earlier that simultaneous inoculation of slow and fast isolates had no effect on the incubation time of the shorter-incubation-time strain (20). However, the long-incubation-time isolate prevailed if it was inoculated several weeks prior to the short-incubation-time isolate. Kimberlin & Walker (34) determined that the competition was dependent on the presence of infectious scrapie agent in the original inoculum of the long-incubation-time agent. The usual interpretation of these experiments is that there is either competition for the "scrapie-replication site" or that the interference occurs at the level of genomic interactions. That the strain of mouse used for the inoculum is of no apparent consequence has been used as evidence for scrapie genetic information.

At least one additional interpretation can be offered. If new scrapie prions arise from specific posttranslational modifications of the normal host-encoded cellular protein (PrP^c), several modifications may be possible, and could explain some forms of "strain" behavior. If it is further assumed that the specific modification is mediated by the scrapie protein (PrP^{Sc}) itself, then in cases of sequential inoculation the relatively small number of scrapie prions present in the second inoculation would have no proportional effect on the prion-protein pool. If this were the case, then the source of the original inoculum, and its specific modification, might play a critical role in the eventual outcome of the infection. These experiments will only be clarified when more information is available about the nature of the posttranslational modifications and how PrP acts as a template.

CONCLUSIONS

It has been obvious for several years that the host plays a very significant role in determining the pattern of infection and incubation time for slow infections of the nervous system. What has not been as clear is the contribution made by genetic information associated with the agent itself. The most quantitative of the biological properties asserted to be a useful indicator of the genetic autonomy of the agent is the incubation time of infection. This marker has been repeatedly used as the basis of scrapie-strain differentiation (19). Those who hold this position most strongly point out that the usefulness of this property requires careful control over the mouse strain and the passage history of the scrapie agent. Under their highly controlled conditions, incubation time appears to be a stable property of the *agent-host combination*, but not a stable property of the agent strain alone. This observation is now fully explained by the results from the transgenic mouse work described earlier (59). The information derived from various studies cited above paints a fairly compelling picture that scrapie, CJD, GSS, and kuru are most probably all "genetic" diseases and that the transmissible factor is host encoded. This conclusion is derived from a variety of studies, but the most compelling are those involving gene transfer in transgenic mice. This work demonstrates that the gene for the prion protein, PrP, is responsible for: (*a*) the host range of the infecting material; (*b*) the incubation time; (*c*) the pathology of the brain and the deposition of amyloid; and, (*d*) the host range of the resulting transmissible agent. Coupled with the observations that anti-PrP antibodies can "neutralize" the transmissible agent and that infectivity can be greatly enriched by affinity chromatography of detergent-lipid-protein complexes on anti-PrP columns (25–27), this conclusion leaves very little possibility of a transmissible conventional virus, especially a retroviral-like agent as postulated by some (48). Furthermore, the known physical properties of the scrapie and CJD agents such as resistance to formalin and glutaraldehyde are not consistent with a retroviral etiology. What is far from clear, and what will consume research efforts and stimulate controversy in the future, is the mechanism whereby a normal cellular protein can be transformed into a transmissible agent. Furthermore, as reviewed above, some evidence apparently supports the existence of strains of some type. While it is possible to imagine a mechanism of strain generation from the modification of endogenous proteins, it is a process without precedence in cell biology.

To date no new information describes the role of the host genes that act in addition to *Prn-p*. *Pid-1* exerts a significant influence on the incubation time, but the exact identity of the gene and its role remain obscure. It is very likely that this gene influences some aspect of the cell surface and might, therefore, interact with the PrP protein. Alternatively, Pid-1 and the other loci that show a minor influence may well be involved in the movement of PrP within the infected animal and may be necessary for targeting PrP to the appropriate site in the reticuloendothelial system or brain.

The focus of work on the genetics of the slow infections over the next few years will undoubtedly be on the resolution of the details of the *Prn-p* gene that influence susceptibility, host range, and incubation time. Assuredly, the transgenic mouse will be one of the core tools and many preconceived ideas of cell biology will be modified.

ACKNOWLEDGMENTS

The work reported here was supported by research grants from the National Institutes of Health (AG 02132 and NS 14069). I also thank many collaborators, especially H. Amyx, J. Bockman, G. Carlson, S. Prusiner, D. Smeltzer, and J. Watson.

Literature Cited

1. Aldhous, P. 1990. Antelopes die of "mad cow" disease. *Nature* 344:183
2. Baker, H. F., Ridley, R. M., Crow, T. J. 1985. Experimental transmission of autosomal dominant spongiform encephalopathy: does infectious agent originate in human genome? *Br. Med. J.* 291:299–302
3. Basler, K., Oesch, B., Scott, M., Westaway, D., Wachli, M., et al. 1986. Scrapie and cellular PrP isoforms are encoded by the same chromosomal gene. *Cell* 46:417–28
4. Bockman, J. M., Kingsbury, D. T. 1988. Immunological analysis of host and agent effects on Creutzfeldt-Jakob disease and scrapie prion proteins. *J. Virol.* 62:3120–27
5. Bockman, J. M., Kingsbury, D. T., McKinley, M. P., Bendheim, P. E., Prusiner, S. B. 1985. Creutzfeldt-Jakob disease prion proteins in human brains. *New Engl. J. Med.* 312:73–78
6. Bockman, J. M., Prusiner, S. B., Tateishi, J., Kingsbury, D. T. 1987. Immunoblotting of Creutzfeldt-Jakob disease prion proteins—Host species-specific epitopes. *Ann. Neurol.* 21:589–95
7. Brinster, R. L., Chen, H. Y., Trumbauer, M. E., Yagle, M. K., Palmiter, R. D. 1985. Factors affecting the efficiency of introducing foreign DNA into mice by micro-injecting eggs. *Proc. Natl. Acad. Sci. USA* 82:4438–42
8. Brown, P., Cathala, F., Raubertas, R. F., Gajdusek, D. C., Castaigne, P. 1987. The epidemiology of Creutzfeldt-Jakob disease: conclusion of a 15-year investigation in France and review of the world literature. *Neurology* 37:895–904
9. Bruce, M. E., Dickinson, A. G. 1979. Biological stability of different classes of scrapie agent. See Ref. 54a, 2:71–86

10. Bruce, M. E., Dickinson, A. G. 1987. Biological evidence that scrapie has an independent genome. *J. Gen. Virol.* 68:79–89
11. Carlson, G. A., Goodman, P. A., Lovett, M., Taylor, B. A., Marshall, S. T., et al. 1988. Genetic polymorphism of the mouse prion gene complex: control of scrapie incubation time. *Mol. Cell. Biol.* 8:5528–40
12. Carlson, G. A., Kingsbury, D. T., Goodman, P. A., Coleman, S., Marshall, S. T., et al. 1986. Prion protein and scrapie incubation time genes are linked. *Cell* 46:503–11
13. Carlson, G. A., Westaway, D., DeArmond, S. J., Peterson-Torchia, M., Prusiner, S. B. 1990. Primary structure of prion protein may modify scrapie isolate properties. *Proc. Natl. Acad. Sci. USA* 86:7475–79
14. Carp, R. I., Moretz, R. C., Natelli, M., Dickinson, A. G. 1987. Genetic control of scrapie: incubation period and plaque formation in I mice. *J. Gen. Virol.* 68:401–7
15. Chesebro, B., Race, R., Wehrly, K., Nishio, J., Bloom, M., et al. 1985. Identification of a scrapie prion protein-specific mRNA in scrapie infected and uninfected brain. *Nature* 315:331–33
15a. Court, L., Cathala, F., eds. 1983. *Unconventional Viruses and the Central Nervous System*. Paris: Masson
16. DeArmond, S. J., McKinley, M. P., Barry, R. A., Braunfeld, M. B., McColloch, J. R., Prusiner, S. B. 1985. Identification of prion amyloid filaments in scrapie-infected brain. *Cell* 41:221–35
17. Dickinson, A. G. 1976. Scrapie in sheep and goats. In *Slow Virus Diseases of Animals and Man*, ed. R. H. Kimberlin, pp. 209–41. Amsterdam: North Holland

18. Dickinson, A. G., Fraser, H. 1977. Scrapie: pathogenesis in inbred mice: an assessment of host and response involving many strains of agent. In *Slow Virus Infections of the Central Nervous System*, ed. V. ter Meulen, M. Katz, pp. 3–14. New York: Springer-Verlag

19. Dickinson, A. G., Fraser, H. G. 1979. An assessment of the genetics of scrapie in sheep and mice. See Ref. 54a, 1:367–86

20. Dickinson, A. G., Outram, G. W. 1979. The scrapie replication site hypothesis and its implications for pathogenesis. See Ref. 54a, 2:13–31

21. Dickinson, A. G., Meikle, V. M. H. 1969. A comparison of some biological characteristics of the mouse-passaged scrapie agents, 22A and ME7. *Genet. Res.* 13:213–25

22. Dickinson, A. G., Meikle, V. M. H. 1971. Host-genotype and agent effects in scrapie incubation: change in allelic interaction with different strains of agent. *Mol. Gen. Genet.* 112:73–79

23. Dickinson, A. G., Meikle, V. M. H., Fraser, H. G. 1968. Identification of the gene which controls the incubation period of some strains of scrapie in mice. *J. Comp. Pathol.* 78:293–99

24. Diener, T. O. 1987. PrP and the nature of the scrapie agent. *Cell* 49:719–21

25. Gabizon, R., McKinley, M. P., Groth, D. F., Kenaga, L., Prusiner, S. B. 1988. Properties of scrapie prion liposomes. *J. Biol. Chem.* 263:4950–55

26. Gabizon, R., McKinley, M. P., Groth, D. F., Prusiner, S. B. 1988. Immunoaffinity purification and neutralization of scrapie prion infectivity. *Proc. Natl. Acad. Sci. USA* 85:6617–21

27. Gabizon, R., McKinley, M. P., Prusiner, S. B. 1987. Purified prion proteins and scrapie infectivity copartition into liposomes. *Proc. Natl. Acad. Sci. USA* 84:4017–21

28. Gadjuseck, D. C. 1977. Unconventional viruses and the origin and disappearance of kuru. *Science* 197:943–60

29. Gibbs, C. J. Jr., Gajdusek, D. C., Amyx, H. 1979. Strain variation in the viruses of Creutzfeldt-Jakob disease and kuru. See Ref. 54a, 2:87–110

30. Gordon, W. S. 1966. In *ARS 91-53 USDA Report of Scrapie Seminar*, pp. 53–67. Washington, DC: US Dep. Agric.

31. Hadlow, W. J., Prusiner, S. B., Kennedy, R. C., Race, R. E. 1980. Brain tissue from persons dying of Creutzfeldt-Jakob disease causes scrapie-like encephalopathy in goats. *Ann. Neurol.* 8:628–31

32. Hsiao, K., Baker, H. F., Crow, T. J., Poulter, M., Qwen, F., et al. 1989. Linkage of a prion protein missense variant to Gerstmann-Straussler syndrome. *Nature* 338:342–45

33. Hunter, N., Hope, J., McConnell, I., Dickinson, A. G. 1987. Linkage of the scrapie-associated fibril protein (PrP) gene and *sinc* using congenic mice and restriction fragment length polymorphism analysis. *J. Gen. Virol.* 68:2711–16

34. Kimberlin, R. H., Walker, C. A. 1985. Competition between strains of scrapie depends on the blocking agent being infectious. *Intervirology* 23:74–81

35. Kingsbury, D. T., Amyx, H. L., Gibbs, C. J. Jr. 1983. Biophysical properties of the Creutzfeldt-Jakob disease agent. See Ref. 15a, pp. 125–37

36. Kingsbury, D. T., Carlson, G. A., Prusiner, S. B. 1987. Genetic control of prion replication. See Ref. 54b, pp. 315–29

37. Kingsbury, D. T., Kasper, K. C., Stites, D. P., Watson, J. D., Hogan, R. N., Prusiner, S. B. 1983. Genetic control of scrapie and Creutzfeldt-Jakob disease in mice. *J. Immunol.* 131:491–96

38. Kingsbury, D. T., Watson, J. D. 1983. Genetic control of the incubation time of mice infected with a human agent of spongiform encephalopathy. See Ref. 15a, pp. 138–44

39. Kitamoto, T., Ogomori, K., Tateishi, J., Prusiner, S. B. 1987. Formic acid pretreatment enhances immunostaining of cerebral and systemic amyloids. *Lab. Invest.* 57:230–36

40. Kretzschmar, H. A., Stowring, L. E., Westaway, D., Stubblebine, W. H., Prusiner, S. B., DeArmond, S. J. 1986. Molecular-cloning of a human prion protein cDNA. *DNA* 5:315–24

41. Lacy, E., Roberts, S., Evans, E. P., Burtenshaw, M. D., Costantini, F. D. 1983. A foreign b-globin gene in transgenic mice: integration at abnormal chromosomal positions and expression in inappropriate tissues. *Cell* 34:343–58

42. Locht, C., Chesebro, B., Race, R., Keith, J. M. 1986. Molecular cloning and complete sequence of prion protein cDNA from mouse brain infected by the scrapie agent. *Proc. Natl. Acad. Sci. USA* 83:6372–76

43. Lowenstein, D. H., Butler, D. A., Westaway, D., McKinley, M. P., DeArmond, S. J., Prusiner, S. B. 1990. Three hamster species with different scrapie incubation times and neuropathological features encode distinct prion proteins. *Mol. Cell. Biol.* 10:1153–63

44. Manuelidis, E. E., Gorgacz, E. J., Manuelidis, L. 1978. Interspecies transmission of Creutzfeldt-Jakob disease to Syrian hamster with reference to clinical syndromes and strains of agent. *Proc. Natl. Acad. Sci. USA* 75:3432–36

45. Manuelidis, L., Sklaviadis, T., Manuelidis, E. E. 1987. Evidence suggesting that PrP is not the infectious agent in Creutzfeldt-Jakob disease. *EMBO J.* 6:341–47

46. Masters, C. L., Gajdusek, D. C., Gibbs, C. J. Jr. 1981. Creutzfeldt-Jakob disease virus isolations from the Gerstmann-Straussler syndrome. *Brain* 104:559–88

47. McGourty, C. 1989. Bovine encephalopathy: Cattle disease set for cure. *Nature* 338:102

48. Murdoch, G. H., Sklaviadis, T., Manuelidis, E. E., Manuelidis, L. 1990. Potential retroviral RNAs in Creutzfeldt-Jakob disease. *J. Virol.* 64:1477–86

49. Oesch, B., Westaway, D., Wachli, M., McKinley, M. P., Kent, S. B. H., et al. 1985. A cellular gene encodes scrapie PrP 27-30 protein. *Cell* 40:735–46

50. Parry, H. B. 1983. *Scrapie Disease in Sheep*, pp. 1–192. New York: Academic

51. Pattison, I. H. 1965. Experiments with scrapie with special reference to the nature of the agent and the pathology of the disease. In *Slow, Latent and Temperate Virus Infections*, ed., D. C. Gajdusek, C. J. Gibbs, Jr., M. Alpers. Bethesda, MD: *NINDB Monogr.*

52. Prusiner, S. B. 1982. Novel proteinaceous infectious particles cause scrapie. *Science* 216:134–44

53. Prusiner, S. B. 1987. Prions and neurodegenerative diseases. *New Engl. J. Med.* 317:1571–81

54. Prusiner, S. B. 1989. Scrapie prions. *Annu. Rev. Microbiol.* 43:345–74

54a. Prusiner, S. B., Hadlow, W. J., eds. 1979. *Slow Transmissible Diseases of the Central Nervous System*, Vols. 1, 2. New York: Academic

54b. Prusiner, S. B., McKinley, M. P., eds. 1987. *Prions: Novel Infectious Pathogens Causing Scrapie and Creutzfeldt-Jakob Disease.* New York: Academic

55. Prusiner, S. B., McKinley, M. P., Bowerman, K. A., Bolton, D. C., Bendheim, P. E., et al. 1983. Scrapie prions aggregate to form amyloid-like birefringent rods. *Cell* 35:349–58

56. Prusiner, S. B., McKinley, M. P., Groth, D. F., Bowerman, K. A., Mock, N. I., et al. 1981. Scrapie agent contains a hydrophobic protein. *Proc. Natl. Acad. Sci. USA* 78:6675–79

57. Robakis, N. K., Sawh, P. R., Wolfe, G. C., Rubenstein, R., Carp, R. I., Innis, M. A. 1986. Isolation of a cDNA clone encoding the leader peptide of prion protein and expression of the homologous gene in various tissues. *Proc. Natl. Acad. Sci. USA* 83:6377–81

58. Roberts, G. W., Lofthouse, R., Brown, R., Crow, T. J., Barry, R. A., Prusiner, S. B. 1986. Prion-protein immunoreactivity in human transmissible dementias. *New Engl. J. Med.* 315:1231–33

59. Scott, M., Foster, D., Mirenda, C., Serban, S., Coufal, F., et al. 1990. Transgenic mice expressing hamster prion protein produce species-specific scrapie infectivity and amyloid plaques. *Cell* 59:847–57

60. Tateishi, J., Hikita, K., Kitamoto, T., Nagara, N. 1987. Experimental Creutzfeldt-Jakob disease: induction of amyloid plaques in rodents. See Ref. 54b, pp. 451–65

61. Tateishi, J., Kitamoto, T., Hashiguchi, H., Shii, H. 1988. Gerstmann-Straussler-Scheinker disease: immunohistochemical and experimental studies. *Ann. Neurol.* 24:35–40

62. Tateishi, J., Ohta, M., Koga, M., Sato, Y., Kuroiwo, Y. 1978. Transmission of chronic spongiform encephalopathy with kuru plaques from humans to small rodents. *Ann. Neurol.* 5:581–84

63. Walker, A. S., Inderlied, C. B., Kingsbury, D. T. 1983. Chemical and physical stability of the K. Fu. strain of Creutzfeldt-Jakob disease virus. *Am. J. Public Health* 73:661–66

64. Westaway, D., Carlson, G. A., Prusiner, S. B. 1989. Unraveling prion diseases through molecular genetics. *Trends Neurosci.* 12:221–27

65. Westaway, D., Goodman, P. A., Mirenda, C. A., McKinley, M. P., Carlson, G. A., Prusiner, S. B. 1987. Distinct prion proteins in short and long scrapie incubation period mice. *Cell* 51:651–62

66. Westaway, D., Prusiner, S. B. 1986. Conservation of the cellular gene encoding the scrapie prion protein. *Nucleic Acids Res.* 14:2035–44

67. Wu, Y., Brown, W. T., Dobkin, C., Devine-Gage, E., Merz, P., et al. 1987. A PvuII RFLP detected in the human prion protein (PrP) gene. *Nucleic Acids Res.* 15:3191

Annu. Rev. Genet. 1990. 24:133–70

THE LDL RECEPTOR LOCUS IN FAMILIAL HYPERCHOLESTEROLEMIA: Mutational Analysis of a Membrane Protein

Helen H. Hobbs, David W. Russell, Michael S. Brown, and Joseph L. Goldstein

Departments of Molecular Genetics and Internal Medicine, University of Texas Southwestern Medical Center, Dallas, Texas 75235

KEY WORDS: human biochemical genetics, lipoprotein receptors, cholesterol metabolism, exon shuffling, LDL receptor mutations, *Alu* elements—repetitive elements, recombination

CONTENTS

133

0066-4197/90/1215-0133$02.00

INTRODUCTION

The classic studies of the inborn errors of metabolism have taught us much about the properties of enzymes, the mechanisms of their mutation, and the metabolic consequences that follow. Another class of proteins, the cell surface receptors, has begun to yield its secrets to the same type of genetic analysis. Discovered more recently than the enzymes, the receptors are more difficult to study experimentally because they are membrane glycoproteins that reside in a hydrophobic environment and resist standard purification techniques.

Important advances in the understanding of cell surface receptors are now emerging from analysis of naturally occurring mutations that disrupt their function in human diseases. The most extensively studied system involves mutations in the low density lipoprotein (LDL) receptor in patients with familial hypercholesterolemia (FH). More than 150 different mutant alleles at the LDL receptor locus distort receptor function in informative ways. This review attempts to catalogue these mutations and to focus the light that they shed on receptor genes and their protein products.

LDL RECEPTOR LOCUS

Historical Perspective

The LDL receptor is a cell surface glycoprotein that regulates plasma cholesterol by mediating endocytosis of LDL, the major cholesterol transport protein in human plasma (8). Mutations in the LDL receptor gene cause familial hypercholesterolemia (FH), a common disease that affects about 1 in 500 people in most populations (30). Individuals heterozygous for an LDL receptor mutation express half the normal number of functional receptors on their cell surface. Their cells bind, internalize, and degrade plasma LDL at half the normal rate. This produces a twofold elevation in plasma LDL-cholesterol concentration (300 to 500 mg/dl). The excess plasma LDL-cholesterol deposits in tendons and arterial walls, forming tendon xanthomas and atherosclerotic plaques (30).

The rare FH homozygotes (about 1 per million people) have a more severe disease than heterozygotes. They have mutations in both LDL receptor genes, which can either be identical (true homozygotes) or different (compound heterozygotes). These individuals express few to no functional LDL receptors on their cell surfaces, their clearance of plasma LDL is markedly impaired, and their plasma LDL-cholesterol level rises dramatically (600 to 1200 mg/dl). FH homozygotes display a pathognomonic skin lesion, the cutaneous xanthoma, that invariably appears by age 4. They frequently die of heart attacks before age 20 (30). The clinical characteristics of FH were reviewed in the *Annual Review of Genetics* in 1979 (29) and more recently in ref. 30.

The LDL receptor was identified in 1973 in studies comparing cholesterol metabolism in cultured fibroblasts of normal and FH-homozygote subjects (reviewed in ref. 8). The cholesterol that normal fibroblasts require for growth can be synthesized intracellularly from acteyl CoA or taken up from plasma LDL. When grown in the presence of serum containing LDL, normal cells synthesize low amounts of cholesterol, which is reflected by a low activity of 3-hydroxy-3-methylglutaryl CoA reductase (HMG CoA reductase), the rate-limiting enzyme in cholesterol synthesis. When cells are grown in lipoprotein-deficient serum, their HMG CoA reductase activity rises, and cholesterol synthesis increases. In contrast, FH-homozygote fibroblasts maintain the same high level of HMG CoA reductase in the presence or absence of lipoproteins, owing to their inability to take up LDL. The LDL receptor was purified in 1982, its cDNA was cloned shortly thereafter by Russell et al (97) and Yamamoto et al (121), and its gene was isolated and characterized in 1985 by Südhof et al (106). These advances laid the groundwork for the molecular analysis of the mutations underlying FH.

LDL Receptor Gene: Relation of Exons to Protein Domains

The human LDL receptor mRNA, 5.3 kilobases (kb) in length, encodes a protein of 860 amino acids (121). About half of the mRNA constitutes a long 3' untranslated region that contains three copies of the *Alu* family of middle repetitive DNAs (121). The human LDL receptor gene, located on the distal short arm of chromosome 19 (p13.1–p13.3) (78), spans 45 kb, and is divided into 18 exons and 17 introns (106). Many of the exons share an evolutionary history with exons of other genes, an observation that suggested that the LDL receptor gene was assembled by exon shuffling (see last section). Figure 1 shows how the exons of the gene correlate with the functional domains of the mature protein (106).

Exon 1 encodes a short 5' untranslated region and 21 hydrophobic amino acids that comprise the signal sequence. This sequence is cleaved from the protein during translocation into the endoplasmic reticulum (ER), leaving a mature protein of 839 amino acids. *Exons 2–6* encode the ligand-binding domain, which is made up of seven repeats of ~ 40 amino acids each. These repeats show a strong resemblance to sequences in several proteins of the complement cascade (82, 106). Each repeat is encoded by a single exon except the third, fourth, and fifth repeats, which are joined in one exon (106). The seven repeats each contain six cysteine residues that form three intrarepeat disulfide bonds. The COOH-terminal end of each repeat contains a negatively charged triplet, Ser-Asp-Glu, that is important for ligand binding. The receptor binds two apolipoproteins, apo B-100 and apo E, that have completely different sequences. Both possess short segments that are rich in basic amino acids (81), and these are believed to bind to the receptor's negatively charged triplets (32). Apo B-100 is a huge 514-kilodalton (kd)

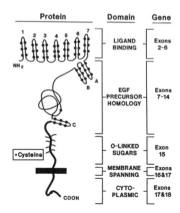

Protein	Domain	Gene

Figure 1 Domain structure of the human LDL receptor protein and its relation to the exon organization of the gene. The domains of the 839-amino acid mature protein are shown at the left and the corresponding exons encoding the protein domains at the right. Exon 1 (not shown) encodes the 21-amino acid signal sequence, which is cleaved from the mature protein during synthesis in the endoplasmic reticulum.

glycoprotein, a single copy of which constitutes the sole protein of LDL. Apo E is a 34-kd protein that occurs in multiple copies on a wide variety of lipoproteins, including chylomicrons, very low density lipoproteins (VLDL) and their remnants, and high density lipoproteins (HDL). Apo E is especially abundant on particles called β-migrating very low density lipoprotein (β-VLDL), which are remnants of chylomicrons and VLDL that accumulate in the plasma of cholesterol-fed rabbits and dogs (81).

Exons 7–14 of the human LDL receptor gene encode a 400-amino acid sequence that is 33% identical to a portion of the human epidermal growth factor (EGF) precursor gene (106, 108). This region includes three growth factor repeats, which are 40-amino acid cysteine rich sequences that differ from the cysteine-rich sequences in the ligand-binding domain. The first two growth factor repeats, designated A and B, are contiguous and separated from the third (C) by a 280-amino acid sequence that contains five copies of a conserved motif (YWTD) repeated once each 40–60 amino acids. The EGF precursor homology domain is required for the acid-dependent dissociation of lipoproteins from the receptor in the endosome during receptor recycling. It also serves to position the ligand-binding domain so that it can bind LDL on the cell surface (15).

Exon 15 encodes 58 amino acids that are enriched in serine and threonine residues, many of which serve as attachment sites for O-linked carbohydrate chains (14). Absence of this exon has no significant functional consequence in cultured hamster fibroblasts (14). *Exon 16 and the 5'-end of exon 17* encode 22 hydrophobic amino acids that comprise the membrane-spanning domain.

The *remainder of exon 17 and the* 5'-*end of exon 18* encode the 50 amino acids that make up the cytoplasmic domain. This domain is important in the localization of the receptor in coated pits on the cell surface (9, 16, 17, 67). The remainder of exon 18 specifies the 2.6-kb 3' untranslated region of the mRNA.

The amino acid sequence of the LDL receptor from six species (human, cow, rabbit, hamster, rat, and the toad *Xenopus laevis*) has been determined by cDNA cloning and DNA sequence analysis (66, 96a, 120, 121; authors' unpublished data). Overall, the protein is highly conserved. The most conserved region is the cytoplasmic domain, which is more than 86% identical among all six species. Next in order of conservation is the EGF precursor domain (70–86% identity), followed by the ligand-binding domain (69–78% identity), and the transmembrane domain (46–62% identity). Two regions of the receptor show minimal primary sequence conservation among the six species; these are the signal sequence and the O-linked sugar domain. Even though the linear order of amino acids in the latter differs widely among species, this sequence is invariably enriched in threonines and serines that are glycosylated.

LDL Receptor Promoter

The 5'-flanking region of the LDL receptor gene contains most if not all of the *cis*-acting DNA sequences responsible for the ubiquitous and regulated expression of the gene in animal cells (31, 107). Within 200 basepairs (bp) of the initiator methionine codon are three imperfect direct repeats of 16 bp each, two A/T-rich sequences, and a cluster of mRNA initiation sites, all of which function in transcription. This architecture has been maintained in the promoters of rat, hamster, and primate LDL receptor genes (authors' unpublished data).

Standard methods of promoter analysis, including hybrid gene construction, transfection into cultured cells, DNAase I footprinting, and linker-scanner mutagenesis, have been used to dissect the roles of the 5' flanking sequences in the expression and regulation of the LDL receptor gene (reviewed in ref. 31). Two of the direct repeat sequences (numbers 1 and 3) interact with the general factor Sp1 to promote transcription. These sequences in themselves are not sufficient for high-level expression. They require the contribution of the third direct repeat (number 2), which contains a conditional-positive sterol regulatory element, termed SRE-1, that is 10 nucleotides in length (103). In the absence of sterols this element synergizes with the two Sp1 sites to promote transcription. The SRE-1 loses its activity in the presence of sterols (31, 103). Through this regulatory mechanism, cells suppress receptor mRNA (97) and develop a corresponding decrease in cell surface LDL receptors (8) when they are cultured in the presence of sterols.

To date, no mutations in individuals with FH have been mapped to the transcriptional regulatory elements. This paucity may be attributed to the restricted length of this segment (200 bp) in contrast to the much longer protein coding sequence. In part, the lack of promoter mutations may arise from redundancy among the promoter elements, so that only a few nucleotides are indispensable.

LDL Receptor Polymorphisms

Prior to the cloning of the gene, the only genetic marker for the LDL receptor was a protein polymorphism in the third component of complement (C3) loosely linked to FH with a recombination fraction of 0.25 (29). After the LDL receptor cDNA was cloned, multiple restriction fragment length polymorphisms (RFLPs) were identified (Table 1). The polymorphisms span the entire LDL receptor locus; six are located in exons and one of these (*Stu*I in exon 8) results in an amino acid substitution that has no apparent effect on receptor function (73). Ten of these RFLPs were used to deduce haplotypes at the receptor locus in 20 Caucasian American pedigrees (73). A total of 123 haplotypes were identified, each varying in frequency from 0.8% to 29.3%. Because of linkage disequilibrium between many of the RFLP sites, the cumulative analysis of five selected sites (shown by the asterisks in Table 1) is almost as informative as the analysis of all 10 sites (heterozygosity index of 84% vs 86%).

Haplotype analysis can be used to follow the segregation of the LDL receptor gene in families with hypercholesterolemia to establish or confirm the diagnosis of FH. Efforts to use the RFLPs to identify common LDL receptor alleles that affect the plasma cholesterol level in the general population have thus far not been revealing.

LDL RECEPTOR MUTATIONS CAUSING FAMILIAL HYPERCHOLESTEROLEMIA

One Locus with Multiple Alleles

Over the past 17 years, fibroblast cultures from 149 FH homozygotes from 134 unrelated families have been analyzed in our laboratory. These cell strains are designated as the Dallas collection. Approximately 45% of the FH homozygotes in the Dallas collection are true homozygotes as determined by haplotype analysis. The remaining 55% are compound heterozygotes with two different mutant alleles. The 134 unrelated FH homozygotes in the Dallas collection could have as many as 183 mutant alleles as revealed by the calculations shown in Table 2. The true number of alleles is probably at the upper range of this estimate since we have seen only three examples in which the same mutation was encountered in FH homozygotes from unrelated

Table 1   Allele frequencies of 22 RFLPs at the LDL receptor locus listed in order from 5' to 3'[a]

Restriction site[b]	Location	Size of bands, kb	Allele frequencies, %	Population	Reference
*Rsa*I	5' FR	3.2/1.8	75/25[c]	American (n = 130)[d]	36
*Stu*I	5' FR	20/12.5	86/14	American (n = 28)	35
*Bcl*I	5' FR	8.7/5.5	22/78	American (n = 18)	34
*Sac*I/*Kpn*I	5' FR	2.05/2	50/50	American (n = 10)	34
*Apa*I	5' FR	4.5/4.2	4/96	American (n = 40)	34
*Pvu*II	5' FR	8.5/4.9	76/24	American (n = 90)	36
*Bsm*I	5' FR	18/13	17/83	American (n = 123)	73
*Apa*LI	Intron 3	12.0/7.2	95/5	German (n = 72)	26
*Taq*I	Intron 4	1.7/0.6	67/33	Japanese (n = 67)	118
			87/13	American (n = 20)	g
*Sph*I	Intron 6	3.7/3.2	41/59	American (n = 123)	73
*Stu*I	Exon 8	17/15	6/94	South African (n = 60)	61
				American (n = 123)	73
				English (n = 154)	111
*Hinc*II[e]	Exon 12	0.133/0.098	55/45	American (n = 10)	74
*Ava*II (+ *Xba*I)[h]	Exon 13	1.9/1.7	57/43	American (n = 123)	73
*Spe*I	Intron 15	21/13	92/8	American (n = 123)	73
*Apa*LI (+ *Bam*HI)[h]	Intron 15	9.4/2.8	49/51	American (n = 123)	73
			48/52	English (n = 66)	111
*Pvu*II	Inron 15	16.5/14	75/25	English (n = 147)	111
			76/24	American (n = 123)	73
			65/35	Afrikaner (n = 65)	6
			60/40	Italian (n = 112)	13
*Msp*I	Exon 18	0.47/0.44	23/73	German (n = 91)	25
*Nco*I	Exon 18	13/3.4	36/64	South African (n = 52)	60
			28/72	American (n = 123)	73
			26/74	English (n = 117)	111
(dTA)n[f]	Exon 18	0.106–0.114		American (n = 27)	122
*Pst*I	3' FR	3.4/2.85	60/40	German (n = 35)	24
			55/45	American (n = 123)	73
*Bst*EII	3' FR	33/22	75/25	European (n = 20)	109
*Apa*LI (+ *Bam*HI)[h]	3' FR	17/13	27/73	American (n= 123)	73

[a] *Abbreviations:* 5' FR, 5' flanking region; 3' FR, 3' flanking region; kb, kilobases.
[b] The asterisk (*) denotes the five RFLP sites that, when analyzed together, have a heterozygosity index of 84%.
[c] Allele frequency: larger band/smaller band.
[d] The number in parenthesis denotes the number of chromosomes studied. All populations studied were Caucasian.
[e] RFLP detected by polymerase chain reaction.
[f] Length polymorphism detected by polymerase chain reaction. Heterozygosity index, 48%.
[g] G. Zuliani, H. H. Hobbs, unpublished observations.
[h] RFLP is best detected by digestion with the indicated second enzyme.

families (discussed below). In certain ethnic populations specific LDL receptor mutations have achieved a high frequency via a founder effect. Examples include the French Canadians, Afrikaners, and Christian Lebanese (discussed below). In these populations the number of different mutant alleles is much less than in outbred populations.

Table 2   Numerology of LDL receptor mutations among 134 FH homozygotes in the Dallas collection

61 True homozygotes	=	61 Possible alleles
73 Compound heterozygotes	=	146 Possible alleles
		207 Total possible alleles
		−24 Redundant alleles[a]
		183 Probable alleles

[a] Identicial mutant alleles observed in more than one FH homozygote. Most of these redundant alleles are found in inbred populations (see text).

Twenty-four of the 183 mutant alleles in the Dallas collection (13%) have major structural rearrangements detectable by Southern blotting. This group includes three insertions and 21 deletions. The 87% of mutations that have no overt structural rearrangement are predominantly point mutations or small inframe deletions. To date, we have not identified any mutations that primarily disrupt RNA splicing, but this most likely reflects a selection bias in that we have concentrated primarily on mutations that alter the structure of the receptor protein.

Classification of Mutations

LDL receptor mutations can be divided into five classes based on their phenotypic effects on the protein (Figure 2). *Class 1 mutations* fail to produce immunoprecipitable protein (null alleles). *Class 2 mutations* encode proteins that are blocked, either partially or completely, in transport between the ER and the Golgi complex (transport-defective alleles). *Class 3 mutations* encode proteins that are synthesized and transported to the cell surface, but fail to bind LDL normally (binding-defective alleles). *Class 4 mutations* encode proteins that move to the cell surface and bind LDL normally, but are unable to cluster in clathrin-coated pits and thus do not internalize LDL (internalization-defective alleles). *Class 5 mutations* encode receptors that bind and internalize ligand in coated pits, but fail to discharge the ligand in the endosome and fail to recycle to the cell surface (recycling-defective alleles).

Forty-two LDL receptor mutations have been characterized in sufficient detail (i.e. analysis of cellular LDL uptake, receptor biosynthesis, receptor transport, and gene structure) to allow classification according to the above scheme. These 42 mutant alleles, which include 33 from the Dallas collection and nine others reported in the literature as of April 1, 1990, are summarized in Table 3.

CLASS 1 MUTATIONS: NULL ALLELES   Twenty of the 134 FH-homozygote fibroblast strains in the Dallas collection (15%) produce no immunoprecip-

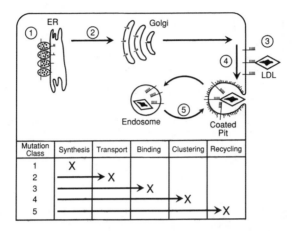

Figure 2   Five classes of mutations at the LDL receptor locus. These mutations disrupt the receptor's synthesis in the endoplasmic reticulum (ER), transport to the Golgi, binding of apolipoprotein ligands, clustering in coated pits, and recycling in endosomes. Each class is heterogenous at the DNA level. See Table 3 for a description of specific mutant alleles in each class.

itable LDL receptor protein and thus have two null alleles of the Class 1 type (44). These 40 null alleles include at least 15 different mutations, of which seven have been characterized at the DNA level (Table 3). Six of the 20 strains with the Class 1 phenotype produce no LDL receptor mRNA. Of these, five are homozygous for the *FH French Canadian-1* allele, a deletion of ~ 15 kb that includes exon 1 and the promoter (42). The sixth mRNA-negative strain is from a compound heterozygote in which one allele contains a 6-kb deletion that removes the promoter and exon 1 *(FH Denver-1)*. The mutation in the partner allele responsible for the absence of mRNA has not been defined, but it does not involve the promoter region.

The majority of null alleles produce an mRNA that is normal in size, but reduced in concentration. Only two of these alleles have been characterized, *FH Turkey* and *FH Nashville* (Table 3). Both generate nonsense codons—in exons 2 and 8, respectively. The absence of receptor in these cells may be due to a rapid turnover of the mRNA as occurs with some nonsense mutations at the β-globin locus (114) or to accelerated degradation of the truncated receptor protein. Two other null alleles, *FH Italy-1* and *FH Potenza*, have premature termination codons due to deletions involving all or part of exons 13, 14, and 15; both produce truncated mRNAs and no detectable protein. Several uncharacterized null alleles in the Dallas collection produce an mRNA of abnormal size even though there is no obvious structural rearrangement in the gene. These alleles may harbor mutations that interfere with mRNA splicing.

In the Dallas collection of 20 FH homozygotes with the Class 1 phenotype,

Table 3 Mutations at the LDL receptor locus that produce familial hypercholesterolemia (FH)[a]

| Allele designation[b] | Patient designation[c] | Molecular lesion | | | | Comments | Reference |
		Type	Size	Location	Alteration of DNA		
CLASS 1: NULL ALLELES							
*FH Denver-1	26	Deletion	5 kb	5' FR to In 1	Deletion of promoter + Ex 1	Deletion joint not sequenced; compound heterozygote; "A" allele	44
*FH French Canadian-1	49, 549, 859, 808, 896	Deletion	~ 15 kb	5' FR to In 1	Deletion of promoter + Ex 1	Deletion joint not sequenced	42
*FH Turkey	573, 941	Nonsense		Ex 2	C → T; Gln_{12} → Stop		e
*FH Nashville	431	Insertion	4 bp	Ex 8	Insertion of GGGT; frameshift → stop		e
*FH Italy-1	651	Deletion	4 kb	In 12 to In 14	Deletion of Ex 13 + 14; frameshift → stop	Deletion joint not sequenced; ? same as FH London-1	44
FH London-1	TD	Deletion	4 kb	In 12 to In 14	Deletion of Ex 13 + 14; frameshift → stop	Alu-Alu recombination (see Figure 5); ? same as FH Italy-1	46
*FH Potenza	381	Deletion	5 kb	Ex 13 to In 15	Deletion of Ex 14, 15, + part of 13	Alu-Exon recombination; compound heterozygote; "A"allele (see Figure 5)	70
*FH Portugal	132	Insertion	> 5 kb	ND	ND	Complex rearrangement	44

CLASS 2A: TRANSPORT-DEFECTIVE ALLELES—NO PROCESSING

FH Saint Omer	375	Missense		Ex 11	$G \rightarrow A$ $Gly_{525} \rightarrow Asp$		e
FH Genoa	850	Missense		Ex 11	$G \rightarrow A$ $Gly_{528} \rightarrow Asp$		e
FH Naples	429	Missense		Ex 11	$G \rightarrow T$ $Gly_{544} \rightarrow Val$		23
FH French Canadian-2	47	Missense		Ex 14	$G \rightarrow A$ $Cys_{646} \rightarrow Tyr$		76
FH Lebanese[d]	264, 550, 738, 786, 794, 868	Nonsense		Ex 14	$G \rightarrow A$ $Cys_{650} \rightarrow Stop$		71

CLASS 2B: TRANSPORT-DEFECTIVE ALLELES—SLOW PROCESSING

FH Cape Town-1	TT	Deletion	6 bp	Ex 2	Deletion of GCGATG; deletion of $Asp_{26}Gly_{27}$		75
FH Mexico	525	Missense		Ex 4	$G \rightarrow A$ $Glu_{207} \rightarrow Lys$	Reduced β-VLDL binding; same as *FH French Canadian-3*	e
FH French Canadian-3	787	Missense		Ex 4	$G \rightarrow A$ $Glu_{207} \rightarrow Lys$	Reduced β-VLDL binding; same as *FH Mexico*	76
FH-Piscataway	563	Deletion	3 bp	Ex 4	Deletion of GGT; deletion of Gly_{197}		f
FH Puerto Rico	848	Missense		Ex 4	$C \rightarrow T$ $Ser_{156} \rightarrow Leu$	Near normal β-VLDL binding	45
FH Afrikaner-1	12	Missense		Ex 4	$C \rightarrow G$ $Asp_{206} \rightarrow Glu$	Same as *FH Maine*	77
FH Maine	101	Missense		Ex 4	$C \rightarrow G$ $Asp_{206} \rightarrow Glu$	Same as *FH Afrikaner-1*	e
FH Denver-2	692	Missense		Ex 6	$G \rightarrow A$ $Asp_{283} \rightarrow Asn$	Normal β-VLDL binding	e
FH Zambia	MM	Missense		Ex 14	$C \rightarrow T$ $Pro_{664} \rightarrow Leu$		59, 104

Table 3 (Continued)

Allele designation[b]	Patient designation[c]	Molecular lesion				Comments	Reference
		Type	Size	Location	Alteration of DNA		
CLASS 3: BINDING-DEFECTIVE ALLELES							
FH Tonami-2		Deletion	10 kb	In 1 to In 3	Deletion of Ex 2 & 3	Deletion joint not sequenced	55
FH French Canadian-5		Deletion	5 kb	In 1 to In 3	Deletion of Ex 2 & 3	Deletion joint not sequenced	80
*FH St. Louis[d]	295	Insertion	14 kb	In 1 to In 8	Duplication of Ex 2 to 8	Alu-Alu recombination (see Figure 5); "A" allele	69
*FH-Paris-2	833	Insertion	> 5 kb	In 2 to In 5	Duplication of Ex 2 to 5	compound heterozygote; "A" allele	e
*FH French Canadian-4	883	Missense		Ex 3	T → G Trp$_{66}$ → Gly		76
*FH Paris-1[d]	626	Deletion	0.8 kb	In 4 to In 5	Deletion of Ex 5	Near normal β-VLDL binding; compound heterozygote; "A" allele; ? same as FH London-2.	41
FH London-2	PD, JA	Deletion	~ 1 kb	In 4 to In 5	Deletion of Ex 5	Deletion joint not sequenced; ? same as FH Paris-1	47
*FH Leuven[d]	359	Deletion	4 kb	In 6 to In 8	Deletion of Ex 7 & 8	Deletion joint not sequenced; compound heterozygote; "A" allele; ? same as FH Cape Town-2.	96
FH CapeTown-2	RI, TD	Deletion	2.5 kb	In 6 to In 8	Deletion of Ex 7 & 8	? same as FH Leuven	38

CLASS 4A: INTERNALIZATION-DEFECTIVE ALLELES—NO SECRETION

*FH Bahrain[d]	682	Nonsense		Ex 17	G → A Trp$_{792}$ → Stop		67
*FH Paris-3[d]	763	Insertion	4 bp	Ex 17	Insertion of GAAA; frameshift → stop		67
*FH Bari	JD	Missense		Ex 18	A → G Tyr$_{807}$ → Cys	Compound heterozygote; "B" allele; same as FH Syria	16
*FH Syria	864	Missense		Ex 18	A → G Tyr$_{807}$ → Cys	Same as FH Bari(JD)	[f]
*FH Rochester[d]	274	Deletion	5.5 kb	In 15 to Ex 18	Deletion of Ex 16, 17 & part of 18	Alu-Alu recombination; (see Figure 5); compound heterozygote; "A" allele.	72
*FH Osaka-1[d]	781	Deletion	7.8 kb	In 15 to Ex 18	Deletion of Ex 16, 17 & part of 18	Alu-Alu recombination; (see Figure 5).	68
FH Helsinki[d]		Deletion	8 kb	In 15 to Ex 18	Deletion of Ex 16, 17 & part of 18		1

CLASS 5: RECYCLING-DEFECTIVE ALLELES

FH Osaka-2[d]		Deletion	12 kb	In 6 to In 14	Deletion of Ex 7 to 14	Alu-Alu recombination (see Figure 5); near normal β-VLDL binding	85
*FH Algeria	646	Missense		Ex 9	G → A Ala$_{410}$ → Thr		[g]
*FH Afrikaner-2	724	Missense		Ex 9	G → A Val$_{408}$ → Met		77
*FH Kuwait	742	Missense		Ex 10	G → A Val$_{502}$ → Met		[g]

[a] *Abbreviations:* kb, kilobases; bp, base pairs; ND, not determined; 5' FR, 5' flanking region; In, intron; Ex, exon. An asterisk (*) refers to mutant alleles characterized in Dallas.
[b] Refers to geographic origin of FH patient.
[c] Refers to identification of FH patient in previous publications.
[d] Mutant receptor protein with abnormal molecular weight.
[e] E. Leitersdorf, H. H. Hobbs, unpublished observations. For each mutant allele (except nonsense mutations), the identified mutation was reproduced in an expressible LDL receptor cDNA and transfected into simian COS cells. The mutant phenotype was verified by biosynthetic studies of the transfected cells.
[f] J. Cali, D. W. Russell, unpublished observations.
[g] G. Zuliani, H. H. Hobbs, unpublished observations.

every mutation characterized at a molecular level is unique to a single family with two exceptions, *FH French Canadian-1* (occurrence in five homozygotes) and *FH Turkey*. The latter allele was found in two reputedly unrelated Turkish FH homozygotes who have the same surname. Inasmuch as they are homozygous for the same mutation and have identical haplotypes, they presumably share a common ancestor.

The total number of null alleles among the 183 FH alleles in the Dallas collection cannot be determined precisely, since 55% of the FH homozygotes are compound heterozygotes. If compound heterozygote cells produce receptor protein, we cannot readily determine whether both alleles produce defective receptors or whether one of the alleles is null. The occurrence of null alleles must be relatively rare, because 80% of the Class 1 homozygotes are true homozygotes by RFLP analysis. Even if the multiple *French Canadian-1* alleles are excluded from the sample, then 75% are true homozygotes. This contrasts with the FH homozygotes in Classes 2–5 in which only 40% are true homozygotes. The apparent rarity of null alleles creates a situation in which the simultaneous inheritance of two of them requires some degree of consanguinity.

CLASS 2 MUTATIONS: TRANSPORT DEFECTIVE ALLELES   The rate of transport of the LDL receptor from the ER to the Golgi can be measured in pulse-chase experiments in which fibroblasts are incubated with $[^{35}S]$methionine, and the receptor is isolated by immunoprecipitation and SDS polyacrylamide gel electrophoresis (112). The normal receptor is synthesized in the ER as a partially glycosylated precursor that migrates with an apparent molecular mass of 120 kd. When the receptor reaches the Golgi, the N- and O-linked sugars are processed so that the receptor migrates with an apparent mass of 160 kd (32, 91). In normal fibroblasts this transport and processing occurs within 60 min after synthesis. In Class 2 fibroblasts this processing is markedly delayed or abolished. The defective receptor does not leave the ER, and its apparent molecular mass remains at 120 kd (32, 91). Eventually, it is degraded without ever having reached the cell surface (23).

The block in transport caused by Class 2 mutations results from either missense mutations or short inframe deletions that alter the amino acid sequence of the receptor, leading to complete or partial misfolding. These findings suggested that cells possess a fail-safe mechanism that detects misfolded proteins and prevents their transport from the ER to the Golgi (120). A similar conclusion has been reached from the study of viral lipid envelope proteins, which follow the same transport pathway (27).

A total of 71 (or 53%) of the 134 FH homozygotes in the Dallas collection have at least one Class 2 allele. Since null alleles are relatively rare (see above), the vast majority of these cells presumably have two Class 2 alleles

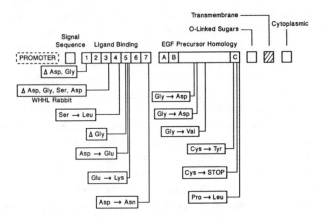

Figure 3   Class 2 LDL receptor mutations. A schematic diagram of the domain structure of the LDL receptor protein is shown at the top. All of the class 2 mutations either obliterate or slow the transport of newly synthesized LDL receptors from the endoplasmic reticulum to the plasma membrane (see Table 3).

that are indistinguishable at the level of analysis employed. Thus, the Class 2 alleles probably account for at least half of all mutant alleles at the LDL receptor locus in the Dallas collection.

Thirteen different Class 2 alleles (12 human and 1 rabbit) have been characterized at the DNA level (Figure 3). The defects cluster in the exons that encode two domains, the ligand-binding and the EGF precursor homology domains. Five of the mutations completely block receptor transport (Table 3) (32, 112). These are designated Class 2A alleles, as opposed to Class 2B alleles, which produce proteins that are transported at a detectable, but markedly reduced rate. All five of the Class 2A alleles harbor mutations in the EGF precursor homology domain. Two of these are in exon 14, which encodes the third cysteine-rich growth factor repeat (repeat C) (Figure 3). One of these mutations, *FH Lebanese,* is frequent in FH patients from Lebanon and Syria (71). It is a transversion (C→A) that produces a stop codon at position 660, which normally encodes the third of six cysteine residues in repeat C. The mutant gene produces a reduced amount of receptor protein with an apparent molecular mass of 100 kd that lacks the O-linked sugar, membrane-spanning, and cytoplasmic domains. The truncated protein is apparently misfolded, and it fails to reach the Golgi (23). Another Class 2A allele, *FH French Canadian-2,* also involves a cysteine codon in repeat C. A missense mutation converts this cysteine to a tyrosine (76).

Class 2A defects are also produced by missense mutations in the YWTD-containing region between repeats B and C (Figure 3). All three of these *(FH Saint Omer, FH Genoa,* and *FH Naples)* involve glycine (Table 3). Because of its small side chain, glycine is often found at space-restricted regions of proteins that do not tolerate other amino acids. In *FH Saint Omer* and *FH Genoa*, aspartic acid is substituted for Gly_{525} or Gly_{528}, respectively, and in *FH Naples* a valine replaced Gly_{544}. These bulky amino acids must interfere with the folding of the LDL receptor, as has been reported for glycine substitutions in procollagen α-1 and α-2 (92). The localization of all five Class 2A mutations to the EGF precursor homology domain suggests that this domain has a compact structure that easily unfolds if only a single amino acid is changed. This observation is consistent with the fact that the EGF precursor homology domain is highly conserved in evolution: this region of the human LDL receptor is 84% identical to the same region in the receptor of the *Xenopus laevis* (authors' unpublished data).

The class 2B phenotype (slow transport to the Golgi) has been traced to seven different mutations. Six occur in the cysteine-rich repeats of the ligand binding domain. Deletion of an entire cysteine-rich repeat does not impair transport to the cell surface (41), but deletion of part of a repeat, as in *FH Cape Town-1* (75) and *FH Piscataway* (authors' unpublished data), retards transport. A similar situation is observed for the *FH WHHL* allele, which causes hypercholesterolemia in the Watanabe-heritable hyperlipidemic rabbit (discussed below). All three of these mutations are short in-frame deletions that change the spacing between highly conserved cysteine residues, presumably interfering with disulfide bond formation and preventing correct folding.

Four Class 2B alleles *(FH Mexico, FH Puerto Rico, FH Afrikaner-1/FH Maine,* and *FH Denver-2)* have missense mutations that replace one of the highly conserved amino acids in the negatively charged Ser-Asp-Glu triplet, which is conserved at the COOH-terminus of all ligand-binding repeats (Figure 3). This triplet is thought to be located on the surface of the protein where it interacts with positively charged residues in the receptor-binding region of apo E and apo B (22, 32). Even the conservative substitution of a glutamic acid for the aspartic acid in this triplet in the fifth repeat results in defective transport *(FH Afrikaner-1/FH Maine)* (77).

Several human inborn errors of metabolism in addition to FH are caused by mutations that produce transport-defective proteins. These mutations involve genes for three membrane proteins—leukocyte adhesion protein (58), sucrase-isomaltase (79), and the insulin receptor (2)—and for three secretory proteins—α-1-antitrypsin (12), pro-α-1 collagen (92), and pro-α-2 collagen (92). Together with observations on the LDL receptor, these findings suggest that a transport-defective phenotype may frequently underlie mutations involving cell surface or secreted proteins. This high frequency is attributable to the ability of the cell to detect small regions of misfolding within a protein and thereby to interdict its movement to the cell surface.

CLASS 3 MUTATIONS: BINDING DEFECTIVE ALLELES   Approximately 25% of the 134 FH homozygote fibroblasts in the Dallas collection produce proteins that are transported to the cell surface, but fail to bind LDL normally. As with Class 2 mutants, it is difficult to estimate precisely the overall frequency of Class 3 alleles, since approximately 73% of the fibroblasts with this phenotype are compound heterozygotes, and we have not determined whether the partner allele is a Class 1, Class 5, or another Class 3 allele. Nine Class 3 mutations have been characterized (Table 3). Seven of these mutations involve substitutions or rearrangements in the cysteine-rich repeats that comprise the ligand-binding domain. The other two are deletions in the adjacent EGF precursor homology domain.

Each of the seven cysteine-rich ligand binding repeats (Figure 1) is encoded by a single exon except repeats 3, 4, and 5, which are encoded by exon 4 (106). The splice junctions for all of these exons are in frame such that if any one of them is deleted, the translational reading frame is preserved, and the resultant protein is transported normally. Extensive deletional analysis by in vitro mutagenesis has revealed that each repeat makes an independent contribution to ligand binding (95). Deletion of repeat 1 has no effect on binding of either LDL or β-VLDL (115). Deletion of any other single repeat impairs LDL binding by up to 95%, but does not impair the binding of β-VLDL. The sole exception is repeat 5, whose deletion reduces β-VLDL binding by 60% (95). The more stringent binding requirements of the receptor for LDL vs β-VLDL creates a situation in which an FH individual may have a selective ability to remove β-VLDL (or IDL) but not LDL from the circulation (discussed below).

Among the Class 3 mutations, *FH Tonami-2* has a deletion that removes the first and second ligand-binding repeats (55). When this deletion is produced in vitro, the transfected receptor binds β-VLDL normally, but it binds only 71% of the normal amount of LDL (22). Deletions of other ligand-binding repeats have selective effects on the ability of the receptor to bind LDL and β-VLDL. For example, in *FH Paris-1* binding repeat 6 is deleted, and the mutant LDL receptor binds β-VLDL with normal affinity but fails to bind LDL (41).

In vitro mutagenesis studies have shown that replacement of a single conserved amino acid in a ligand-binding repeat leads to the same functional abnormality as does deletion of the entire repeat (95). Thus, when the isoleucine adjacent to the third cysteine residue in each of the seven repeats was individually changed by in vitro mutagenesis to an aspartic acid, the resultant receptors had the same binding characteristics as did receptors lacking each of those repeats (95). *FH Puerto Rico* and *FH Mexico*, two slowly processed Class 2B alleles, have missense mutations in the highly conserved Ser-Asp-Glu sequence in repeats 4 and 5, respectively. The *FH*

Puerto Rico receptor ($Ser_{156} \rightarrow$ Leu) fails to bind LDL but binds β-VLDL with near normal affinity (45), a phenotype identical to that seen when the entire fourth repeat is deleted in vitro (95). The *FH Mexico* receptor ($Glu_{207} \rightarrow$ Lys) binds reduced amounts of β-VLDL as well as LDL, as was seen when repeat 5 was deleted in vitro (95). These findings suggest that missense mutations selectively abolish the function of single repeats, which is consistent with the concept that each repeat folds independently.

Normal binding of LDL (but not β-VLDL) requires part of the EGF precursor homology domain as well as the ligand-binding domain (15, 22). If the A repeat or the A plus B repeats are deleted (as in *FH Leuven*), β-VLDL binding is maintained, but LDL binding is markedly reduced (22). Deletion of repeat B alone has no effect on binding of either ligand (22). If the entire EGF precursor homology domain is deleted, as in *FH Osaka-2* (85), then the receptor is unable to bind LDL but continues to bind β-VLDL (15). If such a truncated receptor is denatured in SDS, electrophoresed under nonreducing conditions, and transferred to nitrocellulose, it recovers its ability to bind LDL (15). These observations suggest that the EGF precursor homology domain plays a role in allowing the ligand-binding domain to gain access to LDL on the cell surface.

Two large insertions within the ligand-binding domain of the LDL receptor, *FH St. Louis* and *FH Paris-2*, reduce LDL binding. The better characterized of the two mutations, *FH St. Louis*, encodes a receptor that is 50-kd larger than normal (apparent molecular mass of 210 vs 160 kd on SDS gels) (69). The elongated protein results from a 14-kb duplication of exons 2–8 (discussed below) and contains 18 instead of the usual 9 contiguous cysteine-rich repeats. Seven of these duplicated repeats are derived from the ligand-binding domain, and two are part of the EGF precursor homology region. Remarkably, the elongated receptor is transported normally to the cell surface where it binds reduced amounts of both LDL (69) and β-VLDL (authors' unpublished data). Thereupon, it undergoes efficient internalization and recycling (69). Thus, the only effect of the extensive duplication is a reduction in ligand binding.

CLASS 4 MUTATIONS: INTERNALIZATION-DEFECTIVE ALLELES   Electron microscopy studies show that the Class 4 receptors are distributed diffusely over the surface of the cell rather than being concentrated in coated pits (32). They therefore cannot carry bound LDL into the cell. This rare class of mutations is of considerable historical interest because it provided the earliest evidence that cell surface receptors must cluster in clathrin-coated pits in order to carry ligands into cells (8). The early biochemical, genetic, and ultrastructural studies of these mutations were reviewed in 1979 (29).

Six different internalization-defective alleles have been characterized through cloning. All have mutations that alter the 50-amino acid cytoplasmic domain (1, 16, 67, 68, 72). The mutations have been subclassified into two groups, based on whether the mutation involves the cytoplasmic domain alone or the adjacent membrane-spanning region as well.

Two internalization-defective mutations, *FH Bahrain* and *FH Paris-3* (Table 3), have premature stop codons that leave the membrane-spanning region intact, but remove all but the first 2 or 6 of the normal 50 amino acids in the cytoplasmic tail, respectively. Another Class 4 mutation, the *J.D.* allele (*FH Bari* allele) has a single bp change that substitutes a cysteine for a tyrosine at position 807 in the cytoplasmic domain (16). The latter observation stimulated a series of in vitro mutagenesis experiments, which revealed that position 807 must be occupied by an aromatic amino acid (tyrosine, phenylalanine, or tryptophan) in order for internalization to occur normally (17).

Tyr_{807} is part of a tetrameric sequence NPVY (Asn-Pro-Val-Tyr) that is totally conserved in LDL receptors from six species (9). A variant of this sequence, NPxY (where x can be any amino acid), is present in multiple copies in the cytoplasmic tail of two other members of the LDL receptor supergene family, GP330 protein (Heymann nephritis antigen) (93) and LDL receptor-related protein (40) (discussed below). Extensive in vitro mutagenesis experiments confirmed the importance of the NPxY sequence as a signal for directing the LDL receptor to coated pits (9). An NPxY sequence is also present in the cytoplasmic domains of 13 other cell surface receptors, including several with tyrosine kinase activity (EGF, c-erb-B/*neu*, insulin, IGF-1) and the β-subunits of the integrin receptors (9). Whether this sequence plays a role in directing these proteins to coated pits is unknown.

The second subclass of internalization-defective alleles produce truncated receptors that lack the membrane-spanning domain as well as the cytoplasmic tail. Most of these receptors are secreted from the cell, but approximately 10% remain adherent to the cell membrane where they bind LDL but do not internalize it, thus giving rise to the internalization-defective phenotype (Table 3). Three deletion mutations—*FH Rochester* (72), *FH Osaka-1* (68), and *FH Helsinki* (1)—have this phenotype. All of the deletions extend from intron 15 to the noncoding region of exon 18, but each has a different endpoint (Figure 4, deletions 23–25). In each case, the truncated intron 15 is not removed by splicing. Translation of the mRNA continues into intron 15, and this produces an abnormal COOH-terminal sequence that contains 55 novel amino acids (1, 68, 72). This sequence includes a cluster of 14 hydrophobic amino acids that may constitute a pseudotransmembrane domain (68) that anchors some of the receptors to the cell membrane. Inasmuch as this novel sequence lacks the NPxY signal, it is unable to direct the receptor to coated pits and hence the lack of internalization.

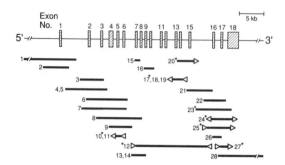

Figure 4   Large deletions in the LDL receptor gene in individuals with FH. Exons, shown in hatched boxes, are separated by introns, which are drawn to approximate scale. The extent of each deletion is shown below the gene map. The functional consequences of deletions no. 1 *(FH French Canadian-1)*; 2 *(FH Denver-1)*; 5 *(FH Tonami-2)*; 10, 11 *(FH Paris-1, FH London-2)*; 12 *(FH Osaka-2)*; 13, 14 *(FH Leuven, FH Cape Town-2)* 17, 19 *(FH Italy-1, FH London-1)*; 20 *(FH Potenza)*; 23 *(FH Helsinki)*; 24 *(FH Rochester)*, and 25 *(FH Osaka-1)* are described in Table 3. The functional consequences of deletions no. 3 (ref. 80), 4 (ref. 119), 6–9 (ref. 65), 15 (ref. 47), 16 (ref. 110), 18 (ref. 65), 21 (ref. 56), 22 (ref. 119), 26 (ref. 113), 27 (S. Langlois, personal communication), and 28 (authors' unpublished data) have not been determined. An asterisk (*) denotes those mutations where the deletion joint has been sequenced. Arrowheads denote involvement of an *Alu* repeat with its orientation indicated by the direction of the arrowhead (see Figure 5).

CLASS 5 MUTATIONS: RECYCLING-DEFICIENT ALLELES   In addition to its role in facilitating the binding of LDL, the EGF precursor homology domain mediates the acid-dependent dissociation of receptor and ligand in the endosome, an essential event for receptor recycling (15). This function was discovered by Davis et al (15), who deleted the EGF precursor homology domain by in vitro mutagenesis of an expressible LDL receptor cDNA. The truncated LDL receptor protein bound and internalized β-VLDL, but it failed to release the ligand in the acidic environment of the endosome. The receptor was then degraded, apparently because it was unable to return to the surface in an unoccupied state (15). A naturally occurring mutation *(FH Osaka-2)* with the same deletion and phenotype was recently described by Miyake et al (85).

Several missense mutations within the EGF precursor homology domain in FH homozygotes produce a phenotype similar to that observed when the entire domain is deleted. *FH Afrikaner-2* has a single amino acid substitution in the YWTD region, which separates growth factor repeat B from repeat C. The protein reaches the cell surface (though more slowly than normal), but is then rapidly degraded (77). Two other single amino acid substitutions in the YWTD region, *FH Algeria* and *FH Kuwait,* reduce the amount of mature 160 kd protein (G. Zuliani, H. H. Hobbs, unpublished data). These reduced amounts of receptor are likely to result from increased degradation, but this has not been demonstrated. Inasmuch as the Class 5 mutations can produce a phenotype that superficially resembles the Class 3 mutations in terms of deficient LDL binding, Class 5 mutations may be considerably commoner than currently believed.

LDL Receptor Mutations Involving Alu Repeats

Many deletions and insertions in the LDL receptor gene have arisen because of recombination between ubiquitously located repetitive *Alu*-type elements. The human LDL receptor gene has three tandem copies of *Alu* sequences in the 3' untranslated region of the mRNA (121). These *Alu* sequences are present in LDL receptor mRNAs from chimpanzees and gorillas, but not baboons or lower mammals, suggesting that the insertion arose during evolution of the primates (43).

Alu repeats are the predominant "middle repetitive" DNA sequences in mammals. The human genome contains about 910,000 copies (50). *Alu* sequences are distributed throughout all chromosomes, occurring mostly in intergenic regions and in introns (53), but occasionally in the mRNA as in the LDL receptor. They are thought to have arisen by duplication and recombination of a small number of founder repeats (7). Through cloning and sequencing of multiple members of the *Alu* family, Deininger et al (18) derived a consensus organization for the typical repeat. Each *Alu* sequence is approximately 300 bp in length and is composed of two tandem repeats referred to as the left and right arms (Figure 5). *Alu* repeats can be transcribed by RNA polymerase III, which recognizes sequence blocks present in each arm (A and

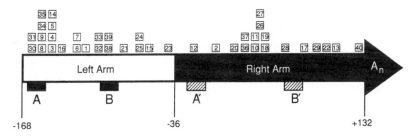

Figure 5   Human mutations involving *Alu* repeats. A consensus *Alu* repeat is depicted with left and right direct repeats and the A and B blocks of the promoter. The deduced breakpoints, representing 25 different mutations in 12 different loci and 40 different *Alu* repeats, are numbered: LDL receptor, 1/2 (ref. 72), 3 (ref. 70), 4/5 (ref. 41), 6/7 (ref. 68), 8/9 (ref. 69), 10/11 (ref. 46), 12/13 (ref. 85); β-globin cluster, 14, 15, 16, and 17 (refs. 39, 52); α-globin cluster, 18/19, 20, and 21 (ref. 88); *c-abl/bcr*, 22 and 23 (ref. 37); C1 complement inhibitor 24/25 and 26/27 (ref. 105); α₂-plasmin inhibitor, 28/29 (ref. 84); apolipoprotein B, 30/31 (ref. 48); X-Y pseudoautosomal region, 32/33 (ref. 94); glycophorin, 34/35 (ref. 63); β-hexosaminidase-α, 36/37 (ref. 87); adenosine deaminase, 38/39 (ref. 83); lipoprotein lipase, 40 (ref. 19). Mutations involving *Alu* sequences in the same orientation are: 4/5, 6/7, 8/9, 10/11, 18/19, 24/25, 26/27, 30/31, 32/33, 34/35, 36/37, and 38/39. Mutations involving *Alu* sequences in opposite orientations are: 1/2, 12/13, and 18/29. Mutations involving a single *Alu* sequence are: 3, 14, 15, 16, 17, 20, 21, 22, 23, and 40.

B, Figure 5). These sequences act as a bipartite promoter that resembles those of the 5S ribosomal RNA genes and tRNA genes (53). The large number of *Alu* repeats in the human genome, coupled with their capacity to be transcribed into RNA, has led to the idea that these sequences might serve as sites for genome rearrangements (7). This hypothesis has received its strongest support from the almost invariant finding of *Alu* sequences at the junctions of large rearrangements in FH genes (41, 46, 68–70, 72, 85).

Figure 4 shows the location of all large deletions in the LDL receptor gene reported as of April 1, 1990. Of these 28 deletions, 8 have been sequenced, and 7 involve an *Alu* repeat at one or both mutation endpoints. The first *Alu*-mediated mutation in the LDL receptor gene, characterized by Lehrman et al (72), involved recombinatioin between two *Alu* sequences oriented in opposite directions, an event that presumably occurred by intrastrand recombination. The mutation is a deletion of 5 kb that extends from intron 15 to exon 18 (*FH Rochester*, Table 3). Sequence analysis of the cloned deletion joint revealed that an *Alu* repeat in intron 15 recombined with an oppositely oriented *Alu* repeat in exon 18 (Figure 4). Inverted repeats in the region surrounding the deletion joint had sufficient complementarity to form an intrastrand double stem-loop structure that may have facilitated the recombination (72). An additional deletion mutation in the LDL receptor gene (*FH Osaka-2*, Table 3 and Figure 4) (85) and a mutation in the alpha$_2$-plasmin inhibitor gene (84) also involve recombination between oppositely oriented *Alu* sequences. The location of the mutation endpoints for these rearrangements as well as those in other human mutations involving *Alu* repeats is shown in Figure 5.

A second type of recombination event, i.e. between *Alu* repeats in the same orientation, appears to be the most prevalent. This type of mutation (*FH Paris-1*, Table 3) was initially characterized by Hobbs et al (41) in a patient with a 0.8-kb deletion that removes exon 5 (Figure 4). Presumably, homology between *Alu* sequences in introns 4 and 5 led to a mispairing of chromosome 19 chromatids during meiosis, followed by an unequal crossover. The two expected products of this recombination would be the observed deletion of exon 5 (Figure 4) and an LDL receptor gene bearing a duplication of this exon. Although this precise duplication has not been observed, we have characterized another duplication caused by an *Alu*-mediated unequal crossover (69). This mutation, *FH St. Louis* (Table 3), duplicates 14 kb spanning exons 2–8 (see above section). Several additional deletions in the LDL receptor gene, as well as in other genes, have occurred via recombination between *Alu* sequences in the same orientation (Figure 4 and 5). This collection includes an interchromosomal *Alu*-mediated crossover between the X and Y sex chromosomes (94). Thus, *Alu* sequences can contribute to both intrachromosomal and interchromosomal gene recombinations.

Alu-mediated recombination can also involve only a single *Alu* sequence. This type of rearrangement was the first reported human mutation involving a middle repetitive sequence and was found by Jagadeeswaran et al (52) in a subject with hereditary persistence of fetal hemoglobin. Since this time, at least nine rearrangements involving single *Alu* sequences have been isolated in several human genes (Figure 5), including one in the LDL receptor gene (*FH Potenza*, Table 3, and Figure 4).

An important question is whether the preponderance of *Alu*-mediated recombinations in the LDL receptor gene is caused by an unusually large number of *Alu* sequences in this gene. To date, approximately 25% of the 45-kb human LDL receptor gene has been sequenced, and 28 *Alu* repeats have been identified, indicating a minimum average distribution of 1 *Alu* sequence per 1.6 kb of gene. In the human genome one *Alu* repeat occurs every 3 kb. Therefore, the LDL receptor gene may have twice as many *Alu* repeats within its borders as the average region of the genome. In the globin genes, more *Alu*-mediated recombination events have been described in the α-globin locus where the distribution of *Alu* repeats is 1 per 2 kb (88) than in the β-globin locus where the frequency is 1 per ~ 10 kb (10, 39). Thus, it seems likely that the number of rearrangements in a region of DNA may be proportional to the number of *Alu* sequences, which may act as hotspots for recombination (51, 105).

Recurrent Mutations

Three LDL receptor mutations have been identified in two unrelated individuals from different countries (Figure 6). To determine whether these represent descendants from a common ancestor or recurrent mutations, we performed receptor haplotype analysis. Previous studies of the β-globin locus have shown that most mutations are in linkage disequilibrium with a set of RFLPs that form a haplotype (90). Unless there has been recombination within a locus, a mutation will remain associated with the haplotype of the allele on which the mutation originally arose. If identical mutations arose by independent events, they should be found on chromosomes with different haplotypes.

The *FH Afrikaner 1* and *FH Maine* Class 2B alleles have the same base pair substitution (GAC → GAG) in codon 206 at the 3' end of exon 4 (Figure 6 and Table 3). Haplotype analysis disclosed an identical restriction pattern at 9 of the 10 RFLP sites: only the 3' most site (*ApaLI*) was different. Therefore, the two individuals, an Afrikaner from South Africa and an American of British descent, probably share an ancestor, with a recent crossover event at the 3' end of the gene explaining the single difference in the haplotype. Since about 5% of the original settlers in South Africa were British, one would

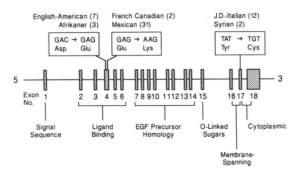

Figure 6   Recurrent LDL receptor mutations in individuals with FH. A schematic diagram of the LDL receptor gene and the relation of the coding regions (exons 1–18) to the protein domains are shown in the bottom half. The site of three recurrent mutations, each occurring in two FH individuals of different ethnic background, is shown above the gene structure. The number in parenthesis denotes the haplotype of the LDL receptor gene on which the mutation arose (73).

predict that this mutation should be found in Britain, but to date it has not been reported.

Two internalization-defective alleles, *J. D.* (or *FH Bari*) and *FH Syria*, have the same mutation, TAT → TGT (Tyr → Cys) at codon 807 in exon 17, which encodes the cytoplasmic domain (Figure 6 and Table 3). Their LDL receptor haplotypes are identical in 9 out of 10 sites; only the 5'-most RFLP site is different (*Bsm*I). It is likely that the Italian and Syrian subjects share an ancestor, and a single crossover event at the 5' end of the LDL receptor gene produced the different restriction pattern at the *Bsm*I site.

The *FH Mexico* and *FH French Canadian-3* alleles have the same base-pair substitution in codon 207 at the 3' end of exon 4, GAG → AAG (Figure 6 and Table 3). However, in this case haplotype analysis showed a difference in restriction pattern on both sides of the substitution at four polymorphic sites. This mutation certainly occurred on two separate occasions. The mutation is a cytosine (C) to thymidine (T) transition at a CpG dinucleotide, a frequent cause of mutations in the human genome (11). At the LDL receptor locus, 8 of the 18 (or 44%) of the characterized base-pair substitutions involve C to T substitutions at CpG dimers.

Population Genetics

In most populations of the world, the frequency of heterozygous FH is ~ 0.2% (30). However, in certain populations the frequency of FH is much higher because a founder effect has caused one or more high-frequency mutations (102). The highest frequency of FH in the world is in the Afrikaner population of South Africa, where the prevalence of FH is fivefold higher than in the European population from which it originated (101). The present-day Afrikaners are descended from ~ 2,000 original settlers, mostly from

Holland, Germany, and France, who emigrated to the Cape of South Africa in the 17th and 18th centuries. In the 19th century they moved into the interior of the country, the Transvaal, where they remained largely isolated from the surrounding populations. The fertility rate was high, and the population grew dramatically to its present size of ~ 3 million. In addition to FH, Afrikaners have a high frequency of several other genetic disorders, owing to founder effects (5).

Among 12 unrelated FH homozygotes of Afrikaner descent, 96% of the 24 mutant alleles had either of two missense mutations, designated *Afrikaner-1* and *Afrikaner-2* (Table 3) (77). The overall frequencies of these mutations in Afrikaners are presently being determined, but it can be expected that these two mutations are the reasons for the high frequency of FH in this population. Seftel et al (100) recently reported a strikingly elevated frequency of FH heterozygotes (1 in 67) in the Jewish population in Johannesburg, South Africa. This population is largely descended from ~ 40,000 Jews who came to South Africa from Lithuania between 1880 and 1910. Like the general white South African population, the Johannesburg Jews have a high incidence of coronary heart disease. The molecular nature of the mutation in the South African Jewish population is not yet known. If it differs from the Afrikaner alleles, this would create the remarkable situation of three different mutations becoming enriched in South Africa. This finding would raise the possibility that the enrichment is caused by a Darwinian selection that favors the heterozygous state in this region of the world. The explanation for such selection, if it exists, is obscure.

The frequency of FH heterozygotes in the French Canadian population has been estimated to be 1 in 270 based on the number of FH homozygotes identified in Quebec Province in 1981 (86), and an even higher frequency (1 in 154) was found in the northeastern region of the province. The French Canadians have a high frequency of several other genetic disorders (64), all attributed to founder effects. The 5.3 million modern French Canadians are descended from about 7,000 French settlers who emigrated to Quebec Province between 1608 and 1763 (64), founding an agrarian population that has remained genetically isolated.

As discussed above, five apparently unrelated French Canadian FH homozygotes in the Dallas collection were shown to be homozygous for a large deletion in the 5' end of the LDL receptor gene that produced a Class 1 phenotype (42). In only one case did the parents give evidence of known consanguinity. These findings suggested that this mutation, *French Canadian-1*, was prevalent in the French Canadian population. The hypothesis was supported when the mutation was found in 59% of 130 French Canadian FH heterozygotes from Montreal (42, 76). We have also identified two FH heterozygotes living in the United States (San Francisco and Boston) with the

French Canadian-1 allele. Both were subsequently found to have ancestors who lived in the same region of Quebec (54).

Four additional mutations, *French Canadians 2–5*, have been identified in this population, but each is much less common than the *French Canadian-1* deletion (76, 78). Together, the five mutations comprise 76% of the mutant LDL receptor alleles in French Canadians. Within both the French Canadian and Afrikaner populations, it is now feasible to screen for FH at the DNA level and detect the mutation in the majority of individuals who are identified as candidates because of high blood cholesterol levels.

Historically, the Christian Lebanese population has played an important role in delineating the genetics of FH. Khachadurian (57) noted a high frequency of FH homozygotes in this population. His careful clinical studies first distinguished between heterozygotes and homozygotes and demonstrated the increased severity of the disease in the latter. The Dallas collection contains six FH homozygotes with a Christian Lebanese or Syrian ancestry. All were found to be homozygous for the *FH Lebanese* allele (71; authors' unpublished observations), which has not yet been identified in any other population. The exact frequency of the mutation in Lebanon has not been determined, but the high frequency of homozygosity suggests that it is common (102).

The general European population, like the North American population, has a plethora of different LDL receptor mutations that cause FH. One exception is the Finnish population, where a single mutation, *FH Helsinki*, is found in at least one third of unrelated Finnish individuals with heterozygous FH (1). The prevalence of FH in the Finnish population has not been determined.

Do Receptor Mutations Correlate with Clinical Expression?

CLINICAL VARIABILITY   In individuals with homozygous FH, the degree of elevation of plasma LDL and the severity of coronary atherosclerosis vary, even among individuals who are homozygous for the same mutation, indicating that other genes play important roles. On average, however, individuals with one or more receptor alleles that produce receptors that retain some function, have a lower level of plasma cholesterol, are more responsive to therapy, and have less aggressive coronary atherosclerosis than individuals whose genes produce totally defective receptors (30).

The dramatic clinical variability in FH homozygotes with the same mutation is exemplified by two individuals, both homozygous for the *French Canadian-1* allele, which abolishes production of receptor mRNA and protein (42). The first, FH 49, did not develop symptoms of coronary atherosclerosis until age 17. Despite treatment with multiple lipid-lowering medications and ileal bypass, her cholesterol level remained between 600 and 1200 mg/dl.

Nevertheless, she lived until age 33. Although she had longstanding and severe angina pectoris, she did not die of a myocardial infarction, but rather of malignant melanoma (42). This clinical picture contrasts strikingly with that the other, FH 549, who died suddenly at age three of an apparent myocardial infarction. At autopsy severe three-vessel coronary atherosclerotic disease was seen (42).

VARIABILITY CAUSED BY ALTERED LIGAND-RECEPTOR INTERACTIONS   Among individuals with different LDL receptor mutations, one source of variability stems from the differential effects of certain mutations on the ability of the receptor to bind lipoproteins containing apo B-100 or apo E. These mutations disrupt the ligand binding and EGF precursor homology domains of the receptor so as to abolish binding of LDL, without impairing binding of apo E-containing lipoproteins, such as β-VLDL. As discussed below, they exert different effects upon LDL metabolism than do the mutations that abolish binding of both ligands.

Normal LDL metabolism begins with the triglyceride-rich particle, VLDL, secreted by the liver. In capillaries, the triglycerides of VLDL are removed by lipoprotein lipase, forming a denser, more cholesterol-rich, apo E-containing particle, termed intermediate density lipoprotein (IDL). Approximately 50% of this IDL is cleared by hepatic LDL receptors, which recognize apo E with high affinity (33). Even when receptors are normal, however, some IDL remains in the plasma, where it is converted into LDL (33).

In FH homozygotes with an LDL receptor mutation that blocks binding of both IDL and LDL, the synthesis of LDL is increased, since the precursor IDL cannot be cleared efficiently by the liver (33). The degradation of LDL is also decreased, owing to the defective receptors. As a result of this combined overproduction and undercatabolism, LDL rises to high levels (8, 33). In contrast, if the mutant LDL receptor retains the ability to bind IDL, but not LDL, the production of LDL is not increased. The major abnormality is a delayed removal of LDL from plasma, and the overall effect is a more moderate elevation in plasma LDL.

Experimentally, it is difficult to obtain sufficient IDL to perform binding studies with human cells. As a substitute for IDL, we generally use β-VLDL (22), which is a mixture of lipoproteins of d < 1.006 g/ml obtained from the plasma of cholesterol-fed rabbits (81). β-VLDL contains particles that are equivalent to IDL, as well as chylomicron remnants, which also bind to LDL receptors by virtue of their apo E content.

Clinical studies of a few FH individuals who have LDL receptors that retain the ability to bind β-VLDL suggest that these mutations are indeed associated with relatively mild disease. The proband of the O. family (99) is homozygous for the *FH-Denver 2* allele, which has a missense mutation in the aspartic acid residue of the Ser-Asp-Glu sequence in repeat 7 of the ligand-binding domain. The mutation produces a Class 2B LDL receptor protein that reaches the cell surface slowly and can bind β-VLDL but not LDL (Table 3). LDL-turnover studies performed in the proband and her obligate heterozygous parents demonstrated a reduced fractional catabolic rate of the injected radiolabeled LDL, but no increase in LDL production (4). This is in marked contrast to the results with other FH homozygotes in whom LDL production is increased (30). There is a striking absence of symptomatic coronary atherosclerosis in the FH heterozygotes in the *FH-Denver 2* family, and both heterozygous parents have only a modest elevation in plasma cholesterol (89).

VARIABILITY CAUSED BY AN LDL-LOWERING GENE   Studies of one FH family, the P. family, have provided evidence that some of the variability in clinical expression of FH may be attributed to a dominant LDL-lowering gene that suppresses the effect of receptor mutations (45). The mutation in the P. family produces a mutant LDL receptor that can bind β-VLDL, but not LDL. This allele, *FH-Puerto Rico,* is present in homozygous form in one individual and in heterozygous form in 18 relatives. The mutant allele has a missense mutation that substitutes leucine for serine in the Ser-Asp-Glu sequence of the fourth ligand-binding repeat (Table 3).

Six heterozygotes in the P. family, including the proband's mother, have concentrations of plasma LDL that fall below the 90th percentile of normal despite their LDL receptor mutations. Genetic analysis suggests segregation of an autosomal dominant gene whose effect is to lower the plasma LDL level. This putative suppressor gene is not linked to the LDL receptor locus itself or to the genes for apo B-100 or apo E as determined by RFLP analysis (45). Efforts are now underway to identify the gene responsible for the LDL-lowering effect.

LDL RECEPTOR MUTATIONS IN ANIMALS

WHHL Rabbit

A rabbit model for FH, the Watanabe-heritable hyperlipidemic (WHHL) rabbit, was discovered at Kobe University and reported in 1980 (116). Homozygous WHHL rabbits are markedly hypercholesterolemic from the time of birth, and they develop severe atherosclerosis by two to three years of age (33). The mutation is a 12-base pair in-frame deletion that removes four amino acids from the fourth ligand-binding repeat of the LDL receptor (120) (Figure 2). The protein has a Class 2B phenotype and reaches the cell surface at a markedly reduced rate (99). The few receptors that reach the surface retain the ability to bind β-VLDL but not LDL. The WHHL rabbits have proved useful in dissecting the role of the LDL receptor in vivo (for review, see ref. 33).

Rhesus Monkey

A second animal model for FH has been described recently by Scanu et al (98) in a pedigree of rhesus monkeys with hypercholesterolemia inherited in an autosomal dominant pattern. Fibroblasts from the hypercholesterolemic monkeys showed a half-normal level of LDL receptor activity, protein, and mRNA. The responsible mutation, a base-pair substitution that produces a stop codon at amino acid 284, was identified by selectively amplifying and sequencing each exon of the monkey LDL receptor gene with oligonucleotide primers homologous to the human sequences (49).

LDL RECEPTOR SUPERGENE FAMILY

Proteins in the LDL receptor supergene family share three distinct sets of sequences (Table 4). One set, which forms the ligand-binding domain of the receptor (Figure 1), consists of complement repeats, which are about 40 amino acids long, contain six cysteines, and possess the negatively charged Ser-Asp-Glu triplet (82, 106). The second set of shared sequences are the growth factor repeats, which are also about 40 amino acids long and contain six cysteines, but differ from the complement repeats in cysteine spacing and in the lack of Ser-Asp-Glu sequences (20, 106). The growth factor repeats can be divided into two types, designated A and B, that differ from each other in the spacing of the cysteines, but are thought to have a common evolutionary origin (20). All three of the growth factor repeats in the LDL receptor are of the B type. The EGF precursor has both types of growth factor repeats, whereas EGF itself is type A.

The third set of shared sequences are those that were originally observed to separate clusters of growth factor repeats in the LDL receptor and the EGF precursor (106, 108). These spacer sequences, which are encoded by five contiguous exons, contain the sequence Tyr-Trp-Thr-Asp (YWTD) repeated approximately every 40 to 60 amino acids (one YWTD motif per exon) (40, 108). For this reason, this shared sequence is designated as the YWTD region.

Group 1 proteins in the LDL receptor supergene family are those that contain all three sets of shared sequences (Table 4 and Figure 7). Group 2 proteins contain growth factor repeats and YWTD sequences, but they lack complement repeats. Group 3 proteins contain complement repeats with or without growth factor repeats, but no YWTD sequences. Group 4 proteins, by far the largest group, contain only growth factor repeats (Table 4).

The three Group 1 proteins (Table 4) resemble each other most closely in structure as well as in function. Each of these is a cell surface transmembrane protein, and each contains the cytoplasmic NPxY sequence that is necessary for coated-pit internalization (9, 40, 62, 93). Two of these proteins, the LDL receptor and LRP, are known to bind and internalize plasma lipoproteins.

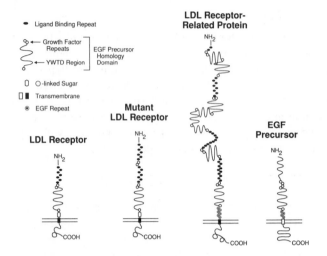

Figure 7   Similarities in shared exons and transmembrane orientation of the LDL receptor and several members of the LDL receptor supergene family (see Table 4). The mutant LDL receptor refers to *FH St. Louis* (see Table 3).

LRP differs functionally from the LDL receptor in that it only binds lipoproteins that have been enriched in vitro with excess apo E (3, 62). GP330 is a protein tha resides in coated pits on the surface of epithelial cells in the renal glomerulus and elsewhere. It is a putative target for an autoantibody that circulates in rats with Heymann-type autoimmune nephritis (93). Whether GP330 binds lipoproteins is currently unknown.

Table 4   The LDL receptor supergene family

		Shared sequences (number)		
	Complement	Growth factor repeat[a]		
Gene product	repeat (1 exon)	Type A (1 exon)	Type B (1 exon)	YWTD region (5 contiguous exons)
GROUP 1—Complement + Growth Factor + YWTD				
LDL receptor	7	0	3	1
LRP	31	6	16	8
GP 330[b]	>13	—	>4	>1
GROUP 2—Growth Factor + YWTD				
EGF precursor	0	4	5	2
Entactin	0	0	6	1
GROUP 3—Complement +/− Growth Factor				
Complement factors C7, C8α, C8β, and C9	1	1	0	0
Complement control factor 1	2	0	0	0

Table 4   (*Continued*)

	Shared sequences (number)			
GROUP 4—Growth Factor Alone				
Clotting Cascade				
Factors VII, IX, and X	0	1	1	0
Proteins C and Z	0	1	1	0
Factor XII	0	2	0	0
Protein S	0	4	0	0
Tissue plasminogen activator, urokinase	0	1	0	0
Thrombospondin	0	3	0	0
Cell Surface Receptors				
TGF-α precursor	0	1	0	0
Lymphocyte homing receptor	0	1	0	0
ELAM-1, GMP-140	0	1	0	0
Thrombomodulin	0	0	6	0
Adhesive Glycoproteins				
Tenascin	0	13	0	0
Cartilage proteoglycan core protein	0	1	0	0
Developmental Proteins				
Notch	0	36	0	0
Slit	0	7	0	0
Delta	0	9	0	0
lin-12	0	13	0	0
glp-1	0	10	0	0
μEGF	0	21	0	0
SpEGF2	0	0	4	0
Exogastrulation peptides of sea urchin	0	0	1	0
Others				
Phospholipase A_2	0	1	0	0
Vaccinia virus-19K	0	1	0	0
Plasmodium-25K	0	0	3	0
Uromodulin	0	0	2	0
Complement factors Cl_s and Cl_r	0	0	1	0

[a] Classification of growth factor repeats according to Doolittle (20).
[b] Complete amino acid sequence not available.

The other groups of proteins in the LDL receptor supergene family do not seem to share any global functions. They are not known to bind lipoproteins, at least with high affinity. The only universal feature is that the cysteine-rich repeat sequences always function in the extracellular milieu, either in external domains of cell surface proteins or in secreted proteins. Many of these proteins engage in protein-protein interactions (21, 117), some of which may be mediated by the shared regions. Like the LDL receptor (32), many of them bind calcium (21). The growth factor repeats in the LDL receptor catalyze an acid-dependent reaction, namely, the dissociation of ligands from the receptor in the endosome (15). This dissociation appears to be mediated by a conformational change in the growth factor region that occurs at about pH 5 (15). It would be of interest to determine whether other proteins that contain growth factor repeats also undergo acid-dependent conformational changes.

The most likely explanation for the widespread sharing of sequences among members of the LDL receptor supergene family relates to exon shuffling (28) in which functional protein domains are reused during evolution. In the shared cysteine-rich repeats, each of these complex structures must fold so that each of the cysteines in a given repeat can form a proper disulfide bond. These bonds are important because they endow the protein with great stability. The LDL receptor, for example, can be boiled in SDS and retain its ability to bind LDL as long as the disulfide bonds remain intact (32). Once nature has evolved such a stable structure, it has a high chance of using the same motif in other proteins through exon duplication and shuffling. The likelihood of such an event must be much greater than the likelihood of evolving a similar structure de novo (28). Once such an exon is incorporated into a gene, it may undergo limited substitutions, which allow the domain to acquire a new function that is vital and unique to that protein.

The genomic rearrangements that underlie exon shuffling in the LDL receptor supergene family may be continuing at the present time, owing to the recombinational activity of the *Alu* elements (discussed above). Figure 7 shows the structure of a mutant LDL receptor that is produced by the *FH St. Louis* allele and compares it with several other members of the LDL receptor supergene family. As discussed above, this mutant gene has undergone an *Alu*-mediated duplication of 14 kb, which reiterates exons 2–8 (69). Although the resulting mutant protein has two complete ligand-binding domains, it binds LDL and β-VLDL poorly. In contrast to the earlier duplications that gave rise to the seven repeats in the ligand-binding domain of the wild-type LDL receptor, this new rearrangement is deleterious, and so would be expected to disappear through natural selection. It is possible to envision another duplication that might improve or expand the function of the LDL receptor, especially given that different combinations of cysteine-rich repeats mediate binding of different apolipoprotein ligands to the LDL receptor (95).

A catalog of LDL receptor mutations in FH subjects is being assembled through the ongoing genetic studies that are reviewed in this article. These studies not only shed light on the structural requirements for the function of an endocytotic cell surface receptor, but they also provide a glimpse of evolution in action.

ACKNOWLEDGMENTS

The original research described in this review was supported by grants from the National Institutes of Health (HL 20948) and the Perot Family Foundation.

Literature Cited

1. Aalto-Setala, K., Helve, E., Kovanen, P. T., Kontula, K. 1989. Finnish type of low density lipoprotein receptor gene mutation (FH-Helsinki) deletes exons encoding and carboxy-terminal part of the receptor and creates an internalization-defective phenotype. *J. Clin. Invest.* 84:499–505

2. Accili, D, Frapier, C., Mosthaf, L., McKeon, C., Elbein, S. C., et al. 1989. A mutation in the insulin receptor gene that impairs transport of the receptor to the plasma membrane and causes insulin-resistant diabetes. *EMBO J.* 8:2509–17

3. Beisiegel, U., Weber, W., Ihrke, G., Herz, J., Stanley, K. K. 1989. The LDL-receptor-related protein, LRP, is an apolipoprotein E-binding protein. *Nature* 341:162–64

4. Bilheimer, D. W., East, C., Grundy, S. M., Nora, J. J. 1985. II. Clinical studies in a kindred with a kinetic LDL receptor mutation causing familial hypercholesterolemia. *Am. J. Med. Genet.* 22:593–98

5. Botha, M. C., Beighton, P. 1983. Inherited disorders in the Afrikaner population of southern Africa. Part I. Historical and demographic background, cardiovascular, neurological, metabolic and intestinal conditions. *S. Afr. Med. J.* 64:609–12

6. Brink, P. A., Steyn, L. T., Bester, A. J., Steyn, K. 1986. Linkage disequilibrium between a marker on the LDL receptor and high cholesterol levels. *S. Afr. Med. J.* 70:80–82

7. Britten, R. J., Baron, W. F., Stout, D. B., Davidson, E. H. 1988. Sources and evolution of human *Alu* repeated sequences. *Proc. Natl. Acad. Sci. USA* 85:4770–74

8. Brown, M. S., Goldstein, J. L. 1986. A receptor-mediated pathway for cholesterol homeostasis. *Science* 232:34–47

9. Chen, W.-J., Goldstein, J. L., Brown, M. S. 1990. NPXY, a sequence often found in cytoplasmic tails, is required for coated pit-mediated internalization of the low density lipoprotein receptor. *J. Biol. Chem.* 265:3116–123

10. Collins, F. S., Weissman, S. M. 1984. The molecular genetics of human hemoglobin. *Prog. Nucleic Acid Res. Mol. Biol.* 31:315–462

11. Cooper, D. N., Youssoufian, H. 1988. The CpG dinucleotide and human genetic disease. *Hum. Genet.* 78:151–55

12. Crystal, R. G. 1990. The α_1-antitrypsin gene and its deficiency states. *Trends Genet.* 5:411–17

13. Daga, A., Fabbi, M., Mattioni, T., Bertolini, S., Corte, G. 1988. Pvu II polymorphism of low density lipoprotein receptor gene and familial hypercholesterolemia. Study of Italians. *Arteriosclerosis* 8:845–50

14. Davis, C. G., Elhammer, A., Russell, D. W., Schneider, W. J., Kornfeld, S., et al. 1986. Deletion of clustered O-linked carbohydrates does not impair function of low density lipoprotein receptor in transfected fibroblasts. *J. Biol. Chem.* 261:2828–38

15. Davis, C. G., Goldstein, J. L., Südhof, T. C., Anderson, R. G. W., Russell, D. W., Brown, M. S. 1987. Acid-dependent ligand dissociation and recycling of LDL receptor mediated by growth factor homology region. *Nature* 326:760–65

16. Davis, C. G., Lehrman, M. A., Russell, D. W., Anderson, R. G. W., Brown, M. S., Goldstein, J. L. 1986. The J. D. mutation in familial hypercholesterolemia: Amino acid substitution in cytoplasmic domain impedes internalization of LDL receptors. *Cell* 45:15–24

17. Davis, C. G., van Driel, I. R., Russell, D. W., Brown, M. S., Goldstein, J. L. 1987. The LDL receptor: Identification of amino acids in cytoplasmic domain required for rapid endocytosis. *J. Biol. Chem.* 262:4075–82

18. Deininger, P. L., Jolly, D. J., Rubin, C. M., Friedmann, T., Schmid, C. W. 1981. Base sequence studies of 300 nucleotide renatured repeated human DNA clones. *J. Mol. Biol.* 151:17–33

19. Devlin, R. H., Deeb, S., Brunzell, J., Hayden, M. R. 1990. Partial gene duplication involving exon-Alu interchange results in lipoprotein lipase de-

ficiency. *Am. J. Hum. Genet.* 46:112–19

20. Doolittle, R. F. 1985. The genealogy of some recently evolved vertebrate proteins. *Trends Biochem. Sci.* 10:233–37

21. Esmon, C. T. 1989. The roles of protein C and thrombomodulin in the regulation of blood coagulation. *J. Biol. Chem.* 264:4743–46

22. Esser, V., Limbird, L. E., Brown, M. S., Goldstein, J. L., Russell, D. W. 1988. Mutational analysis of the ligand binding domain of the low density lipoprotein receptor. *J. Biol. Chem.* 263:13282–90

23. Esser, V., Russell, D. W. 1988. Transport-deficient mutations in the low density lipoprotein receptor: Alterations in the cysteine-rich and cysteine-poor regions of the protein block intracellular transport. *J. Biol. Chem.* 263:13276–81

24. Funke, H., Klug, J., Frossard, P., Coleman, R., Assmann, G. 1986. Pst I RFLP close to the LDL receptor gene. *Nucleic Acids Res.* 14:7820

25. Geisel, J., Weisshaar, B., Oette, K., Mechtel, M., Doerfler, W. 1987. Double Msp I RFLP in the human LDL receptor gene. *Nucleic Acids Res.* 15:3943

26. Geisel, J., Weisshaar, B., Oette, K., Doerfler, W. 1988. A new *Apa* LI restriction fragment length polymorphism in the low density lipoprotein receptor gene. *J. Clin. Chem. Clin. Biochem.* 26:429–33

27. Gething, M.-J., McCammon, K., Sambrook, J. 1986. Expression of wild-type and mutant forms of influenza hemagglutinin: The role of folding in intracellular transport. *Cell* 46:939–50

28. Gilbert, W. 1985. Genes-in-pieces revisited. *Science* 228:823–24

29. Goldstein, J. L., Brown, M. S. 1979. The LDL receptor locus and the genetics of familial hypercholesterolemia. *Annu. Rev. Genet.* 13:259–89

30. Goldstein, J. L., Brown, M. S. 1989. Familial hypercholesterolemia. In *The Metabolic Basis of Inherited Disease,* ed. C. R. Scriver, A. L. Beaudet, W. S. Sly, D. Valle, pp. 1215–50. New York: McGraw-Hill. 6th ed.

31. Goldstein, J. L., Brown, M. S. 1990. Regulation of the mevalonate pathway. *Nature* 343:425–30

32. Goldstein, J. L., Brown, M. S., Anderson, R. G. W., Russell, D. W., Schneider, W. J. 1985. Receptor-mediated endocytosis: Concepts emerging from the LDL receptor system. *Annu. Rev. Cell Biol.* 1:1–39

33. Goldstein, J. L., Kita, T., Brown, M. S.

1983. Defective lipoprotein receptors and atherosclerosis: Lessons from an animal counterpart of familial hypercholesterolemia. *New Engl. J. Med.* 309:288–95

34. Hegele, R. A., Emi, M., Nakamura, Y., Lalouel, J.-M., White, R. 1989. Three RFLPs upstream of the low density lipoprotein receptor (LDLR) gene. *Nucleic Acids Res.* 17:470

35. Hegele, R. A., Emi, M., Nakamura, Y., Lalouel, J.-M., White, R. 1989. A StuI RFLP upstream of the low density lipoprotein receptor (LDLR) gene. *Nucleic Acids Res.* 17:1786

36. Hegele, R. A., Emi, M., Nakamura, Y., Lalouel, J.-M., White, R. 1988. RFLPs upstream of the low-density lipoprotein receptor (LDLR) gene. *Nucleic Acids Res.* 16:7214

37. Heisterkamp, N., Stam, K., Groffen, J., de Klein, A., Grosveld, G. 1985. Structural organization of the *bcr* gene and its role in the Ph¹ translocation. *Nature* 315:758–61

38. Henderson, H. E., Berger, G. M. B., Marais, A. D. 1988. A new LDL receptor gene deletion mutation in the South African population. *Hum. Genet.* 80:371–74

39. Henthorn, P. S., Smithies, O., Mager, D. L., 1990. Molecular analysis of deletions in the human β-globin gene cluster: Deletion junctions and locations of breakpoints. *Genomics* 6:226–37

40. Herz, J., Hamann, U., Rogne, S., Myklebost, O., Gausepohl, H., Stanley, K. K. 1988. Surface location and high affinity for calcium of a 500-kd liver membrane protein closely related to the LDL-receptor suggest a physiological role as lipoprotein receptor. *EMBO J.* 7:4119–27

41. Hobbs, H. H., Brown, M. S., Goldstein, J. L., Russell, D. W. 1986. Deletion of exon encoding cysteine-rich repeat of low density lipoprotein receptor alters its binding specificity in a subject with familial hypercholesterolemia. *J. Biol. Chem.* 261:13114–20

42. Hobbs, H. H., Brown, M. S., Russell, D. W., Davignon, J., Goldstein, J. L. 1987. Deletion in the gene for the low-density-lipoprotein receptor in a majority of French Canadians with familial hypercholesterolemia. *New Engl. J. Med.* 317:734–37

43. Hobbs, H. H., Lehrman, M. A., Yamamoto, T., Russell, D. W. 1985. Polymorphism and evolution of *Alu* sequences in the human low density lipoprotein receptor gene. *Proc. Natl. Acad. Sci. USA* 82:7651–55

44. Hobbs, H. H., Leitersdorf, E., Goldstein, J. L., Brown, M. S., Russell, D. W. 1988. Multiple crm⁻ mutations in familial hypercholesterolemia. Evidence for 13 alleles, including four deletions. *J. Clin. Invest.* 81:909–17

45. Hobbs, H. H., Leitersdorf E., Leffert, C. C., Cryer, D. R., Brown, M. S., Goldstein, J. L. 1989. Evidence for a dominant gene that suppresses hypercholesterolemia in a family with defective low density lipoprotein receptors. *J. Clin. Invest.* 84:656–64

46. Horsthemke, B., Beisiegel, U., Dunning, A., Havinga, J. R., Williamson, R., Humphries, S. 1987. Unequal crossing-over between two alu-repetitive DNA sequences in the low-density-lipoprotein-receptor gene. A possible mechanism for the defect in a patient with familial hypercholesterolaemia. *Eur. J. Biochem.* 164:77–81

47. Horsthemke, B., Dunning, A., Humphries, S. 1987. Identification of deletions in the human low density lipoprotein receptor gene. *J. Med. Genet.* 24:144–47

48. Huang, L.-S., Ripps, M. E., Korman, S. H., Deckelbaum, R. J., Breslow, J. L. 1989. Hypobetalipoproteinemia due to an apolipoprotein B gene exon 21 deletion derived by Alu-Alu recombination. *J. Biol. Chem.* 264:11394–400

49. Hummel, M., Li, Z., Pfaffinger, D., Neven, L., Scanu, A. M. 1990. Familial hypercholesterolemia in a rhesus monkey pedigree: Molecular basis of LDL receptor deficiency. *Proc. Natl. Acad. Sci. USA* 87:3122–26

50. Hwu, H. R., Roberts, J. W., Davidson, E. H., Britten, R. J. 1986. Insertion and/or deletion of many repeated DNA sequences in human and higher ape evolution. *Proc. Natl. Acad. Sci. USA* 83:3875–79

51. Hyrien, O., Debatisse, M., Buttin, G., de Saint Vincent, B. R. 1987. A hotspot for novel amplification joints in a mosaic of *Alu*-like repeats and palindromic A+T-rich DNA. *EMBO J.* 6:2401–8

52. Jagadeeswaran, P., Tuan, D., Forget, B. G., Weissman, S. M. 1982. A gene deletion ending at the midpoint of a repetitive DNA sequence in one form of hereditary persistence of fetal haemoglobin. *Nature* 296:469–70

53. Jelinek, W. R., Schmid, C. W. 1982. Repetitive sequences in eukaryotic DNA and their expression. *Annu. Rev. Biochem.* 51:813–44

54. Jomphe, M., Bouchard, G., Davignon, J., De Braekeleer, M., Gradie, M., et al. 1988. Familial hypercholesterolemia in French-Canadians: Geographical distribution and centre of origin of an LDL-receptor deletion mutation. *Am. J. Hum. Genet.* 43(Suppl):861a

55. Kajinami, K., Fujita, H., Koizumi, J., Mabuchi, H., Takeda, R., et al. 1990. Genetically determined mild type of familial hypercholesterolemia including normocholesterolemic patients: FH-Tonami-2. *Circulation* 80(Suppl.): II–278

56. Kajinami, K., Mabuchi, H., Itoh, H., Michishita, I., Takeda, M., et al. 1988. New variant of low density lipoprotein receptor gene FH-Tonami. *Arteriosclerosis* 8:187–92

57. Khachadurian, A. K. 1964. The inheritance of essential familial hypercholesterolemia. *Am. J. Med.* 37:402–7

58. Kishimoto, T. K., Hollander, N., Roberts, T. M., Anderson, D. C., Springer, T. A. 1987. Heterogeneous mutations in the β subunit common to the LFA-1, Mac-1, and p150,95 glycoproteins cause leukocyte adhesion deficiency. *Cell* 50:193–202

59. Knight, B. L., Gavigan, S. J. P., Soutar, A. K., Patel, D. D. 1989. Defective processing and binding of low-density lipoprotein receptors in fibroblasts from a familial hypercholesterolaemic subject. *Eur. J. Biochem.* 179:693–98

60. Kotze, M. J., Langenhoven, E., Dietzsch, E., Retief, A. E. 1987. A RFLP associated with the low-density lipoprotein receptor gene (LDLR). *Nucleic Acids Res.* 15:376

61. Kotze, M. J., Retief, A. E., Brink, P. A., Welch, H. F. H. 1986. A DNA polymorphism (RFLP) in the human low-density lipoprotein (LDL) receptor gene. *S. Afr. Med. J.* 70:77–79

62. Kowal, R. C., Herz, J., Goldstein, J. L., Esser, V., Brown, M. S. 1989. Low density lipoprotein receptor-related protein mediates uptake of cholesteryl esters derived from apoprotein E-enriched lipoproteins. *Proc. Natl. Acad. Sci. USA* 86:5810–14

63. Kudo, S., Fukuda, M. 1989. Structural organization of glycophorin A and B genes: Glycophorin B gene evolved by homologous recombination at *Alu* repeat sequences. *Proc. Natl. Acad. Sci. USA* 86:4619–23

64. Laberge, C. 1966. Prospectus for genetic studies in the French Canadians, with preliminary data on blood groups and consanguinity. *Bull. Johns Hopkins Hosp.* 118:52–68

65. Langlois, S., Kastelein, J. J. P., Hayden, M. R. 1988. Characterization

of six partial deletions in the low-density-lipoprotein (LDL) receptor gene causing familial hypercholesterolemia (FH). *Am. J. Hum. Genet.* 43:60–68

66. Lee, L. Y., Mohler, W. A., Schafer, B. L., Freudenberger, J. S., Byrne-Connolly, N., et al. 1989. Nucleotide sequence of the rat low density lipoprotein receptor cDNA. *Nucleic Acids Res.* 17:1259–60

67. Lehrman, M. A., Goldstein, J. L., Brown, M. S., Russell, D. W., Schneider, W. J. 1985. Internalization-defective LDL receptors produced by genes with nonsense and frameshift mutations that truncate the cytoplasmic domain. *Cell* 41:735–43

68. Lehrman, M. A., Russell, D. W., Goldstein, J. L., Brown, M. S. 1987. Alu-Alu recombination deletes splice acceptor sites and produces secreted low density lipoprotein receptor in a subject with familial hypercholesterolemia. *J. Biol. Chem.* 262:3354–61

69. Lehrman, M. A., Goldstein, J. L., Russell, D. W., Brown, M. S. 1987. Duplication of seven exons in LDL receptor gene caused by Alu-Alu recombination in a subject with familial hypercholesterolemia. *Cell* 48:827–35

70. Lehrman, M. A., Russell, D. W., Goldstein, J. L., Brown, M. S. 1986. Exon-Alu recombination deletes 5 kilobases from the low density lipoprotein receptor gene, producing a null phenotype in familial hypercholesterolemia. *Proc. Natl. Acad. Sci. USA* 83:3679–83

71. Lehrman, M. A., Schneider, W. J., Brown, M. S., Davis, C. G., Elhammer, A., et al. 1987. The Lebanese allele at the low density lipoprotein receptor locus: Nonsense mutation produces truncated receptor that is retained in endoplasmic reticulum. *J. Biol. Chem.* 262:401–10

72. Lehrman, M. A., Schneider, W. J., Südhof, T. C., Brown, M. S., Goldstein, J. L., Russell, D. W. 1985. Mutation in LDL receptor: Alu-Alu recombination deletes exons encoding transmembrane and cytoplasmic domains. *Science* 227:140–46

73. Leitersdorf, E., Chakravarti, A., Hobbs, H. H. 1989. Polymorphic DNA haplotypes at the LDL receptor locus. *Am. J. Hum. Genet.* 44:409–21

74. Leitersdorf, E., Hobbs, H. H. 1988. Human LDL receptor gene: HincII polymorphism detected by gene amplification. *Nucleic Acids Res.* 16:7215

75. Leitersdorf, E., Hobbs, H. H., Fourie, A. M., Jacobs, M., van der Westhuyzen, D. R., Coetzee, G. A. 1988.

Deletion in the first cysteine-rich repeat of low density lipoprotein receptor impairs its transport but not lipoprotein binding in fibroblasts from a subject with familial hypercholesterolemia. *Proc. Natl. Acad. Sci. USA* 85:7912–16

76. Leitersdorf, E., Tobin, E. J., Davignon, J., Hobbs, H. H. 1990. Common low-density lipoprotein receptor mutations in the French Canadian population. *J. Clin. Invest.* 85:1014–23

77. Leitersdorf, E., Van Der Westhuyzen, D. R., Coetzee, G. A., Hobbs, H. H. 1989. Two common low density lipoprotein receptor gene mutations cause familial hypercholesterolemia in Afrikaners. *J. Clin. Invest.* 84:954–61

78. Lindgren, V., Luskey, K. L., Russell, D. W., Francke, U. 1985. Human genes involved in cholesterol metabolism: Chromosomal mapping of the loci for the low density lipoprotein receptor and 3-hydroxy-3-methylglutaryl-coenzyme A reductase with cDNA probes. *Proc. Natl. Acad. Sci. USA* 82:8567–71

79. Lloyd, M. L., Olsen, W. A. 1987. A study of the molecular pathology of sucrase-isomaltase deficiency. A defect in the intracellular processing of the enzyme. *New Engl. J. Med.* 316:438–42

80. Ma, Y., Betard, C., Roy, M., Davignon, J., Kessling, A. M. 1989. Identification of a second "French Canadian" LDL receptor gene deletion and development of a rapid method to detect both deletions. *Clin. Genet.* 36:219–28

81. Mahley, R. W., Innerarity, T. L., Weisgraber, K. H., Rall, S. C. Jr., Hui, D. Y., et al. 1986. Cellular and molecular biology of lipoprotein metabolism: Characterization of lipoprotein receptor-ligand interactions. *Cold Spring Harbor. Symp.* 51:821–27

82. Marazziti, D., Eggertsen, G., Fey, G. H., Stanley, K. K. 1988. Relationships between the gene and protein structure in human complement component C9. *Biochemistry* 27:6529–34

83. Markert, M. L., Hutton, J. J., Wigington, D. A., States, J. C., Kaufman, R. E. 1988. Adenosine deaminase (ADA) deficiency due to deletion of the ADA gene promoter and first exon by homologous recombination between two Alu elements. *J. Clin. Invest.* 81:1323–27

84. Miura, O., Sugahara, Y., Nakamura, Y., Hirosawa, S., Aoki, N. 1989. Restriction fragment length polymorphism caused by a deletion involving Alu sequences within the human α_2-plasmin inhibitor gene. *Biochemistry* 28:4934–38

85. Miyake, Y., Tajima, S., Funahashi, T., Yamamoto, A. 1989. Analysis of a re-cycling-impaired mutant of low density lipoprotein receptor in familial hyper-cholesterolemia. *J. Biol. Chem.* 264: 16584–90

86. Moorjani, S., Roy, M., Gagne, C., Davignon, J., Brun, D., et al. 1989. Homozygous familial hypercholesterol-emia among French Canadians in Quebec Province. *Arteriosclerosis* 9:211–16

87. Myerowitz, R., Hogikyan, N. D. 1987. A deletion involving Alu sequences in the β-hexosaminidase α-chain gene of French Canadians with Tay-Sachs dis-ease. *J. Biol. Chem.* 262:15396–99

88. Nicholls, R. D., Fischel-Ghodsian, N., Higgs, D. R. 1987. Recombination at the human α-globin gene cluster: se-quence features and topological con-straints. *Cell* 49:369–78

89. Nora, J. J., Lortscher, R. M., Spangler, R. D., Bilheimer, D. W. 1985. I. Famil-ial hypercholesterolemia with "normal" cholesterol in obligate heterozygotes. *Am. J. Med. Genet.* 22:585–91

90. Orkin, S. H., Kazazian, H. H. Jr. 1984. The mutation and polymorphism of the human β-globin gene and its surround-ing DNA. *Annu. Rev. Genet.* 18:131–71

91. Pathak, R. K., Merkle, R. K., Cum-mings, R. D., Goldstein, J. L., Brown, M. S., Anderson, R. G. W. 1988. Im-munocytochemical localization of mutant low density lipoprotein receptors that fail to reach the Golgi complex. *J. Cell. Biol.* 106:1831–41

92. Prockop, D. J., Constantinou, C. D., Dombrowski, K. E., Hojima, Y., Kad-ler, K. E., et al. 1989. Type I Pro-collagen: The gene-protein system that harbors most of the mutations causing osteogenesis imperfecta and probably more common heritable disorders of connective tissue. *Am. J. Med. Genet.* 34:60–67

93. Raychowdhury, R., Niles, J. L., McCluskey, R. T., Smith, J. A. 1989. Autoimmune target in Heymann nephri-tis is a glycoprotein with homology to the LDL receptor. *Science* 244:1163–66

94. Rouyer, F., Simmler, M.-C., Page, D. C., Weissenbach, J. 1987. A sex chromosome rearrangement in a human XX male caused by Alu-Alu recombina-tion. *Cell* 51:417–25

95. Russell, D. W., Brown, M. S., Gold-stein, J. L. 1989. Different combina-tions of cysteine-rich repeats mediate binding of low density lipoprotein recep-tor to two different proteins. *J. Biol. Chem.* 264:21682–88

96. Russell, D. W., Lehrman, M. A., Südhof, T. C., Yamamoto, T., Davis, C. G., et al. 1987. The LDL receptor in familial hypercholesterolemia: Use of human mutations to dissect a membrane protein. *Cold Spring Harbor Symp. Quant. Biol.* 51:811–19

96a. Russell, D. W., Schneider, W. J., Yamamoto, T., Luskey, K. L., Brown, M. S., Goldstein, J. L. 1984. Domain map of the LDL receptor: Sequence homology with the epidermal growth factor precursor. *Cell* 37:577–85

97. Russell, D. W., Yamamoto, T., Schneider, W. J., Slaughter, C. J., Brown, M. S., Goldstein, J. L. 1983. cDNA cloning of the bovine low density lipoprotein receptor: Feedback regula-tion of a receptor mRNA. *Proc. Natl. Acad. Sci. USA* 80:7501–5

98. Scanu, A. M., Khalil, A., Neven, L., Tidore, M., Dawson, G., et al. 1988. Genetically determined hypercholester-olemia in a rhesus monkey family due to a deficiency of the LDL receptor. *J. Lipid Res.* 29:1671–81

99. Schneider, W. J., Brown, M. S., Gold-stein, J. L. 1983. Kinetic defects in the processing of the low density lipoprotein receptor in fibroblasts from WHHL rab-bits and a family with familial hyper-cholesterolemia. *Mol. Biol. Med.* 1: 353–67

100. Seftel, H. C., Baker, S. G., Jenkins, T., Mendelsohn, D. 1989. Prevalence of familial hypercholesterolemia in Johan-nesburg Jews. *Am. J. Med. Genet.* 34:545–47

101. Seftel, H. C., Baker, S. G., Sandler, M. P., Forman, M. B., Joffe, B. I., et al. 1980. A host of hypercholesterolaemic homozygotes in South Africa. *Br. Med. J.* 281:633–36

102. Slack, J. 1979. Inheritance of familial hypercholesterolemia. *Atheroscler. Rev.* 5:35–66

103. Smith, J. R., Osborne, T. F., Goldstein, J. L., Brown, M. S. 1990. Identification of nucleotides responsible for enhancer activity of sterol regulatory element in low density lipoprotein receptor gene. *J. Biol. Chem.* 265:2306–10

104. Soutar, A. K., Knight, B. L., Patel, D. D. 1989. Identification of a point muta-tion in growth factor repeat C of the low density lipoprotein-receptor gene in a patient with homozygous familial hyper-cholesterolemia that affects ligand bind-ing and intracellular movement of recep-tors. *Proc. Natl. Acad. Sci. USA* 86:4166–70

105. Stoppa-Lyonnet, D., Carter, P. E., Meo, T., Tosi, M. 1990. Clusters of

intragenic *Alu* repeats predispose the human C1 inhibitor locus to deleterious rearrangements. *Proc. Natl. Acad. Sci. USA* 87:1511–55

106. Südhof, T. C., Goldstein, J. L., Brown, M. S., Russell, D. W. 1985. The LDL receptor gene: A mosaic of exons shared with different proteins. *Science* 228:815–22

107. Südhof, T. C., Russell, D. W., Brown, M. S., Goldstein, J. L. 1987. 42-bp element from LDL receptor gene confers end-product repression by sterols when inserted into viral TK promoter. *Cell* 48:1061–69

108. Südhof, T. C., Russell, D. W., Goldstein, J. L., Brown, M. S., Sanchez-Pescador, R., Bell, G. I. 1985. Cassette of eight exons shared by genes for LDL receptor and EGF precursor. *Science* 228:893–95

109. Steyn, L. T., Pretorius, A., Brink, P. A., Bester, A. J. 1987. RFLP for the human LDL receptor gene (LDLR): Bst EII. *Nucleic Acids Res.* 15:4702

110. Taylor, R., Bryant, J., Gudnason, V., Sigurdsson, G., Humphries, S. 1989. A study of familial hypercholesterolaemia in Iceland using RFLPs. *J. Med. Genet.* 26:494–98

111. Taylor, R., Jeenah, M., Seed, M., Humphries, S. 1988. Four DNA polymorphisms in the LDL receptor gene: their genetic relationship and use in the study of variation at the LDL receptor locus. *J. Med. Genet.* 25:653–59

112. Tolleshaug, H., Hobgood, K. K., Brown, M. S., Goldstein, J. L. 1983. The LDL receptor locus in familial hypercholesterolemia: Multiple mutations disrupt transport and processing of a membrane receptor. *Cell* 32:941–51

113. Top, B., Koeleman, B. P. C., Leuven, J. A. G., Havekes, L. M., Frants, R. R. 1990. Rearrangements in the LDL receptor gene in Dutch familial hypercholesterolemic patients. Presence of a common 4 kb deletion. *Atherosclerosis.* In press

114. Trecartin, R. F., Liebhaber, S. A., Chang, J. C., Furbetta, M., Angius, A., Cao, A. 1981. B° Thalassemia in Sardinia is caused by a nonsense mutation. *J. Clin. Invest.* 68:1012–17

115. van Driel, I. R., Goldstein, J. L., Südhof, T. C., Brown, M. S. 1987. First cysteine-rich repeat in ligand-binding domain of low density lipoprotein receptor binds Ca^{2+} and monoclonal antibodies, but not lipoproteins. *J. Biol. Chem.* 262:17443–49

116. Watanabe, Y. 1980. Serial inbreeding of rabbits with hereditary hyperlipidemia (WHHL-rabbit). Incidence and development of atherosclerosis and xanthoma. *Atherosclerosis* 36:261–68

117. Xu, T., Rebay, I., Fleming, R. J., Scottgale, T. N., Artavanis-Tsakonas, S. 1990. The *Notch* locus and the genetic circuitry involved in early *Drosophila* neurogenesis. *Genes Dev.* 4:464–75

118. Yamakawa, K., Okafuji, T., Iwamura, Y., Yuzawa, K., Satoh, J., et al. 1988. *Taq*I polymorphism in the LDL receptor gene and a *Taq*I 1.5-kb band associated with familial hypercholesterolemia. *Hum. Genet.* 80:1–5

119. Yamakawa, K., Takada, K., Yanagi, H., Tsuchiya, S., Kawai, K., et al. 1989. Three novel partial deletions of the low-density lipoprotein (LDL) receptor gene in familial hypercholesterolemia. *Hum. Genet.* 82:317–21

120. Yamamoto, T., Bishop, R. W., Brown, M. S., Goldstein, J. L., Russell, D. W. 1986. Deletion in cysteine-rich region of LDL receptor impedes transport to cell surface in WHHL rabbit. *Science* 232:1230–37

121. Yamamoto, T., Davis, C. G., Brown, M. S., Schneider, W. J., Casey, M. L., et al. 1984. The human LDL receptor: A cysteine-rich protein with multiple Alu sequences in its mRNA. *Cell* 39:27–38

122. Zuliani, G., Hobbs, H. H. 1990. Dinucleotide repeat polymorphism at the 3' end of the LDL receptor gene. *Nucleic Acids Res.* In press

Annu. Rev. Genet. 1990. 24:171–87

GENETICS OF ATHEROSCLEROSIS

C. F. Sing[1] and P. P. Moll[1,2]

Departments of Human Genetics[1] and Epidemiology,[2] University of Michigan, Ann Arbor, Michigan 48109

KEY WORDS: coronary artery disease, quantitative traits, genetic architecture, genetics of common diseases

CONTENTS

INTRODUCTION

Atherosclerosis is a disease of the walls of the aorta and large arteries. This disorder is thought to be initiated by injury to the intimal layer of cells that line the lumen of the blood vessel. Progression of disease is characterized by infiltration of lipids into the vessel wall and the formation of fibrous tissue called an atheromatous plaque. Abnormal metabolic processes involving a number of cell types are believed to be involved in determining plaque growth (20, 77). Although there is evidence that this disease begins early in life,

171

0066-4197/90/1215-0171$02.00

clinical symptoms of atherosclerosis do not usually occur until over half of the lumen becomes obstructed (occluded) by the plaque, typically in the fifth and sixth decades (49). The nature of clinical manifestations depends on the blood vessel that is occluded. Interference of blood flow by an occlusion of a coronary artery may cause chest pain (angina pectoris) because of a decrease in the supply of oxygen-rich blood to the heart muscle. Total blockage of an artery may cause death (myocardial infarction) of the area of the heart muscle supplied by the blocked artery. Prolonged blockage of one or more of the coronary arteries often causes death of the individual. Peripheral atherosclerosis may involve any of the major vessels outside the heart. Occlusion of the femoral artery may lead to pathological effects on the lower extremities while atherosclerotic disease of the arteries leading to the brain may result in stroke.

Diseases of the heart constitute the leading cause of death in North America and most countries of Central and Western Europe (31). For this reason, much of the research to understand atherosclerosis has focused on coronary artery disease (CAD). Most investigators believe that CAD has a multifactorial etiology involving many genetic and environmental factors. However, very little is known about how gene products interact with effects of environmental exposures to influence interindividual differences in site of initiation, age of onset, rate of progression, and degree of disease severity. Population research discussed in this review relies on measurements of blood and presence of disease determined by noninvasive clinical methods. Only a fraction of the genetic variation involved in CAD can be revealed by such studies because only a few of the pathways that link gene variation to variation in characteristics of CAD are represented. New, more realistic measures of initiation, progression, and severity of CAD in the living are required to reveal the complete array of genes involved in influencing this common disease. Progress thus far merely serves to convince how complex the genetics of CAD may be when all of the processes that are involved are understood.

HIERARCHICAL ORGANIZATION OF CAUSES OF CORONARY ARTERY DISEASE (CAD)

CAD aggregates in families but does not segregate as if it were determined by a single Mendelian gene (4, 5). This state of affairs is expected when the risk of developing clinically defined disease is determined by the levels of many intermediate quantitative traits whose distributions are each influenced by the segregation of many genes and environmental factors (85, 89, 91, 92). Candidate intermediate traits include cell and tissue characteristics and measures of biochemical and physiological processes that are involved in the pathobiology of CAD (11, 20). In the population at large there is a continuous relationship between quantitative levels of many of these traits and risk of developing disease (76). There is no combination of levels for which risk is totally absent or CAD an absolute certainty.

Because of the complexity of the etiological relationships between genetic variation and variation in risk of CAD introduced by the continuously distributed intermediate traits, we expect there to be few examples of single gene variations that predict risk of CAD directly in the population at large. Thus far, only a few rare single gene mutations have been described that, when homozygous, predict CAD with certainty (14). Simply, mapping between mutational change at the DNA level and measures of CAD at the clinical level is not one-to-one for the majority of those with the disease. Such a complex relationship between causes and outcomes is expected when etiology involves the interaction of many factors at multiple levels in a hierarchical system of causation (51, 79, 83). To reduce the research problem to manageable dimensions, most geneticists have focused on the study of causes of interindividual variability of intermediate quantitative traits that have been identified by epidemiological studies as being predictors of variation in risk of disease in the population at large (86).

INTERMEDIATE QUANTITATIVE RISK FACTOR TRAITS FOR GENETIC STUDIES OF CAD

The role of plasma lipids in the determination of CAD risk has dominated CAD research for over 50 years. Very early investigations of the atherosclerotic plaque that revealed deposits of cholesterol esters focused attention on metabolism of plasma lipids (62). Subsequently, epidemiological studies (37) have established that after gender, age, and smoking, total plasma cholesterol level measured on clinically normal individuals is one of the strongest predictors of subsequent development of CAD. These same epidemiological studies identified casual blood pressure and fasting plasma levels of cholesterol associated with high density lipoprotein (HDL) and low density lipoprotein (LDL) particles as additional predictors of CAD risk. Although appropriate epidemiological studies have not been carried out, an association between elevated levels of plasma apo B, the major protein fraction of the LDL particle, and presence of CAD is generally accepted (16).

The establishment of other candidate intermediate traits has been deterred by the expense of proper epidemiological studies and bounds set by methods of measurement. For instance, the involvement of thrombosis and thrombolysis in the occlusion of a diseased coronary artery has long been appreciated (27, 62). However, only very recently have epidemiological studies of normal variability in quantitative levels of plasma fibrinogen, a molecule involved in the control of blood clotting, established this trait as a predictor of onset of clinical symptoms of CAD (50). Research into the genetic control of normal variability of measures of hemostasis in the population at large has been slowed by lack of appropriate methods of measurement (48).

Discovery that the apo (a) protein, found attached to LDL particles in plasma, has a structure similar to plasminogen, which has a central role in breakdown of blood clots, suggests possibilities for interaction between lipid metabolism and hemostasis in the determination of risk of CAD (81, 100). These observations further serve to convince that a particular gene may influence a network of intermediate traits that are measures of biochemical and physiological processes that establish an individual's risk of CAD. Complexity of genotype-phenotype relationships in CAD is not appreciated by the majority working in the field. No studies have yet attempted the formidable challenge of distinguishing pleiotropic effects of a gene from the effects of gene-by-gene and gene-by-environment interaction.

Numerous biochemical and physiological processes may act as links between genetic and environmental variation and variation in risk to CAD (20, 60, 77). Little effort has focused on how genetic mechanisms directly involved in the pathobiology of the atherosclerotic plaque might contribute to the development of CAD. Recent studies that have defined scavenger-cell receptors involved in lowering injurious levels of oxidized cholesterol illustrate the potential for genetic analysis of a basic process that may be central to the development of CAD (15, 40). It is likely that the blood vessel itself plays a key role in mediating the interaction between circulating lipid and clotting factors and host-specific metabolic processes, all of which are undoubtedly under genetic control. Unfortunately, biological measures of plaque initiation and development and of properties of blood vessel structures in the intact human are not currently available for genetic studies of CAD risk. Most genetic research to characterize the predictors of interindividual variability in risk of CAD in the population at large is likely to continue to rely on those measures of lipid metabolism, blood pressure regulation or hemostasis that are accessible from blood.

Genetic studies of the intermediate biological traits hypothesized to be risk factors for CAD have been reviewed in detail (12, 13, 32, 33, 36, 46, 85, 86). In this brief review we focus on the strategies that are being used and give selected examples to illustrate progress in understanding the genetics of intermediate traits. We close with a discussion of the challenge we face in relating allelic variation in genes that influence these intermediate traits to risk of CAD in the population at large.

GENETIC ARCHITECTURE OF QUANTITATIVE RISK FACTOR TRAITS

A primary goal of geneticists involved in CAD research is to understand the genetic architecture responsible for the distribution of each of the quantitative risk factor traits in the population at large (85, 86). The genetic architecture of a trait in a particular population is defined by the number of genes involved in regulating the metabolic processes that determine the phenotypic expression of the trait, the number of alleles of each gene and their relative frequencies, the impact of each allele on the level and variability of the trait and the impact of each allele on the relationship of the trait with the other risk factors that are involved in the development of CAD. Measurements of genes responsible for the genetic architecture of a risk factor trait are expected to be useful in predicting initiation, progression, and severity of CAD in a particular individual or family in a specific reference population.

Information about the genetic architecture of interindividual differences in levels of major risk factors has accumulated slowly. Epidemiological studies have clearly established that variation in levels of most measures of lipid metabolism, blood pressure regulation, and hemostasis that are predictors of CAD aggregate in families (26, 30, 101). Subsequent biometrical genetic studies have clearly established that genetic differences are involved in determining interindividual variation of every one of these biological traits (1, 6, 25, 26, 36, 53, 55, 58, 61, 69–71, 86, 94, 97). Only a few of the responsible genes have been identified and characterized.

What can we expect for the genetic architecture of risk factors such as blood pressure and plasma levels of total cholesterol, HDL cholesterol, and fibrinogen? A genetic architecture that involves many genetic loci each having polymorphic alleles that contribute relatively small effects is not unreasonable. The level of each important risk factor trait is the consequence of regulation of complex metabolic pathways (27, 43, 52) by the products of many genes. At least 200 genes may be involved in the regulation of lipid uptake in the gut, metabolism in the plasma, liver and peripheral cells, and removal from the body. A large fraction of these genes is expected to be polymorphic in every human population (78). Because the distribution of every major risk factor is unimodal in the population at large, the effects of a single locus are not readily distinguishable. The goodness-of-fit of models that have evolved from Fisher's work (24) suggests to some, but certainly not all, geneticists that the genetic architecture of these traits involves many common alleles, each with small effects that combine additively. In addition, we expect the genetic architecture to involve only a few loci with rare alleles that are associated with very large effects. An alternative belief that the genetic architecture of CAD, and hence the intermediate risk factor traits, is characterized by many rare "inborn errors of metabolism" (2) serves as a useful null hypothesis.

Studies by Brown & Goldstein (14) of individuals with very high total plasma cholesterol levels who carry a rare single gene mutation that determines defective cell surface receptors for LDL particles have provided a definitive link between inherited molecular variation and interindividual differences in one of the primary biological risk factors for CAD. Population studies have consistently associated the common inherited isoforms of apolipoprotein (apo) E (9, 10, 21, 38, 88) and the apo(a) molecule (7) with a significant fraction of interindividual variation in total plasma cholesterol and triglycerides. Molecular mechanisms that explain the effects of the apo E polymorphism are better understood (47) than they are for the apo(a) polymorphism (100). Even less is known about how inherited variation in the proteins that control clot formation and degradation acts to influence interindividual differences in levels of intermediate traits like fibrinogen in apparently healthy individuals (34). Efforts to obtain molecular evidence that a single gene can be responsible for interindividual differences in casual blood pressure levels have been unsuccessful (87, 101).

At present the complete genetic architecture is not known for any quantitative trait in any species (3). Because the effort being spent is great, one of the biological risk factor traits for CAD may be among the first to have its genetic architecture defined. We next review progress in using the top-down and bottom-up research strategies to achieve this goal.

STRATEGIES USED IN THE STUDY OF GENETIC ARCHITECTURE OF QUANTITATIVE RISK FACTOR TRAITS

Top-Down, an Unmeasured Genotype Approach

A top-down strategy begins with a biometrical analysis to establish the impact of unmeasured genetic and environmental variation on interindividual phenotypic variability. Schull & Weiss (80) give an overview. Primary questions are: (a) what is the contribution of the segregation of unmeasured loci to the variance of the trait among individuals who define an interbreeding population? (b) is there statistical evidence for a single locus with alleles that have large effects on the levels of the trait? Top-down studies have included a wide range of sampling designs that include twins, nuclear families, extended pedigrees, families with adoptees, and family sets (56). While all of these designs have been used to address the first question, some are not appropriate for assessing the impact of a single locus. Statistical strategies include those that only consider polygenes (path analysis and variance component estimation) and those that consider the additional role of alleles at a single locus that have large effects (complex segregation analysis).

Path analysis, a form of linear structural equation model, was first introduced by Wright (103) as a general theory for explaining measured phenotypic variation in terms of variation in unmeasured causes. It is used in human genetics to estimate the relative contribution of unmeasured polygenes using covariances computed from samples of different types of relatives (e.g.

spouses, parents and offspring, and siblings) (45, 72). However, when pedigree data are available, using only sample covariances wastes information. Elston & Stewart (23) introduced an algorithm for using the likelihood of a genetic model that includes phenotypic information and the genetic relationships among individuals in a set of pedigrees. Lange et al (42) extended the algorithm for estimation of the component of variance attributable to unmeasured polygenes. Estimates of path coefficients and variance components have provided similar inferences about trait variation when the same biological models are used. In most cases the complex segregation analysis model used to search for the existence of allelic variations at a single locus having major effects has included the simultaneous estimation of the effects of polygenes (41, 59, 98). Most applications of these biometrical analyses assume that the Hardy-Weinberg law is true for all loci and there is no epistasis, no genotype by environment interaction and dominance may exist only for the single locus having major effects. The validity of this restrictive set of conditions cannot be tested unless one employs a bottom-up strategy that involves measures of candidate genes.

SELECTED TOP-DOWN EXAMPLES   Studies of plasma levels of total cholesterol, LDL-cholesterol (C) and apo B illustrate the study of three related traits using the traditional top-down approaches. LDL-C is of interest because it is a better predictor of disease risk than total cholesterol (17). Quantitative levels of plasma apolipoproteins are of interest because they are measures of gene products involved in the metabolism of the cholesterol and triglyceride-rich lipoprotein particles. For example, one molecule of apo B combines with cholesterol to form the LDL particle. Apo B also acts as the ligand for binding the LDL particle to a cell surface receptor protein.

Studies of total plasma cholesterol variability after adjustment for concomitants (such as gender, age, height, and weight) suggest that polygenes explain between 49 and 64% of interindividual variability (25, 35, 55, 61, 70, 71, 94). Several studies have shown that a single locus having large effects also contributes to cholesterol variability (57, 58, 84, 102). In all studies the frequency of an allele associated with elevated cholesterol levels is rare and accounts for a relatively small percentage of the total genetic variance. Estimates of the contribution of polygenes (25, 35, 61, 70, 71) and a single gene locus having large effects (58) to LDL-C variability are similar to those estimated for total cholesterol. This is expected since most of the plasma cholesterol is contained in the LDL particle. Studies of apo B suggest that when polygenes alone are considered, between 50 and 70% of variability, adjusted for age and gender, is genetic (25, 65). Studies that considered both a single locus and polygenes to explain interindividual variation of apo B, after adjustment for age and gender, concluded that the contribution of alleles at a single locus to total phenotypic variance (43–54%) is more than the contribution of polygenes (15–21%) (29, 65). Recent studies of fibrinogen suggest that the clotting traits will be similarly influenced by genetic variation. Hamsten et al (26) estimated that 51% of the plasma fibrinogen variance is associated with segregation of polygenes. To date, no studies of the contribution of a single locus with large allelic effects have been reported for this trait.

A few studies have incorporated measurements of covariates into the biometrical analysis of a quantitative risk factor (28, 57). In one example, Perusse et al (67) found evidence for the effects of a single locus on the increase of systolic blood pressure with age. This genetic effect was gender specific. Comparison with an analysis of age- and gender-adjusted data suggested that some genetic factors that influence interindividual differences in level of a trait at a particular age may be different from the gene, or genes, that influence change in level as the individual ages.

PROSPECTS FOR THE TOP-DOWN APPROACH   The strength of the top-down approach is its utility as a statistical tool to estimate the relative role of genetic factors and to test the null hypothesis that no single gene exists that determines large phenotypic effects. Once the existence of a single locus with large allelic effects has been established, the next step is to identify pedigrees segregating for this locus. Information about the number of individuals available from each pedigree and their genetic relationships to one another will determine the power of a proposed linkage study to find the responsible gene (68). Once it is established that the pedigrees have the potential to be informative, the next step is to select candidate regions of the genome to include in the search for the responsible gene. Once there is evidence for linkage to a marked region, the individuals carrying mutations need to be identified so that DNA responsible for the major effects can be characterized. This "reverse genetics" approach to identifying responsible DNA changes has been successful for cystic fibrosis (75), a discrete Mendelian phenotype. The search for a single locus that determines a major effect on a quantitative trait remains an unconquered challenge of major importance to establishing the utility of the biometrical genetic analysis in human genetics. In reviewing the complexities of using linkage analysis to find genes responsible for a binary phenotype, Ott (64) makes clear the problems involved in applying the methodology to a quantitative phenotype.

In summary, a top-down, biometrical study can give statistical evidence for the action of genetic factors and provide a strategy for searching for the genes responsible for large effects. It cannot provide definitive information about the nature of gene-gene interactions, genotype-by-environment interactions or heterogeneity of the penetrance function among genotypes, all of which are important aspects of the genetic architecture of a quantitative risk factor trait.

Bottom-Up: a Measured Genotype Approach

A complete definition of genetic architecture of a trait requires the identification and measurement of the allelic variations that contribute to the distribution of the trait in the population at large. The bottom-up strategy for unraveling genetic architecture of a quantitative trait has involved three approaches. The first centers on a search for candidate genes using randomly selected polymorphic marker loci that have no a priori reason to be involved in the metabolic or physiological processes that determine the trait of interest. Marker loci at known locations throughout the genome may be typed by either DNA restriction fragment polymorphisms (RFLPs) or inherited protein polymorphisms. The second uses discrete measures of DNA variation in a gene that is a candidate for being involved in determination of the trait of interest. A third approach employs discrete measures of a protein product that define allelic differences in a candidate gene.

The first and second strategies depend on linkage (family studies) or linkage disequilibrium (population studies) between a marker and a mutation determining a phenotypic effect. One can locate the gene using these statistical procedures but cannot define the mutations responsible for phenotypic effects. The third is limited to detection of mutations in coding regions of a gene that are reflected in measurable protein differences, usually charge changes detected by electrophoresis. This type of polymorphism measures only one class of mutations. No information about mutations in regulatory or enhancer sequences is provided. In fact, the DNA sequence change that causes the amino acid substitution may not be responsible for the observed association between electrophoretic variation in the protein and trait variation but simply be in linkage disequilibrium with the responsible sequence change. Hence, in all three strategies DNA sequence data is required to characterize the mutational event responsible for statistical associations between measures of genetic variation and phenotypic variation. It is an arduous task to prove that a particular DNA change is responsible for a particular phenotypic effect (44).

SELECTED BOTTOM-UP EXAMPLES   An application of the first strategy identified 4 of 12 polymorphic blood and serum markers as being significantly associated with levels of blood cholesterol (93). A random marker approach to mapping quantitative trait loci in humans cannot be expected to be as productive as it has been in plants when experimental data were available (22). We agree with Neel (63) that for humans an approach that focuses on measures of candidate genes has greater potential for success. For most quantitative risk factors of CAD, information about the metabolic processes that determine the phenotype provides guidance to selection of these candidate genes.

Studies of RFLP variation in genes involved in lipid metabolism have detected a number of significant associations (33). As expected for linked markers, the association between a particular RFLP and level of a particular trait has not been consistent across populations (19). Studies that use protein isoforms to mark genotypic variation of candidate genes are limited to the subset that are polymorphic. Most work has involved the genes coding for apo E (21, 38, 90), apo (a) (100), apo AIV, and apo H (38, 90). Even though allelic effects across studies have been more consistent than for RFLPs, without knowledge of responsible DNA change, estimates of true genetic effects will also be a function of the distance between the protein marker and the responsible mutation (18).

Pederson & Berg (66) have presented data that suggest nonadditivity of effects of variations of the LDL receptor gene and effects of common alleles of the gene coding for plasma apo E on the levels of plasma cholesterol. A study to characterize the predictors of interindividual variation in plasma apo E levels by Sing et al (90), summarized in Table 1, serves to illustrate just how complex the genetic architecture of an intermediate quantitative risk trait that links genetic variation to variation in risk of CAD might be. Males and females (ages 26 to 63), sampled to represent the population at large, are considered separately. An association between allelic variation in three candidate genes coding for apolipoproteins and variation in plasma apo E levels depends on plasma levels of triglycerides, total cholesterol, and HDL cholesterol in males but not females (A vs B for each gender in Table 1). Allelic variations in unlinked loci coding for apo H and apo AIV have epistatic effects on the plasma level of the apo E gene product in males but not females. Very different sets of predictors explain interindividual differences in males compared to females. Only a small fraction of the population variance of plasma apo E is explained by polymorphic structural variation in the apo E gene measured by the apo E isoforms. Studies that do not consider gender-specific effects of concomitants like age and body size as well as genotypic effects, or genotype-by-genotype interaction, are likely to provide only weak inferences about genetic architecture of quantitative risk factors for CAD.

Other studies further indicate how complex the relationship between genetic variation and variation in quantitative risk factors for CAD might be. Using twin data, Berg et al (4) have found that different RFLP genotypes at loci coding for apolipoproteins B, AI, and AII are associated with different magnitudes of interindividual variability in plasma cholesterol levels. Gender, age, and genotype at the apo E locus influence interindividual variability of lipid traits and the correlations between traits that are measures of lipid metabolism (10, 73, 74, 95). One of the few studies of genotype-by-

Table 1   Percent of population variance of plasma apo E levels associated with measured concomitants and apo isoforms estimated from data collected by the Rochester Family Heart Study. Columns A and B present results without and with inclusion of plasma lipids and lipoproteins, respectively.

Source[1]	Males		Females	
	A	B	A	B
Age, Ht and Wt	2.3	2.3	4.0	4.0
Triglycerides, Total-C and HDL-C	—	66.9	—	16.8
Genotypes				
apo E	11.6	3.5	14.0	16.2
apo H and AIV	6.0	2.9	0.0	0.6
Unexplained	80.1	24.4	82.0	62.4

[1]The reduction in error mean square associated with fitting each source (after fitting the preceding source) is taken as the estimate of the variance due to the source.

environment effects on lipid traits in humans by Kaprio et al (39) further argues that simple analyses of marginal genotype effects may lead to erroneous inferences about genetic architecture. Smoking decreases apo AII levels in subjects with the apo H 2/2 genotype, but not in subjects with other apo H genotypes.

PROSPECTS FOR THE BOTTOM-UP APPROACH   Measurements of candidate genes can provide strong inferences about gene-by-gene interaction, genotype-by-environment (measured by age, body size, life style, diet, etc) interaction and the impact of genotype on trait variance and correlation between traits. The bottom-up approach using observational data on humans presents several unresolved problems in study design and analysis. If the genetic architecture of an intermediate trait involves many genes with small effects and a few genes with large effects, few families will be segregating for exactly the same subset of genes (96). Furthermore, if the effects of an allele depend on the effects of alleles at other loci and/or exposures to particular environments, the sample of individuals with the same relationship between genotype and phenotype will be even smaller. Since most populations are probably made up of small subdivisions, each representing a different relationship between genotype and phenotype, no subdivision is likely to be large enough to provide a statistically meaningful sample. Equally challenging is the modeling problem. At present no statistical model is available that enables one to analyze simultaneously the relationship between levels in the hierarchy represented by measures of genotype, measures of intermediate traits, and clinical measures of CAD. Such a model is needed to address the question of whether measured genetic information contributes to prediction of disease beyond that provided by the intermediate quantitative traits that link genetic and environmental variation to variation in risk. We have suggested elsewhere (92) that there may not be an equation that incorporates all of the known etiological pathways that link genotype and endpoint. Finding a simple, useful approach to modeling diseases such as CAD that realistically represents known biological complexities will test the mettle of the most experienced researchers.

A Hybridization of Strategies

A biometrical strategy can serve to (a) reject the null hypothesis of no genetic variation for a quantitative trait, (b) identify traits with sufficient genetic variation to justify a bottom-up study and (c) identify subsets of families with different genetic and environmental etiologies (54, 67). The bottom-up strategy provides a means of testing assumptions about pleiotropy, epistasis, homogeneity of variance and genotype-by-environment interaction made when applying the models used in the top-down strategy. All of these effects are likely to be an important part of the genetic architecture of most quantitative traits. A combination of these strategies offers the possibility for successive partitioning of the genetic architecture as new measures of candidate genes become available. In one example, Boerwinkle & Sing (8) simultaneously estimated the frequency and effects of the measured genotypes of the gene coding for apo E and the variance attributable to unmeasured polygenes. The percent of interindividual variance associated with genotypes determined by the common apo E alleles was 8% while the contribution of unmeasured polygenes was 56%. In another study, the information about the apo E polymorphism was incorporated into a complex segregation analysis of plasma apo B levels. The effects of a single, unmeasured locus and unmeasured polygenes were estimated before and after adjusting apo B levels for the differences among the means of measured apo E genotypes (65). It was concluded that significant effects of the common alleles of the apo E gene on apo B levels act independently of allelic effects of the unmeasured single locus (65). A combination of the top-down and bottom-up strategies offers possibilities for determining which combination of measured genes explains the polygenic component of variation in quantitative intermediate risk factor traits.

CHALLENGES FOR FUTURE RESEARCH

A need to understand the genetic architecture of the common chronic diseases like coronary artery disease is emerging as a primary objective of medicine (2, 60, 82, 99). Geneticists will be asked to determine which candidate genes are involved in influencing risk of disease in a particular individual, a particular family, and a particular population (89). Research reviewed here suggests that the task is enormous. Simple one-to-one relationships between allelic variations and clinical endpoints are not expected. Collaborations with molecular biologists, biochemists, physiologists, and clinicians will be required to understand the network of causation between genotype and clinical endpoints through quantitative intermediate traits. As Wright (104) so aptly stated,

"The task of science is not complete until it has followed phenomena through all levels of the hierarchy, up and down as far as possible, and after obtaining the best statistical description at each, has tied them all together."

The search for the pathways of causation up and down the hierarchy between the molecular gene and clinical disease has just begun for the complex multifactorial diseases. Research on CAD holds promise for revealing fundamental knowledge about the relationships between levels in this hierarchy. To fully accomplish this ambitious goal will require (a) noninvasive measures of initiation, development, and severity of disease, (b) new measures of intermediate traits that predict disease and disclose the role of genetic variation, (c) statistical models for evaluating how genetic variation in the population at large contributes to variation in measures of disease through intermediate quantitative traits and (d) methods for applying these models to identify subgroups of individuals and families with a homogeneous relationship between genotype, intermediate traits, and disease endpoint. In the search for solutions to these difficult problems, a struggle will certainly arise between those who wish to make general inferences about the impact of a particular allele on a trait in the population and those who seek to make inferences about the consequences of interaction of the effects of the allele with the effects of other alleles and exposures to environments in special individuals and families. We will argue that a valid prediction of the impact of an allele on the phenotype of a particular individual or family cannot be made without knowledge about the range of effects of the gene in the Mendelian population in which these individuals dwell.

ACKNOWLEDGMENTS

We are grateful to our colleagues Martha Haviland, Jim Neel, Tim Rebbeck, Sharon Reilly, and Kim Zerba for helpful discussions about the content and organization of this manuscript. We especially thank Debbie Theodore for her unshakable dedication to providing the best possible secretarial assistance. This research was supported in part by NIH grants HL30428, HL39107, and HL24489.

Literature Cited

1. Annest, J. L., Sing, C. F., Biron, P., Mongeau, J. G. 1979. Familial aggregation of blood pressure and weight in adoptive families. II. Estimation of relative contributions of genetic and common environmental factors to the blood pressure correlations between family members. *Am. J. Epidemiol.* 110:492–503

2. Baird, P. A. 1990. Genetics and health care: A paradigm shift. *Perspect. Biol. Med.* 33:203–13

3. Barton, N. H., Turelli, M. 1989. Evolutionary quantitative genetics: How little do we know? *Annu. Rev. Genet.* 23:337–70

4. Berg, K. 1983. Genetics of coronary heart disease. In *Progress in Medical Genetics*, ed. A. G. Steinberg, A. G. Bearn, A. G. Motulsky, B. Childs, 5:35–90. Philadelphia: Saunders

5. Berg, K. 1987. See Ref. 6, pp. 14–27

6. Bock, G., Collins, G. M., eds. 1987. *Molecular Approaches to Human Polygenic Disease*, Ciba Found. Symp. 130. Chichester: Wiley. 274 pp.

7. Boerwinkle, E., Menzel, H. J., Kraft, H. G., Utermann, G. 1989. Genetics of the quantitative Lp(a) lipoprotein trait. III. Contribution of Lp(a) glycoprotein phenotypes to normal lipid variation. *Hum. Genet.* 82:73–78

8. Boerwinkle, E., Sing, C. F. 1987. The use of measured genotype information in the analysis of quantitative phenotypes in man. III. Simultaneous estimation of the frequencies and effects of the *apolipoprotein E* polymorphism and residual polygenetic effects on cholesterol, betalipoprotein and triglyceride levels. *Ann. Hum. Genet.* 51:211–26

9. Boerwinkle, E., Utermann, G. 1988. Simultaneous effects of the apolipoprotein E polymorphism on apolipoprotein E, apolipoprotein B, and cholesterol metabolism. *Am. J. Hum. Genet.* 42:104–12

10. Boerwinkle, E., Visvikis, S., Welsh, D., Steinmetz, J., Hanash, S. M., Sing, C. F. 1987. The use of measured genotype information in the analysis of quantitative phenotypes in man. II. The role of the apolipoprotein E polymorphism in determining levels, variability and covariability of cholesterol, betalipoprotein and triglycerides in a sample of unrelated individuals. *Am. J. Med. Genet.* 27:567–82

11. Braunwald, E., ed. 1988. *Heart Disease: A Textbook of Cardiovascular Medicine*, 3rd ed. Philadelphia: Saunders. 1900 pp.

12. Breslow, J. L. 1988. Apolipoprotein genetic variation and human disease. *Physiol. Rev.* 68:85–132

13. Breslow, J. L. 1989. Genetic basis of lipoprotein disorders. *J. Clin. Invest.* 84:373–80

14. Brown, M. S., Goldstein, J. L. 1986. A receptor-mediated pathway for cholesterol homeostasis. *Science* 232:34–47

15. Brown, M. S., Goldstein, J. L. 1990. Scavenging for receptors. *Nature* 343:508–9

16. Brunzell, J. D., Sniderman, A. D., Albers, J. J., Kwiterovich, P. O. Jr. 1984. Apoproteins B and A-I and coronary artery disease in humans. *Arteriosclerosis* 4:79–83

17. Castelli, W. P., Garrison, R. J., Wilson, P. W., Abbott, R. D., Kalousdian, S., et al. 1986. Incidence of coronary heart disease and lipoprotein cholesterol levels. The Framingham Study. *J. Am. Med. Assoc.* 256:2835–38

18. Chakraborty, R., Boerwinkle, E. 1989. Effects of multiple markers on variation of a quantitative trait: Power analysis with measured genotype information. *Am. J. Hum. Genet.* 45(4):A236 (Abstr.)

19. Cooper, D. N., Clayton, J. F. 1988. DNA polymorphism and the study of disease associations. *Hum. Genet.* 78:299–312

20. Davignon, J., DuFour, R., Cantin, M. 1983. Atherosclerosis and hypertension. In *Hypertension, Physiopathology and Treatment*, ed. J. Genest, O. Kuchel, P. Hamet, M. Cantin pp. 810–52. New York: McGraw-Hill. 2nd ed.

21. Davignon, J., Gregg, R. R., Sing, C. F. 1988. Apo E polymorphism and atherosclerosis. *Arteriosclerosis* 8:1–21

22. Edwards, M. D., Stuber, C. W., Wendel, J. F. 1987. Molecular-marker-facilitated investigations of quantitative-trait loci in maize. I. Numbers, genomic distribution and types of gene action. *Genetics* 116:113–25

23. Elston, R. C., Stewart, J. 1971. A general model for the genetic analysis of pedigree data. *Hum. Hered.* 21:523–42

24. Fisher, R. A. 1918. The correlation between relatives on the supposition of Mendelian inheritance. *Trans. R. Soc. Edin.* 52:399–433

25. Hamsten, A., Iselius, L., Dahlen, G., de Faire, U. 1986. Genetic and cultural

inheritance of serum lipids, low and high density lipoprotein cholesterol and serum apolipoproteins A-I, A-III and B. *Atherosclerosis* 60:199–208

26. Hamsten, A., Iselius, L., de Faire, U., Blomback, M. 1987. Genetic and cultural inheritance of plasma-fibrinogen concentration. *Lancet* 2:988–91

27. Handin, R. I., Loscalzo, J. 1988. See Ref. 11, pp. 1758–81

28. Hanis, C. L., Sing, C. F., Clark, W. R., Schrott, H. G. 1983. Multivariate models for human genetic analysis: Aggregation, coaggregation, and tracking of systolic blood pressure and weight. *Am. J. Hum. Genet.* 35:1196–210

29. Hasstedt, S. J., Wu, L., Williams, R. R. 1987. Major locus inheritance of apolipoprotein B in Utah pedigrees. *Genet. Epidemiol.* 4:67–76

30. Heuch, I., Namboodiri, K. K., Green, P. P., Kaplan, E. B., Laskarzewski, P., et al. 1985. A multivariate analysis of familial associations of lipoprotein levels in the lipid research clinics collaborative family study: I. Familial correlation and regression analyses. *Genet. Epidemiol.* 2:283–300

31. Higgins, M. W., Luepker, R. V., eds. 1989. *Trends and Determinants of Coronary Heart Disease Mortality: International Comparisons. Int. J. Epidemiol.* 18, No. 3(Suppl. 1) 232 pp.

32. Hopkins, P. N., Williams, R. R. 1989. Human genetics and coronary heart disease: A public health perspective. *Annu. Rev. Nutr.* 9:303–45

33. Humphries, S. E. 1988. DNA polymorphisms of the apolipoprotein genes—their use in the investigation of the genetic component of hyperlipidaemia and atherosclerosis. *Atherosclerosis* 72:89–108

34. Humphries, S. E., Cook, M., Dubowitz, M., Stirling, Y., Meade, T. W. 1987. Role of genetic variation at the fibrinogen locus in determination of plasma fibrinogen concentrations. *Lancet* 1:1452–55

35. Iselius, L. 1979. Analysis of family resemblance for lipids and lipoproteins. *Clin. Genet.* 15:300–6

36. Iselius, L. 1988. Genetic epidemiology of common diseases in humans. In *Proc. 2nd Int. Conf. Quant. Genet.*, ed. B. S. Weir, E. J. Eisen, M. M. Goodman, G. Namkoong, pp. 341–52. Sunderland, MA: Sinauer

37. Kannel, W. B., Gordon, T. 1970. Section 26. Some characteristics related to the incidence of cardiovascular disease and death: Framingham Study, 16-year

follow-up. Washington, DC: US Gov. Print. Off.

38. Kaprio, J., Ferrell, R. E., Kottke, B. A., Kamboh, M. I., Turner, S. T., et al. 1990. Effect of polymorphisms in apolipoprotein E, A-IV and H on quantitative traits related to risk for cardiovascular disease. *Arteriosclerosis.* In press

39. Kaprio, J., Ferrell, R. E., Kottke, B. A., Sing, C. F. 1989. Smoking and reverse cholesterol transport: Evidence for gene-environment interaction. *Clin. Genet.* 36:266–68

40. Kodama, T., Freeman, M., Rohrer, L., Zabrecky, J., Matsudaira, P., et al. 1990. Type I macrophage scavenger receptor contains α-helical and collagen-like coiled coils. *Nature* 343:531–35

41. Lalouel, J. M., Rao, D. C., Morton, N. E., Elston, R. C. 1983. A unified model for complex segregation analysis. *Am. J. Hum. Genet.* 35:816–26

42. Lange, K., Westlake, J., Spence, M. A. 1976. Extensions to pedigree analysis III. Variance components by the scoring method. *Ann. Hum. Genet., London* 39:485–91

43. Laragh, H. H., Brenner, B. M., eds. 1990. *Hypertension: Pathophysiology, Diagnosis, and Management.* Vol. 1. New York: Raven. 1382 pp.

44. Laurie-Ahlberg, C. C., Stam, L. F. 1987. Use of *P*-element mediated transformation to identify the molecular basis of naturally occurring variants affecting *Adh* expression in *Drosophila melanogaster. Genetics* 115:129–40

45. Li, C. C. 1975. *Path Analysis: A Primer.* Pacific Grove, CA: Boxwood

46. Lusis, A. J. 1988. Genetic factors affecting blood lipoproteins: The candidate gene approach. *J. Lipid Res.* 29:397–429

47. Mahley, R. W. 1988. Apolipoprotein E: cholesterol transport protein with expanding role in cell biology. *Science* 240:622–30

48. Mann, K. G. 1989. Factor VII assays, plasma triglyceride levels, and cardiovascular disease risk. *Arteriosclerosis* 9:783–84

49. McGill, H. C. Jr. 1979. See Ref. 97, pp. 27–49

50. Meade, T. W., Mellows, S., Brozovic, M., Miller, G. J., Chakrabarti, R. R., et al. 1986. Haemostatic function and ischaemic heart disease: Principal results of the Northwick Park Heart Study. *Lancet* 2:533–37

51. Miller, J. G. 1978. *Living Systems.* New York: McGraw-Hill

52. Miller, N. E., Lewis, B., eds. 1981.

Lipoproteins, Atherosclerosis and Coronary Heart Disease. Amsterdam: Elsevier/North-Holland. 214 pp.

53. Moll, P. P., Harburg, E., Burns, T. L., Schork, M. A., Özgoren, F. 1983. Heredity, stress and blood pressure, a family set approach: The Detroit project revisited. J. Chronic Dis. 36:317–28

54. Moll, P. P., Michels, V. V., Weidman, W. H., Kottke, B. A. 1989. Genetic determination of plasma apolipoprotein AI in a population-based sample. Am. J. Hum. Genet. 44:124–39

55. Moll, P. P., Powsner, R., Sing, C. F. 1979. Analysis of genetic and environmental sources of variation in serum cholesterol in Tecumseh, Michigan. V. Variance components estimated from pedigrees. Ann. Hum. Genet. 42:343–54

56. Moll, P. P., Sing, C. F. 1979. See Ref. 97, pp. 307–42

57. Moll, P. P., Sing, C. F., Lussier-Cacan, S., Davignon, J. 1984. An application of a model for a genotype-dependent relationship between a concomitant (Age) and a quantitative trait (LDL Cholesterol) in pedigree data. Genet. Epidemiol. 1:301–14

58. Morton, N. E., Gulbrandsen, C. L., Rhoads, G. G., Kagan, A., Lew, R. 1978. Major loci for lipoprotein concentrations. Am. J. Hum. Genet. 30: 583–89

59. Morton, N. E., MacLean, C. J. 1974. Analysis of family resemblance. III. Complex segregation of quantitative traits. Am. J. Hum. Genet. 26:489–503

60. Motulsky, A. G. 1984. See Ref. 97, pp. 541–48

61. Namboodiri, E. B., Kaplan, E. B., Heuch, I., Elston, R. C., Green, P. P., et al. 1985. The collaborative lipid research clinics family study: Biological and cultural determinants of familial resemblance for plasma lipids and lipoproteins. Genet. Epidemiol. 2:227–54

62. National Heart, Lung, and Blood Institute. 1987. Forty Years of Achievement in Heart, Lung, and Blood Research. Nat. Inst. Health. 118 pp.

63. Neel, J. V. 1990. The "bottom up" approach to multifactorial inheritance: Some problems. In Genetic Analysis of Complex Disease, ed. E. Lander, M. A. Chakravarti, J. Witkowski. Cold Spring Harbor, NY: Cold Spring Harbor Press. In press

64. Ott, J. 1990. Invited editorial: Cutting a Gordian knot in the linkage analysis of complex human traits. Am. J. Hum. Genet. 46:219–21

65. Pairitz, G., Davignon, J., Mailloux, H.,

Sing, C. F. 1988. Sources of interindividual variation in the quantitative levels of apolipoprotein B in pedigrees ascertained through a lipid clinic. Am. J. Hum. Genet. 43:311–21

66. Pedersen, J. C., Berg, K. 1989. Interaction between low density lipoprotein receptor (LDLR) and apolipoprotein E (apoE) alleles contributes to normal variation in lipid level. Clin. Genet. 35: 331–37

67. Perusse, L., Moll, P. P., Sing, C. F. 1990. Complex segregation analysis of systolic blood pressure using a genotype- and gender-specific age regression model. Am. J. Hum. Genet. In press

68. Ploughman, L. M., Boehnke, M. 1989. Estimating the power of a proposed linkage study for a complex genetic trait. Am. J. Hum. Genet. 44:543–51

69. Rao, D. C., Elston, R. C., Kuller, L. H., Feinleib, M., Carter, C., et al., eds. Genetic Epidemiology of Coronary Heart Disease: Past, Present, and Future. New York: Liss. 575 pp.

70. Rao, D. C., Morton, N. E., Glueck, C. J., Laskarzewski, P. M., Russell, J. M. 1983. Heterogeneity between populations for multifactorial inheritance of plasma lipids. Am. J. Hum. Genet. 35:468–83

71. Rao, D. C., Morton, N. E., Gulbrandsen, C. L., Rhoads, G. G., Kagen, A., et al. 1979. Cultural and biological determinants of lipoprotein concentrations. Ann. Hum. Genet. 42:467–77

72. Rao, D. C., Province, M. A., Wette, R., Glueck, C. J. 1984. See Ref. 69, pp. 193–212

73. Reilly, S., Kottke, B., Ferrell, R., Sing, C. 1989. Apolipoprotein E genotype and gender influence phenotypic variation and covariation of lipids and apolipoproteins in the population at large. Am. J. Hum. Genet. 45(4):A247 (Abstr.)

74. Reilly, S. L., Kottke, B. A., Sing, C. F. 1990. The effects of generation and gender on the joint distributions of lipid and apolipoprotein phenotypes in the population at large. J. Clin. Epidemiol. In press

75. Rommens, J. M., Iannuzzi, M. C., Kerem, B.-S., Drumm, M. L., Melmer, G., et al. 1989. Identification of the cystic fibrosis gene: Chromosome walking and jumping. Science 245:1059–65

76. Rose, G. 1987. See Ref. 6, pp. 247–56

77. Ross, R. 1988. See Ref. 11, pp. 1135–52

78. Roychoudhury, A. K., Nei, M. 1988. Human Polymorphic Genes. World Distribution. New York: Oxford Univ. Press. 393 pp.

79. Salthe, S. N. 1985. *Evolving Hierarchical Systems.* New York: Columbia Univ. Press. 329 pp.
80. Schull, W. J., Weiss, K. M. 1980. Genetic epidemiology: Four strategies. *Epidemiol. Rev.* 2:1–18
81. Scott, J. 1989. Thrombogenesis linked to atherogenesis at last? *Nature* 341:22–23
82. Scriver, C. R. 1984. An evolutionary view of disease in man. *Proc. R. Soc. Lond.* B220:273–98
83. Simon, H. A. 1962. The architecture of complexity. In *Proc. Am. Philos. Soc.* 106(6):467–82
84. Simpson, J. M., Brennan, P. J., McGilchrist, C. A., Blacket, R. B. 1981. Estimation of environmental and genetic components of quantitative traits with application to serum cholesterol levels. *Am. J. Hum. Genet.* 33:293–99
85. Sing, C. F., Boerwinkle, E. 1987. See Ref. 6, pp. 99–127
86. Sing, C. F., Boerwinkle, E., Moll, P. P., Templeton, A. R. 1988. Characterization of genes affecting quantitative traits in humans. See Ref. 36, pp. 250–69
87. Sing, C. F., Boerwinkle, E., Turner, S. T. 1986. Genetics of primary hypertension. *Clin. Exper. Theory Pract.* A8(4&5):623–51
88. Sing, C. F., Davignon, J. 1985. Role of the apolipoprotein E polymorphism in determining normal plasma lipid and lipoprotein variation. *Am. J. Hum. Genet.* 37:268–85
89. Sing, C. F., Kaprio, J., Perusse, L., Moll, P. P. 1990. Genetic differences in risk of disease within and between populations. *World Rev. Nutr. Diet, Genetic Variation and Nutrition,* ed. A. P. Simopoulous, B. Childs. Basel, Switzerland: Karger
90. Sing, C. F., Kottke, B. A., Kamboh, M. I., Ferrell, R. E. 1990. Genetic architecture of interindividual variation of plasma apolipoprotein E levels: I. Evidence for epistatic effects of polymorphic alleles at unlinked genes coding for apolipoproteins E, H and AIV. *Arteriosclerosis.* In press
91. Sing, C. F., Moll, P. P. 1989. See Ref. 31, pp. S183–95
92. Sing, C. F., Moll, P. P. 1990. Strategies for unravelling the genetic basis of coronary artery disease. In *From Phenotype to Gene in Common Disorders,* ed. K. Berg, N. Retterstol, S. Refsum. Copenhagen, Denmark: Munksgaard A/S Int.
93. Sing, C. F., Orr, J. D. 1976. Analysis of genetic and environmental sources of variation in serum cholesterol in Tecumseh, Michigan. III. Identification of genetic effects using 12 polymorphic genetic blood marker systems. *Am. J. Hum. Genet.* 28:453–64
94. Sing, C. F., Orr, J. D. 1978. Analysis of genetic and environmental sources of variation in serum cholesterol in Tecumseh, Michigan. IV. Separation of additive polygene from common environmental effects. *Am. J. Hum. Genet.* 30:491–504
95. Sing, C. F., Reilly, S., Ferrell, R., Kottke, B. 1989. Apolipoprotein E genotype and gender influence the regression of lipid and apolipoprotein levels on concomitants in the population at large. *Am. J. Hum. Genet.* 45(4):A249 (Abstr.)
96. Sing, C. F., Reilly, S., Moll, P. P. 1990. Family studies of common chronic diseases: Potentials, limitations and paradoxes. In *Genetics of Cellular, Individual, Family and Population Variability.* New York: Oxford Univ. Press. In press
97. Sing, C. F., Skolnick, M., eds. 1979. *Genetic Analysis of Common Diseases: Applications to Predictive Factors in Coronary Disease.* New York: Liss. 749 pp.
98. Thompson, E. A. 1986. *Pedigree Analysis in Human Genetics.* Baltimore/London: Johns Hopkins Univ. Press
99. Trimble, B. K., Smith, M. E. 1977. The incidence of genetic disease and the impact on man of an altered mutation rate. *Can. J. Genet. Cytol.* 19:375–85
100. Utermann, G. 1989. The mysteries of lipoprotein(a). *Science* 246:904–10
101. Ward, R. 1990. See Ref. 43, pp. 81–100
102. Williams, W. R., Lalouel, J. M. 1982. Complex segregation analysis of hyperlipidemia in a Seattle sample. *Hum. Hered.* 32:24–36
103. Wright, S. 1921. Correlation and causation. *J. Agric. Res.* 20:557–85
104. Wright, S. 1964. Biology and the philosophy of science. In *Process and Divinity: The Hartshorne Festschrift,* ed. W. R. Reese, E. Freeman, pp. 101–25. La Salle, IL: Open Court

Annu. Rev. Genet. 1990. 24:189–213

FRAMESHIFT MUTATION: DETERMINANTS OF SPECIFICITY

L. S. Ripley

Department of Microbiology and Molecular Genetics, New Jersey Medical School, University of Medicine and Dentistry of New Jersey, Newark, New Jersey 07103-2714

KEY WORDS: mutagen, DNA, spontaneous mutation, mutational mechanisms, deletion mutation

CONTENTS

INTRODUCTION

A frameshift mutation is a genetically detectable consequence of DNA metabolism gone awry. Perturbed DNA and/or perturbed enzyme behavior are generally believed to be responsible. However, even "frequent" mutations are inherently rare. This characteristic often limits the experimental approaches that can be used to identify mutational mechanisms. A mutational model predicts the existence of specific, often rare, intermediates of mutation. Tests of these predictions support a model when they reveal the site-specific presence of a hypothesized intermediate and behavior of the intermediate consistent with mutation. A model for frameshift mutation can often be hypothesized from knowledge of the DNA sequence and contexts of the mutants and the sequence-specific behavior of enzymes believed to be in-

189

0066-4197/90/1215-0189$02.00

volved in mutation. Further tests may contrast the characteristics of frameshift-prone DNA sequences and contexts with those not prone to mutation. A successful model accounts for both the specific DNA sequence change observed and the sequence-specific pattern of high and low mutation frequencies. This review focuses on several hypotheses for frameshift mutagenesis, drawing attention to the DNA sequence changes they predict, the role of enzymology in frameshift specificity, and improved understanding of some critical DNA intermediates.

Crucial to the development of a frameshift model is a highly resolved view of frameshift specificity and frequency. This view is most easily obtained when the genetic assay specifically identifies frameshift mutants in diverse DNA sequences (87). Additions or deletions of 3N ± 1 bases to coding regions in DNA are readily detected by their frameshifting effects on translation (15). Thus, selection for gene inactivation leads to the recovery of frameshifts along with other mutations such as bases substitutions, insertions, and deletions. Frameshifts can be more specifically selected as revertants of existing frameshifts. Many frameshift assays do not limit reversion to a few nucleotide positions. For example, in genes for proteins with N-terminal regions that are insensitive to amino acid sequence changes, a frameshift at one site can be reverted by frameshifts of the opposite sign at many alternative sites (18, 33, 60, 81, 89). This selection combines the selectivity of a reversion assay with the power of a forward assay to examine numerous sites simultaneously. Mutational spectra obtained in forward or in multisite reversion assays can be particularly informative when the mutants are found frequently at some sites, but rarely at the remaining sites. Such a "saturated" spectrum permits an evaluation of the frequent frameshift sites (hotspots) not only on the basis of what characteristics they share, but also by the absence of those characteristics at other detectable frameshift sites. When the hotspots represent a small fraction of the genetically detectable sites, the non hotspots represent the majority of the data. However, these data can be relied upon only when the spectrum is saturated for the mutational class of interest. Unsaturated spectra appear frequently in the literature when only a few mutations are sequenced or when the specificity of interest (e g. a particular kind of frameshift) constitutes a small fraction of the sequenced mutants. Improved assays addressing detectability issues have been employed (58, 84, 93, 98, 101).

A REPERTOIRE OF FRAMESHIFT MODELS

The ability of a frameshift model to account for the site-specific frequency and specificity of mutagenesis is an important measure of its success. As the number of sequenced frameshift mutations has increased, it has become apparent that more than one model is required to explain them and, thus, that multiple mechanisms contribute to frameshift mutation (107). As the repertoire of frameshift models grows, it becomes increasingly obvious that different mechanisms sometimes predict the same DNA sequence changes. In such instances, DNA sequencing alone will not specifically identify the relevant model and more subtle tests of the mechanism are required.

Additions or Deletions Mediated by a Single Misalignment

Frameshifts often occur in repeated DNA sequences. When the mutations in these sequences are the deletion or the duplication of the repeating unit of the sequence, the correlation and specificity are explained by a model proposed by Streisinger and co-workers (106). In the model (Figure 1), a locally out-of-register alignment of a DNA primer relative to the template is extended as though it were normally aligned. This leads to deletions when the extra base or bases lie in the template strand, or to duplications when the extra bases lie in the primer strand. This model predicts that misalignments between repeats will lead to the concerted deletion or duplication of DNA sequences lying between them. An example is a frameshift hotspot in the T4 *rII*B gene (81, 89). Misaligned pairing can convert the ATTGGCtgATTGGC sequence to ATTGGC or to ATTGGCtgATTGGCtgATTGGC. However, a 2-bp deletion in this sequence, ATTGGCtg–TGGC, is not accounted for by this misalignment model, even though it is flanked by TG repeats.

DNA misalignments between direct repeats can be readily described as intermediates of either recombination or replication. Recombinational mechanisms have been frequently evoked for misalignments operating over very large chromosomal distances. These misalignments produce large duplications that are usually unstable unless maintained by selection or deletions that can be detected only when the deleted DNA is nonessential. Multiple mech-

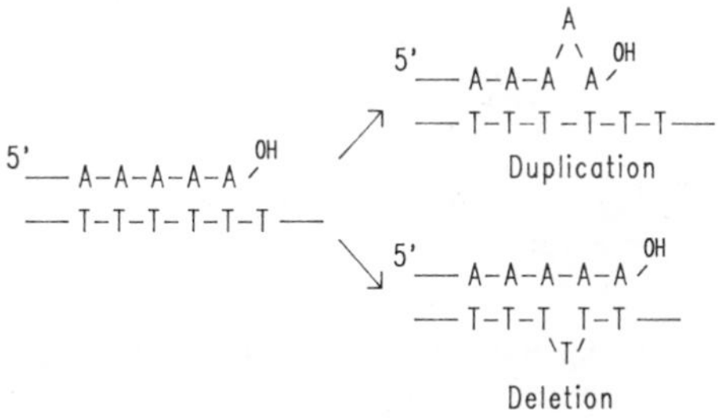

Figure 1   Misalignment permitted by the tandem repeat of A:T. An additional base in the primer strand relative to its template produces a duplication upon further synthesis, while an extra base in the template strand produces a deletion.

anisms undoubtedly contribute to these mutations (2, 34, 35) but are beyond the scope of this review.

MONOTONIC RUNS   Monotonic runs are tandem repeats of a single base pair and have frequently been found to be sites of spontaneous and mutagen-induced frameshifts (e.g. 12, 81, 105, 107). The misalignment model predicts duplications and deletions of a base pair in the run. Six consecutive A:T pairs are the most frequent sites of spontaneous mutation in the bacteriophage T4 *rII*AB cistron (7). Spontaneous frameshifts occur more frequently in longer than in shorter runs (81). However, on average, spontaneous frameshifts at "runs" of two consecutive base pairs are not more frequent than at two single base pairs that cannot misalign (89). Site-specific variation of frameshift frequency in runs is high, at least tenfold in the 6-bp runs in T4. The molecular basis of this variation is unknown. However, the variation is differentially sensitive to experimental perturbation (92, 93), and its existence points to the inadvisability of testing frameshift models in a single DNA context. No carefully controlled study yet defines the relative mutability of A:T and G:C runs.

Misalignments within monotonic runs of at least 3 bp, in theory, could produce 2-bp deletions or duplications. The misalignment model predicts an intermediate with a 2-base bulge in one strand. Although few 2-bp frameshifts have been sequenced, they are not usually found in mononucleotide runs. Multiple mechanisms probably contribute to 2-bp frameshifts and some do not depend on misalignment (18, 28, 88, 89).

Physical studies of the structure of the bulges predicted by misalignment have only been attempted for single bases. Early model building suggested that an extra base could be readily accommodated with B-DNA helix with only local distortions (27). Crystallographic and NMR studies of oligonucleotides indicate that the extra base can be accommodated by a helix in extra- or intrahelical positions (41, 43, 44, 67, 77). The effects of DNA sequence on extra base structure are not yet predictable but are important. For example, G and C bulges examined by ^{1}H NMR in a G:C-run inbedded in an otherwise complementary oligonucleotide behave differently. The C bulge occupied all available sites within the C-run with approximately equivalent probability, while the G bulge primarily occupied the central positions of the G-run (112, 113). It is not yet clear whether the bulges studied by NMR would mimic those hypothesized to mediate misalignment frameshifts in runs. In the model, mutagenesis requires bulges when the run is at a DNA end rather than when bounded on both sides by base pairs that are not part of the run. The potential importance of the identity of bounding base pairs on these structures has not been reported. On the other hand, the tested structure should be an excellent model for the behavior of the frameshift heterozygote, an intermediate that is subject to mismatch repair in *Escherichia coli* (101). An in vivo experiment did not reveal differential repair of a bulged pyrimidine and purine (22), but could only have detected large differences in net repair.

TANDEM REPEATS   The duplication of a repeating unit longer than 1 bp offers a robust test of the misalignment model prediction that additions are templated. For example, the nontemplated addition of one nucleotide to the DNA in a monotonic run might create a duplication consistent with the misalignment model, despite the absence of misaligned pairs. Frameshifts can be induced in vitro by nucleotide additions by terminal-transferase (56). Even DNA polymerase I of $E.$ $coli$ may add untemplated bases at DNA ends (13, 14), although in vitro reactions do not frequently lead to 1-bp duplications (18). A frameshift hotspot in the $lacI$ gene of $E.$ $coli$ supports the templating prediction of the misalignment model (25). The wild-type sequence $(CTGG)_3$ is mutated repeatedly to $(CTGG)_4$, as predicted by templated addition.

Support for the misalignment model has been sought by examining the effect of the length or base content of the repeat on mutation frequency. Misalignments might be more stable when mediated by more base pairs or by G:C base pairs compared to A:T base pairs. This prediction is very difficult to test rigorously. Increased length offers not only improved stability, but also increased opportunity for DNA discontinuities at more sites to initiate misalignments. Comparing different sequences complicates both tests. In $lacI,$ for example, $(CTGG)_4 \rightarrow (CTGG)_3$ is more frequent than is $(CTGG)_3 \rightarrow (CTGG)_2$, a result consistent with both increased opportunities for misalignment and increased stabilization. In T4, deletions that were promoted by one pair of G:C-rich repeats were more frequent than those promoted by one pair of identically placed A:T-rich repeats, a result consistent with increased stability (104). However, a change in the DNA context adjacent to the repeats did not influence deletion frequencies involving the G:C-rich repeats, but did increase deletions involving the A:T-rich repeats to a level similar to that of the G:C-rich repeats. Many alternative explanations, including the nature of the perturbation caused by the context effect and the sequence-specific characteristics of the particular repeats tested, can be offered to explain the results (104). Those operating in vivo are not yet identified.

Repeats need not be fully complementary to promote deletions. This has been convincingly demonstrated in $E.$ $coli$ (1), and results consistent with this idea continue to mount (3). For example, an unexpectedly large increase in deletion frequency ($>10^4$-fold) was observed in bacteriophage T7 when deletions between repeats of 5 bp and 10 bp were compared (80). However, the 10-bp repeat, but not the 5-bp repeat, can be imperfectly extended to provide a 21/29 $\{[5 + 10 + 6]/[8 + 10 + 11]\}$ homology in the misaligned configuration, a factor anticipated to contribute to increased deletion frequency.

It is becoming increasingly clear that the frequency of mutations arising from misalignments between a pair of direct repeats is influenced by the number of competing misalignments possible in the sequence. Evidence for competing misalignment has been seen in plasmids (3). Opportunities for competing misalignment to account for a bias between duplications and deletions at the *lacI* hotspot have been discussed (100). However, an alternative general hypothesis has been presented to explain a similar bias observed in bacteriophage T4 (89).

It has been hypothesized that palindromic sequences in association with repeated sequences could increase the frequency with which deletions between repeats are seen. These DNA misalignments have been proposed to account for spontaneous deletions and excision of some transposons (e.g. 1, 3, 8, 45, 91, 104). Subtly distinct models have been proposed in which palindromic alignments exactly juxtapose centers of the repeats, tandemly align the repeats, or bring the repeats somewhat closer together (Figure 2).

Some deletions that can be explained by the ability of a palindrome to exactly juxtapose the endpoints (Figure 2A) occur in sequences in which no other model currently offers an explanation (32). However, such juxtapositioning may also play a role in mutations that might otherwise be attributed to misalignment between direct repeats. An examination of spontaneous *lacI* deletion sequences revealed that an unexpectedly large fraction of those deletions that could be explained by misaligned pairing between repeats, could alternatively be explained by palindromic juxtapositioning of the endpoints. In the sequences examined, only 6% of repeats (5 examples) allowed both models to predict identical deletion endpoints (32). Nonetheless, these examples accounted for 40% of the deletions and 3 of the 5 examples were represented in the sample of only 12 mutants sequenced. Two of the three examples are extensively palindromic (S74 and S23) and have been found repeatedly in subsequent reports of *lacI* deletion sequences (e.g. 26, 100). The ultimate accumulation of enough spontaneous *lacI* deletion

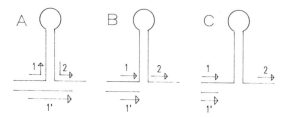

Figure 2 Misalignment between tandem repeats improved by a palindrome. In each case the deletion is produced by misaligning arrow 1' with arrow 2. Arrow 1 is a repeat of arrow 2; arrow 1' is the normal complement of arrow 1. In A, the palindrome brings arrow 2 to precisely the misaligned position that produces a deletion (37). In B, the palindrome brings arrow 2 adjacent to the misaligned position (50). In C the palindrome brings arrow 2 closer (1).

sequences to provide a saturated deletion spectrum should allow a clearer view of the potential association of palindromic juxtapositioning of deletion endpoints with high frequency of deletions.

Recent reviews (8, 24, 45) have detailed how deletions of transposons Tn5 and Tn10 from a host may occur by processes that resemble spontaneous deletion. These deletions exemplify the loss of DNA flanked by repeats and containing palindromic sequences (Figure 2B). It is well established that the deletions do not require transposase. Deletion frequencies are modulated by position in the chromosome and by mutant host proteins, including single strand binding (SSB) protein, recBCD (endonuclease, exonuclease, helicase), and mutD (a mutator defective in the exonucleolytic proof-reading subunit of DNA polymerase III and in DNA mismatch repair (99)). Transposon deletions do not require recA, implying no requirement for the protein-mediated misalignments typical of recombination. However, other spontaneous deletions can be influenced by recA (1, 3).

Whether misalignments that produce deletions occur primarily during replication or during recombination is not yet clear. Recombination is capable of mediating deletions in T4 (104), but the relative contribution to mutation by intra- versus intergenomic misalignments is unknown, as is the contribution of replication and recombination to intragenomic misalignments. In vitro polymerization reactions do produce deletions by misaligned pairing between tandem repeats (18, 57, 76). The propensity to produce deletions is polymerase specific (76), although the characteristic of the polymerase responsible is not yet known. In vitro studies suggest that DNA sequence plays an important role in polymerization-mediated deletions (57, 76). For example, repeats of similar base content and length can mediate deletions at strikingly different frequencies.

The potential contribution of additional characteristics of DNA structure to misalignment mutagenesis has been suggested. Particular interest in the possible contribution of Z-DNA structure to mutation has been excited by the observation that DNAs that can assume Z-DNA conformation are often subject to high frequencies of deletion (e.g. 28, 39). The speculation that Z-DNA conformations may be associated with recombination intermediates (47), coupled with the fact that Z-DNA sequences are routinely examined in dinucleotide tandem repeats capable of misaligned pairing, has led to consideration of the possibility that the high deletion frequency in Z-DNA-containing plasmids is due to recombination-mediated misaligned pairing. Unfortunately, control experiments have rarely demonstrated to what degree "high" rates of deletion may reflect selection for deleted genotypes in plasmids rather than high rates of mutation formation. Since misalignment models predict duplications as well as deletions, if both types of mutations are not recovered, models will have to account for this by some additional aspect of the process. In one study both duplications and deletions of the 2-bp repeat were found (28), consistent with misalignment. The mutations are not dependent on *recA*. The potential role(s) for the unusual DNA structures in promoting mutations in repeats await further definition. Alternating TC or TG tracts that can adopt triplex or Z-DNA configurations may also be prone to unequal chromatid exchange (111). The possibility that unusual DNA structures are subject to higher frequencies of aberrant processing during replication, recombination, or repair is intriguing. It is too soon to rule out the possibility that these structures may define mutational intermediates that are different from those predicted by the misalignment model.

MUTAGEN-INDUCED FRAMESHIFTS   Mutagen treatments often produce a fraction of mutants predicted by misalignments between repeats. Mutagens that strongly stimulate frameshifts specifically usually induce frameshifts in monotonic or alternating base runs (12, 105). Although misaligned pairing in the runs can explain the mutant sequences, it rarely explains context specificities. Why is the spontaneous *lacI* frameshift hotspot $(CTGG)_3$ not a hotspot for mutagen-induced frameshifts? Why do duplication and deletion preferences differ so strikingly among the G:C-run frameshifts induced by ICR-191 (12)?

Advances in the definition of site-specific positions of DNA modification, coupled with the analysis of the consequence of single DNA modifications of a defined type at a specific site, and the identification of the requirement for specific enzymatic processes have improved our view of frameshift mutation induced by covalent modification of DNA. For aflatoxin, sequences of preferred adduction correlate well to mutational specificity (84, 97, 98). In contrast, for 2-(N-acetoxy-N-acetylamino) fluorene (AAF), this simple correlation failed (29). The AAF result is due to at least two factors: multiple mechanisms and site-specific differences in mutagenic specificity of a lesion (11, 48). The AAF-induced frameshifts occur by at least two mechanisms: one produces frameshifts in monotonic runs and depends on *recA* and *umuDC;* the other produces frameshifts in alternating G:C runs and does not require *recA* or *umuDC*, although it does require some other *lex*-regulated function. Site-specific modification of a GGCGCC hotspot leads to frequent mutation when the third G is modified, but not when either of the first two Gs is modified. The absence of mutation as a consequence of modification at the second G suggests that misalignment is not the mechanism, or that sequence context may be very important. Structural studies in this system are likewise intriguing. In crystals, the oligonucleotides bearing the mutation-prone wild-type sequence pack in an usually close, sequence-specific manner (108). This close packing might be considered an analogue of the close approach of DNA duplexes in recombination. Perhaps the structural properties of G:C-rich DNA sequences offer a particular challenge to DNA alignment fidelity (57, 84, 104).

Future studies of mutagen-induced frameshifts are likely to provide detailed insights into frameshift mutation, some of which may be applicable to "spontaneous" mutation (85). They will surely extend the repertoire of frameshift models beyond misalignment.

Mutations Requiring Multiple Misalignments

Mutations arising by misalignment between direct repeats require one DNA misalignment. The remaining misalignment models invoke a two-step misalignment process: first, the mutant sequence is created in a misaligned strand, then the strand moves to a second position in the DNA. The requirement for two perturbed alignments makes the significant frequency of mutant sequences implicating multiple misalignment somewhat unexpected and suggests that DNA misalignments are a major challenge to the fidelity of DNA metabolism.

Mutations clearly attributable to two-step misalignment models have been seen in quasi-palindromic or quasi-repeated DNA sequences and adjacent to monotonic runs. Mutations produced by two-step misalignments can be deletions, insertions, and/or base substitutions, depending on the characteristics of the DNA sequence. These models uniformly predict that the sequence change at the site of mutation is the product of a DNA template and not the product of misincorporation or nontemplated incorporation of nucleotides.

QUASI-PALINDROMES   Spontaneous mutations in a variety of organisms and mutations produced by in vitro DNA synthesis reactions can arise from DNA synthesis in misaligned palindromic sequences. These misalignments can occur by either inter- or intrastrand misaligned pairing of the primer terminus. DNA synthesis at the site of the misalignment incorporates the nucleotides needed to account for the DNA sequence changes of the mutant and to realign the DNA back to the template (86). The repeated recovery of a predicted frameshift with complicated sequence changes (17) is an argument for a templated source for these mutations. As mutant sequencing continues, the data suggest that these mutations arise ubiquitously and that the general outline of the misalignment model is correct. However, just as with misalignments between direct repeats, the molecular details of these processes are still being discovered.

Quasi-palindromes can be predicted to produce mutations through two types of misalignments (76, 86). In one, the single-strand complementarity of the palindromic portions of the sequence permits fold-back or hairpin formation. Processing of this structure as if it were typical double-stranded DNA generates the mutant sequence (Figure 3A). In the other misalignment (Figure 3B), the interstrand complementarity of palindromic sequence permits DNA strand-switches that produce mutations that can be identical to those produced by the intrastrand misalignments. Although the interstrand-switching reaction

A

```
3'C-A-C-A-T-C-G-T-G-G-A-A-A-G-A-A-T-C-G-T-C-T-T-G-G-C-C-G-A-A-C 5'
5'G-T-G-T-A-G-C-A-C-C-T- T-T-C-T-T-A
                    OH  A-A-G-A-C-G
```

```
3'C-A-C-A-T-C-G-T-G-G-A-A-A-G-A-A-T-C-G-T-C-T-T-G-G-C-C-G-A-A-C 5'
5'G-T-G-T-A-G-C-A-C-C-T-T-T-T-C-T-T-A
                    OH  T-C-G-T-G-G-A-A-A-G-A-C-G
```

```
3'C-A-C-A-T-C-G-T-G-G-A-A-A-G-A-A-T-C-G-T-C-T-T- G-G-C — C-G-A-A-C
5'G-T-G-T-A-G-C-A-C-C-T-T-T-T-C-T-T-A-G-C-A-G-A-A-A-G-G-T-G-C-T OH
```

Sequence Change   CCG ⟶ AGGT

B

```
3'C-A-C-A-T-C-G-T-G-G-A-A-A-G-A-A-T-C-G-T-C-T-T-G-G-C-C-G-A-A-C 5'
5'G-T-G-T-A-G-C-A-C-C-T- T-T-C-T-T-A-G C-A-G-A-A-C-C-G-G-C-T-T-G 3'
                    OH  A-A-G-A  C      A  G'
                              C    T
5' G-T-G-T-A-G-C-A-C-C-T-T-T-C-T  T
```

```
3'C-A-C-A-T-C-G-T-G-G-A-A-A-G-A-A-T-C-G-T-C-T-T-G-G-C-C-G-A-A-C 5'
5'G-T-G-T-A-G-C-A-C-C-T- T-T-C-T-T-A-G C-A-G-A-A-C-C-G-G-C-T-T-G 3'
                  OH  T-C-G-T-G-G-A-A-A-G-A  C      A  G'
                                        C    T
5' G-T-G-T-A-G-C-A-C-C-T-T-T-T-C-T  T
```

```
3'C-A-C-A-T-C-G-T-G-G-A-A-A-G-A-A-T-C-G-T-C-T-T- G-G-C — C-G-A-A-C
5'G-T-G-T-A-G-C-A-C-C-T-T-T-T-C-T-T-A-G-C-A-G-A-A-A-G-G-T-G-C-T OH
```

C

```
                                    T-C
                                  A     G
                                  A-T
                                  G-C
                                  A-T
                                  A-T
3'C-A-C-A-T-C-G-T-G-G-A       G-G-C-C-G-A-A-C 5'
5'G-T-G-T-A-G-C-A-C-C-T-  T  A  C-C-G-G-C-T-T-G 3'
                            T  A
                            T  A
                            C  G
                         OH T  A
5' G-T-G-T-A-G-C-A-C-C-T-T-T-C-T-T-A-G-C  A
```

```
3'C-A-C-A-T-C-G-T-G-G-A-A-A-G-A-A-T-C-G-T-C-T-T-G-G-C-C-G-A-A-C 5'
5'G-T-G-T-A-G-C-A-C-C-T-T-T-T-C-T-G-C-T-A-A-G-A-A OH
```

Sequence Change   TAGC → GCTA

is illustrated between two strands of the same duplex, switching might involve a separate duplex. In theory, strand-switching models may describe either recombination or replication intermediates. Essentially nothing is yet known about potential recombination processes. In vitro polymerization reactions do permit the study of polymerization misalignments (76).

It is difficult to define which misalignments occur in vivo. In most cases, both intra- or interstrand misalignments predict the same mutant sequences. However, interstrand misalignments do predict one class of sequence changes not expected from intrastrand misalignments. An example (Figure 3C) illustrates the ability of intrastrand misalignments to produce an inversion of nonpalindromic DNA in the "loop of a hairpin". Mutant sequences with such inversions (inversions of loops of 1 base are G:C→C:G or A:T→T:A transversions) would specifically implicate interstrand misalignments. Among sequenced complex mutations recovered from CHO cells, partial inversions of loop sequences have been found coupled to deletions (37) or insertions (79). However, it is not yet clear whether these mutations are templated; each example is unique.

Nearly 200 mutant sequences arising from misalignments in quasi-palindromic DNA during in vitro polymerization reactions have been characterized (76). This partially saturated spectrum permits the identification of some determinants of misalignment mutagenesis specificity. These studies were done using "ordinary" DNA containing only modestly quasi-palindromic sequences and thus, it is hoped, will define determinants likely to be operating in many DNA sequences subject to genetic analysis.

In this in vitro study only one DNA strand served as a template and DNA synthesis began at a defined site. Thus, strand-switching is required to account for any mutation that arises from misaligned synthesis at a position "ahead" of the site of the mutation after realignment. The template is provided by a displaced strand after complete synthesis around the circular template or by another duplex (76). However, these strand-switch mutations are in a minority when compared to mutations produced by misalignment to a template "behind" the point of mutation, where either fold-back or strand-switch misalignments may operate. Because no quasi-palindromic misalignments to the template DNA were detected, most were probably intramolecular, fold-back misalignments.

←

Figure 3 Alternative misalignments of quasi-palindromes. Panel A shows an intra-molecular misalignment producing the same DNA change as the intermolecular misalignment in Panel B: CCG→AGGT in the misaligned strand. Panel C shows a different strand-switch involving the same quasi-palindrome. This misalignment produces an inversion of the sequence between the two palindromic segments: TAGC→GCTA in the switching strand. Misalignments of this type correspond to misalignments "ahead" of the site of mutation in vitro (76) and described in the text.

DNA polymerase plays a major role in defining the site-specific frequency of misalignment mutagenesis in vitro. DNA polymerase I of *E. coli,* and its large fragment derivative lacking the 5' → 3' exonuclease domain, differ substantially from one another with respect to specificity even though the overall frequency of mutation arising by misalignments is similar. In contrast, -1-bp frameshifts that appear to arise by a mechanism not depending on misalignment are unaffected (76).

In vitro, it is predicted that a major opportunity for DNA misalignment would be provided by DNA discontinuities associated with nonprocessive polymerization. An attempt to correlate polymerase pausing to misalignment specificity revealed that most hotspots for misalignment are at pause sites. However, differential polymerase pausing does not account for misalignment specificity differences (C. Papanicolaou & L. S. Ripley, submitted). These same studies demonstrate that differential opportunities for misalignments are an important specificity feature of misalignment mutagenesis. Sequences that could be predicted to lead to many different undetectable frameshifts are represented at significantly lower frequencies in the spectrum than those in which most misalignments can be detected.

The in vitro studies give a clearer view that enzymatic specificities are important but do not directly address to what degree DNA structure is important to the frequency of particular mutations. In vitro studies of the extrusion, structure, and stability of cruciforms in supercoiled DNA indicate that nearly all aspects of the sequence are likely to be important (61, 68). Loop sequences play a particularly important, though not yet predictable, role in cruciform formation and stability (9, 68, 114, 119). It may be that polymerases and other enzymes detect sequence-dependent structural differences. Nothing is known about strand-switch structures.

The demonstration that the Pol I polymerase of *Saccharomyces cerevisiae* produces misalignment mediated mutations in quasi-palindromes in vitro suggests that it would be very interesting to examine that enzyme complex with respect to yeast sequences known to undergo such mutations in vivo (31, 33). In the bacteriophage T4 *rII*B system, the *ts*L98 polymerase background has been the only genetic background in which quasi-palindrome-mediated frameshifts have been recovered to date (18; L. S. Ripley, unpublished data). Comparison of the characteristics of this mutant with the wild-type DNA polymerase may ultimately provide direct information about the role of DNA polymerase in vivo in promoting misalignment mutation.

Mutations occurring in quasi-palindromes in vivo depend on modest amounts of sometimes imperfect homology (17, 33). Similarly, in vitro misalignments examined involve small, usually imperfect homologies of 4 to 8 bp (76). Within this size range the length of the homology and G:C/A:T content do not by themselves predict mutation frequency. This might be less true of long palindromes.

In vivo, a T4 t-RNA gene and an end of IS2 were subject to a significant
rate of mutation involving multiple palindromic misalignments (62). Interestingly, quasi-palindromic misalignments have rarely been reported in
shuttle-vector assays employing the *supF* tRNA suppressor (31, 51). Consideration of the specific mutant sequences detected suggested that site-
specific DNA cleavage in addition to the structural opportunities for misalignments might play an important role (62). Although no enzymatic role for
endoVII of T4 (gp 49) has been experimentally implicated in mutagenesis,
this enzyme displays in vitro specificity for the base of hairpin stems and
sequence preference for cleavage (46). Alternative enzymes that could be
considered might include RNA-processing enzymes. An endonuclease involved in mRNA splicing in T4 does mediate site-specific transposition (6,
82).

Further description of mutational misalignments in vivo mediated by quasi-
palindromes depends significantly on defining the enzymes and structures
responsible for the mutations. Intriguing prospects are offered by the potential
for unusual processing of palindromic intermediates by endonucleases
responsible for the resolution of recombinational intermediates. In vitro these
enzymes cleave both 3- and 4-armed palindromic structures and display DNA
sequence preferences characteristic of the enzyme (20, 21, 23, 40, 75). The
demonstration that the presence of palindromes in yeast DNA profoundly
influences local gene conversion (69), strongly implicates disruption of normal DNA metabolism by palindromes in a manner that may promote not only
the formation but also the recovery of mutation in these sequences. Yeast is
prone to mutation that is mediated by quasi-palindromic templates (31, 33),
and encodes at least three different endonucleases that specifically cleave
palindromic structures (40).

Less is known about the potential role of quasi-palindromic misalignments
as a mechanism for templated DNA lesion bypass. Mutant sequences consistent with the unusual alignments of quasi-palindromes have been recovered
after in vivo mutagenesis by alkylating agents (30, 33), cross-linking
mutagens (115; T. Cebula, personal communication), and gamma-rays (110).
Closely spaced lesions in opposite DNA strands may produce alternative
kinds of DNA rearrangements (49).

QUASI-REPEATS   Frameshifts attributable to multiple DNA alignments between quasi-repeats have also been found. Figure 4 illustrates an example
from the bacteriophage T4 *rII* system. Frameshifts consistent with this mechanism occur frequently in *E. coli* (100). The examples from *E. coli* involve
nearby short homologies. These provide an explanation for the most frequent
spontaneous frameshifts in wild-type cells, but not in cells defective for
mismatch repair (101). Examples of mutant sequences that can be explained
by this mechanism also occur in cultured cells (37). Deletions may also

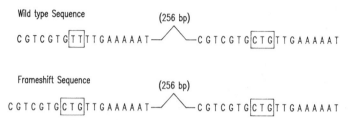

Figure 4   The complex frameshift sequence change TT→CTG, in the sequence at the left can be accounted for the sequence to the right that is 256 bp away. Mutations would be mediated by misaligned pairing between complementary portions of the left and right sequences followed by local processing and realignment to the usual position.

sometimes be explained by this mechanism (32). An intriguing opportunity for interstrand misalignments of this sort to participate in mutation is offered by examples of mutations of yeast iso-1-cytochrome *c* from DNA sequences derived from the iso-2-cytochrome *c* gene that lies on a different chromosome (33).

In vivo, frameshift mutations arising by this mechanism are strongly increased in T4 by the DNA polymerase mutant *ts*L141 (92; unpublished data). This mutant polymerase is less processive than the wild-type (59), a characteristic anticipated to provide increased opportunity to initiate misalignments and to interrupt misaligned processing that might otherwise result in deletion between imperfect repeats. Details of the frameshift process are not yet known, but deletions between the imperfect repeats in Figure 4 occur in vivo (81). In vitro, the Klenow fragment of DNA polymerase I of *E. coli* may occasionally produce frameshifts by this mechanism (18). But these mutations are comparatively rare and have not been examined further.

INTERCONVERSION   OF   MUTATIONAL   INTERMEDIATES   DNA   misalignments can convert one mutational intermediate to another. The interconversions that link base substitutions and frameshifts have been called "dislocations" (55). A base substitution consistent with the transient formation of a deletion intermediate converted by realignment to produce a templated base substitution is shown in Figure 5A. Linked base-substitution and frameshift hotspots consistent with interconversions are produced frequently by eukaryotic polymerase β and by HIV reverse transcriptase in vitro (5, 53, 58). The conversion of frameshifts to base substitutions is strongly supported by the demonstration that a change in the identity of the adjacent nucleotide produced a large increase in base substitutions of precisely the type predicted (58). Nonetheless, base substitutions of the previous type still occurred, albeit

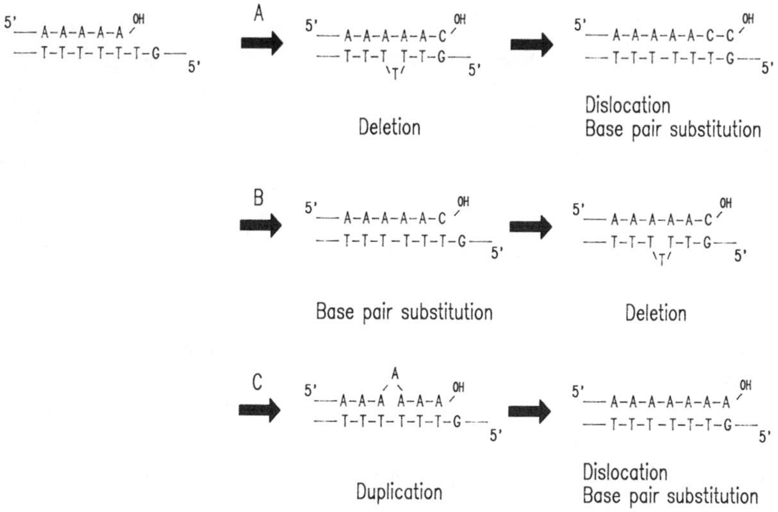

Figure 5   Dislocation is a two-step process. As shown in panel A the initial frameshift intermediate is extended by polymerization to a position 1 bp past the run. Rearrangement of the primer converts the frameshift intermediate into a base substitution intermediate, having a predictable sequence change at a predictable site. Alternative themes of two-step processes and their predictions are shown in B and C. In B, the two steps illustrated in A are reversed; a base substitution occurs first. This substitution is not templated, but is due to misincorporation. The specific identity of this base permits rearrangement to produce a deletion intermediate. In C, duplication is followed by rearrangement of the template/primer to give a templated base substitution 1 bp after the run. The reverse order of events would predict a duplication by dislocation after base substitution. Although the illustrated mutations are associated with a monotonic run, no specificity prediction of dislocation requires the run. Rearrangements between particular base substitution and frameshift intermediate can be described for any sequence.

at a lower frequency, and frameshift frequencies dropped. It may be that the converse reaction is responsible for some frameshifts. The production of a frameshift by dislocation of the base-substitution intermediate is shown in Figure 5B. Consistent with the possibility of the converse reaction, constructed base-substitution mismatches can be successfully converted to frameshifts by polymerase I of yeast (57), an enzyme lacking 3' exonuclease activity. However, conversions of base substitutions to frameshifts were unsuccessful when a mismatched substrate was presented to DNA polymerase I of *E. coli* (Klenow fragment) (J. G. de Boer & L. S. Ripley, unpublished data).

Dislocation is firmly associated with monotonic T runs (58). However, nothing about dislocation demands a run. Interconversion demands only a frameshift hotspot and/or a base-substitution hotspot appropriately situated so that misalignment after the transient formation of one mutational intermediate is converted into the other mutational intermediate. Like other misalignments considered above, the DNA structure and DNA polymerase characteristics (50, 52) can be anticipated to influence site-specific misalignment probabilities.

Interconversion also displays no intrinsic sequence requirement for a deletion frameshift. Additions can also be converted to base substitutions by dislocation (Figure 5C). Moreover, the reversal of the intermediate illustrated in Figure 5C describes the potential ability of misalignment to account for an addition frameshift by dislocation of a base-substitution intermediate. The recovery of 1-bp duplications after in vitro polymerization by HIV reverse transcriptase (5) offers an excellent opportunity to test whether dislocation can generate duplications. Other polymerases examined so far produce no large frequency of duplications, even in monotonic sequences, despite the predictions of the misalignments model that both duplications and deletions should be expected. In instances where frameshift hotspots are not at monotonic runs, but in other sequences (17, 57, 76), dislocation is predicted to be associated with these other sequences.

Frameshifts at Nicks: A Role for Misligation

The mutation models considered to this point hypothesize that mutations are initiated by DNA misalignments. In these models, DNA nicks may potentiate the site-specific frequency of misalignment but production of the mutant sequence demands misalignment before DNA processing. It has now been demonstrated that DNA nicks are associated with other mutational events that do not involve initial misalignments.

The characteristics of many acridine-induced frameshifts in T4 support misalignment models for frameshift mutations (106, 107). This turned out to be true because these sequences lie in repeats. Other acridine-induced frameshifts in T4 do not lie in repeats and thus cannot be accounted for by the misalignment model (74, 88). Analysis of the dichotomy has revealed that both classes of frameshifts can be explained by an alternative, single mechanism (88, 90). The mutations are uniformly found to be either deletions or templated additions (i.e. duplications). These are initiated at a free 3' OH end that is created by acridine-induced DNA cleavage mediated by the T4-encoded type II topoisomerase (Figure 6). Templated polymerization accounts for the duplication; deletions are believed to depend on 3' exonuclease. Final incorporation of the sequence requires aberrant religation of the processed 3' end with the unchanged 5' end. A topoisomerase defect abolishes acridine-induced mutagenesis (90). DNA sequence changes that alter the cleavage specificity of topoisomerase alter mutagenic specificity in precise accord with the model (M. Masurekar et al, submitted).

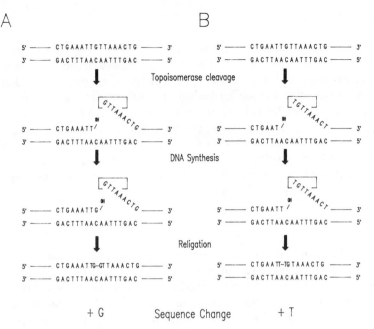

A   B

+ G   Sequence Change   + T

Figure 6   The model illustrated accounts for the frequency and specificity of frameshifts induced by acridines at a hotspot for mutation in the T4 *rII*B gene. In panel A, the wild-type sequence is preferentially cleaved by topoisomerase at the position indicated. Duplications at this site are consistent with a templated addition at the 3' end made available by this cleavage. Aberrant religation, perhaps mediated by the topoisomerase, indicated by the box over the unprocessed 5' end of the DNA, is necessary for mutation. Panel B illustrates the preferred cleavage site in a mutated DNA sequence. +T rather than +G mutations were preferentially recovered in this DNA sequence. A number of DNA sequences have been tested. Each is consistent with the predictions of the model and the specificity of topoisomerase cleavage.

     In the model, aberrant religation of duplication or deleted DNA is an important step whose details are not yet known. Identification of the enzymatic and sequence determinants involved in this process may provide some insight into the potential for aberrant processing of DNA nicks as a general mode of frameshift production. A substantial number of spontaneous frameshift mutations sequences occur in DNA sequences in which misalignment cannot account for the mutant sequences (89). Perhaps a fraction of these arise by related types of aberrant processing and religation of DNA nicks produced in other ways.

     As reviewed by Ehrlich (24), illegitimate recombination can lead to the covalent joining of DNA sequences in a number of ways. Some of these have previously implicated topoisomerases in chromosomal rearrangements, both in vitro and in vivo. The mutants and mechanism described here are different. The mutants have sequences templated at the site of the DNA break and no joining of previously distant sequences is required.

Aberrant processing of DNA nicks at replication origins of plasmids leads to deletions (4, 24, 64–66). The detailed characterization of alternative deletions suggests that more than a single mechanism operates (64). In some cases, misalignment involving limited complementarities appears to be sufficient to account for deletion sequences, whereas in other cases more unexpected types of DNA rejoining reactions may be responsible. As has been emphasized (4, 118), the characteristics of the deletion produced may also be important in influencing apparent frequencies of these mutations, which are generally subject to selection. It may be that *E. coli* has genes that can suppress some aspect of such mutagenesis (118).

OUTLOOK

Many frameshifts and some other mutations are clearly produced by mechanisms involving broken and/or misaligned DNA intermediates. Increasingly, models for converting these intermediates to mutations are able not only to offer explanations for the observed mutations, but also to provide directly testable hypotheses. Nonetheless, frameshift sequences continue to present a panoply of characteristics, only some of which can be accounted for by the models described to date. Thus, it is not likely that the full repertoire of frameshift mechanisms operating in vivo has been identified.

An important role for misalignments is supported not only by the high frequency of mutations predicted by the misalignment of repeated sequences, but by the impressive frequency of mutations involving multiple misalignments. Despite the apparent importance of misalignments, some frameshift mutations, consistent with misalignments in monotonic runs, are instead due to the processing of site-specific nicks. Moreover, under certain circumstances nontemplated additions and quasi-palindromically templated additions may produce these specificities (14, 91). The overlapping predictions of these models in monotonic runs suggests that it may be unwise to automatically assume that all mutations lying in monotonic runs arise by misalignment.

The demonstration that in vitro DNA polymerization reactions produce many different DNA sequence changes by misalignment, but almost never produce +1-bp duplications in monotonic runs, suggests that close attention to examples of this specificity in vivo and in vitro may be mechanistically rewarding. Many frameshifts produced during in vitro polymerization do not appear to be due to misalignments. That these still exhibit polymerase-specific, sequence-specific characteristics suggests that DNA structures distinct from misalignments may be involved in frameshift mutation. Possibly these account for some aspects of mutagen-induced frameshift specificity, and for some unexplained spontaneous frameshift specificities.

The mechanism by which other, intriguing mutant sequences arise still need to be defined:

1. Many spontaneous deletions recovered from cultured cells. These have endpoints that are within shorter repeats than those typically found in prokaryotes or in vitro, and often have added or missing bases that could be due to further processing (16, 19, 37, 38, 70, 71, 109). Perhaps some combination of misalignment and nick-processing models could explain these, but novel mechanisms may be involved. In vitro studies of eukaryotic ligation processes may provide a model for some processing (78, 95, 96). More sequences will be needed to better define whether processing is templated or otherwise directed.

2. Hot and warm spots for deletions in prokaryotes that are not explained by misalignments (e.g. 102, 117). These deletions might represent a variation on the theme of nick processing and rejoining, but no enzymatic determinants have been identified. Some deletions correlate to DNA sequences that are near matches to the Chi sequence that stimulates endonucleolytic cleavage by recBCD (103), but involvement of this enzyme has not been reported. In T4, warm spots for 2-bp deletions are seen in vivo that are not accounted for by misalignments (89). In vitro polymerization leads to 2-bp deletions that are not predicted by misalignment (18).

In vitro approaches can often separately address the multiple determinants of in vivo mutational specificity. This method may also identify novel mechanisms not initially obvious from in vivo spectra. Spontaneous frameshifts in vivo depend on DNA polymerase (92, 93, 99; L. S. Ripley, unpublished data). DNA polymerization in vitro produces mutations (5, 54, 57, 76). Studies to correlate the frameshift specificities are underway. It is widely expected that the DNA sequence specificity of mutations will reflect in part the characteristics of the enzymology. However, underlying these characteristics may well lie DNA sequence-specific structure, which, if investigated, would contribute to an improved understanding of DNA structure-protein interactions. The detailed examination of DNA-sequence specific behavior of polymerases are providing intriguing clues (e.g. 63, 72, 73, 83, 85) to specificity. Incorporation of information from physical models of DNA and polymerase structure is becoming possible (42, 110, 113; L. S. Ripley et al, submitted).

Understanding mutational specificity at the level of the perturbed enzyme behavior and/or DNA structure provides a method for evaluating the differences in mutational spectra observed when alternative genotypes or organisms are compared (10, 36, 94, 116). This perspective can provide a unifying framework within which to evaluate the fidelity perturbations most relevant to man.

ACKNOWLEDGMENTS

My thanks to Z. Humayun for his thoughtful comments and suggestions. Unpublished work from this laboratory cited in this manuscript was supported by grants from the National Institutes of Health and the American Cancer Society. Jeanette Shipman, Nettie Cleveland, and Tracy Mann prepared the manuscript. Figures were prepared using Computer Graphics software provided by the Academic Computing Center of the University of Medicine and Dentistry of New Jersey.

Literature Cited

1. Albertini, A. M., Hofer, M., Calos, M. P., Miller, J. H. 1982. On the formation of spontaneous deletions: the importance of short sequence homologies in the generation of large deletions. *Cell* 29: 319–28

2. Anderson, P., Roth, J. 1981. Spontaneous tandem genetic duplications in *Salmonella typhimurium* arise by unequal recombination between rRNA(*rrn*) cistrons. *Proc. Natl. Acad. Sci. USA* 78:3113–17

3. Balbinder, E., MacVean, C., Williams, R. E. 1989. Overlapping direct repeats stimulate deletions in specially designed derivatives of plasmid pBR325 in *Escherichia coli*. *Mutat. Res.* 214:233–52

4. Ballester, S., Lopez, P., Espinosa, M., Alonso, J. C., Lacks, S. A. 1989. Plasmid structural instability associated with pC194 replication functions. *J. Bacteriol.* 171:2271–77

5. Bebenek, K., Abbotts, J., Roberts, J. D., Wilson, S. H., Kunkel, T. A. 1989. Specificity and mechanism of error-prone replication by human immunodeficiency virus-1 reverse transcriptase. *J. Biol. Chem.* 264:16948–56

6. Bell-Pederson, D., Quirk, S. M., Aubrey, M., Belfort, M. 1989. A site-specific endonuclease and co-conversion of flanking exons associated with the mobile *td* intron of phage T4. *Gene* 82:119–26

7. Benzer, S. 1961. On the topography of the genetic fine structure. *Proc. Natl. Acad. Sci. USA* 47:403–15

8. Berg, D. E. 1989. Transposon Tn*5*. In *Mobile DNA*, ed. D. E. Berg, M. M. Howe, pp. 185–210. Washington, DC: Am. Soc. Microbiol. 972 pp.

9. Blommers, M. J. J., Walters, A. L. I.,

Haasnoot, C. A. G., Aelen, J. M. A., van der Marel, G. A., et al. 1989. Effects of base sequence on the loop folding in DNA hairpins. *Biochemistry* 28:7491–98

10. Bredberg, A., Kraemer, K. H., Seidman, M. M. 1986. Restricted ultraviolet mutational spectrum in a shuttle vector propagated in *xeroderma pigmentosum* cells. *Proc. Natl. Acad. Sci. USA* 83: 8273–77

11. Burnouf, D., Koehl, P., Fuchs, R. P. P. 1989. Single adduct mutagenesis: strong effect of the position of a single acetylaminofluorene adduct within a mutation hot spot. *Proc. Natl. Acad. Sci. USA* 86:4147–51

12. Calos, M. P., Miller, J. H. 1981. Genetic and sequence analysis of frameshift mutations induced by ICR-191. *J. Mol. Biol.* 153:39–66

13. Clark, J. M. 1988. Novel non-templated nucleotide addition reactions catalyzed by procaryotic and eucaryotic DNA polymerases. *Nucleic Acids Res.* 16: 9677–86

14. Clark, J. M., Joyce, C. M., Beardsley, G. P. 1987. Novel blunt-end addition reactions catalyzed by DNA polymerase I of *Escherichia coli*. *J. Mol. Biol.* 198:123–27

15. Crick, F. H. C., Barnett, L., Brenner, S., Watts-Tobin, R. J. 1961. General nature of the genetic code for proteins. *Nature* 192:1227–32

16. de Boer, J. G., Drobetsky, E. A., Grosovsky, A. J., Mazur, M., Glickman, B. W. 1989. The Chinese hamster *aprt* gene as a mutational target. Its sequence and an analysis of direct and inverted repeats. *Mutat. Res.* 226:239–44

17. de Boer, J. G., Ripley, L. S. 1984.

Demonstration of the production of frameshift and base-substitution mutations by quasipalindromic DNA sequences. *Proc. Natl. Acad. Sci. USA* 81:5528–31

18. de Boer, J. G., Ripley, L. S. 1988. An *in vitro* assay for frameshift mutations: Hotspots for deletions of 1 bp by Klenow-fragment polymerase share a consensus DNA sequence. *Genetics* 118:181–91

19. de Jong, P.J., Grosovsky, A. J., Glickman, B. W. 1988. Spectrum of spontaneous mutation at the *APRT* locus of Chinese hamster ovary cells: An analysis at the DNA sequence level. *Proc. Natl. Acad. Sci. USA* 85:3499–503.

20. de Massy, B., Weisberg, R. A., Studier, F. W. 1987. Gene *3* endonuclease of bacteriophage T7 resolves conformationally branched structures in double-stranded DNA. *J. Mol. Biol.* 193:359–76

21. Dickie, P., McFadden, G., Morgan, A. R. 1987. The site-specific cleavage of synthetic Holliday junction analogs and related branched DNA structures by bacteriophage T7 endonuclease I. *J. Biol. Chem.* 262:14826–36

22. Dohet, C., Wagner, R., Radman, M. 1986. Methyl-directed repair of frameshift mutations in heteroduplex DNA. *Proc. Natl. Acad. Sci. USA* 83:3395–97

23. Duckett, D. R., Murchie, A. I. H., Diekmann, S., von Kitzing, E., Kemper, B., Lilley, D. M. J. 1988. The structure of the Holliday junction, and its resolution. *Cell* 55:79–89

24. Ehrlich, S. D. 1989. Illegitimate recombination in bacteria. See Ref 8, pp. 799–832

25. Farabaugh, P. J., Schmeissner, U., Hofer, M., Miller, J. H. 1978. Genetic studies on the *lac* repressor. VII. On the molecular nature of spontaneous hotspots in the *lacI* gene of *Escherichia coli*. *J. Mol. Biol.* 126:847–63

26. Fix, D. F., Burns, P. A., Glickman, B. W. 1987. DNA sequence analysis of spontaneous mutation in a polAl strain of *Escherichia coli* indicates sequence-specific effects. *Mol. Gen. Genet.* 202:267–72

27. Fresco, J. R., Alberts, B. M. 1960. The accommodation of noncomplementary bases in helical polyribonucleotides and deoxyribonucleic acids. *Proc. Natl. Acad. Sci. USA* 46:311–21

28. Freund, A.-M., Bichara, M., Fuchs, R. P. P. 1989. Z-DNA-forming sequences are spontaneous deletions hot spots. *Proc. Natl. Acad. Sci. USA* 86:7465–69

29. Fuchs, R. P. P. 1984. DNA binding spectrum of the carcinogen *N*-acetoxy-*N*-2-acetylaminofluorene significantly differs from the mutation spectrum. *J. Mol. Biol.* 177:173–80

30. Gentil, A., Margot, A., Sarasin, A. 1986. 2-(*N*-acetoxy-*N*-acetylamino)fluorene mutagenesis in mammalian cells: Sequence-specific hot spot. *Proc. Natl. Acad. Sci. USA* 83:9556–60

31. Giroux, C. N., Mis, J. R. A., Pierce, M. K., Kohalmi, S. E., Kunz, B. A. 1988. DNA sequence analysis of spontaneous mutations in the *SUP4-o* gene of *Saccharomyces cerevisiae*. *Mol. Cell. Biol.* 8:978–81

32. Glickman, B. W., Ripley, L. S. 1984. Structural intermediates of deletion mutagenesis: A role for palindromic DNA *Proc. Natl. Acad. Sci. USA* 81:512–16

33. Hampsey, D. M., Ernst, J. F., Stewart, J. W., Sherman, F. 1988. Multiple base-pair mutations in yeast. *J. Mol. Biol.* 201:471–86

34. Hoffmann, G. R. 1985. Genetic duplications in bacteria and their relevance for genetic toxicology. *Mutat. Res.* 150:107–17

35. Hoffmann, G. R., Freemer, C. S., Parente, L. A. 1989. Induction of genetic duplications and frameshift mutations in *Salmonella typhimurium* by acridines and acridine mustards: Dependence on covalent binding of the mutagen to DNA. *Mol. Gen. Genet.* 218:377–83

36. Hsia, H. C., Lebkowski, J. S., Leong, P.-M., Calos, M. P., Miller, J. H. 1989. Comparison of ultraviolet irradiation-induced mutagenesis of the *lacI* gene in *Escherichia coli* and in human 293 cells. *J. Mol. Biol.* 205:103–13

37. Ikehata, H., Akagi, T., Kimura, H., Akasaka, S., Kato, T. 1989. Spectrum of spontaneous mutations in a cDNA of the human *hprt* gene integrated in chromosomal DNA. *Mol. Gen. Genet.* 219:349–58

38. Ingle, C. A., Drinkwater, N. R. 1989. Mutational specificities of l'-acetoxysafrole, *N*-benzoyloxy-*N*-methyl-4-aminoazobenzene, and ethyl methanesulfonate in human cells. *Mutat. Res.* 220:133–42

39. Jaworski, A., Blaho, J. A., Larson, J. E., Shimizu, M., Wells, R. D. 1989. Tetracycline promoter mutations decrease non-*B* DNA structural transitions, negative linking differences and deletions in recombinant plasimids in *Escherichia coli*. *J. Mol. Biol.* 207:513–26

40. Jensch, F., Kosak, J., Seeman, N. C., Kemper, B. 1989. Cruciform cutting en-

donucleases from *Saccharomyces cerevisiae* and phage T4 show conserved reactions with branched DNAs. *EMBO J.* 8:4325–34

41. Joshua-Tor, L., Rabinovich, D., Hope, H., Frolow, F., Appella, E., Sussman, J. L. 1988. The three-dimensional structure of a DNA duplex containing looped out bases. *Nature* 334:82–84

42. Joyce, C. M. 1989. How DNA travels between the separate polymerase and 3'-5'-exonuclease sites of DNA polymerase I (Klenow fragment). *J. Biol. Chem.* 264;10858–66

43. Kalnik, M. W., Norman, D. G., Swann, P. F., Patel, D. J. 1989. Conformation of adenosine bulge-containing deoxytridecanucleotide duplexes in solution. Extra adenosine stacks into duplex independent of flanking sequence and temperature *J. Biol. Chem.* 264:3702–12

44. Kalnik, M. W., Norman, D. G., Zagorski, M. G., Swann, P. F., Patel, D. J. 1989. Conformational transitions in cytidine bulge-containing deoxytridecanucleotide duplexes: Extra cytidine equilibrates between looped out (low temperature) and stacked (elevated temperature) conformations in solution. *Biochemistry* 28:294–303

45. Kleckner, N. 1989. Transposon Tn*10*. See Ref. 8, pp. 227–68

46. Kleff, S., Kemper, B. 1988. Initiation of heteroduplex-loop repair by T4-encoded endonuclease VII *in vitro*. *EMBO J.* 7:1527–35

47. Kmiec, E. B., Angelides, K. J., Holloman, W. K. 1985. Left-handed DNA and the synaptic pairing reaction promoted by *Ustilago rec1* protein. *Cell* 40:139–45

48. Koffel-Schwartz, N., Fuchs, R. P. P. 1989. Genetic control of AAF-induced mutagenesis at alternating GC sequences: An additional role for *recA*. *Mol. Gen. Genet.* 215:306–11

49. Kokontis, J. M., Vaughan, J., Harvey, R. G., Weiss, S. B. 1988. Illegitimate recombination induced by benzo[a]pyrene diol epoxide in *Escherichia coli*. *Proc. Natl. Acad. Sci. USA* 85:1043–46

50. Korn, D., Fisher, P. A., Wang, T. S.-F. 1983. Enzymological characterization of human DNA polymerases-α and β. In *New Approaches in Eukaryotic DNA Replication*, ed. A. M. de Recondo, pp. 17–55. New York: Plenum Publ. Corp

51. Kraemer, K. H., Seidman, M. M. 1989. Use of *supF*, an *Escherichia coli* tyrosine suppressor tRNA gene, as a mutagenic target in shuttle-vector plasmids. *Mutat. Res.* 220:61–72

52. Kumar, A., Widen, S. G., Williams, K. R., Kedar, P., Karpel, R. L., Wilson, S. H. 1990. Studies of the domain structure of mammalian DNA polymerase-β. Identification of a discrete template binding domain. *J. Biol. Chem.* 265:2124–31

53. Kunkel, T. A. 1985. The mutational specificity of DNA polymerase-β during *in vitro* DNA synthesis. Production of frameshift, base substitution, and deletion mutations. *J. Biol. Chem.* 260:5787–96

54. Kunkel, T. A. 1986. Frameshift mutagenesis by eucaryotic DNA polymerase *in vitro*. *J. Biol. Chem.* 261:13581–87

55. Kunkel, T. A., Alexander, P. S. 1986. The base substitution fidelity of eucaryotic DNA polymerases. Mispairing frequencies, site preferences, insertion preferences, and base substitution by dislocation. *J. Biol. Chem.* 261:160–66

56. Kunkel, T. S., Gopinathan, K. P., Dube, D. K., Snow, E. T., Loeb, L. A. 1986. Rearrangements mediated by terminal transferase. *Proc. Natl. Acad. Sci. USA* 83:1867–71

57. Kunkel, T. A., Hamatake, R. K., Motto-Fox, J., Fitzgerald, M. P., Sugino, A. 1989. Fidelity of DNA polymerase I and DNA polymerase I-DNA primase complex from *Saccharomyces cerevisiae*. *Mol. Cell. Biol.* 9:4447–58

58. Kunkel, T. A., Soni, A. 1988. Mutagenesis by transient misalignment. *J. Biol. Chem.* 263:14784–89

59. Lo, K. Y., Bessman, M. J. 1976. An antimutator deoxyribonucleic acid polymerase. I. Purification and properties of the enzyme. *J. Biol. Chem.* 251:2475–79

60. Lorenzetti, R., Cesareni, G., Cortese, R. 1983. Frameshift mutations induced by an *Escherichia coli* strain carrying a mutator gene, *mutD5*. *Mol. Gen. Genet.* 192:515–16

61. Lu, M., Guo, Q., Seeman, N. C., Kallenbach, N. R. 1989. DNase I cleavage of branched DNA molecules. *J. Biol. Chem.* 264:20851–54

62. McClain, W. H. 1988. Specific duplications fostered by a DNA structure containing adjacent inverted repeat sequences. *J. Mol. Biol.* 204:27–40

63. Mendelman, L. V., Petruska, J., Goodman, M. F. 1990. Base mispair extension kinetics. Comparison of DNA polymerase alpha and reverse transcriptase. *J. Biol. Chem.* 265:2338–46

64. Michel, B., D'alencon, E., Ehrlich, S. D. 1989. Deletion hot spots in chimeric *Escherichia coli* plasmids, *J. Bacteriol.* 171:1846–53

65. Michel, B., Ehrlich, S. D. 1986. Illegitimate recombination at the replication origin of bacteriophage M13. *Proc. Natl. Acad. Sci. USA* 83:3386–90

66. Michel, B., Ehrlich, S. D. 1986. Illegitimate recombination occurs between the replication origin of the plasmid pC194 and a progressing replication fork. *EMBO J.* 5:3691–96

67. Miller, M., Harrison, R. W., Wlodawer, A., Appella, E., Sussman, J. L. 1988. Crystal structure of 15-mer DNA duplex containing unpaired bases. *Nature* 334:85–86

68. Murchie, A. I. H., Lilley, D. M. J. 1987. The mechanism of cruciform formation in supercoiled DNA: initial opening of central basepairs in salt-dependent extrusion. *Nucleic Acids Res.* 15:9641–54

69. Nag, D. K., White, M. A., Petes, T. D. 1989. Palindromic sequences in heteroduplex DNA inhibit mismatch repair in yeast. *Nature* 340:318–20

70. Nalbantoglu, J., Hartley, D., Phear, G., Tear, G., Meuth, M. 1986. Spontaneous deletion formation at the *aprt* locus of hamster cells: the presence of short sequence homologies and dyad symmetries at deletion termini. *EMBO J.* 5:1199–204

71. Nalbantoglu, J., Phear, G., Meuth, M. 1987. DNA sequence analysis of spontaneous mutations at the *aprt* locus of hamster cells. *Mol. Cell. Biol.* 7:1445–49

72. Nevinsky, G. A., Veniaminova, A. G., Levina, A. S., Podust, V. N., Lavrik, O. I. 1990. Structure-function analysis of mononucleotides and short oligonucleotides in the priming of enzymatic DNA synthesis. *Biochemistry* 29:1200–7

73. Ng, L., Weiss, S. J., Fisher, P. A. 1989. Recognition and binding of template-primers containing defined abasic sites by *Drosophila* DNA polymerase α holoenzyme. *J. Biol. Chem.* 264:13018–23

74. Owen, J. E., Schultz, D. W., Taylor, A., Smith, G. R. 1983. Nucleotide sequence of the lysozyme gene of bacteriophage T4: analysis of mutations involving repeated sequences. *J. Mol. Biol.* 165:229–48

75. Panayotatos, N., Fontaine, A. 1987. A native cruciform DNA structure probed in bacteria by recombinant T7 endonuclease. *J. Biol. Chem.* 262:11364–68

76. Papanicolaou, C., Ripley, L. S. 1989. Polymerase-specific differences in the DNA intermediates of frameshift mutagenesis. *In vitro* synthesis errors of

Escherichia coli DNA polymerase I and its large fragment derivative. *J. Mol. Biol.* 207:335–53

77. Patel, D. J., Kozlowski, S. A., Marky, L. A., Rice, J. A., Broka, C., et al. 1982. Extra adenosine stacks into the self-complementary d(CGCAGAAT-TCGCG) duplex in solution. *Biochemistry* 21:445–51

78. Pfeiffer, P., Vielmetter, W. 1988. Joining of nonhomologous DNA double strand breaks *in vitro*. *Nucleic Acids Res.* 16:907–24

79. Phear, G., Armstrong, W., Meuth, M. 1989. Molecular basis of spontaneous mutation at the *aprt* locus of hamster cells. *J. Mol. Biol.* 209:577–82

80. Pierce, J. C., Masker, W. 1989. Genetic deletions between directly repeated sequences in bacteriophage T7. *Mol. Gen. Genet.* 217:215–22

81. Pribnow, D., Sigurdson, D. C., Gold, L., Singer, B. S., Napoli, C., et al. 1981. rII cistrons of bacteriophage T4: DNA sequence around the intercistronic divide and positions of genetic landmarks. *J. Mol. Biol.* 149:337–76

82. Quirk, S. M., Bell-Pedersen, D., Belfort, M. 1989. Intron mobility in the T-even phages: high frequency inheritance of group I introns promoted by intron open reading frames. *Cell* 56:455–65

83. Ramesh, N., Shouche, Y. S., Brahmachari, S. K. 1986. Recognition of B and Z forms of DNA by *Escherichia* DNA polymerase I. *J. Mol. Biol.* 193:635–8

84. Refolo, L. M., Bennett, C. B., Humayun, M. Z. 1987. Mechanisms of frameshift mutagenesis by aflatoxin b1-2, 3-dichloride. *J. Mol. Biol.* 193:609–36

85. Revich, G. G., Ripley, L. S. 1990. Effects of proflavin and photoactivated proflavin on the template function of single-stranded DNA. *J. Mol. Biol.* 211:63–74

86. Ripley, L. S. 1982. Model for the participation of quasi-palindromic DNA sequences in frameshift mutation. *Proc. Natl. Acad. Sci. USA* 79:4128–32

87. Ripley, L. S. 1990. Mechanisms of gene mutation. In *Genetic Toxicology; A Treatise*, ed. A. P. Li, R. H. Heflich, pp. 13–40 NJ: Telford Press,

88. Ripley, L. S., Clark, A. 1986. Frameshift mutations produced by proflavin in bacteriophage T4. *Proc. Natl. Acad. Sci. USA* 83:6954–58

89. Ripley, L. S., Clark, A., de Boer, J. G. 1986. Spectrum of spontaneous frameshift mutations. Sequences of bac-

teriophage T4 *rII* gene frameshifts. *J. Mol. Biol.* 191:601–13

90. Ripley, l. S., Dubins, J. S., de Boer, J. G., DeMarini, D. M., Bogerd, A. M., Kreuzer, K. N. 1988. Hotspot sites for acridine-induced frameshift mutations in bacteriophage T4 correspond to sites of action of the T4 type II topoisomerase. *J. Mol. Biol.* 200:665–80

91. Ripley, L. S., Glickman, B. W. 1983. The unique self-complementarity of palindromic sequences provides DNA structural intermediates for mutation. *Cold Spring Harbor Symp. Quant. Biol.* 47:851–61

92. Ripley, L. S., Glickman, B. W., Shoemaker, N. B. 1983. Mutator versus antimutator activity of a T4 DNA polymerase mutant distinguishes two different frameshifting mechanisms. *Mol. Gen. Genet.* 189:113–17

93. Ripley, L. S., Shoemaker, N. B. 1983. A major role for bacteriophage T4 DNA polymerase in frameshift mutagenesis. *Genetics* 103:353–66

94. Romac, S., Leong, P., Sockett, H., Hutchinson, F. 1989. DNA base sequence changes induced by ultraviolet light mutagenesis of a gene on a chromosome in Chinese hamster ovary cells. *J. Mol. Biol.* 209:195–204

95. Roth, D. B., Chang, X.-B., Wilson, J. H. 1989. Comparison of filler DNA at immune, nonimmune and oncogenic rearrangements suggests multiple mechanisms of formation. *Mol. Cell. Biol.* 9:3049–57

96. Roth, D. B., Wilson, J. H. 1986. Nonhomologous recombination in mammalian cells: Role for short sequence homologies in the joining reaction. *Mol. Cell. Biol.* 6:4295–304

97. Sahasrabudhe, S., Sambamurti, K., Humayun, M. Z. 1989. mutagenesis by aflatoxin in M13 DNA: base-substitution mechanisms and the origin of strand bias. *Mol. Gen. Genet.* 217:20–25

98. Sambamurti, K., Callahan, J., Luo, X., Perkins, C. P., Jacobsen, J. S., Humayun, M. Z. 1988. Mechanisms of mutagenesis by a bulky DNA lesion at the guanine N7 position. *Genetics* 120:863–73

99. Schaaper, R. M. 1989. *Escherichia coli* mutator *mut*D5 is defective in the *mut*HLS pathway of DNA mismatch repair. *Genetics* 121:205–12

100. Schaaper, R. M., Danforth, B. N., Glickman, B. W. 1986. Mechanisms of spontaneous mutagenesis: an analysis of the spectrum of spontaneous mutation in the *Escherichia coli lacI* gene. *J. Mol. Biol.* 189:273–84

101. Schaaper, R. M., Dunn, R. L. 1987. Spectra of spontaneous mutations in *Escherichia coli* strains defective in mismatch correction: The nature of *in vivo* DNA replication errors. *Proc. Natl. Acad. Sci. USA.* 84:6220–24

102. Sedwick, W. D., Brown, O. E., Glickman, B. W. 1986. Deoxyuridine misincorporation causes site-specific mutational lesions in the *lacI* gene of *Escherichia coli. Mutat. Res.* 162:7–20

103. Shpakovski, G. V., Akhrem, A. A., Berlin, Y. A. 1988. Structural bases of a long-stretched deletion: completing the λ *plac5* DNA primary structure. *Nucleic Acids Res.* 16:10199–212

104. Singer, B. S., Westlye, J. 1988. Deletion formation in bacteriophage T4. *J. Mol. Biol.* 202:233–43

105. Skopek, T. R., Hutchinson, F. 1984. Frameshift mutagenesis of lambda prophage by 9-aminoacridine, proflavin and ICR-191. *Mol. Gen. Genet.* 195:418–23

106. Streisinger, G., Okada, Y., Emrich, J., Newton, J., Tsugita, A., et al. 1966. Frameshift mutations and the genetic code. *Cold Spring Harbor Symp. Quant. Biol.* 31:77–84

107. Streisinger, G., Owen, J. E. 1985. Mechanisms of spontaneous and induced frameshift mutation in bacteriophage T4. *Genetics* 109:633–59

108. Timsit, Y., Westhof, E., Fuchs, R. P. P., Moras, D. 1989. Unusual helical packing in crystals of DNA bearing a mutation hot spot. *Nature* 341:459–62

109. Tindall, K. R., Stankowski, L. F. 1989. Molecular analysis of spontaneous mutations at the *gpt* locus in Chinese hamster ovary (AS52) cells. *Mutat. Res.* 220:241–53

110. Tindall, K. R., Stein, J., Hutchinson, F. 1988. Changes in DNA base sequence induced by gamma-ray mutagenesis of lambda phage and prophage. *Genetics* 118:551–60

111. Weinreb, A., Collier, D. A., Birshtein, B. K., Wells, R. D. 1990. Left-handed Z-DNA and intramolecular triplex formation at the site of an unequal sister chromatid exchange. *J. Biol. Chem.* 265:1352–59

112. Woodson, S. A., Crothers, D. M. 1987. Proton nuclear magnetic resonance studies on bulge-containing DNA oligonucleotides from a mutational hot-spot sequence. *Biochemistry* 26:904–12

113. Woodson, S. A., Crothers, D. M. 1988. Preferential location of bulged guanosine internal to a G:C Tract by 1H NMR. *Biochemistry* 27:436–45

114. Xodo, L. E., Manzini, G., Quadrifoglio, F., van der Marel, G. A., van

Boom, J. H. 1988. Oligodeoxynucleotide folding in solution: loop size and stability of B-hairpins. *Biochemistry* 27:6321–26

115. Yatagai, F., Glickman, B. W. 1986. Mutagenesis by 8-methoxypsoralen plus near-UV treatment: analysis of specificity in the *lacI* gene of *Escherichia coli*. *Mutat. Res.* 163:209–24

116. Yatagai, F., Glickman, B. W. 1990. Specificity of spontaneous mutation in the *lacI* gene cloned into bacteriophage M13. *Mutat. Res.* 243:21–28

117. Yatagai, F., Horsfall, M. J., Glickman,

B. W. 1987. Defect in excision repair alters the mutational specificty of PUVA treatment in the *lacI* gene of *Escherichia coli*. *J. Mol. Biol.* 194:601–7

118. Yi, T.-M., Stearns, D., Demple, B. 1988. Illegitimate recombination in an *Escherichia coli* plasmid: modulatin by DNA damage and a new bacterial gene. *J. Bacteriol.* 170:2898–903

119. Zheng, G., Sinden, R. R. 1988. Effect of base composition at the center of inverted repeated DNA sequences on cruciform transitions in DNA. *J. Biol. Chem.* 263:5356–61

Annu. Rev. Genet. 1990. 24:215–48

GENETIC ANALYSIS OF PROTEIN EXPORT IN *ESCHERICHIA COLI*

Peter J. Schatz and Jon Beckwith

Department of Microbiology and Molecular Genetics, Harvard Medical School, Boston, Massachusetts, 02115

KEY WORDS: protein, secretion, signal sequences, sec genes, translocation.

CONTENTS

INTRODUCTION

Protein secretion has been an active field of study in both prokaryotes and eukaryotes for over ten years. Two major approaches have been used to

215

0066-4197/90/1215-0215$02.00

analyze the features of specific proteins that promote their export and to determine the nature of the secretion machinery. (*a*) In vitro studies in eukaryotic systems revealed a complex apparatus required for the entry of proteins into the lumen of the rough endoplasmic reticulum (RER). (*b*) Genetic studies in *Escherichia coli* defined components of exported proteins necessary for their translocation and cellular factors comprising the secretion machinery. Only recently have prokaryotic in vitro systems and genetic approaches to eukaryotic secretion been developed.

In this review, we describe mainly the genetic studies carried out in *E. coli*. These studies have led to a more precise picture of the important features of signal sequences and the role of the mature protein in export. They have also led to the definition of a set of *sec* genes that code for essential components of the export machinery. In many cases, the role of the *sec* genes has been shown by both in vivo and in vitro studies. More recent genetic studies in yeast are beginning to yield a picture similar in some respects to that seen in *E. coli*. The amount of genetic data that has accumulated is now quite substantial. In certain areas, we have briefly summarized these data. For more detail, we refer the reader to a recent collection of reviews that goes into considerable depth on many of these subjects (139).

GENETICS OF EXPORT SIGNALS IN CELL ENVELOPE PROTEINS

In vivo Selections for Export-Defective Proteins

The existence of signal sequences in secreted proteins in both prokaryotes and eukaryotes suggested an important role for these small peptides in the process of membrane translocation (12, 57, 59). Subsequent compilation of such sequences for a variety of *E. coli* cell envelope proteins revealed several common features (91). Signal sequences range in size from 15 to 26 amino acids. They have short hydrophilic amino-termini that invariably contain at least one positively charged amino acid. The hydrophilic amino-terminus is followed by a highly hydrophobic sequence of at least eight amino acids. The carboxy-terminus of the signal sequence is hydrophilic and contains information necessary for the cleavage of the peptide from the mature protein. Genetics has been used to approach several questions raised by the existence of the signal sequence: (*a*) Is it necessary for the export process? (*b*) Is it sufficient to specify export of a protein or is there essential information in the mature sequence of such proteins? (*c*) What role do its different portions play in the export process?

Genetic selections that allowed the isolation of mutants defective for the export of specific *E. coli* proteins employed gene fusions coding for a hybrid protein, consisting of the cytoplasmic enzyme β-galactosidase fused to one or another cell envelope protein. Because β-galactosidase cannot be fully exported across the cytoplasmic membrane the hybrid protein becomes imbedded in the membrane and, when produced at high levels, jams the cell's export machinery, leading to cell death (130). Fusions of β-galactosidase to the maltose-inducible proteins LamB and maltose binding protein (MBP, the product of the *malE* gene) result in a sensitivity to the presence of maltose in the growth medium (61). The lethal effects of the fusion protein are dependent on the amount of the protein produced. For instance, the amount of a hybrid alkaline phosphatase-β-galactosidase protein produced in a partly constitutive (*phoR*) strain is lethal when the fusion gene is on a multicopy plasmid but not when it is present in single copy (92).

Since the lethality associated with the production of these hybrid proteins is due to their entry into the export pathway, mutations that prevent such entry relieve the lethal phenotype. Thus, with *malE-lacZ* and *lamB-lacZ* fusions, selection for resistance to the lethal-inducing effects of maltose yields mutations in the signal sequence of the exported protein. The finding that defects in the signal sequence block entry into the translocation pathway demonstrates that it is a primary signal for the initiation of the process.

In addition to jamming the export machinery, the membrane location of the hybrid protein interferes with the proper assembly of β-galactosidase and results in a hybrid protein with low β-galactosidase activity. The strains are phenotypically Lac$^-$. This property has been seen with fusions of β-galactosidase to several different cell envelope proteins, including integral proteins of the cytoplasmic membrane. Mutations that cause the internalization of the hybrid protein to the cytoplasm allow proper assembly of β-galactosidase into an active enzyme. Therefore, selection for Lac$^+$ derivatives of the Lac$^-$ fusion strains is a direct selection for a change in location of the protein and hence for mutants defective in secretion.

Using either the Lac$^+$ selection or the maltose-resistance selection, mutations altering four different exported proteins, LamB (42), MBP (8, 46), alkaline phosphatase (93, 94) and MalM (120), have been obtained. In nearly all cases, these mutations cause a significant block in translocation of the fusion and, after recombination, the corresponding wild-type protein as well. Altogether, 57 mutants have been isolated that had changes in the sequence of one or another of these proteins. The striking result is that, in all but one case, the component of the exported protein that is mutationally altered is the hydrophobic core of the signal sequence. (With one of the *malM* mutations, no sequence alteration was found in the signal sequence, and the mutant was not studied further.)

These findings could be taken to imply that there are no features, other than the hydrophobic core of the signal sequence, that are required for protein export. However, there could be biases to the genetic selections. For instance, the selections demand internalization of the hybrid protein. If components of the protein are involved in export steps later than the actual insertion in the membrane, mutations affecting them might not have the desired phenotype. At this point, there is no evidence for such components except cleavage sites for the signal peptidases. Another possibility is some redundancy in export information in a specific protein. Then, either a double mutation or a deletion would be required to observe the appropriate phenotype in these selections. Studies described below suggest that the positively charged amino acids at the amino-terminus of the signal sequence can fit this category. A third explanation, that the selection is too stringent, is not the case. The Lac$^+$ selection is quite sensitive, since only a small amount of β-galactosidase has to be internalized to allow growth on lactose. Signal sequence mutations have been detected in this selection that allow the export of 30% of the normal amounts of alkaline phosphatase when recombined into a wild-type *phoA* gene. Clearly, the selection is not demanding strong mutants.

Signal Sequence Mutations Obtained by in vitro Mutagenesis

One of the most interesting questions to arise out of the selections described above relates to the role of the positively charged amino acids at the amino-terminus of the signal sequence. Every *E. coli* cell envelope protein with a cleavable signal sequence has at least one lysine or arginine in this region and most have more than one (149). This conserved feature suggests an important role for these amino acids. Yet, the in vivo selections never yielded mutations that altered these amino acids. One explanation for this apparent paradox is that one lysine or arginine is sufficient for export so that alteration of only one of them in a signal sequence will not block export. To test this hypothesis mutations that altered these residues were made in vitro in lipoprotein (58, 147), LamB (9, 52) and MBP (108). The results show that eliminating the positive charges and reducing the net charge to neutral does not eliminate export. In some cases, it reduces the efficiency of export, slowing down its kinetics. Mutations that result in a net negative charge in this region have stronger effects, but, in most cases, still allow significant amounts of export. Even when the net charge in this region of preMBP is changed to -2, substantial export and processing occurs in pulse-chase experiments. It must be concluded that these amino acids are not essential for export, but that they do improve its efficiency. However, it is hard to rule out the essentiality of a positive charge at the amino-terminus, since, as von Heijne (149) points out, the first amino acid of the signal sequence may have a free amino group. (Whether the formyl group is removed from this amino acid in exported proteins has never been established.)

One surprising finding with mutations that remove these positively charged amino acids is that they often result in a reduction in the production of the exported protein. These results have raised the possibility that secretion and translation are coupled through this portion of the signal sequence. According to this explanation, a slowing down of secretion would then feed back to slow down translation. This possibility has been considered seriously, since in the eukaryotic in vitro secretion system, a translation-secretion coupling has been shown (152). However, the search for such a coupling in prokaryotes has revealed none. These results make more likely other explanations for the reduction in expression in mutants of the positively charged amino acids. First, a prediction of the secretion-translation coupling model is that signal sequence mutations should reverse the synthesis block by preventing the entry of the mutant proteins into the export pathway. This is not the case (9, 108). Second, in vitro studies indicate that the mutations are directly blocking the initiation of translation (P. J. Bassford Jr., personal communication). This block may occur because the altered base sequence has affected a primary or secondary structure of the messenger-RNA necessary for efficient initiation.

The carboxy-terminal hydrophilic portion of the signal sequence contains information necessary for recognition and processing by signal peptidase. No mutations affecting this region were detected in the in vivo selections, which required internalization of hybrid proteins. Since early results had shown that processing was not necessary for export, this is not a surprising finding (159). Mutations have been obtained in this region by in vitro mutagenesis, and many of them do affect processing (47, 70, 71, 106). In some cases, the protein is exported, but is anchored to the cytoplasmic membrane by the uncleaved signal sequence.

Another approach to determining the important features of signal sequences is to clone random DNA sequences in front of an exportable protein and determine which kinds of sequences are sufficient to permit export. This has been done using a β-lactamase vector in *E. coli* (190) and an invertase vector in the yeast *Saccharomyces cerevisiae* (66). In both cases, a large fraction (15–20%) of random sequences permitted export. These sequences all contained a hydrophobic stretch of amino acids apparently of sufficient length to promote export. The results provide dramatic support for the previous deduction that there is no sequence specificity to the hydrophobic core of the signal sequence and that any hydrophobic sequence of the appropriate length could act as a signal. Further, in the prokaryotic system, in contrast to the yeast studies, most of the random sequences that functioned in export also had a net positive charge preceding the hydrophobic core. These findings are consistent with the observation that signal sequences of eukaryotic proteins do not necessarily have such positively charged amino acids.

The Role of Sequences of the Mature Protein in Export

There is no evidence that the mature portion of exported proteins contains information required to promote translocation. However, the nature of the mature sequence is important. First, alterations of the sequence at the very beginning of the mature protein can severely interfere with export (84, 137, 138, 168). Several laboratories have shown that the presence of a net positive charge in this region can reduce export by as much as 50-fold. The inhibitory positive charges must be relatively close to the cleavage site, perhaps within the first five amino acids. Summers et al (138) have shown that by moving an arg-lys pair 14 amino acids away from the junction with the signal sequence, the defect in export is eliminated. Further, while two arginines in the first and third positions of the mature portion of an artificial alkaline phosphatase construct interfere with export, a tandemly arranged pair of arginines at positions 24 and 25 present in the wild-type protein obviously do not cause a problem (14, 84). Although Summers et al (137) have reported that with β-lactamase, arginines but not lysines interfere with export, Li et al (84) showed that while lysine is less deleterious than arginine, it still has an effect. This difference in the effects of arginine and lysine is probably related to the lower pKa of lysine.

Von Heijne (150) and Knowles and collaborators (137) have suggested that positively charged amino acids at the amino-terminus of the mature protein cannot enter the lipid bilayer, thus blocking translocation. It is assumed in this explanation that the protein is actually entering the lipid bilayer rather than a proteinaceous pore favored in some models (133). A second explanation suggests that the signal sequence orients itself in the membrane according to the charge distribution around it and the electrochemical gradient across the membrane. Considering that gradient, the more positively charged end of the signal sequence would be on the inside of the cell and the less positively charged portion on the periplasmic face of the membrane. This is the presumed normal orientation of the signal sequence that leads to protein export. However, in the altered constructs, the polarity has been reversed, or at least reduced, so that the signal sequence might reverse its orientation in the membrane, in this way blocking export. Attempts in our laboratory to reverse the effects of the charges at the beginning of the mature sequence by putting additional positive charges at the amino-terminus of the signal sequence have failed (84; L. Stevens, J. Beckwith, unpublished results). Addition of up to three extra lysines to the alkaline phosphatase constructs had no effect.

It is perhaps surprising that in selections for export-defective mutants, none were detected that introduced positive charges at the amino-terminus of the mature protein. It is not clear that such alterations would have the same effect on the fusion proteins as did mutations in the signal sequence itself. Further, mutations that generate new codons for the appropriate amino acids may be rare, and might not be possible with single base alterations, given the DNA sequence of regions of this sort and the nature of the genetic code.

A second proposed role for the mature sequence in export is related to protein folding. As described below (see section on *secB*), mutations that interfere with the folding of maltose binding protein can affect the export process under some conditions (27, 86). In addition, evidence suggests that the chaperone SecB interacts with a portion of the mature maltose binding protein to maintain it in a secretion-competent state (25, 72; L. L. Randall, T. B. Topping, S. J. S. Hardy, submitted).

Folding properties may also explain why some cytoplasmic proteins cannot be exported when they are preceded by a signal sequence. The best-studied case of this phenomenon involves β-galactosidase fusions. Genetic studies on the basis of the nonexportability of β-galactosidase gave results consistent with a model in which the folding of portions of the protein in the cytoplasm prevented its complete transfer through the membrane (81). Further support for this explanation comes from the finding that overproduction of the GroEL chaperone protein appears to facilitate transfer of a LamB-β-galactosidase hybrid into the membrane (G. J. Phillips, T. J. Silhavy, submitted). Although β-galactosidase cannot be exported, other cytoplasmic proteins are transferred through the membrane (137, 138, 140). However, no quantitation has been done to determine the efficiency of export.

The studies on translocation of both cell envelope and cytoplasmic proteins raise the possibility that genetic selections can be used to study the folding process. The MBP mutant that affects folding is an example of the potential of such an approach. Mutations that permit the export of normally cytoplasmic proteins may also be ones that alter the folding pathway (81).

Improving Secretion by Genetics

Alteration of a signal sequence can lead to more efficient export. Goldstein et al (51) have shown that increasing the hydrophobicity of the OmpA signal sequence can enhance the rate of export in a fused protein. However, such enhancement may very much depend both on the nature of the starting signal sequence and of the protein whose export it is initiating.

GENETICS OF THE SECRETORY APPARATUS

Loci with a Well-Established Role in Secretion

A number of genes have a well-established role in secretion, including *secA (prlD), secB, secD, secE (prlG), secF, secY (prlA)*, as well as two genes for leader peptidases (*lep* and *lspA*). These genes have been identified and analyzed by various genetic and biochemical techniques. A number of other genes and proteins that may have a role in secretion are also discussed below. In several cases, we know something about the biochemical activity of the products of genes first identified by mutation. In addition, the in vivo function of biochemically identified proteins has been confirmed in a

Table 1   Genetic loci involved in secretion

Gene	Map Position (min)	Phenotype	Gene Product
secA(prlD)	2.5	ts or cs lethal secretion defect, suppressor of signal-sequence mutations	102-kd peripheral membrane protein, ATPase
secB	81	secretion defect for some proteins, null not lethal on minimal medium	16.6-kd cytoplasmic protein, antifolding factor
secD	9.5	cs lethal secretion defect	67-kd inner-membrane protein
secF	9.5	cs lethal secretion defect	35-kd inner-membrane protein
secE(prlG)	90	cs lethal secretion defect, suppressor of signal sequence mutations	13.6-kd inner-membrane protein
secY(prlA)	72	ts or cs lethal secretion defect, suppressor of signal sequence mutations	49-kd inner-membrane protein
lep	55.5	lethal defect in signal sequence processing	36-kd inner-membrane protein signal (leader) peptidase I
lspA	0.5	lethal defect in signal sequence processing of lipoproteins, globomycin resistance	18-kd inner-membrane protein, signal peptidase II

number of cases through analysis of the genes that code for them. It is the correspondence between the genetic and biochemical analysis that makes the argument for a direct role in secretion for a given gene/protein very strong. A summary of the loci implicated in *E. coli* secretion is given in Table 1.

Approaches for Identifying Secretion Genes

For *E. coli* secretion, the direct approach of looking for conditional lethal mutations that affected secretion was not a success (131). Only one mutation in the secretory apparatus has been reported from a simple mutagenize and screen procedure (63). Successful selections for secretion mutants relied on demanding an alteration in the way that the mutant cell recognizes export signals in portions of secreted proteins. The mutant cells localize the proteins differently than do the wild type, thus giving an indication of an alteration in some component of the export apparatus.

SUPPRESSORS OF SIGNAL SEQUENCE MUTATIONS   With the identification of the signal sequence as necessary for protein export, Emr et al (41) selected mutations that affected the secretory apparatus as extragenic suppressors of signal sequence mutations. Using strains bearing a *lamB* gene deleted for the hydrophobic core of the signal sequence, they selected for transport of maltodextrins (Dex$^+$) and, therefore, export of LamB protein (maltoporin). The mutations recovered were dominant, which can be explained readily since the secretory apparatus has gained a new function, the ability to recognize a defective signal sequence.

Almost all of the mutations isolated by Emr et al (41) mapped to a gene that they named *prlA*, with only single alleles of two other loci called *prlB* and *prlC*. More recently, Stader et al (135) used a different *lamB* signal sequence mutation (a charge introduced in the hydrophobic core) to isolate additional alleles of *prlA*, as well as mutations in *prlD* and *prlG* (see below). Bankaitis & Bassford used a similar selection with a deletion in the signal sequence of MBP to identify alleles of prlA as well as a mutation they called *prlD1* (6).

The signal sequence suppressors in the *prlA* gene are the strongest known. They suppress signal sequence mutations in genes coding for a variety of periplasmic and outer-membrane proteins, including *lamB*, *malE*, and *phoA*, to allow export of the proteins to their normal location (6, 40, 41, 94, 121). The *prlA* alleles are generally most efficient suppressors of defects in the hydrophobic core of signal sequences. They do not efficiently suppress defects in the amino terminal hydrophilic domain of the signal sequence (108, 134). This preference for one class of signal sequence mutations has been interpreted as evidence that the *prlA (secY)* gene product interacts with the hydrophobic core (108; see below). It is also possible that the spectrum of *prlA* mutations is limited by the fact that the known alleles were isolated as suppressors of defects in the hydrophobic core of signal sequences (6, 41, 121, 135). The lack of allele specificity towards various hydrophobic core alleles suggests that *prlA* mutations may simply relax the specificity of the translocation apparatus. This interpretation is supported by recent evidence that export of alkaline phosphatase totally lacking a signal sequence can be partially restored by *prlA* mutations (A. I. Derman & J. Beckwith, unpublished data).

The *prlD* locus is a victim of one of the many nomenclature problems that plague the secretion field. The original *prlD1* allele is a suppressor of a deletion in the hydrophobic core of the signal sequence of MBP (6). Several additional alleles (*prlD2* to *prlD5*) arose as suppressors of *malE14-1*, which specifies a charge change in the hydrophobic core of MBP (121). The *prlD1* allele maps near but not in the *secA* gene (6, 48). It is a fairly weak suppressor of signal-sequence defects, but does show allele specificity. It also causes severe growth defects in the presence of some but not all *prlA* alleles. In spite of attempts to interpret the significance of these interactions (6), no further characterization of *prlD1* has been reported. The *prlD2* to *prlD5* alleles, on the other hand, have been cloned and sequenced (48). They are alleles of the *secA* gene, which codes for a hydrophilic peripheral membrane/cytoplasmic protein (102; see below). Although they arose as suppressors of changes in the hydrophobic core of the MBP signal sequence, they can suppress defects in the amino terminal hydrophilic (generally positively charged) domain (108). Puziss et al (108) speculated that "SecA interacts more strongly with the hydrophilic segment of the signal." It is also possible that SecA interacts with the mature portion of secreted proteins or even some other component of the secretion machinery, and thus is able to suppress defects in multiple regions of the signal sequence.

To clear up the *prlD* nomenclature problem, we suggest that the original allele *prlD1* be given a new name and that *prlD2* to *prlD5* retain their names. The *prlD* alleles isolated by Stader et al (135) have not been mapped precisely.

The mechanism of suppression by the other two genes identified by Emr et al, *prlB* and *prlC* (41), is not known. The only known *prlB* mutation is a deletion in the *rbsB* gene that codes for the periplasmic ribose binding protein (131). It suppresses certain signal sequence mutations in *lamB* but not in *malE* (40). Unlike the other *prl* suppressors, it does not cause an increase in the amount of signal sequence processing (40). Suppression probably occurs by an indirect mechanism, perhaps because of an interaction between the shortened *rbsB* gene product and some component of the secretion machinery.

Alleles of *prlC* weakly suppress signal-sequence mutations in the hydrophobic core region of both *lamB* and *malE* (40, 145, 146). Biochemical analysis reveals, however, that although processing of mutant precursor proteins is almost as efficient as the wild type, very little mature protein is exported from the cell (146). This small amount of export is dependent on proper functioning of the *secA* and *secB* genes (see below), but revealingly, the processing is not (146). Since processing of precursors normally requires *secA* and *secB*, this result implies that the processing observed in *prlC* strains is not part of the normal export pathway. Whether *prlC* is part of another export pathway or allows suppression by some bypass mechanism remains to be determined. The ability to isolate nonlethal amber mutations in *prlC* (146) strongly suggests that it does not have an essential role in protein export.

By refining the suppressor selection to allow the use of point mutations, Stader et al (135) used a weaker signal sequence mutation *(lamB14D)* to identify a new locus, *prlG,* as well as additional alleles that mapped to the *prlA* and *prlD* loci. The *prlG* mutations weakly suppressed signal-sequence defects in *lamB* and *malE,* and mapped to a region of the chromosome containing the *secE* gene (115, 125; see below). Subsequent cloning and sequencing confirmed that *prlG* and *secE* are the same gene (P. J. Schatz, K. Bieker, T. J. Silhavy, & J. Beckwith, in preparation). Thus in three cases, *prlA (secY), prlD (secA),* and *prlG (secE),* the same gene yields both alleles that suppress signal sequence mutations and alleles with a conditional-lethal secretion defect (see below). This correspondence is strong evidence that these genes have a direct role in secretion.

The ability of several genes to yield signal sequence suppressors suggests that there is not a simple rate-limiting step in translocation of defective precursors that is catalyzed by a single protein. Rather it suggests that several proteins function in close association, and alteration of a subset of them can alleviate the rate limitation imposed by a defective signal sequence.

LOCALIZED MUTAGENESIS   Localized mutagenesis of the region of the chromosome containing *prlA* (41) led to the discovery of a mutation *(ts215)* that caused the accumulation of precursors of exported proteins (64). A piece of DNA coding for a protein called Y complemented this mutation, and the gene was named *secY.* Subsequent analysis showed that the *ts215* mutation was an amber allele in the *rplO* gene that was polar on the immediately downstream *secY* reading frame. It was temperature-sensitive for growth due to the presence of a temperature-sensitive amber suppressor in the strain (62). Shiba et al (129) subsequently used localized mutagenesis to identify a mutation in the *secY* reading frame that was temperature-sensitive for growth and had a secretion defect. Alleles of *prlA* are changes in the same reading frame (123, 134), and the dual names have persisted.

USE OF EXPORTED β-GALACTOSIDASE FUSION PROTEINS   As described above, fusions between exported proteins and β-galactosidase are enzymatically inactive because they become jammed in the membrane. Oliver & Beckwith (101) used this localization phenotype of a MBP-β-galactosidase fusion to isolate mutations that allowed the strain to become Lac$^+$. In addition to mutations that affect the signal sequence of the normally secreted portion of the protein, mutations that lead to a change in the interaction of the fusion protein with the secretion apparatus are recovered. Assuming that the secretion apparatus was essential, Oliver & Beckwith searched among the Lac$^+$ candidates for those that had simultaneously acquired a conditional-lethal phenotype. The sensitivity of the selection described earlier allowed detection of strains with slight defects at the permissive temperature of 30°. When shifted to the nonpermissive temperature of 42°, the *secA* mutants resulting from this search caused lethality and a stronger secretion defect.

By using the same scheme, but without requiring conditional-lethal mutations, Kumamoto & Beckwith isolated alleles of a different gene called *secB* (73). Like the *secA* mutations, these alleles caused the accumulation of the precursors of several exported proteins. Unlike *secA*, the *secB* alleles did not cause temperature-dependent lethality. In fact, null alleles of *secB* are viable on certain media and appear to cause a detectable secretion defect for only a subset of exported proteins (74).

Alleles at the *secD* locus were recovered in a selection analogous to that used to identify *secA* but using *phoA-lacZ* and *lamB-lacZ* fusions (50). These alleles caused cold-sensitive lethality and a pleiotropic secretion defect. Although the selection was very similar to that used to obtain *secA* and *secB*, no alleles of those genes were detected. This bias in the genes identified by very similar selections remains unexplained, especially since both *secA* and *secD* alleles affect the secretion of the products of all of the genes used to construct the fusions (*malE, phoA*, and *lamB*). One possible reason for this phenomenon is that some combinations of *sec* alleles and fusion genes are lethal to the cell. For example, introduction of a *secAts* or *secYcs* mutation into a strain carrying a *phoA-lacZ* fusion causes severe instability of the strain (E. Brickman & J. Beckwith, unpublished observations). In any case, this selectivity reinforces the utility of using several different approaches to find mutations that affect a single process.

The second general phenotype associated with fusions of secreted proteins to β-galactosidase is overproduction lethality due to obstruction of the secretion apparatus. Kiino & Silhavy (69) used this phenotype to select mutations in *prlF*. The phenotype of the *prlF1* mutant is the opposite of that of *sec* mutants. The *prlF* allele allows more efficient export of the hybrid protein from the cytoplasm, thus the resulting strain is Lac⁻. This ability of *prlF1* to increase the efficiency of export extends to suppression of mutations that cause a defect in the hydrophilic region of signal sequences (54, 108). Subsequent analysis revealed that *prlF1* is a small deletion in an autoregulatory domain of its protein product. The resulting overexpression of the mutant protein, or overexpression of the wild-type protein, is sufficient to allow suppression of lethality from fusion protein overexpression (68). Since a null mutation in *prlF* had no detectable effect on cell growth or overproduction lethality, it is clear that it has no essential role in protein export. Its suppression probably depends on some effect of PrlF on other components of the secretion apparatus or on the hybrid protein itself (68).

SELECTIONS BASED ON *secA* REGULATION   The observation that SecA is regulated according to the secretion needs of the cell (see below) led to the design of a very general procedure for identifying secretion mutants. At the time, the secA protein was known to be induced by a number of conditions that inhibit protein export, including the presence of *secA, secD,* or *secY* mutations or the expression of a fusion protein that jams the export apparatus (50, 102, 116). These observations led Riggs et al (115) to design a procedure to detect export mutants based on overexpression of a *secA-lacZ* fusion protein. The selection was based on the reasoning that mutants that over-expressed *secA* might have defects in a variety of genes necessary for protein export.

The mutants were either selected on lactose or raffinose minimal medium or isolated by screening on indicator plates (115). The first two selections yielded mostly mutations linked to the fusion gene or to *secA,* the two loci expected to give rise to the strongest overexpression of the fusion. The screening on indicator plates, on the other hand, gave rise to alleles of a wide variety of genes in this study and in a subsequent one (125). These included a number of *secA* and *secY* alleles, as well as alleles linked to *secD.* Subsequent analysis of these *secD* linked alleles has shown that there are two adjacent secretion genes at the locus, now called *secD* and *secF,* which were both represented by a number of mutations (C. Gardel, K. Johnson, A. Jacq & J. Beckwith, submitted). Riggs et al also discovered a new secretion gene, *secE,* originally identified by a cold-sensitive mutation. As expected, no *secB* alleles arose in the selection, because *secB* mutations do not cause over-expression of *secA* (116).

The *secA* expression selection is the most general that has been used to date. Since the signal for induction of *secA* is unknown, the basis for this generality is unclear. That inducing signal may be the best known indicator of a block in the export pathway. Another advantage to this selection is its sensitivity. The indicator plates can detect a very small increase in β-galactosidase activity, so small perturbations in the protein export pathway are apparent.

GENES WHOSE PRODUCTS WERE IDENTIFIED BIOCHEMICALLY   Two en-zymes are known in *E. coli* that remove signal peptides from proteins during the export process. Wickner and his colleagues purified signal peptidase I or leader peptidase from detergent extracts of membranes (171). They sub-sequently cloned, sequenced, and mapped the gene coding for this enzyme, which was named *lep* (33, 132, 158). Several groups have demonstrated that *lep* is essential for cell growth, presumably because its product cleaves the precursors of a wide variety of periplasmic and outer membrane proteins. (31, 32, 55).

A second enzyme, signal peptidase II, has evolved specifically to cleave the precursors of several lipoproteins of *E. coli* (162). This essential enzyme is inhibited by the antibiotic globomycin in vitro (144), and mutants resistant to globomycin map to the gene coding for the enzyme (164, 165). The gene, *lspA,* has been cloned by complementation of the globomycin-resistant mutation (163) or because overproduction of the enzyme leads to increased globomycin resistance (143). The gene maps to minute 0.5 on the *E. coli* chromosome (113, 165).

Note that none of the previously described selections detected alleles in either of these signal peptidase genes. We can rationalize this result because failure to remove the signal sequence does not inhibit membrane translocation, but does inhibit release of the protein into the periplasmic space (31, 70). Thus the selections, which are targeted to early steps in membrane translocation, would miss genes involved in the late release step.

False Leads

One of the attempts to devise new strategies for obtaining mutations in genes coding for components of an export machinery involved the use of a strategy developed by Jarvik & Botstein (65). In protein export, the approach begins with a known *sec* mutation from which suppressors are isolated. The principle underlying this approach is that suppressors will often be mutations that alter another gene coding for a protein that interacts with the first gene product. Ideally, the suppressor would work by restoring a functional protein-protein interaction. Early studies with suppressors appeared very promising. In one case, a mutation that appeared to map in the *prlA/secY* gene suppressed a *secAts* mutation (15). In another case, we mapped a suppressor mutation to a gene we termed *secC,* which coded for a protein subunit of the ribosome (45). Initially, we thought we had identified a ribosomal protein that interacted with the export apparatus. However, as we and others isolated more and more suppressors, it became clear that any mutation that slowed down protein synthesis would suppress many *sec* mutations (82, 100, 128). It appears that slowing down the process allows time for the partially defective secretion machinery to handle exported proteins. Subsequently, sequencing showed that the suppressor mutation thought to map to *prlA/secY* was not in this gene, and may well have altered a nearby gene for a ribosomal protein (T. Baba, A. Jacq, E. Brickman, et al, submitted). It also seems likely that the *secC* mutation exerted its effect by altering the rate of protein synthesis. The suppressors had revealed nothing about the mechanism of secretion.

Another false lead depended in part on the attempt to draw close analogies between protein secretion in eukaryotes and prokaryotes. The in vitro secretion system developed in eukaryotes is dependent on signal recognition particle, a complex of 6 proteins and a 7S RNA. Early studies with antibody to SecA, one of the *E. coli* proteins that is part of the machinery, suggested that the 6S RNA of *E. coli* might be complexed with SecA (H. Liebke, personal communication). However, a null mutation of the *E. coli ssr* gene. coding for 6S RNA, had no effect on protein export or cell viability (83).

Other Candidate sec Genes

A different approach to finding genes involved in secretion is to seek homologs to *sec* genes of other organisms. Poritz et al (107) have pointed out structural and sequence similarities between the 7S RNA of mammalian SRP and the 4.5S RNA of *E. coli* (the product of the *ffs* gene). However, conditional-lethal mutations in *ffs* affect protein synthesis, but not protein export (16). Two genes of *E. coli* have been detected that share sequence similarity with the 54 kd-protein subunit of SRP. These are *ffh* (fifty-four homolog), of unknown function, and *ftsY,* a gene in an operon of cell division genes (10, 117). The product of the former gene shares extensive similarity with the 54 kd protein. Currently, reverse genetics is being used to test whether any of these genes are truly involved in the export process. These studies represent an interesting test of the extent to which shared sequences can be used to infer shared function. It seems possible that the impressive sequence similarity in these cases represents common evolutionary origin but not a common role in protein secretion.

Another case of potentially interesting homology involves the *dnaK* gene, which shares sequences with a heat-shock gene of yeast that participates in protein export (34). Again *dnaK* (or *dnaJ*) mutations do not appear to affect protein export in *E. coli* (B. Bukau & K. Johnson, personal communication). However, genes other than *secB* (see below) may code for products that affect the folding of exported proteins in the cytoplasm. One might suspect this from the fact that *secB* mutations have detectable effects on only a subset of exported proteins. Perhaps the others use different factors to maintain themselves in an export-competent state. Candidates for such proteins are the GroEL protein (13, 78) and trigger factor (28). A mutation in the *groEL* gene has a slight but significant effect on the export of β-lactamase. Recently, it has been shown that the GroEL protein retards folding of β-lactamase (A. Pluchthun, personal communication). No other proteins tested in vivo or in vitro were affected by GroEL. Trigger factor was detected in an in vitro system because of its ability to maintain OmpA in an export-competent state. There is no genetic evidence that this protein plays any role in export in vivo. It could be that this protein affects protein folding in vivo, but for some other cellular process than protein export. The finding that the affinity of trigger factor for an exported protein is considerably less than that of SecB is consistent with this explanation (79). Recent genetic analysis of trigger factor suggests that it may not be involved in secretion. A null mutation of the gene for trigger factor is viable and does not affect export of OmpA, the protein with which it was studied in vitro (B. Guthrie & W. Wickner, personal communication).

Study of Translocation in vitro

Preparations containing inverted membrane vesicles and cytoplasmic extracts from *E. coli* that efficiently translocate precursor proteins have been developed to allow the biochemical study of secretion (98, 114). These systems require energy and can operate co- or posttranslationally (19, 98). ATP is a necessary energy requirement but the proton motive force also contributes to the efficiency of translocation (20–23, 98, 141, 160, 161, 166, 167). These observations are consistent with biochemical studies of protein translocation in intact cells (for review see 111). The role of various genetically identified components in the functioning of these in vitro systems has been studied to establish links between the genetic and biochemical analysis of protein translocation (see below).

Properties of Genes and Proteins

secA (prlD)   The original *secA* alleles cause temperature-sensitive lethality and a secretion defect that grows progressively worse with increasing time at nonpermissive temperatures (101). The *secA* gene codes for a 102-kd peripheral membrane/cytoplasmic protein that is essential for cell growth (102, 103). Schmidt et al determined the sequence of the gene and of a number of mutant alleles (127). It is cotranscribed with the upstream gene *X* (7, 127) and the downstream gene *mutT* (M. Schmidt & D. Oliver, personal communication).

The SecA protein plays a central role in the mechanism of protein translocation. It is essential for the functioning of in vitro translocation systems (17, 67). More specifically, SecA is required for functional binding of precursor proteins to membrane vesicles (29). It has ATPase activity that is stimulated in the presence of membranes with functional SecY and a translocation-competent precursor protein (30, 85). If this stimulatable ATPase activity is destroyed, the protein will not catalyze translocation (85). High concentrations of SecA will overcome defects of in vitro systems caused by defective SecY in membranes (43) or by a lack of proton motive force (160).

In addition to the two previously described phenotypes of *secA (prlD)* mutants, other genetic properties of *secA* emphasize its crucial role in secretion. Levels of expression of *secA* can change the effectiveness of *prlA* suppressors of signal sequence mutations (104). Evolution has also selected *secA* as a gene whose product is induced substantially by a variety of defects in *E. coli* protein export (see below).

For years, geneticists have wished for a specific inhibitor of protein export as an aid in isolating mutants and for biochemical studies. In fact, one has existed for many years but has gone unnoticed until recently. Mutants resistant to low levels of azide (azi[r]) were first isolated in 1950 (80). They map in the 2.5 minute position on the *E. coli* chromosome and are in fact alleles of *secA* (D. Oliver, K. Dolan, R. Cabelli, personal communication). Low millimolar levels of azide act as a fairly specific inhibitor of protein export, presumably by inhibiting the *secA* ATPase. Although azide undoubtedly inhibits other enzymes in *E. coli*, the most sensitive essential protein is *secA*, raising the possibility of using the azide block as a physiological marker for the *secA* catalyzed stage of secretion. This marker should allow analysis of the secretion pathway by various biochemical and physiological means. Although the existing azi[r] mutants do not exhibit a secretion block, these mutants or new ones may have revealing phenotypes related to protein export.

secB   The *secB* protein's biochemical role in secretion is probably the best understood. Both point mutations and null alleles of *secB* are viable and only have strong effects on a subset of exported proteins (73, 74). The null mutants are only viable on minimal but not rich medium, suggesting a greater need for secB at higher rates of growth. The gene codes for a protein of 16.6 kd that is soluble and found mostly in the cytoplasm (75, 77, 156, 157).

One of the proteins whose export is strongly affected by *secB* is the periplasmic maltose binding protein, MBP (73, 74). Study of MBP in *secB*+ cells reveals a competition between folding and export. Rapid folding of preMBP in the cytoplasm apparently results in a conformation that is incompatible with translocation across the membrane (109). Amino acid changes in the mature part of the protein that suppress defects in the signal sequence decrease the rate of folding of MBP, thus presumably giving the precursor more opportunity to interact with the secretion machinery (27, 86). Mutations in *secB* slow the rate of export of MBP and also increase the amount of folded, translocation-incompetent preMBP in the cytoplasm (76). The defect in MBP export in *secB*⁻ cells was suppressed by amino acid changes that slowed the rate of folding of mature MBP (25), or by changes in the signal sequence that increase the rate of export (26). The model that emerged from these studies is that SecB binds to the preMBP during or soon after its synthesis and stabilizes an export-competent conformation until the protein associates with later components of the secretion apparatus.

Several lines of evidence support the proposed antifolding role of SecB. Purified SecB enhances the activity of in vitro secretion systems and inhibits folding of precursor proteins (75, 157). A soluble component previously identified as important for activity of an in vitro system is a tetramer of SecB subunits (153, 156). SecB binds precursor proteins in vitro and in vivo (72, 79, 155; L. L. Randall, T. B. Topping & S. J. S. Hardy, submitted, J. B. Weiss & P. J. Bassford Jr., submitted). There is some disagreement about whether SecB binds to the mature part of the protein or to the signal sequence. Genetic experiments with fusion proteins (49) or expression of deletion derivatives of MBP to titrate SecB (25) indicated that SecB binds to the mature portion of MBP. On the other hand, Watanabe & Blobel (155) interpreted their data to suggest that SecB binds to the signal sequence of preMBP. Further work has shown, however, that SecB does bind directly to mature MBP (L. L. Randall, T. B. Topping & S. J. S. Hardy, submitted, J. B. Weiss & P. J. Bassford, Jr., submitted). Apparently, one role of the signal sequence is to retard folding of the mature portion of MBP to allow SecB to bind (86, 87, 105, 110, 142).

Since SecB is presumably reused many times, it fits the definition of a molecular chaperone (39). It joins a growing list of proteins that bind other proteins to facilitate folding, membrane transit, or assembly of large complexes (119). As in other membrane-translocation systems, the proper conformation of E. coli precursors is crucial to the translocation reaction (38). Since some proteins do not require SecB for export, it is possible that other chaperones function in E. coli secretion (see below), or that these proteins fold slowly enough to engage other components of the export apparatus without help.

secY (prlA)   Originally identified by *prlA* alleles that suppressed signal sequence mutations, *secY* alleles can also cause temperature- or cold-sensitive lethality and a block in protein translocation (41, 115, 125, 129). The *secY (prlA)* gene is in the *spc* operon, downstream from ten ribosomal protein genes, and codes for a 49-kd gene product very rich in hydrophobic amino acids (18). As predicted from the sequence, SecY is an integral inner-membrane protein (1, 60). Based on fusions of secY to alkaline phosphatase (88, 89), and protease sensitivity assays, Akiyama & Ito proposed that SecY has ten transmembrane segments that span the inner membrane (2, 3). The sequence alterations caused by a number of *secY* and *prlA* mutations are known (63, 123, 129, 134).

The SecY protein is required for proper functioning of in vitro translocation systems. Vesicles prepared from a *secYts* mutant are temperature-sensitive for translocation activity (5, 44). Antibodies against SecY will also inhibit the in vitro system (154), but interpretation of this experiment is complicated by the absence of a control for nonspecific inhibition by any antibody that binds the inner membrane. Two lines of biochemical evidence suggest an interaction between SecY and SecA. First, excess SecA will compensate for the defect in membranes from a *secYts* mutant (43). Second, functional SecY is necessary for activity of the SecA ATPase (85). Cells carrying a *secYts24* mutant are extremely sensitive to the presence of a MBP-β-galactosidase fusion or to overproduction of normal secreted proteins, consistent with a central role for

SecY in the export pathway (63). Further genetic evidence for the role of SecY (PrlA) in protein translocation is discussed below.

The finding of homologs of the *secY (prlA)* gene in both the gram-positive bacterium *Bacillus subtilis* and in the archaebacterium *Methanococcus vannielii* provide an evolutionary argument for its importance (4, 136). The similarity of *secY (prlA)* genes in these two divergent species extends to the location of the gene at the end of a ribosomal protein operon and to the predicted structure of the protein. Further work will be required to test for a role for these *secY* homologs in secretion.

secE (prlG)   Like *secY (prlA)*, *secE (prlG)* can mutate to a cold-sensitive, secretion-defective state or to suppress signal sequence mutations (115, 125, 135). It maps near a cluster of ribosomal and transcription genes and is the first gene in an operon with *nusG*, a gene for an antitermination factor (37).

The sequence of SecE predicts a 13.6-kd protein with three hydrophobic segments long enough to span a lipid bilayer (125). Alkaline phosphatase fusion analysis and cell fractionation confirms that SecE is an integral inner-membrane protein in *E. coli* (125). The SecE protein has a small region of sequence similarity to SecY, but the significance of this observation is unclear.

Several mutant alleles have been sequenced (P. J. Schatz, K. L. Bieker, T. J. Silhavy, J. Beckwith, in preparation). Interestingly, both of the known strong cold-sensitive alleles of *secE* cause base changes expected to decrease the efficiency of translation of the mutant mRNAs and not to alter the *secE* protein. The cold-sensitivity caused by these changes implies either that the cell requires more SecE at lower temperature or that translation of the mutant mRNA is cold sensitive. Three *prlG* alleles, on the other hand, make changes in the protein expected to cause disruptions in its normal structure, consistent with their dominant suppression ability.

secD AND *secF*   Mutations at the *secD* locus cause a cold-sensitive secretion defect (50, 115, 125). No *prl* type signal sequence suppressors occur at this locus. Originally thought to define a single gene, the *secD*-linked alleles in fact occur in two adjacent genes called *secD* and *secF* (C. Gardel, K. Johnson, A. Jacq, J. Beckwith, submitted), as shown by complementation experiments with derivatives of a *secDF* plasmid. Sequencing reveals that *secD* and *secF* genes code for 67-kd and 35-kd polypeptides, respectively, both of which have several potential membrane-spanning sequences (C. Gardel, K. Johnson, A. Jacq, J. Beckwith, submitted). Alkaline phosphatase fusion analysis indicates that they are integral inner-membrane proteins (K. Johnson, J. Beckwith, in preparation). SecD and SecF are similar to each other in sequence and predicted structure. Both have large periplasmic domains of unknown, but intriguing, function.

SIGNAL PEPTIDASES   The *lep* gene codes for a 36-kd protein required for the removal of signal peptides from most periplasmic and outer-membrane proteins in *E. coli*. The gene has been sequenced and the protein analyzed biochemically (97, 158). These studies, along with an alkaline phophatase fusion analysis (124), indicate that Lep is an integral inner-membrane protein that spans the membrane twice, with a large domain in the periplasm. The *lep* gene shares an operon with an upstream gene, *lepA*, which codes for a 76-kd protein of unknown function (90).

The *lspA* gene is necessary for the cleavage of signal peptides from a variety of lipoproteins. It codes for an 18-kd product that appears to be an integral membrane protein with four membrane-spanning stretches (56, 169). The *lspA* gene is downstream from *ileS*, the gene for isoleucine-tRNA-synthetase, in an operon with three other genes of unknown function (95).

HAS GENETICS ALREADY DEFINED THE EXPORT APPARATUS?

Genetics has permitted the identification of six *sec* genes. Several of these genes code for proteins that are directly involved in the export process. Recently developed selections are yielding mutations repeatedly in the same genes. These results raise the possibility that all of the genes involved in the process have been detected. The reasons why this might not be the case seem limited: (*a*) Mutations in genes that carry out similar functions to *secB* would only be detected if selections were done using fusions of the appropriate protein; (*b*) Small genes, presenting small targets for mutation, might easily be missed; (The fact that in the screening procedure using the *secA-lacZ* fusion, only one mutation was picked up in the *secE* gene, which codes for a 13.6-kd protein, illustrates this point). (*c*) Mutations in late steps of the export process might not give the phenotypes used in the selection and screening procedures developed so far. For instance, final steps in export may involve periplasmic proteins. This possibility is raised particularly by the finding that mutations in a gene coding for a protein thought to be in the lumen of the RER of yeast have a significant effect on secretion (see below). In the secretion process, the lumen of the RER is analogous to the bacterial periplasm.

Aside from these qualifications, it appears to us that most, if not all of the genes required for protein export have been identified. If we are correct, then the apparatus for protein export in *E. coli* may be considerably simpler than that defined in the mammalian in vitro system. First, the only truly soluble factors known are those involved in the folding of exported proteins. Of these, only SecB is clearly established by both in vivo and in vitro studies as a protein that plays a broad role in the export process. It is not clear whether the SecA protein function is carried out mainly at the membrane level, in the cytoplasm or as a factor that shuttles back and forth between the two compartments. The remaining proteins, SecD, SecE, SecF and PrlA/SecY, are all integral proteins of the cytoplasmic membrane. These proteins may constitute a complex corresponding to both a receptor and a pore or to a facilitator of translocation through the membrane.

PHILOSOPHY OF THE GENETIC APPROACH

Before the in vitro studies of the last several years, there was no direct evidence that any of the *sec* genes truly coded for a component of the export machinery. At various times, it has been suggested that they might code for enzymes involved in lipid biosynthesis, in membrane potential, or even in protein synthesis. What gives a geneticist the confidence to continue pursuing the approaches such as those decribed here without direct evidence for an important role of the genes? Several points are worth noting. First, the selections developed were quite specific. Direct selection for the alteration in location of a protein from the cell envelope to the cytoplasm is the most direct way of looking for mutations affecting the process of protein export. Given past experience, a geneticist has a fair degree of confidence that such selections will yield mutations in genes central to the process under study. Second, a large portion of the *E. coli* map is already defined. A number of the genes on this map code for enzymes of fatty acid and lipid biosynthesis. The fact that none of the *sec* genes were identical with previously identified *E. coli* genes, including these latter ones, suggested that a new pathway was being identified. Third, mutations in certain of the *sec* genes can lead to either a defect in secretion or to suppression of signal sequence mutations. The existence of mutations in the same gene with both positive and negative effects on secretion reinforces the presumption that these genes are central to the export process.

ANALYSIS OF FUNCTION AND PATHWAYS USING GENETICS

Once a number of genes involved in a given pathway are identified, a geneticist generally attempts to order the functions of the genes using epistasis tests. Epistasis analysis relies on drawing conclusions about pathways from the phenotypes of double mutants made by combining mutations that cause blocks at different steps. In the secretion field, this sort of analysis has been limited by the lack of good assays for various stages of the translocation process so that the specific "functions" of secretion gene products can be assigned. Nevertheless, several genetic experiments have revealed information about the function and order of action of various secretion genes.

As soon as two *sec* mutations were identified, they were combined to make a *secA, secB* double mutant. The synergistic effects apparent in the double mutant suggested that the genes might be involved in the same pathway (73). In addition, *secB* was epistatic to *secA* when analyzed for an effect of *secA* on synthesis of MBP (74). In a formal genetic sense, this result places *secB* before *secA* in a sequential biochemical pathway. Later genetic and biochemical evidence (see above) confirmed this hypothesized order of function.

An elegant series of genetic experiments has been done to analyze the function and interactions of the SecY (PrlA) and SecE (PrlG) proteins. Bieker & Silhavy (11) analyzed the component of the secretion machinery jammed by a LamB-β-galactosidase fusion protein. Since the fusion protein becomes stuck in a transmembrane orientation (112, 151), they reasoned that the SecY (PrlA) membrane protein was a likely site for the blockage. Consistent with this hypothesis, overproduction of SecY (PrlA) relieved the phenotype. To gain additional evidence for a direct interaction between the fusion protein and SecY (PrlA), they examined fusion proteins with defective signal sequences that do not jam the secretion machinery of a prlA$^+$ strain. When expressed in a strain with a *prlA4* signal sequence suppressor, however, the lethal effects of the fusion protein returned. This restored lethality was recessive in a *prlA$^+$/prlA4* diploid condition, implying that the defective fusion protein could block only the PrlA4 protein, leaving the wild-type PrlA to carry on export of normal proteins. This explanation was strongly supported by the specific blockage of export of an intact LamB protein (with a partially defective signal sequence) dependent on the suppressor in a *prlA$^+$/prlA4* strain expressing the fusion, at the same time that normal proteins were unaffected.

The ability to block export specifically mediated by a PrlA4 suppressor protein, while allowing wild-type PrlA to continue functioning, has been exploited to identify other proteins that may be part of the jammed membrane complex (K. L. Bieker & T. J. Silhavy, submitted). The approach relies on the hypothesis that proteins that are part of the jammed complex should be titrated by it. If they are present at functional excess, no effect will be observed. If their functional levels are lowered, the titration could cause a return of a secretion defect as the jammed complex absorbs the functional subunits. Experiments to test this hypothesis, using *secA* and *secE* mutations to lower levels of functional protein, indicate that SecE is probably part of the jammed complex but that SecA is not. A jammed complex analogous to the one involving PrlA4 can be produced at PrlG1, but with an interesting difference. While the protein jammed at PrlA4 is processed by leader pepti-

dase, the same protein jammed at PrlG1 is not. Interpretation of this difference in terms of a molecular pathway of protein translocation is still difficult,
but the evidence for an interaction between SecY (PrlA) and SecE (PrlG)
seems fairly strong. Additional evidence for a SecE - SecY interaction comes
from the observation of allele-specific synthetic lethality between *prlA4* and
prlG1 (K. L. Bieker & T. J. Silhavy, submitted) and the biochemical
identification of a SecY-SecE protein complex (L. Brundage, J. Hendrick, E.
Schiebel, A. Driessen, & W. Wickner, personal communication).

One recurring theme of genetic analysis of complex systems such as
secretion concerns the interpretation of genetic interactions in terms of functional interactions between gene products. In the case cited above, a *prlA4*
prlG1 double mutant is dead. While death is clearly a strong phenotype, what
does it tell us about the interaction between the PrlA and PrlG proteins? Some
relevant examples come to mind. For instance, *polA recA* double mutants of
E. coli are dead (96). Extensive biochemical characterization of these two
proteins, however has not revealed a direct physical interaction between
them. Both are involved in DNA metabolism, but to interpret the lethality
caused by a double mutation in terms of a protein-protein interaction is clearly
not warranted. A similar example concerns several recently isolated allleles of
secE, originally identified as cold-sensitive lethals. Further characterization
showed that they only caused cold-sensitive growth when the tetracycline-
resistance protein was expressed at the same time (125). Both gene products
are integral membrane proteins, but again, the cause of the interaction is
unclear. In the *prlA4 prlG1* case other lines of evidence discussed above
demonstrate an interaction. In addition the lethality is allele-specific. Eight
other combinations of mutations did not cause a problem (K. L. Bieker & T.
J. Silhavy, submitted). Although the concept of allele-specificity has been
overused in the secretion field (see discussion by Randall et al; 111), in this
case it is probably meaningful. Sequencing of these two mutations in fact
reveals that both produce changes in the outer half of membrane-spanning
stretches of these two proteins (123, 134; P. J. Schatz, K. L. Bieker, T. J.
Silhavy, J. Beckwith, in preparation). The *prlA4 prlG1* synthetic lethality is
probably due to direct physical contact between the proteins. The distinguishing feature of these various examples is not the strength of the initial genetic
interaction, but the additional genetic and biochemical evidence that exists in
each case.

REGULATION OF THE SECRETION MACHINERY

E. coli has evolved a means of responding to conditions that interfere with the
protein export process. When protein export is blocked by various means, the

secA gene is depressed at least tenfold (101). This depression is seen at permissive and nonpermissive temperatures with mutations in all *sec* genes except for *secB* (115, 116). This regulatory mechanism may allow the bacteria to compensate for the export block by increasing the amount of a rate-limiting component of the translocation machinery. Certain of the other *sec* genes have been examined for their response to a block in secretion. The *secE* and the *prlA (secY)* genes are apparently not derepressed by a secretion block (P. Schatz & J. Beckwith, unpublished results; Y. Akiyama, C. Ueguchi, K. Ito, unpublished results; T. Silhavy, personal communication). In contrast, the *secD* operon is regulated. When a transcriptional *secD-lacZ* fusion is combined with a *secAts* mutation, the amount of *lacZ* expression increases about tenfold at the nonpermissive temperature (C. Gardel & J. Beckwith, unpublished results). However, other *sec* mutations that block secretion do not affect *lacZ* expression in this fusion.

Studies on the mechanism of *secA* regulation (126; M. G. Schmidt & D. B. Oliver, personal commuication) indicate that the regulation operates at a posttranscriptional level and is mediated by a site between the *X* and *secA* genes. It has been suggested that the SecA protein may act as a translational repressor of its own synthesis. When the secretion process is blocked, the SecA protein might become fully engaged with the secretion machinery, thus causing a relief of the translational repression. It is curious that the regulation of the *secD* operon appears to be at the transcriptional level and is only seen in a *secA* mutant. The mechanism of regulation appears to be different in this case.

The regulation of *secA* may explain one puzzling phenomenon associated with *secA* mutants. A large number of *secAts* mutants were tested for the rapidity with which secretion shuts down after a shift to the nonpermissive temperature of 42°. In all cases, it took 2–3 hours before secretion was fully blocked. The derepression of *secA* that occurs after the temperature-shift might increase the concentration of SecA sufficiently to overcome the secretion defect temporarily.

Most of the secretion genes described above are in operons. *secA*, *secE*, *secY*, *lep* and *lspA* are in operons with seemingly unrelated genes, including factors involved in translation, transcription and DNA replication. So far, only *secD* and *secF* are part of a "secretion operon." Since these genes probably are required under all conditions of cell growth, there may be less need for tight negative regulations. The secretion genes could be considered to be part of the housekeeping apparatus for localization of routinely synthesized proteins. As such, they may be regulated by global regulatory signals that respond to growth rate or starvation. Such controls would be in addition to the depression of the *secA* gene and the *secDF* operon seen when secretion is blocked.

GENETICS OF THE YEAST SECRETION MACHINERY

Two approaches have been used to obtain mutations affecting the secretory machinery of *Saccharomyces cerevisiae*. In the first, mutant cells are sought that increase the density of cells in a Ludox gradient because of the internalization of cell envelope proteins (99). This enrichment procedure detects mutants all along the complex secretion pathway. Mutations have been found that block transfer of proteins between various secretory organelles. As a result, this very productive selection is overwhelmed by mutants in steps after entry into the rough endoplasmic reticulum. No mutations were detected that altered components of the machinery involved in the early steps of secretion, other than a glycosylation mutant.

A second approach is based on the selection developed in *E. coli* using gene fusions. In the yeast, the analogous selection to the internalization of β-galactosidase demanded the internalization of histidinol dehydrogenase (35). Strains were constructed so that the cells could only use histidinol as a histidine source when a hybrid protein containing histidinol dehydrogenase fused to a signal sequence was internalized to the cytoplasm. This selection has yielded mutations in three genes, *SEC61, SEC62,* and *SEC63* (36, 118). Curiously, mutations in the *SEC63* gene also alter localization of proteins to the nucleus (122). All three of these genes code for membrane proteins and one, *SEC61,* codes for a protein sharing extensive sequence similarity with the *E. coli prlA/secY* product (C. Sterling & R. Schekman, personal communication). The fact that the selections have revealed genes coding only for membrane proteins raises the possibility that the yeast-secretion machinery is also simpler than that found in mammalian cells. Since signal recognition particle contains six proteins and a 7S RNA molecule, there should have been significant genetic targets for mutagenesis in this selection, if yeast has an SRP homolog. On the other hand, a gene coding for a homolog of one of the mammalian SRP subunits has been detected, although its function is not known (53).

In another analogy to the *E. coli* system, factors that maintain proteins in a secretion-competent state have been implicated in the process in yeast (24, 34). Mutations that result in depletion of heat-shock proteins of the 70 kd family cause accumulation of precursors of exported proteins. They also interfere with the import of proteins into the mitochondria.

Another gene that plays a role in secretion in *S. cerevisiae* is *KAR2* (148). The *KAR2* gene codes for a protein that is highly homologous to the BiP protein found in the lumen of the endoplasmic reticulum of mammalian cells. BiP is thought to act within the lumen of the RER to facilitate the folding of multimeric proteins and to scavenge misfolded proteins. *KAR2* mutations have pleiotropic effects, blocking secretion of proteins into the RER as well as blocking nuclear fusion. This is the first example of a protein which is required for the translocation process being located on the opposite side of the membrane from where translocation is initiated.

A WORKING MODEL FOR PROTEIN EXPORT

The evidence cited above leads to the following model for protein export in *E. coli*. As the precursor protein emerges from the ribosome, or soon thereafter, SecB or another chaperone protein may bind to it to prevent folding into an export-incompetent conformation. The dependence of various precursor proteins on a chaperone varies widely due to differences in their folding properties. Some may not require chaperones at all. The precursor-SecB complex then interacts with SecA either in the cytoplasm or on the inner surface of the membrane. The speed with which this interaction takes place determines whether export is cotranslational or posttranslational, but not whether it occurs. The success of the export reaction probably depends most critically on the export-competent conformation of the precursor. It remains to be determined whether SecB and other chaperones directly interact with SecA or function only to maintain proper folding of the precursor. The signal sequence probably has two roles in these early steps. First the signal sequence of some proteins, for example MBP, retards folding sufficiently to allow SecB to bind. Second, the signal sequence interacts with SecA and other components in the membrane to initiate the translocation reaction.

The component in the membrane with which SecA and the precursor interact is likely to be an integral membrane complex containing SecY and SecE. In a reaction mediated by the signal sequence, SecA and the membrane complex initiate translocation. The proposed interaction between SecA and SecY is supported by both genetic and biochemical evidence cited above. A complex of SecY and SecE is suggested by the titration experiments, the synthetic lethal interaction, and the biochemical experiments described previously. Later steps in translocation probably involve the SecD and SecF proteins. There are two reason for placing these two proteins later in the pathway. First, no suppressors of defective signal sequences have been detected in these genes. Thus, mutations in these genes cannot easily affect the interaction of the signal sequence with the translocation apparatus. Second, both have large periplasmic domains (K. Johnson, J. Beckwith, in preparation) that may function in late steps in translocation or in release of the cleaved precursor into the periplasm. The signal peptidases also act late in translocation to remove the signal sequence and allow release of exported proteins from the inner membrane.

Now that many of the components of the secretion apparatus have been identified in *E. coli*, future work should concentrate on genetic and biochemical analysis of details of the molecular mechanism by which they catalyze the passage of proteins through the membrane.

ACKNOWLEDGMENTS

We thank the many colleagues who communicated results prior to publication and we thank the members of the lab for critically reading the manuscript. J. B. was supported by a grant from the National Institutes of Health. P. J. S. was supported by a Damon Runyon-Walter Winchell Cancer Research Fund Fellowship (DRG-985).

Literature Cited

1. Akiyama, Y., Ito, K. 1985. The SecY membrane component of the bacterial protein export machinery: Analysis by new electrophoretic methods for integral membrane proteins. *EMBO J.* 4:3351–56

2. Akiyama, Y., Ito, K. 1987. Topology analysis of the SecY protein, an integral membrane protein involved in protein export in *Escherichia coli*. *EMBO J.* 6:3465–70

3. Akiyama, Y., Ito, K. 1989. Export of *Escherichia coli* alkaline phosphatase attached to an integral membrane protein, SecY. *J. Biol. Chem.* 264:437–42

4. Auer, J., Lechner, K., Bock, A. 1989. Gene organization and structure of two transcriptional units from *Methanococcus* coding for ribosomal proteins and elongation factors. *Can. J. Microbiol.* 35:200–4

5. Bacallao, R., Crooke, E., Shiba, K., Wickner, W., Ito, K. 1986. The secY protein can act post-translationally to promote bacterial protein export. *J. Biol. Chem.* 261:12907–10

6. Bankaitis, V. A., Bassford, P. J. Jr. 1985. Proper interaction between at least two components is required for efficient export of proteins to the *Escherichia coli* cell envelope. *J. Bacteriol.* 161:169–78

7. Beall, B., Lutkenhaus, J. 1987. Sequence analysis, transcriptional organization, and insertional mutagenesis of the *envA* gene of *Escherichia coli*. *J. Bacteriol.* 169:5408–15

8. Bedouelle, H., Bassford, P. J. Jr., Fowler, A. V., Zabin, I., Beckwith, J., et al. 1980. The nature of mutational alterations in the signal sequence of the maltose binding protein of *Escherichia coli*. *Nature* 285:78–81

9. Benson, S. A., Hall, M. N., Rasmussen, B. A. 1987. Signal sequence muta-tions that alter coupling of secretion and translation of an *Escherichia coli* outer membrane protein. *J. Bacteriol.* 169:4686–91

10. Bernstein, H. D., Poritz, M. A., Strub, K., Hoben, P. J., Brenner, S., et al. 1989. Model for signal sequence recognition from amino-acid sequence of 54K subunit of signal recognition particle. *Nature* 340:482–86

11. Bieker, K. L., Silhavy, T. J. 1989. PrlA is important for the translocation of exported proteins across the cytoplasmic membrane of *Escherichia coli*. *Proc. Natl. Acad. Sci. USA* 86:968–72

12. Blobel, G., Dobberstein, B. 1975. Transfer of proteins across membranes, I. Presence of proteolytically processed and unprocessed nascent immunoglobulin light chains on membrane-bound ribosomes of murine myeloma. *J. Cell Biol.* 67:835–51

13. Bochkareva, E. S., Lissin, N. M., Girshovich, A. S. 1988. Transient association of newly synthesized unfolded proteins with the heat-shock GroEL protein. *Nature* 336:254–57

14. Bradshaw, R. A., Cancedda, F., Ericsson, L. H., Newmann, P. A., Piccoli, S. P., et al. 1981. Amino acid sequence of *Escherichia coli* alkaline phosphatase. *Proc. Natl. Acad. Sci. USA* 78:3473–78

15. Brickman, E. R., Oliver, D. B., Garwin, J. L., Kumamoto, C. A., Beckwith, J. 1984. The use of extragenic suppressors to define genes involved in protein export in *Escherichia coli*. *Mol. Gen. Genet.* 196:24–27

16. Brown, S. 1987. Mutations in the gene for EF-G reduce the requirement for 4.5S RNA in the growth of *E. coli*. *Cell* 49:825–33

17. Cabelli, R. J., Chen, L., Tai, P. C., Oliver, D. B. 1988. SecA protein is required for secretory protein translocation

into *E. coli* membrane vesicles. *Cell* 55:683–92

18. Cerretti, D. P., Dean, D., Davis, G. R., Bedwell, D. M., Nomura, M. 1983. The *spc* ribosomal protein operon of *Escherichia coli*: sequence and cotranscription of the ribosomal protein genes and a protein export gene. *Nucleic Acids Res.* 11:2599–16

19. Chen, L., Rhoads, D., Tai, P. C. 1985. Alkaline phosphatase and OmpA protein can be translocated posttranslationally into membrane vesicles of *Escherichia coli*. *J. Bacteriol.* 161:973–80

20. Chen, L., Tai, P. C. 1985. ATP is essential for protein translocation into *Escherichia coli* membrane vesicles. *Proc. Natl. Acad. Sci. USA* 82:4384–88

21. Chen, L., Tai, P. C. 1986. Effects of nucleotides on ATP-dependent protein translocation into *Escherichia coli* membrane vesicles. *J. Bacteriol.* 168:828–32

22. Chen, L., Tai, P. C. 1986. Roles of H^+-ATPase and proton motive force in ATP dependent protein translocation in vitro. *J. Bacteriol.* 167:389–92

23. Chen, L., Tai, P. C. 1987. Evidence for the involvement of ATP in co-translational protein translocation. *Nature* 328:164–66

24. Chirico, W. J., Waters, M. G., Blobel, G. 1988. 70K heat shock related proteins stiumlate protein translocation into microsomes. *Nature* 332:805–10

25. Collier, D. N., Bankaitis, V. A., Weiss, J. B., Bassford, P. J. Jr. 1988. The antifolding activity of *secB* promotes the export of the *E. Coli* maltose-binding protein. *Cell* 53:273–83

26. Collier, D. N., Bassford, P. J. Jr. 1989. Mutations that improve export of maltose-binding protein in secB⁻ cells of *Escherichia coli*. *J. Bacteriol.* 171:4640–47

27. Cover, W. H., Ryan, J. P., Bassford, P. J. Jr., Walsh, K. A., Bolinger, J., et al. 1987. Suppression of a signal sequence mutation by an amino acid substitution in the mature portion of the maltose-binding protein. *J. Bacteriol.* 169:1794–800

28. Crooke, E., Wickner, W. 1987. Trigger factor: a soluble protein that folds proOmpA into a membrane-assembly-competent form. *Proc. Natl. Acad. Sci. USA* 84:5216–20

29. Cunningham, K., Lill, R., Crooke, E., Rice, M., Moore, K., et al. 1989. SecA protein, a peripheral membrane protein of the *Escherichia coli* plasma membrane, is essential for the functional binding and translocation of proOmpA. *EMBO J.* 8:955–59

30. Cunningham, K., Wickner, W. 1989. Specific recognition of the leader region of precursor proteins is required for the activation of the translocation ATPase of *Escherichia coli*. *Proc. Natl. Acad. Sci. USA* 86:8630–34

31. Dalbey, R. E., Wickner, W. 1985. Leader peptidase catalyzes the release of exported proteins from the outer surface of the *Escherichia coli* plasma membrane. *J. Biol. Chem.* 260:15925–31

32. Date, T. 1983. Demonstration by a novel genetic technique that leader peptidase is an essential enzyme of *Escherichia coli*. *J. Bacteriol.* 154:76–83

33. Date, T., Wickner, W. 1981. Isolation of the *Escherichia coli* leader peptidase gene and effects of leader peptidase overproduction in vivo. *Proc. Natl. Acad. Sci. USA* 78:6106–10

34. Deshaies, R. J., Koch, B. D., Werner-Washburn, M., Craig, E. A., Schekman, R. 1988. A subfamily of stress facilitates translocation of secretory and mitochondrial precursor polypeptides. *Nature* 332:800–5

35. Deshaies, R. J., Schekman, R. 1987. A yeast mutant defective at an early stage in import of secretory protein precursors into the endoplasmic reticulum. *J Cell Biol.* 105:633–45

36. Deshaies, R. J., Schekman, R. 1989. *SEC62* encodes a putative membrane protein required for protein translocation into the yeast endoplasmic reticulum. *J. Cell Biol.* 109:2653–64

37. Downing, W. L., Sullivan, S. L., Gottesman, M. E., Dennis, P. P. 1990. Sequence and transcriptional pattern of the essential *Escherichia coli secE-nusG* operon. *J. Bacteriol.* 172:1621–27

38. Eilers, M., Schatz, G. 1988. Protein unfolding and the energetics of protein translocation across biological membranes. *Cell* 52:481–83

39. Ellis, J. 1987. Proteins as molecular chaperones. *Nature* 328:378–79

40. Emr, S. D., Bassford, P. J. Jr. 1982. Localization and processing of outer membrane and periplasmic proteins in *Escherichia coli* strains harboring export specific suppressor mutations. *J. Biol. Chem.* 257:5852–60

41. Emr, S. D., Hanley-Way, S., Silhavy, T. J. 1981. Suppressor mutations that restore export of a protein with a defective signal sequence. *Cell* 23:79–88

42. Emr, S. D., Hedgpeth, J., Clement, J.-M., Silhavy, T. J., Hofnung, M. 1980. Sequence analysis of mutations that prevent export of lambda receptor, an *Escherichia coli* outer membrane protein. *Nature* 285:82–85

43. Fandl, J. P., Cabelli, R., Oliver, D, Tai, P. C. 1988. SecA suppresses the temperature-sensitive secY24 defect in protein translocation in *Escherichia coli* membrane vesicles. *Proc. Natl. Acad. Sci. USA* 85:8953–57

44. Fandl, J. P., Tai, P. C. 1987. Biochemical evidence for the secY24 defect in *Escherichia coli* protein translocation and its suppression by soluble cytoplasmic factors. *Proc. Natl. Acad. Sci. USA* 84:7448–52

45. Ferro-Novick, S., Honma, M., Beckwith, J. 1984. The product of gene secC is involved in the synthesis of exported proteins in *Escherichia coli*. *Cell* 38:211–17

46. Fikes, J. D., Bankaitis, V. A., Ryan, J. P., Bassford, P. J. Jr. 1987. Mutational alterations affecting the export competence of a truncated but fully functional maltose-binding protein signal peptide. *J. Bacteriol.* 169:2345–51

47. Fikes, J. D., Bassford, P. J. Jr. 1987. Export of unprocessed precursor maltose-binding protein to the periplasm of *Escherichia coli* cells. *J. Bacteriol.* 169:2352–59

48. Fikes, J. D., Bassford, P. J. Jr. 1989. Novel secA alleles improve export of maltose-binding protein synthesized with a defective signal peptide. *J. Bacteriol.* 171:402–09

49. Gannon, P. M., Li, P., Kumamoto, C. A. 1989. The mature portion of *Escherichia coli* maltose-binding protein (MBP) determines the dependence of MBP on SecB for export. *J. Bacteriol.* 171:813–18

50. Gardel, C., Benson, S. A., Hunt, J., Michaelis, S., Beckwith, J. 1987. secD, a new gene involved in protein export in *Escherichia coli*. *J. Bacteriol.* 169:1286–90

51. Goldstein, J., Lehnhardt, S., Inouye, M. 1990. Enhancement of protein translocation across the membrane by specific mutations in the hydrophobic region of the signal peptide. *J. Bacteriol.* 172:1225–31

52. Hall, M. N., Gabay, J., Schwartz, M. 1983. Evidence for a coupling of synthesis and export of an outer membrane protein in *Escherichia coli*. *EMBO J.* 2:15–19

53. Hann, B. C., Poritz, M. A., Walter, P. 1989. *Saccharomyces cerevisiae* and *Schizosaccharomyces pombe* contain a homologue to the 54-kD subunit of the signal recognition particle that in *S. cerevisiae* is essential for growth. *J. Cell Biol.* 109:3223–30

54. Iino, T., Sako, T. 1988. Inhibition and resumption of processing of the stapholkinase in some *Escherichia coli prlA* suppressor mutants. *J. Biol. Chem.* 263:19077–82

55. Inada, T., Court, D. L., Ito, K., Nakamura, Y. 1989. Conditionally lethal amber mutations in the leader peptidase gene of *Escherichia coli*. *J. Bacteriol.* 171:585–87

56. Innis, M. A., Tokunaga, M., Williams, M. E., Loranger, J. M., Chang, S.-Y., et al. 1984. Nucleotide sequence of the *Escherichia coli* prolipoprotein signal peptidase (*lsp*) gene *Proc. Natl. Acad. Sci. USA* 81:3708–12

57. Inouye, H., Beckwith, J. 1977. Synthesis and processing of an *E. Coli* alkaline phosphatase precursor in vitro. *Proc. Natl. Acad. Sci. USA* 74:1440–44

58. Inouye, S., Soberon, X., Franceschini, T., Nakamura, K., Inouye, M. 1982. Role of positive charge on the amino-terminal region of the signal peptide in protein secretion across the membrane. *Proc. Natl. Acad. Sci. USA* 79:3138–41

59. Inouye, S., Wang, S., Sekizawa, J., Halegoua, S., Inouye, M. 1977. Amino acid sequence for the peptide extension on the prolipoprotein of the *Escherichia coli* outer membrane. *Proc. Natl. Acad. Sci. USA* 74:1004–8

60. Ito, K. 1984. Identification of the secY (prlA) gene product involved in protein export in *Escherichia coli*. *Mol. Gen. Genet.* 197:204–8

61. Ito, K., Bassford, P. J. Jr., Beckwith, J. 1981. Protein localization in *E. coli*. Is there a common step in the secretion of periplasmic and outer membrane proteins? *Cell* 24:707–14

62. Ito, K., Cerretti, D. P., Nashimoto, H., Nomura, M. 1984. Characterization of an amber mutation in the structural gene for ribosomal protein L15, which impairs the expression of the protein export gene, secY, in *Escherichia coli*. *EMBO J.* 3:2319–24

63. Ito, K., Hirota, Y., Akiyama, Y. 1989. Temperature-sensitive sec mutants of *Escherichia coli*: inhibition of protein export at the permissive temperature. *J. Bacteriol.* 171:1742–43

64. Ito, K., Wittekind, M., Nomura, M., Shiba, K., Yura, T., et al. 1983. A temperature-sensitive mutant of *E. coli* exhibiting slow processing of exported proteins. *Cell* 32:789–97

65. Jarvik, J., Botstein, D. 1975. Conditional-lethal mutations that suppress genetic defects in morphogenesis by altering structural proteins. *Proc. Natl. Acad. Sci. USA* 72:2738–42

66. Kaiser, C. A., Preuss, D., Grisafi, P.,

Botstein, D. 1987. Many random sequences functionally replace the secretion signal sequence of yeast invertase. *Science* 235:312–17

67. Kawasaki, H., Matsuyama, S.-I., Sasaki, S., Akita, M., Mizushima, S. 1989. SecA protein is directly involved in protein secretion in *Escherichia coli*. *FEBS Lett.* 242:431–34

68. Kiino, D. R., Phillips, G. J., Silhavy, T. J. 1990. Increased expression of the bifunctional protein PrlF suppresses overproduction lethality associated with exported β-galactosidase hybrid proteins in *Escherichia coli*. *J. Bacteriol.* 172:185–92

69. Kiino, D. R., Silhavy, T. J. 1984. Mutation prlF1 relieves the lethality associated with export of β-galactosidase hybrid proteins in *Escherichia coli*. *J. Bacteriol.* 158:878–83

70. Koshland, D., Sauer, R. T., Botstein, D. 1981. Diverse effects of mutations in the signal sequence on the secretion of β-lactamase in *Salmonella typhimurium*. *Cell* 30:903–14

71. Kuhn, A., Wickner, W. 1985. Conserved residues of the leader peptide are essential for cleavage by leader peptidase. *J. Biol. Chem.* 260:15914–18

72. Kumamoto, C. A. 1989. *Escherichia coli* SecB protein associates with exported protein precursors in vivo. *Proc. Natl. Acad. Sci. USA* 86:5320–24

73. Kumamoto, C. A., Beckwith, J. 1983. Mutations in a new gene, *secB*, cause defective protein localization in *Escherichia coli*. *J. Bacteriol.* 154:254–60

74. Kumamoto, C. A., Beckwith, J. 1985. Evidence for specificity at an early step in protein export in *Escherichia coli*. *J. Bacteriol.* 163:267–74

75. Kumamoto, C. A., Chen, L., Fandl, J., Tai, P. C. 1989. Purification of the *Escherichia coli secB* gene product and demonstration of its activity in an in vitro protein translocation system. *J. Biol. Chem.* 264:2242–49

76. Kumamoto, C. A., Gannon, P. M. 1989. Effects of *Escherichia coli secB* mutations on pre-maltose binding protein conformation and export kinetics. *J. Boil. Chem.* 263:11554–58

77. Kumamoto, C. A., Nault, A. K. 1989. Characterization of the *Escherichia coli* protein-export gene *secB*. *Gene* 75:167–75

78. Kusukawa, N., Yura, T., Ueguchi, C., Akiyama, Y., Ito, K. 1989. Effects of mutations in heat-shock genes *groES* and *groEL* on protein export in *Escherichia coli*. *EMBO J.* 8:3517–21

79. Lecker, S., Lill, R., Ziegelhoffer, T.,

Georgopoulos, C., Bassford, P. J. Jr., et al. 1989. Three pure chaperone proteins of *Escherichia coli*—SecB, trigger factor and GroEL—form soluble complexes with precursor proteins in vitro. *EMBO J.* 8:2703–9

80. Lederberg, J. 1950. The selection of genetic recombinations with bacterial growth inhibitors. *J. Bacteriol.* 59:211–15

81. Lee, C. A., Beckwith, J. 1986. Suppression of growth and protein secretion defects in *Escherichia coli secA* mutants by decreasing protein synthesis. *J. Bacteriol.* 166:878–83

82. Lee, C., Li, P., Inouye, H., Brickman, E. R., Beckwith, J. 1989. Genetic studies on the ability of β-galactosidase to be translocated across the *E. coli* cytoplasmic membrane. *J. Bacteriol.* 171:4609–16

83. Lee, C. A., Fournier, M. J., Beckwith, J. 1985. The 6S RNA of *Escherichia coli* is not essential for growth or protein secretion. *J. Bacteriol.* 161:1156–61

84. Li, P., Beckwith, J., Inouye, H. 1988. Alteration of the amino-terminus of the mature sequence of a periplasmic protein can severely effect protein export in *Escherichia coli*. *Proc. Natl. Acad. Sci. USA* 85:7685–89

85. Lill, R., Cunningham, K., Brundage, L. A., Ito, K., Oliver, D., et al. 1989. SecA protein hydrolyzes ATP and is an essential component of the protein translocation ATPase of *Escherichia coli*. *EMBO J.* 8:961–66

86. Liu, G., Topping, T. B., Cover, W. H., Randall, L. L. 1988. Retardation of folding as a possible means of suppression of a mutation in the leader sequence of an exported protein. *J. Biol. Chem.* 263:14790–93

87. Liu, G., Topping, T. B., Randall, L. L. 1989. Physiological role during export for the retardation of folding by the leader peptide of maltose-binding protein. *Proc. Natl. Acad. Sci. USA* 86:9213–17

88. Manoil, C., Beckwith, J. 1985. TnphoA: A transposon probe for protein export signals. *Proc. Natl. Acad. Sci. USA* 82:8129–33

89. Manoil, C., Beckwith, J. 1986. A genetic approach to analyzing membrane protein topology. *Science* 233:1403–8

90. March, P. E., Inouye, M. 1985. Characterization of the *lep* operon of *Escherichia coli*. *J. Biol. Chem.* 260:7206–13

91. Michaelis, S., Beckwith, J. 1982. Mechanism of incorporation of cell envelope proteins in *Escherichia coli*. *Annu. Rev. Microbiol.* 36:435–65

92. Michaelis, S., Guarente, L., Beckwith, J. 1983. In vitro construction and characterization of *phoA-lacZ* gene fusions in *Escherichia coli*. *J. Bacteriol.* 154:356–65

93. Michaelis, S., Hunt, J., Beckwith, J. 1986. Effects of signal sequence mutations on the kinetics of alkaline phosphatase export to the periplasm in *Escherichia coli*. *J. Bacteriol.* 167:160–67

94. Michaelis, S., Inouye, H., Oliver, D., Beckwith, J. 1983. Mutations that alter the signal sequence of alkaline phosphatase in *Escherichia coli*. *J. Bacteriol.* 154:366–74

95. Miller, K. W., Bouvier, J., Stragier, P., Wu, H. C. 1987. Identification of the genes in the *Escherichia coli iles-lsp* operon. *J. Biol. Chem.* 262:7391–97

96. Monk, M., Kinross, J. 1972. Conditional lethality of *recA* and *recB* derivatives of a strain of *Escherichia coli* K-12 with a temperature-sensitive deoxyribonucleic acid polymerase I. *J. Bacteriol.* 109:971–78

97. Moore, K. E., Miura, S. 1987. A small hydrophobic domain anchors leader peptidase to the cytoplasmic membrane of *Escherichia coli*. *J. Biol. Chem.* 262: 8806–13

98. Muller, M., Blobel, G. 1984. In vitro translocation of bacterial proteins across the plasma membrane of *Escherichia coli*. *Proc. Natl. Acad. Sci. USA* 81: 7421–25

99. Novick, P., Field, C., Schekman, R. 1980. Identification of 23 complementation groups required for posttranslational events in the yeast secretory pathway. *cell* 21:205–15

100. Oliver, D. B. 1985. Identification of five new essential genes involved in the synthesis of a secreted protein in *Escherichia coli*. *J. Bacteriol.* 161:285–91

101. Oliver, D. B., Beckwith, J. 1981. *E. coli* mutant pleiotropically defective in the export of secreted proteins. *Cell* 25:2765–72

102. Oliver, D. B., Beckwith, J. 1982. Regulation of a membrane component required for protein secretion in *Escherichia coli*. *Cell* 30:311–19

103. Oliver, D. B., Beckwith, J. 1982. Identification of a new gene (*secA*) and gene product involved in the secretion of envelope proteins in *Escherichia coli*. *J. Bacteriol.* 150:686–91

104. Oliver, D. B., Liss, L. R. 1985. *prlA*-mediated suppression of signal sequence mutations is modulated by the *secA* gene product of *Escherichia coli* K-12. *J. Bacteriol.* 161:817–19

105. Park, S., Liu, G., Topping, T. B.,

Cover, W. H., Randall, L. L. 1988. Modulation of folding pathways of exported proteins by the leader sequence. *Science* 239:1033–35

106. Pollitt, S., Inouye, S., Inouye, M. 1986. Effect of amino acid substitutions at the signal peptide cleavage site of the *Escherichia coli* major outer membrane lipoprotein. *J. Biol. Chem.* 261:1835–37

107. Poritz, M. A., Strub, K., Walter, P. 1988. Human SRP RNA and *E. coli* 4.5S RNA contain a highly homologous structural domain. *Cell* 55:4–6

108. Puziss, J. W., Fikes, J. D., Bassford, P. J. Jr. 1989. Analysis of mutational alterations in the hydrophilic segment of the maltose-binding protein signal peptide. *J. Bacteriol.* 171:2303–11

109. Randall, L. L., Hardy, S. J. S. 1986. Correlation of competence for export with lack of tertiary structure of the mature species: a study in vivo of maltose-binding protein in *E. coli*. *Cell* 46:921–28

110. Randall, L. L., Hardy, S. J. S. 1989. Unity of function in the absence of consensus in sequence: role of leader peptides in export. *Science* 243:1156–59

111. Randall, L. L., Hardy, S. J. S., Thom, J. R. 1987. Export of protein: a biochemical view. *Annu. Rev. Microbiol.* 41:507–41

112. Rasmussen, B. A., Bankaitis, V. A., Bassford, P. J. Jr. 1984. Export and processing of MalE-LacZ hybrid proteins in *Escherichia coli*. *J. Bacteriol.* 160:612–17

113. Regue, M., Remenick, J., Tokunaga, M., Mackie, G. A., Wu, H. C. 1984. Mapping of the lipoprotein signal peptidase gene (*lsp*). *J. Bacteriol.* 158:632–35

114. Rhoads, D. B., Tai, P. C., Davis, B. D. 1984. Energy-requiring translocation of the OmpA protein and alkaline phosphatase of *Escherichia coli* into inner membrane vesicles. *J. Bacteriol.* 159:63–70

115. Riggs, P. D., Derman, A. I., Beckwith, J. 1988. A mutation affecting the regulation of a *secA-lacZ* fusion defines a new *sec* gene. *Genetics* 118:571–79

116. Rollo, E. R., Oliver, D. B. 1988. Regulation of the *Escherichia coli secA* gene by protein secretion defects: analysis of *secA*, *secB*, *secD*, and *secY* mutants. *J. Bacteriol.* 170:3281–82

117. Romisch, K., Webb, J., Herz, J., Prehn, S., Frank, R., et al. 1989. Homology of 54K protein of signal-recognition particle, docking protein and two *E. coli* proteins with putative GTP-binding proteins. *Nature* 340:478–81

118. Rothblatt, J. A., Deshaies, R. J., Sanders, S. L., Daum, G., Schekman, R. 1989. Multiple genes are required for proper insertion of secretory proteins into the endoplasmic reticulum in yeast. *J. Cell Biol.* 109:2641–52

119. Rothman, J. E. 1989. Polypeptide chain binding proteins: catalysts of protein folding and related processes in cells. *Cell* 59:591–601

120. Rousset, J.-P., Gilson, E., Hofnung, M. 1986. *malM*, a new gene of the maltose regulon in *Escherichia coli* K12. II. Mutations affecting the signal peptide of the MalM protein *J. Mol. Biol.* 191: 313–20

121. Ryan, J. P., Bassford, P. J. Jr. 1985. Post-translational export of maltose-binding protein in *Escherichia coli* strains harboring *malE* signal sequence mutations and either *prl*+ or *prl* suppressor alleles. *J. Biol. Chem.* 260:14832–37

122. Sadler, I., Chiang, A., Kurihara, T., Rothblatt, J., Way, J., et al. 1989. A yeast gene important for protein assembly into the endoplasmic reticulum and the nucleus has homology to DnaJ, an *Escherichia coli* heat shock protein. *J. Cell Biol.* 109:2665–75

123. Sako, T., Iino, T. 1988. Distinct mutation sites in *prlA* suppressor mutant strains of *Escherichia coli* respond either to suppression of signal peptide mutations or to blockage of staphylokinase processing. *J. Bacteriol.* 170:5389–91

124. San Millan, J. L., Boyd, D., Dalbey, R., Wickner, W., Beckwith, J. 1989. Use of *phoA* fusions to study the topology of the *Escherichia coli* inner membrane protein leader peptidase. *J. Bacteriol.* 171:5536–5541

125. Schatz, P. J., Riggs, P. D., Jacq, A., Fath, M. J., Beckwith, J. 1989. The *secE* gene encodes an integral membrane protein required for protein export in *E. coli*. *Genes Dev.* 3:1035–44

126. Schmidt, M. G., Oliver, D. B. 1989. SecA protein autogenously represses its own translation during normal protein secretion in *Escherichia coli*. *J. Bacteriol.* 171:643–49

127. Schmidt, M. G., Rollo, E. E., Grodberg, J., Oliver, D. B. 1988. Nucleotide sequence of the *secA* gene and *secA(Ts)* mutations preventing protein export in *Escherichia coli*. *J. Bacteriol.* 170: 3404–14

128. Shiba, K., Ito, K., Yura, T. 1984. Mutation that suppresses the protein export defect of the *secY* mutation and causes cold-sensitive growth of *Es-*

cherichia coli. *J. Bacteriol.* 160:696–701

129. Shiba, K., Ito, K., Yura, T., Cerretti,D. P. 1984. A defined mutation in the protein export gene within the *spc* ribosomal protein operon of *Escherichia coli:* isolation and characterization of a new temperature-sensitive *secY* mutant. *EMBO J.* 3:631–36

130. Silhavy, T. J., Beckwith, J. 1985. Use of *lac* fusions for the study of biiological problems. *Microbiol. Revs.* 49:398–418

131. Silhavy, T. J., Benson, S. A., Emr, S. D. 1983. Mechanisms of protein localization. *Microbiol. Revs.* 47:313–44

132. Silver, P., Wickner W. 1983. Genetic mapping of the *Escherichia coli* leader (signal) peptidase gene (*lep*): a new approach for determining the map position of a cloned gene. *J. Bacteriol.* 154:569–72

133. Singer, S. J., Maher, P. A., Yaffe, M. P. 1987. On the translocation of proteins across membranes. *Proc. Natl. Acad. Sci. USA* 84:1015–19

134. Stader, J., Benson, S. A., Silhavy, T. J. 1986. Kinetic analysis of *lamB* mutants suggests the signal sequence plays multiple roles in protein export. *J. Biol. Chem.* 261:15075–80

135. Stader, J., Gansheroff, L. J., Silhavy, T. J. 1989. New suppressors of signal-sequence mutations, *prlG*, are linked tightly to the *secE* gene of *Escherichia coli*. *Genes Dev.* 3:1045–52

136. Suh, J.-W., Boylan, S. A., Thomas, S. M., Dolan, K. M., Oliver, D. B., et al. 1990. Isolation of a *secY* homologue from *Bacillus subtilis:* Evidence for a common protein export pathway in Eubacteria. *Mol. Microbiol.* 4:305–14

137. Summers, R. G., Harris, C. R., Knowles, J. R. 1989. A conservative amino acid substitution, arginine for lysine, abolishes export of a hybrid protein in *Escherichia coli*. *J. Biol. Chem.* 264:20082–88

138. Summers, R. G., Knowles, J. R. 1989. Illicit secretion of a cytoplasmic protein into the periplasm of *Escherichia coli* requires a signal peptide plus a portion of the cognate secreted protein. *J. Biol. Chem.* 264:20074–81

139. Tai, P. C., ed. 1990. Protein export in bacteria: structure, function and molecular genetics. *J. Bioenerget. Biomemb.* 22:209–491

140. Takahara, M., Sagai, H., Inouye, S., Inouye, M. 1988. Secretion of human superoxide dismutase in *Escherichia coli*. *Biotechnology* 6:195–98

141. Tani, K., Shiozuka, K., Tokuda, H., Mizushima, S. 1989. In vitro analysis of the process of translocation of OmpA across the *Escherichia coli* cytoplasmic membrane. *J. Biol. Chem.* 264:18582–88

142. Thom, J. R., Randall, L. L. 1988. Role of the leader peptide of maltose-binding protein in two steps of the export process. *J. Bacteriol.* 170:5654–61

143. Tokunaga, M., Loranger, J. M., Wu, H. C. 1983. Isolation and characterization of an *Escherichia coli* clone overproducing prolipoprotein signal peptidase. *J. Biol. Chem.* 258:12102–5

144. Tokunaga, M., Tokunaga, H., Wu, H. C. 1982. Post-translational modification and processing of *Escherichia coli* prolipoprotein in vitro. *Proc. Natl. Acad. Sci. USA* 79:2255–9

145. Trun, N. J., Silhavy, T. J. 1987. Characterization and in vivo cloning of *prlC*, a suppressor of signal sequence mutations in *Escherichia coli* K12. *Genetics* 116:513–21

146. Trun, N. J., Silhavy, T. J. 1989. *PrlC*, a suppressor of signal sequence mutations in *Escherichia coli*, can direct the insertion of the signal sequence into the membrane. *J. Mol. Biol.* 205:665–76

147. Vlasuk, G. P., Inouye, S., Ito, H., Itakura, K., Inouye, M. 1983. Effects of the complete removal of basic amino acid residues from the signal peptide on secretion of lipoprotein in *Escherichia coli*. *J. Biol. Chem.* 258:7141–48

148. Vogel, J. P., Misra, L. M., Rose, M. D. 1990. Loss of BiP/GRP78 function blocks translocation of secretory proteins in yeast. *J. Cell Biol.* In press

149. von Heijne, G. 1984. Analysis of the distribution of charged residues in the N-terminal region of signal squences: implications for protein export in prokaryotic and eukaryotic cells. *EMBO J.* 3:2315–18

150. von Heijne, G. 1086. Net N-C charge imbalance may be important for signal sequence function in bacteria. *J. Mol. Biol.* 192:287–90

151. Voorhout, W., de Kroon, T., Leunissen-Bijvelt, J., Verkleij, A., Tommassen, J. 1988. Accumulation of LamB-LacZ hybrid proteins in intracytoplasmic membrane-like structures in *Escherichia coli* K12. *J. Gen. Microbiol.* 134:599–604

152. Walter, P., Blobel, G. 1981. Translocation of proteins across the endoplasmic reticulum. III. Signal recognition protein (SRP) causes signal sequence-dependent and site-specific arrest of chain elongation that is released by microsomal membrances. *J. Cell Biol.* 91:557–61

153. Watanabe, M., Blobel, G. 1989. Binding of a soluble factor of *Escherichia coli* to preproteins does not require ATP and appears to be the first step in protein export. *Proc. Natl. Acad. Sci. USA* 86:2248–52

154. Watanabe, M., Blobel, G. 1989. Site specific antibodies against the PrlA (SecY) protein of *Escherichia coli* inhibit protein export by interfering with plasma membrane binding of preproteins. *Proc. Natl. Acad. Sci. USA* 86:1895–99

155. Watanabe, M., Blobel, G. 1989. SecB functions as a cytosolic signal recognition factor for protein export in *E. coli*. *Cell* 58:695–705

156. Watanabe, M., Blobel, G. 1989. Cytosolic factor purified from *Escherichia coli* is necessary and sufficient for the export of a preprotein and is a homotetramer of SecB. *Proc. Natl. Acad. Sci. USA* 86:2728–32

157. Weiss, J. B., Ray, P. H., Bassford, P. J. Jr. 1988. Purified SecB protein of *Escherichia coli* retards folding and promotes membrane translocation of the maltose-binding protein in vitro. *Proc. Natl. Acad. Sci. USA* 85:8978–82

158. Wolfe, P. B., Wickner, W., Goodman, J. M. 1983. Sequence of the leader peptidase gene of *Escherichia coli* and the orientation of leader peptidase in the bacterial envelope. *J. Biol. Chem.* 258:12073–80

159. Wu, H. C., Hou, C., Lin, J. J. C., Yem, D. W. 1977. Biochemical characterization of a mutant lipoprotein of *Escherichia coli*. *Proc. Natl. Acad. Sci. USA* 74:1388–92

160. Yamada, H., Matsuyamas, S., Tokuda, H., Mizushima, S. 1989. A high concentration of SecA allows proton motive force-independent translocation of a model secretory protein into *Escherichia coli* membrane vesicles. *J. Biol. Chem.* 264:18577–81

161. Yamada, H., Tokuda, H. Mizushima, S. 1989. Proton motive force-dependent and independent protein translocation revealed by an efficient in vitro assay system of *Escherichia coli*. *J. Biol. Chem.* 264:1723–28

162. Yamada, H., Yamagata, H., Mizushima, S. 1984. The major outer membrane lipoprotein and new lipoproteins share a common signal peptidase that exists in the cytoplasmic membrane of *Escherichia coli*. *FEBS Lett.* 166:179–182

163. Yamagata, H., Daishima, K., Mizushi-

ma, S. 1983. Cloning and expression of a gene coding for the prolipoprotein signal peptidase of *Escherichia coli. FEBS Lett.* 158:301–4

164. Yamagata, H., Ippolito, C., Inukai, M., Inouye, M. 1982. Temperature-sensitive processing of our membrane lipoprotein in an *Escherichia coli* mutant. *J. Bacteriol.* 152:1163–68

165. Yamagata, H., Taguchi, N., Daishima, K., Mizushima, S. 1983. Genetic characterization of a gene for prolipoprotein signal peptidase in *Escherichia coli. Mol. Gen. Genet.* 192:10–14

166. Yamane, K., Ichihara, S., Mizushima, S. 1987. In vitro translocation of protein across *Escherichia coli* membrane vesicles requires both the proton motive force and ATP. *J. Biol. Chem.* 262:2358–62

167. Yamane, K., Matsuyama, S.-I., Mizushima, S. 1988. Efficient in vitro translocation into *Escherichia coli* membrane vesicles of a protein carrying an uncleavable signal peptide. *J. Biol. Chem.* 263:5368–72

168. Yamane, K., Mizushima, S. 1988. Introduction of basic amino acid residues after the signal peptide inhibits protein translocation across the cytoplasmic membrance of *Escherichia coli. J. Biol. Chem.* 263:19690–96

169. Yu, F., Yamada, H., Daishima, K., Mizushima, S. 1984. Nucleotide sequence of the *lspA* gene, the structural gene for lipoprotein signal peptidase of *Escherichia coli. FEBS Lett.* 173:264–68

170. Zhang, Y., Broome-Smith, J. K. 1989. Identification of amino acid sequences that can function as translocators of β-lactamase in *Escherichia coli. Mol. Microbiol.* 3:1361–69

171. Zwizinski, C., Wickner, W. 1980. Purification and characterization of leader (signal) peptidase from *Escherichia coli. J. Biol. Chem.* 255:7973–77

Annu. Rev. Genet. 1990. 24:249–74

BACTERIAL CELL DIVISION

Piet A. J. de Boer, William R. Cook, and Lawrence I. Rothfield

Department of Microbiology, University of Connecticut Health Center, Farmington, Connecticut 06032

KEY WORDS: nucleoid, membrane, *E. coli,* septation, bacterial chromosome

CONTENTS

INTRODUCTION

Bacterial cell division is a classic problem in subcellular differentiation—a complex structure, the division septum, is formed at a specific location within a cell in a process that must be temporally and topologically coordinated with other cellular events such as chromosome replication and nucleoid segregation. This complicated process is carried out with remarkable fidelity, as shown by the high precision of placement of the division septum at midcell and the vanishingly low production of chromosomeless cells during normal growth (67). To accomplish this, the cell must solve three basic problems:

249

HOW to differentiate the division site and to carry out the circumferential and coordinated invagination of the multiple cell envelope layers that form the septum; WHERE to place the septum; WHEN to undergo septal invagination. In addition, it is likely that a positive mechanism exists to account for the equipartition of chromosomes into daughter cells, defining a fourth problem that is closely linked to the division process. Little is known about the molecular mechanisms responsible for these important cellular processes.

The application of genetics to the cell division problem began in 1968 with the ground-breaking work of Hirota and his collaborators (44), who identified a large number of mutants of *Escherichia coli*, that failed to septate when grown at elevated temperature. During the succeeding years many additional division mutants have been identified. Of particular interest are mutations that block septum formation without any detectible defects in DNA synthesis or in other aspects of cellular metabolism. As a result, the cells form long nonseptate filaments. These were, in turn, divided into two general groups by Hirota and his colleagues—*fts* mutants (for *t*emperature-*s*ensitive *f*ilamentation) where the filaments contain nucleoids that appear to be regularly distributed along the length of the cell, and *par* mutants (for *par*tition) which were thought to be defective in chromosome segregation because of the aberrant distribution of DNA within the cells and the formation of anucleate cells. The use of the *fts* group and related mutants to identify proteins involved in the division process and to probe the biogenetic history of the division apparatus, and the use of "partition" mutants to probe the mechanism of chromosome segregation are discussed further in the body of this review.

For many years the developmental history of the division apparatus was difficult to study because division-related events that occur before the onset of septal ingrowth could not be identified. In recent years, however, early differentiation events at the future division site have become accessible to study. This, in turn, has led to the idea that the division site has a developmental history that begins long before septal invagination. In addition, evidence indicates that the residual division sites at the cell poles continue to function after cell separation, suggesting that the history of the division site does not end with septation and cell separation. The details of the differentiation process have recently been reviewed (14). In this review we develop this theme further by treating the genetics of cell division as a problem in developmental genetics.

The review is divided into three main parts. First, we address the question of HOW the cell forms the division apparatus and carries out the septation process. We outline the developmental pathway that leads to biogenesis of the division apparatus and to formation of the division septum, as deduced from studies of cells at different stages of the cell cycle. We then examine the evidence that permits various division genes to be assigned to specific stages of the developmental sequence. Second, we discuss the problem of how the cell decides WHERE to place the division site. Third, we describe gene products that act operationally as activators and inhibitors of the division process, which may help to elucidate how the cell decides WHEN to divide.

We have not attempted to discuss all cell division genes nor all aspects of the cell division problem. For a discussion of other aspects of cell division and a more complete description of other genes that affect the division process or related cell-cycle events the reader is referred to other recent reviews (14, 19a, 27, 28, 83).

HOW

Developmental History of the Division Site

Until the early 1980s the only observable landmark of the division process was the formation of a visible septum. The window of observation then was extended to earlier stages in the development of the division site through the identification of a division-related organelle, the periseptal annular apparatus, that appears at the future division site before the onset of septal ingrowth (3, 62). This has provided an entry into preseptation stages of the division process, yielding information on the process of differentiation and site-specific localization of the division apparatus.

The periseptal annuli are differentiated regions of the cell envelope that extend around the circumference of the cell at the site of cell division, forming two concentric rings that flank the division site (Figure 1). The annuli were discovered and have been defined by serial section electron microscopy (16, 62). Each annulus is a continuous zone of adhesion in which the inner membrane is closely associated with the murein-outer membrane layer of the cell envelope (Figure 1). The annuli appear at the future division site before septum formation begins and they define a cell envelope domain, the periseptal domain, within which the septum is later assembled.

Because of their microscopic appearance, the periseptal annuli are postulated to segregate the periseptal domain from the remainder of the cell envelope (62). This view was supported by fluorescence photobleaching studies in which fluorescent proteins were introduced into the periplasmic space (31). These studies showed that regions of the periplasm that corresponded to future division sites or to old division sites at the cell poles were not in free communication with the remainder of the periplasm, implying compartmentalization of the periplasm at these sites. This presumably allows the cell to establish and maintain proteins and other molecules required for septum formation at the proper location by preventing them from diffusing to other regions of the cell envelope.

The positions of the annuli and their biogenetic precursors can be visualized

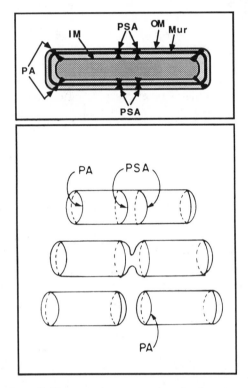

Figure 1 Periseptal annuli. Diagrammatic representation of periseptal annuli as seen in cross-section (upper panel) and as reconstructed from serial section electron micrographs (lower panel, adapted from (75)). PSA, periseptal annuli; PA, polar annulus; IM, inner membrane; Mur, murein (peptidoglycan); OM, outer membrane. The lightly stippled area corresponds to the periplasmic space.

in phase contrast micrographs of plasmolyzed cells because of the presence of localized regions of plasmolysis that are bounded by the annular adhesion zones. The localized plasmolysis bays therefore act as markers for the locations of the annular structures (16) and have been used to study their sequential formation and localization as cells progress through the division cycle (15).

These studies indicated that annuli were already in place at the midpoint of the youngest cells in the population. Further studies suggested that the midcell annuli are the starting point of a cycle in which three stages can be identified (Figure 2): (*a*) *genesis* of new periseptal annuli by replication of the midcell annuli present in the newborn cells; (*b*) *lateral displacement* of the nascent annuli toward the two poles; (*c*) *displacement arrest* when the structures have reached their proper locations at 1/4 and 3/4 cell lengths. The annuli at 1/4 and

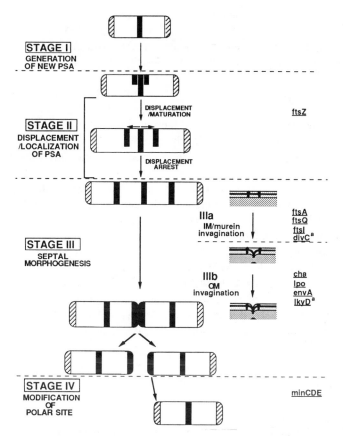

Figure 2 Developmental pathway of the *E. coli* division site. Stages of development are indicated on the left. Genes that affect progression through the individual stages are indicated on the right. The periplasmic compartment that is sequestered by the developing and mature annuli is indicated in black. Inactivation of the polar compartment is indicated by cross-hatching. For abbreviations see Figure 1.

[a]Genes of *Salmonella typhimurium*.

3/4 cell lengths are then retained during division to become the structures that are already in place at the midpoint of the newborn cells. Light and electron microscopic studies have suggested that the nascent annuli continue to mature during the displacement phase by growing progressively around the circumference of the cylinder. Generation of new division sites therefore appears to be a cyclical process in which new sites are not formed de novo at midcell but rather arise from preexisting sites formed during the preceding division cycle.

The mechanism responsible for the lateral displacement is unknown. Evidence conflicts as to whether zonal cell envelope growth occurs near midcell

(11, 21, 78, 87, 92). Therefore, the attractive possibility that the nascent annuli are pushed apart by insertion of new cell envelope material remains unproven. Indeed, another mechanism might exist in which the annular structures are "pulled" or "pushed" toward the two poles by as yet unrecognized cellular elements.

At the appropriate time in the cycle, the septum is formed between the two central periseptal annuli by the circumferential invagination of the three cell envelope layers (inner membrane, murein, and outer membrane), leading finally to separation of the two daughter cells. Genetic studies have indicated that septal invagination can be further dissociated into individual steps (see below).

After cell separation each newborn cell inherits one of the old midcell annuli, which remains as a permanent polar annulus at the end of the daughter cell (61, 62). Evidence to be discussed below suggests that the polar domain between the polar annulus and the cell pole may continue to function after cell separation so that the developmental history of the division site extends beyond the cell cycle in which the site is used for septum formation.

Based on this information, we have divided the developmental cycle into four major stages, as schematically indicated in Figure 2, to provide a framework to classify division mutants. We assume that future research will define additional substages within these major groups, as already shown for Stage III.

Assignment of Genes to Different Stages of the Developmental Pathway

As noted above, thermosensitive mutants (*fts* mutants) have been identified that fail to form a septal crosswall when grown under nonpermissive conditions. In several cases, the stage of the division block has been more precisely defined by analyzing the locations and appearance of annuli along the filaments. Additional genes that affect the differentiation process probably remain to be identified. Studies of their phenotypes may well define additional intermediate steps in the developmental pathway. Such studies are based on the usual assumption (not without risk) that a recessive temperature-sensitive mutation that results in an apparent block at one stage of a developmental sequence is likely to identify a gene product that normally plays a positive role in the transition to the next stage.

STAGE I MUTANTS   These mutants should fail to form new annuli at nonpermissive temperatures. In the simplest case the mutant cells should grow as filaments that lack periseptal annuli. Annuli generated prior to the switch to nonpermissive conditions would be used to support septum formation during the first one or two cell cycles at the nonpermissive temperature and should therefore disappear from the body of the cell. Polar annuli should continue to be present since they were produced before the temperature upshift.

No Stage I mutants have yet been identified. These would be of great interest since their phenotype is expected to give clues as to the mode of generation of the annuli. However, Stage I mutants will be difficult to identify if annulus formation and displacement turns out to be an obligatory part of the cell elongation mechanism. This might be the case, for example, if the nascent annuli acted as the site of insertion of new cell envelope material during their displacement from midcell to their final positions at the cell quarters, thereby playing an essential role in cell elongation. In this case, mutants blocked in genesis of periseptal annuli would not form filaments and therefore would not be recognized as cell division mutants. They would instead have to be sought within the large group of nonfilamenting conditional lethal mutants.

STAGE II MUTANTS   Stage II mutants should accumulate incomplete structures that are arrested at intermediate stages of annulus development and/or should fail to arrest the displacement process when the annuli reach their proper locations.

Recent studies of the classical *ftsZ*84 mutant (60) grown at 42° suggest that the product of the *ftsZ* gene acts at this developmental stage (W.R.C. & L.I.R., unpublished observations). The annuli in the FtsZ⁻ cells were not clustered at regular intervals but were randomly distributed along the length of the cell, as indicated by analysis of bay positions in the nonseptate filaments. In addition, maturation of the annual structures appeared to be retarded, as shown by the accumulation of incomplete bays; these appear to represent annuli that extend only part way around the cylinder in serial section electron micrographs. These results contrast with those obtained from studies of FtsA⁻ filaments (described below) in which annulus localization and maturation appear normal. The block in the *ftsZ*(Ts) cells did not appear to interfere with formation of new annulus progenitors since there was no decrease in the total number of plasmolysis bays per unit cell length. The accumulation of "immature" annuli together with their failure to localize at the proper positions suggests that the FtsZ protein acts during Stage II of the developmental pathway. The failure to localize the incomplete structures suggests a close, perhaps obligatory, coupling between annulus maturation and annulus localization, as would be expected if displacement arrest required completion of development of the periseptal annuli. Other aspects of FtsZ function are discussed later.

STAGE III MUTANTS

1. Mutants blocked at the Stage II–III transition   Mutants of this class should be capable of forming and localizing the periseptal annuli at potential division sites but unable to progress through the subsequent stages of septal morphogenesis. Several *fts* mutants are blocked at this point, as illustrated by studies of a thermosensitive *ftsA* mutant (77).

In contrast to the results in FtsZ⁻ filaments, studies of FtsA⁻ cells showed annular structures clustered at unit cell lengths along the filaments, at positions where septa would have formed if not for the division block, i.e. at 1/8, 1/4, 3/8, 1/2, . . . 7/8 cell lengths. When septation was permitted to resume by shifting back to permissive temperature, septa appeared at the same locations, confirming that the annuli did in fact correspond to potential cell division sites. Similar results were obtained with an *ftsQ* mutant of *E. coli* (W.R.C. & L.I.R., unpublished observations) and a *divC* mutant of *Salmonella typhimurium* (16). These mutants are all able to generate and localize periseptal annuli but are unable to initiate septal ingrowth. They therefore are blocked at the Stage II–III transition, suggesting that the FtsA, FtsQ, and DivC proteins play a role in the initial stage of septal ingrowth (Stage IIIa in Figure 2).

The *ftsI* gene product [penicillin-binding protein 3] probably also acts at this stage. (The *ftsI* gene has also been variously called *pbpB* and *sep*.) When *ftsI*(Ts) mutants are grown at nonpermissive temperature they form long nonseptate filaments. In addition, antibiotics that bind specifically to FtsI also block septum formation (80). The high affinity of the FtsI protein for β-lactam antibiotics and the report that FtsI has murein transglycosylase and transpeptidase activity (49) suggest that the protein plays a role in murein synthesis or remodeling. It therefore is likely to play a role in the formation of septal murein and is tentatively assigned to Stage IIIa in the present scheme. It is not yet known whether annuli are regularly spaced along the FtsI⁻ filaments, as would be predicted if this were correct.

Evidence indicates that incorporation of new septal murein is confined to a localized site of membrane-murein association at the leading edge of the ingrowing septum (61, 91). This may prove to be the site of localization or activation of the FtsI protein.

Genetic experiments have suggested that FtsI interacts with the *rodA* gene product that plays a role in maintaining the rod shape of the cell (5), and possibly also with the *ftsA* gene product (86). These interactions may play a role in the switch from exclusively lateral to preferentially septal murein synthesis, which seems to occur at the onset of septum formation (66).

Other studies (4, 84) also support the idea that the *ftsA*, *ftsQ* and *ftsI* gene products act after FtsZ in the developmental sequence. These studies showed the presence of broad, flattened constrictions at regular intervals along the length of *ftsA*, *ftsQ*, and *ftsI* but not *ftsZ* filaments. The broad constrictions do not resemble normal septa and are not visible in thin section electron micrographs where early septal ingrowth can be readily detected in populations of wild-type cells. The appearance and location of these constrictions suggests an alteration of cell envelope organization at the division site that occurs after localization of the periseptal annuli, perhaps to prepare for septal invagination. This could be related to the local changes in murein hydrolase activity at potential division sites that had previously been suggested from studies of murein sacculi (79). Consistent with the studies of annulus development discussed above, these studies indicated that FtsZ acts at an earlier stage in the division pathway than FtsA, FtsQ, and FtsI.

2. Mutations that block division after the onset of septal morphogenesis Several genes have been identified that affect the process of septal morphogenesis itself. The mutant phenotypes are characterized by uncoupling of the coordinate ingrowth of the three layers of the normal division septum. In all cases the mutant cells contain abnormal septal crosswalls that consist of inner membrane and murein but which lack the outer-membrane layer of the normal septum (illustrated as Stage IIIb in Figure 2). This leads to the formation of chains of unseparated cells in which the cytoplasm is segregated into cell-sized units by inner membrane-murein crosswalls while the outer membrane remains as a bridge that holds the filaments together. This group of genes includes *IkyD* of *S. typhimurium* (90) and the *cha* (27) and *envA* genes (70, 71) of *E. coli*. In addition, an *lpo* mutant of *E. coli* that fails to synthesize murein-lipoprotein (34) shows a similar phenotype when grown in low-Mg^{++} medium. Because these mutations dissociate the ingrowth of outer membrane from the ingrowth of murein and inner membrane, the affected gene products are assigned a role in the last stage of morphogenesis, the stage of ingrowth of outer membrane (Stage IIIb).

No comparable mutations have been described in which invagination of septal inner membrane occurs in the absence of murein ingrowth. The expected phenotype of these mutants would depend on the mechanism by which the inner membrane component of the septum is assembled. The inner membrane components of the septum might be recruited from molecules that were randomly inserted into inner membrane along the length of the cell and then diffused laterally into the periseptal domain. Alternatively, the inner membrane components of the septum might be assembled locally by insertion into the inner membrane in the septal region. Therefore, mutants blocked in murein ingrowth without a parallel block in inner membrane proliferation would be expected to contain redundant inner membrane located either at the blocked septation sites or diffusely along the body of the filaments, depending on whether assembly occured locally or diffusely.

STAGE IV   Evidence that the residual site that remains at the new cell pole continues to function, and the identification of gene products that act during Stage IV will be discussed later.

WHERE

Work of the past few years has indicated that the correct placement of the septum at the midpoint of the cell involves two stages: (*a*) differentiation of future division sites, as indicated by the development and localization of periseptal annuli (described above); (*b*) selection of the proper midcell site by preventing septation at other potential division sites within the same cell. In addition, it has been suggested that the placement of the division site may be directly affected by the location of the nucleoids.

Selection of the Proper Division Site

Studies of minicell mutants have provided evidence that placement of the septum at midcell requires an additional stage of site selection distinct from the process that localizes the periseptal annuli at midcell. Although minicell mutants have been described in other bacteria (33), the present discussion is limited to *E. coli.*

The most characteristic feature of minicell mutants is the placement of a large percentage of all septation events at one of the cell poles instead of at midcell. The polar divisions lead to the formation of small chromosomeless cells (minicells) with the approximate dimensions of two polar caps (33). The total number of septation events per unit increase in cell mass appears similar to that of wild-type cells, suggesting that the mutant cells have a choice between dividing at the normal midcell site or at one of the two poles (85). As a result, the population consists of a mixture of anucleate minicells and chromosome-containing cells that are heterogeneous in length, including many short filaments. Although the minicells are devoid of chromosomal DNA, plasmids (especially high copy number plasmids) readily segregate into the minicells. Therefore, since it is relatively easy to physically separate the small minicells from the larger chromosome-containing cells, minicell mutants have been extensively used as experimental tools to identify plasmid-encoded gene products (33, 74).

The first isolated minicell mutant of *E. coli* (1) was initially thought to contain two distinct mutations (*minA* and *minB*) that were both required to cause the minicell phenotype (33). Later work failed to confirm the existence of the *minA* gene and a single mutation is clearly responsible for the minicell phenotype of the original mutant (20). Genetic complementation analysis (authors' unpublished observations) and DNA sequence analysis (C. Labie &

J. P. Bouché, personal communication) have mapped the mutation to the *minD* gene of the *minB* operon (see below) and it is now termed *minD1*.

In the early 1970s it was proposed that the normal role of the minicell locus is to inactivate potential division sites at the cell poles (2, 85). Assuming that the *minD1* mutation resulted in loss of the inactivation function, this role nicely explained the occurrence of polar septation events in the mutant cells.

Two possible origins have been proposed for the polar sites. It was first suggested that the polar sites represented an early stage in the development of the normal division site that later appears at midcell (2). In this view, the *min* system would repress function of the potential division sites until they were translocated from cell poles to the current position at midcell. The studies of the periseptal annuli described above later indicated that the precursor of the future division site is generated at midcell rather than at the poles, making this explanation unlikely. It was later proposed that the potential division sites at the poles were remnants of the site that had been present at midcell before the preceding septation event (85). This idea was supported by the finding that each new pole inherits one of the two periseptal annuli that originally flanked the division septum (62). As discussed above, the polar annulus segregates a polar periplasmic domain that is directly derived from the original periseptal domain and that might reasonably be expected to include elements required for septum formation.

More recently, detailed studies of the *minB* locus have corroborated the view that the role of the minicell locus is to block septum formation at potential division sites at the cell poles. These studies (22, 23) have shown that *minB* is an operon consisting of three genes, *minC*, *minD*, and *minE*, and that expression of all three genes is required for the correct placement of the division septum. The correct placement pattern is accomplished by the combined action of a division inhibitor, dependent on the expression of *minC* and *minD*, and a topological specificity factor coded for by *minE*, as schematically illustrated in Figure 3.

Expression of *minC* and *minD* in the absence of *minE* causes a global block in septum formation, resulting in the formation of nonseptate filaments. The MinCD-induced septation block lacks topological specificity since it affects internal division sites as well as polar division sites. The *minE* gene product modifies the MinCD-mediated septation block as shown by the prevention of MinCD-induced filamentation and the restoration of the wild-type septation pattern when *minE* is expressed at normal levels. Interestingly, expression of *minE* at high levels results in the classical minicell phenotype. Thus, minicell formation can be caused either by loss of MinC or MinD, or by overexpression of MinE. According to the most straightforward interpretation of these results (Figure 3) the role of the *minE* gene product is to give topologic specificity to the MinCD division inhibitor. Normal levels of MinE prevent

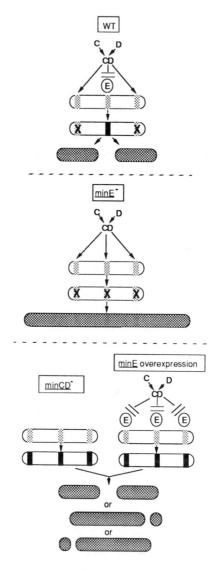

Figure 3 Model of minCDE action. Diagrammatic representation of a model describing the proposed roles of the *minCDE* gene products in the selection of the correct site of septum formation. Light gray areas within cells represent potential division sites. Sites capable of supporting septum formation are indicated in black. Sites that have been inactivated by the MinCD division inhibition mechanism are indicated by the crosses (X). For details, see text. (Adapted from (23))

the inhibitor from blocking division at the site at midcell without interfering with the ability of the inhibitor to block septation at polar sites. At high levels of expression, MinE prevents the MinCD division inhibitor from blocking septation at polar sites as well as at midcell sites, explaining the induction of the minicell phenotype by high levels of MinE. The final result therefore depends on the ratio of MinE to MinCD. It is not known whether MinE can give topological specificity to any other elements that are related to the division process.

Many questions concerning the site-specific inactivation of potential division sites by the *mincDE* system remain: (*a*) What is the nature of the MinCD-mediated division inhibitor? Is one or both of the two proteins the actual inhibitor or do MinC and MinD activate another molecule that is responsible for the septation block? (*b*) What is the direct molecular target of the division inhibition reaction? (*c*) How does MinE give topological specificity to the division inhibitor? Does MinE interact with the potential division sites in the cell envelope, as indicated for simplicity in Figure 3, or does MinE interact with the division inhibitor itself, acting as a pilot peptide to direct the inhibitor to polar sites rather than midcell sites? (*d*) What is the difference between the "old" sites at the poles and the "new" sites at midcell that can be recognized by the topological specificity factor? The answers are still unknown although evidence relevant to several of these points is discussed further below.

The Relation between Nucleoid Partition and Division Site Placement

Under normal conditions, the placement of the division site at midcell automatically positions the new septum between the daughter chromosomes formed during the preceding round of DNA replication. (For further information the reader is referred to: 14, 19, 26, 41, 76, 92.)

1. Is there an active mechanism that moves daughter chromosomes away from midcell, thereby partitioning them on either side of the potential division site at midcell?   Although passive diffusion is unlikely to be the sole mechanism of nucleoid partition in view of the extremely low production of chromosomeless cells under normal circumstances, this possibility cannot be dismissed out of hand. The study of mutants that show aberrant chromosome segregation has provided some pertinent evidence.

Most "partition" mutants map in genes coding for proteins with topoisomerase activity [*parA(gyrB)* (54); *parC* (55; J. Kato, personal communication); *parD(gyrA)* (47); *parE* (J. Kato, personal communication)]. In these cases the DNA is located in a single large mass at midcell and anucleate cells are produced by septum formation in the nucleoid-free regions of the cell. The apparent defect in nucleoid separation in gyrase mutants reflects a requirement for DNA gyrase to decatenate the linked DNA molecules that are formed during replication (29, 81). Thus, decatenation of the progeny chromosomes defines the first required event before movement of the daughter chromosomes to both halves of the cell.

Perhaps more relevant to the partition process itself are several mutations that Hiraga and his colleagues (41, 42) recently described. The mutants were identified by an elegant genetic approach based on the ability of a plasmid-borne reporter gene to partition into chromosomeless cells and thereby escape transcriptional repression by a chromosomally encoded repressor.

There is no apparent defect in nucleoid decatenation in the best-studied of these mutants *(mukA1)* since cells that give rise to anucleate progeny contain two visible nucleoids in one half of the cell, leaving the other half devoid of chromosomal material. Septation in the chromosome-free half of the cell then gives rise to normal-sized cells that lack DNA. The defect therefore is likely to affect the partition process itself.

2. Does the positioning of the potential division site directly affect the location of the daughter nucleoids?   The most widely favored model to explain chromosome segregation remains that of Jacob et al (50) in which the segregation of daughter chromosomes is mediated by a direct attachment of the chromosome to specific cell envelope sites. Cell envelope growth between the sites would then push the attached chromosomes apart into the halves of the cell. If this model is correct, the periseptal annuli would be attractive candidates for the cell envelope attachment sites since the newly generated annuli that appear on either side of the midcell annuli in newborn cells are displaced toward the two poles and are destined irrevocably for different daughter cells (Figure 2). If the annuli were used as chromosome attachment sites, mutants in which annulus localization was abnormal would also show an aberrant distribution of nucleoids. It will therefore be important to determine whether nucleoid distribution is abnormal in the filaments resulting from inactivation of the *ftsZ* gene product, which causes a random rather than clustered distribution of nascent annuli.

Minicell mutants may also have a defect in chromosome segregation in addition to their obvious inability to partition nucleoids into the minicells (51, 64a). The roles of the individual *min* gene products in this phenomenon remain to be defined.

3. Does the position of the nucleoids directly affect the location at which the new septum is formed.   Circumstantial evidence indicates that the presence of chromosomal material exerts a local veto on septum formation in the region of the nucleoid. This is illustrated in situations where chromosome replication or decatenation is prevented, leading chromosomal material to accumulate at midcell. Septation does not occur at these sites, resulting in the formation of filaments, each containing a central chromosomal mass (39, 43, 46, 52, 65, 73). It is not known whether the local inhibition of septation occurs at the stage of initiation of septal ingrowth or at an earlier stage of the developmental pathway.

Whatever the mechanism, a negative control on septation in these regions would prevent septal closure from exerting a guillotine effect on the unsegregated chromosome, thereby avoiding fragmentation of the genome. The apparent prevention of septation at midcell in these situations is not associated with a block in septation at other sites, as shown by the formation of chromosomeless cells. Thus, the placement of the division septum does not require that the site be flanked by two nucleoids.

More controversial than the local inhibition of septation in the immediate vicinity of the nucleoid is the suggestion that the position of the nucleoid is the only factor determining where the septum is formed (26, 46, 65, 92, 93). In this view, the entire chromosome-free region of the cell is equally competent to form a septum. Considerable evidence argues against this model in its simplest form (for further discussion, see 14, 19, 40, 76). Nevertheless, although nucleoid location is unlikely to be the only factor involved in septal placement, the idea that the position of the nucleoid can influence the probability of septation at different sites within the nucleoid-free region of the cell calls for further study.

WHEN

A number of genes are known whose expression can alter the frequency of septation by either specifically inhibiting or activating cell division. The division inhibitors and activators are of interest because such molecules may play a role in regulating the timing of division during the normal division cycle as well as the frequency of septation events in response to changes in growth conditions or to cell damage.

Division Inhibitors

Several genes are known whose expression leads to a global inhibition of septation. Some are chromosomal whereas others reside on plasmid or phage genomes. We focus on chromosomally encoded division inhibitors (for information on the phage and plasmid-encoded inhibitors see: 38, 69).

MINC-DEPENDENT DIVISION INHIBITION SYSTEMS   The only known division inhibition system that operates during normal growth and that plays a role in the normal pattern of cell division is the MinC/MinD system. As described above, the division inhibition system acts, together with the *minE* gene product, to inactivate polar division sites and thus ensures that division is restricted to the proper site at midcell. Evidence that the MinC and MinD proteins play different roles in the site-specific inhibition reaction has come from study of another division inhibitor, DicB.

The *dicB* gene is part of an operon located at 34.9 units on the genetic map (6–8). The DicB protein is a 7kd peptide formerly called DicB$_S$ (12). [The original suggestion that a second peptide product (DicB$_L$) can be produced by the DicB gene (12) now appears incorrect (30)]. Expression of the operon is controlled by two upstream repressor genes, *dicA* and *dicC*, so that *dicB* is normally not expressed. Physiological conditions that can lead to derepression have not yet been identified. Expression of *dicB*, either in a repressor mutant or when placed under control of another promoter, leads to a prompt general block in septation, resulting in the formation of long filaments.

A functional relationship between *dicB* and the *minCD* division inhibition system was first suggested when some mutations that suppress *dicB*-induced filamentation were observed also to result in a minicell phenotype and to map in or near the *minB* locus (56). It was subsequently shown that DicB-induced filamentation requires the *minC* gene product, but not MinD (24). Consistent with this observation, Labie & Bouché (personal communication) have shown that the *minB*-suppressors of DicB-induced filamentation were point mutations in *minC*. The MinC/MinD and the MinC/DicB division inhibition reactions are both functional in the absence of the *recA* and *sfiA* gene products (24) and therefore are not part of the SOS reaction (discussed below).

DicB and MinD show no significant sequence homologies, suggesting that DicB does not merely act as a MinD homolog in formation of the division inhibitor. Moreover, the MinC/DicB and MinC/MinD division inhibition reactions can be distinguished by their different responses to the *minE* gene product. Whereas the MinC/MinD-mediated division block is modified by low levels of expression of *minE*, MinC/DicB-mediated division inhibition is unaffected by even high levels of *minE* expression (24).

MinC therefore is the common element in a class of division inhibition systems that is distinct from the well-studied SfiA-mediated inhibition system. From these observations a model was proposed (24) in which MinC is the effector of the division inhibition process while MinD and DicB act as independent activators of the MinC-mediated inhibition mechanism. The activator proteins in turn determine the other specific properties of the system. Since MinC is the common element in the two systems, it is likely to be the component that interacts with the division machinery or that plays a role in the formation of another molecule that could be the actual division inhibitor. Similarly, since MinD imparts sensitivity to MinE, the MinD protein may well be the component that interacts with MinE. It is not known whether any other proteins can also serve as activators of the MinC-mediated division inhibition reaction.

DICF   Recent experiments have shown that inhibition of septation can also result from transcription of a small RNA *(dicF)* from an untranslated region that lies upstream of *dicB* (9, 30). The *dicF*-mediated effect does not require the DicB protein or components of the *minCDE* operon.

SFIA   The *sfiA* gene (also called *sulA*) is one of the set of genes that are derepressed as part of the SOS response to DNA damage (28, 37, 88). Expression of *sfiA* leads to inhibition of septum formation. SfiA does not play a significant role in regulation of the division process in unperturbed growing cells since *sfiA⁻* mutants grow and divide normally (45). The SfiA-mediated division block is rapidly reversible (64) and is therefore presumably useful to the cell by temporarily stopping division until the DNA damage has been repaired.

SFIC   A second division inhibitor induced in some strains as part of the SOS response is the product of the *sfiC* gene. This gene is part of excisable element e14 (63) located at 25 units of the genetic map (10). e14 has many properties of a defective prophage and is present in some but not all strains of *E. coli*. Upon induction of the SOS response, e14 is excised from the chromosome and expression of *sfiC* is induced, leading to a general inhibition of septation. Interestingly, although *sfiC* expression depends on RecA, thereby placing it in the group of SOS genes, it is not directly regulated by LexA (18), thereby distinguishing it from many other members of the SOS system.

Division Activators

FTSZ   The *ftsZ* gene is part of a large cluster of genes involved in cell growth and cell division that are located near two units on the chromosome of *E. coli* (28, 48, 57). FtsZ is generally considered an essential component of the cell division machinery based on the observation that growth of a conditional mutant *(ftsZ84)* at elevated temperature leads to the formation of long nonseptate filaments. In addition, attempts to disrupt the chromosomal gene by insertional mutagenesis have so far been unsuccessful (57).

That the *ftsZ* gene product is an activator of the division process is most strongly supported by the observation that overexpression of *ftsZ* in wild-type cells leads to an increased number of septation events per increase in cell mass (89). Consistent with the view of FtsZ as an activator of division is the ability of high concentrations of FtsZ to prevent the filamentation that is induced by many division inhibitors. These include SfiA (59), MinC/MinD (24), MinC/DicB (24), and DicF (J. P. Bouché, personal communication). The induction of minicell formation by FtsZ overproduction in wild-type cells (17, 89) also reflects the ability of the FtsZ protein to overcome the MinC/MinD block, in this case at polar division sites. The relation between the ability of over-expression of *ftsZ* to increase the number of septation events at midcell and poles, and the apparent defect in maturation and localization of periseptal annuli in the *ftsz*84 mutant, remains to be defined.

FtsZ has been proposed to be the direct target of the SfiA division inhibition system (53, 58). This was originally suggested by the observation that a group of mutations (*sfiB, sulB*) that confer resistance to SfiA-mediated division inhibition were actually mutations in *ftsZ* (58). The ability of FtsZ to counter-act SfiA-induced filamentation is also consistent with this view. In addition, high rates of expression of *ftsZ* greatly reduced the rate of degradation of the SfiA protein in maxicells, suggesting that the FtsZ and SfiA proteins might interact (53). These observations are consistent with the idea that FtsZ is the immediate target of the SfiA protein, but more direct evidence is needed to establish this important point. It has also been suggested that FtsZ is the target of SfiC, based on the observation that *ftsZ(sfiB)* alleles that confer resistance to SfiA also conferred resistance to SfiC-mediated division inhibition (18, 63).

The molecular mechanism of FtsZ action is not known. FtsZ could be an integral component of the developing division site, playing a direct role in the events that lead to septum formation. On the other hand, the protein could play a regulatory role, affecting the activity or amount of essential com-ponents of the division machinery without itself being located at the division site.

The FtsZ protein is recovered in the cell envelope fraction of maxicells (53) in which large amounts of the protein are synthesized. However, at this time (April 1990) it is not known whether the FtsZ protein is located in the cell envelope in intact normal cells, which would be consistent with its being an integral part of the machinery, or whether it is a cytosolic protein that may play an indirect role in regulating the developmental pathway. Such informa-tion, as well as direct biochemical evidence of possible interactions between FtsZ and other proteins, should be forthcoming soon and will permit a more informed assessment of the role of the FtsZ protein in the division process.

The similarities in response of the SfiA, MinC/MinD, MinC/DicB and DicF division inhibitors to *ftsZ* overexpression, and the ability of *ftsZ(sfiB)* mutations to suppress the division inhibition activity of SfiA (35), SfiC (18), and MinC/MinD (E. Bi & J. Lutkenhaus, personal communication) is strik-ing. This could mean that all inhibitor proteins can interact directly with the FtsZ protein or that they all produce a common inhibitor molecule that interacts with FtsZ. However, alternative models are also possible.

In addition, as noted below, other gene products mimic some or all of these effects of FtsZ. One example is the reported ability of *ftsA* and *ftsQ* mutations to modify the sensitivity of cells to SfiA (25). More strikingly, two recently identified genes resemble *ftsZ* in the ability of their products both to increase the frequency of septation events in wild-type cells and to counteract the effects of the division inhibitors.

SDIA   The *sdiA* gene (for *s*uppressor of *d*ivision *i*nhibition) is located at 42 map units (X. Wang, L. Rothfield, P. de Boer, unpublished results). The wild-type *sdiA* gene was identified by its ability to confer resistance to MinC/MinD-mediated division inhibition when present in multiple copies per cell. Elevated rates of expression of *sdiA* also result in suppression of the filamentation that is induced by expression of MinC/DicB and SfiA. In wild-type cells *sdiA* expression results in the formation of shorter than normal cells and in the production of minicells. In all these properties *sdiA* resembles *ftsZ*. In addition, *sdiA* expression suppresses the filamentation that results from growth of the *ftsZ84* mutant under nonpermissive conditions. These results could be explained if the 28kd *sdiA* gene product acted to increase the amount or activity of FtsZ or another "limiting" component of the division system, or if SdiA were capable of substituting for such a factor. The *sdiA* gene product is not itself an essential component of the normal division machinery since a chromosomal deletion of the gene does not lead to division inhibition.

CFCA1   The *cfcA* gene, located at 79.2 units, was identified by isolating a recessive mutation *(cfcA1)* that suppressed SfiA-mediated division inhibition (68). In addition, the mutation significantly suppressed the filamentation phenotype of the *ftsZ84*(Ts) mutant and, in wild-type cells, caused an increase in septation events. In these respects the mutation mimicked the effects of *ftsZ* or *sdiA* overexpression. The *cfcA*$^+$ gene product may thus be a negative regulator of the amount or activity of FtsZ or SdiA. However, the mutation apparently did not result in minicell formation, as would be expected if increased FtsZ or SdiA activity were responsible for the *cfcA1* phenotype. This may indicate that CfcA is a negative regulator of a different, unidentified activator of cell division to which the MinC/MinD division inhibitor is relatively resistant.

Roles of the Division Inhibitors and Activators

As discussed above, the balance between negative effectors and positive effectors of septation determines whether or not septation occurs (illustrated in Figure 4).

The existence of a number of proteins that operate in *E. coli* as activators and inhibitors of division can be compared with the situation in eukaryotic

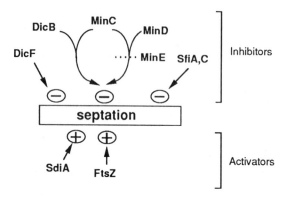

Figure 4   Activators and inhibitors of cell division (Adapted from (24))

cells, in which the timing of entry into the mitotic phase of the cell cycle is controlled by a network of activating and inhibitory gene products (72). The concentration of some of these gene products varies cyclically during the cell cycle. This regulatory network operates by modulating the phosphorylation and dephosphorylation of key proteins that trigger a multitude of cellular changes associated with mitosis. At present, the possible functional relationships between division inhibitors and activators in *E. coli* are less well understood and it is not known whether any division-related gene products behave cyclically or are controlled by a phosphorylation cascade. Future research should indicate whether in these regards meaningful parallels exist between prokaryotic and eukaryotic cells.

CURRENT STATE AND FUTURE PROSPECTS

Nearly all work on bacterial cell division during the past three decades has been performed on intact cells, using morphological, physiological, and genetic approaches. The morphological and physiological studies, for the most part outside this review, have provided a large body of information on the physical parameters and physiological regulation of the overall process. The genetic studies have identified many genes and gene products that affect the division process, but other important division-related genes probably remain to be identified. In this regard, a major challenge for the immediate future is to distinguish between gene products that regulate the differentiation events that lead to septum formation from those that play a direct role in the differentiation events themselves.

Much information has emerged from these studies. However, little is known about the molecular mechanisms responsible for any stage of the division process. The failure to extend these studies to the biochemical level reflects the formidable problems involved in studying a biological problem— the division of one cell into two cells—that cannot be directly assayed in broken cells.

Because of the complexity of the problem, the main hope of understanding the molecular events responsible for cell division lies in dissecting the overall process into smaller bite-sized pieces that can then be studied in isolation from the overall process. As a step toward this goal our major theme has been that cell division is a developmental process in which individual stages in the life history of the division site can be recognized. A number of gene products have now been identified that affect individual stages of the developmental sequence. Much work remains with this approach, which will likely identify other mutations that define additional steps in the developmental sequence.

An important step will be the identification of the cellular locations of the gene products that have already been identified. The questions are clear and the requisite tools are now available or obtainable. Three types of information are needed: (*a*) Is the gene product cytosolic or cell envelope-associated? (*b*) If the latter, is it associated with inner membrane, periplasm, periseptal annuli, murein, or outer membrane? (*c*) Is the gene product randomly located along the long axis of the cell or is it localized at poles, midcell, or at other potential division sites? Information on the first question is emerging from the identification of the proteins in cell fractionation studies (13, 36, 53, 80, 82; J. Pla, A. Dopazo, M. Vicente, personal communication). Nevertheless, since the questions concern cellular topology, definitive information will require morphologic approaches such as immunoelectron microscopy. This information is essential before the functions of the gene products can be meaningfully deduced.

An understanding of the molecular organization of the division site and of the functions of the division-related gene products will also require knowledge of how division-related proteins interact with other cell components. Considerably more information about protein-protein interactions can probably be obtained from the classic genetic approach of identifying allele-specific suppressor mutants. However, the combined use of biochemical and immunological methods such as protein fractionation, covalent crosslinking methods, and immunoaffinity approaches should provide more direct evidence of molecular interactions of the division-related proteins.

Finally, meaningful biochemical advance has been seriously impeded by the inability to isolate the division sites themselves. This difficult problem will be more tractable after the cell localization studies discussed above have identified the most suitable protein markers for the sites. This will warrant considerable experimental effort since isolation of the division sites or of structures associated with the sites will be an important and necessary step toward defining the molecular architecture of the division apparatus and developing in vitro assays for the activities of the division-related proteins. In this review we have emphasized the events that transpire at the division site itself. It is, of course, impossible to fully describe the cell division process without also understanding the interrelationships between the division process and other essential cellular processes that must be temporally coordinated with division site differentiation and septum formation, such as DNA replication, chromosome segregation, and cell elongation. Understanding the basis of the coordination of these complex events remains a major challenge.

ACKNOWLEDGMENTS

We thank the many colleagues who have informed us of their unpublished observations and given permission to include them in this review. Studies from the authors' laboratory were supported by grants from the National Science Foundation and the National Institutes of Health.

Literature Cited

1. Adler, H. I., Fisher, W. D., Cohen, A., Hardigree, A. A. 1967. Miniature *Escherichia coli* cells deficient in DNA *Proc Natl. Acad. Sci. USA* 57:321–26
2. Adler, H. I., Hardigree, A. A. 1972. Biology and radiobiology of minicells. In *Biology and Radiobiology of Anucleate Systems. Bacteria and Animal Cells,* ed. S. Bonotto, R. Kirchmann, R. Goutior, J.-R. Maisin, 1:51–66. New York: Academic
3. Anba, J., Bernadac, A., Pages, J., Lazdunski, C. 1984. The periseptal annulus in *Escherichia coli. Biol. Cell.* 50:273–78
4. Begg, K. J., Donachie, W. D. 1985. Cell shape and division in *Escherichia coli:* experiments with shape and division mutants. *J. Bacteriol.* 163:615–22
5. Begg, K. J., Spratt, B. G., Donachie, W. D. 1986. Interaction between membrane proteins PBP3 and RodA is required for normal cell shape and division in *Escherichia coli. J. Bacteriol.* 167:1004–8
6. Béjar, S., Bouché, F., Bouché, J.-P. 1988. Cell division inhibition gene *dicB* is regulated by a locus similar to lambdoid bacteriophage immunity loci. *Mol. Gen. Genet.* 212:11–19
7. Béjar, S., Bouché, J.-P. 1985. A new dispensable genetic locus of the terminus region involved in control of cell division in *Escherichia coli. Mol. Gen. Genet.* 201:146–50
8. Béjar, S., Cam, K., Bouché, J.-P.
1986. Control of cell division in *Escherichia coli.* DNA sequence of *dicA* and of a second gene complementing mutation *dicA2, dicC. Nucleic Acids Res.* 14:6821–33
9. Bouché, F., Bouché, J.-P. 1989. Genetic evidence that DicF, a second division inhibitor encoded by the *Escherichia coli dicB* operon, is probably RNA. *Mol. Microbiol.* 3:991–94
10. Brody, H., Greener, A., Hill, C. W. 1985. Excision and reintegration of *Escherichia coli* K-12 chromosomal element e14. *J. Bacteriol.* 161:1112–17
11. Burman, L. G., Raichler, J., Park, J. T. 1983. Evidence for diffuse growth of the cylindrical portion of the *Escherichia coli* murein sacculus. *J. Bacteriol.* 155:983–88
12. Cam, K., Béjar, S., Gil, D., Bouché, J.-P. 1988. Identification and sequence of division inhibitor gene *dicB* suggests expression from an internal inframe translation start. *Nucleic Acids Res.* 16:6327–38
13. Chon, Y., Gayda, R. 1988. Studies with FtsA-LacZ protein fusions reveal FtsA located inner-outer membrane junctions. *Biochem. Biophys. Res. Commun.* 152: 1023–30
14. Cook, W. R., de Boer, P. A. J., Rothfield, L. I. 1989. Differentiation of the bacterial cell division site. *Int. Rev. Cyto.* 118:1–31
15. Cook, W. R., Kepes, F., Joseleau-Petit, D., MacAlister, T. J., Rothfield, L. I.

1987. A proposed mechanism for the generation and localisation of new division sites during the division cycle of *Escherichia coli*. *Proc. Natl. Acad. Sci. USA* 84:7144–48

16. Cook, W. R., MacAlister, T. J., Rothfield, L. I. 1986. Compartmentalization of the periplasmic space at division sites in Gram-negative bacteria. *J. Bacteriol.* 168:1430–38

17. Corton, J. C., Ward, J. E., Lutkenhaus, J. 1987. Analysis of cell division gene *ftsZ* (*sulB*) from Gram-negative and Gram-positive bacteria. *J. Bacteriol.* 169:1–7

18. D'Ari, R., Huisman, O. 1983. Novel mechanism of cell division inhibition associated with the SOS response in *Escherichia coli*. *J. Bacteriol.* 156:243–50

19. D'Ari, R., Maguin, E., Bouloc, P., Jaffé, A., Robin, A., et al. 1990. Aspects of cell cycle regulation. *Res. Microbiol.* 141:9–16

19a. D'Ari, R., Nanninga, N. 1990. Cell shape and division in *Escherichia coli*. *Res. Microbiol.* 141:1–156

20. Davie, E., Sydnor, K., Rothfield, L. I. 1984. Genetic basis of minicell formation in *Escherichia coli* K-12. *J. Bacteriol.* 158:1202–3

21. Davison, M. T., Garland, P. B. 1983. Immunochemical demonstration of zonal growth of the cell envelope of *Escherichia coli*. *Eur. J. Biochem.* 130: 589–97

22. de Boer, P. A. J., Crossley, R. E., Rothfield, L. I. 1988. Isolation and properties of *minB*, a complex genetic locus involved in correct placement of the division site in *Escherichia coli*. *J. Bacteriol.* 170:2106–12

23. de Boer, P. A. J., Crossley, R. E., Rothfield, L. I. 1989. A division inhibitor and a topological specificity factor coded for by the minicell locus determine proper placement of the division septum in *E. coli*. *Cell* 56:641–49

24. de Boer, P. A. J., Crossley, R. E., Rothfield, L. I. 1990. Central role of the *Escherichia coli minC* gene product in two different cell division-inhibition systems. *Proc. Natl. Acad. Sci. USA* 87: 1129–33

25. Descoteaux, A., Drapeau, G. R. 1987. Regulation of cell division in *Escherichia coli* K-12: Probable interactions among proteins FtsQ, FtsA, and FtsZ. *J. Bacteriol.* 169:1938–42

26. Donachie, W. D., Begg, K. J. 1990. Genes and the replication cycle of *Escherichia coli*. *Res. Microbiol.* 141:64–75

27. Donachie, W. D., Begg, K. J., Sulli-

van, N. F. 1984. Morphogenes of *Escherichia coli*. In *Microbial Development*, ed. R. Losick, L. Shapiro, pp. 27–62. Cold Spring Harbor, NY: Cold Spring Harbor Lab.

28. Donachie, W. D., Robinson, A. C. 1987. Cell division: Parameter values and the process. See Ref. 67a, pp. 1578–93

29. Drlica, K. 1984. Biology of bacterial deoxyribonucleic acid topoisomerases. *Microbiol. Rev.* 48:273–89

30. Faubladier, M., Cam, K., Bouché, J.-P. 1990. The *Escherichia coli* cell division inhibitor dicF-RNA of the *dicB* operon: evidence for its generation *in vivo* by transcription termination and by RNase III and RNase E-dependent processing. *J. Mol. Biol.* 212:461–71

31. Foley, M., Brass, J. M., Birmingham, J., Cook, W. R., Garland, P. B., et al. 1989. Compartmentalization of the periplasm at cell division sites in *Escherichia coli* as shown by fluorescence photobleaching experiments. *Mol. Microbiol.* 3:1329–36

32. Deleted in proof

33. Frazer, A. C., Curtiss, R. III. 1975. Production, properties and utility of bacterial minicells. *Curr. Top. Microbiol. Immunol.* 69:1–84

34. Fung, J., MacAlister, T. J., Rothfield, L. I. 1978. Role of murein-lipoprotein in morphogenesis of the bacterial division septum; phenotypic similarity of *lkyD* and *lpo* mutants. *J. Bacteriol.* 133: 1467–71

35. George, J., Castellazzi, M., Buttin, G. 1975. Prophage induction and cell division in *E. coli* III. Mutations *sfiA* and *sfiB* restore division in *tif* and *lon* strains and permit the expression of mutator properties of *tif*. *Mol. Gen. Genet.* 140:309–32

36. Gill, D. R., Salmond, G. P. C. 1987. The *E. coli* cell division proteins, FtsY, FtsE and FtsX are inner-membrane associated. *Mol. Gen. Genet.* 210:504–8

37. Gottesman, S. 1989. Genetics of proteolysis in *Escherichia coli*. *Annu. Rev. Genet.* 23:163–98

38. Greer, H. 1975. The *kil* gene of bacteriophage lambda. *Virology* 66:589–604

39. Helmstetter, C. E., Pierucci, O., Weinberger, M., Holmes, M., Tang, M. 1979. Control of cell division in *Escherichia coli*. In *The Bacteria*, ed. J. R. Sokatch, L. N. Ornston, pp. 517–79. New York: Academic

40. Hiraga, S. 1990. Discussion. *Res. Microbiol.* 141:140

41. Hiraga, S. 1990. Partitioning of nucleoids. *Res. Microbiol.* 141:50–56
42. Hiraga, S., Niki, H., Ogura, T., Ichinose, C., Mori, H., et al. 1989. Chromosome partitioning in *Escherichia coli*: Novel mutants producing anucleate cells. *J. Bacteriol.* 171:1496–505
43. Hirota, Y., Ricard, M., Shapiro, B. 1971. The use of thermosensitive mutants of *E. coli* in the analysis of cell division. In *Biomembranes*, ed. L. A. Manson, pp. 13–31. New York: Plenum
44. Hirota, Y., Ryter, A., Jacob, F. 1968. Thermosensitive mutants of *E. coli* affected in the process of DNA synthesis and cellular division. *Cold Spring Harbor Symp. Quant. Biol.* 33:677–93
45. Huisman, O., Jacques, M., D'Ari, R., Caro, L. 1983. Role of the *sfiA*-dependent cell division regulation system in *Escherichia coli*. *J. Bacteriol.* 153:1072–74
46. Hussain, K., Begg, K. G., Salmond, G. P. C., Donachie, W. D. 1987. *ParD*: a new gene coding for a protein required for chromosome partitioning and septum localization in *Escherichia coli*. *Mol. Microbiol.* 1:73–81
47. Hussain, K., Elliott, E. J., Salmond, G. P. C. 1987. The ParD⁻ mutant of *Escherichia coli* also carries a *gyrA*ₐₘ mutation. The complete sequence of *gyrA*. *Mol. Microbiol.* 1:259–73
48. Ishino, F., Jung, H. K., Ikeda, M., Doi, M., Wachi, M., et al. 1989. New mutations *fts-36*, *lts-33*, and *ftsW* clustered in the *mra* region of the *Escherichia coli* chromosome induce thermosensitive cell growth and division. *J. Bacteriol.* 171: 5523–30
49. Ishino, F., Matsuhashi, M. 1981. Peptidoglycan synthetic activities of highly purified penicillin-binding protein 3 in *Escherichia coli*: a septum-forming reaction sequence. *Biochem. Biophys. Res. Commun.* 101:905–11
50. Jacob, F., Brenner, S., Cuzin, F. 1963. On the regulation of DNA replication in bacteria. *Cold Spring Harbor Symp. Quant. Biol.* 28:329–48
51. Jaffe, A., D'Ari, R., Hiraga, S. 1988. Minicell-forming mutants of *Escherichia coli*: Production of minicells and anucleate rods. *J. Bacteriol.* 170:3094–101
52. Jaffé, A., D'Ari, R., Norris, V. 1986. SOS-independent coupling between DNA replication and cell division in *Escherichia coli*. *J. Bacteriol.* 165:66–71
53. Jones, C., Holland, I. B. 1985. Role of the SulB (FtsZ) protein in division inhibition during the SOS response in *Escherichia coli*: FtsZ stabilizes the in-

hibitor SulA in maxicells. *Proc. Natl. Acad. Sci. USA* 82:6045–49
54. Kato, J., Nishimura, Y., Suzuki, H. 1989. *Escherichia coli parA* is an allele of the *gyrB* gene. *Mol. Gen. Genet.* 217:178–81
55. Kato, J.-l., Nishimura, Y., Yamada, M., Suzuki, H., Hirota, Y. 1988. Gene organization in the region containing a new gene involved in chromosome partition in *Escherichia coli*. *J. Bacteriol.* 170:3962–77
56. Labie, C., Bouché, F., Bouché, J.-P. 1989. Isolation and mapping of *Escherichia coli* mutations conferring resistance to division inhibition protein DicB. *J. Bacteriol.* 171:4315–19
57. Lutkenhaus, J. 1990. Regulation of cell division in *E. coli*. *Trends Genet.* 6:22–25
58. Lutkenhaus, J. F. 1983. Coupling of DNA replication and cell division: *sulB* is an allele of *ftsZ*. *J. Bacteriol.* 154: 1339–46
59. Lutkenhaus, J., Sanjanwala, B., Lowe, M. 1986. Overproduction of FtsZ suppresses sensitivity of *lon* mutants to division inhibition. *J. Bacteriol.* 166:756–62
60. Lutkenhaus, J. F., Wolf-Watz, H., Donachie, W. D. 1980. Organization of genes in the *ftsA-envA* region of the *Escherichia coli* genetic map and identification of a new *fts* locus (*ftsZ*). *J. Bacteriol.* 142:615–20
61. MacAlister, T. J., Cook, W. R., Weigand, R., Rothfield, L. I. 1987. Membrane-murein attachment at the leading edge of the division septum: a second membrane-murein structure associated with morphogenesis of the Gram-negative bacterial division septum. *J. Bacteriol.* 169:3945–51
62. MacAlister, T. J., MacDonald, B., Rothfield, L. I. 1983. The periseptal annulus; an organelle associated with cell division in gram-negative bacteria. *Proc. Natl. Acad. Sci. USA* 80:1372–76
63. Maguin, E., Brody, H., Hill, C. W., D'Ari, R. 1986. SOS-associated division inhibition gene *sfiC* is part of excisable element e14 in *Escherichia coli*. *J. Bacteriol.* 168:464–66
64. Maguin, E., Lutkenhaus, J., D'Ari, R. 1986. Reversibility of SOS-associated division inhibition in *Escherichia coli*. *J. Bacteriol.* 166:733–38
64a. Mulder, E., El'Bouhali, M., Pas, E., Woldringh, C. L. 1990. The *Escherichia coli minB* mutation resembles *gyrB* in defective nucleoid segregation and decreased supercoiling of plasmids. *Mol. Gen. Genet.* 221:87–93

65. Mulder, E., Woldringh, C. L. 1989. Actively replicating nucleoids influence positioning of division sites in *Escherichia coli* filaments forming cells lacking DNA. *J. Bacteriol.* 171:4303–14

66. Nanninga, N., Wientjes, F. B., de Jonge, B. L. M., Woldringh, C. L. 1990. Polar cap formation during cell division in *Escherichia coli*. *Res. Microbiol.* 141:103–18

67. Nanninga, N., Woldringh, C. L. 1985. Cell growth, genome duplication, and cell division. In *Molecular Cytology of Escherichia coli*, ed. N. Nanninga, pp. 259–318. London: Academic

67a. Neidhardt, F. C., Ingraham, J. L., Low, K. B., Magasanik, B., Schaechter, M., Umbarger, H. E., eds. 1987. *Escherichia coli and Salmonella typhimurium: Cellular and Molecular Biology*. Washington, DC: Am. Soc. Microbiol.

68. Nishimura, A. 1989. A new gene controlling the frequency of cell division per round of DNA replication in *Escherichia coli*. *Mol. Gen. Genet.* 215:286–93

69. Nordström, K., Austin, S. J. 1989. Mechanisms that contribute to the stable segregation of plasmids. *Ann. Rev. Genet.* 23:37–69

70. Normark, S., Boman, H. G., Bloom, G. D. 1971. Cell division in a chain-forming *envA* mutant of *Escherichia coli* K12. *Acta Pathol. Microbiol. Scand.* 79:651–64

71. Normark, S., Boman, H. G., Matsson, E. 1969. Mutant of *Escherichia coli* with anamolous cell division and ability to decrease episomally and chromosomally mediated resistance to ampicillin and several other antibiotics. *J. Bacteriol.* 97:1334–42

72. Nurse, P. 1990. Universal control mechanism regulating onset of M-phase. *Nature* 344:503–8

73. Orr, E., Fairweather, N. F., Holland, I. B., Pritchard, R. H. 1979. Isolation and characterization of a strain carrying a conditional lethal mutation in the *cou* gene of *Escherichia coli* k12. *Mol. Gen. Genet.* 177:103–12

74. Reeve, J. N. 1984. Synthesis of bacteriophage and plasmid-encoded polypeptides in minicells. In *Advanced Molecular Genetics*, ed. A. Puhler, K. N. Timmis, pp. 212–23. Berlin: Springer-Verlag

75. Rothfield, L. I., Cook, W. R. 1988. Periseptal annuli: organelles involved in the bacterial cell division process. *Microbiol. Sci.* 5:182–85

76. Rothfield, L. I., de Boer, P., Cook, W. R. 1990. Localization of septation sites. *Res. Microbiol.* 141:57–63

77. Rothfield, L. I., MacAlister, T. J., Cook, W. R. 1986. Murein-membrane interactions in cell division. In *Bacterial Outer Membranes as Model Systems*, ed. M. Inouye, pp. 247–75. New York: Wiley

78. Ryter, A., Hirota, Y., Schwarz, U. 1973. Process of cellular division in *Escherichia coli*. Growth pattern of *E. coli* murein. *J. Mol. Biol.* 78:185–95

79. Schwarz, U., Asmus, A., Frank, J. 1969. Autolytic enzymes and cell division of *Escherichia coli*. *J. Mol. Biol.* 41:419–29

80. Spratt, B. G. 1975. Distinct penicillin binding proteins involved in the division, elongation and shape of *Escherichia coli* K12. *Proc. Natl. Acad. Sci. USA* 82:2999–3003

81. Steck, T. R., Drlica, K. 1984. Bacterial chromosome segregation: Evidence for DNA gyrase involvement in decatenation. *Cell* 36:1081–88

82. Storts, D. R., Aparicio, O. M., Schoemaker, J. M., Markovitz, A. 1989. Overproduction and identification of the *ftsQ* gene product, an essential cell division protein in *Escherichia coli* K-12. *J. Bacteriol.* 171:4290–97

83. Taschner, P. E. M. 1988. *Genetic and morphological analysis of cell division in* Escherichia coli. PhD thesis. Univ. Amsterdam, Holland

84. Taschner, P. E. M., Huls, P. G., Pas, E., Woldringh, C. L. 1988. Division behavior and shape changes in isogenic *ftsZ, ftsQ, ftsA, pbpB*, and *ftsE* cell division mutants of *Escherichia coli* during temperature shift experiments. *J. Bacteriol.* 170:1533–40

85. Teather, R. M., Collins, J. F., Donachie, W. D. 1974. Quantal behaviour of a diffusible factor which initiates septum formation at potential division sites in *Escherichia coli*. *J. Bacteriol.* 118:407–13

86. Tormo, A., Ayala, J. A., de Pedro, M. A., Aldea, M., Vicente, M. 1986. Interaction of FtsA and PBP3 proteins in the *Escherichia coli* septum. *J. Bacteriol.* 166:985–92

87. Verwer, R. W. H., Nanninga, N. 1980. Pattern of meso-DL-2,6-diaminopimelic acid incorporation during the division cycle of *Escherichia coli*. *J. Bacteriol.* 144:327–36

88. Walker, G. C. 1987. The SOS response of Escherichia coli. See Ref. 67a, pp. 1346–57

89. Ward, J. E., Lutkenhaus, J. 1985. Over-

production of FtsZ induces minicell formation in *E. coli. Cell* 42:941–49

90. Weigand, R. A., Vinci, K. D., Rothfield, L. I. 1976. Morphogenesis of the bacterial division septum: a new class of septation-defective mutants. *Proc. Natl. Acad. Sci. USA* 73:1882–86

91. Wientjes, F. B., Nanninga, N. 1989. Rate and topography of peptidoglycan synthesis during cell division in *Escherichia coli:* Concept of a leading edge. *J. Bacteriol.* 171:3412–19

92. Woldringh, C. L., Huls, P., Pas, E., Brakenhoff, G. J., Nanninga, N. 1987. Topography of Peptidoglycan Synthesis during Elongation and Polar Cap Formation in a Cell Division Mutant of *Escherichia coli* MC4100. *J. Gen. Microbiol.* 133:575–86

92a. Woldringh, C. L., Mulder, E., Valkenburg, J. A. C., Wientjes, F. B., Zaritzky, A., et al. 1990. Role of the nucleoid in the toporegulation of division. *Res. Microbiol.* 141:39–49

93. Woldringh, C. L., Valkenburg, J. A. C., Pas, E., Taschner, P. E. M., Huls, P., et al. 1985. Physiological and geometrical conditions for cell division in *Escherichia coli. Ann. Inst. Pasteur Microbiol.* 136A:131–38

Annu. Rev. Genet. 1990. 24:275–303

CELL-SPECIFIC GENE EXPRESSION IN PLANTS

Janice W. Edwards and Gloria M. Coruzzi

The Laboratory of Plant Molecular Biology, The Rockefeller University, New York, New York 10021-6399

KEY WORDS: plant transformation, agricultural biotechnology, β-glucuronidase (GUS), tissue-specific gene expression, histochemistry

CONTENTS

0066-4197/90/1215/0275$02.00

INTRODUCTION

The structural complexity of the plant body results from variation in the form and function of cells and also from differences in the manner of combination of cells into tissues and tissue systems.

Katherine Esau, 1953.

Cell-specific gene expression studies in plants are indebted to early botanical monographs in which the cytological structure and function of organs were defined. These studies showed that in many cases plant organs are not sharply delimited, and a number of cell types are present in more than one plant organ. In the absence of strict organ distinction in plants, it is appropriate for biological questions to be addressed at the cellular level

Previous characterizations of plant cells were based on morphological and physiological criteria. With the advent of "reporter" gene systems and plant transformation technology, cells can now be defined at the molecular level by the genes that they express. Cell-specific gene expression studies have provided a more accurate definition of the function of known cell types, and have shown that tissues appearing to be morphologically homogeneous are often composed of different cell types. The ability to detect the expression of an individual gene in single cells has also elucidated differences between individual members of multigene families, and has supplied clues to the identity of genes of unknown function.

Plant genes expressed in a tissue-specific fashion have been identified genetically (homeotic genes) (184) or by their differential expression in various plant organs. More recently, tissue-specific promoters have been identified using "reporter gene trapping" systems (101,144). "Reporter genes" such as chloramphenicol acetyl transferase (CAT) and β-glucuronidase (GUS) have facilitated the detection of tissue- and cell-specific gene expression in plants. Because organogenesis is repeated in plants, gene expression can be monitored within organs of a single plant over its entire life-span. Since plant cells differentiate in response to external signals, plant growth conditions have a significant effect on gene expression (19) and the pattern of cell-specific expression.

The literature on tissue-specific gene expression in plants is extensive and incorporates studies of large multigene families in many species. This review encompasses many cell-specific gene expression studies that have provided insight into the function of plant cells and is not intended to catalog all gene studies in plants. Since cell-specific gene expression studies in plants have had a large impact on agricultural biotechnology, relevant examples have been included in this review.

GENE EXPRESSION IN VEGETATIVE ORGANS

Leaf-Specific Gene Expression

As the major photosynthetic organ of the plant, leaves contain numerous gene products that are produced primarily in a leaf-specific manner and regulated by light or photosynthetic metabolism. To understand the significance of leaf-specific gene expression at the cellular level, it is important to consider the specialized functions of different leaf cell-types in C3 and C4 plants. Two carbon fixation pathways (C3 and C4) exist in plants and are named according to the number of carbon atoms in the first fixation product [for review see (50)]. Photosynthesis in C3 plants requires only one photosynthetic cell type whereas most C4 plants contain Kranz leaf anatomy, which comprises two photosynthetically active cell types, bundle sheath (BS) cells and mesophyll (M) cells, that cooperate in a multistep scheme of CO_2 fixation.

C4 PLANTS   In C4 plants, CO_2 is first fixed into a C4 acid in M cells. Subsequent decarboxylation and refixation steps occur in BS cells where the O_2-sensitive ribulose-1,5-bisphosphate carboxylase (RUBISCO) enzyme is sequestered. The cell-specific distribution of C4 metabolic enzymes in BS and M cells is regulated by gene expression. In situ hybridization and cell-separation experiments have shown that genes for (RUBISCO) and NADP-malic enzyme (NADP-ME) are expressed specifically in BS cells in light-grown C4 plants. In contrast, other C4 genes encoding phosphoenolpyruvate carboxylase (PEPCase), and NADP-malate dehydrogenase (NADP-MDH) are expressed specifically in mesophyll cells [for review see (140)]. Another C4 pathway enzyme, pyruvate, Pi dikinase (PPdK) is expressed at high levels in M cells and at lower levels in BS cells (5).

In situ hybridization studies have demonstrated that cell-specific gene expression in BS and M cells is committed very early in development. For example, RUBISCO large and small subunit genes are expressed in a ring surrounding the provascular cells. This occurs prior to the discernible differentiation of BS cells in developing maize leaves (108). NADP-MDH mRNA accumulation shows a similar pattern of expression at a slightly greater distance from the developing vein (108). C4 gene expression is also regulated spatially within maize leaves. In foliar leaves in which veins are closely spaced, M cells develop in a C4 manner (109). In contrast, in husk leaves, where veins are separated by as many as twenty cells, M cells develop as conventional C3 chlorenchyma (109). This distinction is reflected by gene expression since RUBISCO mRNA is expressed in M cells distal from veins in husks whereas vein-proximal M cells express PEPCase, PPdK, and NADP-MDH (109).

Light changes the cell-specific expression patterns of C4 genes. The distribution of RUBISCO protein and mRNA in BS and M cells differs in dark-versus light-grown C4 plants (109, 173). In dark-grown plants RUBISCO gene expression occurs predominantly in M cells. Upon illumination, RUBISCO gene expression switches to BS cell. At the same time, light induces the expression of other C4 genes in M cells (140). Thus, these studies indicate that light plays a role both in positively and negatively influencing cell-specific gene expression.

C3 PLANTS   In C3 plants, the nuclear genes encoding the small subunit of RUBISCO (rbcS) are regulated by light and are expressed predominantly in a leaf-specific manner. The light-regulated expression of rbcS and other plant genes has been reviewed recently (69), therefore, this review concentrates on studies concerning the leaf-specific expression of these plant genes. Studies using in situ immunofluorescence have demonstrated that rbcS gene expression occurs in several distinct photosynthetic cell types; leaf mesophyll cells, guard cells of the leaf epidermis, and chlorenchymal cells of the midrib (4). The rbcS-3A gene is a highly expressed member of the large rbcS multigene family in pea (35,58). The rbcS-3A promoter is able to confer leaf-specific and light-dependent expression on the CAT reporter gene in transgenic tobacco (57). Recent promoter-manipulation experiments in transgenic plants have shown that some of the cis-acting elements of the rbcS gene responsible for conferring either light-regulation (105) or organ-specificity (4) are distinct. Other photosynthetic genes such as those encoding chlorophyll a/b binding protein (Cab) and a component of the oxygen-evolving complex of photosystem II (ST-LS1) are also expressed in a light-regulated manner in leaves (57, 177–179, 188). In addition to its expression in photosynthetic leaves, the ST-LS1 promoter can be induced in roots that are exposed to light (189). These results reveal that the organ-specific expression of ST-LS1 can be altered by external factors (i.e. light). In addition to photosynthetic genes, vegetative storage protein genes [for review see (186)], and genes involved in specialized processes such as leaf abscission (196), or trichome development (79), display predominantly leaf-specific expression patterns.

The cell-specific expression of glutamine synthetase (GS) in leaves has provided an explanation for the existence of multiple isoenzymes in plants. Studies using GS promoter-GUS reporter gene fusions have revealed that pea nuclear genes for chloroplast and cytosolic GS are expressed in different cell types in transgenic tobacco (52). The promoter for chloroplast GS2 directs GUS expression within photosynthetic cells (e.g. palisade parenchyma, chlorenchyma of the midrib and stem, and in photosynthetic cotyledons) (52). This cell-specific expression pattern is consistent with previous molecular and genetic studies (51, 212) and indicates that chloroplast GS2 plays a role in the reassimilation of photorespiratory ammonia. For cytoslic GS, histochemical analysis of transgenic plants reveals that one gene *(GS3A)* is expressed specifically in phloem cells of leaves, stems, roots, and abundantly in cotyledons of germinating seedlings (52). This cell-specific expression pattern indicates that the cytosolic GS3A isoform plays a role in generating glutamine for transport. Taken together, the nonoverlapping, cell-specific expression patterns of the GS genes indicate that the chloroplast GS2 and cytosolic GS3A isoforms perform unique metabolic functions.

Stem-Specific Gene Expression

Several plant genes are expressed abundantly in stems, although few are expressed exclusively in this organ. The known plant genes that are expressed predominantly in stems include those encoding cell wall proteins and other metabolic enzymes of the plant vasculature. The organ-specific expression of genes in the hypocotyl is also reviewed in this section.

CELL WALL PROTEIN GENES   Three classes of cell wall proteins exist in dicotyledonous plants: hydroxyproline-rich glycoproteins (HRGPs), the glycine-rich proteins (GRPs), and proline- or hydroxyproline-rich proteins (PRPs) [for review see (27)]. A common feature is the presence of a repeating basic amino acid motif. Each class of cell wall protein is encoded by a small multigene family in several plant species (32, 84, 85, 96, 97, 123). In the soybean *PRP* gene family, there is a complex pattern of organ-specific and developmentally regulated expression for each of the three *PRP* genes studied (84, 85). Individual *PRP* mRNAs are expressed differentially in the apical, elongating, and mature regions of the hypocotyl (84, 85).

Hydroxyproline-rich glycoproteins (HRGPs) are major structural components of the plant cell wall, and are thought to play a role in defense against pathogens (32, 175). In soybean seed coats, HRGPs have been localized to the palisade epidermal and hourglass cells (26), but little else is known about the cell-specific expression of this class of structural proteins.

The cell-specific expression of genes encoding the glycine-rich cell wall proteins (GRPs) of *Phaseolus* (96, 97) and *Petunia* (30, 31) has been examined. Antibodies to a *Phaseolus* GRP fusion protein have revealed that GRPs are cell wall proteins (96) and that GRPs in young hypocotyls, ovaries, and seed coats are associated with the protoxylem cells (97). Another gene family encoding putative cell wall proteins in pea *(S2* and *P4)* is expressed differentially in the stems and pods (216).

A technique that employs plant tissue printing onto nitrocellulose (26) has been used to detect the tissue-specific distribution of small auxin-induced RNAs (SAURs) in seedlings (125). These studies have shown that SAURs accumulate primarily in the hypocotyl within the cortex, epidermis, and pith tissue. This expression pattern is induced but conserved within the same regions in seedlings treated with the auxin analoge, 2, 4-dichlorophen-oxyacetic acid (125). Gravitrophic changes alter the tissue distribution of SAURs, and the redistribution of SAURs in seedlings that have been switched from a vertical to a horizontal position shows a strong correlation with hypocotyl cell extension (125).

VASCULATURE-SPECIFIC EXPRESSION OF BIOSYNTHETIC ENZYMES
Several genes encoding biosynthetic enzymes are expressed in a vascula-ture-specific manner. Among these are the genes encoding S-adenosylmethio-nine synthetase (*sam-1*) (151) and the cytosolic isoform of glutamine syn-thetase (*GS3A*) (52). S-Adenosylmethionine synthetase catalyzes the biosyn-thesis of S-adenosylmethionine (AdoMet) and plays a role in the biosynthesis of polyamines and ethylene (217). Histochemical analysis of transgenic *Ara-bidopsis* plants demonstrates that the *sam-1* promoter (151) directs GUS gene expression primarily in the vascular tissues (xylem and phloem), sclerenchy-ma, and root cortex. The high expression of the *sam-1* gene in xylem and phloem tissues may be related to the role that AdoMet plays in lignification (81). A pea gene for cytosolic GS (*GS3A*) is expressed exclusively in phloem (52) (see previous section).

Root-Specific Gene Expression

Roots provide mechanical support, function in nutrient absorption, serve as storage organs in certain species, and are the primary contact point for soil microbes. Tap and lateral roots are composed of several different cell types: epidermis, cortex, stele, apical, meristem, and root cap. Through differential screening, genes have been identified which are expressed to high levels in roots, however, none were expressed exclusively in this organ (55). Cell-specific gene expression in roots has been examined for a gene encoding a tobacco cell wall protein, *HRGPnt3* (95). Histochemical analysis of transgen-ic tobacco revealed that the *HRGPnt3* promoter directed transient GUS expression in the pericycle and endodermis, specifically in a small subset of cells destined to form the lateral root tip (95). This transient cell-specific expression pattern suggests that the *HRGPnt3* cell wall protein may play a role in the penetration of the cortex and epidermis of the main root (95).

In legumes, root-specific genes play a role in plant-*Rhizobium* interactions. Root-hair cells, which are the site of *Rhizobium* attachment and penetration, contain mRNAs that are expressed specifically in response to *Rhizobium* inoculation (70, 166). Transgenic experiments have shown that host range is determined, at least in part, by lectins expressed in legume roots. Transgenic roots of white clover that express a pea seed lectin gene acquire the ability to be nodulated by the pea group-specific *Rhizobium leguminosarum* (43). While these results concern the artificial expression of a seed lectin in legume roots, root-specific lectin genes may occur in legumes as they do in a nonlegume (barley) (114).

A recently identified plant transcription factor (TGAla) that is active predominantly in roots has sequence homology to animal and yeast transcription factors (92). Expression of the plant transcription factor (TGAla) is ten-fold higher in roots compared to leaves (92) and the DNA sequence element to which TGAla binds confers root-specific expression to a heterologous *rbcS* gene (63, 106).

Biotechnological Applications of Gene Expression in Vegetative Organs

The study of the cell-specific expression patterns of plant genes in vegetative organs has numerous applications to agricultural biotechnology. For example, the analysis of the organ- and cell-specific expression patterns of glutamine synthetase (GS) genes has application to the engineering of herbicide-resistant plants (146). Because different GS isoenzymes are expressed in different cell types (52), it may be necessary to express herbicide resistant forms of GS in both mesophyll and phloem cells to confer high-level herbicide resistance. Plant promoters that direct vascular-specific gene expression [e.g. *GS3A* (52), *sam-1* (151), *GRP* (97)] can be used to express products that protect against phloem- or xylem-borne pathogens. The trichome-specific promoter (*gl-1*) (79) could be used to express novel antipest products in trichomes, cells that are involved in protection against preying insects (61).

GENE EXPRESSION IN REPRODUCTIVE ORGANS

A flower is a complex organ comprising several types of tissues. The majority of flowering plants are hermaphroditic, and develop flowers that contain stamen (filament plus anther containing pollen) and a pistil (stigma, style, and ovary containing ovules). Pollen that lands on the stigma germinates and the single-celled pollen tube grows between the files of cells in the pistil to reach the ovule where fertilization occurs. The complex processes of flower development and plant reproductive biology are beginning to be dissected at the molecular level.

Genes Controlling Fertilization

In many species, self-fertilization can be blocked by a genetically controlled mechanism termed self-incompatibility (SI), whereby the pistil rejects pollen derived from the same plant. The SI phenotype is determined by the genetic constitution of the haploid pollen grain (gamete) in gametophytic plants (e.g. Solanaceae), and by the diploid genotype of the parent (sporophyte) in sporophytic plants (e.g. Cruciferae). The distinguishing morphological difference between these two SI systems (gametophytic and sporophytic) is the region of the pistil in which pollen tube growth is arrested. In *Nicotiana alata*, which is gametophytic, SI pollen germinates, but subsequent pollen tube growth is arrested in the style. In *Brassica*, which is sporophytic, SI pollen fails to germinate and/or penetrate the papillar epidermal cells of the stigma [for reviews see (49, 138)].

SI is controlled by a complex multiallelic locus (*S-locus*) and genes encoding S-locus-specific glycoproteins (SLSGs) have been cloned from both *N. alata* and *Brassica oleracea* (2, 137). In *N. alata*, the *S-locus* genes are expressed in stigma throughout the pathway of pollen tube growth in the style, and also in the placental epidermis of the ovary, reflecting the path of pollen tube growth toward the ovules (2, 34). In contrast, in *Brassica*, the *S-locus* genes are expressed exclusively in the papillar cells of the stigma (139). Thus, the cell-specific expression pattern of *S-locus* genes correlates with the site and timing of pollen tube inhibition in each SI system (2, 139).

Transgenic experiments have demonstrated that while flower-specific expression of an *S-locus* gene from *Brassica* is superficially maintained in a heterologous host, its cell-specific expression pattern is determined by the host environment. In particular, the *S-locus* gene from *Brassica*, which is normally expressed exclusively in the papillar cells of the style, is expressed throughout the stylar-transmitting tissue in transgenic *Nicotiana tabacum* (131). The introduction of this *S-locus* gene from self-incompatible *Brassica* into self-compatible *N. tabacum* did not alter the compatibility phenotype of the host (131). This is not suprising since the *S-locus* genes from *Nicotiana* and *Brassica* are not similar (49), and since it is not known how the SLSGs function to prevent self-fertilization.

It has been postulated that the SLSGs function in a receptor-ligand fashion to block pollen tube growth. While this model is still operative for *Brassica*, it has recently been shown that the *S-locus* gene products of *N. alata* encode ribonucleases, which are postulated to prevent self-fertilization by entering the pollen tube cell and causing cytotoxicity (126). Although the nature of gametophytic SI indicates that the S-gene must be expressed in the pollen as well as in the style, only trace amounts of ribonuclease were detected in pollen (126). Future studies on the nature of the *S-locus* gene products expressed in the pollen should elucidate the molecular mechanism of SI in *N. alata*.

Stamen- and Pistil-Specific Gene Expression

Stamen-specific genes have been identified that are expressed in the anther in sporophytic cells (e.g. tapetal cells) (72, 181) or in gametophytic cells (e.g. pollen) (75, 187). For a set of anther-specific genes in tomato, the cell-specific expression in sporophytic or gametophytic cells varies during development, and transcripts for these genes increase during gametogenesis and reach a maximum in mature pollen grains (199, 200). At flower maturity, transcripts were also detected in the epidermal and endothecial cell layers of the anther wall (199, 200). That these stamen-specific genes are involved in pollen development and pollen tube growth is supported by studies demonstrating their expression at the tip of a growing pollen tube, and their lack of expression in male-sterile flowers (200). A gene family (P2) has been identified in *Oenothera organensis* whose members are expressed at high levels only in mature pollen and in pollen tubes (23). This finding, plus the homology of P2 proteins to polygalacturonase, suggests that the P2 gene products may function to depolymerize pectin during pollen germination and pollen tube growth (23).

For pistil-specific genes, their expression in the stigma, style, or ovaries at certain stages of development has provided insight into their possible function (46, 65, 72). For example, one pistil-specific gene is expressed exclusively in the stylar-transmitting tissue just prior to fertilization, suggesting that this gene product may play a role in regulating pollen tube growth (65).

In certain homeotic plant mutants, morphogenesis of male and female organs is aberrant (18, 54, 185), and in two cases, the affected genes have been cloned. In *Antirrhinum majus*, the deficiens mutation (*defA*) causes conversion of the male organ into an abnormal female organ and petals into sepaloid leaves. The *defA* gene has recently been cloned by differential screening and shown to encode a putative transcription factor (185). The *defA* gene is expressed exclusively in flowers, but its expression is not restricted to a particular floral organ (185). In *Arabidopsis*, the agamous mutation (*Ag*) causes the reproductive organs of the flower to be replaced by petals and sepals. The *Ag* gene has recently been cloned by T-DNA tagging and has been shown to encode a putative transcription factor. The *Ag* gene is expressed exclusively in flowers, specifically in the stamens, carpels, and also in their progenitor cells (E. Meyerowitz, personal communication).

Petal-Specific Gene Expression

Among the genes identified with high expression in petals are structural and regulatory genes involved in pigment production. The most common flower pigments are flavonoids synthesized via the phenylpropanoid and flavonoid biosynthetic pathways (48). Some of the best-studied genes for enzymes along these pathways are phenylalanine ammonia lyase (PAL), chalcone synthase (CHS), chalcone isomerase (CHI), and dihydroflavanol-4-reductase (DFR), which are all encoded by multigene families [for a list of cloned genes see (130)]. Although members of these gene families are also expressed in vegetative organs where their expression is linked to the production of lignins and plant defense compounds, the discussion here is limited to members expressed in flowers.

For the PAL gene family of *Phaseolus*, only one member is expressed at high levels in petals (PAL2) (115). The *PAL2* promoter was able to direct high-level expression of GUS in transgenic tobacco flowers, specifically in the pink region of the corolla (15, 116). Of the 12 members of the CHS gene family of *Petunia*, only two (*CHS-A* and *CHS-J*) are expressed in floral tissues, specifically in petals and anthers (99, 100). In pea, the two CHS genes (*CHS1* and *CHS3*) expressed in petals are also expressed in roots where phenylpropanoid biosynthetic genes function in disease resistance and nodulation (76). In situ studies have localized CHS gene expression to epidermal cells in petals of tobacco (142). The CHI genes of *Petunia* are both expressed in flowers, although the *CHI-A* gene is expressed in the limb and tube of the flower, while the *CHI-B* gene is expressed exclusively in immature anthers (205). mRNA for DFR accumulates in *Petunia* flowers, first in the anther, and subsequently in the limb and tube, mirroring the timing of development of these organs (10).

These studies have begun to uncover the molecular mechanisms mediating phenylpropanoid gene expression in floral organs. CHS, CHI, and DFR mRNAs accumulate coordinately during limb, tube, and anther development (205). That differences in the expression of these genes occur in distinct floral organs is supported by the finding that the CHI-A gene uses alternate promoters in limb and tube versus in mature pollen grains (204). That floral-specific transcription factors exist is supported by the identification of two regulatory loci (*a* and *a2*) that specifically affect the expression of CHS genes in flowers but not in roots (76).

Another gene with petal-specific expression is enolpyruvylshikimate-3-phosphate synthase (EPSPs), an enzyme involved in the biosynthesis of phenylalanine that feeds into the phenylpropanoid pathway. In *Petunia*, EPSPs gene expression is more abundant in petals and anthers than in leaves, suggesting a link between EPSPs gene expression and pigment production (66). Further analysis demonstrated that EPSPs gene expression is activated later than that of the CHS and CHI genes, even though EPSPs functions earlier in the pathway of pigment production (11). Therefore EPSPs expression is not consonant with pigment production, and the reason for high-level EPSPs gene expression in *Petunia* flowers remains unknown (11). Transgenic experiments have demonstrated that the EPSPs promoter directs a complex pattern of flower-specific expression within specific cells of the petal, tube, ovules, and pollen in *Petunia* (11).

Biotechnological Applications of Floral-Specific Gene Expression

One important practical application of genes expressed specifically in floral organs involves the creation of male sterile plants for the production of hybrid seed and for genetic studies. Conditional male sterile lines would facilitate genetic complementation studies, especially in plants with small flowers (e.g. *Arabidopsis*) where hand emasculation is tedious and requires flower microdissection. Male sterile flowers were recently created using an anther-specific promoter (72) to direct the expression of a ribonuclease in transgenic plants (R. Goldberg, personal communication). The notion that ribonucleases may be effective agents for preventing pollen formation, germination, or pollen tube growth is supported by the finding that the S-locus gene products in *N. alata* encode ribonucleases (126). The study of flower-specific gene expression also has applications to the manipulation of flower color in the production of ornamental flowers [for review see (129)]. For example, a new flower color was produced when a maize DFR gene that encodes an enzyme with a novel substrate specificity was introduced into *Petunia* (127). Inhibition of flower color can be achieved by antisense CHS mRNA production (202), or surprisingly, by the over expression of CHS or DFR in plants via a mechanism described as "co-repression" (136, 203).

GENE EXPRESSION IN SEEDS

In flowering plants, seed development is initiated after two fertilization events occur. One fertilization involves one pollen nucleus and an egg that produces the diploid zygote (embryo). A second genetically identical pollen nucleus unites with two polar nuclei of the embryo sac to give rise to the triploid endosperm. The expression of many seed-specific genes is confined to the embryo or endosperm tissues and this expression has been examined at the cellular level in a large number of plant species.

Embryo-Specific Gene Expression

SEED-STORAGE PROTEIN GENES   A common feature of globulin-storage protein gene expression is that it is predominantly restricted to embryonic tissues of the dicot seed. Embryo-specific gene expression has been demonstrated for a large number of seed-storage protein genes including; β-conglycinin (6, 9, 20, 28, 135), glycinin (141), legumins (8, 53, 174), vicilins (86), helianthinin (17, 91), β-phaseolin (24), and a 12S-storage protein of *Arabidopsis* (148). The correct expression of many of the genes from legumes in transgenic tobacco (6, 8, 28, 53, 174) indicates that the DNA sequences and *trans*-acting factors controling seed-specific gene expression are conserved between these distantly related phyla. This observation is particularly noteworthy since mature legume seeds (without endosperm) are strikingly different from tobacco seeds (with endosperm). Thus, endosperm tissue is not required for correct expression of seed-storage protein genes within the embryo. Similarities in the *cis*-acting regulatory regions of genes encoding seed-storage proteins (53, 67, 174), as well as in genes of different seed-protein types (24, 89, 91), indicate that a common transcriptional mechanism may regulate embryo-specific gene expression.

OTHER EMBRYO-SPECIFIC GENES   Although the function of seed-storage proteins in supplying carbon and nitrogen to the germinating seedling is well established, the function of several other embryo-specific seed proteins (e.g. lectins, proteinase inhibitors) is not yet fully understood. In addition to possibly serving as storage proteins, these proteins have also been implicated as defense proteins that are synthesized in response to wounding. Because these proteins appear to have multiple functions in the seed, the regulation of their genes may be complex.

Lectins represent a class of proteins that function as seed-storage proteins and also have the distinctive property of binding carbohydrates. Although the function of lectins in plants is not yet fully understood, they are widely used as tools in medical cell biology. Phytohemaglutenin (PHA) is a lectin-storage protein in seeds of *Phaseolus vulgaris* and soybean. Transgenic studies in tobacco have shown that *Phaseolus* and soybean lectin genes are expressed predominantly in the embryo (143, 210) and that the 5' upstream region of a *Phaseolus* PHA gene (*lec2*) is sufficent to direct correct temporal and spatial expression (155). The conservation of a 63-bp repeat containing a core consensus sequence in the promoters of several PHA genes implicates these sequences in gene regulation (44, 209, 210).

Protease inhibitors expressed in seeds have been proposed to function as storage proteins, regulators of endogenous proteinases, or factors that protect plants from insect attack (82, 161). Individual members of the Kunitz trypsin inhibitor (KTi) gene family are expressed in the embryo in both soybean and transgenic tobacco (88, 152). Transgenic studies have shown that the expression of each KTi gene in the cotyledon or axis varies at different times during embryogenesis (152). The cell-specific expression of soybean KTi genes in the cotyledons is temporally regulated in developing seeds of transgenic tobacco (88, 152). The KTi mRNAs accumulate progressively from the outside to inside of each cotyledon in a "wave-like" pattern as embryogenesis proceeds (152). Both the timing and extent to which expression occurs is specific for each KTi gene. The observation that a similar KTi mRNA accumulation pattern is observed in somatic embryos indicates that this wave-like pattern of expression is not due to the influence of surrounding nonembryonic tissue (152). The mechanisms responsible for the wave like pattern of KTi mRNA accumulation have not yet been established, but may be the result of transcriptional activation of KTi genes due to a gradient of regulatory substances (i.e. hormones) (152).

Endosperm-Specific Gene Expression

SEED-STORAGE PROTEIN GENES   Zeins are endosperm-specific seed-storage proteins that are encoded by a large multigene family in maize (104). Although efficient transformation systems have not been established for monocotyledonous plants, the presence of endosperm tissue in tobacco seeds has made tobacco a model transgenic system for the study of endosperm-specific genes from cereals. For example, when a zein promoter was fused to the GUS reporter gene and introduced into transgenic tobacco, endosperm-specific expression was conserved (167). A comparison of zein promoter sequences reveals a 15-bp consensus sequence (22) that binds a nuclear factor (120). This zein consensus "box" is similar to a conserved element in the promoters of other endosperm-specific genes (i.e. cereal storage proteins (60) and maize sucrose synthase (215)). However, when deletions of a zein promoter were assayed in transgenic tobacco, the entire 15-bp sequence was not required for endosperm-specific gene expression (124).

Genetics has played an important role in elucidating the mechanisms behind cell-specific gene expression in maize. Mutants exist in which the synthesis of zeins is reduced [for review see (132)]. The *opaque-2* (*O2*) gene in maize is involved in the differential transcription of zein genes (98) and has been cloned by transposon tagging (133, 169). DNA sequence comparison of the *O2* gene with mammalian oncogenes (e.g. *v-jun, v-fos*) and a yeast transcription factor (*GCN4*) reveals a strong similarity in the leucine zipper DNA binding domain (77, 168). The ability of the O2 protein to bind to the promoter of a zein gene has been demonstrated in an in vitro binding assay (168).

Wheat storage proteins are divided into glutenins and gliands that together form the gluten complex responsible for the bread-making properties of wheat. Glutenin storage protein genes of wheat are expressed in an endosperm-specific manner. A high molecular weight (HMW) glutenin genomic clone from wheat is expressed correctly within the endosperm of transgenic tobacco (156). The promoters for a wheat HMW glutenin and low molecular weight (LMW) glutenin gene are individually sufficient for endosperm-specific expression of the chloramphenicol acetyl transferase (CAT) reporter gene (29).

Hordeins, the prolamines of barley, are structurally related to the wheat prolamines. Barley hordein accumulates in an endosperm-specific manner and recently it has been demonstrated that the 5' upstream sequences of a hordein gene of barley confer endosperm-specific expression to the CAT reporter gene (122).

PROTEINASE INHIBITOR GENES   The bifunctional α-amylase/-subtilisin inhibitor (BASI) is a member of the trypsin (Kunitz) inhibitor family that is specific for bacterial serine proteases (218) and for the endogenous α-amylase 2 isozyme of germinating barley (134). mRNA for this "double-headed" inhibitor is expressed in the starchy endosperm during wheat seed development (112). However, during germination, the BASI mRNA is expressed in aleurone layers of wheat seeds where its accumulation is enhanced by abscisic acid and is abolished by gibberellic acid (112). This pattern of "tissue-switching" suggests that BASI serves two separate functions: protection of the seed against bacterial infection during seed development, and modulation of germination events by controlling α-amylase activity during germination (112).

Aleurone-Specific Gene Expression

The aleurone is a specialized layer of the cereal seed that is derived developmentally from the outer region of the endosperm. Enzymes within the aleurone layer become functional after the rehydration of the desiccated seed. Enzymes that are specifically expressed in the aleurone layer break down the endosperm tissue to provide nourishment for the developing seedling. These aleurone-specific hydrolases are synthesized in response to plant hormones (gibberellins) produced by the embryo during germination. Several genes have been identified that are expressed predominantly within the aleurone, the best characterized are the α-amylase genes.

Three classes of α-amylase genes (α-Amy1-3) have been identified in wheat that are differentially expressed in seeds (7, 111). The α-Amy1 and αAmy2 genes are active in the aleurone layer of germinating wheat grains and show different temporal patterns of expression during germination. Individual α-Amy2 genes are expressed in both the gibberellin-responsive aleurone of the germinating grain and earlier in the developing grain (87). The α-Amy3 genes are unique in that they are expressed only in the developing grain (111). Because no tissue in tobacco seeds is analogous to aleurone, oat aleurone protoplasts have been used to analyze the transient expression of the α-Amy2 promoter (86). In the oat protoplast transient-expression system, gibberellins specifically control expression of the α-Amy2 promoter (86). Gibberellins are also important for activity of a protein factor that binds to a 500-bp sequence in a rice α-amylase gene (145).

Biotechnological Applications of Seed-Specific Gene Expression

Storage proteins are a major component of the seed and are of considerable economic and nutritional importance. Since cereal grains are typically poor in lysine, modifications in the genes encoding major storage proteins can increase their nutritional quality. Similarly, a modification in the methionine and cysteine content of dicotyledonous seed-storage proteins can increase their nutritional value. The ability of seeds to store proteins stably over long periods of time makes them a good system for the production of pharmaceutical products. Recently, a seed-storage protein gene was used as a molecular vehicle to produce foreign peptides in plants. A chimeric gene consisting of a 2S albumin seed-storage gene of *Arabidopsis* and a mammalian neuropeptide gene has been expressed in transgenic *Arabidopsis* and *Brassica napus* plants (207). In transformed plants carrying the chimeric construct, large amounts of the neuropeptide were obtained specifically from the seeds (207). The elasticity of flour is a trait important in bread making and is a function of the high molecular weight glutenin seed-storage protein. Modified glutenin genes could be introduced into wheat and used to produce flours with greater elasticity (56). Another area of seed biotechnology involves enhancing the protection of crop seeds against insect pests; this would greatly reduce the losses in yeild and the need for pesticides. For example, the cowpea trypsin inhibitor gene confers resistance against insect pests when introduced into transgenic tobacco (82).

GENE EXPRESSION IN FRUITS

The understanding of genes involved in fruit ripening has important agricultural implications. Fruit ripening is a complex, developmentally regulated process that depends, in part, on the plant hormone, ethylene (118). Several genes encoding fruit-specific mRNAs have been cloned from tomato (121, 180). Although many of these mRNAs encode products of unknown function, a member of this group whose function has been identified is the polygalacturonase (PG) gene (74, 180). PG functions to partially solubilize the pectin fraction of the cell wall and is therefore thought to play a role in fruit softening. PG mRNA accumulation increases during fruit ripening due to the transcriptional activation of the gene (42), and increases coordinately with the rise in ethylene production in the fruit (40). PG gene expression is different in three different ripening-impaired mutants, rin, nor, and Nr (41). This suggests that at least three distinct molecular mechanisms contribute to the overall expression of the PG gene, and to fruit ripening in general.

Ethylene plays an important role in fruit ripening and a gene encoding the rate-limiting enzyme of ethylene biosynthesis (ACC synthase; 1-aminocyclopropane-1-carboxylate synthase) has been cloned from zucchini fruits (165). Separate studies have focused on other genes that are induced in response to ethylene (21, 118, 119). The expression of a fruit-specific mRNA, E8, is responsive to ethylene (118, 119). A 4.4-Kb DNA sequence of the E8 gene required for both ethylene-responsive and fruit-specific gene expression in transgenic tomato plants specifically interacts with a DNA binding protein activity (33, 38).

Biotechnological Applications of Fruit-Specific Gene Expression

Fruit softening and fruit texture are the primary targets in attempts to modify fruit quality and have commercial importance for both processing and fresh-market tomato varieties. Fruit-specific genes can now be used as molecular tools to modify the fruit-ripening process. Recent studies have focused on the use of antisense mRNA for the fruit-specific PG gene to reduce PG enzyme levels in plants with the aim of reducing the rate of ripening and increasing the shelf-life of tomato fruits (103, 182, 183). Surprisingly, the expression of PG antisense mRNA in transgenic tomato plants had no effect on the onset of ripening or fruit softening (182, 183). The results of these transgenic studies suggest that PG is not the primary determinant of fruit softening. However, since PG is extremely abundant in ripe fruit, residual PG whose synthesis is not affected by the PG antisense mRNA may be sufficient for the observed fruit ripening. The recent isolation of the gene encoding ACC synthase (165) provides the potential to manipulate endogenous ethylene production and consequently to deter fruit ripening. It may be possible to reduce ethylene synthesis by inhibiting ACC synthase expression with antisense RNA. Further experiments are required to adequately address the role of PG and ACC synthase in fruit ripening and to effectively modify crop plants to affect ripening charcteristics.

GENE EXPRESSION IN SPECIALIZED ORGANS

Tuber-Specific Gene Expression

Tubers are specialized storage organs derived from stems. The tuber-bearing members of the Solanaceae form a small group in comparison to the family. Most species in the subsection *Potatoe* can be crossed with relative ease, suggesting their recent evolutionary divergence within the genus *Solanum* (78). This has been borne out by molecular studies demonstrating that genes expressed exclusively in potato tubers can be activated in nontuber-bearing solanaceous species by metabolic signals, as detailed below.

PATATIN GENES   Up to 40% of the soluble protein of potato tubers is represented by a family of immunologically identical glycoproteins called patatin, whose exact function remains unknown (149). While the abundance of patatin in tubers suggests that it serves as a storage protein, patatin also possesses enzymatic activities corresponding to lipid acyl hydrolase and esterase, which may play protective roles (3, 158). The characterized patatin genes fall into two classes (I and II) that are distinguished by their expression patterns and by a small sequence difference in their 5' noncoding regions (14, 128, 154, 159, 198). Class I patatin accounts for 99% of patatin mRNA found in tubers, and is not normally expressed in leaves, roots, or stems (128). Class II patatin accounts for only 1% of the patatin mRNA in tubers and is also detected at low concentrations in roots (153). When the isolated promoters for class I and class II patatins were used to direct the expression of GUS or CAT reporter genes in tuber-bearing transgenic potato plants, additional distinctions between these two classes of patatin genes were observed. GUS expression directed by the class I patatin promoter was 100–1000-fold higher in tubers than in leaves, stems, or roots, and histochemical analysis reveals that GUS expression occurs in the parenchymal cells of the tuber (157, 214). In transgenic experiments using the class II patatin promoter-CAT fusions, the class II patatin promoter was expressed at low levels in tubers and roots of transgenic potato (197). Further analysis of these fusions revealed that the low level of expression of the class II patatin promoter in tubers is actually due to high-level expression in a few defined cells in the pericyclic ring and the phellem, as revealed by histochemical GUS analysis (102). Thus, the ability to monitor cell-specific expression has enabled the discrimination of a promoter that functions at low levels in many cells, from one that functions at high levels in only a few specialized cells.

Although tuberization is always accompanied by patatin gene expression, patatin can in some cases be induced to accumulate in nonstorage tissues. Accumulation of patatin was high in stems and petioles when carbon sinks (e.g. tubers and axillary buds) were removed (150). These early results suggested that patatin could accumulate in other organs in the absence of morphological tuberization, and that photosynthase accumulation may play a role in inducing patatin expression (150). This suggestion has been confirmed by recent transgenic analyses which demonstrate that the expression of class I patatin could be induced by sucrose in leaves and stems in both tuber-bearing and nontuber-bearing solanaceous species (157, 213, 214). By contrast, the expression of the patatin class II gene is unaffected by the concentrations of sucrose (102). These studies suggest that the mechanism for high-level expression of class I patatin gene in potato tubers has evolved from one that functions in other organs and also in nontuber-bearing solanaceous species.

PROTEINASE INHIBITOR II GENES   Storage organs of plants such as tubers, seeds, and fruits often contain considerable amounts of proteins that inhibit proteolytic enzymes. One such is proteinase inhibitor II (PI-II), which inhibits trypsin and chymotrypsin. The idea that PI-II protein functions as a defense against microbial or animal pathogens (160) has been corroborated by studies in which over-expression of PI-II in tobacco retarded the growth of hornworm larve, compared to control untransformed plants (90). While the PI-II gene has been a paradigm for the study of wound-inducible gene expression in several organs (1, 93, 94, 147, 163, 164, 192), its discussion here is limited to expression in tubers.

RNA analyses have shown that the expression of the PI-II gene is tuber-specific in nonwounded potato plants, and accumulation in leaves and stems occurs only in response to wounding (163). Wounding of tubers represses the expression of PI-II in tubers (163), but causes a systemic induction of PI-II gene expression in leaves and stems (162). In transgenic studies, the promoter for the PI-II gene of potato directed constitutive expression of GUS which was localized to the parenchymal cells associated with vascular bundles (93), in stolon and tubers of nonwounded potato plants. While GUS expression in leaves, stems, and petioles was low in nonwounded plants, wounding induced high-level GUS expression in cells associated with the vascular tissue in these organs (93). The vascular-associated localization of GUS expression in both tubers of nonwounded plants and in organs of wounded plants supports the theory that a signal produced by wounding is transmitted to nonwounded parts of the plant via the vascular system (93).

Nodule-Specific Gene Expression

The association of *Rhizobium* with leguminous plants results in the formation of a novel organ, the nitrogen-fixing root nodule. While *Rhizobium* genes involved in this process have been identified genetically, plant genes involved in this process have, for the most part, been identified by their differential expression in nodules versus roots of legumes such as soybean (64, 113, 172, 208), pea (73), *Phaseolus* (25), and alfalfa (45, 206). The term "nodulin," defined to include only those plant genes expressed exclusively in nodules, implies that legumes possess a unique set of genes sequestered for the sole purpose of nodule formation and metabolism. However, more sensitive methods of detection (e.g. PCR, GUS) and the examination of other organs (e.g. leaves, stems, flowers, seeds) have shown that genes formerly regarded as nodulins are expressed in other organs (12, 59, 166), and can be assigned a nonsymbiotic counterpart, even in nonlegumes (16). Several nodulins encode structural proteins (e.g. peribacteroid membrane proteins) or proteins involved in nodule metabolism (e.g. leghemoglobin, glutamine synthetase, sucrose synthase, uricase) [for a list of cloned nodulin genes see (39)].

EARLY NODULINS   "Early" nodulins (ENOD) are expressed at early stages of root nodule development well before the onset of nitrogen-fixation. Within a few hours after inoculation with *Rhizobium,* bacteria begin to enter root-hair cells through an infection thread (71). Two ENOD genes of pea (*ENOD 12* and *ENOD 2*) both encode cell wall proteins and are expressed similarly during nodule development. However, recent analysis of their cell-specific expression patterns has distinguished these genes (62, 166, 201). In situ analysis revealed that the *ENOD 12* gene is expressed in root-hair and cortex cells in which infection threads are present, in cells preparing for infection thread passage, and in the nodule primordium where bacteria are released (166). While this suggests that the ENOD 12 protein is involved in infection thread formation in nodules, ENOD 12 has a function in nonsymbiotic organs as well (stems and flowers) (166). In situ analysis shows that *ENOD 2* is expressed exclusively in the cells surrounding the central infected zone of the nodule, suggesting that this cell wall protein plays a role in creating an oxygen barrier to protect oxygen-sensitive nitrogenase of *Rhizobium* (201).

LATE NODULINS   Leghemoglobin (Lb) is present in large amounts in the cytoplasm of the infected nodule cells (208) where it serves as an oxygen carrier. Several Lb promoters confer developmentally correct, nodule-specific expression in transgenic legumes (37, 190, 191). While the Lb genes have been the paradigm for genes expressed exclusively in nodules, recent studies have uncovered their counterparts in nonlegumes. Hemoglobin (Hb) genes found in nodulating and nonnodulating nonlegumes (e.g. *Parasponia andersonii* and *Trema tomentosa*) are expressed in a root-specific fashion (16, 107). The Hb promoter of *P. andersonii* confers root-specific expression to a CAT reporter gene in transgenic tobacco (110). Thus, hemoglobin may be involved in the respiratory metabolism of root cells of all species, but has evolved for high-level expression in nodules.

Plant genes involved in nitrogen assimilation and transport also play a critical role in nodule metabolism. Ammonia assimilated into glutamine via glutamine synthetase (GS), is transported predominantly as asparagine in alfalfa and pea, and as ureides in soybean and *Phaseolus*. For both *Phaseolus* and pea, expression of the nuclear gene for plastid GS2 is slightly higher in nodules than roots (117, 193, 194). However, GS genes that are most highly expressed in nodules are cytosolic GS genes (12, 36, 47, 68, 83, 171, 194, 211). In *Phaseolus,* the GS gene formerly classified as nodule-specific (GSn1) (36, 68) is also expressed in stems, petioles, and in green cotyledons (12). In transgenic *Lotus corniculatus,* GUS expression directed by the GSn1 promoter is 180-fold higher in nodules compared to roots and leaves, and as histochemical analysis revealed, restricted to the infected cells of the central nodule core (59). Ammonia assimilated by GS in the infected cells is converted to ureides for transport in uninfected cells, as determined by studies of cell-specific expression of a nodule-specific uricase II in soybean nodules (13). Thus, these molecular results support biochemical models for the cellular localization of these nitrogen metabolic enzymes in nodules (170).

Organ-specific expression studies in pea have shown that the genes for GS and asparagine synthetase (AS), which are highly expressed in nodules, are also highly expressed in cotyledons of germinating seeds (194, 195, 211) where large amounts of nitrogen also must be mobilized. Transgenic studies have shown that the promoter for the cytosolic GS gene most highly expressed in nodules and cotyledons of pea (*GS3A*) is able to direct high-level GUS expression in cotyledons of tobacco, specifically within the phloem cells (52). This phloem-specific expression suggests that the GS3A isoform functions in the synthesis of glutamine for intercellular nitrogen transport. It will be interesting to determine whether the high-level expression of the pea *GS3A* gene in nodules is also correlated with phloem-specific expression. It is noteworthy that the promoter for cytosolic GS of *P. vulgaris* (GS-R1), which is expressed lower in nodules than roots (68), directs expression of GUS in the vascular and cortex regions of mature nodules in transgenic *Lotus corniculatus* (59).

CONCLUDING REMARKS

The ability to monitor gene expression in situ in single cells has bridged the gap between plant molecular biology and botany. The study of cell-specific gene expression in plants has contributed to the understanding of plant cell structure and function by identifying the precise cell types in which metabolic reactions occur. While many of the genes covered in this review are expressed predominantly in one plant organ, in most cases these genes are also active in other organs. Therefore, a comprehensive understanding of plant gene regulation and function should encompass a study of gene expression in pertinent cell types throughout the plant.

ACKNOWLEDGMENTS

We thank our colleagues who have shared preprints and other unpublished results with us during the preparation of this manuscript. We also thank members of the Laboratory of Plant Molecular Biology at Rockefeller University for critically reading the manuscript. This work was supported by NIH Grant GM 32877 and DOE Grant DEFGO-289ER-140-34. J.W.E. is the recipient of an NSF Fellowship in Plant Molecular Biology DCB-8710630. This review is based on literature available before April 1, 1990.

Literature Cited

1. An, G., Mitra, A., Choi, H. K., Costa, M. A., An, K., et al. 1989. Functional analysis of the 3' control region of the potato wound inducible proteinase inhibitor II gene. *Plant Cell* 1:115–122
2. Anderson, M. A., McFadden, G. I., Bernatsky, R., Atkinson, A., Orpin, T., et al. 1989. Sequence variability of three alleles of the self-incompatibility gene of *Nicotiana alata. Plant Cell* 1: 483–91
3. Andrews, D. L., Beames, B., Summers, M. D., Park, W. D. 1988. Characterization of the lipid acyl hydrolase activity of the major potato *(Solanum tuberosum)* tuber protein, patatin, by cloning and abundant expression in a baculovirus vector. *Biochem. J.* 252:199–206
4. Aoyagi, K., Kuhlemeier, C., Chua, N.-H. 1988. The pea *rbcS-3A* enhancer-

like element directs cell-specific expression in transgenic tobacco. *Mol. Gen. Genet.* 213:179–85
5. Aoyagi, K., Nakamoto, H. 1985. Pyruvate, Pi dikinase in bundle sheath strands as well as in mesophyll cells in maize leaves. *Plant Physiol.* 78:661–64
6. Barker, S. J., Harada, J. J., Goldberg, R. B. 1988. Cellular localization of soybean storage protein messenger RNA in transformed tobacco seeds. *Proc. Natl. Acad. Sci. USA* 85:458–62
7. Baulcombe, D. C. Huttly, A. K., Martienssen, R. A., Barker, R. F., Jarvis, M. G. 1987. A novel wheat alpha-amylase gene. *Mol. Gen. Genet.* 209: 33–40
8. Baumlein, H., Muller, A. J., Schiemann, J., Helbing, D., Manteuffel, F., Wobus, U. 1987. A legumin

B gene of *Vicia faba* is expressed in developing seeds of transgenic tobacco. *Biol. Zentralbl.* 106:569–75

9. Beachy, R. N., Chen, Z.-L., Horsch, R. B., Rogers, S. G., Hoffmann, N. J., Fraley, R. T. 1985. Accumulation and assembly of soybean B-conglycinin in seeds of transformed petunia plants. *EMBO J.* 4:3047–53

10. Beld, M., Martin, C., Huits, H., Stuitje, A. R., Gerats, A. G. M. 1989. Flavonoid synthesis in *Petunia hybrida*: Partial characterization of dihydroflavanol-4-reductase. *Plant Mol. Biol.* 13:491–502

11. Benfey, P. N., Chua, N.-H. 1989. Regulated genes in transgenic plants. *Science* 244:174–81

12. Bennett. M. J., Lightfoot, D. A., Cullimore, J. V. 1989. cDNA sequence and differential expression of the gene encoding the glutamine synthetase gamma polypeptide of *Phaseolus vulgaris* L. *Plant Mol. Biol.* 12:553–65

13. Bergmann, H., Preddie, E., Verma, D. P. S. 1983. Nodulin-35: A subunit of a specific uricase (uricase II) induced and localized in the uninfected cells of soybean nodules. *EMBO J.* 2:2333–39

14. Bevan, M., Barker, R., Goldsbrough, A., Jarvis, M., Kavanagh, T., Iturriaga, G. 1986. The structure and transcription start site of a major tuber protein patatin. *Nucleic Acids Res.* 14:4625–38

15. Bevan, M., Shufflebottom, D., Edwards, K., Jefferson, R., Schuch, W. 1989. Tissue- and cell-specific activity of a phenylalanine ammonia lyase promoter in transgenic plants. *EMBO J.* 8:1899–906

16. Bogusz, D., Appleby, C. A., Landsmann, J., Dennis, E., Trinick, M. J., Peacock, J. 1988. Functional haemoglobin genes in non-nodulating plants. *Nature* 331:178–80

17. Boque, M. A., Vonder Haar, R. A., Nuccio, M. L., Griffing, L. R., Thomas, T. L. 1990. Developmentally regulated expression of a sunflower 11S seed protein gene in transgenic tobacco. *Mol. Gen. Genet.* In press

18. Bowman, J. L., Yanofsky, M. F., Meyerowitz, E. 1988. *Arabidopsis thaliana*: A review. *Oxf. Surv. Plant Mol. Cell Biol.* 5:57–87

19. Braam, J., Davis, R. W. 1990. Rain-, wind-, and touch-induced expression of calmodulin and calmodulin-related genes in *Arabidopsis*. *Cell* 60:357–64

20. Bray, B., Naito, S., Beachy, R. N. 1987. Expression of the B-subunit of β-conglycinin in seeds of transgenic plants. *Planta* 172:364–70

21. Broglie, K., Biddle, P., Cressman, R., Broglie, R. 1989. Functional analysis of DNA sequences responsible for ethylene regulation of a bean chitinase gene in transgenic tobacco. *Plant Cell* 1:599–607

22. Brown, J. W. S., Wandelt, C., Feix, G., Neuhaus, G., Schweiger, H. G. 1986. The upstream regions of zein genes: Sequence analysis and expression in the unicellular green algae *Acetabularia*. *Eur. J. Cell Biol.* 42:161–70

23. Brown, S. M., Crouch, M. L. 1990. Characterization of a gene family abundantly expressed in *Oenothera organensis* pollen that shows sequence similarity to polygalacturonase. *Plant Cell* 2:263–74

24. Bustos, M. M., Guiltinan, M. J., Jordano, J., Begum, D., Kalkan, F. A., Hall, T. C. 1989. Regulation of beta-glucuronidase expression in transgenic tobacco plants by an A/T rich *cis*-acting sequence found upstream of a French bean beta-phaseolin gene. *Plant Cell* 1:839–53

25. Campos, F., Padilla, J., Vazquez, M., Ortega, J. L., Enriquez, C., Sanchez, F. 1987. Expression of nodule-specific genes in *Phaseolus vulgaris* L. *Plant Mol. Biol.* 9:521–32

26. Cassab, G. I., Varner, J. E. 1987. Immunocytolocalization of extensin in developing soybean seed coats by immunogold-silver staining and by tissue printing on nitrocellulose paper. *J. Cell Biol.* 105:2581–88

27. Cassab, G. I., Varner, J. E. 1988. Cell wall proteins. *Annu. Rev. Plant Physiol.* 39:321–53

28. Chen, Z.-L., Naito, S., Nakamara, I., Beachy, R. N. 1989. Regulated expression of genes encoding soybean Beta conglycinins in transgenic plants. *Dev. Genet.* 10:112–22

29. Colot, V., Robert, L. S., Kavanagh, T. A., Bevan, M. W., Thompson, R. D. 1987. Localization of sequences in wheat endosperm protein genes which confer tissue-specific expression in tobacco. *EMBO J.* 6:3559–64

30. Condit, C. M., Meagher, R. B. 1986. A gene encoding a novel glycine-rich structural protein of petunia. *Nature* 323:178–81

31. Condit, C. M., Meagher, R. B. 1987. Expression of a gene encoding a glycine-rich protein in petunia. *Mol. Cell. Biol.* 7:4273–79

32. Corbin, D. R., Sauer, N., Lamb, C. J. 1987. Differential regulation of a hydroxyproline-rich glycoprotein gene

family in wounded and infected plants. *Mol. Cell. Biol.* 7:4337–44

33. Cordes, S., Deikman, J., Margossian, L. J., Fischer, R. L. 1989. Interaction of a developmentally regulated DNA binding factor with sites flanking two different fruit ripening genes from tomato. *Plant Cell* 1:1025–34

34. Cornish, E. C., Pettitt, J. M., Bonig, I., Clarke, A. E. 1987. Developmentally controlled, tissue-specific expression of a gene associated with self-incompatibility in *Nicotiana alata. Nature* 326:99–102

35. Coruzzi, G., Broglie, R., Edwards, C., Chua, N.-H. 1984. Tissue-specific and light-regulated expression of a pea nuclear gene encoding the small subunit of ribulose-1, 5-bisphosphate carboxylase. *EMBO J.* 3:1671–79

36. Cullimore, J. V., Gebhardt, C., Saarelainen, R., Miflin, B. J., Idler, K. B., Barker, R. F. 1984. Glutamine synthetase of *Phaseolus vulgaris* L.: Organspecific expression of a multigene family. *J. Mol. Appl. Genet.* 2:589–99

37. De Bruijn, F. J., Felix, G., Grunenberg, B., Hoffman, H. J., Metz, B., et al. 1989. Regulation of plant genes specifically induced in nitrogen-fixing nodules: Role of *cis*-acting elements and *trans*-acting factors. *Plant Mol. Biol.* 13:319

38. Deikman, J., Fischer, R. L. 1988. Interaction of a DNA binding factor with the 5'-flanking region of an ethyleneresponsive fruit ripening gene from tomato. *EMBO J.* 7:3315–20

39. Delauney, A. J., Verma, D. P. 1990. Cloned nodulin genes for symbiotic nitrogen fixation. *Plant Mol. Biol. Rep.* 279:285

40. DellaPenna, D., Alexander, D. C., Bennett, A. B. 1986. Molecular cloning of tomato fruit polygalacturonase: analysis of polygalacturonase mRNA levels during ripening. *Proc. Natl. Acad. Sci. USA* 83:6420–24

41. DellaPenna, D., Kates, D. S., Bennett, A. B. 1987. Polygalacturonase gene expression in Rutgers, *rin, nor* and *Nr* tomato fruits. *Plant Physiol.* 85:502–7

42. DellaPenna, D., Lincoln, J. E., Fischer, R. L., Bennett, A. B. 1989. Transcriptional analysis of polygalacturonase and other ripening associated genes in Rutgers, *rin, nor,* and *Nr* tomato fruit. *Plant Physiol.* 90:1372–77

43. Diaz, C. L., Melchers, L. S., Hooykaas, P. J. J., Lugtenberg, B. J. J., Kijne, J. W. 1989. Root lectin as a determinant of host-plant specificity in the *Rhizobium*-legume symbiosis. *Nature* 338:579–81

44. Dickinson, C. D., Evans, R. P., Nielson, N. C. 1988. TY repeats are conserved in the 5' flanking regions of legume seed-protein genes. *Nucleic Acids Res.* 16:371–80

45. Dickstein, R., Bisseling, T., Reinhold, V. N., Ausubel, F. M. 1988. Expression of nodule specific genes in alfalfa root nodules blocked at an early stage of development. *Genes Dev.* 2:677–87

46. Drews, G. N., Goldberg, R. B. 1989. Genetic control of flower development. *Trends Genet.* 5:256–61

47. Dunn, K., Dickstein, R., Feinbaum, R., Burnett, B. K., Peterman, T. K., et al. 1988. Developmental regulation of nodule-specific genes in alfalfa root nodules. *Mol. Plant-Micr. Interact.* 1: 66–74

48. Ebel, J., Hahlbrock, K. 1982. Biosynthesis. In *The Flavonoids,* ed. J. B. Harborne, T. J. Mabry, p. 641. London: Chapman & Hall

49. Ebert, P. R., Anderson, M. A., Bernatzky, R. Altschuler, M., Clarke, A. E. 1989. Genetic polymorphism of self-incompatibility in flowering plants. *Cell* 56:255–62

50. Edwards, G. E., Walker, D. A. 1983. C3, C4: *Mechanisms, and Cellular and Environmental Regulation of Photosynthesis.* Oxford: Blackwell Sci.

51. Edwards, J. W., Coruzzi, G. M. 1989. Photorespiration and light act in concert to regulate the expression of the nuclear gene for chloroplast glutamine synthetase. *Plant Cell* 1:241–48

52. Edwards, J. W., Walker, E. L., Coruzzi, G. M. 1990. Cell-specific expression in transgenic plants reveals non-overlapping roles for chloroplast and cytosolic glutamine synthetase. *Proc. Natl. Acad. Sci. USA* 87:3459–63

53. Ellis, J. R., Shirast, A. H., Hepher, A., Yarwood, J. N., Gatehouse, J. A., et al. 1988. Tissue-specific expression of a pea legumin gene in seeds of *Nicotiana plumbaginifolia. Plant Mol. Biol.* 10: 203–14

54. Estelle, M. A., Somerville, C. 1986. The mutants of *Arabidopsis. Trends Genet.* 2:89–93

55. Evans. I. M., Swinhoe, R., Gatehouse, L. N., Gatehouse, J. A., Boulter, D. 1988. Distribution of root messenger RNA species in other vegetative organs of pea *Pisum sativum* L. *Mol. Gen. Genet.* 214:153–57

56. Flavell, R. B., Goldsbrough, A. P., Robert, L. S., Schnick, D., Thompson, R. D. 1989. Genetic variation in wheat

HMW glutenin subunits and the molecular basis of breadmaking quality. *Biotechnology* 7:1281–85

57. Fluhr, R., Kuhlemeier, C., Nagy, F., Chua, N.-H. 1986. Organ-specific and light-induced expression of plant genes. *Science* 232:1106–12

58. Fluhr. R., Moses, P., Morelli, G., Coruzzi, G., Chua, N.-H. 1986. Expression dynamics of the pea *rbcS* multigene family and organ distribution of the transcripts. *EMBO J.* 5:2063–71

59. Forde, B. G., Day, H. M., Turton, J. F., Shen, W.-J., Cullimore, J. V., Oliver, J. E. 1989. Two glutamine synthetase genes from *Phaseolus vulgaris* L. display contrasting developmental and spatial patterns of expression in transgenic *Lotus corniculatus* plants. *Plant Cell* 1:391–401

60. Forde, B. G., Heyworth, A., Pywell, J., Kreis, M. 1985. Nucleotide sequence of a B1 hordein gene and the identification of possible upstream regulatory elements in endosperm storage protein genes from barley, wheat and maize. *Nucleic Acids Res.* 13:7327–39

61. Fox, J. L. 1988. Plant biotechnology—yet more novelties. *Biotechnology* 6:865–68

62. Franssen, H. J., Nap, J. P., Gloudeman, T., Stiekema, W., van Dam, H., et al. 1987. Characterization of a cDNA for nodulin-75 of soybean: A gene product involved in early stages of root nodule development. *Proc. Natl. Acad. Sci. USA* 84:4495–99

63. Fromm, H., Katagiri, F., Chua, N.-H. 1989. An octopine synthase enhancer element directs tissue-specific expression and binds ASF-1, a factor from tobacco nuclear extracts. *Plant Cell* 1:997–84

64. Fuller, F., Kunstner, P. W., Nguyen, T., Verma, D. P. 1983. Soybean nodulin genes: analysis of cDNA clones reveals several major tissue-specific sequences in nitrogen-fixing root nodules. *Proc. Natl. Acad. Sci. USA* 80:2594–98

65. Gasser, C. S., Budelier, K. A., Smith, A. G., Shah, D. M., Fraley, R. T. 1989. Isolation of tissue-specific cDNAs from pistils. *Plant Cell* 1:15–24

66. Gasser, C. S., Winter, J. A., Hironake, C. M., Shah, D. M. 1988. Structure, expression, and evolution of the 5 enolpyruvylshikimate-3-phosphate synthase genes of *Petunia* and tomato. *J. Biol. Chem.* 263:4280–87

67. Gatehouse, J. A. Evans, I. M., Croy R. R. D., Boulter, D. 1986. Differential expression of genes during legumin seed

development. *Philos. Trans. R. Soc. London Ser. B* 314:343–53

68. Gebhardt, C., Oliver, J. E., Forde, B. G., Saarelinen, R., Miflin, B. J. 1986. Primary structure and differential expression of glutamine synthetase genes in nodules, roots and leaves of *Phaseolus vulgaris. EMBO J.* 5:1429–35

69. Gilmartin, P. M., Sarokin, L., Memelink, J., Chua, N.-H. 1990. Molecular light switches for plant genes. *Plant Cell* 2:369–78

70. Gloudemans, T., Bhuvaneswari, T. V., Moerman, M., van Brussel, T., van Kammen, A., Bisseling, T. 1989. Involvement of *Rhizobium leguminosarum* nodulation genes in gene expression in pea root hairs. *Plant Mol. Biol.* 12:157–67

71. Gloudemans, T., Bisseling, T. 1989. Plant gene expression in early stages of *Rhizobium*-legume symbiosis. *Plant Sci.* 65:1–14

72. Goldberg, R. B. 1988. Plants: Novel developmental processes. *Science* 240:1460–67

73. Govers, F., Gloudemans, T., Moerman, M., van Kammen, A., Bisseling, T. 1985. Expression of plant genes during the development of pea root nodules. *EMBO J.* 4:861–67

74. Grierson, D., Tucker, G. A., Keen, J., Ray, J., Bird, C. R., Schuch, W. 1986. Sequencing and identification of a cDNA clone for tomato polygalacturonase. *Nucleic Acids Res.* 14:8595–603

75. Hanson, D. D., Hamilton, D. A., Travis, J. L., Bashe, D. M., Mascarenhas, J. P. 1989. Characterization of a pollen-specific complementary DNA clone from *Zea mays* and its expression. *Plant Cell* 1:173–80

76. Harker, C. L., Ellis, T. H. N., Coen, E. S. 1990. Identification and genetic regulation of the chalcone synthase multigene family or pea. *Plant Cell* 2:185-94

77. Hartings, H., Maddaloni, M., Lazzaroni, N., Di Fonzo, N., Motto, M., et al. 1989. The O2 gene which regulates zein deposition in maize indosperm encodes a protein with structural homologies to transcriptional activators. *EMBO J.* 8:2795–801

78. Hawkes, J. G. 1979. Evolution and polyploidy in potato species. In *The Biology and Taxonomy of the Solanaceae*, ed. J. G. Hawkes, R. N. Lester, A. D. Skelding, pp. 637–645. London: Academic

79. Herman, P., Marks, M. D. 1989. Trichome development in *Arabidopsis thaliana*. II. Isolation and complementa-

tion of the *GLABROUS1* gene. *Plant Cell* 1:1051–55

80. Higgins. T. J. V., Newbigin, E. J., Spencer, D., Llewellyn, D. J., Craig, S. 1988. The sequence of a pea vicilin gene and its expression in transgenic tobacco plants. *Plant Mol. Biol.* 11:683–95

81. Higuchi, T. 1981. Biosynthesis of lignin. In *Plant Carbohydrates* II, *Encyclopedia of Plant Physiology;* ed. W. Tanner, F. A. Loewus, pp. 194–224. Berlin: Springer-Verlag

82. Hilder, B. A., Gatehouse, A. M. R., Sheerman, S. E., Barker, R. F., Boulter, D. 1987. A novel mechanism of insect resistance engineered into tobacco. *Nature* 330:160–63

83. Hirel, B., Bouet, C., King, B., Layzell, B., Jacobs, F., Verma, D. P. S. 1987. Glutamine synthetase genes are regulated by ammonia provided externally or by symbiotic nitrogen-fixation. *EMBO J.* 6:1167–71

84. Hong, J. C., Nagao, R. T., Key, J. L. 1987. Characterization and sequence analysis of a developmentally regulated putative cell wall protein gene isolated from soybean. *J. Biol. Chem.* 262: 8367–76

85. Hong, J. C., Nagao, R. T., Key, J. L. 1989. Developmentally regulated expression of soybean proline-rich cell wall protein genes. *Plant Cell* 1:937–44

86. Huttly, A. K., Baulcombe, D. C. 1989. A wheat α-Amy2 promoter is regulated by gibberellin in transformed oat aleurone protoplasts. *EMBO J.* 8:1907–13

87. Huttly, A. K., Martienssen, R. A., Baulcombe, D. C. 1988. Sequence heterogeneity and differential expression of the alpha-Amy-2 gene family in wheat. *Mol. Gen. Genet.* 214:232–40

88. Jofuku, K. D., Goldberg, R. B. 1989. Kunitz trypsin inhibitor genes are differentially expressed during the soybean life cycle and in transformed tobacco plants. *Plant Cell* 1:1079–93

89. Jofuku, K. D., Okamuro, J. K., Goldberg, R. B. 1987. Interaction of an embryo DNA binding protein with a soybean lectin gene upstream region. *Nature* 328:734–36

90. Johnson, R., Narvaez, J., An, G., Ryan, C. 1989. Expression of proteinase inhibitors I and II in transgenic tobacco plants: Effects on natural defense against *Manduca sexta* larvae. *Proc. Natl. Acad. Sci. USA* 86:9871–75

91. Jordano, J., Almoguera, C., Thomas, T. L. 1989. A sunflower heliathinin gene upstream sequence ensemble contains an enhancer and sites of nuclear protein interaction. *Plant Cell* 1:855–66

92. Katagiri, F., Lam, E., Chua, N.-H. 1989. Two tobacco DNA Binding proteins with homology to the nuclear factor CREB. *Nature* 340:727–30

93. Keil, M., Sanchez-Serrano, J. J., Willmitzer, L. 1989. Both wound-inducible and tuber-specific expression are mediated by the promoter of a single member of the potato proteinase inhibitor II gene family. *EMBO J.* 8:1323–30

94. Keil, M., Sanchez-Serrano, J., Schell, J., Willmitzer, L. 1990. Localization of elements important for the wound-inducible expression of a chimeric potato proteinase inhibitor II-CAT gene in transgenic tobacco plants. *Plant Cell* 2: 61–70

95. Keller, B., Lamb, C. J. 1990. Specific expression of a novel cell wall hydroxyproline-rich glycoprotein gene in lateral root initiation. *Genes Dev.* 3:1639–46

96. Keller, B., Sauer, N., Lamb, C. J. 1988. Glycine-rich cell wall proteins in bean gene structure and association of the protein with the vascular system. *EMBO J.* 7:3625–34

97. Keller, B., Templeton, M. D., Lamb, C. J. 1989. Specific localization of a plant cell wall glycine-rich protein in protoxylem cells of the vascular system. *Proc. Natl. Acad. Sci. USA* 86:1529–33

98. Kodrzycki, R. Boston, R. S., Larkins, R. A. 1989. The opaque-2 mutation of maize differentially reduces zein gene transcription. *Plant Cell* 1:105–14

99. Koes, R. E., Spelt, C. E., Reif, H. J., van den Elzen, P. J. M., Veltkamp, E., Mol, J. N. M. 1986. Floral tissue of *Petunia hybrida* (V30) expresses only one member of the chalcone synthase multigene family. *Nucleic Acids Res.* 14:5229–39

100. Koes, R. E., Spelt, C. E., van den Elzen, P. J. M., Mol, J. N. M. 1989. Cloning and molecular characterization of chalcone synthase multigene family of *Petunia hybrida*. *Gene* 81:245–57

101. Koncz, C., Martini, N., Mayerhofer, R., Koncz-Kalman, Z., Korber, H., et al. 1990. High frequency T-DNA mediated gene tagging in plants. *Proc. Natl. Acad. Sci. USA* 86:8467–71

102. Koster-Topfer, M., Frommer, W. B., Rocha-Sosa, M., Rosahl, S., Schell, J., Willmitzer, L. 1989. A class II patatin promoter is under developmental control in both transgenic potato and tobacco plants. *Mol. Gen. Genet.* 219:390–96

103. Kramer, M., Sheehy, R. E., Hiatt, W.

R. 1989. Progress towards the genetic engineering of tomato fruit softening. *Trends Biotechnol.* 7:191–94

104. Kriz, A. L., Boston, R. S., Larkins, B. A. 1987. Structural and transcriptional analysis of DNA sequences flanking genes that encode 19 kilodalton zeins. *Mol. Gen. Genet.* 207:90–98

105. Kuhlemeier, C., Strittmatter, G., Ward, K., Chua, N.-H. 1989. The pea *rbcS-3A* promoter mediates light-responsiveness but not organ-specificity. *Plant Cell* 1: 471–78

106. Lam, E., Benfey, P. N., Gilmartin, P. M., Fang, R. X., Chua, N. H. 1989. Site-specific mutations alter in vitro factor binding and change promoter expression pattern in transgenic plants. *Proc. Natl. Acad. Sci. USA* 86:7890–94

107. Landsman, J., Dennis, E., Higgins, T. J., Appleby, C. A., Kortt, A. A., Peacock, J. 1986. Common evolutionary origin of legume and non-legume plant haemoglobins. *Nature* 324:166–68

108. Langdale, J. A., Rothermel, B. A., Nelson, T. 1988. Cellular pattern of photosynthetic gene expression in developing maize leaves. *Genes Dev.* 2:106–15

109. Langdale, J. A., Zelitch, I., Miller, E., Nelson, T. 1988. Cell position and light influence C4 verses C3 patterns of photosynthetic gene expression in maize. *EMBO J.* 7:3643–51

110. Lansmann, J., Llewellyn, D., Dennis, E. S., Peacock, W. J. 1988. Organ regulated expression of the *Parasponia andersonii* hemoglobin gene in transgenic tobacco plants. *Mol. Gen. Genet.* 214: 68–73

111. Lazarus, B. M., Baulcombe, D. C., Martienssen, R. A. 1985. Alpha-amylase genes of wheat are two multigene families which are differentially expressed. *Plant Mol. Biol.* 5:13–24

112. Leah, R., Mundy, J. 1989. The bifunctional alpha amylase-subtilisin inhibitor of barley: Nucleotide sequence and patterns of seed-specific expression. *Plant Mol. Biol.* 12:673–82

113. Legocki, R. P., Verma, D. P. S. 1980. Identification of nodule-specific host proteins (nodulins) involved in the development of the *Rhizobium*-legume symbiosis. *Cell* 20:153–63

114. Lerner, D. R., Raikhel, N. V. 1989. Cloning and characterization of root-specific barley lectin. *Plant Physiol.* 91: 124–29

115. Liang, X., Dron, M., Cramer, C., Dixon, R. A., Lamb, C. J. 1989. Differential regulation of pheylanine ammonia lyase genes during plant development

and by environmental cues. *J. Biol. Chem.* 264:14486–92

116. Liang, X., Dron, M., Schmid, J., Dixon, R. A., Lamb, C. J. 1989. Developmental and environmental regulation of a phenylalanine ammonia lyase-beta glucuronidase gene fusion in transgenic tobacco plants. *Proc. Natl. Acad. Sci. USA* 86:9284 –88

117. Lightfoot, D. A., Green, N. K., Cullimore, J. V. 1988. The chloroplast located glutamine synthetase of *Phaseolus vulgaris* L: Nucleotide sequence, expression in different organs and uptake into isolated chloroplast. *Plant Mol. Biol.* 11:191–202

118. Lincoln, J. E., Fischer, R. L. 1988. Diverse mechanisms for the regulation of ethylene-inducible gene expression. *Mol. Gen. Genet.* 212: 71–75

119. Lincoln, J. E., Fischer, R. L. 1988. Regulation of gene expression by ethylene in wild-type and *rin* tomato *(Lycopersicon esculentum)* fruit. *Plant Physiol.* 88:370–74

120. Maier, R.-G., Brown, J. W. S., Toloczyki, C., Feix, G. 1987. Binding of a nuclear factor to a consensus sequence in the 5' flanking region of zein genes from maize. *EMBO J.* 6:17–22

121. Mansson, P.-E., Hsu, D., Stalker, D. 1985. Characterization of fruit-specific cDNAs from tomato. *Mol. Gen. Genet.* 200:356–61

122. Marris, C., Gallois, P., Copley, J., Kreis, M. 1988. The 5' flanking region of a barley B hordein gene controls tissue- and developmental-specific CAT expression in tobacco plants. *Plant Mol. Biol.* 10:359–66

123. Mason, H. S., Guerrero, F. D., Boyer, J. S., Mullet, J. E. 1988. Proteins homologous to leaf glycoproteins are abundant in stems of dark-grown soybean seedlings. Analysis of proteins and cDNAs. *Plant Mol. Biol.* 11:845–56

124. Matzke, A. J. M., Stoger, E. M., Schernthaner, J. P., Matzke, M. A. 1990. Deletion analysis of a zein gene promoter in transgenic tobacco plants. *Plant Mol. Biol.* 14:323–32

125. McClure, B. A., Guilfoyle, T. J. 1989. Tissue print hybridization. A simple technique for detecting organ- and tissue-specific gene expression. *Plant Mol. Biol.* 12:517–24

126. McClure, B. A., Haring, V., Ebert, P. R., Anderson, M. A., Simpson, R. J., et al. 1990. Style self-incompatibility gene products of *Nicotiana alata* are ribonucleases. *Nature* 342:955–57

127. Meyer, P., Heidmann, I., Forkmann,

G., Saedler, H. 1987. A new petunia flower color generated by transformation with a maize gene. *Nature* 330:677–78

128. Mingnery, C. A., Pikaard, C. S., Park, W. D. 1988. Molecular characterization of the patatin multigene family of potato. *Gene* 62:27–44

129. Mol, J., Stuitje, A., Gerats, A., van der Krol, A., Jorgensen, R. 1989. Saying it with genes: Molecular flower breeding. *Trends Biotechnol.* 7:148–53

130. Mol, J. N., M., Stuitje, T. R., Gerats, A. G. M., Koes, R. E. 1988. Cloned genes of phenylpropanoid metabolism in plants. *Plant Mol. Bio. Rep.* 6:274–79

131. Moore, H. M., Nasrallah, J. B. 1990. A *Brassica* self-incompatibility gene is expressed in the stylar transmitting tissue of transgenic tobacco. *Plant Cell* 2:29–38

132. Motto, M., Di Fonzo, N., Hartings, S. H., Maddaloni, M., Salamini, F., et al. 1990. Regulatory genes affecting maize storage protein synthesis. *Oxf. Surv. Plant Mol. Cell Biol.* 6:87–114

133. Motto, M., Maddaloni, M., Panziani, G., Brembilla, M., Marotta, R., et al. 1988. Molecular cloning of the o2-m5 allele of *Zea mays* using transposon marking. *Mol. Gen. Genet.* 212:488–94

134. Mundy, J., Svendsen, I., Hejgaard, J. 1983. Barley alpha-amylase/subtilisin inhibitor. I+II. Isolation and characterization. *Carlsberg Res. Commun.* 48:81–94

135. Naito, S., Dube, P. H., Beachy, R. N. 1988. Differential expression of conglycinin A' and B subunits in transgenic plants. *Plant Mol. Biol.* 11:109–21

136. Napoli, C., Lemieux, C., Jorgensen, R. 1990. Introduction of a chalcone synthase transgene into Petunia results in reversible cosuppression of expression of the transgene and its homologue. *Plant Cell* 2:279–89

137. Nasrallah, J. B., Kao, T. H., Chen, C. H., Goldberg, M. L., Nasrallah, M. E. 1987. Amino acid sequence of glycoproteins encoded by three alleles of the *S-locus* of *Brassica oleracea*. *Nature* 326:617–19

138. Nasrallah, J. B., Nasrallah, M. E. 1989. The molecular genetics of self-incompatibility in *Brassica*. *Annu. Rev. Genet.* 23:121–39

139. Nasrallah, J. B., Yu, S. M., Nasrallah, M. E. 1988. Self-incompatibility genes of *Brassica oleracea*: Expression, isolation, and structure. *Proc. Natl. Acad. Sci. USA* 85:5551–55

140. Nelson, T., Langdale, J. A. 1989. Patterns of leaf development in C4 plants. *Plant Cell* 1:3–13

141. Nielsen, N. C., Dickenson, C. D., Cho, T. J., Thanh, V. H., Scallon, B. J., et al. 1989. Characterization of glycinin gene family in soybean. *Plant Cell* 1: 313–28

142. Okamuro, J. K., Goldberg, R. B. 1989. Regulation of plant gene expression: General principles. In *The Biochemistry of Plants*, ed. A. Marcus 15:1–83. New York: Academic

143. Okamuro, J. K., Jofuku, K. D., Goldberg, R. B. 1986. Soybean seed lectin gene and flanking non-seed protein genes are developmentally regulated in transformed tobacco plants. *Proc. Natl. Acad. Sci. USA* 83:8240–44

144. Ott, R. W., Ren, L., Chua, N.-H. 1990. A bi-directional enhancer cloning vehicle for higher plants. *Mol. Gen. Genet.* In press

145. Ou-Lee, T.-M., Turgeon, R., Wu, R. 1988. Interaction of a gibberellin-induced factor with the upstream region of an alpha-amylase gene in rice aleurone tissue. *Proc. Natl. Acad. Sci. USA.* 85:6366–69

146. Oxtoby, E., Hughes, M. A. 1990. Engineering herbicide tolerance into crops. *Trends Biotechnol.* 8:61–65

147. Palm, C. J., Costa, M. A., An, G., Ryan, C. 1990. Wound-inducible nuclear protein binds DNA fragments that regulate a proteinase inhibitor II gene from potato. *Proc. Natl. Acad. Sci. USA* 87:603–7

148. Pang, P. P. Pruitt, R. E., Meyerowitz, E. M. 1988. Molecular cloning, genomic organization, expression and evolution of seed storage protein genes of *Arabidopsis thaliana*. *Plant Mol. Biol.* 11:805–20

149. Park, W. D. 1983. Tuber proteins of potato—A new and surprising molecular system. *Plant Mol. Biol. Rep.* 1:61–66

150. Pavia, E., Lister, R. M., Park, W. D. 1983. The major tuber proteins of potato can be induced to accumulate in stems and petioles. *Plant Physiol.* 71:161–68

151. Peleman, J., Boerjan, W., Engler, G., Seurinck, J., Botterman, J., et al. 1989. Strong cellular preference in the expression of housekeeping gene of *Arabidopsis thaliana* encoding S-Adenosylmethionine synthetase. *Plant Cell* 1:81–93

152. Perez-Grau, L., Goldberg, R. B. 1989. Soybean protein genes are regulated spatially during embryogenesis. *Plant Cell* 1:1095–109

153. Pikaard, C. S., Brusca, J. S., Hannapel, D. J., Park, W. D. 1987. The two classes of genes for the major tuber protein patatin are differentially expressed

in tubers and roots. *Nucleic Acids Res.* 15:1979–94

154. Pikaard, C. S., Mignery, G. A., Ma, D. P., Stark, V. J., Park, W. D. 1986. Sequence of two apparent pseudogenes of the major tuber protein patatin. *Nucleic Acids Res.* 14:5564–66

155. Riggs, C. D., Voelker, T. A., Chrispeels, M. J. 1989. Cotyledon nuclear proteins bind to DNA fragments harboring regulatory elements of phytohemagglutinin genes. *Plant Cell* 1:609–21

156. Robert, L. S., Thompson, R. D., Flavell, R. B. 1989. Tissue-specific expression of a wheat high molecular weight glutenin gene in transgenic tobacco. *Plant Cell* 1:569–78

157. Rocha-Sosa, M., Sonnewald, U., Frommer, W., Stratmann, M., Schell, J., Willmitzer, L. 1989. Both developmental and metabolic signals activate the promoter of a class 1 patatin gene. *EMBO J.* 8:23–29

158. Rosahl, S., Schell, J., Willmitzer, L. 1987. Expression of a tuber-specific storage protein in transgenic tobacco plants: Demonstration of an esterase activity. *EMBO J.* 6:1155–59

159. Rosahl, S., Schmidt, R., Schell, J., Willmitzer, L. 1986. Isolation and characterization of a gene from *Solanum tuberosum* encoding patatin, the major storage protein of potato tubers. *Mol. Gen. Genet.* 203:214–20

160. Ryan, C. A. 1977. Proteolytic enzymes and their inhibitors in plants. *Annu. Rev. Plant Physiol.* 24:173–96

161. Ryan, C. A. 1981. Proteinase inhibitors. In *Biochemistry of Plants* ed. A. Marcus, pp. 351–70. New York: Academic

162. Sanchez-Serrano, J., Keil, M., Pena-Cortes, H., Rocha-Sosa, M., Willmitzer, L. 1987. Wound-induced expression of proteinase inhibitor II in potato and transgenic tobacco plants. In *Plant Gene Systems and Their Biology*, ed. J. L. Key, L. McIntosch, pp. 331–38. New York: Liss

163. Sanchez-Serrano, J., Schmidt, R., Schell, J., Willmitzer, L. 1986. Nucleotide sequence of proteinase inhibitor II encoding cDNA of potato (*Solanum tuberosum*) and its mode of expression. *Mol. Gen. Genet.* 203:15–20

164. Sanchez-Serrano, J. J., Keil, M., O'Connor, A., Schell, J., Willmitzer, L. 1987. Wound induced expression of a potato proteinase inhibitor II gene in transgenic tobacco plants. *EMBO J.* 6:303–6

165. Sato, T., Theologis, A. 1989. Cloning the mRNA encoding 1-aminocyclopropane-1-carboxylate synthase, the key enzyme for ethylene biosynthesis in plants. *Proc. Natl. Acad. Sci. USA* 86:6621–25

166. Scheres, B., van de Wiel, C., Zalensky, A., Horvath, B., Spaink, H., et al. 1990. The ENOD12 gene product is involved in the infection process during the pea-*Rhizobium* interaction. *Cell* 60: 281–94

167. Scherntaner, J. P., Matzke, M. A., Matzke, A. J. M. 1988. Endosperm-specific activity of a zein gene promoter in transgenic tobacco plants. *EMBO J.* 7:1249–56

168. Schmidt, R. J., Burr, F. A., Aukerman, M. J., Burr, B. 1990. Maize regulatory gene opaque-2 encodes a protein with a "leucine-zipper" motif that binds DNA. *Proc. Natl. Acad. Sci. USA* 87:46–50

169. Schmidt, R. J., Burr, F. A., Burr, B. 1987. Transposon tagging and molecular analysis of the maize regulatory locus opaque-2. *Science* 238:960–63

170. Schubert, K. R. 1986. Products of biological nitrogen fixation in higher plants: Synthesis, transport, and metabolism. *Annu. Rev. Plant Physiol.* 37:539–75

171. Sengupta-Gopalan, C., Pitas, J. W. 1986. Expression of nodule specific glutamine synthetase genes during nodule development in soybeans. *Plant Mol. Biol.* 7:189–99

172. Sengupta-Gopalan, C., Pitas, J. W., Thompson, D. V., Hoffman, L. M. 1986. Expression of host during root-nodule development in soybeans. *Plant Mol. Biol.* 7:189–99

173. Sheen, J.-Y., Bogorad, L, 1985. Differential expression of the ribulose bisphosphate carboxylase large subunit gene in bundle sheath and mesophyll dells of developing maize leaves is influenced by light. *Plant Physiol.* 79:1072–76

174. Shirsat, A., Wilford, N., Croy, R., Boulter, D. 1989. Sequences responsible for the tissue specific promoter activity of a pea legumin gene in tobacco. *Mol. Gen. Genet.* 215:326–31

175. Showalter, A. M., Bell, J. N., Cramer, C. L., Bailey, J. A., Varner, J. E., Lamb, C. J. 1986. Accumulation of hydroxyproline-rich glycoprotein mRNAs in response to fungal elicitor and infection. *Proc. Natl. Acad. Sci. USA* 82:6551–55

176. Sijmons, P. C., Dekker, B. M. M., Schrammeijer, B., Verwoerd, T. C., van den Elzen, P. J. M., Hoekema, A. 1990. Production of correctly processed human serum albumin in transgenic plants. *Biotechnology* 8:217–21

177. Simpson, J., Schell, J., Van Montagu,

M., Herrera-Estrella, L. 1986. Light-inducible and tissue-specific pea lhcp gene expression involves an upstream element combining enhancer and silencer-like properties. *Nature* 323: 551–54

178. Simpson, J., Timko, M. R., Cashmore, A. R., Schell, J., Van Montagu, M., Herrera-Estrella, L. 1985. Light-inducible and tissue-specific expression of a chimeric gene under control of the 5' flanking sequence of a pea chlorophyll a/b-binding protein gene. *EMBO J.* 4:2723–29

179. Simpson, J., Van Montagu, M., Herrera-Estrella, L. 1986. Photosynthesis-associated gene families: Differences in response to tissue-specific and environmental factors. *Science* 233:34–38

180. Slater, A., Maunders, M. J., Edwards, K., Schuch, W., Grierson, D. 1985. Isolation and characterization of cDNA clones for tomato polygalacturonase and other ripening-related proteins. *Plant Mol. Biol.* 5:137–47

181. Smith, A. G., Hinchee, M. A., Horsch, R. B. 1987. Cell and tissue specific expression localized by in situ hybridization in floral tissues. *Plant Mol. Biol. Rep.* 5:237–41

182. Smith, C. J. S., Watson, C. F., Morris, P. C., Bird, C. R., Seymour, G. B., et al. 1990. Inheritance and effect on ripening of antisense polygalacturonase genes in transgenic tomatoes. *Plant Mol Biol.* 14:369–79

183. Smith, C. J. S., Watson, C. F., Ray, J., Bird, C. R., Morris, P. C., et al. 1988. Antisense RNA inhibition of polygalacturonase gene expression in transgenic tomatoes. *Nature* 334:724–26

184. Somerville, C. 1989. *Arabidopsis* blooms. *Plant Cell* 1:1131–35

185. Sommer, H., Beltran, J. P., Huijser, P., Pape, H., Lonnig, W. E., et al. 1990. *Deficiens*, a homeotic gene involved in the control of flower morphogenesis in *Antirrhinum majus*: The protein shows homology to transcription factors. *EMBO J.* 9:605–13

186. Staswick, P. E. 1990. Novel regulation of vegetative storage protein genes. *Plant Cell* 2:1–6

187. Stinson, J. R., Eisenberg, A. J., Willing, R. P., Pe, M. E., Hanson, D. D., Mascarenhas, J. P. 1987. Genes expressed in the male gametophyte of flowering plants and their isolation. *Plant Physiol.* 83:442–47

188. Stockhaus, J., Schell, J., Willmitzer, L. 1989. Identification of enhancer elements in the upstream region of the nu-

clear photosynthetic gene ST-LS1. *Plant Cell* 1:805–13

189. Stockhaus, J., Schell, J., Willmitzer, L. 1989. Correlation of the expression of the nuclear photosynthetic gene *ST-LS1* with the presence of chloroplasts. *EMBO J.* 8:2445–51

190. Stougaard, J., Marcker, K. A. 1986. Nodule specific expression of a chimaeric soybean leghaemoglobin gene in transgenic *Lotus corniculatus*. *Nature* 321:669–74

191. Stougaard, J., Petersen, T. E., Marcker, K. A. 1987. Expression of a complete soybean leghemoglobin gene in root nodules of transgenic *Lotus corniculatus*. *Proc. Natl. Acad. Sci. USA* 84: 5754–57

192. Thornburg, R. W., An, G., Cleveland, T. E., Johnson, R., Ryan, C. A. 1987. Wound-inducible expression of a potato inhibitor II-chloramphenicol acetyltransferase gene fushion in transgenic tobacco plants. *Proc. Natl. Acad. Sci. USA* 84:744–48

193. Tingey, S. V., Tsai, F.-Y, Edwards, J. W., Walker, E. L., Coruzzi, G. M. 1988. Chloroplast and cytosolic glutamine synthetase are encoded by homologous nuclear genes which are differentially expressed in vivo. *J. Biol. Chem.* 263:9651–57

194. Tingey, S. V., Walker, E. L., Coruzzi, G. M. 1987. Glutamine synthetase genes of pea encode distinct polypeptides which are differentially expressed in leaves, roots and nodules. *EMBO J.* 6:1–9

195. Tsai, F.-Y, Coruzzi, G. M. 1990. Dark-induced and organ-specific expression of two asparagine synthetase genes in *Pisum sativum*. *EMBO J.* 9:323–32

196. Tucker, M. L., Sexton, R., Del Compillo, E., Lewis, L. M. 1988. Bean abscission cellulase characterization of a complementary DNA clone and regulation of gene expression by ethylene and auxin. *Plant Physiol.* 88:1257–62

197. Twell, D., Ooms, G. 1987. The 5' flanking DNA of a patatin gene directs tuber-specific expression of a chimeric gene in potato. *Plant Mol. Biol.* 9:365

198. Twell, D., Ooms, G. 1988. Structural diversity of the patatin gene family in potato cv. Desiree. *Mol. Gen. Genet.* 212:325–36

199. Twell, D., Wing, R., Yamaguchi, J., McCormick, S. 1989. Isolation and expression of an anther specific gene from tomato. *Mol. Gen. Genet.* 217:240–245

200. Ursin, V. M., Yamaguchi, J., McCormick, S. 1989. Gametophytic and sporophytic expression of anther-

specific genes in developing tomato anthers. *Plant Cell* 1:727–36

201. van de Wiel, C., Scheres, B., Franssen, H., van Lierop, M. J., van Lammeren, A., et al. 1990. The early nodulin transcript *ENOD2* is located in the nodule parenchyma (inner cortex) of pea and soybean root nodules. *EMBO J.* 9:1–7

202. van der Krol, A. R., Lenting, P. E., Veenstra, J., van der Meer, I. M., Koes, R. E., et al. 1988. An antisense chalcone synthase gene in transgenic plants inhibits flower pigmentation. *Nature* 333:866–69

203. van der Krol, S., Mur, L., Beld, M., Mol, J. N. M., Stuitje, A. R. 1990. Flavonoid genes in *Petunia hybrida:* Addition of a limited number of gene copies may lead to a collapse of gene expression. *Plant Cell* 2:291–99

204. Van Tunen, A. J., Hartman, S. A., Mur, L. A., Mol, J. N. M. 1989. Regulation of chalcone flavanone isomerase *CHI* gene expression in *Petunia hybrida:* the use of alternative promoters in corolla, anthers and pollen. *Plant Mol. Biol.* 12:539–52

205. Van Tunen, A. J., Koes, R. E., Spelt, C. E., van der Krol, A. R., Stuitje, A. R., Mol, J. N. M. 1988. Cloning of the two chalcone flavanone isomerase genes from *Petunia hybrida* coordinate light-regulated and differential expression of flavonoid genes. *EMBO J.* 7:1257–64

206. Vance, C. P., Boylan, K. L. M., Stade, S., Somers, D. A. 1985. Nodule-specific proteins in alfalfa (*Medicago sativa* L.). *Symbiosis* 1:69–84

207. Vandekerckhove, J., Damme, J. F., Lijsebettens, M., Botterman, J., De-Block, M., et al. 1989 Enkephalins produced in transgenic plants using modified 2S seed storage proteins. *Biotechnology* 7:929–32

208. Verma, D. P. S., Fortin, M. G., Stanley, J., Mauro, V. P., Purohit, S., Morrison, N. 1986. Nodulins and nodulin genes of *Glycine max:* A perspective. *Plant Mol. Biol.* 7:51–55

209. Voelker, T. A., Staswick, P., Chrispeels, M. J. 1986. Molecular analysis of

two phytohemagglutinin genes and their expression in *Phaseolus vulgaris* cv. Pinto, a lectin-deficient cultivar of the bean. *EMBO J.* 5:3075–82

210. Voelker, T. A., Sturm, A. J., Chrispeels, M. J. 1987. Differences in expression between two seed lectin alleles obtained from normal and lectin-deficient beans are maintained in transgenic tobacco. *EMBO J.* 6:3571–77

211. Walker, E. L., Coruzzi, G. M. 1989. Developmentally regulated expression of the gene family for cytosolic glutamine synthetase in *Pisum sativum. Plant Physiol.* 91:702–8

212. Wallsgrove, R. M., Turner, J. C., Hall, N. P., Kendally, A. C., Bright, S. W. J. 1987. Barley mutants lacking chloroplast glutamine synthetase-biochemical and genetic analysis. *Plant Physiol.* 83:155–58

213. Wenzler, H., Mignery, G., Fisher, L., Park W. 1989. Sucrose regulated expression of a chimeric potato tuber gene in leaves of transgenic tobacco plants. *Plant Mol. Biol.* 13:347–54

214. Wenzler, H. C., Mignery, G. A., Fisher, L. M., Park, W. D. 1989. Analysis of a chimeric class I patatin-GUS gene in transgenic potato plants: High level expression in tubers and sucrose-inducible expression in cultured leaf and stem explants. *Plant Mol. Biol.* 12:41–50

215. Werr, W., Frommer, W.-B, Maas, C., Starlinger, P. 1985. Structure of the sucrose synthase gene on chromosome 9 of *Zea mays* S. L. *EMBO J.* 4:1373–80

216. Williams, M. E., Mundy, J., Kay, S. A., Chua, N.-H. 1990. Differential expression of two related organ-specific genes in pea. *Plant Mol. Biol.* 14:765–74

217. Yang, S. F., Hoffman, N. E. 1984. Ethylene biosynthesis and its regulation in higher plants. *Annu. Rev. Plant Physiol.* 35:155–89

218. Yoshikawa, M., Iwasaki, T., Fujii, M., Oogaki, M. 1976. Isolation and some properties of a subtilisin inhibitor from barley. *J. Biochem.* 79:765–73

Annu. Rev. Genet. 1990. 24:305–26

IN VIVO SOMATIC MUTATIONS IN HUMANS: MEASUREMENT AND ANALYSIS

Richard J. Albertini, Janice A. Nicklas, J. Patrick O'Neill and Steven H. Robison

Genetics Laboratory, University of Vermont, Burlington, Vermont 05401

KEY WORDS: *hprt*, mutation in vivo, human T-lymphocyte, Lesch-Nyhan, gene sequencing

CONTENTS

0066-4197/90/1215-0305$02.00

INTRODUCTION

There are many reasons to study somatic cell gene mutation arising in vivo in humans. Mutations in critical genes almost certainly underlie pathological processes such as malignant transformations as well as some aspects of normal in vivo cell differentiation. Furthermore, in vivo somatic mutations of indicator genes provide realistic markers for genetic toxicology in that mutagen metabolism and delivery to genetic targets are automatically assessed, and individual differences in mutagen susceptibility may be determined. Somatic mutations in indicator genes may also be correlated with disease outcomes in the individual studied. If this becomes possible, a laboratory test result will substitute for a disease state for making human risk assessments. Studies in this area, therefore, have both biological and toxicological relevance.

The mouse "spot-test" was the first useful assay for quantitating *in vivo* somatic cell gene mutations in mammals (84). There were early attempts to study somatic mutations arising in vivo in humans. Three rare variant types of cells—two in red blood cells (RBCs) and one in white blood cells—were described between 1958 and 1975 (12, 87, 88, 99). Assays for these variants were initially proposed as measures of in vivo somatic mutations in humans. However, all eventually failed because it could never be demonstrated that the phenotypic changes were due to gene mutations. Similar variant phenotypes could have resulted from nongenetic causes, i. e. there were "phenocopies" (11, 76, 77).

This situation has changed over the past 15 years. There are now four systems for detecting in vivo gene mutations in four genes in human somatic cells, two in RBCs and two in T-lymphocytes. Each assay has its own unique characteristics and may yield its own patterns of mutation, either related to the cell type or the gene studied.

THE CELL SYSTEMS

Red Blood Cells

Red blood cells are abundant in human blood. Mature RBCs have lost their nuclei, an event that occurs after the last of four cell divisions of an earlier, committed bone-marrow precursor cell (proerythroblast). Proerythroblasts, in turn, arise through progressive differentiation and division of earlier committed and uncommitted bone marrow cells, of which the earliest is a "multipotent" stem cell, whose progeny produce both cells that begin the differentiation process and cells that renew this stem-cell compartment (31, 83). Somatic mutations recognized in RBCs can arise in any of these nucleated precursor cells.

Quantitatively relating RBC mutants to underlying mutational events can be complex because of the extended in vivo differentiation process (87). If the in vivo mutations arise in multipotent stem cells, one factor that determines the frequency of mutants in peripheral blood at any time is the number of stem cells producing that population of RBCs. Timing of the mutation event relative to the number of succeeding cell divisions will cause variation in the measured frequency. An advantage of *stem-cell* mutations for human mutagenicity monitoring is memory, i.e. variant RBCs will continually be produced. Mutations may also arise in *committed* red cell precursors. In such cases, the quantitative relationship between mutations and RBC variants is more direct. However, these mutations will only transiently produce variants because committed precursors are lost with time. Therefore, the "advantage" of mutations arising in committed precursors is close temporal relationship between the appearance of RBC variants and mutagen exposures and less variability in measurements.

Finally, because RBCs have no nucleus, there is no possibility for molecular analyses of variants. Systems using these cells must rely on indirect methods to avoid phenocopies.

T-lymphocytes

T-lymphocytes are nucleated white blood cells that are widely dispersed in peripheral blood, lymph nodes, and most body tissues. Mature T-lymphocytes arise from stem cells in the bone marrow and migrate to the thymus, where they differentiate.

One characteristic of the human in vivo T-cell population is its heterogeneity (reviewed in 26). It differs in this regard from most cell populations, and certainly from those used for in vitro mutagenicity studies. At the extreme, each T-cell acquires a uniqueness as it differentiates in the thymus in terms of its ability to recognize a specific antigen. This progress, which is achieved by rearrangements of T-cell receptor (TCR) genes, occurs early in life, and may be almost complete in humans shortly after adolescence. There are at least four TCR genes (termed α, β, δ, γ) that are located on chromosome 14q (α and δ) (13, 49), on chromosome 7q (β) (46), and on chromosome 7p (γ) (64). These genes rearrange their variable, joining, diversity, and constant regions in vivo in each cell to confer the uniqueness of the TCRs found in specifically reactive, mature T-lymphocytes. Once rearranged, the TCR-gene pattern is permanent and persists throughout the life of the T-cell and in all of its clonal descendants. Because these rearrangement patterns are fixed in most T-cells when the thymus is active, they provide one marker of the in vivo independence of T-cell mutant isolates obtained from an individual. There is probably little addition to the mature T-cell population from bone marrow during adult life in humans. Therefore, somatic mutations recognized in differentiated T-lymphocytes in adults most likely arise directly in these cells.

Most T-cells are in an arrested G_0 stage of the cell cycle in vivo at any given time. However, periodically, probably in response to antigen stimulus, these cells undergo a burst of cell division to generate progeny helper, suppressor or effector cells and also long-lived (memory) cells that maintain this clone for the specific immunological memory of the individual.

THE GENETIC SYSTEMS

Genes Scored in Red Cells

Somatic mutations currently being studied in RBCs arise in one of the hemoglobin genes, or in the glycophorin-A (GPA) gene. The hemoglobin genes exist as two sets of linked loci on chromosome 11p and chromosome 16p (27, 28) and the GPA gene is on chromosome 4q (52).

The use of hemoglobins to study in vivo somatic mutations was initially proposed as a means to avoid the phenocopy problem (87). Hemoglobin is an abundant tetrameric protein in red blood cells that is made up of two heterodimers. In humans, most postnatal hemoglobin is hemoglobin A (HbA), consisting of two α and two β chains. Mutations that produce *altered* (rather than absent) hemoglobin polypeptide chains are specific and unambiguous. The best known example is mutation of the β gene (chromosome 11p) that produces "sickle-cell" hemoglobin (HbS). This mutation is an A $\rightarrow$ T base change that causes the amino acid valine to be substituted for glutamine at position 6 of the β polypeptide. The β gene spans 2 kb and contains 3 exons; however, the effective target size for the HbS mutation is a single base pair.

GPA is a glycosylated cell-surface protein (approximately 10^5 molecules per red blood cell) that carries the M and N blood group antigens (34, 35). The M and N proteins, which differ by two non-adjacent amino acids, are codominantly expressed and represent the only two alleles of the GPA locus. The allele frequencies for M and N are approximately equal in all populations thus far studied, resulting, therefore, in 50% of individuals being M/N heterozygotes. The GPA gene spans 44 kb and has 4 exons (52). In principle, this is a "large-target" for mutation. However, because immunological selection for variants is employed, the effective target size will be a function of antibody used and the protein structure required for recognition.

Genes Scored in T-lymphocytes

The somatic mutations being studied in T-lymphocytes arise in the gene for hypoxanthine-guanine phosphoribosyltransferase (HPRT enzyme; *hprt* gene) or in one of the genes for the human leukocyte antigens (HLA antigens; HLA genes). The *hprt* gene is on chromosome Xq (43); the HLA genes are on chromosome 6p (17).

The HPRT enzyme is constitutive but dispensable in mammalian cells. It functions in the purine salvage pathway where it phosphoribosylates its normal substrates, as well as cytotoxic purine analogues. Cells lacking this enzyme are resistant to these analogues (3, 90).

The *hprt* gene spans 44 kb and contains 9 exons (78). Therefore, it is a "large-target" for mutation. However, the location of *hprt* on the X-chromosome means that this gene is either actually (in males) or functionally (in females) haploid in all human cells. This raises the issue of equal recoverability of all classes of mutants, i.e. because of possible codeletions of essential genes, will mutations resulting from large deletions involving *hprt* be recoverable as viable mutants?

The HLA complex includes several linked loci containing two classes of genes encoding cell surface recognition (restriction) molecules of importance in immune responses (17). The class I loci contain the HLA-A, HLA-B and HLA-C genes, plus a large number of as yet undefined genes. For class I genes, there are at least 23 alleles of HLA-A, 47 alleles of HLA-B, and 8 alleles of HLA-C that are codominantly expressed. Mutational loss of an antigen specified by one allele can be readily detected in heterozygotes. In addition, gene conversion or mitotic recombination can easily be studied with this autosomal gene assay in contrast to *hprt*. Thus far, only mutations in two alleles of HLA-A genes have been studied, i.e. HLA-A2 and HLA-A3 (47, 56).

The HLA-A gene spans 5 kb and contains 7 exons (17). This, in principle, provides a "large-target" for the loss mutations scored. However, since mutants are assayed by immunological methods, it is difficult to define target size as for GPA.

THE ASSAYS

HbS

The detection of RBCs with this altered hemoglobin has recently been refined through the use of automated image analysis to screen large numbers of cells (92, 97). This assay employs treatment of RBCs fixed on slides with fluorescent anti-HbS polyclonal antibodies. The frequency of rare fluorescent cells defines the HbS-variant frequency (Vf).

GPA

This assay employs highly specific anti-M and anti-N antibodies linked with green and red fluorophors, respectively. These label the wild-type RBCs from M/N heterozygous individuals with double fluorescence. The rare RBCs that have lost either fluorescence can be enumerated by use of a single or a dual beam flow cytometer (14–16, 53). The GPA-variant frequency (Vf) can be determined for the loss of either antigen and defines the hemizygous (simple loss) type of variants (MØ, NØ). Another class of GPA variants shows loss of expression of one antigen (either M or N) with double expression of the other (MM, NN). These are interpreted as products of gene conversion or somatic crossing-over and define the homozygous GPA variant frequency (Vf).

hprt

The *hprt* T-lymphocyte assays are based on the cytotoxicity of 6-thioguanine (TG), which is mediated by the HPRT enzyme (90). Rare lymphocytes that have lost HPRT activity are TG resistant (TG^r). There are two different assays based on this selection system.

The first assay uses autoradiography to detect ^{3}H-thymidine incorporation in variant cells that are able to overcome TG inhibition of first-round phytohemagglutinin (PHA) stimulated DNA synthesis in vitro (91). A recent modification employs 5-bromodeoxyuridine (BrdU) incorporation followed by Hoechst 33258 dye staining to score the TG^r cells by fluorescence differences (75). These short-term assays are useful for population studies because they are simple, relatively inexpensive, and have the potential for automation (89). The frequency of cells labeled in the presence of TG defines the *hprt* variant frequency (Vf).

The second assay employs cell cloning to measure the frequency of TG^r lymphocytes and has the advantage of allowing recovery and further analysis of individual clones (5, 62, 72). Peripheral blood T-lymphocytes are cultured in limiting dilution in microtiter dish wells in the presence or absence of TG. The wells also contain a source of growth factor (interleukin-2), phytohemagglutinin, and X-irradiated feeder cells. Cloning efficiencies are determined from the Poisson relationship ($P_0 = e^{-x}$). Growing colonies can be isolated and propagated in vitro to confirm their mutant phenotype. In addition, molecular analysis of the TCR gene-rearrangement patterns allows one to determine whether two mutants are siblings or the result of independent mutation events. This allows calculation of the actual *hprt* mutation frequency that underlies the measured TG^r mutant frequency (reviewed in 69). The exact nature of the *hprt* mutation can also be defined by Southern blotting and DNA sequencing.

HLA

This assay is based on earlier studies employing Epstein-Barr virus transformed B-lymphocytes in vitro (38, 50, 51, 68, 79, 80). Detection of mutants involves immunoselection using specific anti-HLA monoclonal antibodies plus complement. This T-lymphocyte assay also employs cell cloning in microtiter dishes as described for the *hprt* cloning assay (47, 56). The frequency of cells that survive the immunoselection defines the HLA-mutant frequency (Mf). This assay also allows further analysis of the mutant colonies, and makes possible comparison of the Mf at an autosomal locus (HLA) with that at an X-linked locus *(hprt)*.

QUANTITATIVE MEASUREMENTS OF Vf OR Mf IN HUMAN SOMATIC CELLS

Quantitative results of 20 independent studies of normal nonmutagen-exposed adults or newborns (placental cord blood samples) are summarized in Table 1.

Table 1 Frequency values for normal humans

Genetic assay	Endpoint	# of samples	Mean frequency	Reference
Newborn values				
hprt	Mf	45	0.6×10^{-6}	57
	Mf	66[a]	1.6×10^{-6}	55
	Mf	9	0.4×10^{-6}	44
	Mf	5	1.4×10^{-6}	93
Adult values				
HbS	Vf	5	3.8×10^{-8}	92
GPA	Vf	54	10.9×10^{-6} (MØ, NØ)	48
		—[b]	10×10^{-6} (MØ)	54
		—	11×10^{-6} (NØ)	54
		—	10×10^{-6} (MM)	54
		—	17×10^{-6} (NN)	54
hprt	Vf	26 (82 assays)	8.7×10^{-6}	8
		18	6.9×10^{-6}	75
		8	1.9×10^{-6}	10
	Mf	12	3.1×10^{-6}	85
		17	3.4×10^{-6}	39
		117	3.7×10^{-6}	94
		14	3.7×10^{-6}	44
		12	4.7×10^{-6}	21
		6 (68 assays)	5.5×10^{-6}	73
		27 (115 assays)	6.5×10^{-6}	8
		35	7.2×10^{-6}	35
		12	9.5×10^{-6}	61
HLA	Mf	21	30×10^{-6}	47

[a] A high risk patient population
[b] # of samples not listed in reference

Results where the mutational basis of the scored variants cannot be verified are given as variant frequencies (Vfs); results where mutants can be characterized are given as mutant frequencies. The relative rank order of these Vf or Mf values is HbS Vf < GPA Vf = *hprt* Vf = *hprt* Mf < HLA Mf, and is partially explained by the relative size of the genetic mutational targets. These and other quantitative results are considered according to assay systems.

HbS

The HbS assay has been employed to measure Vfs in only a single study to date (92). The mean Vf for five adult samples was 3.8×10^{-8}. Such a low frequency reflects the fact the HbS can only arise from an A→T transversion at a single base. This automated assay allows an analysis of large numbers of cells and the sensitivity may be increased by the simultaneous use of several antibodies specific for other mutant hemoglobins.

GPA

The Vf GPA results are expressed as hemizygous variants (gene loss, MØ or NØ) and homozygous variants (recombinants, MM or NN) (48, 54). The mean Vfs for normal adults show rather wide interindividual (but not intraindividual) variations, with frequencies of hemi- and homozygous variants being similar. The overall mean Vf for normal adults is approximately 10×10^{-6}. Age-relationship analysis has revealed a linear correlation of GPA Vf with age, with an increase of approximately 2% per year.

Samples from patients with DNA-repair defects have also been analyzed. The Xeroderma Pigmentosum homo- and heterozygotes showed normal values (W. Bigbee, personal communication), which may be explained by the fact that there is probably little exposure of RBC precursors in the bone marrow to ultraviolet light. By contrast, Ataxia Telangietasia homozygotes showed elevated frequencies of hemizygous (gene loss) variants while AT heterozygotes had normal values (15). An increased sensitivity to ionizing radiations would explain the AT homozygotes' increase. The largest increases in Vf values were found in Bloom Syndrome (BS) patients (54) although BS heterozygotes showed normal values.

Transient Vf increases were observed in cancer patients receiving adjuvant chemotherapy regimens after a lag of 1–3 weeks (16). The transient nature of the increases indicates that the GPA mutations probably arose in committed RBC precursors rather than in stem cells. Two studies of ionizing radiation-exposed atomic bomb survivors have been reported (reviewed in 53). There were large interindividual variations and persistence of elevated Vfs 40 years after exposure, which are consistent with the occurrence of mutations in stem cells. Employing an updated exposure-dose estimate, induced rates per gray using bone-marrow dose estimates were 40 variants/10^6RBC/gray for hemizygous (gene loss) variants and 20 variants/10^6RBC/gray for homozygous (recombinant) variants. These GPA Vf values correlated with both estimated exposure dose and measured chromosome aberration frequency. It is suggested that the combination of the two measures may allow better dose estimate of putative radiation exposure.

hprt

Detection of *hprt* mutant T-lymphocytes was the first in vivo somatic mutation assay to become operational (1, 4, 6) and, therefore, this sytem has the largest data base.

The background Vfs for normal adults determined by the autoradiography assay, using cryopreserved cells, are listed in Table 1. An age-related increase of approximately 0.26×10^{-6} per year (5% per year based on a Vf value of 5×10^{-6}) has been found (8). Increases in the *hprt* Vfs have been reported in cancer patients treated with chemo- or radiotherapy (2, 9, 10, 30, 45) or in individuals exposed in radiation accidents (74).

Either the autoradiography or the BrdU *hprt* Vf assay should be useful for rapid population monitoring studies with the development of automated image-analysis methods for the rapid screening of cell samples (89). The quantitative nature of the assay has been validated by the cell-cloning assay (8, 29). The disadvantage of these short-term assays, however, is that variant cells cannot be isolated for molecular studies. This disadvantage led to the development of the cloning assay.

Several laboratories have studied "background" or "spontaneous" *hprt* Mfs in normals with the T-cell cloning assay (Table 1). The range of values for adults is 3.1–9.5×10^{-6} and the mean ($\pm$SD) value for a total of 252 individuals is $5.3 (\pm 2.7) \times 10^{-6}$. A clear age-related increase in Mf has been reported and is most pronounced for newborns. The simple linear regression of adult values on age shows an increase of 1.7–5% per year (21, 92, 94).

Studies of individuals with DNA-repair disorders have shown increased *hprt* Mfs in both XP and AT homozygotes (20, 71, 92). Heterozygotes for both genetic disorders showed normal frequencies.

As with the short-term assays, chemo- and radiotherapy in cancer patients caused significant rises in Mfs as did exposure to external beam-ionizing radiation (30, 59–61, 65). A study of Hiroshima atomic bomb survivors showed an increased Mf in this group with a significant dose-response relationship (40). However, the slope was not nearly as steep as that seen with the GPA assay (53). This undoubtedly reflects the different differentiation stages of the mutational target cells, i. e. stem cells for mutations detected in RBCs, and mature T-cells for mutations detected in T-lymphocytes (however, see ref. 41 for an exception).

HLA

The T-lymphocyte HLA assay has been developed and used by a single laboratory to date (47, 56). The mean Mf for normal adults (30×10^{-6}) is approximately threefold higher than the mean found with the *hprt* T-lymphocyte Mf assay (Table 1).

RELATIONSHIPS BETWEEN MEASURED MUTANT FREQUENCIES AND PROBABLE MUTATION FREQUENCIES

These four assays for mutations in human somatic cells allow quantitative measurement of variant (HbS, GPA, *hprt*) or mutant (*hprt*, HLA) frequencies. The T-lymphocyte assays allow study of the stability of the mutant phenotype and of the mutational change in the gene since the growing mutant colonies can be isolated and propagated in vitro. The *hprt* mutants can be tested for the stability of the TG^r phenotype, assayed for HPRT enzyme activity and analyzed for *hprt* gene changes. The HLA mutants can be tested for the stability of the immunoselection phenotype and also analyzed for HLA gene changes.

The unique advantage of T-lymphocytes is the heterogeneity of their T-cell receptors that confer the antigen specificity to the mature T-cell. These receptors are the products of rearranged TCR genes. These different TCR-gene rearrangements are readily recognized on Southern blots after appropriate restriction enzyme digestion of genomic DNAs. A collection of mutant colonies from a given individual can be characterized as to each colony's TCR-gene rearrangement pattern. Pairwise or groupwise comparisons allow the definition of truly independent mutants (resulting from unique mutations), and the recognition of pre- or intrathymic in vivo mutations. The scheme for interpreting these comparisons has been fully described (7, 69, 70). The ability to determine the actual mutation frequency responsible for the measured Mf by exclusion of replicate clones derived from in vivo proliferation of an original mutant cell (which would inflate the mutation frequency) allows these T-cell assays to yield truly quantitative measures of *mutation,* and not simply estimates.

Studies of isolated mutants from normal adults (Table 2), atomic bomb survivors and radioimmunotherapy patients have demonstrated that approximately 90% of the *hprt* mutants have unique TCR-gene rearrangement patterns (40, 65, 66). This argues that, in these groups, the measured Mf does closely approximate the mutation frequency. However, some individuals with elevated *hprt* mutant frequencies do show discordances between Mf and mutation frequency (7, 70). Here, TCR-gene rearrangement pattern analysis demonstrate large TCR-gene defined mutant sets. One individual showed a progressively rising Mf over four years with 98% of these mutants showing the same TCR-gene rearrangement pattern. Here, the measured mutant frequency ($>200 \times 10^{-6}$) greatly overestimated the mutation frequency ($15–20 \times 10^{-6}$). Even individuals with "normal" mutant frequencies show TCR-defined mutant sets (Table 2; 66). The most significant was a set of nine mutants with identical TCR-gene rearrangement patterns in a collection of 141 mutants from one individual (Table 2, Individual E). Although this set does not change the mutant-mutation relationship significantly, knowledge of the existence of such a group of *hprt* sibling mutants is essential for defining the spectrum of in vivo somatic mutations in humans (see below). This information is absolutely necessary in order to differentiate sibling mutant isolates from legitimate mutational "hot spots".

MOLECULAR ANALYSES OF IN VIVO SOMATIC MUTATIONS IN HUMANS

The largest data base of in vivo molecular changes has been accumulated for mutations of the *hprt* gene.

Hprt Mutations

Mutations in the *hprt* gene can be studied by Southern blots and by DNA sequencing analyses. The latter method can employ *hprt* mRNA converted to

Table 2 Use of TCR gene-rearrangement patterns to define mutation frequency in normal adults

	Individual		
	D	E	F
# of mutants	90	141	95
# of different TCR gene patterns	82	120	85
% unique TCR gene patterns	91%	85%	89%
hprt mutant frequency	4.8×10^{-6}	4.7×10^{-6}	5.6×10^{-6}
"Conservative" *hprt* mutation frequency	4.4×10^{-6}	4.0×10^{-6}	5.0×10^{-6}
# of TCR-gene defined mutant sets	5 doublets 1 quadruplet	9 doublets 1 quadruplet 1 nonamer	3 doublets 2 triplets 1 quadruplet

a cDNA by reverse transcriptase and amplified by the polymerase chain reaction (PCR) methodology (19, 22–25, 32, 33, 36, 37, 81, 82, 86, 100) or exon-specific PCR amplification of genomic DNA (36). *Hprt* somatic mutations occurring in vivo in T-lymphocytes can be compared with in vivo *hprt* germinal mutations, as seen in Lesch-Nyhan individuals. The *hprt* system, therefore, has this additional advantage of allowing somatic-germinal comparisons.

LARGE DELETIONS AND REARRANGEMENTS   Southern blot studies allow definition of large structural alterations such as deletions and rearrangements. The results are listed in Table 3 for both the somatic and germinal in vivo *hprt* mutations. Simple deletions, i.e. loss of exon-containing fragments, are divided into partial and total *hprt* deletions. Complex changes include changes in fragment molecular weight, new fragments, or apparent amplification of specific exon fragments. Although the percentages in somatic and germinal mutations are similar, the distribution of the types of alterations is different, with more complex changes seen in the somatic mutations. This may simply reflect the small number of germinal mutations reported, but it is possible that many complex alterations, such as restriction fragment molecular weight changes are the result of chromosome translocations that are lethal as germinal mutations but viable as somatic mutations in lymphocytes. Karyotype analyses have not been performed with deletion mutations but only with some that show no Southern blot changes (42, 63).

Somatic deletion mutations from one study (48/326; Table 2) have been analyzed for breakpoint distribution within the *hprt* gene (66). The number of

Table 3 Large structural alterations in the *hprt* genes of somatic and germinal mutants

Fraction with changes	Simple deletions		Complex changes	Reference
	Partial	Total		
In vivo somatic				
0.57 (12/21)	2	0	10	96
0.09 (15/164)	1	4	10	67
0.17 (15/90)	14	0	1	18
0.13 (14/105)	5	0	9	40
0.14 (48/326)	27	8	13	66
Total 0.15 (104/706)	49	12	43	
In vivo germinal				
0.08 (2/24)	0	1	1	101
0.18 (5/28)	2	1	2	98
Total 0.13 (7/53)	4[a]	1[b]	2[c]	

[a] includes 2 newly described individuals (36)
[b] same individual in both studies
[c] one individual was the same in both studies

breakpoints is proportional to intron size, as expected for a random distribution. Assuming that extragenic breakpoints occur at the same frequency per kb of DNA as do intragenic breakpoints, a deletion size of approximately 100 kb involving the *hprt* gene can be recovered as a viable somatic mutation. Additional studies show that large deletions can be recovered at the *hprt* locus as detected by codeletion of anonymous probes that hybridize to genomic sequences of chromosome Xq. Study of partial (extending 5' or 3' outside the gene) and total *hprt* deletion mutants using the probes DXS10, DXS86, DXS79, and DXS177 has allowed the sequence order of these probes on the X-chromosome relative to *hprt* to be established. This order is DXS79–5' *hprt* 3'–DXS86–DXS10–DXS177 (J. A. Nicklas et al, unpublished).

Although the distribution of deletion breakpoints in adult somatic *hprt* mutations appears to be random, a nonrandom spectrum is found in newborns with deletions of exons 2 and 3 predominating (55, 58). Combined TCR/*hprt* gene analysis showed several mutants from individual newborns with the same *hprt* alteration but with different TCR-gene rearrangement patterns. These results suggest that *hprt* mutations in the fetus frequently occur in pre- or intrathymic cells, prior to the time of TCR-gene rearrangement.

POINT MUTATIONS   Approximately 85% of the background in vivo somatic *hprt* mutations in adults occur as "point" mutations, i.e. mutations in which the molecular change is not detected on Southern blots. Both in vivo somatic mutations (81, 82) and germinal mutations (19, 22–25, 32, 33, 36, 37, 86) have been sequenced. Twenty-five of 60 somatic and 24 of 45 germinal mutations have been single base substitutions (42 and 53%, respectively). The specific base changes found in the mutant *hprt* mRNA sequences are listed in Table 4a for each type of transition and transversion mutation, and then grouped as base-pair alterations in Table 4b. Comparison of the distributions for somatic and germinal mutations show little difference, with 48 and 46% of the mutations being transitions, respectively. The distribution of these single base substitution mutations within the *hprt* gene are shown in Figure 1. The sites of somatic (top) and germinal (bottom) mutations are designated by vertical lines, with the lengths of the lines indicating the number of mutations at that base.

The remainder of the *hprt* molecular changes found by DNA sequencing of somatic and germinal mutations are listed in Table 5. Again, the distributions of changes are similar with the exception of mutations thought to involve mRNA splice-site alterations. Single exon loss mutations are presumably alterations of intron or exon splice sites. The frequency of these mutations is higher in the somatic mutations than in the germinal (20 vs 4). The 15 somatic single-exon loss mutations involve four different exons, three involve exon 4, two exon 5, three exon 7, and seven exon 8, while the three germinal mutants

Table 4a Number of base substitution mutations in the *hprt* gene of in vivo germinal and in vivo somatic mutants

	Transitions				Transversions								Total
	G→A	C→T	A→G	T→C	G→T	C→A	A→C	T→G	G→C	C→G	A→T	T→A	
In vivo germinal	4	3	1	3	2	2	1	2	1	3	1	1	24
In vivo somatic	7	1	2	2	2	2	0	1	3	0	0	5[a]	25

[a] includes 4 mutants which may have resulted from a single prethymic mutation (66, 81)

Table 4b Data in Table 4a listed as % of total mutations

	Transitions		Transversions			
	GC→AT	AT→GC	GC→TA	AT→CG	GC→CG	AT→TA
In vivo germinal	29%	17%	17%	12%	17%	8%
In vivo somatic	32%	16%	16%	4%	12%	20%

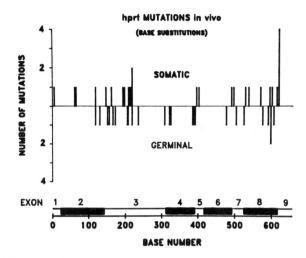

Figure 1   Distribution of single base substitutions in the *hprt* gene in in vivo somatic (top) and germinal (bottom) mutations. The relative placements of the nine exons are indicated by the horizontal bars above the base number scale.

are one each of exons 2, 7, and 8. The high frequency of exon 8 mutations suggest a susceptible or weak splice site involving this exon. The higher frequency of splice-site mutations in somatic cells may reflect unique events in these cells.

HLA Mutations

HLA-mutant T-cell isolates have also been studied for DNA changes. Complete deletions of the HLA gene were reported for approximately 30% of the in vivo background mutants derived from adults (95). Some deletions were quite large as demonstrated by the associated phenotypic loss of the HLA-B antigen coded by the allele in *"cis"* with the deleted HLA-A allele (approximately 1000 kb distant). Importantly, the most recent molecular analyses of in vivo derived HLA-loss mutants by Southern blot studies show that approximately 30% result from mitotic recombination (A. A. Morley, personal communication). Therefore, this important class of in vivo somatic events is detectable in the HLA assay.

CONCLUSIONS

If the in vivo somatic mutations in neutral genes, as detected by current assays, represent human in vivo somatic mutations generally, certain conclusions can be drawn from available data. The obvious ones are that such in

Table 5   Other sequenced mutations in the *hprt* gene of in vivo germinal and in vivo somatic mutants

	Base subst.	Small Δ	Insertions (1bp)	Splice sites			Complex changes	No RNA	TOTAL
				Single exon loss	Intron inclusion[a]	Exon exclusion[b]			
In vivo germinal	24	10	2	3	0	1	2[c]	3	45
In vivo somatic	25	9	2	15	2	3	1[d]	3	60

[a] part of 5' intron sequence added into the mRNA presumably due to mutational inactivation of the usual splice site and use of a new cryptic site into the intron or a mutation caused a new more efficacious splice site in the intron
[b] part of 5' exon sequence missing in the mRNA presumably due to mutational inactivation of the usual splice site and use of a new cryptic site farther into the exon or a mutation caused a new more efficacious splice site in the exon
[c] adjacent 2 bp change (GA→AT) and 2bp changes 1bp apart (TTG→GTA)
[d] 5bp deletion replaced by 6bp insertion

vivo events do occur to produce mutant progeny of measurable frequencies, that exposures to mutagens increase measured Vfs and Mfs, that there is an age effect, and that DNA repair is important in maintaining the genetic integrity of human somatic cells in vivo. There is also a relationship between the frequencies of mutant cells and the effective sizes of the genetic targets in which mutations occur. Finally, in vivo somatic mutations arise by a variety of mechanisms, including large deletions and gene rearrangements (detectable on Southern blots), smaller deletions and insertions, and single base changes represented by transitions and transversions, including those that produce exon losses secondary to splice-site changes. Somatic mutations can also arise in vivo by processes related to somatic crossing-over at autosomal genes.

In addition, other conclusions have been surprising and point to unique aspects of somatic cells in vivo. One conclusion is that the differentiation stage of the cell, or the developmental stage of the individual, has relevance in terms of the biological consequences of the mutation. Another is that cell division may be an important determinant of in vivo somatic mutation.

Somatic mutations arising in stem cells during any stage of an individual's development will differ from mutations arising in mature, nondividing, cells in adults. This is because stem cells produce many progeny during the life of the individual. Therefore, somatic mutations arising in such target cells will result in many mutant progeny.

The relationship of cell division to in vivo somatic mutations arising in mature cells in adults was covered above in the context of the *hprt* T-cell assay. One individual with a massive T-cell clonal amplification was discussed. Molecular analysis of hprt mutants from this individual revealed the same *hprt* mutation in most mutant isolates of the TCR-defined clonal set.

However, some isolates showed secondary mutations in addition to the primary mutation (R. J. Albertini et al, unpublished). This illustrates how unremitting in vivo cell division can create "space" for sequential gene mutations to accumulate in descendants of the same cell. If this is one route to malignant transformation, perhaps the evolution of cancer occurs by "spontaneous" mutations, with the role of some environmental carcinogens being primarily to induce cell division.

The ability to investigate in vivo somatic mutations in humans has advanced a great deal in the past fifteen years. Insights have been obtained that give a somewhat different perspective than that obtained from in vitro studies. The assays described, and new ones to be developed, will allow continued progress in this field. The utility of these assays for basic investigations seems assured.

The precise role of human in vivo somatic mutation assays in genetic toxicology remains to be defined. Although the assays are almost certainly useful as population indicators of genotoxicant exposures, their utility for detecting individual exposures or for making individual health risk assessments is unknown. Perhaps definitions of "induced" mutational spectra, added to determinations of Mfs, will provide better assessments of genotoxic hazards. A major motivation for pursuing studies of in vivo somatic mutations has always been to develop methods for human mutagenicity monitoring that translate to meaningful health benefits for the individuals being monitored.

ACKNOWLEDGMENTS

The authors wish to thank Drs. M. Ammenheuser, W. Bigbee, J. Cole, R. Gibbs, M. Hakoda, R. Jensen, D. Manchester, K. Messing, A. Morley, L. Recio, T. Skopek, A. Tates, and their colleagues for supplying data and collaborative support; M. Falta, T. Hunter, M. Lippert and L. Sullivan for performing laboratory experiments; I. Gobel for preparing this manuscript. This work was supported by NCI R01 CA 30688, NCI R01 CA39543, NCI 5 P01 CA 43791, and DOE FG02 87 60502. DOE support does not constitute an endorsement of the views expressed in this paper.

Literature Cited

1. Albertini, R. J. 1982. Studies with T-lymphocytes: An approach to human mutagenicity monitoring. In *Banbury Report: Indicators of Genotoxic Exposure*, ed. B. A. Bridges, B. E. Butterworth, I. B. Weinstein, 13:393–412. Cold Spring Harbor: Cold Spring Harbor Laboratory 580 pp.

2. Albertini, R. J. 1985. Somatic gene mutations in vivo as indicated by the 6-thioguanine-resistant T-lymphocytes

in human blood. *Mutat. Res.* 150:411–22

3. Albertini, R. J. 1985. The use of human T-lymphocytes to monitor mutations in human populations. In *New Approaches in Toxicity Testing and Their Application in Human Risk Assessments.* ed. A. P. Li, pp. 51–66. New York: Raven. 280 pp.

4. Albertini, R. J., Allen, E. F., Quinn, A. S., Albertini, M. R. 1981. Human

somatic cell mutation: in vivo variant lymphocyte frequencies determined by 6-thioguanine resistant lymphocytes. In *Population and Biological Aspects of Human Mutation.* ed. E. B. Hook, J. H. Potter, pp. 235–63. New York: Academic. 435 pp.

5. Albertini, R. J., Castle, K. L., Borcherding, W. R. 1982. T-cell cloning to detect the mutant 6-thioguanine-resistant lymphocytes present in human peripheral blood. *Proc. Natl. Acad. Sci. USA* 79:6617–21

6. Albertini, R. J., Gennett, I. N., Lambert, B., Thilly, W. G., Vrieling, H. 1989. Mutation at the *hprt* locus; Report of a workshop. *Mutat. Res.* 216:65–88

7. Albertini, R. J., Nicklas, J. A., Sullivan, L. M., Hunter, T. C., O'Neill, J. P. 1987. *Hprt* mutation in vivo in human T-lymphocytes. In *Banbury Reports 28: Mammalian Cell Mutagenesis,* ed. M. Moore, D. Tindall, F. DeMarini, F. de-Serres, 28:139–48. Cold Spring Harbor: Cold Spring Harbor Laboratory. 385 pp.

8. Albertini, R. J., Sullivan, L. S., Berman, J. K., Green, C. J., Stewart, J. A., et al. 1988. Mutagenicity monitoring in humans by autoradiographic assay for mutant T-lymphocytes. *Mutat. Res.* 204:481–92

9. Ammenheuser, M. M., Ward, J. B., Au, W. W., Belli, J. A. 1989. A prospective study comparing 6-thioguanine-resistant variant frequencies with chromosome aberration frequencies in lymphocytes from radiotherapy and chemotherapy patients. *Environ. Mol. Mutagen.* 14:9

10. Ammenheuser, M. M., Ward, J. B. Jr., Whorton, E. B. Jr., Killian, J. M., Legator, M. S. 1988. Elevated frequencies of 6-thioguanine resistant lymphocytes in multiple sclerosis patients treated with cyclophosphamide: A prospective study. *Mutat. Res.* 204: 509–20

11. Atwood, K. C., Petter, F. J. 1961. Erythrocyte automosaicism in some persons of known genotype. *Science* 134: 2100–2

12. Atwood, K. C., Scheinberg, S. L. 1958. Somatic variation in human erythrocyte antigens. *J. Cell. Comp. Physiol.* 52: 97–123

13. Baer, R., Boehm, T., Yssell, H., Spits, H., Rabbitts, T. H. 1988. Complex rearrangements within the human J-delta-C-delta/J alpha-C alpha locus and aberrant recombination between J alpha segments. *EMBO J.* 7:1661–68

14. Bigbee, W. L., Langlois, R. G., Stanker, L. H., Vanderlaan, M., Jensen, R.

H. 1990. Flow cytometric analysis of erythrocyte populations in Tn syndrome blood using monoclonal antibodies to glycophorin A and the Tn antigen. *Cytometry* 14:261–71

15. Bigbee, W. L., Langlois, R. G., Swift, M., Jensen, R. H. 1989. Evidence for an elevated frequency of in vivo somatic cell mutations in ataxia telangiectasia. *Am. J. Hum. Genet.* 44:402–8

16. Bigbee, W. L., Wryobek, R. G., Langlois, R. G., Jensen, R. H., Everson, R. B. 1990. The effect of chemotherapy on the in vivo frequency of glycophorin A "null" variant erythrocytes. *Mutat. Res.* 240:165–75

17. Bodmer, W. F. 1984. The HLA system, 1984. In *Histocompatibility Testing 1984,* ed. E. D. Albert, M. P. Baur, W. R. Mayr, pp. 11–22. Berlin: Springer Verlag. 764 pp.

18. Bradley, W. E. C., Gareau, J. L. P., Seifert, A. M., Messing, K. 1987. Molecular characterization of 15 rearrangements among 90 human in vivo somatic mutants shows that deletions predominate. *Mol. Cell Biol.* 7:956–60

19. Cariello, N. F., Scott, J. K., Kat, A. G., Thilly, W. G., Keohavong, P. 1988. Resolution of a missense mutant in human genomic DNA by denaturing gradient gel electrophoresis and direct sequencing using in vitro DNA amplification: HPRTMunich. *Am. J. Hum. Genet.* 42:726–34

20. Cole, J., Arlett, C. F., Green, M. H. L., Harcourt, S. A., Priestley, A., et al. 1988. Comparative human cellular radiosensitivity: II. The survival following gamma-irradiation of unstimulated T-lymphocytes, T-lymphocyte lines, lymphoblastoid cell lines and fibroblasts from normal donors, from ataxia telangiectasia patients and ataxia-telangiectasia heterozygotes. *Int. J. Radiat. Biol.* 54:929–43

21. Cole, J., Green, M. H. L., James, S. E., Henderson, L., Cole, H. 1988. A further assessment of factors influencing measurements of thioguanine-resistant mutant frequency in circulating T-lymphocytes. *Mutat. Res.* 204:493–507

22. Davidson, B. L., Palella, T. D., Kelley, W. N. 1988. Human hypoxanthine-guanine phosphoribosyltransferase: a single nucleotide substitution in cDNA clones isolated from a patient with Lesch-Nyhan syndrome (HPRTMontreal). *Gene* 68:85–91

23. Davidson, B. L., Pashmforoush, M., Kelley, W. N., Palella, T. D. 1988. Genetic basis of hypoxanthine guanine phosphoribosyltransferase deficiency in

a patient with the Lesch-Nyhan syndrome (HPRTFlint). *Gene* 63:331–36

24. Davidson, B. L., Pashmforoush, M., Kelley, W. N., Palella, T. D. 1989. Human hypoxanthine-guanine phosphoribosyltransferase deficiency: The molecular defect in a patient with gout (HPRTAshville). *J. Biol. Chem.* 264:502–25

25. Davidson, B. L., Tarle, S. A., Palella, T. D., Kelley, W. N. 1989. Molecular basis of hypoxanthine-guanine phosphoribosyltransferase deficiency in ten subjects determined by direct sequencing of amplified transcripts. *J. Clin. Invest.* 84:342–46

26. Davis, M. M., Kappler, J. 1988. *The T-cell Receptor.* New York: Liss. 410 pp.

27. Deisseroth, A., Nienhuis, A., Lawrence, J., Giles, R., Turner, P., et al. 1978. Chromosomal localization of human beta-globin gene on human chromosome 11 in somatic cell hybrids. *Proc. Natl. Acad. Sci. USA* 75:1456–60

28. Deisseroth, A., Nienhuis, A., Turner, P., Velez, R., Anderson, W. F., et al. 1977. Localization of the human alpha-globin structural gene to chromosome 16 in somatic cell hybrids by molecular hybridization assay. *Cell* 12:205–18

29. Dempsey, J. L., Morley, A. A. 1983. Evidence that thioguanine-resistant lymphocytes detected by autoradiography are mutant cells. *Mutat. Res.* 119:203–11

30. Dempsey, J. L., Seshadri, R. S., Morley, A.A. 1985. Increased mutation frequency following treatment with cancer chemotherapy. *Cancer Res.* 45:2873–77

31. Erslev, A. J., Weiss, L. 1983. Structure and function of the marrow. In *Hematology*, ed. W. J. Williams, E. Beutler, A. J. Erslev, M. A. Lichtman, pp. 75–82. New York: McGraw-Hill. 3rd ed. 1728 pp.

32. Fujimori, S., Davidson, B. L., Kelley, W. N., Palella, T. D. 1989. Identification of a single nucleotide change in the hypoxanthine-guanine phosphoribosyltransferase gene (HPRTYale) responsible for Lesch-Nyhan syndrome. *J. Clin. Invest.* 83:11–13

33. Fujimori, S., Hidaka, Y., Davidson, B. L., Palella, T. D., Kelley, W. N. 1988. Identification of a single nucleotide change in a mutant gene for hypoxanthine-guanine phosphoribosyltransferase (HPRTAnnArbor). *Hum. Genet.* 79:39–43

34. Furthmayer, H. 1977. Structural analysis of a membrane glycoprotein: glycophorin A. *J. Supramol. Struct.* 7:121–34

35. Gahmberg, C. G., Jakinen, M., Andersson, L. C. 1979. Expression of the major red cell sialoglycoprotein, glycophorin A, in the human leukemic cell line K562, *J. Biol. Chem.* 254:7442–48

36. Gibbs, R. A., Nguyen, P. -N., Edwards, A., Civitello, A. B., Caskey, C. T. 1990. Multiplex DNA deletion detection and exon sequencing of hypoxanthine phosphoribosyltransferase gene in Lesch-Nyhan families. *Genomics.* In press

37. Gibbs, R. A., Nguyen, P.-N., McBride, L. J., Koepf, S. M., Caskey, C. T. 1989. Identification of mutations leading to the Lesch-Nyan syndrome by automated direct DNA sequencing of in vitro amplified cDNA. *Proc. Natl. Acad. Sci. USA* 86:1919–23

38. Gladstone, P., Fueresz, L., Pious, D. 1982. Gene dosage and gene expression in the HLA region: Evidence from deletion variants. *Proc. Natl. Acad. Sci. USA* 79:1235–39

39. Hakoda, M., Akiyama, M., Kyoizumi, S., Kobuke, K., Awa, A. A., et al. 1988. Measurement of in vivo HGPRT-deficient mutant cell frequency using a modified method for cloning human peripheral blood T-lymphocytes. *Mutat. Res.* 197:161–69

40. Hakoda, M., Hirai, Y., Kyoizumi, S., Akiyama, M. 1989. Molecular analyses of in vivo *hprt* mutant T cells from atomic bomb survivors. *Environ. Mol. Mutagen.* 13:25–33

41. Hakoda, M., Hirai, Y., Shimba, H., Kusunoki, Y. 1989. Cloning of phenotypically different human lymphocytes originating from a single stem cell. *J Exper. Med.* 169:1265–76

42. He, S. M., Holmberg, K., Lambert, B., Einhorn, N. 1989. *Hprt* mutations and karyotype abnormalities in T-cell clones from healthy subjects and melphalan-treated ovarian carcinoma patients. *Mutat. Res.* 210:353–58

43. Henderson, J. F., Kelley, W. N., Rosenbloom, F. M., Seegmutter, J. E. 1969. Inheritance of purine phosphoribosyltransferase in man. *Am. J. Hum. Genet.* 21:61–70

44. Henderson, L., Cole, H., Cole, J., James, S. E., Green, M. 1986. Detection of somatic mutations in man: Evaluation of the microtitre cloning assay for T-lymphocytes. *Mutagenesis* 1:195–200

45. Huttner, E., Mergner, W., Braun, R., Schoneich, J. 1990. Increased frequency of 6-thioguanine-resistant lymphocytes

in peripheral blood of workers employed in cyclophosphamide production. *Mutat. Res.* 243:101–7

46. Isobe, M., Erikson, J., Emanuel, B. S., Nowell, P. C., Croce, C. M. 1985. Location of gene for beta subunit of human T-cell receptor at band 7q35, a region prone to rearrangements in T-cells. *Science* 228:580–82

47. Janatipour, M., Trainor, K. J., Kutlaca, R., Bennett, G., Hay, J., et al. 1988. Mutations in human lymphocytes studied by an HLA selection system. *Mutat, Res.* 198:221–26

48. Jensen, R. H., Bigbee, W. L., Langlois, R. G. 1987. In vivo somatic mutations in the Glycophorin A locus of human erythroid cells. See Ref. 7.

49. Jones, C., Morse, H. C., Kao, F. -T., Carbone, A., Palmer, E. 1985. Human T-cell receptor alpha-chain genes: Location on chromosome 14. *Science* 228:83–85

50. Kavathas, P., Bach, F. H., DeMars, R. 1980. Gamma ray-induced loss of expression of HLA and glyoxalase I alleles in lymphoblastoid cells. *Proc. Natl. Acad. Sci. USA* 77:4251–55

51. Kavathas, P., DeMars, R., Bach, F. H. 1980. Hemizygous HLA mutants of lymphoblastoid cells: A new cell type for histocompatibility testing. *Hum. Immunol.* 1:317–24

52. Kudo, S., Fukuda, M. 1989. Structural organization of glycophorin A and B genes: Glycoporin B gene evolved by homologous recombination at Alu repeat sequences. *Proc. Natl. Acad. Sci. USA* 86:4619–23

53. Kyoizumi, S., Nakamura, N., Hakoda, M., Awa, A. A., Bean, M. A., et al. 1989. Detection of somatic mutations at the glycophorin A locus in erythrocytes of atomic bomb survivors using a single beam flow sorter. *Cancer Res.* 49:581–88

54. Langlois, R. G., Bigbee, W. L., Jensen, R. H., German, J. 1989. Evidence for increased in vivo mutations and somatic recombination in Bloom's syndrome. *Proc. Natl. Acad. Sci. USA* 86:670–74

55. Lippert, M. J., Manchester, D. K., Hirsch, B., Nicklas, J. A., O'Neill, J. P., et al. 1990. Patterns of *hprt* mutations detected in newborns and elderly by cloning assay. *Environ. Mol. Mutagen.* 15:35

56. McCarron, M. A., Kutlaca, A., Morley, A. A. 1989. The HLA-A mutation assay: Improved technique and normal results. *Mutat. Res.* 225:189–93

57. McGinniss, M. J., Falta, M. T., Sulli-van, L. S., Albertini, R. J. 1990. In vivo *hprt* mutant frequencies in T-cells of normal human newborns. *Mutat. Res.* 240:117–26

58. McGinniss, M. J., Nicklas, J. A., Albertini, R. J. 1990. Molecular analyses of in vivo *hprt* mutations in human T-lymphocytes IV. Studies in newborns. *Environ. Mol. Mutagen.* 14:229–37

59. Messing, K., Bradley, W. E. C. 1985. In vivo mutant frequency rises among breast cancer patients after exposure to high doses of gamma radiation. *Mutat. Res.* 152:107–12

60. Messing, K., Ferraris, J., Bradley, W. E. C., Swartz, J., Seifert, A. M. 1989. Mutant frequency of radiotherapy technicians appears to be associated with recent dose of ionizing radiation. *Health Phys.* 57:537–44

61. Messing, K., Seifert, A. M., Bradley, W. E. C. 1986. In vivo mutant frequency of technicians professionally exposed to ionizing radiation. In *Monitoring of Occupational Genotoxicants*, ed. M. Sorsa, H. Norppa, pp. 87–97. New York: Liss. 250 pp.

62. Morley, A. A., Trainor, K. J., Dempsey, J. L., Seshadri, R. S. 1985. Methods for study of mutations and mutagenesis in human lymphocytes. *Mutat. Res.* 147:363–67

63. Muir, P., Osborne, Y., Morley, A. A., Turner, D. R. 1988. Karyotypic abnormality of the X chromosome is rare in mutant HPRT lymphocyte clones. *Mutat. Res.* 197:157–60

64. Murre, C., Waldman, R. A., Morton, C. C., Bongiovanni, R. F., Waldman, T. A., et al. 1985. Human gamma-chain genes are rearranged in leukemia T-cells and map to the short arm of chromosome 7. *Nature* 316:549–52

65. Nicklas, J. A., Falta, M. T., Hunter, T. C., O'Neill, J. P., Jacobson-Kram, D., et al. 1990. Molecular analyses of in vivo *hprt* mutations in human lymphocytes. V. Effects of total body irradiation secondary to radioimmunoglobulin therapy (RIT) *Mutagenesis*. In press

66. Deleted in proof

67. Nicklas, J. A., Hunter, T. C., Sullivan, L. M., Berman, J. K., O'Neill, J. P., et al. 1987. Molecular analyses of in vivo *hprt* mutations in human T-lymphocytes I. Studies of low frequency "spontaneous" mutants by Southern blots. *Mutagenesis* 2:341–47

68. Nicklas, J. A., Miyachi, J., Taurog, J. D., Wee, S.-L., Chen, L. K. et al. 1984. HLA loss variants of a B27+ lymphoblastoid cell line: Genetic and cellu-

lar characterization. *Hum. Immunol.* 11: 19–30
69. Nicklas, J. A., O'Neill, J. P., Albertini, R. J. 1986. Use of T-cell receptor gene probes to quantify the in vivo *hprt* mutations in human T-lymphocytes. *Mutat. Res.* 173:65–72
70. Nicklas, J. A., O'Neill, J. P., Sullivan, L. M., Hunter, T. C., Allegretta, M., et al. 1988. Molecular analyses of in vivo hypoxanthine-guanine phosphoribosyltransferase mutations in human T-lymphocytes: II. Demonstration of a clonal amplification of *hprt* mutant T-lymphocytes in vivo. *Environ. Mol. Mutagen.* 12:271–84
71. Norris, P. G., Limb, G. A., Hamblin, A. S., Lehmann, A. R., Arlett, C. F., et al. 1989. Immune function, mutant frequency and cancer risk in the DNA repair defective genodermatoses xeroderma pigmentosum, cockaynes syndrome and trichothiodystrophy. *J. Invest. Dermatol.* In press
72. O'Neill, J. P., McGinniss, M. J., Berman, J. K., Sullivan, L. M., Nicklas, J. A., et al 1987. Refinement of a T-lymphocyte cloning assay to quantify the in vivo thioguanine-resistant mutant frequency in humans. *Mutagenesis* 2:87–94
73. O'Neill, J. P., Sullivan, J. P., Booker, J. K., Pornelos, B. S., Falta, M. T., et al. 1989. Longitudinal study of the in vivo *hprt* mutant frequency in human T-lymphocytes as determined by a cell cloning assay. *Environ. Mol. Mutagen.* 13:289–93
74. Ostrosky-Wegman, P., Montero, R., Gomez, M., Cortinas de Nava, C. 1987. 6-Thioguanine resistant T-lymphocyte determination as a possible indicator of radiation exposure. *Environ. Mutagen.* 9:81
75. Ostrosky-Wegman, P., Montero, R. M., Cortinas de Nova, C., Tice, R. R., Albertini, R. J. 1988. The use of bromodeoxyuridine labeling in the human lymphocyte HGPRT somatic mutation assay. *Mutat. Res.* 191:211–14
76. Papayannopoulou, T. H., Brice, M., Stamatoyannopoulos, G. 177. Hemoglobin F synthesis in vitro: Evidence for control at the level of primitive erythroid stem cells. *Proc. Natl. Acad. Sci. USA* 74:2923–27
77. Papayannopoulou, T. H., Nute, P. E., Stamatoyannopoulos, G., McGuire, T. G. 1977. Hemoglobin ontogenesis: Test of the gene exclusion hypothesis. *Science* 197:1215–16
78. Patel, P. I., Nussbaum, R. T., Framson, P. E., Ledbetter, D. H., Caskey, C. T.,

et al. 1984. Organization of the *hprt* gene and related sequences in the human genome. *Som. Cell. Mol. Genet.* 10: 483–93
79. Pious, D., Hawley, P., Forrest, G. 1973. Isolation and characterization of HL-A variants in cultured human lymphoid cells. *Proc. Natl. Acad. Sci. USA* 70:1397–400
80. Pious, D., Soderland, C., Gladstone, P. 1977. Induction of HLA mutations by chemical mutagens in human lymphoid cells. *Immunogenetics* 4:437–88
81. Recio, L., Cochrane, J., Simpson, D., Skopek, T., O'Neill, J. P., et al. 1990. DNA sequence analysis of in vivo *hprt* mutation in human T-lymphocytes. *Mutagenesis.* In press
82. Rossi, A., Thijssen, J., Tates, A., Vrieling, H., Natarajan, A., et al. 1990. Mutations affecting RNA splicing in man are detected more frequently in somatic than in germ cells. Submitted
83. Rundles, R. W. 1983. Chronic mylogenous leukemia. In *Hematology*, ed. W. J. Williams, E. Beutler, A. J. Erslev, M. A. Lichtman. pp. 196–213. New York: McGraw-Hill. 3rd ed. 1728 pp.
84. Russell, L. B., Major, M. J. 1957. Radiation induced presumed somatic mutants in the house mouse. *Genetics* 42:161–75
85. Seifert, A. M., Bradley, W. C., Messing, K. 1987. Exposure of nuclear medicine patients to ionizing radiation is associated with rises in HRRT⁻ mutant frequency in peripheral T-lymphocytes. *Mutat. Res.* 191:57–63
86. Skopek, T. R., Recio, L., Simpson, D., Dallaire, L., Melancon, S. B., et al. 1990. Molecular analyses of a novel Lesch-Nyhan syndrome mutation (*hprt*Montreal) by use of T-lymphocyte cultures. *Hum. Genet.* 85:111–16
87. Stamatoyannopoulos, G., Nute, P., Lindsley, D., Farguhar, M. B., Nakamato, B., et al. 1984. Somatic-cell mutation monitoring system based on human hemoglobin mutants. In *Single Cell Monitoring Systems, Topics in Chemical Mutagenesis*, ed. A. A. Ansari, F. J. de Serres, pp. 1–35. New York: Plenum. 289 pp.
88. Stamatoyannopoulos, G., Wood, W. G., Papayannapoulou, T. N., Nute, P. E. 1975. An atypical form of hereditary persistence of fetal hemoglobin in blacks and its association with sickle cell trait. *Blood* 46:683–92
89. Stark, M. H., Tucker, J. H., Thomson, E. J., Perry, P. E. 1984. An automated image analysis system for the detection

of rare autoradiographically labeled cells in the human lymphocyte HGPRT assay. *Cytometry* 5:250–57

90. Stout, J. T., Caskey, C. T. 1985. *Hprt*: Gene structure, expression, and mutation. *Anni. Rev. Genet.* 19:127–48

91. Strauss, G. H., Albertini, R. J. 1977. 6-Thioguanine resistant lymphocytes in human blood. In *Progress in Genetic Toxicology*, ed. D. Scott, B. A. Bridges, F. H. Sobels, pp. 327–34. Amsterdam: Elsevier. 335 pp.

92. Tates, A. D., Bernini, L. F., Natajaran, A. T., Ploem, J. S., Verwoed, N. P., et al. 1989. Detection of somatic mutations in man: HPRT mutations in lymphocytes and hemoglobin mutations in erythrocytes. *Mutat. Res.* 213:73–82

93. Tatsumi, K., Uchiyama, T., Uchino, H. 1985. Measurement of in vivo mutations in human peripheral T-lymphocytes. *J. Radiat. Res.* 26:89

94. Trainor, K. J., Wigmore, D. J., Chrysostomou, A., Dempsey, J. L., Seshadri, R., et al. 1984. Mutation frequency in human lymphocytes increases with age. *Mech. Ageing Dev.* 27:83–86

95. Turner, D. R., Grist, S. A., Janatipour, M., Morley A. A. 1988. Mutations in human lymphocytes commonly involve gene duplication and resemble those seen in cancer cells. *Proc. Natl. Acad. Sci. USA* 85:3189–93

96. Turner, D. R., Morley, A. A., Haliandros, M., Kutlaca, B., Sanderson, B. J.

1985. In vivo somatic mutations in human lymphocytes frequently result from major gene alterations. *Nature* 315:343–45

97. Verwoerd, N. P., Bernini, L. F., Bonnet, J., Tanke, H. J., Natarajan, A. T., et al. 1987. Somatic cell mutations in humans detected by image analysis of immunofluorescently stained erythrocytes. In *Clinical Cytometry and Histometry*, ed. G. Burger, J. S. Ploem, K. Goerttler, pp. 465–469. San Diego: Academic. 590 pp. 1st ed.

98. Wilson, J., Stout, J., Palella, T., Davidson, B., Kelley, W., et al. 1986. A molecular survey of hypoxanthine-guanine phosphoribosyltransferase deficiency in man. *J. Clin. Invest.* 77:188–95

99. Wood, W. G., Stamatoyannapoulos, G., Lim, G., Nute, P. E. 1975. F-cells in the adult: Normal values and levels in individuals with hereditary and acquired elevations of HbF. *Blood* 46:671–82

100. Yang, J. -L., Maher, V. M., McCormick, J. J. 1989. Amplification of cDNA from the lysate of a small clone of diploid human cells and direct DNA sequencing. *Gene* 83:347–54

101. Yang, T. P., Patel, P. I., Chinault, A. C., Stout, J. T., Jackson, L. G., et al. 1984. Molecular evidence for new mutation at the *hprt* locus in Lesch-Nyhan patients. *Nature* 310:412–14

Annu. Rev. Genet. 1990. 24:327–62

THE COMPARATIVE RADIATION GENETICS OF HUMANS AND MICE

James V. Neel

Department of Human Genetics, University of Michigan Medical School, Ann Arbor, Michigan 48109–0618

Susan E. Lewis

Research Triangle Institute, P. O. Box 12194, Research Triangle Park, North Carolina 27709

KEY WORDS: mutation rates/induced/mouse, mutation rates/induced/man, genetic doubling dose/radiation/man, genetic doubling dose/radiation/mouse, low-level radiation/genetic effect

CONTENTS

0066-4197/90/1215-0327$02.00

INTRODUCTION

The attempt by geneticists to predict the genetic consequences for humans of exposure to ionizing radiation has arguably been one of the most serious social responsibilities they have faced in the past half century. Important for its own sake, this issue also serves as a prototype for the effort to evaluate the ultimate genetic impact on ourselves of other human perturbations of the environment in which our species functions. Recently we (67–69) have been developing the thesis that according to the results of studies on the children of survivors of the atomic bombings, humans may not be as sensitive to the genetic effects of radiation as has been projected by various committees on the basis of data from the most commonly employed paradigm, the laboratory mouse (cf 12, 104, 105). In this review we attempt as detailed a comparison as space permits of the findings on humans and mice, presenting the data in a fashion that will enable those who at certain critical points in the argument wish to make other assumptions, to do so. We argue that a reconsideration that includes *all* the data now available on mice brings the estimate of the doubling dose for mice into satisfactory agreement with the higher estimate based on humans.

Since the concept of a doubling dose is critical in these discussions, a clear definition is needed at the outset. The genetic doubling dose of radiation is the amount of acute or chronic radiation that will produce the same mutational impact on a population as occurs spontaneously each generation. To be societally relevant, that estimate is best expressed in terms of morbidity and mortality. For both spontaneous and induced mutation, the reference is customarily first generation effects. This definition in principle encompasses the impact of the gamut of all possible alterations in DNA, from single nucleotide substitutions and very small insertions and deletions to gain or loss of entire chromosomes. The definition is as simple as the concept is useful. Expressing genetic damage as a doubling dose provides a convenient societal frame of reference.

Unfortunately, for neither mice nor humans is the impact of spontaneous mutation each generation as yet satisfactorily defined. Thus, in the calculation of a doubling dose, the errors inherent in estimating the magnitude of a radiation effect may be further compounded by imprecision in the definition of baseline values. We present radiation in the units employed by the investigator, but for our own studies and analyses present the results in gray (1 Gy = 100 rad) and sieverts (1 Sv = 100 rem). In the experimental settings reviewed herein, one r may be equated to one rad (rem).

A PRÉCIS OF THE HUMAN DATA ON THE GENETIC EFFECTS OF ACUTE RADIATION

This treatment of the human data is confined to the information accumulated in Hiroshima and Nagasaki over the past 40 years, because of their high information content compared to other sets of human data, the relative accuracy of radiation-exposure estimates, and a study design that provides essentially nonoverlapping indicators of radiation effects. The Genetic Program that evolved in Japan has been described in detail on several occasions (4, 39, 64–67, 93), Briefly, in the immediate postwar period in Japan, the economic stringencies were such that a rationing program initiated during the war years to benefit pregnant women was continued. This involved registration of the fact of pregnancy at the completion of the fifth lunar month of gestation. By incorporating the registrants into a genetic study, a *prospective* investigation of pregnancy outcome in Hiroshima and Nagasaki was initiated in February, 1948, that included reference to sex ratio, congenital defect, viability at birth, birth weight, and survival of child during the neonatal period. This more clinical program, which included a physical examination of all newborn infants by a physician, was terminated in February, 1954, but the collection of data on the sex ratio and survival of live-born infants continued, with ascertainment of additional births now through city records. In addition, from the birth registrations in these two cities, the sample for the consideration of these latter two indicators was extended backward in time, to include births between May, 1946, and January, 1948. During the clinical program in Hiroshima, some 62% of all infants who were stillborn or died in the first six days of life were subjected to autopsy. A subset of some 26% of the infants born during this period was reexamined at approximately 9 months of age.

Relatively few survivors beyond 2000 meters from the hypocenter were exposed to radiation from the bombs; the mortality within the 2000 meter zone was high. Accordingly, the original sample of 76,617 children plus subsequent additions contained many more born to parents at distances from the hypocenter > 2000 meters ATB, who seldom received increased radiation from the explosions, than to parents within the 2000 meter radius, whose gonadal doses were relatively large. In 1959, to increase the efficiency of the study of survival, three cohorts were defined from the children born in the two cities since the bombings: The first comprised all children born between May 1, 1946, and December 31, 1958, to parents one or both of whom were within 2000 meters of the hypocenter ATB (proximally exposed); the second was composed of age, sex, and city-matched births to parents one or both of whom were distally exposed to the bombs (> 2500 meters from the hypocenter); and the third, of age, sex, and city-matched children born to parents now residing in Hiroshima and Nagasaki, neither of whom were in the cities ATB. Children born to parents within the 2000–2500 meter radius of distance from the hypocenter are omitted from consideration, because of the exceptional difficulties in estimating the (low) exposures at this distance and the work involved in following children who would contribute relatively little information regarding radiation effects. Setting the earliest date for admission into the cohort at May 1, 1946, effectively eliminates from the sample all children in utero ATB.

These cohorts have been periodically expanded to include matched numbers of recent births, the limiting factor in the expansion being the number of births to the proximally exposed parents. In 1984, some 39 years after the birth of the first children conceived following the bombings, there were only three births to proximally exposed parents and the cohorts were closed, effective January, 1985. The cohort of children born through 1982 (the cut-off date for analyses) to parents receiving > 0.01 Gy of radiation now equals 31,150. In addition to the study of the survival of these and the "control" children, special studies have been mounted on the physical development of these children during their school years and on the incidence of cancer. In 1968, a search for cytogenetic abnormalities in those children in the cohorts over 12 years of age was initiated, and, in 1975, a search was undertaken in these same cohorts for mutation altering the electrophoretic mobility and/or function of a select battery of proteins. For both these studies, the necessary controls were drawn from the children of parents not receiving increased radiation ATB.

The approach to assigning gonadal doses to the survivors underwent a drastic revision during the early 1980s, resulting in the system known as Dose Schedule 1986 (DS86) (81). All the genetic data have now been analyzed on the basis of the new dosage schedules, supplemented, where DS86 dose are not available, by an ad hoc procedure (cf 74). The radiation resulting from the explosions was predominantly gamma, but included a small neutron component. These two components have been separately estimated, and for these exposure levels, the neutron component assigned a Relative Biological Efficiency (RBE) of 20 (cf 37). This permits assigning the total dose in sieverts. The average conjoint parental gonadal exposures of course vary from study to study, as the composition of the parents of the children examined varies, but in general has been between 0.32 and 0.60 Sv; the distribution of conjoint doses is markedly skewed to the right. Given the chronology of this study, the findings must apply almost exclusively to the results of the radiation of spermatogonia and early, immature oocyte.

We turn now to a consideration of the data.

Untoward pregnancy outcome   Because of the interrelatedness of major congenital defect, stillbirth, and neonatal death, we have treated these endpoints as an entity, termed Untoward Pregnancy Outcome (UPO), defined as an outcome resulting in a child with major congenital defect and/or stillborn and/or dying within the first two weeks of life. Between 1948 and 1954, data were collected on these outcomes for a total of 76,617 newborn infants in Hiroshima and Nagasaki, on 69,706 of whom the data were sufficiently complete in all respects to permit inclusion in the analysis (cf 66). A total of 3,498 children were classified as UPOs. The congenital defects considered major are listed in Neel & Schull (66) and Neel (62). Although the data have been analyzed in many ways, we will for this and the subsequent endpoints to be considered present only the results derived from fitting a simple linear dose-response model to the occurrence of the various indicators of radiation-related damage. This model yields a regression coefficient per Sv (β), and an intercept term, α. Table 1 presents the results of the most recent analysis, based on the use of both DS86 and the ad hoc doses (74). The regression of indicator on dose is positive, and so consistent with the hypothesis that genetic damage was produced, but well below the level of significance.

"Prereproductive" deaths among live-born infants (exclusive of those resulting from a malignant tumor)   The mortality data on the three cohorts of live-born children described earlier extend through 1985, when the mean age of the members of the cohorts was 26.2 years. The data are thus relatively complete for death during the prereproductive years. Details of the organization of the study have been presented elsewhere (39, 64). After exclusion of the neonatal deaths included in the analysis of the foregoing section, and cancer deaths (to be discussed in the next section), the three cohorts together include 67,202 individuals, among whom there have been 2,584 deaths. The most recent estimate of the regression of death on parental exposure, employing DS86 doses, is presented in Table 1 (63). Again, the regression is positive, as might be expected if deleterious mutations had been produced, but less than its standard error.

Malignancies in the F_1   A minority of certain cancers of childhood, such as retinoblastoma and Wilms' tumor, result from the presence of one defective gene inherited from a parent plus a second defective allele of this gene resulting from a somatic mutation in the child. Most cancers of childhood, however, seem to be entirely due to somatic cell events occurring in the child. Only the frequency of the former type might be expected to respond to parental radiation. The normal alleles at the loci associated with these "genetic" cancers are usually referred to as "tumor-suppressing" genes. In addition, the occurrence of proto-oncogenes is well established; in principal, radiation could introduce mutations in proto-oncogenes, although thus far there are no clear examples in humans of a genetic (inherited) predisposition to cancer (or any other trait) due to a mutant proto-oncogene. On the thesis that the frequency of tumor-suppressor and proto-oncogenes was probably such that collectively they constituted a significant target for radiation, the incidence of cancer in the populations under consideration was studied, using both death

Table 1 An estimate of the genetic doubling doses that can be excluded at specified confidence levels by the human data. Further explanation in text.

| Trait | Regression | | Observed total background (a) | Mutational contribution background* (b) | Mutational component (%, b ÷ a) | Lower confidence limit (SV) | | |
	βSV	α				99%	95%	90%
UPO	0.00264 ±0.00277	0.03856 ±0.00582	0.0502	0.0017 – 0.0027	3.4–5.4	0.14 – 0.23	0.18 – 0.29	0.21 – 0.33
F_1 Mortality	0.00076 ±0.00154	0.06346 ±0.00181	0.0458	0.0016 – 0.0026	3.5–5.7	0.51 – 0.83	0.68 – 1.10	0.81 – 1.32
F_1 Cancer	−0.00008 ±0.00028	0.00104 ±0.00033	0.0012	0.00002 – 0.00005	2.0–4.0	0.04 – 0.07	0.05 – 0.11	0.07 – 0.15
Sex-chromosome aneuploids	0.00044 ±0.00069	0.00252 ±0.00043	0.0030	0.0030	100	1.23	1.60	1.91
Protein variants	−0.00001 ±0.00001	0.00001 ±0.00001	0.000013	0.000013	100	0.99	2.27	7.41

*per diploid locus

certificates and the cancer registries maintained in Hiroshima and Nagasaki (109).

There were 43 cancers in the 31,150 children born to exposed parents and 49 in the 41,066 children of unexposed parents, for an incidence in the latter of 1.2/1000 persons. As shown in Table 1, the regression on exposure was very small and negative, i.e. counter hypothesis. This was true both for all cancers and for those thought most likely to fit the retinoblastoma model, i.e. retinoblastoma, Wilms' tumor, neuroblastoma, embryonal carcinoma of the testis, and sarcomas (considered an alternative expression of the retinoblastoma allele). None of these latter tumors was in this series associated with a positive family history.

Frequency of balanced structural rearrangements of chromosomes and of sex-chromosome aneuploidy   Cytogenetic studies on the children of survivors were initiated in the 1960s, involving members of the F_1 Mortality Study described earlier (2–4). Two age- and sex-matched samples were established, one drawn from the children of proximally exposed and the other from the children of the distally exposed. To increase participation, no effort was made to obtain the venous blood samples necessary to the study from children under age 13. The study will therefore not supply reliable data on chromosomal abnormalities associated with high childhood mortality (including Down syndrome under the conditions of postwar Japan) but is thought to provide appropriate data on the occurrence of sex-chromosome aneuploidy and balanced structural rearrangements of chromosomes, the latter defined as reciprocal translocations, pericentric inversions, and Robertsonian translocations.

Awa et al (4) detected 19 individuals with sex-chromosome aneuploidy in the 8322 children of proximally exposed parents who were examined, and 25 such individuals in the 7976 children of controls. Individuals with some of these abnormalities are sterile (XXY) and with others may exhibit decreased fertility (XXX); furthermore, XYY and XXX individuals only rarely have similarly affected children (19). Thus a very high proportion of these persons may be presumed to result from primary nondisjunction (i.e. mutation) in a parent, but specific family studies were not undertaken on these individuals. DS86 doses were not available to Awa et al (4); we have now distributed these children according to joint parental gonad exposure and derived the appropriate regressions (67). The regression of incidence on dose (Sv) is 0.00044 ± 0.00069, with an intercept of 0.00252 ± 0.00044 (Table 1).

Awa et al (4) also encountered 18 balanced structural rearrangements in the 8322 children of proximally exposed parents and 25 in the 7976 children of controls. Family studies could be performed on about 60% of the children, and one de novo mutational event was detected in the children of the exposed and one in the controls. With adjustment for the children not subjected to family studies, Awa et al (4) estimate a mutation rate for both groups of 1.0×10^{-4}/gamete/generation, with a 95% confidence interval between 0.2 and 4.5 $\times 10^{-4}$. The number of mutations was not felt sufficient to warrant a regression analysis.

"Point" mutations affecting protein charge and/or function   As used throughout this treatment, the term "point" mutation of necessity encompasses not only specific nucleotide changes but also small insertions/deletions/rearrangements. Between 1972 and 1985, a battery of 30 serum transport and erythrocyte enzyme proteins were screened for the occurrence of mutations altering electrophoretic mobility, and a subset of 11 of these proteins (all enzymes) were screened for mutations resulting in loss of function (65). Three mutations altering electrophoretic mobility were encountered in the equivalent of 667,404 locus product tests on children of proximally exposed persons, and the same number (3) in the equivalent of 466, 881 tests on the children of distally exposed parents. These mutations were teased out of family studies on some 964 rare protein variants that had the potential of resulting from mutation in the parental generation.

The examination of a subset of 60,529 locus products for loss of enzyme activity in the children of proximally exposed parents yielded 26 rare variants, one of which proved to be a newly arisen mutation; no mutations were encountered in the 21 variants identified in the 61,741 determinations on the children of the comparison group. When the data on the mobility and loss-of-function mutations were combined, the mutation rate in the children of proximally exposed was 0.60×10^{-5}/locus/generation, with 95% confidence intervals between 0.2 and 1.5 $\times 10^{-5}$/locus/generation. In the comparison group the rate was 0.64×10^{-5}/locus/generation, with 95% intervals between 0.1 and 1.9 $\times 10^{-5}$/locus/generation. The average conjoint parental exposure for the proximally exposed was 0.41 Sv and the regression of mutation occurrence on conjoint parental gonadal exposure was -0.00001 ± 0.00001/Sv with an intercept of 0.00001 ± 0.00001 (Table 1) (67).

Sex ratio   The relation of sex ratio to parental radiation history in the Japanese experience has been presented on several occasions. With the discovery of sex-chromosome aneuploids in 1959 and recognition of the Lyonization phenomenon in 1961, it became clear that the interpretation of any observed change in the phenotypic sex-ratio at birth would not be simple, and no further analyses have been conducted on sex-ratio since our 1966 publication. The demonstration of no significant difference in the occurrence of sex-chromosome aneuploidy in the two groups, as described earlier, now indicates this phenomenon is not a complication in our findings regarding sex-ratio. Because of Lyonization, the least complicated data on sex-ratio concern the offspring of exposed mother:unexposed father marriages, where the expectation is for a decrease in the frequency of male offspring if the mutation rate had been increased. In our last analysis (92), the regression, based on the previous (T65DR) dose schedule, was $0.0027 \pm 0.0040/Sv$, i.e. insignificantly counter hypothesis. Since we see no way to bring these data into the framework of a doubling-dose estimate (see below), these data have not been reanalyzed or extended with the advent of the DS86 dose system.

Anthropometric studies   Birthweights were obtained on the 69,706 children examined during the clinical phase of the program, and weight, body length, head circumference, and chest circumference obtained on a subset of 18,498 of these infants reexamined between ages 8 and 10 months (66). In addition, the annual school measurements (height, weight, sitting height, and chest circumference) have for a subset of Hiroshima school children been analyzed in relation to parental radiation exposure, using the T65DR schedule (22–26). No significant or suggestive differences existed between the children of proximally exposed and the control series.

THE GENERATION OF AN ESTIMATE OF THE DOUBLING DOSE FROM THE HUMAN DATA

The attempt to derive both the lower limits to the genetic doubling dose(s) compatible with these data and the most probable doubling dose consistent with all the data has presented a number of unique problems recently described in detail (67). On the other hand, an unusual and favorable feature of this study is that all of its various components have been based on a well-defined population subsampled in various, essentially nonoverlapping ways. The result is a multifaceted appraisal of the genetic impact of the bombs on what is essentially a single cohort. We will enumerate below some of the ways in which these studies lack the elegance that can be achieved in an experimental setting. They are, however, based on a single complex population exhibiting the full range of human heterogeneity. Furthermore, the subjects of the study have been born over a sufficiently extended time interval that we begin to approximate the effects on an entire generation rather than on a "window" of access to a very limited period in the reproductive histories of the parents.

We take the position that in the studies in Hiroshima and Nagasaki we are not testing the hypothesis that radiation is mutagenic. It has been so for every properly studied species of prokaryote and eukaryote, and *Homo sapiens* cannot be an exception. The present data are thus the best approximation to the genetic effect of the bombs to be available for some years. To explore their precise implications requires specification for each indicator of the contribution of mutation in the parental generation to the frequency of the trait. For sex-chromosome aneuploids, reciprocal chromosome exchanges, and protein variants, this frequency can and has been directly established by the appropriate family studies. For UPOs, and death and cancer among live-born infants, this is not the case; elsewhere we have attempted to generate the estimate of the contribution of spontaneous mutation in the parental generation to these events that is necessary to derive a doubling dose (67).

Table 1 summarizes the results of this effort. Column 4 presents the incidence of the events to be used in subsequent calculations. For the first three items, it is total incidence; for the last two, where it can be specified, it is mutational incidence. Because of the paucity of balanced chromosomal rearrangement that can be attributed to mutation in the parental generation, this indicator is not incorporated in the calculations to come. The fifth column presents the estimated incidence of the trait as a result of spontaneous parental mutation; for the last two indicators it is of course identical with the entry in column 4. The range reflects the uncertainty in the estimate. A detailed justification of these estimates is presented elsewhere (67). Column 6 attempts to provide perspective by indicating the percent of the entry in column 4 to be attributed to spontaneous parental mutation. Note—contrary to Sankaranarayanan's (90) unfortunate erroneous perception of how the doubling dose has been estimated—that only a small fraction of these three indicators, ranging from 2–5%, is attributed to mutation in the parental generation. Finally, by a standard statistical procedure presented elsewhere (67), we have calculated (cols 7–9) for the various indicators lower confidence limits for the doubling dose at the specified probabilities. The range given for the first three indicators results from the range in the estimated mutational component for each.

Such estimates as these five can be combined only on the assumption that they are drawn from a common pool of estimates, i.e. that the doubling doses for the individual phenomenon are essentially the same. An extensive literature suggests that the true doubling dose for radiation-induced sex-linked chromosome aneuploids and protein variants (resulting from nondisjunction and nucleotide substitutions, respectively) is probably higher than the doubling dose for the other three indicators (resulting predominantly from insertion/ deletion/rearrangement events and unrepaired chromosome breaks). We have therefore pooled the first three and last two estimates of Table 1, using a statistical procedure described elsewhere (67). The resulting lower confidence limits, at the 95% confidence limit, are between 0.63–1.04 Sv for the first three indicators pooled and 2.71 Sv for the last two.

The data of Table 1 can also be used to derive the most probable overall doubling dose suggested by these data. For this we assume the effect on the F_1

of any mutations resulting in protein variants (i.e. point mutations) is subsumed within the UPOs and F_1 mortality. Since we are basically working with independent observations on the same cohort, we have suggested that the doubling dose can be derived by dividing the sum of the first four entries in column 4 (0.00632–0.00835) by the sum of the five regression terms (0.00375), resulting in an estimate of 1.69–2.23 Sv. There is satisfactory agreement between this estimate and the minimal 95% probability estimates.

The most probable estimate cannot be compared directly with any estimate based on acutely irradiated mice because with humans, both sexes enter into the estimate, the average conjoint gonadal exposure being 0.3–0.6 Sv, depending on the precise study. Moreover, the dosage curve is highly skewed to the right, some couples having received an estimated joint gonadal dose as high as 6.0 Sv. The mouse data, on the other hand, are largely derived from gonadal doses to one sex, the male, of 3.0 or 6.0 Gy (=Sv). The difference in radiation exposures can in principle be met by converting both the human and mouse data to equivalent results from low-dose rate, low-LET radiation (LET = linear energy transfer). The many difficulties in extrapolating from acute, relatively high-level radiation effects to the effects of low-level, chronic, or intermittent exposure are detailed in Report 64 of the National Council on Radiation Protection (61), and yet the effort must be made, both to render the man-mouse comparison more valid and to supply guidelines for human exposures. Russell et al (89) have demonstrated with the 7-locus system that high-level, acute radiation exposures are *at least* three times more mutagenic than low-dose pulsed or chronic radiation. Other indicators, notably those on translocation frequencies, have yielded higher dose rate factors (revs. in 98, 106). Following the linear quadratic treatment of the mouse data by Abrahamson & Wolff (1), we suggest applying a dose rate factor of 2 to convert the Hiroshima-Nagasaki data into their low-dose rate, low-LET radiation equivalent. The resulting figure is 3.38–4.46 Sv. This is a conservative estimate, since it does not incorporate three sets of data that showed no hint of an exposure effect, namely, the data on the balanced chromosomal exchanges, the sex-ratio, and physical development. We find it impossible to set limits in the usual statistical sense on this estimate, because of statistical sampling errors and the uncertainty we have described in establishing the contribution of parental spontaneous mutation to the indicators in the control population.

SOME UNUSUAL FEATURES OF THE ESTIMATE OF THE HUMAN DOUBLING DOSE

As we turn to a similar treatment of the accumulated mouse data, the following features of the human data must be kept in mind:

1. None of the studies has yielded a statistically significant difference be-

tween the children of proximally exposed and controls, and the "numerators" for the individual studies are indeed small. On the other hand, collectively these numerators sum to a substantial body of data, based, for the largest study, on approximately 13,395 Sv of parental gonadal exposure.

2. Exposed males and females contribute about equally to the gonadal exposures of the human data. In the absence of significant radiation effects, there is no prospect of effectively distinguishing the contributions of the two sexes. The human estimate is specific to this situation.

3. The human data cover the entire postexposure reproductive spans of the exposed males and females, whether they were aged 40 or aged 1 ATB, and not, as is customary in experimental studies, the period of maximum fertility following the treatment of young adults.

4. The estimates of the mutational contribution to UPOs and prereproductive mortality are time-place specific, in any strict sense valid only for this particular postwar population. In a more favorable environment, the relative and absolute frequency of children dying because of a mutation in the parental generation should be less, and a somewhat different estimate of the doubling dose would be obtained. This fact, plus the nature of the assumptions necessary to obtaining the estimate of the human doubling dose, impart to this estimate a greater error than is typical of an experimental situation.

5. The parents who were not exposed to the explosions of the bombs (i.e. who came to Hiroshima or Nagasaki following the bombings, as released service men, repatriates, or spouses) were slightly younger and had a little more education and somewhat higher occupational ratings than the exposed (39, 64, 66). It is also assumed that the mother had completely recovered from her exposure when she reproduced. To the extent socioeconomic and maternal effects exist, they will inflate the estimate of the genetic effects of the bombs.

PRÉCIS OF THE MOUSE DATA ON THE GENETIC EFFECTS OF ACUTE RADIATION

The data on mice are summarized in the same sequence as the human data. With rare exceptions, only those murine data are cited that involve transmitted genetic effects following single or double acute exposures to radiation, the effects studied in late fetuses or full-term offspring. The justification is that the primary concern in the societal context is with genetic damage that impacts on the population, whereas damage that results in nonfunctional sperm and eggs, or even the early death of zygotes (i.e. early-acting dominant lethals), although of basic biological importance, does not result in the transmitted effects of most concern to society. We emphasize the data most relevant to the totality of human reproduction following exposure to ionizing radiation, namely, those based on the irradiation of spermatogonia and early oocytes. Since we wish to have maximum comparability with the treatment of the human data, we will, insofar as possible, present the findings in terms of doubling dose rather than rate per unit radiation. One dramatic discovery in radiation genetics was the demonstration by W. L. Russell in 1965 (88) that in his 7-locus system no mutations were recovered in the "late" litters of irradiated females. Unfortunately, because female mice rapidly lose fertility after radiation, for most indicators there is relatively little early oocyte data for mice, a deficiency that greatly complicates the calculation of a mouse doubling dose.

Untoward pregnancy outcome   Because parturient mice tend to devour still-born offspring as well as those dying shortly after birth, the acquisition of reliable data on the frequency of congenital defect, stillbirth, or early postnatal death comparable to the human data is difficult. The pioneer study of Charles et al (11), involving what must be termed on the average high-level chronic radiation (daily, lifetime exposure schedules of either 0.1, 0.5, 1.0, or 10.0 r), scored the offspring of irradiated DBA males × unirradiated C57BL females for defects shortly after birth and also on the basis of an autopsy at maturity. Deceased offspring with major defects present at birth had probably been eaten by the mother by the time of scoring, so that the at-birth data are of limited value. With respect to the autopsy results, many of the malformations listed in their report would have been considered as minor (and not included in the analysis) by the standards of the Japanese studies. The control frequency of rare anomalies at autopsy was $0.73 \pm 0.11\%$. Averaged over all experiments, the regression of scored defect on dose was reported to be 2.4×10^{-5} per r. The data appear to indicate a dose-rate effect; it would be conservative to consider this an experiment with chronic radiation.

The data on congenital malformation most nearly comparable to the human data were acquired by scoring fetuses obtained from sacrificing pregnant mice at various stages of gestation. Lyon et al (58) exposed the male offspring of a C3H/HeH × 101/H cross to 600 + 600 r of acute radiation and then mated them, at an interval that ensured the results would reflect spermatogonial radiation, with CBA/H females. The females were sacrificed at about 14 days of gestation and scored for corpora lutea and small and large moles, dead embryos, malformed live embryos and normal live embryos. We will not consider the data on preimplantation losses or on numbers of moles (i.e. early postimplantation losses), with respect to which an excess would result from early-acting dominant lethals and would have no counterpart in the human data, and which in the human societal context have relatively little significance. Of the 1109 control fetuses, 6 were malformed ($0.54 \pm 0.22\%$) and 23 dead ($2.07 \pm 0.43\%$). Of 955 fetuses in the irradiated series, 3 ($0.31 \pm 0.18\%$ were malformed and 25 dead ($2.62 \pm 0.52\%$). There was thus no significant difference between the two series, but the very high doses used in this study are at the level where genetic radiation effects "fall off" in many studies, for complex reasons.

In further experiments, Kirk & Lyon (40, 41) sacrificed at 18 days gestation the offspring of male or female F_1 hybrid C3H/HeH × 101/H mice treated in one experiment with 0, 108, 216, 360, and 504 cGy and in another, with 0 and 500 cGy and scored for moles and living fetuses with gross malformations and dead fetuses (with or without malformation). Since the data on exposed females are based on matings within 28 days of radiation, they presumably reflect the results of radiation of mature or semimature oocytes. Over 50% of the fetuses scored as malformed were "dwarfs" or "runts" (body weight less than 75% of mean of litter mates). Malformations observed in dead fetuses were not scored. There was an increase in defective fetuses with radiation, from values of $1.2 \pm 0.2\%$ in the controls to values of $1.8 \pm 0.3\%$ in the offspring of males receiving an average of 390 r and $3.4 \pm 0.4\%$ in the offspring of females receiving an average of 278 r, the latter increase significant. To meet the possibility that maternal radiation per se was a teratogenic influence, West et al (107, 108) undertook a further series of experiments involving the transplantation of preimplantation embryos after irradiation of C3H/HeH × 101/H females with 3.6 Gy X-rays. These experiments demonstrated that the effect of radiation was truly genetic, as did further experiments with 3.3 or 3.7 Gy of localized (versus whole-body) radiation, which revealed that irradiation that did not involve the ovaries had no effect on the frequency of abnormal embryos.

In other experiments, Nomura (70–73) irradiated male and female ICR mice at doses of 36, 216, and 504 rad, and then scored for anomalies at 19 days gestation and 7 days postpartum. There was a significant effect of radiation in both series (all data combined). Thus, among living fetuses scored on day 19 of gestation, the control value was $0.39 \pm 0.19\%$, whereas following male (spermatogonial) radiation it was $2.20 \pm 0.33\%$ and following female radiation $2.65 \pm 0.52\%$, both these latter two frequencies being significantly different from controls. Likewise, among offspring surviving more than 7 days, the control frequency was $0.12 \pm 0.12\%$, whereas following male radiation the frequency was $0.86 \pm 0.19\%$ and following female radiation, $1.65 \pm 0.31\%$. Again the differences are significant. The average radiation dose for males was about 250 rad and for females, 200 rad (S. Kondo, personal communication). The female data all appear to have been derived from oocytes treated in the follicular stage.

These experiments establish that radiation increases the frequency of congenital defect in the F_1 of irradiated fathers and mothers, and provide empiric justification for employing this indicator in the human studies. Defective and dead fetuses correspond to the first two components of the triad we have termed an UPO. There are, however, several reasons why care must be exercised in comparing observations of this type made in mice with those on humans (see also 18).

To begin with, the mouse fetus at term differs in many respects from the term human fetus. Otis & Brent (75) have suggested a rough equivalence of a term mouse to about a 100-day old human fetus, but there are many asymmetries in the comparison, and a sweeping generalization is impossible. The equivalence would be somewhat less than 100 days in the above-quoted experiments. In the program in Japan, pregnant women were eligible to register at the completion of the fifth lunar month of pregnancy—140 days of gestation—but there was no penalty for nonregistration and many delayed registration a few weeks or even longer after their eligible date. The data on newborn mice thus correspond in some but not all respects to the first half of the second trimester stage of human development, when, for example, although the bulk of the malformed human fetuses with cytogenetic abnormalities have been eliminated, there is still a higher frequency of defect due to karyotypic abnormality than is observed at term (cf 35), and it is now documented that a significant proportion of these will be lost before term (36). Thus, malformed offspring that would have been aborted before the pregnancy came to attention in the study in Japan may still be represented in the mouse series under discussion.

The findings are further rendered noncomparable to the data on humans by what we will term the litter-size effect, of much less importance in the human data. Postnatal mortality is higher in large litters (11) as well as in very small litters (58). This effect for large litters is also present in the prenatal period; in the controls to the experiments just cited, the intrauterine fetal death rate was higher in large than in small litters. In these irradiation experiments, because of the induction of early-acting dominant lethals (identified in the autopsies as "moles"), the average number of fetuses was decreased in those litters where one or the other parent had been irradiated (40, 41, 58, 71). There is thus a major bias in studying the effect of parental radiation on late fetal death, recognized by Lyon et al (58). But the situation is complicated even further because defective but dead fetuses were not scored as defective in these experiments; since defective fetuses may be expected to compete less well *in utero,* the elimination of defective offspring may be higher in the larger control litters than in the experimental material.

Finally, there is the question of the comparability of diagnostic standards for the small mouse fetus and the human infant at term. For instance, even severe club foot or hypospadias might not be recognized in a mouse fetus at

18 or 19 days of gestation. Thus, while these experiments certainly establish the principle that ionizing radiation increases the frequency of congenital defect in the F_1 of treated male and female mice, and provide empiric justification for the use of this indicator in the studies on humans, precise comparisons of the findings in mice and humans are not possible.

"Prereproductive" death among live-born offspring   The murine data most comparable to the human data on mortality prior to the age of reproduction are those on preweaning mortality. Given the complications in interpretation introduced through inbreeding when radiation experiments are continued over many generations, in this section we consider only the results of studies on the F_1. As noted above, a serious problem with the use of this indicator in the polytocous mouse is the negative correlation of survival with litter size, and the smaller mouse litter size in the offspring of irradiated mice. As recognized by the investigators, this fact vitiates many of the observations on parental radiation and offspring survival (79). The various experimental strains used by investigators also differ in mean litter size. Finally, the relative immaturity of the mouse at birth (and differences in maternal care) may result in different postnatal environment-genotype interactions in mice and humans that render precise comparisons difficult.

A representative set of data (S. E. Lewis, unpublished data) is given in Table 2. We note first the variation in control survival rates in the different experiments, a fact that underscores the importance of concurrent controls in studies of this type. Survival is better in the treated than in the control series in four of six experiments. We note further, however, that litter size averages about one animal more in the controls. To test to what extent this is responsible for the *better* survival of the offspring of treated animals observed in four of the experiments, we have conducted a preliminary analysis of the last data set in Table 2, in which each control litter greater than size 5 but less than 11 was randomly matched with a litter in the experimental series, the range 6–10 chosen to optimize the likelihood of survival. This resulted in two subsets of 480 births. Survival in the controls was $97.5 \pm 0.7\%$ and in the treated series, $91.7 \pm 1.3\%$. A much more extensive analysis along these lines will be presented elsewhere; for now it is sufficient to recognize the need for this type of adjustment, either empirically (as here) or statistically, and the possibility it creates for comparable analyses of human and mouse data.

Malignancies in the F_1   Batra & Sridharan (8) reported an increase in leukemia in the offspring of irradiated male mice, but this was not confirmed by Kohn et al (44). More recently, Nomura (70–73) has reported a significant increase in malignant tumors in 8-month-old offspring of both male and female ICR mice following acute radiation exposures of 0, 36, 108, 216, 360,

Table 2   Preweaning mortality in mice in relation to spermatogonial radiation exposure and litter size.[1] Note the impaired survival in the *controls* in four of six experiments.

	No. litters	No. progeny at birth	Birth average litter	No. progeny at weaning	% Survival
Female 300R					
Treated	488	3,265	6.7	2,706	82.9 ± 0.7
Control	51	390	7.6	319	81.2 ± 2.0
Male 300R Spermatogonia					
Treated	536	3,506	6.5	2,701	77.0 ± 0.7
Control	87	584	6.7	413	70.7 ± 1.9
Male 300R Presterile					
Treated	454	2,329	5.1	2,010	86.3 ± 0.7
Control	87	643	7.4	521	81.0 ± 1.6
Male 600R Spermatogonia					
Treated	861	5,569	7.5	4,597	82.5 ± 0.5
Control	175	1,036	5.9	901	87.0 ± 1.0
Male 600R Presterile					
Treated	337	1,291	3.8	761	58.9 ± 1.4
Control	115	784	6.8	529	67.5 ± 1.7
Male 300R × 2 Spermatogonia					
Treated	547	3,755	6.9	3,478	92.6 ± 0.4
Control	102	776	7.6	715	92.1 ± 1.0

[1]S. E. Lewis, unpublished data.

and 504 rad (average dose of 250 rad for males and 200 rad for females). The control data yielded a tumor frequency of 5.29 ± 0.96% whereas the corresponding (consolidated) figure from male exposure was 10.01 ± 0.77%, and from female exposure, 8.74 ± 0.83%. The findings were repeated on a smaller scale in the LT and N5 strains.

The transfer value of these data to the human studies is moot (cf 109). Mouse strains differ widely in their tumor profiles, and Nomura (71) recognized that "the results may be a special property of the mouse strain used." In the experiments with the ICR strain, 87% of the tumors in the offspring of treated mice were scored as papillary adenomas of the lung. In the experiments with LT and N5 mice, 16.0% and 21.0% of the tumors were pulmonary, 25.3% and 22.8% were ovarian tumors, and 5.3% and 3.9% were leukemias. These endpoints had a relatively high frequency in the appropriate controls. These were not tumors observed with any frequency in the human series, but 8 months covers a relatively larger part of a mouse's life span than the fraction of the human span represented in the human series. These differences in the pattern of malignancy in the two species—which will persist even with further follow-up of the human material—render any detailed comparison of the results from the two species suspect.

Furthermore, none of the human tumors based on the genetic mechanism that presumably should be most responsive to radiation (retinoblastoma, Wilms' tumor, sarcomas of various types, and embryonal carcinoma of the testis) were observed in the mouse series. There is no basis for an estimate of the proportions of the control malignancies that result from mutation in the parental generation, an estimate essential to the derivation of a doubling dose. We conclude that these data represent a special circumstance in mouse strains many of which were at one time selected for high frequencies of malignant tumors, and do not believe the data can be used in the calculation of a mouse doubling dose to be compared with the human doubling dose (for further discussion, see 109).

Frequency of structural rearrangements and of sex-chromosome aneuploidy   The subject of sex-chromosome aneuploidy has been comprehensively reviewed by L. Russell (82). Whereas in humans the most commonly encountered abnormalities are the XXY, XYY, and XXX state, in mice the most frequent abnormality is the XO state, and unlike humans with this constitution, these are fertile individuals. Furthermore, XXX mice are extremely rare (17). There are no large-scale cytogenetic surveys of newborn mice comparable to those on humans; the frequencies of sex-chromosome abnormalities have been determined by genetic techniques whereby individuals exhibiting the various types of abnormalities have distinctive phenotypes. To a first approximation the frequency of primary XO and XXY individuals in mice is about 3/1000, very similar to the human total for sex-chromosome aneuploidy, but predominantly due to XOs to which the male parent has not contributed a sex chromosome.

Acute gamma radiation resulted in a clear increase in sex-chromosome anomalies in offspring from males in whom the germ cells were exposed as spermatocytes, spermatids, and spermatozoa, of the order of $2 \times 10^{-5}/r$, but this effect was not observed in offspring resulting from irradiated spermatogonia (Table 7 in 82). The result of the acute radiation of dictyate oocytes as reflected in matings up to 49 days post treatment, based on the occurrence of OX^P offspring, appears more variable but is of the order of $1 \times 10^{-5}/r$ (Table 8 in 82). The collection of data on the effect of *acute* radiation on immature oocytes has been hampered by the extreme sensitivity of these cells to radiation. Two studies in which small doses of *acute* radiation were employed revealed no increased incidence of trisomy in midterm fetuses (60, 103). On the other hand, two studies in which larger doses were delivered as *chronic* radiation revealed either a nonsignificant increase in X-chromosome loss in term mice (82) or a significant increase in hyperhaploidy in metaphase II oocytes (33). These latter findings, while interesting, are of dubious value in establishing a reference point for the situation in Japan.

Spontaneous or induced mutations characterized by structural rearrangements of chromosomes will result in zygotes with either a balanced or unbalanced chromosomal complement. In mice, as in humans, the latter are usually eliminated early in pregnancy but some few survive until the later stages of pregnancy. The cytogenetic studies on the children of survivors in Japan were not initiated until age 13 and so would only very rarely detect children resulting from chromosomally unbalanced zygotes. Such of the latter type of children as survived beyond the fifth gestational month presumably were detected as grossly abnormal newborn infants and their impact was scored in the analysis of malformations. Likewise, in the above-referenced murine studies of Lyon and colleagues, the more-or-less comparable impact of the mutations resulting in chromosomally unbalanced gametes should be reflected in the malformations scored on the 18th gestational day although, as we noted, the differences between the human and mouse fetus at term or near-term renders a direct comparison with the human data problematic.

On the basis of the studies of high, acute doses of X-rays to the C3H/HeH × 101/H strain (21, 58, 94), Lüning & Searle (50) suggested a spermatogonial doubling dose for reciprocal translocations of about 31 r. They reject from this calculation the data of Reddi (78), which showed a much lesser effect of radiation on the CBA strain. The later data of Generoso et al (27), employing male (101 × C3H)F_1 mice, with a spontaneous rate for heritable reciprocal translocation in spermatogonia of 1.8×10^{-4}/gamete/generation and an induced rate of 3.9×10^{-5}/r/gamete, yields a doubling-dose estimate of 4.6 r; this estimate rests on a single mutation in the controls. As noted, the sparse human data were deemed insufficient for the calculation of a doubling dose for mutations resulting in reciprocal translocations, but since the findings were nearly identical in the children of exposed and controls, the human data appear to be inconsistent with the mouse doubling dose reported by Generoso et al (27).

"Point" mutations   Perhaps in recognition of the various complications in the study of the indicator traits thus far reviewed, most studies of the genetic effects of radiation on mice, especially those of recent years, have involved some version of a specific-locus/specific-phenotype test. Again we use the traditional term "point" mutation but recognize that, in addition to radiation-induced specific nucleotide changes [as suggested by the reversibility of radiation-induced mutations in Neurospora (14, 59)] and intragenic insertions/deletions/rearrangements, some of the radiation-induced mutations encompass several loci (83), in which case the term "point" is a misnomer, but we use it for convenience. In this section we draw heavily on the excellent reviews of Searle (96) and Favor (20). The estimates obtained with eight different test systems are summarized in Table 3. *All entries are based primarily on the results of the acute, relatively high-dose (usually 300 or*

Table 3 A summary of the gametic doubling doses for acute, "high-dose" radiation of spermatogonia yielded by the various specific-locus/specific-phenotype systems developed in the laboratory mouse.

System	Data summarized in:	Doubling dose (Gy)	Calculated by:	Origin of treated males
1. Russell 7-locus	16, 96	.44	This paper	101 × C3H
2. Dominant visibles	50	.16	50	various
3. Dominant cataract	20	1.57	This paper	101/El × C3H/El
4. Skeletal malformations	15	.26	50	101
5. Histocompatibility loci	6	>2.60	6	C57BL/6JN
6. Recessive lethals	102	.51 ⎤	50	DBA
	52, 58	.80 ⎬ 1.77	6	C3H/HeH × 101/H
	52, 95	4.00 ⎦	6	CBA, C3H
7. Loci encoding for proteins	This paper	.11	This paper	various
8. Recessive visibles	58	3.89	This paper	C3H/HeH × 101/H
	Av.	1.35		

600r) irradiation of spermatogonia. In assembling this table, we have insofar as possible drawn on doubling-dose estimates already in the literature. Each entry requires some brief discussion.

(a) The multiple, specific-loci test system   The most widely used murine test system has been that developed by W. L. Russell (84). This entails the cross of an irradiated mouse homozygous for the normal alleles at 7 test loci with a test strain homozygous with respect to mutant alleles at the same 7 loci. A mutation is immediately scorable in the F_1. The loci are *a, b, c, p, d, se,* and *s*. Because of its efficiency, the method has been widely used and has led to many important insights in radiation genetics. When the results of various laboratories with single or multiple, acute high doses of spermatogonial radiation are combined, (cf summaries in 16, 96), the mutation rate/locus/r is 2.2×10^{-7} (2,346,687 locus tests), with a control rate of 9.0×10^{-6}/locus/generation (6,417,082 locus tests). This leads to a doubling-dose estimate of 41 r for acute, high-dose spermatogonial radiation.

Female mice treated with doses readily tolerated by males produce a few posttreatment litters but then rapidly become sterile due to the extreme sensitivity of early oocytes to ionizing radiation. Even so, because of the importance of obtaining data on the results of female radiation, a particular effort has been made to obtain such data with this system. The first few litters conceived by female mice following single, acute radiation doses of 50 r (involving mature dictyate oocytes) revealed an induced rate/r of from 2.1 to 5.5×10^{-7}, depending on the dose (1,628,277 locus tests), as contrasted with a spontaneous rate/locus/generation of 4.9×10^{-6} (202,812 locus tests). This yields a doubling dose of 9 to 23 r. However, when breeding of females who had received 50 r continued, no mutations were recovered in the "late" litters of these same females (0 mutations in 547,337 locus tests) (88).

There are major differences between loci in induced rates following acute spermatogonial radiation (96; Table 14). Induced mutations are approximately 18 times more frequent at the s than at the a locus. Unfortunately, only fragmentary data seem to have been published on locus differences in spontaneous mutation rates, insufficient for a proper statistical analysis. However, we note (96; Table 3) that the 3 (b, d, s) loci (among 7) showing the highest spontaneous rates are those with the highest induced rates, and the 3 with the lowest induced rates are those with 0 spontaneous mutations. We return to the great significance of this later.

An alternative 6-locus system was developed by Lyon & Morris (54, 55), consisting of the $a, bp, fz, ln, pa,$ and pe loci. The rate of mutation induction following acute, high-dose spermatogonial radiation was 7.6×10^{-8}/locus/r (149,004 locus tests), approximately one third of the rate in the Russell 7-locus system at the same exposure level. Unfortunately, no mutations were encountered in the 124,614 control locus tests, so that these observations cannot be put into a doubling-dose context.

(b) The "gross" dominant visible test   The observations on the induction of dominant mutations by acute spermatogonial radiation have been well reviewed (96, 99). Since these data are for the most part an incidental by-product of other (specific-locus) studies, they must be placed on a per gamete rather than per locus basis. Searle's analysis (96) reveals a spontaneous rate of 0.8×10^{-5}/gamete and an induced rate of 4.7×10^{-7}/gamete/rad, resulting in a doubling-dose estimate of 17 rad. Although we include this estimate in our summary, the opportunistic use of certain phenotypes suggests that the estimate is based on the more spontaneously mutable loci, and we have just seen that these loci tend to show the higher induced rates (see below also). Otherwise stated, loci with low spontaneous and/or induced rates that failed to yield any mutations would not enter into this estimate.

(c) The dominant cataract system   This approach involves scoring the offspring of male mutagenized mice for lens opacities with the aid of a slit lamp after dilation of the pupils with atropine (47). Animals with such opacities were bred, and only transmitted defects scored. In these studies locus rates are calculated on the assumption that 30 loci may contribute to the phenotype. In fact, the number of loci at which mutation may result in this phenotype is unknown, for mice or humans, and for our purposes it is sufficient to express doubling dose in terms of phenotype. In a recent summary of the data from five different experiments, Favor (20) reported that the phenotypic doubling dose ranged from 0.96 to 2.14 Gy. The unweighted average of these estimates

is 1.57 Gy. That this estimate of the doubling dose is based on a single mutation in the control series imparts to it a wide error.

(d) The skeletal malformation system   Because of the obvious relevance to human disease and because of the potential immediate expression in the F_1 generation, skeletal defects are, like cataract, an attractive mutational endpoint. That dominant mutations resulting in skeletal defects can be identified in the mouse is well documented in M. Green's catalog of known mouse mutations (32). However, dominant skeletal mutations detected through a standard "physical examination" of the mouse in general appear to be relatively rare both as spontaneous (91) and induced (96) occurrences.

The screening of F_1 generation, cleared and stained mouse skeletons have provided a means for intensifying and expanding the search for skeletal mutations (15, 100). The use of these methods has indeed identified skeletal defects that could reasonably be ascribed to treatment of parents with mutagenic agents (15, 100, 101). However, the number of minor defects found with this method, possibly due to developmental accidents, produces a large background noise level that confounds the identification of truly heritable events.

Heritability tests of presumed skeletal defects are especially cumbersome because all F_1 animals to be tested must be bred to assess transmission of these traits, inasmuch as with this method presumptive mutants can only be identified after death. To offset the need for breeding tests, various criteria have been suggested by Selby & Selby (101) to identify skeletal defects due to mutation. The accuracy of these criteria is yet to be rigorosly validated. Selby & Selby (101) apparently did not include controls in their studies. We will therefore accept for this indicator the doubling-dose estimate of 26 r by Lüning & Searle (50), which employed the data of Ehling (15). However, since this estimate is based on a single mutation in the controls, it has a wide error.

(e) The histocompatability system   Bailey & Usama (7; see also 5) developed a system of testing for mutations at loci determining histocompatability, by orthoptic tail-skin transplantation involving graft exchange in a 'reciprocal circle' system that should detect both gain and loss mutations. It was suggested that the system would detect mutation at approximately 30 different loci, but as for the other 'phenotype tests,' we shall only present a phenotypic doubling dose. Bailey & Kohn (6) report that in experiments in which the male parent received 522 rads acutely, spermatogonial mutant frequency was higher among the control group than among the offspring of treated, although the difference was not significant. This finding precludes the calculation of a doubling dose but does permit assigning it a lower limit. They calculate that the spontaneous rate per gamete is $<2.6 \times 10^{-5}$, from which the doubling

dose is estimated to be >260 rads. Similar data published by Godfrey & Searle (28) and later data published by Kohn et al (46) are consistent with this calculation.

(f) The prenatal recessive lethal system   The recessive lethals scored in murine experiments are mostly very early acting, manifesting as moles. The majority of such early-acting lethals could not be recognized in human studies, and their societal impact would come through their heterozygous effects. The murine data are extremely heterogeneous and involved, and cannot be presented in detail in these confines. Very competent investigators have viewed the data in different lights. A basic problem has been to estimate the baseline total gametic rate of mutation to lethals. In general this has been measured indirectly, using assumptions concerning heterozygote disadvantage and the results of inbreeding. Lüning & Searle (50), using the data of Sheridan & Wårdell (102) on 14 generations of male irradiation at 276 r/generation, developed a doubling-dose estimate for this indicator of 51 r (see also 49). On the other hand, Bailey & Kohn (6), using the control data of Lyon (52) and Searle (95) and the radiation data of Lyon et al (58), developed doubling-dose estimates of 80 and 400 rads, respectively, depending on which control figure is used.

Roderick (80), treating DBA/2J males, has described a direct approach to the detection of spontaneous and induced lethals in a chromosomal inversion covering approximately 3% of the mouse genome, but thus far the system does not seem to have been applied to spermatogonial mutation rates. For postspermatogonial cells, however, he adduces, in a small series, a lethal rate per locus of 0.35×10^{-8}/r, based on an estimated 1750 loci in the segment of chromosome 1 being screened. The mutation rate for the same cell stage in the 7-locus system was 45.32×10^{-8} (84); approximately 80% of these mutations were homozygous lethal. As Roderick points out, this is a 100-fold difference, albeit with a large (indeterminate) error. Unfortunately, there were no lethal mutations in Roderick's control series so that a doubling dose cannot be calculated.

(g) The electrophoretic system   In recent years considerable information has accumulated on the frequency in mice of spontaneous and induced mutation that can be detected by the electrophoresis of proteins, either as mobility or absence-of-gene-product variants, or by enzyme activity assays (10, 38, 76, 77). In mice, test crosses can often be designed so that the absence-of-gene-product can be recognized by the absence of an electrophoretic band rather than through the more laborious enzyme assays necessary to the detection of similar variants in humans. In principle, these studies should provide one of the strongest links with the human studies, since there is a substantial overlap in the proteins employed in the two sets of studies. Table 4 summarizes the

Table 4 The frequency of X-ray induced mutations in mouse spermatogonia involving the electrophoretic mobility or loss of physiological function in proteins.

Attribute	Investigator	Strain	Treatment	No. loci	No. observations	Mutations
Protein Charge	W. Pretsch (See also 20)	101/E1 × C3H/E1	0	23	133,676	0
	S. E. Lewis, unpubl.	DBA/2J × C57BL/6J	0	32	1,289,500	1
	J. Peters, unpubl.	C3H/HeH	0	4	1,452	0
	W. Pretsch (See also 20)	101/E1 × C3H/E1	3 + 3 Gy	23	23,069	0
	S. E. Lewis, unpubl.	DBA/2J × C57BL/6J	3	32	91,105	0
			3 + 3	32	115,395	0
			6	32	142,385	0
	J. Peters, unpubl.	C3H/HeH	3 + 3 Gy	4	40,004	0
Enzyme Activity	D. J. Charles & W. Pretsch	101/E1 × C3H/E1	0	12	86,640	0
	S. E. Lewis, unpubl.	DBA/2J × C57BL/6J	0	32	1,289,500	0[a]
	J. Peters, unpubl.	C3H/HeH	0	4	1,452	0
	D. J. Charles & W. Pretsch	101/E1 × C3H/E1	3 + 3 Gy	12	40,656	1[b]
		101/E1 × C3H/E1	5.1 + 5.1 Gy	12	38,244	5[c]
	S. E. Lewis, unpubl.	DBA/2J × C57BL/6J	3	32	40,515	2[a,d]
			3 + 3	32	49,555	2
			6	32	66,784	1
	J. Peters, unpubl.	C3H/HeH	3 + 3 Gy	4	40,004	3

[a] The cross results in heterozygosity at 16 of the loci
[b] This mutation significantly increased specific activity
[c] Of these, 2 increased and 3 decreased specific activity
[d] One of these mutations was detected in a non-heterozygous genotype

data currently available. (We omit reference to publications in which there are no control data.)

Inasmuch as absence-of-gene-product can be reliably detected at only a subset of loci, the data presented in the table are not really descriptive of total mutation rates for these types of variants at the loci concerned. Since, however, the proportion of loci scored for nulls is roughly comparable in the experimental and control series, the data on electrophoretic and null frequencies can be combined in this preliminary evaluation of the doubling doses suggested by this system. For the denominator in the calculation, we use the total number of gene products scored for electrophoretic mutations. The mutation frequency is 0.7×10^{-6}/locus/generation in the controls whereas in the experimental series, the probability of mutation/.01 Gy is 6.3×10^{-8}. The doubling dose (based on the occurrence of a single mutation in the controls) is .11 Gy. We note, by comparison with the data on humans (65), a relatively low spontaneous mutation rate for electromorphs; this observation plays a major role in the low doubling dose. It is perhaps pertinent that some 60%–70% of the control data is derived from female mice, whereas the data on induced rates are derived from spermatogonial radiation. Lyon et al (56) have suggested from specific-locus systems data a control spontaneous mutation rate for females of 2×10^{-6}/locus/generation, approximately one fourth the male rate. A rough correction for the use of female controls would increase this doubling dose estimate by at least a factor of 2.

(h) Recessive visible mutations   In addition to the test systems designed to detect recessive (and dominant) mutations at specific loci, several experiments have been designed to permit the detection of recessive visible mutations at all loci at which such mutations can occur. In a comprehensive experiment involving the male offspring of a C3H/HeH × 101/H cross, Lyon et al (58) recovered one recessive mutation in the equivalent of 142.07 control gametes and six in the equivalent of 142.02 gametes from animals treated with 600 + 600 r acute spermatogonial radiation. They calculate an overall gamete mutation rate to recessive visibles of 7.0×10^{-3} in the controls and 28.2×10^{-3} after radiation. After correction for control frequency, the rate of induction of recessive visible mutations was 1.8×10^{-5}/gamete/r. This leads to a doubling-dose estimate of 3.9 Gy. This estimate is based on a single mutation in the controls and, of course, again implies a large error.

The simple unweighted average of the results of these eight systems is 1.35 Gy. A weighting system and/or the calculation of an error is impossible because the number of loci involved in systems 2, 3, 4, 5, 6, and 8 is unknown. We note, however, that because of the paucity of mutations detected with some of the approaches, the error term should be Poisson in type, i.e. asymmetrical, the upper range at any given probability level exceeding the lower range.

Sex-ratio   In the mouse, the sex-chromosome is about 1/20th of the total genome. Thus, in principle there could be of the order of 1000–2000 genes on the X susceptible to radiation damage, perhaps half of which could yield lethal mutations. The published data on the effect of radiation on the mouse sex-ratio are confusing (reviewed in 29, 30, 57). Most early studies found no effect of radiation of males on the sex-ratio of the F_1 or F_2; in those studies that report findings in the F_2 of irradiated males consistent with the induction of sex-linked lethals, efforts to isolate these presumed lethals have failed (cf 29), leading Lüning & Sheridan (51) to conclude that changes in the sex-ratio are an "unacceptable" way to search for sex-linked recessive lethals. Lyon et al (57) met many of the early problems by developing a strain with an X-chromosome inversion encompassing about 85% of the X-chromosome. Treated (500 + 500 rad) males of the F_1 C3H/HeH × 101/H stock were mated to females heterozygous for the inversion. Two of 536 irradiated and 0/529 control X-chromosome carried a confirmed lethal, corresponding to a rate for recessive lethals of 1.9×10^{-6}/rad/X-chromosome. The authors suggest that this points to a locus rate much below that of the 7-locus system, but in the absence of control mutations, no doubling dose can be calculated.

The murine data most comparable to the least ambiguous aspects of the human data would be the proportion of males in the offspring of irradiated females. We have found no published murine data of this type. Accordingly, we have tabulated the (tertiary) sex-ratio observed at weaning in several relatively small experiments involving female radiation performed in the laboratory of one of the authors (S. E. L.) (Table 5). Note that these results are based on the radiation of mature oocytes. Two observations are noteworthy. First, in the controls (and irradiated) there is an insignificant excess of females ($\chi^2 = 1.658$, d.f. $= 1$, $0.10 < p < 0.20$), whereas in humans at a corresponding stage in development, there is a slight excess of males. Second, the difference between the offspring of control and treated animals is (nonsignificantly) counter-hypothesis in sign ($\chi^2 = 0.124$, d.f. $= 1$, $0.70 < p < 0.80$). Once again, loci on the X-chromosome appear resistant to the induction of lethal mutations.

Table 5 The sex ratio at weaning in the offspring of females treated with 300r and mated during the week following treatment.

	No. litters	No. progeny at birth	Average birth litter size	Females	Males	Total	M/F Ratio	% Survival
Treated	488	3,265	6.7	1,475	1,331	2,806	.902	85.9 ± 0.7
Concurrent control	51	390	7.6	171	148	319	.865	81.8 ± 2.2

Anthropometrics   We have found no analysis of data on mouse body weight or other measurements in the *first* generation following parental radiation that allows for the litter-size effect. Accordingly, there do not seem to be any data suitable for comparison with the measurements on humans.

THE GENERATION OF A DOUBLING DOSE FROM THE MOUSE DATA

Unlike the human situation, in the various experiments with mice it has usually been possible to drive the system under study with sufficient radiation to obtain significant differences between the experimental and control material. This should increase the precision in estimation procedures and provide us in principle with two alternatives in calculating a doubling dose. On the one hand, we can simply combine the results of the various specific-locus/ specific-phenotype studies. This provides a very definite and clear-cut endpoint, which, however, does not readily translate into a morbidity-mortality figure comparable to the basis for the estimation of the human doubling dose. On the other hand, we can attempt to proceed as with the human data. This requires estimating the contribution of mutation in the parental generation to the incidence of malformed and dead fetuses, preweaning mortality, and sex-chromosome aneuploids, with the specific-locus findings represented in the calculation through their heterozygous effects on malformations and fetal deaths and preweaning mortality. While this procedure produces a more relevant figure, it requires assumptions that are probably even softer for mice than for humans.

In the derivations of a mouse gametic doubling dose to be extrapolated to humans, it is essentially the first approach that has been employed in the past. On this basis, various national and international committees have set the doubling dose of acute low-LET gonadal radiation for humans at 30 to 40 r, with limits of 10 to 100 r (reviewed in 12, 105). That estimate appears to be based primarily on the data summarized pictorially by Lüning & Searle in 1971 (50; see also 97), namely, data on semisterility (= reciprocal translocations), the 7-locus system, dominant visible mutations, dominant skeletal mutations, and recessive lethals. The average of these values at that time was 31 r. Our Table 3 in essence extends the approach of Lüning & Searle (50), adding new data but, for consistency in the use of specific-locus/specific-phenotype results, omitting the findings on reciprocal translocations (treated elsewhere in this review). The average of the eight estimates of Table 3 is 1.35 Gy.

As with the human data, these data sets are of variable information content, but none can justifiably be disregarded, and collectively they comprise a formidable body of data. Three of these estimates, however, are based on the occurrence of a single mutation in the controls (cataract, skeletal defects, protein variants). One additional mutation in the controls would double the estimate of the doubling dose for these three outcomes. Thus, this estimate of 1.35 Gy carries a considerable potential for error. For the high doses of radiation employed in these studies, the appropriate dose-rate factor in converting to the effects of low-dose, chronic, or intermittent radiation is no less than 3.0 (86, 87). This would lead to an estimate of the doubling dose for chronic radiation of approximately 4.0 Gy. The data in Table 3 are based on spermatogonial radiation, and are thus a male-oriented figure. When the failure to recover specific-locus mutations in the late litters of irradiated females (88) is considered, that doubling-dose estimate must be increased, to an indeterminate degree (but less than doubled).

The "information content" of the 7-locus test in particular outweighs that of any of the seven other specific-locus/specific-phenotype systems that have been employed, and some will argue a weighting system should be employed in estimating the doubling dose. However, both empirical and theoretical reasons considerations suggest that the component loci of the 7-locus system may be unusually responsive to the mutagenic effects of radiation. In the context of this discussion, it is important to bear in mind that W. L. Russell in his earliest presentations of the results of the system he had developed expressed concern over the "representativeness" of the loci he had selected (84). Since then significant progress has been made on the subject of "representativeness."

To begin with, although the existence of mutable loci has been recognized for many years, it is now clear that loci in general show a wide range in spontaneous and induced mutability. In the mouse, five coat-color loci differed in their spontaneous mutability by a factor of 7 (but the individual estimates carry large errors) (31, 91). Favor (20; Table 5) has emphasized the differing responses to radiation of the 7-locus, the cataract, and the enzyme-activity systems; data to that effect have also been presented in this review. With humans, recently we have suggested a tenfold difference in the mutation rates of loci encoding for serum transport proteins and erythrocyte enzymes, these loci having been selected for study only because of the clarity of resolution of their products with one-dimensional electrophoresis (9). We also suggest that the loci responsible for the cellular proteins visualized with two-dimensional polyacrylamide gel electrophoresis and silver staining, proteins sometimes referred to as 'structural', are as a group less mutable than the loci encoding for erythrocyte and serum transport proteins (63).

Finally, in a recent mutagenization experiment using the TK6 line of lymphocytoid cells and employing ethylnitrosourea as mutagen, the products of 263 loci were scored for the occurrence of mutation resulting in electrophoretic variants, using the two-dimensional gel technique (34). Ten of these 263 loci were known from family-oriented studies on peripheral lymphocytes to be associated with genetic polymorphisms. The induced mutation rate at these 10 loci was 3.6 times greater than at the monomorphic loci, an observation with a probability <0.004. Now, a specific-locus test system such as developed by Russell can only use loci associated with genetic variants. The possibility of a bias towards mutability in selecting the components of the specific-locus system is obvious. On the other hand, there may also be bias towards lesser mutability in some of the other systems; this will presumably become apparent with the passage of time. The least prejudicial procedure at present is simply to average the results of all valid approaches.

The alternative approach to developing an estimate of the doubling dose for the mouse is to attempt to analyze the mouse data in the context of radiation-induced morbidity and mortality, in keeping with the basis for the human estimate. At the outset of this review, it was indeed our intention to proceed in this direction, taking into due consideration the early studies on F_1 characteristics following paternal radiation so well summarized by E. Green (30), as well as the later studies on nondisjunction, congenital malformations, and cancer. Unfortunately, as we have progressed, it has become clear that this is simply not feasible. The three dominant components of the human estimate are UPOs, prereproductive mortality among live-born offspring, and sex-chromosome aneuploids. As we have pointed out, each of the murine equivalents of these indicators differs in such important respects from the human data that a valid comparison does not seem possible. Even if this were not the case, however, there is a further reason for the inability to use the first two of these indicators. For the human data, we can estimate, albeit with considerable error, the contribution of mutation in the parental generation to the indicator; this is the necessary basis for an estimate of a doubling dose. For the mouse, no such estimates exist. Indeed, given the complex backgrounds of the various strains of experimental mice used in radiation experiments, it is not clear a normative figure comparable to the estimates for humans could be developed. Finally, relatively few of the published observations on litter size and survival in the older literature are on F_1. Under the circumstances, then, we must compare a specific-locus/specific-phenotype estimate from the mouse with an estimate for humans based on the presumed genetic component in neonatal and childhood morbidity-mortality and sex-chromosome aneuploids.

SOME UNUSUAL FEATURES OF THE ESTIMATE OF THE MOUSE DOUBLING DOSE (FROM THE STANDPOINT OF EXTRAPOLATION TO HUMANS)

As with the human data, several features must be born in mind concerning the murine data:

1. In the several decades following the advent of the atomic bombs, when major programs on mammalian radiation genetics were funded to provide guidance for human risks, the effort was about evenly distributed between programs oriented towards specific-locus tests and towards such strain characteristics as litter size, preweaning and lifetime mortality, body weight at various stages, and sex-ratio. When the latter programs failed to reveal the expected strain deterioration even after repeated generations of massive radiation (cf 30), whereas the specific-locus test was revealing rates greatly in excess of those encountered in Drosophila (cf 85), a consensus developed that the "strain deterioration" approach—which should reflect the impact of mutation at *all* loci—was too blunt an instrument for the problem (cf 30, 48). At the same time, however, investigators repeatedly suggested that the loci of the specific-locus test of Russell (84) might be unusually sensitive to the genetic effects of radiation (e.g. 42, 45, 48, 80, 84), and also recognized the many difficulties in extrapolating from mice to humans (see 43, 48, 53). The present analysis has attempted to meet those concerns by using all the specific-locus/specific-phenotype data available.

2. The exclusive use of specific-locus/specific-phenotype data in reaching a murine doubling dose entails fewer assumptions than in the human estimate but also reflects a smaller sampling of the total genome.

3. A problem with the mouse data that may never be rectified is lack of data on a variety of genetic endpoints in the late litters of irradiated females. If, for whatever reason, all other specific-locus/specific-phenotype indicators behave as do those in the 7-locus test, then the results of female radiation can be only a fraction of the results of male, and the doubling dose of chronic radiation for a normal (two-sexed) mouse population will be substantially higher than the 4.0 Gy we have estimated, and would exceed the estimate for humans, although, given the looseness of these estimates, probably not "significantly" so.

CONCLUSIONS

Studies of eight indicators in the children of atomic bomb survivors and suitable controls have led us to suggest that the gametic doubling dose for the spectrum of acute gonadal radiation experienced by survivors of the atomic bombings is in the neighborhood of 2.0 Sv. Inasmuch as statistically significant effects were not encountered with any of the indicators, this estimate carries a large but indeterminate error. Additional error may be introduced by the assumptions made concerning the extent to which mutation in the parental generation contributes to some of these indicators. For an extrapolation to the effects of chronic radiation, we employ for these specific exposures a dose-rate factor of 2, resulting in an estimate of 4.0 Sv.

A review of all of the pertinent murine data was undertaken, organized to be as comparable as possible to the human data. Because of the differences in the maturity of mice and humans at birth, plus radiation-induced litter-size effects that, as we have shown, strongly influence pre- and postnatal survival, much of the mouse data cannot be compared directly with the human data. The strongest aspect of the mouse data—which is the weakest aspect of the human data—are the specific-locus, specific-phenotype studies. Averaged, these lead to a doubling-dose estimate for acute spermatogonial radiation of 1.35 Gy. Because of the higher acute doses employed in the mouse experiments than obtain for the human experience, we suggest a dose-rate factor of not less than 3, leading to a doubling-dose estimate for chronic radiation of approximately 4.0 Gy. Although the situation requires that we express the human and mouse estimates in different units, in any practical sense the human and mouse doubling dose estimates for chronic radiation are very similar.

The current general agreement between these estimates, reached by relatively independent approaches, is noteworthy, albeit, as we have repeatedly emphasized, the error in both estimates is large but essentially indeterminate. There is, of course, no reason to expect identical doubling doses in two such different species. Together, however, these estimates suggest that the genetic risk of low-dose rate, low-LET radiation for humans is less than has generally been assumed during the past 30 years.

As noted in the Introduction, the attempt to develop a rounded appraisal of the genetic effects of radiation on either the mouse or human genome represents one of the most complex undertakings of modern genetics. The forthrightness with which we have dealt with certain problems with both the human and murine data should not divert attention from the great progress that has been made. Inasmuch as the general issue of genetic damage from environmental exposures will be with society for some time to come, we hope that this discussion of the problems inherent in the present data, both human and mouse, will be useful in planning the experiments and observations of the future. Space constraints unfortunately do not permit discussion of possible future developments and especially the exciting prospects inherent in the parallel study in the two species of radiation-induced mutation in DNA.

This review was well along when the most recent report of the Committee on the Biological Effects of Ionizing Radiation (13) appeared. With respect to the human data, the Committee treatment was appropriate to the time, but, unfortunately, the Committee did not have access to the most recent analyses-syntheses of the Japanese data, based on the DS86 dose schedule, which, we suggest, place the human estimates on somewhat firmer footings than previously. With respect to the mouse data, where we have analyzed essentially the same findings, our approach in an effort to use the mouse data most appropriate to the human situation leads to a somewhat higher estimate of the doubling dose of chronic radiation for the mouse than the Committee's, and lessens the apparent conflict between the two sets of data that the Committee identifies in its report.

ACKNOWLEDGMENTS

James V. Neel gratefully acknowledges the support of Department of Energy Grant FG02-8760533 and Susan E. Lewis similarly acknowledges the support of NIH Contract No. 1-ES-55078. We are greatly indebted to Dr. Josephine Peters for critiquing the manuscript and making available unpublished data on mutation rates in an electrophoretic system, and to Drs. Michael Shelby and Robert P. Erickson for a critical reading of the manuscript.

Literature Cited

1. Abrahamson, S., Wolff, S. 1976. Reanalysis of radiation-induced specific locus mutations in the mouse. *Nature* 264:715–19
2. Awa, A. A. 1975. Cytogenetic study. *J. Radiat. Res.* 16:75–81 (Suppl.)
3. Awa, A. A., Bloom, A. D., Yoshida, M. C., Neriishi, S., Archer, P. 1968. A cytogenetic survey of the offspring of atomic bomb survivors. *Nature* 218: 367–68
4. Awa, A. A., Honda, T., Neriishi, S., Sufuni, T., Shimba, H., et al. 1987. Cytogenetic study of the offspring of atomic bomb survivors, Hiroshima and Nagasaki. In *Cytogenetics: Basic and Applied Aspects*, ed. G. Obe, A. Basler, pp. 166–83. Heidelberg: Springer-Verlag
5. Bailey, D. W. 1963. Histoincompatability associated with the X chromosome in mice. *Transplantation* 1:70–74
6. Bailey, D. W., Kohn, H. I. 1965. Inherited histocompatability changes in the progeny of irradiated and unirradiated inbred mice. *Genet. Res.* 6:330–40
7. Bailey, D. W., Usama, B. 1960. A rapid method of grafting skin on tails of mice. *Transplant. Bull.* 7:424–25
8. Batra, B. K., Stridharan, B. N. 1964. A study of the progeny of mice descended from X-irradiated females with special reference to the gonads. *Acta Unio Int. Contra Cancrum* 20:1181–86
8a. Baverstock, K. F., Stather, J. W., eds. 1989. *Low Dose Radiation: Biological Bases of Risk Assessment*. London: Taylor & Francis
9. Chakraborty, R., Neel, J. V. 1989. Description and validation of a method for simultaneous estimation of effective population size and mutation rate from human population data. *Proc. Natl. Acad. Sci. USA* 86:9407–11

10. Charles, D. J., Pretsch, W. 1986. Enzyme-activity mutations detected in mice after paternal fractionated irradiation. *Mutat. Res.* 160:243–48
11. Charles, D. R., Tihen, J. A., Otis, E. M., Grobman, A. B. 1960. Genetic effects of chronic x-irradiation exposure in mice. *AEC Res. Devel. Rep. UR-565.* Off. Tech. Serv., Dept. Commerce. 354 pp.
12. Comm. Biol. Effects of Ionizing Radiat. 1980. *The Effects of Populations of Exposure to Low Levels of Ionizing Radiation (BEIR III)*, pp. xv & 524 Washington, DC: Natl. Acad. Press
13. Comm. Biol. Effects of Radiat. 1990. *Health Effects of Exposure to Low Levels of Ionizing Radiation (BEIR V)*, pp. xiii & 421 Washington, DC: Natl. Acad. Press
14. de Serres, F. J. 1958. Studies with purple adenine mutants in *Neurospora crassa*. III. Reversion of X-ray-induced mutants. *Genetics* 43:187–206
14a. de Serres, F. J., Sheridan, W., eds. 1983. *Utilization of Mammalian Specific Locus Studies in Hazard Evaluation and Estimation of Genetic Risk*. New York: Plenum
15. Ehling, U. H. 1966. Dominant mutations affecting the skeleton in offspring of X-irradiated male mice. *Genetics* 54:1381–89
16. Ehling, U. H., Charles, D. J., Favor, J., Graw, J., Kratochvilova, J., et al. 1985. Induction of gene mutations in mice: The multiple endpoint approach. *Mutat. Res.* 150:393–401
17. Endo, A., Watanabe, T. 1989. A case of X-trisomy in the mouse. *Cytogenet. Cell Genet.* 52:98–99
18. Erickson, R. P. 1989. Why isn't a mouse more like a man? *Trends Genet.* 5:1–3

19. Evans, H. S. 1977. Chromosome abnormalities among live-births. *J. Med. Genet.* 14:309–12
20. Favor, J. 1989. Risk estimation based on germ cell mutations in animals. *Proc. 16th Int. Congr. Genet., Genome* 31: 844–52
21. Ford, C. E., Searle, A. G., Evans, E. P., West, B. J. 1969. Differential transmission of translocations induced in spermatogonia of mice by irradiation. *Cytogenetics* 8:447–70
22. Furusho, T., Otake, M. 1978. A search for genetic effects of atomic bomb radiation on the growth and development of the F_1 generation. 1. Stature of 15- to 17-year-old senior high school students in Hiroshima. Hiroshima: *RERF Tech. Rep. 4–78.* 33 pp.
23. Furusho, T., Otake, M. 1978. A search for genetic effects of atomic bomb radiation on the growth and development of the F_1 generation. 2. Body weight, sitting height, and chest circumferences of 15- to 17-year-old senior high school students in Hiroshima. Hiroshima: *RERF Tech. Rep. 5–78.* 16 pp.
24. Furusho, T., Otake, M. 1979. A search for genetic effects of atomic bomb radiation on the growth and development of the F_1 generation. 3. Stature of 12- to 14-year-old junior high school students in Hiroshima. Hiroshima: *RERF Tech. Rep. 14–79.* 26 pp.
25. Furusho, T., Otake, M. 1980. A search for genetic effects of atomic bomb radiation on the growth and development of the F_1 generation. 4. Body weight, sitting height, and chest circumference of 12- to 14-year-old junior high school students in Hiroshima. Hiroshima: *RERF Tech. Rep. 1–80.* 18 pp.
26. Furusho, T., Otake, M. 1985. A search for the genetic effects of atomic bomb radiation on the growth and development of the F_1 generation. 5. Stature of 6- to 11-year-old elementary school pupils in Hiroshima. Hiroshima: *RERF Tech. Rep. 9–85.* 33 pp.
27. Generoso, W. M., Cain, K. T., Cacheiro, N. C. A., Cornett, C. V. 1984. Response of mouse spermatogonial cells to X-ray induction of heritable reciprocal translocations. *Mutat. Res.* 126:177–87
28. Godfrey, J., Searle, A. G. 1963. A search for histocompatibility differences between irradiated sublines of inbred mice. *Genet. Res.* 4:21–29
29. Grahn, D., Leslie, W. P., Verkey, F. A., Lea, R. A. 1972. Determination of the radiation-induced rate for sex-linked

lethals and detrimentals in the mouse. *Mutat. Res.* 15:331–47
30. Green, E. L. 1968. Genetic effects of radiation on mammalian populations. *Annu. Rev. Genet.* 2:87–120
31. Green, E. L., Schlager, G., Dickie, M. M. 1965. Natural mutation rates in the house mouse: plan of study and preliminary estimates. *Mutat. Res.* 2:457–65
32. Green, M. C., ed. 1981. *Genetic Variants and Strains of the Laboratory Mouse,* pp. xvi & 476. New York: Fischer
33. Griffin, C. S., Tease, C. 1988. γ-ray induced numerical and structural chromosome anomalies in mouse immature oocytes. *Mutat. Res.* 202:209–13
34. Hanash, S. M., Boehnke, M., Chu, E. H. Y., Neel, J. V., Kuick, R. D. 1988. Nonrandom distribution of structural mutants in ethylnitrosourea-treated cultured human lymphoblastoid cells. *Proc. Natl. Acad. Sci. USA* 85:165–69
35. Hook, E. B. 1985. The impact of aneuploidy upon public health: mortality and morbidity associated with human chromosome abnormalities. In *Aneuploidy: Etiology and Mechanisms,* ed. V. L. Dellarco, P. E. Voytek, A. Hollaender, pp. 7–33. New York: Plenum
36. Hook, E. B., Topol, B. B., Cross, P. K. 1989. The natural history of cytogenetically abnormal fetuses detected at midtrimester amniocentesis which are not terminated electively: New data and estimates of the excess and relative risk of late fetal death associated with 47, & 21 and some other abnormal karyotypes. *Am. J. Hum. Genet.* 45:855–61
37. Int. Comm. Radiat. Units Meas. 1986. *The Quality Factor in Radiation Protection. Rep. 40.* Bethesda: ICRUM. 32 pp.
38. Johnson, F. M., Lewis, S. E. 1981. Electrophoretically detected germinal mutations induced in the mouse by ethylnitrosourea. *Proc. Natl. Acad. Sci. USA* 78:3138–41
39. Kato, H., Schull, W. J., Neel, J. V. 1966. A cohort-type study of survival in the children of parents exposed to atomic bombings. *Am. J. Hum. Genet.* 18:339–73
40. Kirk, K. M., Lyon, M. F. 1982. Induction of congenital anomalies in offspring of female mice exposed to varying doses of x-rays. *Mutat. Res.* 106:73–83
41. Kirk, K. M., Lyon, M. F. 1984. Induction of congenital malformations in the offspring of male mice treated with x-rays at pre-meiotic and post-meiotic stages. *Mutat. Res.* 125:75–85
42. Kohn, H. I. 1979. X-ray mutagenesis:

Results of the H-test compared with others and the importance of selection and/or repair. *Genetics* 9 2(Suppl. 1 : 1):203–9

43. Kohn, H. I. 1983. Radiation genetics: The mouse's view. *Radiat. Res.* 94:1–9

44. Kohn, H. I., Epling, M. L., Guttman, P. H., Bailey, D. W. 1965. Effect of paternal (spermatogonial) X-ray exposure in the mouse: Lifespan, X-ray tolerance, and tumor incidence of the progeny. *Radiat. Res.* 25:423–34

45. Kohn, H. I., Melvold, R. W. 1976. Divergent x-ray-induced mutation rates in the mouse for H and "7-locus" groups of loci. *Nature* 259:209–10

46. Kohn, H. I., Melvold, R. W., Dunn, G. R. 1976. Failure of x-rays to mutate Class II histocompatibility loci in BALB/c mouse spermatogonia. *Mutat. Res.* 37:237–44

47. Kratochvilova, J. 1981. Dominant cataract mutations detected in offspring of gamma-irradiated male mice. *J. Hered.* 72:302–7

48. Lüning, K. G. 1979. Some problems in the assessment of risks. *Genetics* 92(Suppl. 1 : 1):121–26

49. Lüning, K. G., Eiche, A. 1975. X-ray induced recessive lethal mutations in the mouse. *Mutat. Res.* 34:63–174

50. Lüning, K. G., Searle, A. G. 1971. Estimates of the genetic risks from ionizing radiation. *Mutat. Res.* 12:291–304

51. Lüning, K. G., Sheridan, W. 1971. Changes in sex-proportion: an unacceptable way to estimate sex-linked recessive lethals. *Mutat. Res.* 13:77–83

52. Lyon, M. F. 1959. Some evidence concerning the 'mutational load' in inbred strains of mice. *Heredity* 13:341–52

53. Lyon, M. F. 1983. Problems in extrapolation of animal data to humans. See Ref. 14a, pp. 289–305

54. Lyon, M. F., Morris, T. 1966. Mutation rates at a new set of specific loci in the mouse. *Genet. Res.* 7:12–17

55. Lyon, M. F., Morris, T. 1969. Gene and chromosome mutation after large fractionated or unfractionated radiation doses to mouse spermatogonia. *Mutat. Res.* 8:191–98

56. Lyon, M. F., Phillips, R. J. S., Fisher, G. 1979. Dose-response curves for radiation-induced gene mutations in mouse oocytes and their interpretation. *Mutat. Res.* 63:161–73

57. Lyon, M. F., Phillips, R. J. S., Fisher, G. 1982. Use of an inversion to test for induced X-linked lethals in mice. *Mutat. Res.* 92:217–28

58. Lyon, M. F., Phillips, R. J. S., Searle,

A. G. 1964. The overall rates of dominant and recessive lethal and visible mutations induced by spermatogonial x-irradiation of mice. *Genet. Res.* 5:448–67

59. Malling, H. V., de Serres, F. J. 1973. Genetic alterations at the molecular level in X-ray induced ad-3B mutants of *Neurospora crassa*. *Radiat. Res.* 53:77–87

60. Max, C. 1977. Cytological investigation of embryos in low-dose X-irradiated young and old inbred mice. *Hereditas* 85:199–206

61. Natl. Counc. Radiat. Prot. Meas. 1980. Influence of dose and its distribution in time on dose-response relationships for low-LET radiations. Washington, DC: *NCRP Rep. 64*:vi & 216

62. Neel, J. V. 1958. A study of major congenital defects in Japanese infants. *Am. J. Hum. Genet.* 10:398–445

63. Neel, J. V. 1990. Average locus differences in mutability related to protein "class": A hypothesis. *Proc. Natl. Acad. Sci. USA* 86:9407–11

64. Neel, J. V., Kato, H., Schull, W. J. 1974. Mortality in the children of atomic bomb survivors and controls. *Genetics* 76:311–26

65. Neel, J. V., Satoh, C., Goriki, K., Asakawa, J-I., Fujita, M., et al. 1988. Search for mutations altering protein charge and/or function in children of atomic bomb survivors: Final report. *Am. J. Hum. Genet.* 42:663–76

66. Neel, J. V., Schull, W. J. 1956. The effect of exposure to the atomic bombs on pregnancy termination in Hiroshima and Nagasaki. *Natl. Acad. Sci. Natl. Res. Counc. Publ. 461*. Washington, DC. pp. xvi & 241

67. Neel, J. V., Schull, W. J., Awa, A. A., Satoh, C., Kato, H., et al. 1990. Children of parents exposed to atomic bombs: Estimates of the genetic doubling dose of radiation for humans. *Am. J. Hum. Genet.* 46:1053–72

68. Neel, J. V., Schull, W. J., Awa, A. A., Satoh, C., Otake, M., et al. 1989. The genetic effects of the atomic bombs: Problems in extrapolating from somatic cell findings to risk for children. See Ref. 8a, pp. 42–53

69. Neel, J. V., Schull, W. J., Awa, A. A., Satoh, C., Otake, M., et al. 1989. Implications of the Hiroshima-Nagasaki genetic studies for the estimation of the human "doubling dose" of radiation. *Proc. 16th Int. Congr. Genet., Genome* 31:853–59

70. Nomura, T. 1978. Changed urethan and radiation response of the mouse germ

cell to tumor induction. In *Tumours of Early Life in Man and Animals*, ed. L. Severi, pp. 873–91. Monteluce, Italy: Perugia Quadrenn. Int. Conf. Cancer

71. Nomura, T. 1982. Parental exposure to X-rays and chemicals induces heritable tumours and anomalies in mice. *Nature* 296:575–77

72. Nomura, T. 1983. X-ray-induced germ-line mutation leading to tumors. Its manifestation in mice given urethane postnatally. *Mutat. Res.* 121:59–65

73. Nomura, T. 1986. Further studies on X-ray and chemically induced germ-line alterations causing tumors and malformations in mice. See Ref. 77a, pp. 13–20

74. Otake, M., Schull, W. J., Neel, J. V. 1990. The effects of exposure to the atomic bombing of Hiroshima and Nagasaki on congenital malformations, stillbirths and early mortality among the children of atomic bomb survivors: A reanalysis. *Radiat. Res.* 122:1–11

75. Otis, E. M., Brent, R. 1952. Equivalent ages in mouse and human embryos. *Anat. Res.* 120:33–64

76. Peters, J., Ball, S. T., Andrews, S. J. 1986. The datechain of gene mutations by electrophoresis and their analyses. *Prog. Clin. Biol. Res.* 209B:367–74

77. Pretsch, W. 1986. Protein-charge mutations in mice. *Prog. Clin. Biol. Res.* 209B:383–89

77a. Ramel, C., Lambert, B., Magnussen, J., eds. 1986. *Genetic Toxicology of Environmetal Chemicals, Part B: Genetic Effects and Applied* Mutagenesis. New York: Liss

78. Reddi, D. S. 1965. Radiation induced translocations in mouse spermatogonia. *Mutat. Res.* 2:95

79. Roderick, T. H., ed. 1964. The effects of radiation on the hereditary fitness of mammalian populations. *Genetics* 50: 1213–17

80. Roderick, T. H. 1983. Using inversions to detect and study recessive lethals and detrimentals in mice. See Ref. 14a, pp. 135–67

81. Roesch, W. C., ed. 1987. *US-Japan Reassessment of Atomic Bomb Radiation Dosimetry in Hiroshima and Nagasaki* 1:434. Hiroshima: RERF

82. Russell, L. B. 1976. Numerical sex-chromosome anomalies in mammals: Their spontaneous occurrence and use in mutagenesis studies. In *Chemical Mutagens. Principles and Methods for Their Detection*, ed. A. Hollaender, 4:55–91. New York: Plenum

83. Russell, L. B., Rinchik, E. M. 1987. Genetic and molecular characterization of genomic regions surrounding specific loci of the mouse. *Banbury Rep.* 28: 109–21

84. Russell, W. L. 1951. X-ray induced mutations in mice. *Cold Spring Harbor Symp. Quant. Biol.* 16:327–36

85. Russell, W. L. 1956. Comparison of x-ray-induced mutation rates in Drosophila and mice. *Am. Nat.* 90:69–80

86. Russell, W. L. 1963. The effect of radiation dose rate and fractionation on mutation in mice. In *Repair from Genetic Radiation*, ed. F. H. Sobels, pp. 205–17. London: Pergamon

87. Russell, W. L. 1965. The nature of the dose-rate effect of radiation on mutation in mice. *Proc. Conf. Mech. Dose Rate Effect Radiat. Jpn. J. Genet.* 40:128–40 (Suppl.)

88. Russell, W. L. 1965. Effect of the interval between irradiation and conception on mutation frequency in female mice. *Proc. Natl. Acad. Sci. USA* 54: 1552–57

89. Russell, W. L., Russell, L. B., Kelly, E. M. 1958. Radiation dose rate and mutation frequency. *Science* 128:1546–50

90. Sankaranarayanan, K. 1988. Prevalence of genetic and partially genetic diseases in man and the estimation of genetic risks of exposure to ionizing radiation. *Am. J. Hum. Genet.* 42:651–62

91. Schlager, G., Dickie, M. M. 1967. Spontaneous mutations and mutation rates in the house mouse. *Genetics* 57: 319–30

92. Schull, W. J., Neel, J. V., Hashizume, A. 1966. Some further observations on the sex ratio among infants born to survivors of the atomic bombings of Hiroshima and Nagasaki. *Am. J. Hum. Genet.* 18:328–38

93. Schull, W. J., Otake, M., Neel, J. V. 1981. Genetic effects of the atomic bombs: A reappraisal. *Science* 213: 1220–27

94. Searle, A. G. 1964. Effects of low-level irradiation on fitness and skeletal variation in an inbred mouse strain. *Genetics* 50:1159–78

95. Searle, A. G. 1964. Genetic effects of spermatogonial X-irradiation on productivity of F_1 female mice. *Mutat. Res.* 1:99–108

96. Searle, A. G. 1974. Mutation induction in mice. In *Advances in Radiation Biology*, ed. J. T. Lett, H. I. Adler, A. Zelle, 4:131–207. New York: Academic

97. Searle, A. G. 1977. Radiological protection and assessment of genetic risk. *J. Med. Gen.* 14:307–8

98. Searle, A. G. 1989. Evidence from

mammalian studies on genetic effects of low-level irradiation. See Ref. 8a, pp. 123–38

99. Searle, A. G., Beechey, C. 1986. The role of dominant visibles in mutageneity testing. See Ref. 77a, pp. 511–18

100. Selby, P. B., Selby, P. R. 1977. Gamma-ray induced dominant mutations that cause skeletal abnormalities in mice. I. Plan, summary of results, and discussion. *Mutat. Res.* 43:357–75

101. Selby, P. B., Selby, P. R. 1978. Gamma-ray-induced dominant mutations that cause skeletal abnormalities in mice. II. Description of proved mutations. *Mutat. Res.* 51:199–236

102. Sheridan, W., Wårdell, I. 1968. The frequency of recessive lethals in an irradiated mouse population. *Mutat. Res.* 5:313–21

103. Speed, R. M., Chandley, A. C. 1981. The response of germ cells of the mouse to the induction of non-disjunction by X-rays. *Mutat. Res.* 84:409–18

104. UN Sci. Comm. Effects of At. Radiat. 1977. *Sources and Effects of Ionizing Radiation.* New York: United Nations. 725 pp.

105. UN Sci. Comm. Effects At. Radiat. 1986. *Genetic and Somatic Effects of Ionizing Radiation.* New York: United Nations. 366 pp.

106. Van Buul, P. P. W. 1983. Induction of chromosomal aberrations by ionizing radiation in stem cell spermatogonia of mammals. In *Radiation-induced Chromosome Damage in Man*, ed. T. Ishihara, M. S. Sasaki, pp. 369–400. New York: Liss

107. West, J. D., Kirk, K. M., Goyder, Y., Lyon, M. F. 1985. Discrimination between the effects of X-ray irradiation of the mouse oocyte and uterus on the induction of dominant lethals and congenital anomalies. I. Embryo transfer experiments. *Mutat. Res.* 149:221–30

108. West, J. D., Kirk, K. M., Goyder, Y., Lyon, M. F. 1985. II. Localized irradiation experiments. *Mutat. Res.* 149:231–38

109. Yoshimoto, Y., Neel, J. V., Schull, W. J., Kato, H., Mabuchi, K., et al. 1990. Malignant tumors during the first two decades of life in the offspring of atomic bomb survivors. *Am. J. Hum. Genet.* 46:1041–52

Annu. Rev. Genet. 1990. 24:363–85

PHAGE T4 INTRONS: SELF-SPLICING AND MOBILITY

Marlene Belfort

Wadsworth Center for Laboratories and Research, New York State Department of Health, and School of Public Health, State University of New York, Empire State Plaza, Albany, New York 12201-0509

KEY WORDS: prokaryotic introns, group I introns, intron evolution, intron endonucleases, double-strand-break repair model for intron mobility

CONTENTS

363

And it was this constant dialogue between imagination and experiment that allowed one to form an increasingly fine-grained conception of what is called reality. To this game the bacteriophage was particularly well suited, lending itself to all sorts of experiments: simple, rapid, precise.

François Jacob:
The Statue Within

INTRODUCTION

Prokaryotes were considered to be free of intervening sequences until the discovery in 1984 of an intron in the *td* gene of phage T4 (22). There are now five well-characterized bacteriophage introns, four of which are of the group I self-splicing variety and are the topic of this review, while the fifth is an unspliced intron that is precisely bypassed by the translational apparatus (45; reviewed in 38). Two of the group I introns have recently been shown to be mobile, transferring to intronless cognates of the respective donor genes via a DNA-based mechanism. The RNA processing events that mediate splicing and the DNA transactions associated with the mobility of these composite introns form the basis of this chapter.

A recurring theme throughout this review is the extraordinarily conserved nature of the group I introns in pro- and eukaryotes despite the phylogenetic divergence of their hosts. These similarities extend from intron structure to splicing and mobility functions. Given the availability of recent, comprehensive reviews covering both group I splicing (10, 15–20) and mobility (3, 10, 19, 33, 42, 51, 70, 76), I focus primarily on unusual features of the phage introns and on ways in which they are contributing to our understanding of RNA catalysis and intron inheritance.

THE DISCOVERY OF INTERRUPTIONS IN PHAGE GENES

The intron in the *td* gene, which encodes thymidylate synthase, was discovered in a serendipitous observation that the colinearity rule for the DNA and protein sequences had been violated (22). The two sequences could be reconciled by inferring the presence of a 1016-nt intervening sequence in the gene. The *td* intron turned out to be one of a family of phage introns. An intervening sequence in the T4 *nrd*B gene, encoding the small unit of ribonucleotide reductase, was found independently by comparing the deduced protein sequence with that of the *Escherichia coli* reductase B2 subunit (80) and by a random search for phage introns (40). This latter means of localizing introns was based on the observation that the *td* intron is a typical autocatalytic group I intron that splices via the guanosine-initiated transesterification pathway described below (15, 21, 23, 36). It was therefore possible to screen for introns in phage RNA preparations by labeling them with radioactive GTP

Table 1 Group I phage introns

Gene	Phage	Intron Size (nt)	ORF Product	ORF Function	ORF Loop
td	T2[a], T4, T6, RB3[b]	1016	245 aa	endonuclease	P6
*nrd*B	T4	598	97 aa	unknown	P6
	RB3[b]	1091	269 aa	probable endonuclease	P6
*sun*Y	T2[a], T4	1033	258 aa	endonuclease	P9.1
g31	SPO1	883	174 aa	unknown	P8

[a] Present only in one known T2 derivative, T2W. Other T2 isolates, T2L and T2H, have no known introns (74).

[b] RB3, a member of the T-even phage family, contains the *td* and *nrd*B introns. The *td* intron in RB3 is similar to that in T4, whereas the *nrd*B intron has a 491-nt insertion in the ORF, raising the possibility that the ORF in T4 is vestigial and that the RB3 *nrd*B intron may encode an endonuclease and be mobile (S. Eddy, L. Gold, personal communication).

under self-splicing conditions. With the labeled RNA as hybridization probe, an intron of 598 nt was localized to the *nrd*B gene (40), and another, of 1033 nt, was mapped to a gene termed *sun*Y (*s*plit gene of *un*known function) (40, 78, 79). The map locations of the three T4 introns are shown in Figure 1A and some of their properties are listed in Table 1. Additionally, GTP-labeling revealed a self-splicing group I intron in the DNA polymerase gene, *g31,* of *Bacillus subtilis* phage SPO1 (39). A thought-provoking but unexplained finding is that all three phage intron-containing genes of known function (*td, nrd*B, and *g31*) are involved in DNA metabolism.

OVERALL ARCHITECTURE OF GROUP I PHAGE INTRONS

Figures 1B and C show the *td* gene as the prototypic intron-containing phage gene. The phage self-splicing introns all contain ochre stop codons in frame with the upstream exon sequences (22, 39, 78). The UAA triplet straddles the 5' splice site of the *nrd*B intron, whereas it lies at the 5' end of the *td* intron, and occurs downstream of the 5' splice site of the *sun*Y and *g31* introns (by 5 nt and 36 nt, respectively). Since transcription and translation are coupled, this arrangement results in the capacity of the pre-mRNAs to encode amino-terminal products (NH$_2$TS for the *td* gene) (6; Figure 1C). Although these products are not known to have any biological role, the ochre triplets are thought to function to preserve splicing *in vivo* by preventing ribosomes from entering the critical catalytic core structures of the introns. Translation into the intron has indeed been shown to inhibit splicing of *nrd*B pre-mRNA *in vivo* (B. M. Sjöberg, personal communication).

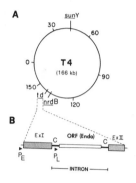

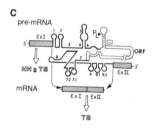

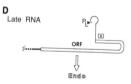

Figure 1   A. Intron-containing loci on the T4 map. B. Linear map of the *td* gene. Exons are shaded and the intron core is represented by a heavy line (C). The early upstream promoter (P_E, precise position not known), and late gp55-dependent promoter that drives ORF expression (P_L) are indicated by arrowheads. C. Secondary structure of *td* pre-mRNA. Numbers in the folded intron core correspond to conserved structure elements (see Figure 3A). The "attenuator" stem that obstructs ORF translation is indicated by a boxed "-". The ORF is shown looped out of the core, except at its 3' end. The pre-mRNA product, NH₂-TS, and exon ligation product, thymidylate synthase (TS), are shown below their respective templates (6). D. The late transcript with a free initiation site ("+") for ORF translation. The *nrd*B and *sun*Y introns have basically similar architecture and regulatory elements (41, 78).

The phage introns consist of two basic components that are functionally distinct and structurally separate, except for about 20 overlapping nucleotides (78). First, approximately 250 nt form the group I RNA secondary structure important in splicing. Second, looped out of this structure is an open reading frame (ORF) that encodes a protein not required for splicing (Figure 1B and C). For the *td* and *sun*Y introns, the ORFs have been shown to encode endonucleases (7, 8, 73, 85), whereas the other ORFs are of unknown function. The T4 *nrd*B ORF appears to be a truncated version of a longer reading frame present in the *nrd*B intron of RB3, a relative of the T-even phages (S. Eddy, L. Gold, personal communication; Table 1).

Both the 5' and 3' ends of the ORFs have unusual features. The translational start sites of the three T4 intron ORFs are separate from the core splicing structures, but are embedded in independent stem-loops that obstruct initiation of translation from the pre-mRNA (41; Figure 1C). ORF expression awaits synthesis of a transcript that is initiated at a T4 late promoter located within the stem-loop. These late RNAs, which are positively regulated by the T4 gp55 transcription factor (49), have their ribosome binding sites available for ORF translation (41; Figure 1D).

The 3' ends of the ORFs all extend about 20 nt into the core structures (78). This arrangement is noteworthy on two accounts. First, it creates the potential for ORF translation from the pre-mRNA to regulate splicing (79; see below). Second, it suggests that the combined core-ORF sequence must have evolved under the dual constraints of RNA catalysis and protein function.

RNA STRUCTURE AND CATALYTIC FUNCTION

Splicing Pathway

Self-splicing of the phage introns follows the group I pathway (21, 23, 36, 39, 40, 79), as first elaborated by Cech and colleagues for the large rRNA intron of *Tetrahymena thermophila* (the *Tet* LSU intron) (Figure 2; reviewed in 15–20). Splicing occurs via a series of transesterification reactions. The first

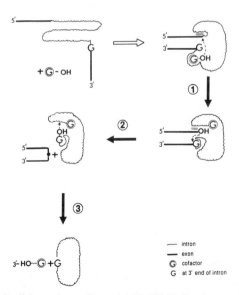

Figure 2   The group I splicing pathway. Guanosine (G-OH) binding (open arrow) is followed by the transesterification reactions (closed arrows; steps 1–3), as described in the text.

transesterification is initiated by free guanosine or a guanosine nucleotide, which binds the ribozyme and, via its 3'-OH, attacks the phosphate at the 5' splice site. This results in guanosine addition to the 5' end of the intron via a 3'–5' phosphodiester bond, and a free 3'-OH at the end of the upstream exon (step 1). In a second transesterification, this latter 3'-OH attacks the phosphate at the 3' splice site, resulting in exon ligation and release of the intron (step 2). Intron cyclization can occur by a third transesterification, in which a conserved G at the 3' intron terminus attacks an internal phosphate near the 5' end, resulting in loss of a 5' intron oligomer (step 3). For the *td* intron, the 5'-terminal guanosine and one additional nucleotide are released (21, 36), as opposed to loss of 15 or 19 nt upon cyclization of the *Tet* LSU intron (1).

Secondary Structure

The transesterification reactions require a characteristic folding of the RNA molecules (12, 18, 28, 57, 61). The secondary structure model for group I introns was based on a comparison of fewer than 10 fungal mitochondrial introns (28, 57, 61). The folding is guided by four conserved sequence elements, P, Q, R and S, and contains nine common pairing elements, P1 through P9 (reviewed in 10, 12). Remarkably, more than 80 group I introns have now been characterized and all, including the phage introns (Figure 3A), conform to this model (18, 25, 26, 78). The T4 introns have additional pairings between P3 and P7 (P7.1 and P7.2) that, together with specific deviations in P, Q, R, and S, are the hallmark of the subgroup IA introns (18, 57, 78; Figure 3). The SP01 *g31* intron also conforms to the basic group IA prototype (39), but it differs from the T4 introns in the presence of a large spacing (69 nt) between P3 and P4, and in the absence of P9.1 and P9.2 pairings. In further contrast to the T4 introns, its ORF is looped out of P8 (39) rather than out of P6 or P9.1 (78; Table 1 and below).

The core elements of the T4 introns are very similar in both sequence and structure, except that the *sun*Y intron lacks a P2 pairing (78; Figure 3A). In contrast, the ORFs are quite different. In addition to being unrelated in sequence, they emanate from different positions in the ribozymes: from the P6 region for *td* and *nrd*B, and from P9.1 for *sun*Y (78). Therefore, it appears that, whereas the core structures have strong ancestral ties, the divergent ORFs may have evolved independently.

Consistent with the functional division of core and ORF, non-directed splicing-defective mutations of the *td* gene are concentrated in the core structure (4, 14, 37, 44; Figure 3A). The structure model has been tested in a number of systems with second-site revertants that concomitantly restore base-pairing and splicing function (10). For the *td* intron, the structure has been verified by compensatory mutations in one short-range (P8) and two long-range pairings (P3 and P6) (Figure 3A). Except for the case of P6 (37), the pseudorevertants were also nondirected; compensatory mutations induced

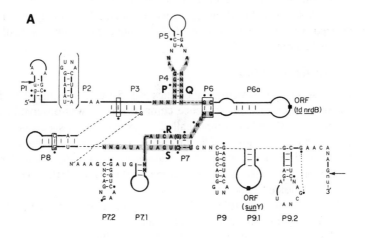

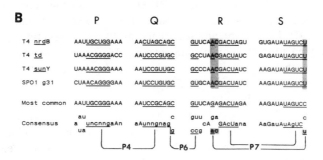

Figure 3   Intron secondary structure. A. Sequence and structure elements conserved among the three T4 introns. Nucleotides common to the three introns are indicated, with conserved spacings between them represented by "N"s and conserved base-pairings indicated by hyphens (78). Dashed lines join adjacent nucleotides, whereas bold lines indicate varying lengths of nonconserved residues. Splice sites are indicated by arrows between the exon (lower case) and intron (upper case) nucleotides, with the structural elements labeled P1 to P9.2 (for conventions see ref. 12). Black boxes represent the ORFs. P2 (bracketed) is absent from *sun*Y. Sites of nondirected splicing-defective mutations in the *td* intron are marked by dots (4, 14, 37), with boxes indicating those pairing elements that have been confirmed by the isolation of compensatory mutations (37; P.S. Chandry, M. Brown, D. Hall & M.B., unpublished data). The conserved G-C pair in P7 that corresponds to a guanine binding site in the *Tet* LSU intron (60) is in parentheses, and the "bulged" nucleotide in P7 is underscored. Conserved sequences, P, Q, R and S are shaded. B. Comparison of conserved sequences for the phage introns. Most common nucleotides at these positions and consensus sequences for a total of 66 group I introns are taken from Cech (18). Consensus nucleotides in upper and lower case letters are conserved for >90% and for 70-90% of 66 group I introns, respectively (18). Variant nucleotides characteristic of group IA introns are shaded (18, 57). Underlined nucleotides are proposed to pair in the secondary structure.

by hydroxylamine, a unidirectional mutagen incapable of reverting the primary hydroxylamine-induced mutagen, were selected by phenotype (5) and characterized, verifying the structure (P. S. Chandry, M. Brown, D. Hall, M. B., unpublished data).

Splice-Site Recognition

As for the secondary structure, the rules for 5' and 3' splice-site (SS) selection were inferred from sequence comparisons. It was observed that a sequence near the 5' end of introns, called the internal guide sequence (IGS), could pair with the 3' residues of the upstream exon (to form element P1) (28, 57, 61) and with 5' nucleotides of the downstream exon (to form element P10) (28; Figures 3 and 4). It was then proposed that P1 and P10 determine 5' SS and 3' SS choice, respectively.

The P1 pairing and its role in 5' SS recognition have been proven by mutations that disrupt the pairing and reduce splicing, and compensatory changes that restore activity (2, 83). The SS typically coincides with a U-G pair, with the U representing a highly conserved 3' base of the upstream exon. Apart from this "wobble" basepair, the P1 sequence is not conserved, leading to the supposition that RNA structure rather than sequence determines 5' SS recognition (31). Loss of splicing specificity results from disruption of the P1 structure by mutation of its exon component (2, 14, 72). For the *td* intron, an exon mutation that disrupts P1 results in use of an upstream cryptic 5' SS 29 nucleotides from the authentic 5' SS. This cryptic SS coincides with an alternative P1-like pairing (14).

Although the IGS was postulated to align the two exons for ligation via P1 and P10 (28), two observations argue against the absolute necessity for P10 in 3' SS selection. First, the pairing can be disrupted without major consequence to splicing (29, 60). Second, some introns have either poor P10 pairings, or none at all. The T4 introns have very weak theoretical P10s of only two or three nucleotides (23, 78). The inadequacy of a 3-nt P10 to specify the 3' SS is striking for the *sun*Y intron, where the 800-nt ORF loop separates the intron core from the downstream exon (62, 78, 79; Figures 3 and 4). Recognition of the 3' SS is faithful despite the ability of the ORF loop to form several potential 3-nt P10 pairings with the IGS, suggesting an additional or independent means of 3' SS selection.

Mutational studies have implicated both the terminal intron residue, an invariant G, and bases 5' to the G in exon ligation (71). Subsequent phylogenetic comparisons have revealed covariation in many group I introns between the second- and third-to-last residues of the intron and two nucleotides of J7/9 (the segment between P7 and P9), termed the 3' splice-site binding sequence (3' SSBS) (11, 62). Mutational analysis in the *Tet* LSU and *sun*Y introns has provided evidence for the importance of the pairing of these two regions (the P9.0 pairing) for exon ligation (13, 60, 62; Figure 4). Thus,

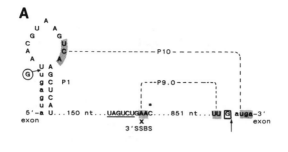

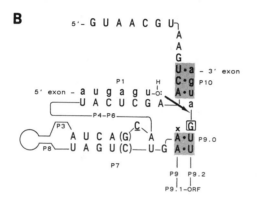

Figure 4 Splice-site selection in the *sun*Y intron. A. Elements of the Pl, P9.0 and P10 pairings. SSs are indicated by arrows, between exon (lower case) and intron sequences (upper case). The guanosine cofactor is circled, and shown attacking the 5' SS in the P1 exon I-intron pairing. The internal guide sequence comprises those intron residues involved in this pairing, plus those that pair with exon II nucleotides to form P10 (shaded). The components of P9.0, the 3' SSBS and two nucleotides preceding the conserved 3' G of the intron (boxed), are also shaded. The 3' element of P7 (P7[3']) is underscored and the 5' nucleotide at the base of P9 is marked by an asterisk. The A in the 3' SSBS that is marked with an X alters 3' SS selection when mutated to a G (62). B. Alignment of the 3' SS. The molecule is shown after 5' SS cleavage with the P9.0 and P10 pairings shaded. All sequences are represented in A, except for the 5' element of P7 (P7[5']), shown here to illustrate the proximity of a predicted guanine binding site (the G-C pair in parentheses) to P9.0. The arrow indicates nucleophilic attack by the 3'-OH of the upstream exon on the 3' SS phosphate.

*sun*Y transcripts truncated within the ORF, 3' to the catalytic core, self-splice to cryptic sites, which are dictated by the presence of a G that is preceded by nucleotide(s) able to pair with the 3' SSBS. That is, in the absence of the natural P9.0 and P10, an alternative P9.0 is both necessary and sufficient to direct exon ligation to cryptic 3' SSs (62). The role of P9.0 in the absence of the natural P10 was proven by changing an A in the 3' SSBS to a G (Figure 4). This mutation resulted in altered specificity of the cryptic 3' SSs, with cleavage occurring after CG rather than after UG residues (62). Experiments with the *Tet* LSU intron specifically address the relative contributions of P9.0

and P10 in 3' SS choice. Mutations that disrupted either P9.0 and P10 had little effect, whereas mutations that disrupted both P9.0 and P10 greatly inhibited 3' SS activity (60). In this case it would appear that P9.0 or P10 contribute to 3' SS selection in a redundant fashion.

The Guanosine-binding Site and Defining the 3-D Structure

Having localized the 3' SSBS, Michel et al identified a proximal, phylogenetically invariant G in P7 as a guanine-binding residue (60). The G of the G264-C311 pair in P7 of the *Tet* LSU intron was shown to bind free guanine via a triple base interaction, and to interact with the 3' terminal G of the intron as well (60). The equivalent base pair in the T4 introns is shown in Figures 3A and 4B. The three-dimensional architecture of the site and interactions of the guanosine cofactor with other residues must also contribute to the specificity of the site. One such candidate is the unpaired ("bulged") nucleotide in P7, which is immediately 5' to the guanine-binding residue (underscored in Figures 3A and 4B) and is invariably a C in group IA introns and an A in all others (18). In support of a role for this residue in the G-binding site, mutation of the bulged C in the *td* intron to U or G, specifically increased K_m for guanosine 20- to 100-fold without appreciably affecting V_{max} (R. Schroeder, M. B. submitted).

The overall architecture of the guanosine-binding site appears similar for all group I introns. The ability of the *Tet* LSU intron to bind arginine, a reflection of the structural homology between the purine ring of guanosine and the guanidino group of arginine (89), is shared by the phage introns (47). A twofold stereoselectivity, or preference for binding L- over D-arginine, is also a common property of the introns. These observations provide further evidence for the spatial uniformity of the guanosine-binding pocket of the phage and eukaryotic introns, despite the sequence heterogeneity of these RNAs (47).

Although a working model has been proposed for the tertiary folding of the *Tet* LSU intron (50), additional studies are required to define the precise three-dimensional structure (reviewed in 19). Input into the tertiary structure is provided by experiments with the phage introns, supporting the existence of a triple-helical stack. The triple helix, proposed on the basis of phylogenetic covariation, is thought to form via P4:J6/7 and J3/4:P6 interactions (59; Figure 5). Mutations in the *sun*Y and *Tet* LSU introns support the P4:J6/7 interaction, while the J3/4:P6 interaction remains to be proven. This triple-helical "scaffold" provides new constraints for 3-D model-building of group I introns.

Physical analyses such as NMR and X-ray crystallography must ultimately go hand-in-hand with mutational and phylogenetic approaches to establish the correct tertiary structure. To this end, the phage introns are well suited because they have by far the smallest core structures of any of the group I self-splicing introns. By deleting the ORF loop, and truncating the intron at

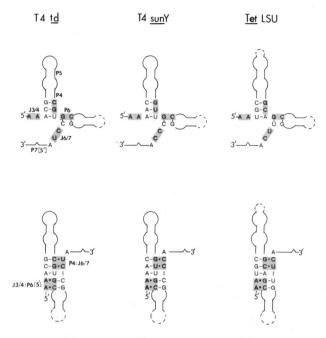

Figure 5 Triple-helical scaffold. Secondary structures (top) show elements (shaded) proposed to form the triple helix (bottom). Dots represent proposed non-Watson-Crick tertiary interactions (59).

the base of P9, a 212-nt *td* ribozyme, which is fully active at the 5' SS, has been generated (75). The *sun*Y intron, which naturally lacks a P2 element, has been similarly reduced by ORF- and 3'-deletions to an active ribozyme of 196 nt (32, 87). Deletion of yet other elements (e.g. P9, P7.1, and P7.2) results in functionally compromised introns with greater than ten-fold reductions in ribozyme activity (32; J. L. G. Salvo & M. B. Belfort, unpublished data). The conformational heterogeneity of such molecules is likely to be greater than that of fully active introns, limiting their usefulness for physical studies. Selectable phenotypes, as exist for the *td* ribozyme (4), may be useful on one hand to generate smaller active mini-introns. On the other hand, the ability to subject larger molecules to physical analyses is anticipated, and these studies should combine with genetic and phylogenetic approaches to solve the 3-D structure of these RNA enzymes.

REGULATORY POTENTIAL OF PHAGE INTRONS

The discovery of introns in phage led to speculation that they may act as control elements. Phage introns were first viewed as devices capable of facilitating expression and temporal regulation of multiple products of the same gene (Figure 1C), given the coupling of transcription, splicing, and translation (6, 22). Second, when group I introns were localized to genes involved in nucleotide metabolism, a global regulatory scheme was proposed in which introns were envisaged as physiological sensors that respond to the concentrations of guanosine, as well as to guanosine derivatives that inhibit splicing (40). Third, the unusual sequence arrangements at the 5' and 3' ends of the ORFs suggested their regulatory potential. Indeed, obstruction of ORF translation from the pre-mRNA appears to serve the dual role of delaying ORF (endonuclease) expression until more opportune times late in infection (3), and of preserving splicing at early times, to allow timely expression of functions required for DNA synthesis (41). Splicing at early times is ensured by preventing ribosomes from approaching the 3' end of the ORF, which overlaps RNA core elements (78; Figure 1C). This was demonstrated by introducing mutations into the sunY stem-loop that precedes the ORF, thereby exposing the ribosome binding site—splicing of sunY pre-mRNA was inhibited in direct relationship to the efficiency of translation of the ORF (M.-Q. Xu & D. A. Shub, in preparation). Whether under some unusual physiological circumstance ORF translation is prematurely and deliberately activated (e.g. see refs. 55, 56) to attenuate splicing is a matter of conjecture.

Despite these attractive regulatory models, it must be recognized that intronless variants of T4 are viable and, indeed, phenotypically indistinguishable from their intron-containing counterparts (7; D. A. Shub, personal communication), casting in doubt a regulatory role for introns under standard laboratory conditions. Whether introns provide a growth advantage under adverse environmental circumstances remains a topic of much speculation.

INTRON MOBILITY

The three T4 introns are not present consistently throughout the T-even phage family (69, 74; S. Eddy, L. Gold, personal communication) despite the highly conserved nature of the td, nrdB, and sunY structural genes. These observations suggested acquisition or loss of introns since the divergence of these phages from a common ancestor (69, 74). Indeed, mobility has been demonstrated for the td and sunY introns (73).

Intron mobility, as first described for the ω intron of the large rRNA gene of *Saccharomyces cerevisiae* (reviewed in 33, 70), is an active process whereby the intron becomes inserted into an allele lacking that intron (for a definition of terms see ref. 34). Although the same basic mobility phenomenon has since been experimentally demonstrated in biological systems as diverse as protist nuclei (66), plant chloroplasts (53), and bacteriophages (73), the DNA-based gene conversion event has thus far been observed only for inheritance of introns by intron-minus alleles of the same or closely related genes (i.e. a "homing" process) rather than by unrelated genetic loci (i.e. a "transposition" process). Nevertheless, the generality of the mobility phenomenon has profound implications for intron evolution, resulting in a recent explosion of reviews on intron mobility (3, 33, 42, 51, 70, 76). I focus here on T4 as a model system for the intron-transfer event, and on a comparison between the mobile introns of T4 and of eukaryotes.

The Double-Strand-Break Repair (DSBR) Model.

Like mobility of the eukaryotic introns, *td* and *sun*Y intron transfer is dependent on expression of the intact intron ORFs (7, 73), which encode site-specific endonucleases that cleave the respective intronless target genes (7, 8, 22a, 73, 85). This double-strand break is believed to initiate the recombination event, as originally proposed in the DSBR model for gene conversion in yeast (68, 82). The DSBR model as the postulated mechanism for intron transfer (24, 90) is depicted for the *td* intron in Figure 6A. Exonucleolytic digestion of the cleaved recipient and alignment of homologous exons (step 1) is followed by 3'-end invasion into the intron-containing duplex. After D-loop formation, the 3' ends act as primers in a repair process that uses the intron-containing strands as template (step 2). Resolution of the Holliday crossovers in the nonrecombinant mode results in two intron-containing duplexes (step 3).

The phage system lends itself favorably to tests of this model. Typically, recipient intron-minus phage infect a cell containing the donor intron flanked by exons on a plasmid; the ORF provides endonuclease in *cis* from the intron donor plasmid, or in *trans* from a compatible plasmid (73). In such crosses, the phage acquires the intron at 10 to 1000 times the frequency observed for endonuclease-independent homologous recombination. Manipulating this basic system resulted in the following findings, all consistent with the DSBR model: (*a*) mobility is absolutely endonuclease-dependent; (*b*) recombination is RNA-independent (i.e. occurs regardless of transcription or splicing proficiency of the intron); (*c*) intron transfer is nonreciprocal; (*d*) intron inheritance is dependent on exon homology (73). This latter finding is consistent with the hypothesis that alignment of homologous exon sequences precedes repair and is supported by recent results indicating the RecA-dependence of the gene conversion process (J. Clyman, M. B., unpublished data).

Exonucleolytic degradation, proposed to precede 3'-end invasion, is evidenced by the coinheritance of flanking exon sequences. In both pro- and eukaryotic systems coconversion frequencies are highest for sequences flanking the intron, and decrease with distance from the intron (7, 48, 66, 84, 90). Interestingly, for the *td* intron, polymorphic exon sequences within 24 nt 5' to the intron are coinherited with a frequency of 100% (7). These results are

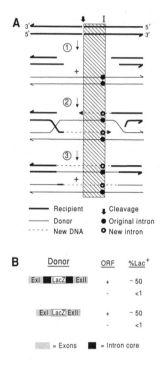

A

Recipient ——— | Cleavage
Donor ——— ● Original intron
New DNA ----- ○ New intron

B   Donor   ORF   %Lac⁺

| ExI | LacZ | ExII | | + | ~50 |
| | | | | - | <1 |

| ExI | LacZ | ExII | | + | ~50 |
| | | | | - | <1 |

▨ = Exons   ■ = Intron core

Figure 6   A. The double-strand-break repair (DSBR) model for intron inheritance. The DSBR model of Szostak et al. (82) is adapted here to the *td* intron, with an eccentric 5' cut site (heavy arrow) upstream of the intron insertion site (I) in the phage recipient. This break is enlarged to a gap, with exonucleolytic degradation extending past the intron insertion site (step 1). After alignment of donor and recipient, 3'-end invasion into the intron-containing donor molecule initiates D-loop formation. Both 3' ends act as primers for donor-directed repair synthesis (step 2). Following ligation, and resolution of the double Holliday cross-overs (only the non-recombinant mode is shown), each duplex harbors an identical copy of the intervening sequence (step 3). The hatching designates obligatory coconversion of exon sequences between the cleavage site and the intron. B. ORF-dependent, core-independent gene conversion. Inheritance of foreign sequences (*lacZ*) from a plasmid donor by intronless recipient phage is expressed as the percentage of Lac⁺ phage among the progeny. The *lacZ* sequences are inherited with equal efficiency when situated within the intron core (top) as when flanked directly by exons (bottom), when the ORF is provided *in trans* (8). Similar results were obtained with *kan*ᴿ as the foreign marker gene (8).

again in accord with the DSBR model, when viewed in the context of the recently mapped *td*-endonuclease cleavage site, located 23–26 nt 5' of the intron (8, 22a; Figure 7). Thus, degradation must proceed from the upstream cleavage site to beyond the intron insertion point for a successful intron transfer event, resulting in obligatory coconversion of exon sequences between the cleavage and insertion sites (Figure 6A).

According to the model, the DNA that encodes the catalytic core of the intron is a passive element in the recombination event, with the sole function of serving as a template for DNA synthesis (Figure 6A). To test this hypothesis, inheritance of foreign sequences (*lacZ* or *Kan*[R]), either contained within the *td* intron or flanked by *td* exons in the absence of intron sequences, was compared. Indeed, foreign sequences were transferred efficiently in an endonuclease-dependent manner regardless of the presence or absence of intron sequences (8; Figure 6B). This result provides strong support for the DSBR model for intron inheritance, while it argues that the ORFs rather than the introns themselves are the primordial mobile elements (70).

Endonuclease-Homing Site Interactions

The homing site of the intron recipient, as defined by Dujon et al (33, 34), comprises the intron insertion site, the endonuclease recognition sequences, and cleavage site. For the *S. cerevisiae* mitochondrial ω and al4α endonucleases (I-SceI and I-SceII, respectively) and the *P. polycephalum* nuclear intron endonuclease (I-PpoI) these sites overlap (24, 30, 65, 84): the long, asymmetric recognition sequence (up to 18 nt) is centered around the insertion site, which is within a single nucleotide of the cleavage site (Figure 7). These highly specific enzymes cleave to generate 3'-OH and 5'-PO_4 ends, with 4-nt 3' extensions. Although the *td* and *sunY* endonucleases (I-TevI and I-TevII, respectively) also yield 3'-OHs and 5'-PO_4s and recognize asymmetric sequences spanning the insertion site, they are sufficiently different to constitute a distinct class of intron-encoded endonucleases (8). First, although both fungal endonucleases have characteristic dodecapeptide (LAGLI-DADG) motifs (46, 61), these are absent from the phage enzymes. Interestingly, I-TevI shares a motif, GIY-N_{10-11}-YIG, with the ORF products of several filamentous fungal introns, some of which are thought to be mobile (26, 58). Second, the phage enzymes appear to have a more relaxed specificity, based on sequence polymorphisms that are tolerated in the homing site (7, 73, 74; Figure 7) and on cleavage promiscuity (D. Bell-Pedersen, M. B., unpublished data). Third, and most strikingly, the cleavage sites are remote from the insertion sites, with I-TevI cleaving 23–26 nt 5' of the insertion site and I-TevII cleaving 13–15 nt 3' (8, 22a; Figure 7). Fourth, I-TevI leaves a 2- or 3-nt 3' extension, and I-TevII a 2-nt 3' extension (Figure 7).

The properties of these phage nucleases raise fascinating questions about their interactions with the DNA target. It will be of interest to determine how these enzymes of ca 30 kd span the residues between their recognition and cleavage sites. Whether they do so in a distance-dependent or sequence-specific manner, critical contacts between the endonucleases and the DNA must be independent of bulky side groups in the major groove, since both unmodified plasmid DNA and glucosylated-hydroxymethylated-C-containing phage DNAs are efficiently cleaved.

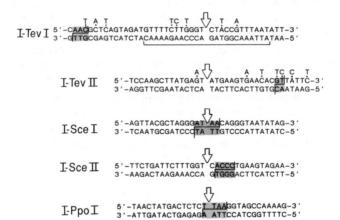

Figure 7 Homing sites. The endonuclease cleavage sites (shaded) and intron insertion sites (arrows) are shown for the phage T4 (I-TevI, *td*; I-TevII, *sun*Y), *S. cerevisiae* (I-SceI, ω; I-SceII, aI4α) and *P. polycephalum* (I-PpoI, rRNA) homing sites (8, 22a, 24, 30, 65, 84). Although the I-TevI cleavages are shown to generate 3-nt extensions, under some experimental conditions the bottom strand is cleaved one nucleotide further 5', to yield 2-nt extensions (8, 22a). Variant nucleotides consistent with efficient intron inheritance, which were either introduced by mutation (above dashes) or are naturally occurring in T2 (above dots), are indicated for the phage homing sites (7, 73, 74). The bar below the *td* sequence demarcates the extent of a synthetic duplex that is specifically recognized by I-TevI (D. Bell-Pedersen, S. Quirk & M.B. in preparation).

EVOLUTIONARY CONSIDERATIONS

Evolution of Mobile Introns

One popular view considers the catalytic RNAs as the primordial elements and the ORFs as more recent additions to the intron (51, 64, 70). The demonstration that the *td* endonuclease can promote inheritance of nonintron sequences contained within homologous exons (8) supports the contention that it is indeed the ORF rather than the intron itself that is the mobile progenitor (70). In such a scenario, the invasive ORFs are maintained within those sequences that ensure viability. Thus, a self-splicing intron provides safe haven for the endonuclease and preserves genetic function, while the endonuclease promotes propagation of the intron core—through this symbiotic relationship the composite intron is perpetuated.

ORF invasion provides a rational basis for very similar catalytic core elements harboring ORFs that are unrelated in sequence and position (10). Examples are provided by the T4 introns (78) and by the *Neurospora* mitochondrial intron NDI, which contains dissimilar ORFs at different positions in closely related species (64). However, the ability of the intron-encoded site-specific endonucleases to then recognize and cleave intronless variants of the intron-containing alleles presents an enigma. A speculative model to account for the ability of the endonuclease to recognize an intronless target to initiate the homing process is presented in Figure 8. The ORF invades the intron, with the success of colonization being dependent on the preservation of some degree of splicing proficiency (step 1). Evolution toward efficient splicing may follow as necessary (e.g. by adaptation of RNA structure elements, or acquisition of maturase-type function (10, 19) by the endonuclease—the latter scenario could account for latent maturase activity of I-SceII (30, 84). Then, based on the assumption that intron transposition can occur by the same basic mechanism as intron homing (33), albeit at extremely low frequency, the endonuclease recognizes and cleaves a natural foreign target sequence, stimulating a rare repair event that results in translocation of the composite intron (step 2). Maintenance of splicing function (i.e. intact P1 and perhaps P10 elements) would be ensured by mechanisms such as exon coconversion compatible with gene function at the new locus, and coadaptation of intron and exon sequences (3, 7, 33). Translocation would position the intron appropriately for the endonuclease to cleave intronless variants at the new locus, thereby promoting efficient homing (step 3). The essence of this model is that the homing introns in modern genomes are translocated descendants of introns from sites at which they acquired endonuclease ORFs.

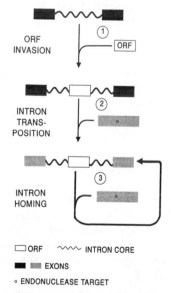

Figure 8   Model for the evolution of mobile introns. The model is described in the text. Step 3 is depicted with a heavy arrow to emphasize the efficiency of homing relative to ORF invasion (step 1) and intron transposition (step 2).

An alternative hypothesis is that after invading the intron (step 1), the endonuclease itself evolves to recognize the intronless target. While both hypotheses are difficult to evaluate experimentally, the phage system is more amenable to tests of the intron transposition model. The phage enzymes have already been shown to cleave at secondary sites in foreign sequence contexts (85; D. Bell-Pedersen, M. B., unpublished data). Defining the limits of exon homology required for intron inheritance via the DSBR pathway will further address the feasibility of ORF-mediated transposition.

Introns as "Recombinogens"

In view of the infectious nature of the introns and because no obvious selective advantage is conferred by their presence, it can be argued that the introns are simply parasitic elements. However, the persistence of the introns despite streamlining of prokaryotic genomes (27) and the degree to which the introns have become adapted to the phage and integrated into the phages' regulatory networks (e.g. ORF codon usage (78), gp 55-dependent ORF expression (41)), suggest that introns are somehow beneficial to their phage hosts (3). Two recent lines of evidence indicate that introns provide homology that promotes ectopic recombination events. First, sequences spontaneously deleted by single crossovers between homologous stretches of the closely linked *td* and *nrd*B introns resulted in the formation of new exon combinations (43). Second, spontaneous exchange of sequences between the unlinked *td* and *sun*Y introns, by double crossovers between homologous regions of the introns, resulted in inter-intron sequence translocation (9). These two observations provide a possible raison d'être for the introns, since they indicate that the introns increase the opportunities for genetic rearrangements in the T-even phages. Enhanced genetic variability could increase the adaptability of the phages to new environments and promote their evolutionary fitness in the wild, without necessarily imparting distinctive laboratory phenotypes.

FUTURE PROSPECTS

Important issues related to RNA and DNA transactions of the introns as well as to intron evolution remain to be addressed. Beyond the self-splicing and physical studies with the phage mini-introns lies the question of whether intron function in vivo is dictated solely by the intrinsic properties of the RNA, or whether accessory factors, encoded by either phage or host, serve to modulate RNA catalysis. The well-characterized *E. coli*/phage system is highly appropriate for exploring extragenic functions that might affect splicing.

Likewise, the phage genetic system promises to aid in defining the gene conversion process associated with intron inheritance. Such issues as recombination mechanism and the role of recombination, replication, and repair functions can be readily addressed. Additionally, endonuclease-DNA interactions and the precise dependency of the recombination process on exon homology can be determined.

Many questions relating to intron evolution are unanswered. Can introns transpose by the basic DSBR mechanism, rather than by an RNA-mediated process (86)? Do introns confer a selective advantage? Might enhancing the opportunities for recombination extend beyond the intron cores providing homologies that promote genetic exchange? Might the endonucleases confer advantage by further boosting an already robust T4 recombination system (63), by generating recombinogenic DNA lesions (67, 81) at secondary sites? Finally, how is "intron balance" (33) maintained and the rampant spread of introns prevented? Although endonuclease promiscuity is likely to be offset by the requirements for exon homology and splicing function in the new environment, the possibility of intron loss should also be considered. Given the precedent in fungal mitochondria for perfect intron deletion via reverse transcription of mRNA and reinsertion into the genome (reviewed in 33, 76), and the discovery of reverse transcriptase in *E. coli* (52, 54), precise intron loss from phage genomes is feasible. Although many evolutionary conundrums are difficult to solve, the phages lend themselves favorably to experimental tests of some of these issues.

ACKNOWLEDGMENTS

I thank Debbie Bell-Pedersen, Mary Bryk, Jonathan Clyman, François Michel, Susan Quirk, Jill Salvo, Renee Schroeder, and David Shub for their helpful comments on the MS and Renee Schroeder for stimulating discussions. I also thank Sean Eddy, Larry Gold, François Michel, David Shub, Britt-Marie Sjöberg and Jack Szostak for providing information prior to publication. Work in my laboratory is supported by grants from NIH (GM39422 and GM44844) and NSF (DMB8502961).

Literature Cited

1. Been, M. D., Cech, T. R. 1985. Sites of circularization of the *Tetrahymena* rRNA IVS are determined by sequence and influenced by position and secondary structure. *Nucleic Acids Res.* 13:8389–408

2. Been, M. D., Cech, T. R. 1986. One binding site determines sequence specificity of Tetrahymena pre-rRNA self-splicing, trans-splicing, and RNA enzyme activity. *Cell* 47:207–16

3. Belfort, M. 1989. Bacteriophage introns: Parasites within parasites? *Trends Genet.* 5:209–16

4. Belfort, M., Chandry, P. S., Pedersen-

Lane, J. 1987. Genetic delineation of functional components of the group I intron in the phage T4 *td* gene. *Cold Spring Harbor Symp. Quant. Biol.* 52:181–92

5. Belfort, M., Ehrenman, K., Chandry, P. S. 1990. Genetic and molecular analysis of RNA splicing in *Escherichia coli*. *Methods Enzymol.* 181:521–39

6. Belfort, M., Pedersen-Lane, J., Ehrenman, K., Chu, F. K., Maley, G. F., et al. 1986. RNA splicing and *in vivo* expression of the intron-containing *td* gene of bacteriophage T4. *Gene* 41:93–102

7. Bell-Pedersen, D., Quirk, S., Aubrey,

M., Belfort, M. 1989. A site-specific endonuclease and co-conversion of flanking exons associated with the mobile *td* intron of phage T4. *Gene* 82:119–26

8. Bell-Pedersen, D., Quirk, S., Clyman, J., Belfort, M. 1990. Intron mobility in phage T4 is dependent upon a distinctive class of endonucleases and independent of DNA sequences encoding the intron core: mechanistic and evolutionary implications. *Nucleic Acids Res.* 18:3763–70

9. Bryk, M., Belfort, M. 1990. Spontaneous shuffling of domains between introns of phage T4. *Nature.* 346:394–96

10. Burke, J. M. 1988. Molecular genetics of group I introns: RNA structures and protein factors required for splicing—a review. *Gene* 73:273–94

11. Burke, J. M. 1989. Selection of the 3' -splice site in group I introns. *FEBS Lett.* 250:129–33

12. Burke, J. M., Belfort, M., Cech, T. R., Davies, R. W., Shweyen, R. J., et al. 1987. Structural conventions for group I introns. *Nucleic Acids Res.* 15:7217–21

13. Burke, J. M., Esherick, J., Burfeind, W. R., King, J. L. 1990. A 3' splice site-binding sequence within the catalytic core of a group I intron. *Nature* 344:80–82

14. Chandry, P. S., Belfort, M. 1987. Activation of a cryptic 5' splice site in the upstream exon of the phage T4 *td* transcript: Exon context, missplicing, and mRNA deletion in a fidelity mutant. *Genes Devel.* 1:1028–37

15. Cech, T. R. 1983. RNA splicing: Three themes with variations. *Cell* 34:713–16

16. Cech, T. R. 1986. The generality of self-splicing RNA: Relationship to nuclear mRNA splicing. *Cell* 44:207–10

17. Cech, T. R. 1987. The chemistry of self-splicing RNA and RNA enzymes. *Science* 236:1532–39

18. Cech, T. R. 1988. Conserved sequence and structures of group I introns: building an active site for RNA catalysis—a review. *Gene* 73:259–71

19. Cech, T. R. 1990. Self-splicing of group I introns. *Annu. Rev. Biochem.* 59:543–68

20. Cech, T. R., Bass, B. L. 1986. Biological catalysis by RNA. *Annu. Rev. Biochem.* 55:599–629

21. Chu, F. K., Maley, G. F., Maley, F. 1987. Mechanism and requirements of *in vitro* RNA splicing of the primary transcript from the T4 bacteriophage thymidylate synthase gene. *Biochemistry* 26:3050–57

22. Chu, F. K., Maley, G. F., Maley, F., Belfort, M. 1984. An intervening sequence in the thymidylate synthase gene of bacteriophage T4. *Proc. Natl. Acad. Sci. USA* 81:3049–53

22a. Chu, F. K., Maley, G., Pedersen-Lane, J., Wang, A-M, Maley, F. 1990. Characterization of the restriction site of a prokaryotic intron-encoded endonuclease. *Proc. Natl. Acad. Sci. USA* 87: 3574–78

23. Chu, F. K., Maley, G. F., West, D. K., Belfort, M., Maley, F. 1986. Characterization of the intron in the phage T4 thymidylate synthase gene and evidence for its self-excision from the primary transcript. *Cell* 45:157–66

24. Colleaux, L., D'Auriol, L., Galibert, F., Dujon, B. 1988. Recognition and cleavage site of the intron-encoded omega transposase. *Proc. Natl. Acad. Sci. USA* 85:6022–26

25. Cummings, D. J., Michel, F., McNally, K. L. 1989. DNA sequence analysis of the 24.5 kilobase pair cytochrome oxidase subunit I mitochondrial gene from *Podospora anserina:* a gene with sixteen introns. *Curr. Genet.* 16:381–406

26. Cummings, D. J., Michel, F., McNally, K. L. 1989. DNA sequence analysis of the apocytochrome *b* gene of *Podospora anserina:* a new family of intronic open reading frame. *Curr. Genet.* 16:407–18

27. Darnell, J. E., Doolittle, W. F. 1988. Speculations on the early course of evolution. *Proc. Natl. Acad. Sci. USA* 83:1271–75

28. Davies, R. W., Waring, R. B., Ray, J. A., Brown, T. A., Scazzocchio, C. 1982. Making ends meet: A model for RNA splicing in fungal mitochondria. *Nature* 300:719–24

29. Davies, R. W., Waring, R. B., Towner, P. 1987. Internal guide sequence and reaction specificity of group I self-splicing introns. *Cold Spring Harbor Symp. Quant. Biol.* 52:165–72

30. Delahodde, A., Goguel, V., Becam, A. M., Creusot, F., Perea, J., et al. 1989. Site-specific DNA endonuclease and RNA maturase activities of two homologous intron encoded proteins from yeast mitochondria. *Cell* 56:431–41

31. Doudna, J. A., Cormack, B. P., Szostak, J. W. 1989. RNA structure, not sequence, determines the 5' splice-site specificity of a group I intron. *Proc. Natl. Acad. Sci. USA* 86:7402–6

32. Doudna, J. A., Szostak, J. W. 1989. Miniribozymes, small derivatives of the *sun*Y intron, are catalytically active. *Mol. Cell. Biol.* 9:5480–83

33. Dujon, B. 1989. Group I introns as

mobile genetic elements: facts and mechanistic speculations—a review. *Gene* 82:91–114

34. Dujon, B., Belfort, M., Butow, R. A., Jacq, C., Lemieux, C., et al. 1989. Mobile introns: definition of terms and recommended nomenclature. *Gene* 82: 115–18

35. Deleted in proof

36. Ehrenman, K., Pedersen-Lane, J., West, D., Herman, R., Maley, F., Belfort, M. 1986. Processing of phage T4 *td*-encoded RNA is analogous to the eukaryotic group I splicing pathway. *Proc. Natl. Acad. Sci. USA* 83:5875–79

37. Ehrenman, K., Schroeder, R., Chandry, P. S., Hall, D. H., Belfort, M. 1989. Sequence specificity of the P6 pairing for splicing of the group I *td* intron of phage T4. *Nucleic Acids Res.* 17:9147–63

38. Gold, L. 1988. Posttranscriptional regulatory mechanisms in *Escherichia coli. Annu. Rev. Biochem.* 57:199–233

39. Goodrich, H. A., Gott, J. M., Xu, M. Q., Scarlato, V., Shub, D. A. A group I intron in *Bacillus subtilis* bacteriophage SPO1. 1988. In *Molecular Biology of RNA*, ed. T. R. Cech, pp. 59–66. New York: Liss

40. Gott, J. M., Shub, D. A., Belfort, M. 1986. Multiple self-splicing introns in bacteriophage T4: evidence from autocatalytic GTP labeling of RNA *in vitro. Cell* 47:81–87

41. Gott, J. M., Zeeh, A., Bell-Pedersen, D., Ehrenman, K., Belfort, M., Shub, D. A. 1988. Genes within genes: Independent expression of phage T4 intron open reading frames and the genes in which they reside. *Genes Devel.* 2: 1791–99

42. Green, M. R. 1988. Mobile RNA catalysts. *Nature* 336:716–18

43. Hall, D. H., Liu, Y., Shub, D. A. 1989. Exon shuffling by recombination between self-splicing introns of bacteriophage T4. *Nature* 340:574–76

44. Hall, D. H., Povinelli, C. M., Ehrenman, K., Pedersen-Lane, J., Chu, F., Belfort, M. 1987. Two domains for splicing in the intron of the phage T4 thymidylate synthase (*td*) gene established by non-directed mutagenesis. *Cell* 48:63–71

45. Huang, W. M., Ao, S.-K., Casjens, S., Orlandi, R., Zeikus, R., et al. 1988. A persistent untranslated sequence within bacteriophage T4 DNA topoisomerase gene 60. *Science* 239:1005–12

46. Hensgens, L. A. M., Bonen, L., De Haan, M., van der Horst, G., Grivell, L. A. 1983. Two intron sequences in yeast mitochondrial COX1 gene: Homology among URF-containing introns and strain-dependent variation in flanking exons. *Cell* 32:379–89

47. Hicke, B. J., Christian, E. L., Yarus, M. 1989. Stereoselective arginine binding is a phylogenetically conserved property of group I self-splicing RNAs. *EMBO J.* 8:3843–51

48. Jacquier, A., Dujon, B. 1985. An intron-encoded protein is active in a gene conversion process that spreads an intron into a mitochondrial gene. *Cell* 41:383–94

49. Kassavetis, G. A., Geiduschek, E. P. 1984. Defining a bacteriophage T4 late promoter: Bacteriophage T4 gene 55 protein suffices for directing late promoter recognition. *Proc. Natl. Acad. Sci. USA* 81:5101–5

50. Kim, S.-H., Cech, T. R. 1987. Three-dimensional model of the active site of the self-splicing rRNA precursor of *Tetrahymena. Proc. Natl. Acad. Sci. USA* 84:8788–92

51. Lambowitz, A. M. 1989. Infectious introns. *Cell* 56:323–26

52. Lampson, B. C., Sun, J., Hsu, M.-Y., Vallejo-Ramirez, J., Inouye, S., Inouye, M. 1989. Reverse transcriptase in a clinical strain of *Escherichia coli:* Production of branched RNA-linked msDNA. *Science* 243:1033–38

53. Lemieux, B., Turmel, M., Lemieux, C. 1988. Unidirectional gene conversions in the chloroplast of *Chlamydomonas* interspecific hybrids. *Mol. Gen. Genet.* 212:48–55

54. Lim, D., Maas, W. K. 1989. Reverse transcriptase-dependent synthesis of a covalently linked, branched DNA-RNA compound in *E. coli* B. *Cell* 56:891–904

55. Macdonald, P. M., Kutter, E., Mosig, G. 1984. Regulation of a bacteriophage T4 late gene, *soc*, which maps in an early region. *Genetics* 196:17–27

56. McPheeters, D. S., Christensen, A., Young, E. T., Stormo, G., Gold, L. 1986. Translational regulation of expression of the bacteriophage T4 lysozyme gene. *Nucleic Acids Res.* 14:5813–26

57. Michel, F., Dujon, B. 1983. Conservation of RNA secondary structures in two intron families including mitochondrial-, chloroplast- and nuclear-encoded members. *EMBO J.* 2:33–38

58. Michel, F., Dujon, B. 1986. Genetic exchanges between bacteriophage T4 and filamentous fungi? *Cell* 46:323

59. Michel, F., Ellington, A. D., Couture,

S., Cherry, M., Szostak, J. W. 1990. Phylogenetic and genetic evidence for base-triples in the catalytic domain of group I introns. *Nature*. In press

60. Michel, F., Hanna, M., Green, R., Bartel, D. P., Szostak, J. W. 1989. The guanosine binding site of the *Tetrahymena* ribozyme. *Nature* 342:391–95

61. Michel, F., Jacquier, A., Dujon, B. 1982. Comparison of fungal mitochondrial introns reveals extensive homologies in RNA secondary structure. *Biochimie* 64:867–81

62. Michel, F., Netter, P., Xu, M.-Q., Shub, D. A. 1990. Mechanism of 3' splice site selection by the catalytic core of the *sun*Y intron of bacteriophage T4: the role of a novel base pairing interaction in group I introns. *Genes Devel.* 4:777–88

63. Mosig, G. 1987. The essential role of recombination in phage T4 growth. *Annu. Rev. Genet.* 21:347–71

64. Mote, E. M., Collins, R. A. 1988. Independent evolution of structural and coding regions in a *Neurospora* mitochondrial intron. *Nature* 332:654–56

65. Muscarella, D.E., Ellison, E. L., Ruoff, B. M., Vogt, V. M. 1990. Characterization of *I-Ppo*, an intron-encoded endonuclease that mediates homing of a group I intron in the rDNA of *Physarum polycephalum*. *Mol. Cell. Biol.* 10:3386–96

66. Muscarella, D. E., Vogt, V. M. 1989. A mobile group I intron in the nuclear rDNA of Physarum polycephalum. *Cell* 56:443–54

67. Nickoloff, J. A., Singer, J. D., Hoekstra, M. F., Heffron, F. 1989. Double-strand breaks stimulate alternative mechanisms of recombination repair. *J. Mol. Biol.* 207:527–41

68. Orr-Weaver, T. L., Szostak, J. W., Rothstein, R. J. 1981. Yeast transformation: A model system for the study of recombination. *Proc. Natl. Acad. Sci. USA* 78:6354–58

69. Pedersen-Lane, J., Belfort, M. 1987. Variable occurrence of the *nrd*B gene in the T-even phages suggests intron mobility. *Science* 237:182–84

70. Perlman, P. S., Butow, R. A. 1989. Mobile introns and intron-encoded proteins. *Science* 246:1106–9

71. Price, J. V., Cech, T. R. 1988. Determinants of the 3' splice site for self-splicing of the *Tetrahymena* pre-rRNA. *Genes Devel.* 2:1439–47

72. Price, J. V., Engberg, J., Cech, T. R. 1987. 5' exon requirement for self-splicing of the *Tetrahymena thermophila* pre-ribosomal RNA and identification of a cryptic 5' splice site in the 3' exon. *J. Mol. Biol.* 196:49–60

73. Quirk, S. M., Bell-Pedersen, D., Belfort, M. 1989 Intron mobility in the T-even phages: high frequency inheritance of group I introns promoted by intron open reading frames. *Cell* 56: 455–65

74. Quirk, S. M., Bell-Pedersen, D., Tomaschewski, J., Ruger, W., Belfort, M. 1989. The inconsistent distribution of introns in the T-even phages indicates recent genetic exchanges. *Nucleic Acids Res.* 17:301–15

75. Salvo, J. L. G., Coetzee, T., Belfort, M. 1990. Deletion tolerance and *trans*-splicing of the T4 *td* intron: Analysis of the P6-L6a region. *J. Mol. Biol.* 211:537–49

76. Scazzocchio, C. 1989. Group I introns: do they only go home? *Trends Genet.* 5:168–72

77. Deleted in proof.

78. Shub, D. A., Gott, J. M., Xu, M.-Q., Lang, B. F., Michel, F., et al. 1988. Structural conservation among three homologous introns of phage T4 and the group I introns of eukaryotes. *Proc. Natl. Acad. Sci. USA* 85:1151–55

79. Shub, D. A., Xu, M.-Q., Gott, J. M., Zeeh, A., Wilson, L. D. 1987. A family of autocatalytic group I introns in bacteriophage T4. *Cold Spring Harbor Symp. Quant. Biol.* 52:193–200

80. Sjöberg, B.-M., Hahne, S., Mathews, C. Z., Mathews, C. K., Rand, K. N., Gait, M. J. 1986. The bacteriophage T4 gene for the small subunit of ribonucleotide reductase contains an intron. *EMBO J.* 5:2031–36

81. Stahl, F. W. 1986. Roles of double-strand breaks in generalized genetic recombination. *Prog. Nucleic Acid Res. Mol. Biol.* 33:169–94

82. Szostak, J. W., Orr-Weaver, T. L., Rothstein, R. J., Stahl, F. W. 1983. The double-strand-break repair model for recombination. *Cell* 33:25–35

83. Waring, R. B., Towner, P., Minter, S. J., Davies, R. W. 1986. Splice-site selection by a self-splicing RNA of *Tetrahymena*. *Nature* 321:133–39

84. Wenzlau, J. M., Saldanha, R. J., Butow, R. A., Perlman, P. S. 1989. A latent intron-encoded maturase is also an endonuclease needed for intron mobility. *Cell* 56:421–30

85. West, D. K., Changchien, L.-M., Maley, G. F., Maley, F. 1989. Evidence that the intron open reading frame of the phage T4 *td* gene encodes a specific endonuclease. *J. Biol. Chem.* 264:10343–46

86. Woodson, S. A., Cech, T. R. 1989. Reverse self-splicing of the Tetrahymena group I intron: implication for the directionality of splicing and for intron transposition. *Cell* 57:335–45

87. Xu, M.-Q., Shub, D. A. 1989. The catalytic core of the *sun*Y intron of bacteriophage T4. *Gene* 82:77–82

88. Deleted in proof.

89. Yarus, M. 1988. A specific amino acid binding site composed of RNA. *Science* 240:1751–58

90. Zinn, A. R., Butow, R. A. 1985. Nonreciprocal exchange between alleles of the yeast mitochondrial 21S rRNA gene: kinetics and the involvement of a double-strand break. *Cell* 40:887–95

Annu. Rev. Genet. 1990. 24:387–407

GENETICS OF EARLY NEUROGENESIS IN *DROSOPHILA MELANOGASTER*

José A. Campos-Ortega and Elisabeth Knust

Institut für Entwicklungsphysiologie, Universität zu Köln, Gyrhofstrase 17, 5000 Köln 41, Federal Republic of Germany

KEY WORDS: genetics of neurogenesis, Drosophila, neurogenic genes, proneural genes

CONTENTS

INTRODUCTION

The origin of cell diversity is one of the major problems in developmental biology. Although the mechanisms controlling the direction and timing of determination and cellular differentiation are still largely unknown, considerable progress has been made in our understanding of the problem. Early

387

0066-4197/90/1215-0387$02.00

neurogenesis in insects, i.e. the processes that lead to the formation of a neural primordium, is an appropriate system in which to study the origin of cell diversity and it has received a great deal of attention. In *Drosophila,* neurogenesis is initiated by the separation of the neural progenitor cells (neuroblasts) from the neurogenic region of the ectoderm. The neuroblasts move inward from the surface to the interior of the embryo, where they build up the primordium of the central nervous system. The remaining cells of the neuroectoderm, with which the presumptive neuroblasts are intermingled prior to their segregation, take on a different developmental fate: they develop as epidermal progenitor cells (epidermoblasts) to give rise to the epidermal sheath of ventral and cephalic regions (32, 65, 69). Thus, each cell within the neuroectoderm apparently must choose between a neural and an epidermal developmental fate.

Observations from embryology (11, 32) suggest that all neuroectodermal cells initially acquire the capability to develop as neuroblasts—in the parlance of developmental biology, neurogenesis appears to be the "primary fate" of the neuroectodermal cells. Two pieces of experimental evidence indicate that interactions among the cells of the neuroectoderm are essential to subsequently deflect a substantial proportion of these cells from the primary neural into the secondary epidermal fate and thus make possible the proper segregation of both cell lineages. First, laser-ablation experiments carried out in grasshoppers have shown that the cells remaining in the neurogenic region after the neuroblasts have segregated are not firmly committed to their fate (24, 62). Under normal circumstances, these cells would develop as epidermoblasts; however, when an adjacent neuroblast is ablated they may adopt the neural fate instead. Second, results of cell transplantations in *Drosophila* show that cells may take on a given fate depending on their position, suggesting that regulatory signals pass between the cells of the neuroectoderm and commit them to either the epidermal or the neural fate (10, 64, 66). The existence of two kinds of signals, one with epidermalizing and the other with neuralizing character, has been inferred from genetic analyses in *Drosophila* (see below) and from cell-transplantation experiments.

Although these latter experiments clearly show that neuroectodermal cells interact with each other, and that cellular interactions determine the acquisition and maintenance of a given developmental fate, the experiments do not show at what precise point during the cells' decision process the crucial interactions take place. We come back to this problem below.

GENES REQUIRED FOR NEUROGENESIS IN *DROSOPHILA*

In *Drosophila,* the correct separation of neural and epidermal progenitor cells is controlled by two groups of genes, the so-called neurogenic genes (52, 53,

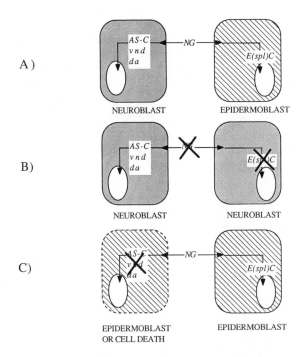

Figure 1   A)   Shows two interacting cells of the neuroectoderm in the wild-type. A large number of genes, i.e. the neurogenic genes *(NG)*, including the genes of the *E(spl)-C*, the genes of the *achaete-scute* complex *(AS-C)*, *ventral nervous system condensation defective (vnd)* and *daughterless,* encode the proteins of a regulatory signal chain that allows the cells to develop either as neuroblasts or as epidermoblasts. The functions encoded by the genes of the *AS-C, vnd,* and *daughterless* are required to regulate the genetic activity of the neuroblasts, those encoded by the various genes of the *E(spl)-C* to regulate the genetic activity of the epidermoblasts. B)   Mutation of one of the neurogenic genes results in the development of all neuroectodermal cells as neuroblasts. C)   Mutation in the genes of the *AS-C, vnd* or *daughterless* results in either the development of additional neuroectodermal cells as epidermoblasts, at the expense of neuroblasts, or in cell death.

56) on the one hand, and on the other, the proneural genes, i.e. the various members of the *achaete-scute* complex *(AS-C)*, the locus *ventral nervous system condensation defective (vnd),* and the second chromosomal *daughterless* gene (6, 16, 36–38, 78, 79; see Figure 1). In the wild-type, the neuroectoderm contains approximately 2000 cells, of which 500 will eventually choose the neural pathway, whereas the remaining 1500 will take on an epidermal fate and develop as epidermoblasts; loss of function of any of the neurogenic genes causes all the neuroectodermal cells to develop as neuroblasts, i.e. neurogenic mutants initiate neurogenesis with 2000 instead of the normal 500 neuroblasts. Consequently, the mutant embryos have a highly hyperplasic nervous system, lack ventral and cephalic epidermis and die.

Thus, the wild-type functions of the neurogenic genes are apparently required to suppress the neural fate in 1500 neuroectodermal cells and allow them to develop as epidermoblasts. Upon mutation, the suppressive effect normally exerted by the neurogenic genes on neurogenesis is relieved, which leads all the neuroectodermal cells to develop as neuroblasts.

The phenotype of loss-of-function mutations in the proneural genes is the opposite of that of neurogenic mutations: embryos lacking the genes of the *AS-C* and/or *vnd* initiate neurogenesis with less than the normal complement of 500 neuroblasts—depending on the mutant, 20–25% of all neuroblasts are missing (38). In addition, during later stages, large numbers of cells degenerate within the neural primordium of all these mutants (5, 36, 38, 78). Consequently, the fully differentiated mutant embryos die with a highly hypoplasic central nervous system. The genes of the *AS-C* and *vnd* are thus required for the commitment of the normal 500 neuroblasts. Since the populations of neuroblasts affected by mutations in different genes do not seem to overlap significantly, *AS-C, vnd,* and probably other as yet unidentified genes, may each control the development of particular sets of neuroblasts. The participation of the gene *daughterless* in neuroblast commitment and segregation is not yet clear. The complement of neuroblasts in *daughterless⁻* mutants is initially normal; however, most of these cells die early during embryonic development (38: M. Brand & J. A. Campos-Ortega, in preparation). However, *daughterless⁻* mutants completely lack the peripheral nervous system, because of defective commitment of the progenitor cells of the sensory organs (16). Interactions with the *AS-C* support the participation of *daughterless* in central neurogenesis (see below).

Poulson (56) called *Notch* (the first neurogenic gene discovered) a "neurogenic" gene, following the convention in *Drosophila* genetics of naming a gene according to the phenotype of the mutation that led to its discovery. Other genes found with the same phenotype were also called "neurogenic" although, as discussed below, their functions are actually required for epidermal development. Since the functions of the genes of the *AS-C, vnd,* and *daughterless* are required for neural development, they have been generically called "proneural" (29, 58). Neurogenic and proneural genes are currently referred to as two different groups, and such a distinction is justified by the phenotype of their mutations: loss-of-function of any neurogenic genes causes neural hyperplasia, and neural hypoplasia in the genes of the subdivision 1B (i.e. *AS-C, vnd*). However, this distinction is in fact artificial, in that, as we discuss in the following section, the products of all these genes are apparently involved in a complex genetic network and together contribute to a single process—the separation of neural and epidermal cell progenitors.

THE NEUROGENIC GENES ARE FUNCTIONALLY INTERRELATED

Abundant evidence from various kinds of genetic analysis supports the assumption that the neurogenic loci are functionally interrelated (13, 22, 61, 72). Due to pleiotropic expression of the genes, a wide variety of phenotypic effects has been observed in the various genotypic combinations studied. However, insofar as their participation in the segregation of neuroblasts from epidermoblasts is concerned, six of the neurogenic loci, i.e. *Notch, almondex, master mind, Delta, neuralized,* and the genes of the *Enhancer of split* complex [*E(spl)-C*] were found to be functionally linked in a chain of epistatic relationships; *big brain* was found to act independently (22). Hence, the function of each of these genes appears to be dependent on that of another member of the group and, consequently, the function of the entire chain is perturbed if any one of the links is missing (Figure 2). Epistatic, or hypostatic, relationships between two genes can be inferred from the effects of various genetic combinations on particular phenotypic traits. If increased wild-type dosage of a gene, for example *Notch,* modifies the phenotype caused by the loss-of-function of another gene, for example *neuralized,* it is likely that the former gene is regulated by the latter. Such considerations allow one to propose formal schemes like that shown in Figure 2. Even though its molecular basis has still to be established, such a scheme is very useful in the design of further experiments.

   In addition to the genetic analysis, results of cell transplantations support the notion that mutations in the different neurogenic genes affect specific parts of a signal chain (67). Mutations in *Notch, almondex, big brain, master mind, Delta,* or *neuralized* behave as if the gene products were required to produce and/or relay a regulatory signal that leads to epidermogenesis (Figure 2). When neuroectodermal cells lacking any of these gene products are trans-

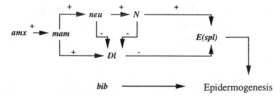

Figure 2   Illustrates the network of relationships of the neurogenic genes. This diagram is based on formal arguments derived from transmission genetics and gives no indication of the molecular level at which the proposed interactions take place. Positive or negative signs reflect the kind of functional influences, i.e. repression or activation, assumed to be exerted by one gene product upon the next one. The data suggest that *E(spl)* provides an epidermalizing signal to the neuroectodermal cells. Although *bib* takes part in the process of neuroblast segregation, its function is apparently independent of the other six genes (from 22, modified).

planted into the neurogenic ectoderm of a wild-type embryo, the mutant cells do not autonomously express their mutant phenotypes; rather they develop as epidermoblasts with the same frequency as wild-type cells. This result could be interpreted to mean that the transplanted mutant cells are still capable of receiving the regulatory signal from the neighboring wild-type cells and processing it to their own genomes. In contrast, neuroectodermal cells lacking the functions of the *E(spl)-C* never adopt an epidermal fate after transplantation into the wild-type neurogenic region, suggesting that this locus is required to receive and/or process the signal.

Evidence for the nonautonomy of *Notch* gene expression derived from the transplantation experiments is still controversial. Under experimental conditions that differ from those of the transplantation assay, cells lacking the *Notch* gene are not able to take on the epidermal fate (23, 35, P. Hoppe & R. Greenspan, submitted). It is important to distinguish between autonomous versus nonautonomous expression of a gene, for the former case may indicate that the *Notch* protein acts as a receptor (see below).

Notch *and* Delta *Encode Transmembrane Proteins with EGF-Like Repeats*

The participation of *Notch* and *Delta* in cell-communication processes is strongly suggested by the primary structure of both proteins. The *Notch* locus (76) encodes a single transcript of about 10.2 kb that is developmentally regulated, but ubiquitously expressed during neurogenesis (3, 4, 34, 43, 80). The *Notch* protein, as deduced from sequences of genomic and cDNA clones (42, 77), comprises 2703 amino acids and shows features typical of a transmembrane protein: two hydrophobic domains, an amino terminal signal peptide and a membrane-spanning domain (Figure 3). Indeed, antibodies raised against different parts of the *Notch* protein confirm its location in the cell membrane (39, 41). A striking feature of the extracellular domain is a tandem array of 36 cysteine-rich motifs, each of about 40 amino acids with striking similarity to the epidermal growth factor (EGF) and to other proteins of vertebrates and invertebrates. In addition, there are three other cysteine-rich repeats, called *Notch* repeats. The putative cytoplasmic domain of about 1000 amino acids contains a putative nucleotide phosphate binding site.

By mapping restriction site polymorphisms, the *Delta* locus has been localized within about 25 kbp of genomic DNA that exhibit a high degree of transcriptional complexity. Three major developmentally regulated, largely overlapping RNAs of 5.4, 4.6, and 3.6 kb, respectively, have been identified as products of the *Delta* gene (49, 70; M. Haenlin, B. Kramatschek & J. A. Campos-Ortega, unpublished data); several other minor RNAs can be detected on Northern blots, but their relationships to *Delta* are not yet understood. Genetic analyses of *Delta* point to considerable genetic complexity

Notch protein

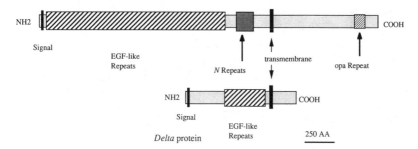

Figure 3 A comparison of the primary structures of the proteins encoded by the genes *Notch* and *Delta*. The *Notch* protein comprises 36 EGF-like repeats. *N* and opa are other repetitive sequences which are not present in the *Delta* protein. The *Delta* protein exhibits a single hydrophobic transmembrane domain and a stretch of nine EGF-like repeats (modified from 42, 49, 70, 77; B. Hassel, M. Haenlin & J. A. Campos-Ortega, unpublished data).

of the locus (2, 71); however, the current results of our transcriptional analysis (M. Haenlin, B. Kramatschek & J. A. Campos-Ortega, in preparation) indicate that *Delta* encodes a single protein (however, see 49, 50).

In contrast to *Notch*, which is ubiquitously transcribed during the segregation of the neuroblasts (34), the transcription of *Delta* is spatially regulated (50, 71). *Delta* RNA is expressed in all regions with neurogenic abilities (e.g. the neurogenic ectoderm, the anlagen of the optic lobes and of the sensory organs, etc). Initially, *Delta* is transcribed in all neuroectodermal cells, but after the segregation of neural and epidermal progenitor cells, *Delta* RNA can be detected in the neuroblasts. Sequencing of cDNA clones complementary to the largest *Delta* RNA revealed a structure of the putative *Delta* protein similar to the one proposed for the *Notch* protein: a transmembrane protein, with a hydrophobic signal sequence and a membrane-spanning domain (49, 71). The extracellular domain contains nine EGF-like repeats, arranged in tandem like those in the *Notch* protein (Figure 3).

Physical Interactions Between the Delta and Notch Proteins?

As mentioned above, there is evidence that the proteins encoded by *Delta* and *Notch* are membrane-bound proteins with EGF-like repeats in their extracellular domains. These EGF-like repeats may well represent essential parts of the cell-communication pathway by mediating direct protein-protein interactions. The participation of EGF-like repeats in specific ligand-receptor interactions has been demonstrated in several cases, for example, the binding of EGF and laminin to their corresponding receptors (31, 48). In the case of *Notch* and *Delta*, genetic mosaic analyses indicate that their products cannot diffuse over long distances (23, 35). Furthermore, results obtained from transplanting single mutant cells into wild-type hosts suggest that membrane-bound proteins are involved in cellular communication. It is, therefore, tempting to speculate that this process is mediated by the EGF-like repeats acting between adjacent cells.

This view is further supported by the data obtained from sequencing some point mutations in the *Notch* locus, namely *split* and several *Abruptex* alleles. The deduced *split* and *Abruptex* proteins each differ from the *Notch* wild-type protein by single amino acid changes in just one of the 36 EGF-like repeats (34, 40). The phenotypes associated with those alleles whose sequences have been determined suggest that the EGF-like repeats of the extracellular domain, although very similar in structure, may perform different functions. One possibility is that they interact with different proteins at different times in development.

Finally, direct physical relationships between the *Delta* and *Notch* proteins are supported by the existence of allele-specific interactions between both genes. Particular alleles of *Delta* are known that suppress the phenotype of the *split* mutation (6).

The structures of the proteins encoded by both *Notch* and *Delta* strongly suggest their participation in cell-communication processes. The *Notch* protein is ubiquitously distributed during neuroblast segregation (39, 41); this, together with other data (9), implies that *Notch* plays a permissive rather than an instructive role in the process, acting, for example, as a kind of general cell-adhesion molecule in cellular interactions. Since no data are available on the distribution of the *Delta* protein, any assessment of its possible function is difficult to make on the basis of its pattern of transcription alone. If the *Delta* protein is present in all neuroectodermal cells during lineage segregation, it would be difficult to think of it as serving instructive functions; such a distribution would be more compatible with cell-adhesive properties similar to *Notch*, which might facilitate more specific interactions. However, the distribution of the *Delta*-RNA, with its preponderant presence in neurogenic territories, both central and peripheral, is striking, and is highly suggestive of an instructive role for *Delta*.

Several Related Functions Are Encoded by the E(spl)-C

Genetic and molecular data indicate that two gene complexes, the *E(spl)-C* (44, 46, 82) and the *AS-C* (27, 29), each consisting of several functionally and structurally related genes, act at either end of the signal chain. Strikingly, both complexes exhibit extensive similarity both in their organization and the structure of the encoded gene products.

Two different aspects of the *E(spl)-C* must be distinguished. One aspect concerns the gene that gave the locus its name and its function in compound eye development; the other concerns the precise composition of the gene complex and its participation in the process of lineage segregation. The gene that gave the *E(spl)-C* its name was defined by a dominant mutation, called *E(spl)^D*, that enhances the phenotype of *split*, in particular the effects of the *split* mutation on compound eye development (45, 75). *Split* is a recessive allele of *Notch*, and the relationships of *split* and *E(spl)^D* are allele-specific; thus, we may assume that during compound eye development the proteins encoded by the two genes interact (12). Work discussed below shows that the gene *E(spl)* corresponds to the transcription unit *m8*. With respect to neuroepidermal lineage separation, the current evidence suggests that the *E(spl)-C* contributes several related, partially redundant, functions to the process (44, 46, 82; H. Schrons, E. Knust & J. A. Campos-Ortega, in preparation), although the composition of the *E(spl)-C* is not yet completely defined. The temporal and spatial expression patterns of the component functions, the presence of aberrant transcripts in mutant embryos, and the results of P element-mediated germ-line transformation experiments indicate that the transcripts *m3, m4, m5, m7, m8, m9,* and *m10* and their encoded proteins may participate in functions of the *E(spl)-C* (44, 46, 57). All these transcription units are uncovered by a deletion of 34 kbp of genomic *DNA* *(Df(3R)E(spl)^{R-A7.1}*; Figure 4), which when homozygous causes a strong but not the most severe form of the neurogenic phenotype. Since deletions

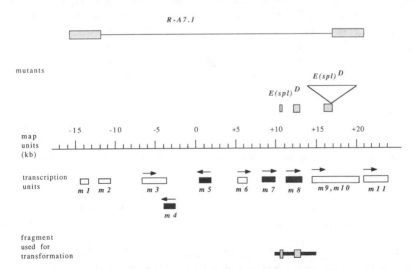

Figure 4 Molecular map of part of the *E(spl)* gene complex (modified from 46). Several mutants have been mapped to the 34–36 kb of genomic DNA deleted in *Df(3R)E(spl)^{R-A7.1}*; transcription units and direction of transcription are indicated beneath the physical map. Below is shown the *E(spl)^D* fragment used for transformation experiments including the *m8* transcription unit of the mutant (deletions are marked stippled) that causes enhancement of the *spl* phenotype (44). *m4, m5, m7,* and *m8* are the transcripts with the same expression pattern. Refer to text.

affecting other genes proximal to the $Df(3R)E(spl)^{R-A7.1}$ cause even more severe neurogenic phenotypes, the $E(spl)$-C very likely includes other genes besides those listed above (46; see below).

The Transcription Unit m8 Is E(spl)

Three major restriction site polymorphisms have been detected in the DNA of the $E(spl)^D$ mutant by Southern blot analysis (46). One is caused by the insertion of a middle repetitive element of about 5 kbp into the $m9$–$m10$ transcription unit; the other two correspond to small deletions in the $m8$ transcription unit (see Figure 4). The more distal deletion causes a reduction in the size of the $m8$ RNA (1.0 kb in the wild-type vs about 0.7 kb in the mutant) that, in addition, is more abundantly transcribed in $E(spl)^D$ embryos. In contrast, the insertion in the $m9$–$m10$ transcription unit does not change the size of these two RNAs, presumably because the insertion is located in an intron. Comparison of the sequence of the genomic DNA of the mutant $m8$ region with the corresponding sequence of the wild-type showed that the larger of the two deletions eliminates the last 56 amino acids in the putative $m8$ protein; in the $E(spl)^D$ mutant, the $m8$ protein is further modified by the addition of nine amino acids to its truncated carboxy-terminus (44).

To establish causal relationships between defects in the $m8$ transcription unit and phenotypic traits associated with the $E(spl)^D$ mutation, P element-mediated transformation experiments have been carried out using mutant and wild-type genes. Transgenic flies were created, with one or two copies of the mutant ($m8^D$) gene, in addition to two copies of the wild-type $m8$ gene. In combination with $split$, the transforming $m8^D$ gene brings about the same enhancement of the spl phenotype as the $E(spl)^D$ mutation itself. This experiment directly demonstrates that the transcription unit $m8$ and the gene $E(spl)$ are identical. Similar germ-line transformation experiments using the mutant $m9$–$m10$ transcription unit, which carries a 5-kbp middle repetitive DNA fragment inserted in an intron, show that this gene does not modify the expression of $split$, demonstrating that $m9$–$m10$ is not $E(spl)$ (44; K. Tietze & E. Knust, in preparation).

The Transcripts m4, m5, m7 and m8: A Molecular Basis For Functional Redundancy?

$m8$ encodes the gene $E(spl)$; in addition, $m8$ is one of four RNAs ($m4$, $m5$, $m7$, and $m8$) that may provide a molecular basis for the functional redundancy found by genetic analysis (82). During early stages of embryogenesis, these four transcripts exhibit nearly identical spatial distributions, as shown by in situ hybridization experiments to embryonic tissue sections (46; E. Knust, F. Grawe & J. A. Campos-Ortega, in preparation). Sequencing of genomic and cDNA clones comprising the coding region of $m4$, $m5$, $m7$, and $m8$ has shown

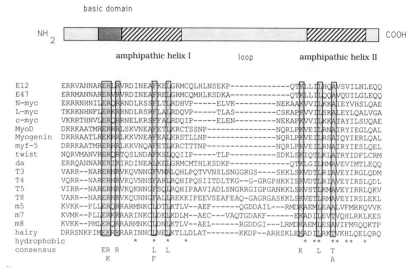

Figure 5   Shows the conserved region, including the helix-loop-helix (HLH) motif, found in many different proteins of mammals and insects. The consensus sequence is indicated below. The extent of the two putative helices, separated by a putative loop region of variable length, is indicated. A basic region is found at the amino terminal side of the proteins. Aminoacids conserved in all proteins are boxed. Refer to text for further details.

that the proteins encoded by *m5, m7,* and *m8* show a high degree of sequence similarity to each other, with three conserved domains, whereas *m4* is different (44). The amino-terminal domains of *m5, m7,* and *m8* exhibit sequence similarity to a region comprising a helix-loop-helix (HLH) motif (55) that is conserved in various proteins of vertebrates (Figure 5), among them members of the *myc* family (47, 51, 74), several proteins involved in muscle development [*MyoD1, myogenin,* and others] (7, 25, 63, 81), and two immunoglobulin enhancer-binding proteins (54). Interestingly, the same HLH motif has been found in other developmentally regulated proteins of *Drosophila,* like *twist* (68), *hairy* (60), and *extra macrochaetae* (26, 28) and, more important for our present purposes, in the proteins encoded by the transcripts *T3, T4, T5,* and *T1a* of the *AS-C* (1, 30, 73) and in the *daughterless* protein (17; see below).

Knust et al (46) pointed to the possibility that the *E(spl)-C* extends beyond the 34 kbp deleted in *Df(3R)E(spl)*$^{R-A7.1}$. At least three additional transcription units have been detected further proximal to the *Df(3R)E(spl)*$^{R-A7.1}$ interval; all exhibit expression patterns similar to those of *m4, m5, m7,* and *m8* and encode HLH proteins (E. Knust, H. Schrons, & J. A. Campos-Ortega, in preparation). Their transcription patterns and sequences strongly suggest that these transcription units are related to *m4, m5, m7,* and *m8* and may therefore be further members of the *E(spl)-C*.

The Transcripts m9 *and* m10

Both RNAs are transcribed from the same transcription unit, and P element-mediated transformation experiments have demonstrated that a genomic fragment comprising the whole coding region of *m9* and *m10* can rescue the lethality of *E(spl)* mutations that cause weak neurogenic phenotypes with incomplete penetrance. However, neither the neurogenic phenotype of the 34-kbp deletion that eliminates the 11 transcripts, or that of a smaller deletion (14 kbp, eliminating transcripts *m6* to *m9–m10*) can be rescued by this fragment (57; H. Schrons, E. Knust & J. A. Campos-Ortega, unpublished data). Since the neurogenic defect caused by these deletions cannot be restored to wild-type by the addition of just the functions encoded by *m9–m10*, the phenotype must be the result of the simultaneous lack of *m9–m10* and some of the other genes contained in the region.

The two RNAs differ only in the lengths of their 3' untranslated regions and the deduced amino acid sequence of the putative protein contains a repeated motif similar to the β-subunit of transducin, a G-protein involved in phototransduction (33).

THE *AS-C* GENES AND *DAUGHTERLESS* ARE MEMBERS OF THE SAME GENE FAMILY AND ENCODE REGULATORY PROTEINS

The *AS-C* includes four genes, *achaete*, *scute*, *lethal of scute*, and *asense* (29), the names being derived from the phenotypic effects of their mutations on bristle development and viability. The genomic DNA of the *AS-C* has been cloned and characterized by J. Modolell and colleagues (14, 15, 30, 58). The complex contains a number of transcription units, of which four, *T5*, *T4*, *T3*, and *T1a* (corresponding to *T8* in (1)) have been identified as corresponding to *achaete*, *scute*, *lethal of scute*, and *asense* functions (Figure 6). Sequence analyses have shown that the proteins encoded by the four *AS-C* genes *T3*, *T4*, *T5*, and *T1a(T8)* are similar to each other, particularly in three domains, a C-terminal acidic domain of 15 amino acids, which may be involved in transcription activation, the HLH motif mentioned above in the discussion of the *E(spl)-C* transcripts, and an N-terminal basic domain (1, 30, 73; Figure 5).

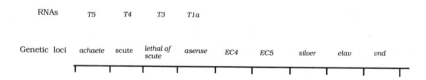

Figure 6 Genetic and transcriptional map of the subdivision 1B (using data from 20, 21, 30, 37; and K. White, personal communication).

The *daughterless* locus has recently been cloned and was found to contain a single transcription unit with two overlapping RNAs of 3.2 and 3.7 kb, respectively (16, 18). Conceptual translation of the corresponding cDNA sequences uncovers the same conserved HLH motif present in the proteins encoded by the *AS-C* transcripts *T3*, *T4*, *T5*, and *T1a*, and in the proteins encoded by the *E(spl)-C* transcripts *m5*, *m7*, and *m8*, among other genes (16; Figure 5).

Helix-Loop-Helix Proteins, Dimer Formation, And DNA Binding

The presence of the HLH motif in several developmentally regulated proteins is striking. Murre et al (54), who found it in two proteins (E_{12} and E_{47}) that bind to the $\kappa E2$ DNA *motif* in the immunglobulin kappa-chain enhancer, showed that proteins with the HLH motif are able to form dimers and bind to specific DNA sequences. Moreover, it has been shown that various members of the HLH protein family, including E_{12}, E_{47}, *MyoD*, *T3*, and *daughterless*, are able to form homo- and heterodimers that specifically bind to DNA in vitro with high affinity (55). This particularly important finding strongly supports the contention that all these proteins function in vivo as transcriptional regulators. A high degree of specificity and complexity in the regulatory functions of the corresponding genes may be achieved through the combination of different proteins to form heterodimers. Furthermore, the formation of heterodimers between the various proteins encoded by the *E(spl)-C* would also explain the functional redundancy found in the genetic analysis of the gene complex. This possibility is currently being tested experimentally.

Genetic interactions between some of the AS-C genes and *daughterless* have recently been described (21), suggesting that these genes are involved in closely related functions. Thus, the recent finding (55) that the *T3* and *daughterless* proteins may form DNA-binding heterodimers corroborates the observations of the genetic analysis.

The AS-C Genes Are Required For Neuroblast Commitment

The spatial distribution of the *T3*, *T4*, and *T5* transcripts in sections of staged embryos as shown by in situ hybridization (8, 59) is similar for all three transcripts and, with respect to *T3* and *T5* in particular, shows a high degree of correlation with the processes of neuroblast segregation and development of sensory organs and the stomatogastric nervous system, i.e. those processes in which the functions of the genes are known to be required. During early neurogenesis, the three transcripts are expressed in partially overlapping clusters of cells within the neuroectoderm. Since their domains of expression partially overlap, some neuroblasts may contain all three RNAs, and possibly their products, whereas other neuroblasts contain only one or two. This pattern of transcription is suggestive of a role for the *AS-C* genes in neuroblast commitment.

Genetic evidence supports this contention. It has been long known that the *AS-C* is required for central neural development (36, 78). Recently, Jiménez & Campos-Ortega (38) have shown that neuroblast commitment is impaired in *AS-C⁻* mutants in that 20–25% of the normal neuroblast complement do not become committed to their fates; similar findings have been made with *vnd* mutants as well. In addition, Dambly-Chaudière & Ghysen (20) describe a correspondence between the deletion of some *AS-C* genes and defects in particular subsets of sensory organs, suggesting specific roles for the *AS-C* genes during development of central and peripheral neural progenitor cells. Cabrera et al (8) proposed that the *AS-C* gene products provide the neuroblasts with specific identities, based on the particular combination of the products of the *AS-C* genes expressed in each neuroectodermal cell.

INTERACTIONS BETWEEN NEUROGENIC AND PRONEURAL GENES

Evidence has been obtained for interactions of the neurogenic genes with the *AS-C* genes and with *daughterless* (5). Double mutants for neurogenic genes and *AS-C* or *daughterless* show that the severity of the phenotype of homozygous neurogenic mutations is considerably reduced if the same genome partially or completely lacks the *AS-C* or *daughterless* functions in homo- or hemizygosity. Additionally, *AS-C⁻* mutations were found to be epistatic to the neurogenic mutations, suggesting that the function of the *AS-C* genes follows that of the neurogenic genes. Genetic interactions deduced from phenotypic changes are certainly difficult to interpret without corresponding biochemical data; however, they do suggest functional interrelationships between the genes involved. Phenotypic interactions may, for example, reflect regulatory interactions at the level of transcription or they may also reflect interactions at the protein level, e.g. heterodimer formation between the HLH motifs encoded by members of the *E(spl)-C*, the *AS-C*, and *daughterless*, similar to those between *T3* and *daughterless*, mentioned above (55).

The distribution of RNA from the *AS-C* genes in neurogenic mutants suggests that at least some of the interactions between neurogenic and *AS-C* genes are likely to involve regulation of the transcription of these genes. Changes in the pattern of transcription of the genes *lethal of scute* and *achaete* (*T3* and *T5*) have been observed in embryos carrying any of several neurogenic mutations (5). In these embryos, the early expression of *T3* and *T5* is indistinguishable from the wild-type (Figure 7). However, at the stage when the neuroblasts normally segregate, RNA from *T3* and *T5* is found in many more cells than in the wild-type. Hence, transcriptional interactions seem to

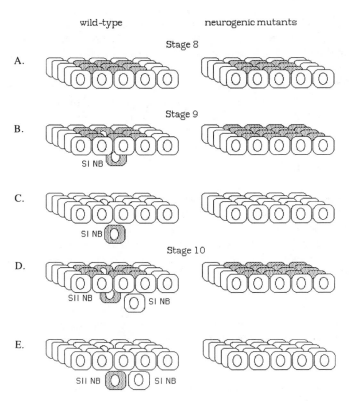

wild-type   neurogenic mutants

Stage 8

A.

Stage 9

B.

SI NB

C.

SI NB

Stage 10

D.

SII NB   SI NB

E.

SII NB   SI NB

Figure 7   Spatial distribution of *T5* transcripts in wild-type (8, 59) and neurogenic mutants (5). An array of 25 neuroectodermal cells is shown at different stages in both cases. In A), before the segregation of the neuroblasts, a cluster of these cells transcribes the *T5* RNA; in B), when the segregation of the SI neuroblasts (SI NB) takes place, one cell segregates as a SI NB and continues transcribing *T5* for some time, whereas the other cells of the group no longer contain *T5* RNA (C). At the end of this stage, in D), transcription of *T5* reappears in another similar cluster of cells, one of which will segregate as a SII NB, which continues transcribing *T5*. (E) In the neurogenic mutants (right panel), the number of cells transcribing *T5* is comparable to the wild-type in young stages, but it increases at the time when SI NB segregation normally occurs; since all neuroectodermal cells develop as NBs, no segregation of lineages occurs in the neurogenic mutants. Transcription of *T5* therefore continues for some time in a large group of neuroectodermal cells. However, temporal aspects of expression are normal.

operate, or at least to become evident, when the segregation of lineages is taking place. In the wild-type, a restriction of *T5* transcription occurs from an initial group of about nine ectodermal cells to one, or a few neuroblasts, as they segregate from the epidermoblasts (8). In neurogenic mutants, this restriction fails to occur; moreover, the total number of *T5* transcribing cells per cluster is larger than nine cells.

All these observations indicate that the neurogenic genes define the normal expression of AS-C genes in that they suppress the transcription (or the accumulation) of RNA from at least T3 and T5 in some of the neuroectodermal cells. This means that from a cluster of several neuroectodermal cells, which in a given region of the wild-type transcribe T3 and T5, only 1–2 will finally become neuroblasts. Since these are the only cells of the cluster in which RNA from those gene is still present, a causal relationship between the presence of RNA of the T3 and T5 genes in a given cell and its development as a neuroblast is very likely.

CONCLUSIONS

One important conclusion to be drawn from the data above is that the proteins encoded by the neurogenic genes, the genes of the AS-C, vnd, and daughterless are functionally interrelated, forming a regulatory network that permits the cells to take on the neural or the epidermal fate. Extensive evidence derived from cell transplantations suggests that cellular interactions are mediated by direct contact between the neuroectodermal cells and not by diffusing substances; such contact is apparently required for regulatory signals to be passed from one cell to another. However, we repeat that these experiments do not allow inferences about when the salient cellular interactions take place. Several pieces of evidence suggest that at least the decision to adopt the neural developmental fate may be taken in a cell-autonomous manner, and that cell communication is subsequently required to permit epidermal development. The best evidence to support this contention is provided by the neurogenic mutants. If the cell-communication process is impaired, as apparently occurs in neurogenic mutants, all cells of the neuroectoderm develop as neuroblasts, suggesting that the primary neurogenic fate of these cells cannot be suppressed. The pattern of transcription of AS-C genes in neurogenic mutants (6) strongly suggests that the cells that normally develop as epidermoblasts are misrouted into neurogenesis because they continue to express the AS-C genes.

The molecular structures of the AS-C and daughterless proteins, together with the phenotypes of their mutants, immediately suggest that these proteins carry out the regulatory functions necessary for development of the neuroblasts. DNA-binding properties have been demonstrated recently for T3 and daughterless (55) and the same is possible for T4, T5, and T1a. Since the HLH motif permits the formation of heterodimers, which bind to DNA with higher affinity than the corresponding homodimers (55), combinations between these proteins are possible

We propose that, at the membrane of neuroectodermal cells, the cellular interactions are mediated by the Delta and Notch proteins, probably at the EGF-like repeats present in the extracellular domains of both proteins. Since we postulate a signal chain, a very appealing possibility is that the interactions between both proteins represent relationships between ligand and receptor. Such a relationship would imply an asymmetrical distribution—or a somehow asymmetrical function—of the two proteins. The *Notch* protein is present in all neuroectodermal cells, whereas the transcription of *Delta* is topologically regulated; however, no data are available yet on the distribution of the *Delta* protein. Another possibility is that both *Delta* and *Notch* are capable of passing signals in both directions, between neighboring cells, as opposed to the unidirectional flow of information implicit in ligand-receptor relationships. Hence, we cannot decide whether the complex assumed to be formed by both proteins plays an instructive or a permissive role in the cell-communication process. The results of transplantations of mutant cells into the neuroectoderm of the wild-type argue against the notion that *Notch* acts as a receptor; however, as already mentioned, the implications of this evidence are still uncertain.

Results of the transplantation of mutant cells into the neuroectoderm of the wild-type were interpreted to mean that the *E(spl)-C* encodes functions related to the reception of the regulatory signals. Although none of the putative products from genes of the *E(spl)-C* whose structure has been determined so far resembles a receptor protein, the proteins encoded by *m5*, *m7*, and *m8*, with the DNA-binding HLH motif (44), are compatible with a function as transcriptional regulators, which fits the expectations for the *E(spl)-C* well. Thus, we assume that the proteins encoded by these latter genes, and probably by other members of the *E(spl)-C* still to be characterized, regulate the specific genetic activities of the neuroectodermal cells that enable them to develop as epidermoblasts.

Thus, we envisage neuroectodermal cell determination as the result of a delicate balance between the activity of two groups of transcriptional regulators, encoded by the proneural genes (*AS-C, daughterless,* perhaps *vnd*) for the neural fate, and by the *E(spl)-C* for the epidermal fate. A possible sequence of events, compatible with the available data, is that all cells in the neuroectoderm express a particular set of proneural genes and initially acquire the competence to develop as neuroblasts (i.e. a primary neurogenic fate). Individual cells would then become committed to the neural fate following stochastic fluctuations in the expression of these latter genes; lateral inhibition mediated by *Notch* and *Delta* would lead to activation of the genes of the *E(spl)-C* and to repression of the proneural genes, allowing development of epidermal progenitor cells.

ACKNOWLEDGMENTS

We thank Dr. Paul Hardy for constructive criticism on the manuscript. The research reported here was supported with grants of the Deutsche Forschungsgemeinschaft (DFG).

Literature Cited

1. Alonso, M. C., Cabrera, C. V. 1988. The *achaete-scute* gene complex of *Drosophila melanogaster* comprises four homologous genes. *EMBO J.* 7:2585–91
2. Alton, A. K., Fechtel, K., Terry, A. L., Meikle, S. B., Muskavitch, M. A. T. 1988. Cytogenetic definition and morphogenetic analysis of *Delta*, a gene affecting neurogenesis in *Drosophila melanogaster*. *Genetics* 118:235–45
3. Artavanis-Tsakonas, S. 1988. The molecular biology of the *Notch* locus and the fine tuning of differentiation in *Drosophila*. *Trends Genet.* 4:95–100
4. Artavanis-Tsakonas, S., Muskavitch, M. A. T., Yedvobnick, B. 1983. Molecular cloning of *Notch*, a locus affecting neurogenesis in *Drosophila melanogaster*. *Proc. Natl. Acad. Sci. USA* 80:1977–81
5. Brand, M., Campos-Ortega, J. A. 1988. Two groups of interrelated genes regulate early neurogenesis in *Drosophila melanogaster*. *Roux's Arch. Dev. Biol.* 197:457–70
6. Brand, M., Campos-Ortega, J. A. 1990. Second site modifiers of the *split* mutation of *Notch* define genes involved in neurogenesis in *Drosophila melanogaster*. *Roux's Arch. Dev. Biol.* 198:275–85
7. Braun, T., Buschhausen-Denker, G., Bober, E., Tannich, E., Arnold, H. H. 1989. A novel human muscle factor related but distinct from *MyoD1* induces myogenic conversion in 10T1/2 fibroblasts. *EMBO J.* 8:701–9
8. Cabrera, C. V., Martinez-Arias, A., Bate, M. 1987. The expression of three members of the *achaete-scute* gene complex correlates with neuroblast segregation in *Drosophila*. *Cell* 50:425–33
9. Cagan, R. L., Ready, D. F. 1989. *Notch* is required for successive cell decisions in the developing *Drosophila* eye. *Genes Dev.* 3:1099–112
10. Campos-Ortega, J. A. 1988. Cellular interactions during early neurogenesis of *Drosophila melanogaster*. *Trends Neurosci* 11:400–5
11. Campos-Ortega, J. A., Hartenstein, V. 1985. *The Embryonic Development of Drosophila melanogaster*. Berlin/Heidelberg/New York/Tokyo: Springer Verlag. 227 p.
12. Campos-Ortega, J. A., Knust, E. 1990. Defective ommatidial cell assembly leads to defective morphogenesis: a phenotypic analysis of the $E(spl)^D$, mutation of *Drosophila melanogaster*. *Roux's Arch. Dev. Biol.* 198:275–85
13. Campos-Ortega, J. A., Lehmann, R., Jimenez, F., Dietrich, U. 1984. A genetic analysis of early neurogenesis in *Drosophila*. In *Organizing Principles of Neural Development*, ed. S. C. Sharma, pp. 129–144. New York/London: Plenum
14. Campuzano, S., Carramolino, L., Cabrera, C. V., Ruiz-Gomez, M., Villares, R., et al. 1995. Molecular genetics of the *achaete-scute* gene complex of *Drosophila melanogaster*. *Cell* 40:327–38
15. Carramolino, L., Ruiz-Gomez, M., Guerrero, M. C., Campuzano, S., Modolell, J. 1982. DNA map of mutations at the *scute* locus of *Drosophila melanogaster*. *EMBO J.* 1:1185–91
16. Caudy, M., Grell, E. H., Dambly-Chaudière, C., Ghysen, A., Jan, L. Y., Jan, Y. N. 1988. The maternal sex determination gene *daughterless* has zygotic activity necessary for the formation of peripheral neurons in *Drosophila*. *Genes Dev.* 2:843–52
17. Caudy, M., Vässin, H., Brand, M., Tuma, R., Jan, L. Y., Jan, L. N. 1988. *daughterless*, a gene essential for both neurogenesis and sex determination in *Drosophila*, has sequence similarities to *myc* and the *achaete-scute* complex. *Cell* 55:1061–67
18. Cronmiller, C., Schedl, P., Cline, T. W. 1988. Molecular characterization of *daughterless*, a *Drosophila* sex determination gene with multiple roles in development. *Genes Dev.* 2:1666–76
19. Dambly-Chaudière, C., Ghysen, A. 1986. The sense organs in the *Drosophila* larva and their relation to the embryonic pattern of sensory organs. *Roux's Arch. Dev. Biol.* 195:222–28
20. Dambly-Chaudière, C., Ghysen, A. 1987. Independent subpatterns of sense organs require independent genes of the *achaete-scute* complex in *Drosophila* larvae. *Genes Dev.* 1:297–306
21. Dambly-Chaudière, C., Ghysen, A., Jan, L. Y., Jan, Y. N. 1988. The determination of sense organs in *Drosophila*: interactions of *scute* with *daughterless*. *Roux's Arch. Dev. Biol.* 197:419–23
22. de la Concha, A., Dietrich, U., Weigel, D., Campos-Ortega, J. A. 1988. Functional interactions of neurogenic genes of *Drosophila melanogaster*. *Genetics* 118:499–508
23. Dietrich, U., Campos-Ortega, J. A. 1984. The expression of neurogenic loci in imaginal epidermal cells of *Drosophi-*

la melanogaster. J. Neurogenet. 1:315–32

24. Doe, C. Q., Goodman, C. S. 1985. Early events in insect neurogenesis. II. The role of cell interactions and cell lineages in the determination of neuronal precursor cells. *Dev. Biol.* 111:206–19

25. Edmondson, D. G., Olson, E. N. 1989. A gene with homology to the myc similarity region of *MyoD1* is expressed during myogenesis and is sufficient to activate the muscle differentiation program. *Genes Dev.* 3:628–40

26. Ellis, H. M., Spann, D. R., Posakony, J. W. 1990. *Extrama crochatae,* a negative regulator of sensory organ development in *Drosophila,* defines a new class of helix-loop-helix proteins. *Cell* 61:27–38

27. Garcia-Bellido, A. 1979. Genetic analysis of the *achaete-scute* system of *Drosophila melanogaster. Genetics* 91:491–502

28. Garrell, J., Modolell, J. 1990. Molecular genetics of the *Drosophila* extramacrochaetae locus, an antagonist of proneural genes that, similarly to these genes, encodes a helix-loop-helix protein. *Cell.* In press

29. Ghysen, A., Dambly-Chaudière, C. (1988). From DNA to form: the *achaete-scute* complex. *Genes Dev.* 2:495–501

30. González, F., Romani, S., Cubas, P., Modolell, J., Campuzano, S. 1989. Molecular analysis of *asense,* a member of the *achaete-scute* complex of *Drosophila melanogaster,* and its novel role in optic lobe development. *EMBO J.* 8:3553–62

31. Graf, J., Iwamoto, Y., Sasaki, M., Martin, G. R., Kleinman, H. K., et al. 1987. Identification of an amino acid sequence in laminin mediating cell attachment, chemotaxis, and receptor binding *Cell* 48:989–96

32. Hartenstein, V., Campos-Ortega, J. A. 1984. Early neurogenesis in wild-type *Drosophila melanogaster. Roux's Arch. Dev. Biol.* 193:308–25

33. Hartley, D. A., Preiss, A., Artavanis-Tsakonas, S. 1988. A deduced gene product from the *Drosophila* neurogenic locus enhancer of split shows homology to mammalian G-protein β subunit. *Cell* 55:785–95

34. Hartley, D. A., Xu, T., Artavanis-Tsakonas, S. 1987. The embryonic expression of the *Notch* locus of *Drosophila melanogaster* and the implications of point mutations in the extracellular EGF-like domain of the predicted protein. *EMBO J.* 6:3407–17

35. Hoppe, P. E., Greenspan, R. J. 1986. Local function of the *Notch* gene for embryonic ectodermal choice in *Drosophila. Cell* 46:773–83

36. Jimenez, F., Campos-Ortega, J. A. 1979. A region of the *Drosophila* genome necessary for CNS development. *Nature* 282:319–12

37. Jimenez, F., Campos-Ortega, J. A. 1987. Genes in subdivision 1B of the *Drosophila melanogaster* X-chromosome and their influence on neural development. *J. Neurogenet.* 4:179–200

38. Jimenez, F., Campos-Ortega, J. A. 1990. Defective neuroblast commitment in mutants of the *achaete-scute* complex and adjacent genes of *Drosophila melanogaster. Neuron* 5:81–89

39. Johansen, K. M., Fehon, R. G., Artavanis-Tsakonas, S. 1989. The *Notch* gene product is a glycoprotein expressed on the cell surface of both epidermal and neuronal precursor cells during *Drosophila* development. *J. Cell Biol.* 109:2427–40

40. Kelley, M. R., Kidd, S., Deutsch, W. A., Young, M. W. 1987. Mutations altering the structure of epidermal growth factor-like coding sequences at the *Drosophila Notch* locus. *Cell* 51:530–48

41. Kidd, S., Baylies, M. K., Gasic, G. P., Young, M. W. (1989). Structure and distribution of the *Notch* protein in developing *Drosophila. Genes Dev.* 3:1113–29

42. Kidd, S., Kelley, M. R., Young, M. W. 1986. Sequence of the *Notch* locus of *Drosophila melanogaster:* Relationship of the encoded protein to mammalian clotting and growth factors. *Mol. Cell. Biol.* 6:3094–108

43. Kidd, S., Lockett, T. J., Young, M. W. 1983. The *Notch* locus of *Drosophila melanogaster. Cell* 34:421–33

44. Klämbt, C., Knust, E., Tietze, K., Campos-Ortega, J. A. 1989. Closely related transcripts encoded by the neurogenic gene complex *Enhancer of split* of *Drosophila melanogaster. EMBO J.* 8:203–10

45. Knust, E., Bremer, K. A., Vässin, H., Ziemer, A., Tepass, U., Campos-Ortega, J. A. 1987. The *Enhancer of split* locus and neurogenesis in *Drosophila melanogaster. Dev. Biol.* 122:262–73

46. Knust, E., Tietze, K., Campos-Ortega, J. A. 1987. Molecular analysis of the neurogenic locus *Enhancer of split* of *Drosophila melanogaster. EMBO J.* 6:4113–23

47. Kohl, N. E., Legouy, E., DePinho, R.

A., Nisen, P. D., Smith, R. K., et al. 1986. Human *N-myc* is closely related in organization and nucleotide sequence to *c-myc*. *Nature* 319:73–77

48. Komoriya, A., Hortsch, M., Meyers, C., Smith, M., Kanety, J., Schlessinger, J. 1984. Biologically active synthetic fragments of epidermal growth factor: localization of a major receptor-binding region. *Proc. Nat. Acad. Sci. USA* 81:1351–55

49. Kopczynski, C. C., Alton, A. K., Fechtel, K., Kooh, P. J., Muskavitch, M. A. T. 1988. *Delta*, a *Drosophila* neurogenic gene, is transcriptionally complex and encodes a protein related to blood coagulation factors and epidermal growth factor of vertebrates. *Genes Dev.* 2:1723–35

50. Kopczynski, C. C., Muskavitch, M. A. T. 1989. Complex spatio-temporal accumulation of alternative transcripts from the neurogenic gene *Delta* during *Drosophila* embryogenesis. *Development* 107:623–36

51. Legouy, E., DePinho, R., Zimmerman, K., Collum, R., Yancopoulos, G., et al. 1987. Structure and expression of the murine *L-myc* gene. *EMBO J.* 6:3359–66

52. Lehmann, R., Dietrich, U., Jimenez, F., Campos-Ortega, J. A. 1981. Mutations of early neurogenesis in *Drosophila. Roux's Arch. Dev. Biol.* 190:226–29

53. Lehmann, R., Jimenez, F., Dietrich, U., Campos-Ortega, J. A. 1983. On the phenotype and development of mutants of early neurogenesis in *Drosophila melanogaster. Roux's Arch. Dev. Biol.* 192:62–74

54. Murre, C., McCaw, P. S., Baltimore, D. 1989. The amphipathic helix-loop-helix: a new DNA-binding and dimerization motif in immunglobulin enhancer binding, *daughterless, MyoD* and *myc* proteins. *Cell* 56:777–83

55. Murre, C., McCaw, P. S., Vaessin, H., Caudy, M., Jan, L. Y., et al. 1989. Interactions between heterologous helix-loop-helix proteins generate complexes that bind specifically to a common DNA sequence. *Cell* 58:537–44

56. Poulson, D. F. 1937. Chromosomal deficiencies and embryonic development of *Drosophila melanogaster. Proc. Natl. Acad. Sci. USA* 23:133–37

57. Preiss, A., Hartley, D. A., Artavanis-Tsakonas, S. 1988. The molecular genetics of *Enhancer of split,* a gene required for embryonic neural development in *Drosophila. EMBO J.* 7:3917–28

58. Romani, S., Campuzano, S., Macagno,

E. R., Modolell, J. 1989. Expression of *achaete* and *scute* genes in *Drosophila* imaginal discs and their function in sensory organ development. *Genes Dev.* 3:997–1007

59. Romani, S., Campuzano, S., Modolell, J. 1987. The achaete-scute complex is expressed in neurogenic regions of *Drosophila* embryos. *EMBO J.* 6:2085–92

60. Rushlow, C. A., Hogan, A., Pinchin, S. M., Howe, K. M., Lardelli, M., Ish-Horowicz, D. 1989. The *Drosophila hairy* protein acts in both segmentation and bristle patterning and shows homology to *N-myc. EMBO J.* 8:3095–103

61. Shepard, S. B., Broverman, S. A., Muskavitch, M. A. T. 1989. A tripartite interaction among alleles of *Notch, Delta* and *Enhancer of split* during imaginal development of *Drosophila melanogaster. Genetics* 122:429–38

62. Taghert, P. H., Doe, C. Q., Goodman, C. S. 1984. Cell determination and regulation during development of neuroblasts and neurones in grasshopper embryos. *Nature* 307:163–65

63. Tapscott, S. J., Davis, R. L., Thayer, M. J., Cheng, P.-F., Weintraub, H., Lassar, A. B. 1988. *MyoD1:* a nuclear phosphoprotein requiring a *myc* homology region to convert fibroblasts to myoblasts. *Science* 242:405–11

64. Technau, G. M., Becker, T., Campos-Ortega, J. A. 1988. Reversible commitment of neural and epidermal progenitor cells during embryogenesis of *Drosophila melanogaster. Roux's Arch. Dev. Biol.* 197:413–18

65. Technau, G. M., Campos-Ortega, J. A. 1985. Fate mapping in wildtype *Drosophila melanogaster.* II. Injections of horseradish peroxidase in cells of the early gastrula stage. *Roux's Arch. Dev. Biol.* 194:196–212

66. Technau, G. M., Campos-Ortega, J. A. 1986. Lineage analysis of transplanted individual cells in embryos of *Drosophila melanogaster.* II. Commitment and proliferative capabilities of neural and epidermal cell progenitors. *Roux's Arch. Dev. Biol.* 195:445–54

67. Technau, G. M., Campos-Ortega, J. A. 1987. Cell autonomy of expression of neurogenic genes of *Drosophila melanogaster. Proc. Natl. Acad. Sci. USA* 84:4500–4

68. Thisse, B., Stoetzel, C., Gorositza-Thisse, C., Perrin-Schmitt, F. 1988. Sequence of the *twist* gene and nuclear localization of its protein in endomesodermal cells of early *Drosophila* embryos. *EMBO J.* 7:2175–83

69. Thomas, J. B., Bastiani, M. J., Bate,

M., Goodman, C. 1984. From grasshopper to *Drosophila:* a common plan for neuronal development. *Nature* 310:203–7

70. Vässin, H., Bremer, K. A., Knust, E., Campos-Ortega, J. A. 1987. The neurogenic locus *Delta* of *Drosophila melanogaster* is expressed in neurogenic territories and encodes a putative transmembrane protein with EGF-like repeats. *EMBO J.* 6:3431–40

71. Vässin, H., Campos-Ortega, J. A. 1987. Genetic analysis of *Delta,* a neurogenic gene of *Drosophila melanogaster. Genetics* 116:433–45

72. Vässin, H., Vielmetter, J., Campos-Ortega, J. A. 1985. Genetic interactions in early neurogenesis of *Drosophila melanogaster. J. Neurogenet.* 2:291–308

73. Villares, R., Cabrera, C. V. 1987. The *achaete-scute* gene complex of *Drosophila melanogaster:* conserved domains in a subset of genes required for neurogenesis and their homology to *myc. Cell* 50:415–24

74. Watt, R., Stanton, L. W., Marcu, K. B., Gallo, R. C., Croce, C. M., Rovera, G. 1983. Nucleotide sequence of cloned cDNA of human c-*myc* oncogene. *Nature* 303:725–28

75. Welshons, W. J. 1956. Dosage experiments with *split* mutants in the presence of an enhancer of *split. Drosophila Inform. Serv.* 30:157–58

76. Welshons, W. J. 1965. Analysis of a gene in *Drosophila. Science* 150:1122–29

77. Wharton, K. A., Johansen, K. M., Xu, T., Artavanis-Tsakonas, S. 1985. Nucleotide sequence from the neurogenic locus *Notch* implies a gene product that shares homology with proteins containing EGF-like repeats. *Cell* 43:567–81

78. White, K. 1980. Defective neural development in *Drosophila melanogaster* embryos deficient for the tip of the X-chromosome. *Dev. Biol.* 80:322–44

79. White, K., DeCelles, N. L., Enlow, T. C. 1983. Genetic and developmental analysis of the locus *vnd* in *Drosophila melanogaster. Genetics* 104:433–88

80. Wright, T. R. F. 1970. The genetics of embryogenesis in *Drosophila. Adv. Genet.* 15:261–395

81. Wright, W. E., Sassoon, D. A., Lin, V. K. 1989. *Myogenin,* A factor regulating mygenesis has a domain homologous to *MyoD. Cell* 56:607–17

82. Ziemer, A., Tietze, K., Knust, E., Campos-Ortega, J. A. 1988. Genetic analysis of *Enhancer of split,* a locus involved in neurogenesis in *Drosophila melanogaster. Genetics* 119:63–74

Annu. Rev. Genet. 1990. 24:409–45

GENERATION OF DIVERSITY IN RETROVIRUSES

Richard A. Katz and Anna Marie Skalka

Institute for Cancer Research, Fox Chase Cancer Center, 7701 Burholme Avenue, Philadelphia, Pennsylvania 19111

KEY WORDS: retrovirus variability, retrovirus recombination, retrovirus evolution, AIDS virus, selective forces on retroviruses

CONTENTS

409

0066-4197/90/1215-0409$02.00

INTRODUCTION AND BACKGROUND

Retroviruses (155, 156) are prototypic retroelements. The common feature of these elements is that their replication cycle includes the transcription of RNA into DNA (reverse transcription). Retroelements have been found in many eukaryotic organisms (e.g. vertebrates, Drosophila, yeast) (135, 151) and more recently in bacteria (139, 143). Retroviruses are distinguished from other retroelements by an extracellular phase, and thus far have only been found in vertebrate species (128, 129).

Retrovirus particles contain two copies of a single-stranded RNA genome that, after infection of the host cell, are copied into double-stranded DNA in a reaction catalyzed by the reverse transcriptase (RT) carried within the particle. The viral DNA product can become stably integrated into the host cell chromosome (thereafter referred to as a "provirus"), from where new RNAs are transcribed by the host RNA polymerase II. The RNAs produced include the viral RNA genome and mRNAs that program the synthesis of progeny virion components. These features are the basis of classification of the Retroviridae family (128, 129). Retroviruses also share features of genetic organization and genetic content. They all contain three essential replicative genes arranged 5' to 3': the first, *gag*, encodes the structural proteins of the virus capsid and core; the second, *pol*, encodes the reverse transcriptase (RT) and integration protein (IN); and the third, *env*, encodes the virus surface glycoproteins required for binding to specific receptor molecules on the surface of cells during subsequent infections.

The retroviruses are a diverse group and can be categorized by several criteria. The Retroviridae taxonomic family is divided into three major sub-families based on the apparent consequences of infection (128, 129); the Oncovirinae (oncoviruses) usually cause malignancies, but also can be nonpathogenic, the Lentivirinae (lentiviruses) are "slow" viruses that cause various diseases; and the Spumavirinae (spumaviruses) cause vacuolization of cells, but no apparent disease. Retroviruses can also be classified according to their morphology and subcellular distribution (128, 129). The A-type onco-viruses produce only intracellular forms, while the C-type only extracellular particles. The extracellular particles produced by the B-, C-, and D-types have distinctive morphologies. Retroviruses have also been grouped according to whether they are passed in the germline of vertebrate species (endoge-nous retroviruses) (19, 123) or replicate by cell-to-cell infection (exogenous viruses). Oncoviruses isolated from mice that are only capable of infecting mice are denoted "ecotropic," while those that can also infect other species are denoted "amphotropic" or "polytropic." Those that only infect heterologous species are denoted "xenotropic."

Reverse transcription was discovered when retroviral particles were shown to contain an enzyme (reverse transcriptase) that can synthesize DNA using an RNA template. It was later found that the oncogenic properties of retroviruses are due to acquisition, or activation, of cellular protooncogenes. More recently, retroviruses have been applied as vectors to deliver foreign genes to the host cell chromosome. In addition, retrovirus research has provided important models for studying mechanisms of gene regulation and has alerted us to the existence of nonviral retroelements. The identification of human retroviral pathogens [e.g. human immunodeficiency viruses (HIVs) and human T-cell leukemia viruses (HTLVs)] (41) has heightened interest in the molecular biology of retroviruses.

Early studies of clonal isolates revealed that retroviruses can display extraordinary genetic diversity when passaged in tissue culture or in animals. It is important to understand the origin of this diversity to elaborate retroviral disease mechanisms, to fully exploit retroviruses as efficient vectors for delivering genes and to determine how retroviruses may evolve, especially "emerging" pathogens, such as HIV. Modern molecular biology has produced powerful tools, such as recombinant DNA technology, monoclonal antibodies and the polymerase chain reaction (PCR) that are useful in such studies.

The purpose of this review is to discuss current knowledge concerning the origins of retroviral diversity. We consider the various mechanisms that contribute to genetic variability and then discuss the selective forces that lead to the loss, maintenance, or increase of particular variants. Variation and selection are dual aspects of this topic. Diversity cannot exist without mechanisms that produce genetic variation. As we discuss below, the retroviruses possess a unique constellation of such mechanisms. On the other hand, little evolution will be apparent in an unchanging environment, i.e. without changes in selective forces. The selective forces that impact on retroviruses are wide-ranging and often ill-defined. Clues to their identity are provided by the *cis*- or *trans*-acting retroviral genomic sequences that display diversity. The manifestations of diversity include the acquisition of cellular genes, the alteration of host range, and evasion of immunological neutralization by the host immune system. We briefly describe the retroviral replication cycle to acquaint the reader with the steps involved in generating variability (e.g. reverse transcription, DNA integration, RNA transcription, virion assembly) and how the functions of virion proteins (e.g. *env*) make them important targets of selective forces.

For further information, we refer the reader to the following reviews: on retroviruses, with details of retroviral replication (18, 115, 144–146); on the genetic variability and evolution of RNA viruses as a group (26, 27, 53, 99, 118, 122, 160); on variability in HIVs (21, 138, 148, 159); and on the evolution and variation of retroviruses and retroelements (4, 20, 29, 133–135, 137, 139, 150, 151).

RETROVIRUS REPLICATION: AN OVERVIEW

The retrovirus infectious cycle consists of two stages (Figure 1). In the early stage, the virus attaches to the host cell, the core passes through the cell membrane into the cytoplasm, and DNA is synthesized by RT using the RNA

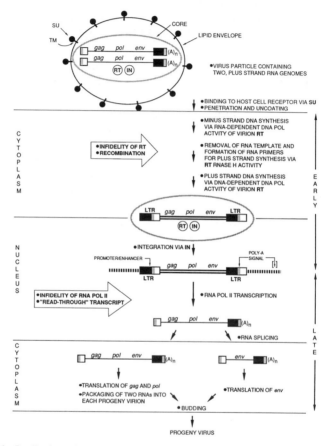

Figure 1 Replication scheme for a "simple" retrovirus. Above; schematic representation of virus particle. Core is shown containing two RNA molecules (thin lines), reverse transcriptase (RT) and the integration protein (IN). U5 and U3 regions are represented by the open and filled boxes, respectively. R regions are indicated by stippled boxes at the ends of the genome. Poly-A tail is indicated [(A)$_n$]. Subcellular sites of replication are indicated in the left margin. Early and late stages are indicated on the right. Middle; linear viral DNA (thick double line) containing flanking LTRs (with R regions) is shown associated within viral core. Cellular DNA is represented by dashed lines. A downstream alternative cellular poly-A addition site is indicated (in brackets) which may be used during oncogene capture (see Figure 3). Large open arrows indicate steps and mechanisms through which genetic changes may occur.

genomes as a template. The viral DNA is delivered to the nucleus where it is integrated (via IN) into the host cell chromosome. In the late stage, the integrated DNA is expressed via transcription and translation of the viral genetic information, thus providing the components of progeny particles. The early stage is dependent on the *pol* gene enzymes (RT, IN) brought into the cell with the virus core. The late stage is dependent primarily on host cell functions: RNA transcription by Pol II, RNA splicing, translation, and post-translational modification of proteins. Mutations may occur during the early and late stages, while recombination takes place during the early stage (Figure 1), as discussed in detail below.

The C-type oncoviruses, avian sarcoma-leukosis viruses (ASLV) and murine leukemia viruses (MuLV), have been studied extensively and serve as prototypes for the replication scheme. The replication cycle described below is a highly generalized summation of results obtained with these two virus groups (reviewed in ref. 144–46). (The references included in the description below document more recent observations). ASLV and MuLV are "simple retroviruses"; lentiviruses, including HIVs, and the HTLVs, encode regulatory genes in addition to the structural genes (*gag, pol,* and *env*) and are therefore considered "complex retroviruses." The HIV and HTLV replication schemes have been reviewed elsewhere (14, 39, 50, 95) and are discussed where relevant in this review.

Early Stage

Retrovirus particles are ca 100 nm in diameter and are composed of a protein and RNA core surrounded by a lipid envelope derived from the host cell's plasma membrane. The envelope contains glycoprotein complexes encoded by the retroviral *env* gene (91). Each complex consists of a transmembrane protein (TM, 15 to 45 kd) that anchors the complex by spanning the lipid bilayer, and a larger surface protein (SU, 65 to 120 kd) that contains the determinants required for recognition of a receptor on the surface of the host cell (30, 31, 90; see ref. 74 for nomenclature for retroviral proteins). The SU/TM complexes associate as multimeric forms on the viral membrane (34). Since the *env* gene products are the outermost viral proteins, they are selectively exposed to certain environmental stimuli, such as immune attack by the host and diverse host cell receptors. Thus, *env* is the most variable retroviral gene (82); this apparently reflects the action of selective forces, as discussed in detail below. The host cell receptors for SU serve essential cellular functions, but are usurped by these viruses as binding sites for entry into the cell (1, 157). The receptor for the HIV SU protein is CD4 (111), a molecule that is found on a subset of T lymphocytes and functions in recognition of antigens. It is thought that the HIV core is deposited in the cell when the virus envelope fuses with the cell membrane. The retroviral core, or some derivative, which includes the viral RNA and several proteins encoded by the *gag* and *pol* genes, is the site of cytoplasmic DNA synthesis (Figure 1) (12). Viral DNA synthesis can be "activated" in vitro by treatment of virus particles with mild detergents that permeabilize the virus envelope and allow entry of deoxynucleotide precursors and essential cations (145, 146).

The core components encoded by *gag* consist of the capsid protein (CA), which forms an icosahedral or rodlike shell and the nucleocapsid protein (NC) that binds to the RNA to form a ribonucleoprotein complex (63). The core components required for DNA synthesis and integration include the *pol* products RT and IN. DNA synthesis uses two single-stranded template RNA genomes (2–10 kb), and a cellular tRNA, which serves as the initial primer for RT. Although the RNA genome is similar in structure to an mRNA (it is of positive polarity, and contains a 5' cap structure and a 3' poly-A tract), it is not translated during the early stage of infection, probably because it is sequestered in the core. Synthesis of the first (minus) strand of DNA initiates near the 5' end of the viral genome using the tRNA primer that is bound to this region through RNA-RNA base pairing. DNA synthesis proceeds toward the 5' end, a distance of 100–300 nucleotides (nt), and this region is denoted "U5". At the immediate 5' terminus of the RNA (upstream of U5) is a sequence, R, of 10 to 200 nt that is repeated at the 3' end of the genome, immediately preceding the poly-A tract (Figure 1). The ribonuclease H (RNaseH) activity of RT removes sequences near the 5' end of the RNA template present as an RNA-DNA, template-product hybrid. This allows the single-stranded R DNA product to hybridize to R sequences on the 3' end of the other RNA molecule in the virion and thus reprime DNA synthesis through the remainder of the RNA genome (92). This processive synthesis by RT between different templates has been referred to as "jumping." A similar mechanism may be exploited during recombination, as discussed below. When this reprimed synthesis of the minus strand extends past the U3 region at the 3' end of the RNA genome (Figure 1), synthesis of plus-strand DNA is initiated from an RNA primer produced by nucleolytic digestion of genomic RNA by the RT RNaseH (79). The position of this RNA primer defines the 5' end of the long terminal repeat (LTR), which has the structure U3-R-U5 and is present at the ends of the viral DNA (Figure 1). Plus-strand synthesis continues using the completed minus strand of DNA as a template to produce the final linear DNA duplex structure: LTR-*gag-pol-env*-LTR (Figure 1). Genetic studies suggest that both RNA molecules within the particle participate in forming one DNA provirus (55, 92). Within the nucleus of the infected cell, circular forms of the DNA, one of which contains tandem LTRs, can also be detected. Studies with MuLV indicate that the linear form is normally used for integration (13). However, other data suggest that the circular form can also be a substrate for integration in some systems (93).

Inspection of the structure and nucleotide sequence of proviral DNAs revealed that two nucleotides are absent from each end as compared to the unintegrated linear forms. A second characteristic of integration is a duplication (4–6 bp) of host DNA sequences flanking the provirus. The duplication is presumed to arise from repair synthesis of a staggered cut in the host DNA target site for integration. The *pol* IN product is required for integration (112, 115) and for cleavage of the viral LTR ends in vivo (107) and in vitro (65); it is also likely to be involved in the staggered cleavage of the cellular DNA.

Late Stage

The integrated provirus is transcribed by host RNA Pol II as a single RNA species that initiates in the 5'-LTR and undergoes polyadenylation at a site encoded in the 3'-LTR (Figure 1). In the nucleus, a portion of this RNA is spliced to produce the *env* subgenomic (and sometimes additional) mRNA(s) (Figure 1). Both full-length and spliced RNAs are transported to the cytoplasm where the full-length transcript serves two functions: *a*) as an mRNA for *gag* and *pol* products, *b*) as genomic RNA to be packaged into progeny virions. The major translation product of the full-length RNA is the *gag* precursor polypeptide (ranging from ca 55 to 75 kd depending on the virus). This precursor is processed after viral assembly by the virus-encoded protease (PR) (116), giving rise to the major core proteins CA and NC. (PR is encoded between *gag* and *pol,* but can be in translational frame with either or can be separate, depending on the virus.) A third processed *gag* protein, the matrix protein (MA), is thought to localize between the capsid and the lipid envelope. Through ribosomal frameshifting or suppression (depending on the virus), the termination codon at the end of *gag* is occasionally by-passed, allowing continued translation of the neighboring reading frame, *pol*. This results in limited synthesis of a *gag-pol* fusion protein, which is the *pol* precursor. This precursor is cleaved by PR to produce RT and IN after viral assembly.

As opposed to the full-length *gag-pol* mRNA, which is translated on free ribosomes, the *env* mRNA is translated on membrane-bound ribosomes as a precursor polypeptide that is directed into the RER secretory pathway via an N-terminal signal peptide. Posttranslational processing occurs in the RER/golgi and includes glycosylation and proteolytic processing at the cell surface by a cellular endopeptidase. The mature SU/TM complex is subsequently found on the cell surface, inserted into the lipid bilayer through the TM domain, as noted above.

The assembly of B-, C- and D-type retrovirus particles initiates at the inner surface of the cell membrane, while maturation to an infectious form usually occurs during or after budding from the cell. The A-type particles bud through intracellular membranes and are not released. Each precursor component of the virus particle contains a specific signal sequence that directs it to the assembly site. The *gag* precursor has an N-terminal signal that directs it to the inner surface of the plasma membrane; the *pol* products are delivered in the form of the *gag-pol* precursor also by virtue of this signal; the *env* products are delivered to the membrane surface by the signal peptide. The viral RNAs contain *cis*-acting signal sequences that may bind to the NC domain of the *gag* precursor, thus allowing selective packaging (63, 76).

After these events, the core components bud from the plasma membrane and acquire the viral envelope containing SU/TM. The PR is then activated and processes the *gag* and *gag-pol* precursors into the mature components. The particle becomes infectious during this processing stage. Maturation also involves formation of a base-paired complex between the two encapsidated genomic RNA molecules. It is not clear what determines that two RNA genomes are packaged per particle. However, this is a highly regulated step and may set the stage for genetic recombination (see Recombination section, below).

GENETIC DIVERSITY IN RNA VIRUSES: GENERAL PRINCIPLES

Retroviruses are distinguished from other RNA (ribo) viruses by their DNA phase of replication (138). However, it is useful to consider retroviruses with other riboviruses since the infidelity of enzymes involved in the replication of all riboviruses is likely central to their diversity. In this section we summarize the approaches to study diversity in riboviruses and review the general principles that relate to their variability; these principles apply directly to retroviruses.

Riboviruses are ubiquitous intracellular parasites that usually encode polymerases that replicate their own RNA genomes. Early studies of eukaryotic RNA viruses (tobacco mosaic virus, Newcastle disease virus, reovirus and vesicular stomatitis virus) revealed that they were prone to rapid variation (26, 122). Genetic variants were usually detected as revertants in stocks of viral mutants. More recently, molecular and sequence analysis of human riboviruses such as influenza and polio has shown nucleotide differences in isolates from separated geographic areas, in variants arising in individuals after vaccination, or by selection in vitro (26, 118, 119). As assayed in tissue culture, riboviruses have mutation rates in the range of 10^{-4} to 10^{-6} per nucleotide per round of replication (26, 118, 160). Retrovirus mutation rates are similar (see below). This contrasts with rates of 10^{-7} to 10^{-11} misincorporations per base pair replicated observed for chromosomal genes (53, 77). Mutations in ribovirus genomes can include nucleotide substitutions and rearrangements (e.g. insertions, deletions, etc.), Most single nucleotide mutations probably reflect misincorporation of nucleotides; in contrast to most cellular DNA polymerases, viral RNA polymerases do not carry editing functions (26, 27, 53, 77).

Mutation rates for riboviruses have been evaluated in several ways (26, 118) by: (*a*) estimation of the frequency of activation or inactivation of a genetic marker after a single infectious cycle; (*b*) direct sequence analysis or physical studies of the RNA after a known number of replication cycles, starting with a cloned virus; (*c*) assessment of the fidelity of the viral polymerases in vitro; and (*d*) measurement of the frequency of appearance of monoclonal antibody-resistant mutants. Each method has its limitations (see ref. 73). For example, some in vitro assays that assess the fidelity of viral enzymes using homopolymers likely result in an overestimate of the mutation rate (118).

Mutation rates of riboviruses (including retroviruses) are so high that a genome cannot be defined. Thus, a ribovirus population is considered to exist in a dynamic equilibrium, or as an "average" of the genetic content of individual genomes. The terms "quasi-species" (27) or "swarm" (138) have been used to describe such mixtures. The heterogeneous population is thought to be advantageous, in that mutant genomes in the population may proliferate rapidly in response to selective conditions (27). Thus, the survival of the virus likely depends on this dynamic equilibrium. A sample of viruses isolated from a single infected animal or individual is defined as an "isolate", a "strain" as a collection of highly related genomes, whereas a "genome" corresponds to the nucleic acid molecule(s) that contains the genetic inheritance of the virus (45).

In replicating ribovirus populations, mutations arise at a high rate and some affect essential genes. However, mutations that inactivate viral proteins or enzymes can be rescued in *trans* through complementation by wild-type viral proteins in mixedly infected cells (a process called pseudotyping or phenotypic mixing). A most dramatic example is the persistence of defective interfering particles (DIs), whose RNAs can contain little more than *cis*-acting replication and packaging signals (53, 56). In addition to lethal mutations, many neutral mutations arise and these can persist in the population (genetic drift). Adaptive mutations are fixed by selection conditions. For riboviruses to be successful, certain criteria must be met (27, 99): (*a*) the average number of mutations per genome cannot be so high that a lethal mutation occurs in every genome; (*b*) sufficient numbers of genomes must be produced to insure the presence of adaptive mutations; and (*c*) the number of mutations required to cause a particular phenotype must be small.

In riboviral genomes (and for molecular evolution in general) the diversity that results from the combination of mutations and selection can be viewed in distinct ways. Since sequences that encode essential functions must be maintained, it has been hypothesized that clustered nucleotide changes in viral structural genes, enzymes or regulatory regions probably identify less critical regions. Indeed, estimates of the evolutionary relationships among proteins of diverse origin, including retroviral enzymes, are based on identification of highly conserved amino acid sequences, which often identify residues that form the enzymatically active site. Such a relationship has been noted in the retroviral PR enzyme (116). Another view suggests that selective forces can act on functional regions and, in some cases these regions may accumulate changes that generate new functions. These theories may not be mutually exclusive, and different mechanisms may operate depending on the nature of the protein and the selective forces.

THE MUTATION RATE OF RETROVIRUSES

Early studies with retroviruses revealed numerous variations detected by genetic and biochemical methods (132). These included changes in antigenicity or host range, and losses of gene products or oncogenic properties. A mutation frequency of 10^{-3} to 10^{-4} misincorporations per genome per infectious cycle was estimated through physical analysis of RNA molecules (22). Thus, it seemed likely that a significant number of mutant genomes were present even in biologically cloned retroviral stocks. The advent of molecular DNA cloning and rapid sequencing methods allowed identification of both silent mutations and those that result in altered phenotypes. Early nucleotide sequence analyses of independent clones of a given retroviral DNA, and direct RNA sequencing, revealed frequent differences among them that included nucleotide substitutions, insertions, and deletions (25, 89). However, because bacterial DNA replication is less error-prone, propagation of these viral sequences as DNA in bacterial plasmids provides a uniform source of "parental" genomes. Introduction of such DNA into a eukaryotic cell (by transfection or microinjection) can program the formation of progeny virions, as would the provirus during a normal infection. In addition, sequences could be introduced into the cloned retroviral DNA to serve as defined markers for analysis of mutation rates and recombination frequencies.

In vivo viral mutation rates are usually expressed as the number of mutations per genome per replication cycle. Therefore, the determination of the mutation rate of retroviruses requires that the exact number of infectious cycles (e.g. DNA→RNA→DNA) be known and that individual genomes be analyzed. This can be difficult since multiple rounds of infection can occur in tissue culture systems or in infected organisms. Two different approaches have been taken to circumvent this difficulty. Dougherty & Temin (32, 33) used a retroviral DNA vector system to measure mutation rates. In general, vector DNAs encode *cis*-acting sequences required for replication (LTRs, RNA packaging signals), but the viral genes are replaced with foreign DNA. A vector RNA transcript that contains the foreign sequences is produced after transfection of a "helper" cell with the cloned vector DNA. Helper cells contain DNA that can express viral structural proteins but no packageable virus RNA. Because it contains packaging sequences, the vector RNA is packaged into a virus particle by complementation with the virus proteins produced in the helper cells. The foreign gene is then delivered, by infection, to a second cell in which the vector RNA is reverse-transcribed, integrated, and efficiently expressed. The vector provirus cannot produce progeny virions in this second cell, because it lacks the genes that specify viral proteins. Thus, the replication cycle is limited to one round (DNA→RNA→DNA). Cells that contain the vector provirus can be selected by expression of the foreign gene. Dougherty & Temin used an avian spleen necrosis virus (SNV) vector encoding the selectable marker, neomycin resistance (neo^r). An amber codon was introduced by a single nucleotide change resulting in neomycin sensitivity (neo^s). The mutation rate at the amber codon site was calculated by the reversion frequency from neo^s to neo^r (32, 33). Nucleotide sequence analyses indicated that the base pair (bp) substitution rate was 2×10^{-5} per bp per replication cycle (33) and the insertion rate 10^{-7} per bp per replication cycle.

In a second approach by Leider et al (73), single-round ASLV infections were performed and RNA expressed from the resultant proviruses was analyzed directly for single nucleotide changes. These changes were detected by enhanced melting of hybrids between wt RNA and progeny RNA that were subjected to gel electrophoresis. In this approach, the authors exploited the resistance of infected cells to superinfection, and thus expanded single cell clones without the occurrence of multiple rounds of infection. A rate of 1.4×10^{-4} mutations per nucleotide per replication cycle was obtained.

Although both experiments described above reveal a high rate of mutation, they do not distinguish between changes incurred during virus RNA synthesis by Pol II after integration or during viral DNA synthesis by RT before integration (see Figure 1). The fidelity of RNA synthesis by Pol II is unknown, but since there is no known editing function, it is likely to be similar to that of viral RNA polymerases. It is clear from the in vitro studies described below that the fidelity of DNA synthesis by RT is low and thus probably contributes significantly to the high in vivo mutation rate. Integrated proviral DNA that replicates as part of the cell chromosome has a much lower mutation rate, presumaly identical to that of the host genes (42).

FIDELITY OF REVERSE TRANSCRIPTASE IN VITRO

Two general approaches have been used to measure the fidelity of RT in vitro (77). An early assay (5) measured the frequency of nucleotide misincorporation using synthetic homo- or copolymer DNA or RNA templates. The second, a genetic assay, analyzed the in vitro misincorporation rate on natural templates (46, 96, 104, 105). In this assay, a single-stranded ϕX174 bacterio-phage DNA serves as a template for DNA synthesis by RT. The phage genome encodes a nonsense mutation that produces a replication defect; during DNA synthesis in vitro, single nucleotide misincorporations can result in reversion of the nonsense mutation. These misincorporations are then scored by production of viable phage.

Early studies by Loeb and coworkers, using the synthetic polymer template assay, showed that the fidelity of ASLV RT was much lower than that of *E. coli* DNA polymerase I (5). The misincorporation rate for RT was 1 per 6,000 nt compared to 1 per 100,000 nt for Pol I. The misincorporation rates for RT were similar regardless of whether an RNA or DNA template was used. Rates of 1 misincorporation per 904 to 1 per 30,000 nt were reported with the ASLV RT using the ϕX174 assay (46, 105), and the reason for this discrep-ancy has been discussed (105). The misincorporation rate for RTs is generally believed to be higher than cellular DNA polymerases because RTs lack proofreading activities (105). However, this conclusion was reached by comparing data from different laboratories rather than by direct experimenta-tion (105).

The high degree of variability in the HIV envelope gene (noted below) has recently prompted investigators to ask whether the HIV RT is more prone to misincorporate nucleotides than ASLV or MuLV RTs. Using nonsense codon reversion assays, the error rate of HIV RT was 1 per 1,700 (104) to 4,000 (96), while the rates for ASLV and MuLV RTs were 1 per 9,000 (96) to 17,000 (104) and 1 per 30,000 (104), respectively. Similar rates for HIV were reported using a homopolymer template primer system (127). Thus, the misincorporation rate of the HIV RT appears to be higher than that of other retroviruses. Recent studies (94) suggest that HIV RT is able to extend primers that contain a mismatched 3' terminus more efficiently than mamma-lian polymerase α. Efficient extension would be required for processive synthesis after misincorporation.

Sequence analysis of mutations produced by MuLV or ASLV RT in vitro have suggested some interesting properties for the enzymes (105). For ex-ample, during extension of a DNA primer containing a terminal-mismatched nucleotide, polymerization may initiate at an upstream site containing a nucleotide complementary to the terminal base of the primer. Thus, interven-ing sequences are deleted. This is not seen using DNA polymerase. RTs also produce deletions in vitro that are not dependent on direct repeats. These, as well as the terminal mismatch-generated deletions may reflect RT's ability to carry out nonprocessive synthesis ("jumping"). Biased misincorporations are also seen with RTs in vitro (105) and in vivo (45), although in the latter case a contribution by RNA Pol II cannot be ruled out.

RECOMBINATION

Mechanisms of Recombination

Recombination between retroviral genomes requires one round of viral replication in which one of each parental RNA genome is packaged into the same virion, producing a "heterozygous particle" (58, 75, 154). After infection of a new host cell, both RNAs are copied (at least in part) into DNA by RT and recombination occurs during this step (Figure 1). Two models for homologous recombination have been proposed for retroviruses (Figure 2).

FORCED COPY-CHOICE MODEL   The forced copy-choice model, described by Coffin (18, 20), proposes that recombinant retroviral DNA molecules are formed by template switching during minus strand DNA synthesis from the viral RNA template (Figure 2A). This model was formulated, in part, to explain how retroviruses can survive the numerous breaks typically observed in virion RNA preparations. Since the breaks are random, if template strand switching occurs at each break, a complete DNA minus strand can be made from fragmented templates. This model can also explain why one virus particle may give rise to only one DNA provirus (55, 92). From a mechanistic standpoint, the forced copy-choice recombination model is made plausible by the observation that RT can "jump" from one RNA to another during extension of R DNA from the 5' to the 3' end of the viral RNA template (18). In this case, the R region provides homology for the jump between RNA templates (see section on Retroviral Replication). It is hypothesized that, as in this jump, the RNaseH activity of RT removes ribonucleotides from the end of the DNA-RNA hybrid to expose sequences that can find their homologues on the second RNA genome and allow repriming of the DNA on this template to promote homologous recombination (18).

STRAND DISPLACEMENT-ASSIMILATION MODEL   This model (60) (Figure 2B), which proposes that recombination occurs by plus strand viral DNA exchange, was formulated to explain the origin of joined viral DNA molecules ("H" structures) observed in the electron microscope after disruption of virions that had been permeabilized to allow reverse transcription (60). The formation of these joined double-stranded DNA molecules is proposed to be facilitated by the generation of free plus strand "tails" as a consequence of strand-displacement synthesis by RT (60, 61). This model is consistent with several biochemical features of retroviral replication. The DNA minus strand copied from the RNA genome starts at a unique primer (tRNA) and is continuous. However, internal plus strand DNA synthesis starts at many sites

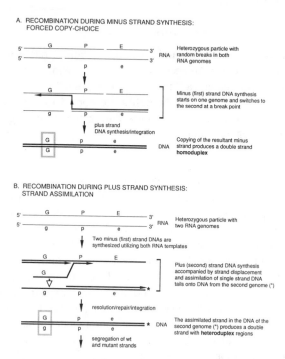

Figure 2   Two models for retroviral recombination (see text for details). RNA is represented by thin lines; DNA as thick lines. Closed arrows indicate direction of DNA synthesis. Viral genes (*gag, pol* and *env* alleles are indicated as G/g, P/p and E/e respectively) are shown to illustrate recombination events. Single recombinations are shown for simplicity, arbitrarily focussing on the *gag* gene (G/g). Multiple crossover events are frequently observed in retroviruses (18, 20). Stippled box highlights the region of recombination. A. Recombination during minus strand synthesis (18, 20). Break in RNA is represented as a gap. "Jumps" in DNA synthesis between RNA templates are thought to occur at RNA breaks, as shown. Forced copy-choice predicts only one DNA product from two RNAs. B. Recombination during plus strand synthesis (60). Plus strand DNA synthesis initiates at internal sites using RNA primers generated by partial RNaseH digestion. The strand assimilation model requires that plus strand DNA synthesis takes place on two templates and predicts formation of a heteroduplex with recombination of plus strands. One DNA is shown for simplicity. Open arrow indicates strand assimilation.

on the minus strand DNA template from RNA primers that are produced by RNaseH digestion of the genomic RNA plus strand. The newly synthesized DNA plus strands are at first short and incomplete, and single-strand gaps are present (60, 61). As plus strand DNA synthesis continues these gaps are filled. However, DNA synthesis does not stop when the 5' end of an adjacent plus strand is encountered. Instead, the 5' end strand is displaced by continued synthesis, forming a single-stranded tail. Such tails (average 2–4 per molecule for ASLV) have been detected in DNA synthesized in both permeabilized virions (60, 61) and in infected tissue culture cells (54). The assimilation model proposes that "H" structures are formed when these free tails invade the DNA being synthesized on the second genome in the virion—perhaps at gaps, which could exist as a consequence of asynchrony of reverse-transcription of the two genomes. Presumably, strand assimilation does not require the participation of a viral protein but, rather, is a stochastic process initiated by the appropriate topological and thermodynamic conditions. Other mechanisms (e.g. branch migration) could be invoked whereby the 3' ends of the transferred plus strand could also be exchanged and plus strand DNA synthesis might then continue on the second (invaded) template.

Each model makes a specific prediction concerning the nature of the product. The product of forced copy-choice must be a homoduplex, because recombination occurs during synthesis of the first DNA (minus) strand (Figure 2A). The strand-assimilation product would be a heteroduplex (Figure 2B). Interpretation of genetic analyses which bear upon this question has been difficult and inconclusive. Goodrich & Duesberg (45a) have reinvestigated this question; they used a sensitive PCR method to amplify products of reverse-transcription in heterozygous retroviral particles that had been permeabilized to permit DNA synthesis in vitro. In this experimental system one of the parental RNA genomes could not be reverse-transcribed efficiently because it lacked 3'-terminal R sequences. Although it is not possible from these data to estimate the frequency of the reaction, analysis of the amplified products suggest that recombination can occur by RNA template switching during minus strand DNA synthesis. Using several markers, and a system in which both parental genomes could be reverse transcribed, Hu & Temin (submitted for publication) have identified patterns of recombination which suggest that both copy-choice and plus strand exchange mechanisms are operative. However, a rigorous test of the possible contribution of a plus-strand DNA exchange reaction requires a system in which both parental molecules can be reverse-transcribed efficiently and the relative frequencies of both homoduplex and heteroduplex recombinant products can be determined (Figure 2).

Frequency of Recombination

A system to study recombination using SNV-based retroviral vectors that contain two selectable genes has been devised by Hu & Temin (55). Two vectors that contained inactivating mutations in either of the selectable markers were introduced into the same cell. Heterozygous particles were formed and the recombination frequency observed in the subsequent provirus was determined to be 2% per kilobase per replication cycle. These experiments also confirm that the production of recombinants requires formation of heterozygous particles. This and another (92) study further suggest that the two virion RNAs give rise to only one DNA copy, although other interpretations are possible.

Recombination between replicating viruses can be readily detected in tissue culture cells. One reason is that cells can be multiply infected with mixtures of parental viruses. However, infection and expression by one provirus results in resistance of the cell to superinfection by a second virus with the same receptor specificity, presumably by blocking or down-regulation of receptors due to the presence of viral *env* protein on the cell surface (152). Thus, for recombination to occur, cells must be infected either by two parental viruses that use different receptors, or simultaneously by parental viruses that use the same receptor. In the latter case, sequential infection could also occur during the lag period before superinfection immunity is established. The frequency of recombination in infected hosts may therefore be limited by these requirements.

Unwanted recombinational events have been detected frequently in retroviral helper systems. As noted above, the use of retroviral vectors requires two retroviral genomes to be coexpressed in a single cell but only the vector genome should be packaged. However, replication-competent viruses can arise in some systems through recombination due to occasional (leaky) packaging of the helper virus genome, with consequent formation of a heterozygous particle. These recombination events are detected when the replication-competent recombinant virus, which has picked up a packaging signal from the vector genome, is amplified in susceptible cells (discussed in ref. 85). This can be a problem in the use of retroviral vectors since the introduction of replication-competent viruses along with the vector is prohibited for gene therapy applications. These difficulties can be resolved by separate expression of *gag-pol* and *env* regions in helper cells (76) or removal of homologies that may promote recombination (85).

Recombination Between Exogenous and Endogenous Viruses

In contrast to exogenous retroviruses, which are transmitted from cell to cell through normal infection, endogenous viruses are perpetuated as chromosomal genes through the germline and thus are present in all cells of an animal (19, 123). Endogenous proviruses can serve as repositories for genetic information that may be acquired by exogenous viruses through recombination. Endogenous viruses may also provide a beneficial function in the host species, as mentioned below. Most endogenous proviruses are replication-defective due to point mutations or rearrangements, but may also contain nondefective *gag, pol,* or *env* gene segments.

The endogenous retroviruses of chickens (108), mice (102), and humans (71) have been studied most thoroughly. The numbers that exist within the germline of a given species vary from one to hundreds, and several types can exist within one species. In most cases, endogenous viruses of a given species have nucleotide sequence similarities with the exogenous viruses that propagate in the same species. Indeed, most endogenous proviruses have been detected by hybridization with exogenous viral nucleic acids. However, some endogenous proviruses may be missed because they lack strong homology to the available exogenous virus probes. Endogenous proviruses can be expressed in specific tissues and at different stages of development, or their expression can be induced in tissue culture cells following treatment with various chemicals. Since the endogenous proviruses often carry mutations, they can give rise to defective or nonpathogenic replication-competent viruses. In some cases, RNA transcripts are produced, but no detectable viral-like proteins are synthesized. Recombination between exogenous and endogenous retroviruses requires expression of the endogenous virus RNA and formation of a heterozygous particle with the exogenous viral RNA genome (154). Certain endogenous retroviral transcripts can be efficiently packaged into particles (pseudotyped); this indicates that the endogenous transcript has packaging signals that are recognized by the proteins of the exogenous virus (63, 113).

Nearly 20 endogenous virus (*ev*) loci that show homologies to the ASLV group of exogenous viruses have been detected and characterized in domestic chickens (19, 108). Certain endogenous proviruses are unique to particular lines of chickens. For example, in line 7 chickens, an infectious, non-pathogenic virus (Rous associated virus-0, RAV-0) is produced spontaneously from the *ev*-2 locus. Most chicken lines contain the defective *ev*-1 provirus at another locus, which can be transcriptionally activated by treatment with 5-azacytidine, but which does not produce infectious particles. Historically, recombination between exogenous and endogenous avian viruses was detected by stable acquisition of a new host range specificity. This phenotype is encoded in the *env* SU acquired from a subset of *ev* loci in certain chicken cells (154). Thus, the potential for recombination between exogenous and endogenous sequences clearly exists, and recombination in nature could conceivably produce a recombinant SU that uses a new host cell receptor (see Table I).

Various strains of laboratory mice contain large numbers of endogenous proviruses belonging to four recognized classes: C-type, B-type, intracisternal A particles (IAPs) (A-type) (67a), and VL30s (19, 102, 123). The B-type and C-type endogenous viruses are related to exogenous mouse viruses, and exogenous-endogenous recombinants have been identified in tissue culture and in mice. Certain mouse strains (e.g. AKR), bred for high tumor incidence, express endogenous C-type viruses from birth. The neonatal mice are viremic, but the viruses produced are initially benign (AKVs). These viruses are ecotropic i.e. they only can infect mouse cells. During the lifetime of the mouse, recombinant viruses appear that have disease potential and cause thymic lymphoma (19, 123, 152). The generation of pathogenic viruses is a stepwise process marked by changes in LTR and *env* sequences that appear to be acquired by recombination with endogenous viruses (19, 20, 102, 123, 124). The pathogenic viruses can frequently be differentiated from AKV by their ability to infect and replicate in nonmurine cells (they are polytropic). This observation is consistent with genetic changes in the LTR (44) and *env* gene (66, 124) that potentially affect cell-specific transcription (11, 15) and *env*-receptor interaction (100), respectively. These changes may allow these viruses to replicate efficiently in the thymus, and cause host cell transformation by activating cellular oncogenes through insertional, or other, mechanisms (123).

Goff and colleagues have described a tissue culture system in which revertants of MuLV deletion mutants are generated by recombination with endogenous retroviral sequences (80). This type of system may be valuable in investigating the mechanisms involved, since the tissue culture cells presumably provide uniform expression of the endogenous virus sequences.

Recombination Between Retroviral and Host Genes

Retroviruses can cause neoplastic disease by at least three mechanisms (144). (*a*) The nonacute oncogenic retroviruses transform cells by affecting the expression of cellular oncogenes (protooncogenes) in the vicinity of their integration site. Since retroviruses integrate at many sites on the host DNA, integration in the vicinity of a protooncogene is a low-frequency event. (*b*) Some leukemia viruses (e.g. HTLV-I,II) express nonhost-related viral products that stimulate cell proliferation (41). (*c*) The acute transforming retroviruses of the oncovirus group carry oncogenes (6, 7), which they have captured from the genomes of their host cells by recombination ("oncogene capture") (43, 49, 126, 136), usually at the expense of some of their own replication genes. Most acute transforming viruses therefore require a replication-competent helper virus for propagation.

Various lines of evidence indicate that the process of oncogene capture begins with the integration of proviruses in the vicinity of a cellular protooncogene (43, 126) (Figure 3). The formation of a chimeric, virus-protooncogene RNA transcript is the presumed next step. The formation of such a transcript requires that 3' processing (cleavage/polyadenylation) occurs at an adjacent 3' cellular signal rather than in the 3' LTR. This could occur in several ways (Figure 3): (*a*) by infrequent skipping of the viral polyadenylation site forming a "read-through" transcript (51); (*b*) by removal of the viral polyadenylation site via RNA splicing between retroviral and downstream cellular sequences present in the read-through transcript (136); (*c*) by deletion of proviral sequences including the downstream LTR at the

DNA level so that transcripts from the provirus can only use a downstream cellular polyadenylation site (43, 126). The next steps would involve removal of introns that might be present in the protooncogene (by RNA splicing) followed by copackaging of the chimeric RNA along with a complete wild-type viral transcript. Packaging of the chimeric transcript presumably occurs by virtue of the 5' viral *cis*-acting packaging signals that are maintained. The mechanism by which the fused transcript acquires the 3' LTR necessary for subsequent replication is unknown. Some evidence suggests that short sequence homologies may facilitate this recombination (8). Thus, the copy-choice mechanism inferred in the experimental system of Goodrich & Duesberg (45a) may apply. The presence of a portion of the 3' poly-A tract of the cellular protooncogene transcript within some transforming virus genomes provides evidence that the recombination can use an RNA intermediate (57, 98).

REARRANGEMENTS

Deletions and duplications can be detected in retroviral genomes. Large (up to thousands of nucleotides) and small (several nucleotides) deletions are generally believed to occur through one of four pathways: (*a*) template slippage (86) or mispriming (20) at direct repeats during reverse transcription; (This usually results in small deletions.) (*b*) homologous recombina-

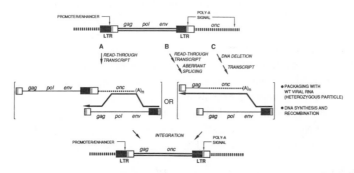

Figure 3   Models for recombination between retroviral and cellular genes: oncogene capture (see text for details). Above; DNA provirus (as designated in Figure 1) is shown integrated upstream of a cellular protooncogene (*onc*). Cellular DNA is shown as thick dashed line. Recombination at the 5' end of the protooncogene is thought to occur by pathways A, B, or C. Middle; RNA and DNA are denoted as thin and thick lines, respectively, as in Figures 1 and 2. Cellular-derived portion of chimeric RNA transcript is indicated by thin dotted line. Copackaging is indicated by brackets. Note that pathway A requires a double recombination event. Bottom; provirus product of recombination. This provirus produces an RNA transcript that can be propagated further by packaging into a replication competent virus.

tion between direct repeats may be the origin of large deletions (8, 20); (c) spurious endonucleolytic action of ASLV IN near the internal border (U5) of the upstream LTR (88) (Figure 1); (d) a pathway that is independent of homologous recombination (147). Duplications have also been observed and are likely the result of errors in reverse transcription (20, 21).

Large deletions that affect viral structural genes and enzymes cause replication defects. These mutant RNA genomes can be perpetuated by complementation by wt virus since the mutant RNAs may retain *cis*-acting replication signals (22). Other large deletions may affect *cis*-acting RNA signals such as those required for RNA packaging (76). These natural packaging mutants were identified by their unique phenotype: particles were formed, but they lacked viral RNA. Large deletions are characteristic of integrated ALV proviruses that activate the neighboring c-*myc* oncogene by promoter insertion (130, 131). This appears to be an obligatory step in c-*myc* activation, but the deletion mechanism and the selective forces involved are unknown. One possibility is that tumor cells that do not express viral antigens are less likely to be destroyed by immune surveillance (130, 131).

ROLE OF SELECTIVE FORCES IN GENERATION OF DIVERSITY

Retrovirus mutations produced by the mechanisms described above generate variability, and selective forces then promote proliferation of variants containing advantageous mutations. In experimental selection, screening conditions are predetermined and can allow one to correlate a specific mutation with a particular phenotype. In an infected organism the selective forces can only be hypothesized. The effects of some, e.g. the host's immune system, can be identified. However, additional, more obscure forces surely exist that contribute to retroviral diversity in ways not yet understood. Selective forces can operate within one infected organism or can be viewed in the broader context of long-term retroviral evolution. In Table I we have summarized ways in which experimental or natural selection may participate in the generation of diversity in retroviruses; a more detailed discussion follows.

Increased and Tissue-Specific Replication

The ability to replicate in different target cells or tissues ("tissue-specific" replication) is a commonly acquired phenotype in retroviruses. This phenomenon can be observed by experimental analysis in tissue culture or in more complex murine animal model systems. At least four regions (LTR, 5' leader, *gag*, and *env*), and possibly others (106), have been implicated in the ability of the viruses to replicate in different tissues and/or cause disease (Table I). In some cases, the pathogenicity of the virus appears to be directly linked to its ability to replicate in certain tissues (20, 123, 130, 131). The selection here is probably for more efficient replication. This may increase pathogenicity through the action of viral gene products or through an increase in the probability of activation of cellular oncogenes by provirus insertion (123).

LTR   Sequence differences in the LTR are associated with different replication efficiencies and/or the ability of the virus to replicate in a certain cell type. In the murine pathogenic viruses, sequence changes in the LTR increase the activity of the U3 transcriptional enhancer (Figure 1) as compared with the benign predecessor (AKV, discussed above) (11, 15, 44). The sequence changes appear to result from recombination with endogenous viruses and/or single base changes (20, 123, 130, 131).

GAG   Ecotropic MuLVs can be classified according to their ability to replicate in NIH-Swiss, BALB/c mouse cells, or both (N-, B- or NB-tropic viruses, respectively) (128). This phenotype is regulated by a host cell gene, and maps to the CA region of *gag*. The biochemical block is at a postpenetration step and appears to affect DNA integration (145, 146). In BALB/c mice, N-tropic viruses are present at low amounts early in life and B-tropic viruses appear later. The source(s) of the B-tropic determinants (specific amino acid changes) appears to be endogenous viruses and these determinants are acquired by recombination (10).

ENV   The SU protein of *env* is involved in recognition of the host cell receptor (30, 31, 90). Thus, selection for increased replication may involve mutations in this region that expand host range. In ASLV, sequence differences in SU between virus subgroups (defined, in part, by host range) reflect the ability to use different receptors present in different chicken flocks (30, 31). This divergence suggests that the viruses have evolved to use new receptors. In mice, recombination between expressed AKV endogenous viruses and other endogenous viruses results in the formation of polytropic pathogenic viruses. These viruses are recombinants that have acquired new SU sequences (as well as LTR mutations) that expand host range as compared to the parent (66, 123). Thus, diversity in this case may also reflect a selection for increased replication.

5' LEADER   Tissue-specific determinants have been mapped to the 5'-noncoding leader region of MuLV (78, 149). MuLV cannot replicate in undifferentiated embryonal carcinoma cells, but host range mutants containing 5'-leader mutations have been isolated.

REPLICATION VARIANTS OF HIV-1   HIV infections, like other animal lentivirus infections (47, 87), are characterized by long latency periods (years in AIDS), degenerative tissue syndromes, and/or immune system dysfunction.

Table 1 Hypothetical role of selective forces in generation of diversity in retroviruses: Potential selective forces, retroviral target regions, and selected genotypic and phenotypic changes

Target region	Potential selective force	Genotypic/phenotypic change	†References
LTR promoter-enhancer	Increased replication	Sequence changes in LTR/enhanced tissue-specific transcription and replication; pathogenicity	11, 15, 44
	Reduced replication or pathogenicity to allow survival of host	Sequence differences between endogenous and exogenous viruses LTRs/endogenous promoter is weaker and virus replicates more slowly	24, 141, 142
5' Noncoding region	Tissue-specific replication	Sequence changes in 5' noncoding region/ability to replicate in undifferentiated cell type	78, 149
Gag	Altered tropism (post penetration)	Amino acid changes in *gag* CA protein/tissue-specific DNA integration	10
Pol	Drug that inhibits RT	Amino acid changes in *pol* gene/drug resistance	69

Env	Attack by the immune system	*Env* sequence changes/neutralization resistance	9, 21, 101, 110
	Tissue specific replication (utilization of a new receptor)	*Env* sequence changes in receptor binding region/ tissue-specific growth; pathogenicity	20, 66, 100, 123
	Ability to replicate in different host species	*Env* sequence changes?/recognition of new receptors?	28
Oncogene	Selective proliferation of transformed host cells	Capture and activation of oncogene/ transformation by virus	6, 7
RNA splice site	Selective proliferation of transformed host cells	*Cis*-acting RNA mutations/increase in splicing of oncogene	86
	Efficient replication	*Cis*-acting RNA mutations/balanced *env* splicing	64
Nonspecific	Attack by the immune system	Inactivation of viral genes by mutations/ replication defect; induction of viral latency?	45, 84

† References discuss or describe phenomena that may be related to the indicated selective force. The list is not comprehensive; some references are representative of a number of similar reports.

HIVs infect cells of the immune system, primarily helper T cells, which express the CD4 cell surface receptor (14, 50, 95). The potential for genetic diversity of HIV (within or between infected individuals) may play a significant role in the disease process, possibly affecting replication rates, tissue tropisms, the ability to evade the immune system, persistence, pathogenesis, cytopathicity (syncytia formation), immune-mediated toxicity, transmissibility, and resistance to antiviral drugs (reviewed in ref. 36, 148, 159) (Table I).

Early studies using restriction enzyme and nucleotide sequence analysis demonstrated that molecular DNA clones from separate isolates of HIV-1 were quite different. Variation was most extensive within the *env* gene; it was manifested as three major variable regions in SU that contained nucleotide substitutions, and deletions and insertions (2, 21, 121, 158). In addition to differences between molecular clones from separate individuals, sequential changes were detected in molecular clones prepared from isolates obtained from the same individual over a period of up to two years (48). However, most isolates from one individual were composed of a predominant strain.

The diversity observed in the *env* gene of HIV-1 is likely a reflection of selective pressure exerted by the human immune system (see below). However, other pressures must also exist, since separate isolates exhibit different phenotypes as gauged by differing criteria. These include replication in fresh T cells (3), macrophages (37), T-cell and astrocytic cell lines (109), syncytium-inducing capacity (140), tissue tropisms (16, 35, 67), and cytotoxicity. Such analyses suggest that, over time, the cytopathic potential and replication rate of the viruses increase. There is also evidence for widening of tissue tropism during the progression of the disease in a single individual (16). Other studies demonstrated that individual genomes within a single isolate possess different growth properties (38, 109). Understanding the origin of these changes may be crucial in understanding the pathogenesis of this virus. One obvious difficulty in assessing the genetic basis for these diverse phenotypes is that a given phenotype may be the result of multiple genetic changes. In addition, the overlapping genes of HIVs may preclude simple genetic analysis (39).

RNA SPLICING   A balance of full-length and spliced viral RNAs is required for efficient virus production (Figure 1) and, therefore, splicing must be regulated such that some full-length RNA remains. ASLV mutants can be selected in tissue culture in which balanced splicing is enforced by sequence changes that result in suboptimal *cis*-acting RNA splicing signals (64). The changes include single nucleotide substitutions as well as deletions and are presumed to arise as a consequence of RT infidelity. Selection for expression of retroviral oncogenes could also be up- or down-modulated by similar changes in response to appropriate selective conditions (86).

Reduced Replication

Diversity may also arise from selection for a reduced rate of replication. In an individual infection this may be manifested as latency; latent viruses may escape detection by the immune system but may subsequently be reactivated. Reduced replication may also be selected for during long-term evolution, giving rise to relatively benign endogenous viruses (see section on Evolution below).

As noted above, a weak but apparently effective immune response appears to occur after the initial HIV-1 infection. However, the virus is not eliminated and eventually reappears years later at high accumulations. It has been proposed that (latent) dormant HIV genomes persist. At least two general models explain HIV latency. In one model it can be hypothesized that viral DNAs are simply not expressed. Activation may subsequently occur after stimulation of host T cells by antigen or cytokines; HIV expression is then regulated by *tat* and *rev* (HIV-encoded *trans*-acting proteins) (23). Virus replication may also be transiently blocked at preintegration or posttranscriptional steps. Putative *trans*-acting negative regulators encoded by HIV-1 could also play a role in repressing expression during latency (23). In a second model, it can be hypothesized that defective DNA genomes persist that may be rescued later (45, 84). In this regard, it was reported that virus-negative, asymptomatic, antibody-positive individuals harbor T cells that contain integrated HIV-1 DNA and also express CD4 receptors on their surface (97). In contrast, CD4 receptors are usually lost in tissue culture cells infected with nondefective HIV. The reason for this discrepancy is not clear, but since receptors are still expressed on the patient's T4 cells, these cells may not be resistant to superinfection. Thus, if defective HIV-1 proviruses are present in these T cells, they could possibly be rescued by recombination with a second HIV with a different defect (17).

Nucleotide sequence analysis of HIV DNAs that were amplified by PCR indicate that considerable numbers of defective proviruses are present in infected individuals (45, 84). In addition, temporal variations during the disease course were observed in the integrated HIV *tat* genes when lymphocyte DNA was analyzed (84). Some *tat* genes, containing nucleotide changes, had reduced activity. When virus isolates were obtained by coculturing these lymphocytes with susceptible cells, the resulting genomes were not representative of the integrated proviruses. Thus, in vitro culturing of virus may select for particular variants (84).

In the ASLV system, the endogenous RAV-0 virus shows a reduced rate of replication compared to exogenous ASLV. This difference can be attributed to sequences located primarily in the U3 region of LTR (141, 142) that affect RNA transcription (24). Such reduced activity may have been selected for in evolution, resulting in benign endogenous viruses that can coexist with the host.

Evasion of the Immune System

Retroviruses of the lentivirus group (47, 87) display antigenic diversity. The *env* gene products are targets for neutralizing antibodies, and diversity appears to reflect selection of *env* variants that can evade the immune system (Table I). The appearance of such lentivirus variants has been documented for visna virus of sheep (120) and equine infectious anemia virus (EIAV) (110). EIA is characterized by episodic illnesses each episode of which, according to virus neutralization studies, is associated with unique antigenic variants. There is a concomitant variation in *env*-coding nucleotides (110). Variations in the *env* protein include altered electrophoretic mobilities and glycosylation patterns. These results suggest that after an initial immune response to EIAV, variants resistant to neutralization can emerge, giving rise to new disease episodes. These cycles repeat over periods of as short as 4–8 weeks. Thus, the neutralizing immune response appears to be a strong selective force for the generation of EIAV *env* variants (Table I).

In AIDS patients, HIV-neutralizing antibodies (9, 83) are weak and likely ineffective during the late stage of the disease (years after infection), although an effective response may occur after the initial infection. The epitopes recognized by neutralizing antibodies are in both conserved and variable regions of SU and therefore some natural neutralizing activities are strain-specific, but most are broadly reactive (52, 103, 117, 153). It may be that clustered changes in SU occur in regions that can withstand variation because they do not disrupt important functions (21). These regions may trigger an immune response that elicits neutralizing antibodies, but variants resistant to neutralization can emerge in which significant amino acid changes have occurred. Not enough is known concerning the three dimensional structure of SU to explain why antibodies to conserved epitopes are not more effective in neutralizing the virus. Some functionally important regions (i.e. the strongly conserved CD4 recognition region) (72) may be partly or entirely inaccessible to the immune system in the intact infectious virion (9). Such an arrangement may have been selected for during HIV evolution. However, antibody-mediated enhancement of HIV infection has also been demonstrated (9). The significance to the disease outcome of this observation is not clear, but suggests that antibody binding may not necessarily be a negative selective pressure for this virus.

In ALV, many natural infections are congenital, producing immunological tolerance to viral antigens (130). As noted above, the most highly variable regions in ALV *env* are involved in receptor recognition. Therefore, where immune elimination may not be a strong selective force, antigenic variation is not a prominent feature of ASLV *env*.

Transformation of the Host Cell

Oncogenic transformation of an infected host cell may provide a temporary selective advantage for a retroviral population (Table I). Nondividing target cells are stimulated to proliferate by expression of the oncogenes carried by transforming retroviruses (6, 7). The provirus is replicated and thus propagated along with the host cell. However, acute transforming viruses that carry oncogenes rapidly kill their hosts and thus are not readily transmitted in nature. Thus, transformation of the host cell provides only a short-term selective advantage for the virus in vivo. Viral oncogenes may also manifest diversity (i.e. point mutations and deletions) that may modulate or be required for their transforming activity (6, 7, 136).

EVOLUTION OF RETROVIRUSES

In tissue culture systems or in infected animals, retroviruses can evolve over periods of weeks to years. In this section we discuss long-term evolution of retroviruses; this time scale ranges from hundreds to perhaps billions of years. In considering retroviral evolution, it is useful to compare retroviral genomes with related elements present in both prokaryotes (139, 143) and eukaryotes (19, 123, 135, 151). These elements include RT-like sequences lacking LTRs (retroposons, e.g. human LINE elements) or RT-like sequences that include LTRs (retrotransposons, e.g. yeast Ty elements). Other elements do not contain RT sequences or LTRs (retrotranscripts, e.g. processed pseudogenes), but appear to have been transposed through RNA intermediates via RT. Classes of retroviruses/retroelements can be distinguished by RT homologies (29).

As noted above, selective forces in molecular evolution may act on functional regions, generating clustered areas of variability or, on the contrary, clustered variations may be restricted to nonessential regions that can withstand changes. The RT gene of retroelements is an example of the latter, in that only small regions of strong homology can be discerned between distantly related RTs (59); one conserved region is likely part of the enzyme active site (62, 70). However, all comparisons of this type must take into account the possibility of convergent evolution.

Two important principles have emerged regarding the evolution of RT and retroviruses:
1. So far, all eukaryotic genes that share strong sequence homology with retroviral RTs appear to be components of recognizable retroelements (4).

However, other RT-related genes may be difficult to identify in eukaryotic genomes because retroelements are so numerous. An RT-like gene has been identified in bacteria that does not appear to be part of a transposable element (139, 143). This may be descended from an ancestral RT gene that existed prior to the divergence of prokaryotes and eukaryotes. However, the possibility that RT was acquired by certain bacterial strains after divergence cannot be excluded. The recently discovered RT activity associated with eukaryotic telomerase may be another descendant of an ancestral RT activity not part of a retrotransposon (114). Additional RT-related genes of this type may be identified through the activities of their gene products rather than by DNA sequence similarity.

2. Non-viral retroelement RTs have diverged considerably in evolution as compared to conserved cellular genes such as cytochrome C. One possible reason is that the RT gene itself is subject to the mutations inherent in reverse transcription as it is transposed as part of the retroelement (29). A striking example has been noted by Doolittle and colleagues (82), who suggested that the protease genes of two human retroviruses (HIV-1 and HTLV-II) had diverged considerably, compared to the divergence of a prototypic eukaryotic cellular protease gene and a hypothetical prokaryotic ancestral cellular protease; the cellular proteases have not been subjected to reverse transcription.

The possible origin(s) of retroviruses has provoked a special interest because retroviral DNA joins with the chromosomal DNA as a normal part of the replication cycle. Thus, cellular DNA seemed like a possible source of retroviral genes. This issue was addressed by Temin in 1970 in his protovirus hypothesis, in which he proposed that retroviruses evolved from mobile DNA genetic elements. This is consistent with the observation that retroviral proviruses share similarities in structure and function with the transposable elements of bacteria, yeast, and Drosophila (134). He further hypothesized that capture of a cellular RT gene by LTR sequences was an early event in the evolution of retroviruses, with *gag* and *env* derived from subsequent captures of cellular genes. The discovery of an RT-like gene in bacteria supports the notion that retroviruses evolved from ancient preexisting cellular genes (139, 143). It can also be argued that retrotransposons (containing only LTRs and RT) might simply represent degenerate retroviruses that no longer encode the genes required for an extracellular existence. A third view contends that if the prebiotic genetic material was RNA, reverse transcription might have been required to formulate DNA-based genetic information; that is, reverse transcriptase might have been one of the first protein enzymes. If so, retroelements might have existed very early in evolution. Some support for this view comes from the characterization of mitochondrial DNA plasmids that encode RT activity (68)

The issue of retroviral evolution is confused by several observations. First, exogenous retroviral sequences continue to enter the germline (19, 123). This makes it difficult to assess whether any individual example represents a retroviral precursor or an acquired retrovirus that has degenerated. However, the presence or absence of common endogenous virus loci in related species indicates whether the insertion occurred pre- or postspeciation (29). Second, considerable circumstantial evidence supports interspecies infection (29). Avian spleen necrosis virus, which is more highly related to murine viruses than other avian retroviruses, may be such a case (82). Finally, it is difficult to trace the origins or dispersion of individual retroviral genes. For example, the RNaseH segment of *pol* is more highly related to RNaseH of *E. coli* than comparisons between RT and *E. coli* polymerase would predict (29, 59). This observation may reflect a particularly strong selection for conservation of RNaseH function as compared to RT. Alternatively, RNaseH may have been acquired by *E. coli* from retroelements (29). Sequence comparisons of this type are also difficult to interpret since distinct selective forces may be acting in different systems. Other sequence comparisons implicate recombination as a facet of retroviral evolution (82) and suggest other mechanisms of gene exchange (81). The apparently recent acquisition by one retrovirus of a protease gene segment from a second retrovirus of a distinct host species supports the notion that new enzymatic components can be acquired (81).

Since HIVs appear to have emerged recently as identifiable pathogens, the question of their origin has elicited particular interest. Where did HIVs come from? The two obvious sources are from humans themselves, or from other primate species found in endemic regions. As discussed by Doolittle (28), recent evidence suggests the latter. Nucleotide sequence analysis indicates that HIV-2 is closely related to a sooty mangabey lentivirus (simian immunodeficiency virus, SIV_{sm}) that does not cause disease in its natural host. It is possible that sooty mangabeys may have been a reservoir for human infection. It appears likely that HIV-2 is a descendant of SIV_{sm}, which was recently transmitted to man, perhaps within the past 50 years (28). Although the origin of HIV-1 is more obscure, it may have independently evolved in a similar manner from SIVs. An alternative model for the origin of HIV-1 suggests that HIV-1 and SIV existed in their respective hosts since these species diverged (millions of years ago) (40).

Retroviruses are unique in that they can evolve in two separate ecological niches—as exogenous infectious agents or as endogenous proviruses. In exogenous retroviruses, as with all pathogens, diversity that affects virulence and transmission must be balanced. Although some selection for high proliferative rate may occur, it is disadvantageous for the virus to kill one host before it can spread to another. On the other hand, germline infection will favor persistence of benign elements. A long-standing question is whether it is advantageous to the host to maintain retroelements (29, 133, 150). The endogenous IAP particles (A-type) of mice, which bud intracellularly, are expressed during early mouse embryo development and in tissue culture cells (19, 67a, 102, 123). New IAP DNA insertions can be detected; the ability to form intracellular particles may make the IAP endogenous elements particularly efficient in re-insertion. The observation that recent C-type endogenous retrovirus insertions have caused mutations in the mouse (125) suggests that these elements may contribute to generation of genetic diversity in this host population (150). Individual chickens that lack endogenous RAV-0-related viruses are not discernibly different from those that do not; however, chickens may harbor other endogenous viruses, not related to ASLV (123). Endogenous retroviruses may provide a subtle selective advantage that is not obvious during the lifetime of an individual animal but may be relevant to evolution in the long term. At the other extreme, endogenous retroviruses may simply be types of "selfish DNAs" that perpetuate themselves without providing any benefit to the host.

SUMMARY AND CONCLUSIONS

Retroviruses are unique in that their propagation includes transfer of genetic information from RNA to DNA. Two enzymes (RT and RNA polymerase II) that participate in their replication process do not encode editing functions and are thus "error prone." Current estimates indicate that up to one nucleotide substitution per genome occurs per retrovirus replication cycle. In addition, rearrangements can occur during reverse transcription. These mutations result in viral populations in which the wild-type sequence can only be defined by consensus. Retroviruses are also unusual among viruses in their high recombination frequency, which is the result of the copackaging, and then reverse transcribing of two different RNA genomes in the same particle. Recombinants are formed during copying of RNA templates into DNA. With the ability to mutate and recombine genetic information at a high rate, retroviral populations are poised to respond to selective forces, which may increase or decrease replication of particular genotypes. Interspecies transmission of retroviruses or retroviral genes also likely plays a role in generating diversity. Endogenous proviruses exist in the germline of many vertebrates; pedigrees indicate that recent infections of the germline have also occurred. Thus, the selective forces that mold the retroviral genome may be opposing: selection for efficient replication as exogenous viruses versus selection for passive replication as endogenous proviruses. The latter may be more advantageous over the course of evolution since the survival of the retrovirus is ensured by survival of the host organism. Segments of endogenous viruses may reappear in exogenous viruses through recombination and thus these endogenous sequences are perpetuated.

ACKNOWLEDGMENTS

We are grateful to John Coffin, Bryan Cullen, Al Knudson, Joe Kulkosky, Marcella McClure, John Taylor and Howard Temin for critical comments on this manuscript. We thank John Coffin and Howard Temin for providing manuscripts prior to publication and other colleagues who provided reprints. We also thank Philip Tsichlis for helpful discussions. We are grateful to Marie Estes and Mary Williamson for their patient and excellent assistance in preparing this manuscript. The authors' research is supported by National Institutes of Health grants CA-06927, CA-48703, RR-05539, a grant from the Pew Charitable Trust, by an appropriation from the Commonwealth of Pennsylvania, and by the W. W. Smith Charitable Trust.

Literature Cited

1. Albritton, L. M., Tseng, L., Scadden, D., Cunningham, J. M. 1989. A putative murine ecotropic retrovirus receptor gene encodes a multiple membrane-spanning protein and confers susceptibility to virus infection. *Cell* 57:659–66
2. Alizon, M., Wain-Hobson, S., Montaignier, L., Sonigo, P. 1986. Genetic variability of the AIDS virus: Nucleotide sequence analysis of two isolates from African patients. *Cell* 46:63–74
3. Asjö, B., Morefeldt-Manson, L., Albert, J., Biberfeld, G., Karlsson, A., et al. 1986. Replicative capacity of human immunodeficiency virus from patients with varying severity of HIV infection. *Lancet* 11:660–62
4. Baltimore, D. 1985. Retroviruses and retrotransposons: The role of reverse transcription in shaping the eukaryotic genome. *Cell* 40:481–82
5. Battula, N., Loeb, L. A. 1974. The infidelity of avian myeloblastosis virus deoxyribonucleic acid polymerase in polynucleotide replication. *J. Biol. Chem.* 249:4086–93
6. Bishop, J. M., Varmus, H. E. 1982. Functions and origins of retroviral transforming genes. See Ref. 155, pp. 999–1108
7. Bishop, J. M., Varmus, H. E. 1985. Functions and origins of retroviral transforming genes. See Ref. 156, pp. 249–356
8. Bizub, D., Katz, R. A., Skalka, A. M. 1984. Nucleotide sequence of noncoding regions in Rous-associated virus-2: comparisons delineate conserved regions important in replication and oncogenesis. *J. Virol.* 49:557–65
9. Bolognesi, D. P. 1989. HIV antibodies and vaccine design. *AIDS* 3 (suppl. 1):S111–18
10. Boone, L. R., Glover, P. L., Innes, C. L., Niver, L. A., Bondurant, M. C., et al. 1988. *Fv-1* N- and B-tropism-specific sequences in murine leukemia virus and related endogenous proviral genomes. *J. Virol.* 62:2644–50
11. Boral, A. L., Okenquist, S. A., Lenz, J. 1989. Identification of the SL3-3 virus enhancer core as a T-lymphoma cell-specific element. *J. Virol.* 63:76–84
12. Bowerman, B., Brown, P. O., Bishop, J. M., Varmus, H. E. 1989. A nucleoprotein complex mediates the integration of retroviral DNA. *Genes Dev.* 3:469–78
13. Brown, P. O., Bowerman, B., Varmus, H. E., Bishop, J. M. 1987. Correct integration of retroviral DNA in vitro. *Cell* 49:347–56
14. Cann, A. J., Karn, J. 1989. Molecular biology of HIV: new insights into the virus life-cycle. *AIDS* 3(Suppl.1):S19–S34
15. Celander, D., Haseltine, W. A. 1984. Tissue-specific transcription preference as a determinant of cell tropism and leukaemogenic potential of murine retroviruses. *Nature* 312:159–62
16. Cheng-Mayer, C., Seto, D., Tateno, M., Levy, J. A. 1988. Biologic features of HIV-1 that correlate with virulence in the host. *Science* 240:80–82
17. Clavel, F., Hoggan, M. D., Willey, R. L., Strebel, K., Martin, M. A., et al. 1989. Genetic recombination of human immunodeficiency virus. *J. Virol.* 63:1455–59
18. Coffin, J. M. 1979. Structure, replication, and recombination of retrovirus genomes: some unifying hypotheses. *J. Gen. Virol.* 42:1–26

19. Coffin, J. M. 1982. Endogenous viruses. See Ref. 155, pp. 1109–203
20. Coffin, J. M. 1990. Genetic variation in retroviruses. In *Applied Virology Research, Virus Variation and Epidemiology*, ed. E. Kurstak, R. G. Marusyk, F. A. Murphy, M. H. V. Van Regenmortel, Vol. 2. In press
21. Coffin, J. M. 1986. Genetic variation in AIDS viruses. *Cell* 46:1–4
22. Coffin, J. M., Tsichlis, P. M., Barker, C. S., Voynow, S. 1980. Variation in avian retrovirus genomes. *Ann. NY Acad. Sci.* 354:410–25
23. Cullen, B. R., Greene, W. C. 1989. Regulatory pathways governing HIV-1 replication. *Cell* 58:423–26
24. Cullen, B. R., Skalka, A. M., Ju, G. 1983. Endogenous avian retroviruses contain deficient promoter and leader sequences. *Proc. Natl. Acad. Sci. USA* 80:2946–50
25. Darlix, J.-L., Spahr, P.-F. 1983. High spontaneous mutation rate of Rous sarcoma virus demonstrated by direct sequencing of the RNA genome. *Nucleic Acids Res.* 11:5953–67
26. Domingo, E., Holland, J. J. 1988. High error rates, population equilibrium, and evolution of RNA replication systems. In *RNA Genetics*, ed. E. Domingo, P. Ahlquist, J. J. Holland, 3:3–36 Boca Raton, Fl: CRC Press
27. Domingo, E., Martínez-Salas, E., Sobrino, F., de la Torre, J. C., Portela, A., et al. 1985. The quasispecies (extremely heterogeneous) nature of viral RNA genome populations: biological relevance—a review. *Gene* 40:1–8
28. Doolittle, R. F. 1989. The simian-human connection. *Nature* 339:338–39
29. Doolittle, R. F., Feng, D.-F., Johnson, M. S., McClure, M. A. 1989. Origins and evolutionary relationshps of retroviruses. *Q. Rev. Biol.* 64:1–30
30. Dorner, A. J., Coffin, J. M. 1986. Determinants for receptor interaction and cell killing on the avian retrovirus glycoprotein gp85. *Cell* 45:365–74
31. Dorner, A. J., Stoye, J. P., Coffin, J. M. 1985. Molecular basis of host range variation in avian retroviruses. *J. Virol.* 53:32–39
32. Dougherty, J. P., Temin, H. M. 1986. High mutation rate of a spleen necrosis virus-based retrovirus vector. *Mol. Cell. Biol.* 168:4387–95
33. Dougherty, J. P., Temin, H. M. 1988. Determination of the rate of base-pair substitution and insertion mutations in retrovirus replication. *J. Virol.* 62:2817–22
34. Einfeld, D., Hunter, E. 1988. Oligomeric structure of a prototype retrovirus glycoprotein. *Proc. Natl. Acad. Sci. USA* 85:8688–92
35. Evans, L. A., McHugh, T. M., Stites, D. P., Levy, J. A. 1987. Differential ability of human immunodeficiency virus isolates to productively infect human cells. *J. Immunol.* 138:3415–18
36. Fenyö, E. M., Albert, J., Åsjö, B. 1989. Replicative capacity, cytopathic effect and cell tropism of HIV. *AIDS* 3(Suppl. 1):S5–S12
37. Fenyö, E. M., Morfeldt-Månson, L., Chiodi, F., Lind, B., von Gegerfelt, A., et al. 1988. Distinct replicative and cytopathic characteristics of human immunodeficiency virus isolates. *J. Virol.* 62:4414–19
38. Fisher, A. G., Ensoli, B., Looney, D., Rose, A., Gallo, R. C., et al. 1988. Biologically diverse molecular variants within a single HIV-1 isolate. *Nature* 334:444–47
39. Franza, R. F., Cullen, B. R., Wong-Staal, F., eds. 1988. *The Control of Human Retrovirus Gene Expression*. Cold Spring Habor, NY: Cold Spring Harbor Lab.
40. Fukasawa, M., Miura, T., Hasegawa, A., Morikawa, S., Tsujimoto, H., et al. 1988. Sequence of simian immunodeficiency virus from African green monkey, a new member of the HIV/SIV group. *Nature* 333:457–61
41. Gallo, R. C., Nerurkar, L. S. 1989. Human retroviruses: Their role in neoplasia and immunodeficiency. *Ann. NY Acad. Sci.* 567:82–94
42. Gojobori, T., Yokoyama, S. 1985. Rates of evolution of the retroviral oncogene of Moloney murine sarcoma virus and of its cellular homologues. *Proc. Natl. Acad. Sci. USA* 82:4198–201
43. Goldfarb, M., Weinberg, R. A. 1981. Generation of novel, biologically active Harvey sarcoma viruses via apparent illegitimate recombination. *J. Virol.* 38:136–50
44. Golemis, E. C., Speck, N. A., Hopkins, N. 1990. Alignment of U3 region sequences of mammalian Type C viruses: Identification of highly conserved motifs and implications for enhancer design. *J. Virol.* 64:534–42
45. Goodenow, M., Huet, T., Saurin, W., Kwok, S., Sninsky, J., Wain-Hobson, S. 1989. HIV-1 isolates are rapidly evolving quasispecies: Evidence for viral mixtures and preferred nucleotide substitutions. *J. Acquired Immune Defic. Syndr.* 2:344–52
45a. Goodrich, D. W., Duesberg, P. H.

1990. Retroviral recombination during reverse transcription. *Proc. Natl. Acad. Sci. USA* 87:2052–56

46. Gopinathan, K. P., Weymouth, L. A., Kunkel, T. A., Loeb, L. A. 1979. Mutagenesis in vitro by DNA polymerase from an RNA tumor virus. *Nature* 278:857–59

47. Haase, A. T. 1986. Pathogenesis of lentivirus infections. *Nature* 322:130–36

48. Hahn, B. H., Shaw, G. M., Taylor, M. E., Redfield, R. R., Markham, P. D., et al. 1986. Genetic variation in HTLV-III/LAV over time in patients with AIDS or at risk for AIDS. *Science* 232:1548–53

49. Hanafusa, H., Halpern, C. C., Buchhagen, D. L., Kawai, S. 1977. Recovery of avian sarcoma virus from tumors induced by transformation-defective mutants. *J. Exp. Med.* 146:1735–47

50. Haseltine, W. 1988. Replication and pathogenesis of the AIDS virus. *J. Acquired Immune Defic. Syndr.* 1:217–40

51. Herman, S. A., Coffin, J. M. 1987. Efficient packaging of readthrough RNA in ALV: Implications for oncogene transduction. *Science* 236:845–48

52. Ho, D. D., Kaplan, J. C., Rackauskas, I. E., Gurney, M. E. 1988. Second conserved domain of gp120 is important for HIV infectivity and antibody neutralization. *Science* 239:1021–23

53. Holland, J., Spindler, K., Horodyski, F., Grabau, E., Nichol, S., VandePol, S. 1982. Rapid evolution of RNA genomes. *Science* 215:1577–85

54. Hsu, T. W., Taylor, J. M. 1982. Single-stranded regions on unintegrated avian retrovirus DNA. *J. Virol.* 44:47–53

55. Hu, W.-S., Temin, H. M. 1990. Genetic consequences of packaging two RNA genomes in one retroviral particle: pseudodiploidy and high rate of genetic recombination. *Proc. Natl. Acad. Sci. USA* 87:1556–60

56. Huang, A. S. 1988. Modulation of viral disease processes by defective interfering particles. See Ref. 26, pp. 195–208

57. Huang, C-C., Hay, N., Bishop, J. M. 1986. The role of RNA molecules in transduction of the proto-oncogene c-*fps*. *Cell* 44:935–40

58. Hunter, E. 1978. The mechanism for genetic recombination in the avian retroviruses. *Curr. Top. Microbiol. Immunol.* 79:295–309

59. Johnson, M. S., McClure, M. A., Feng, D.-F., Gray, J., Doolittle, R. F. 1986. Computer analysis of retroviral *pol* genes: Assignment of enzymatic functions to specific sequences and homologies with nonviral enzymes. *Proc. Natl. Acad. Sci. USA* 83:7648–52

60. Junghans, R. P., Boone, L. R., Skalka, A. M. 1982. Retroviral DNA H structures: Displacement-assimilation model of recombination. *Cell* 30:53–62

61. Junghans, R. P., Boone, L. R., Skalka, A. M. 1982. Products of reverse transcription in avian retrovirus analyzed by electron microscopy. *J. Virol.* 43:544–54

62. Kamer, G., Argos, P. 1984. Primary structural comparison of RNA-dependent polymerases from plant, animal and bacterial viruses. *Nucleic Acids Res.* 12:7269–82

63. Katz, R. A., Jentoft, J. E. 1989. What is the role of the Cys-His motif in retroviral nucleocapsid (NC) proteins? *BioEssays* 11:176–81

64. Katz, R. A., Skalka, A. M. 1990. Control of retroviral RNA splicing through maintenance of suboptimal processing signals. *Mol. Cell. Biol.* 10:696–704

65. Katzman, M., Katz, R. A., Skalka, A. M., Leis, J. 1989. The avian retroviral integration protein cleaves the terminal sequences of linear viral DNA at the in vivo sites of integration. *J. Virol.* 63:5319–27

66. Khan, A. S. 1984. Nucleotide sequence analysis establishes the role of endogenous murine leukemia virus DNA segments in formation of recombinant mink cell focus-forming murine leukemia viruses. *J. Virol.* 50:864–71

67. Koyanagi, Y., Miles, S., Mitsuyasu, R. T., Merrill, J. E., Vinters, H. V., et al. 1987. Dual infection of the central nervous system by AIDS viruses with distinct cellular tropisms. *Science* 236:819–22

67a. Kuff, E. L., Lueders, K. K. 1988. The intracisternal A-particle gene family: Structure and functional aspects. *Adv. Cancer Res.* 51:183–276

68. Kuiper, M. T. R., Lambowitz, A. M. 1988. A novel reverse transcriptase activity associated with mitochondrial plasmids of neurospora. *Cell* 55:693–704

69. Larder, B. A., Kemp, S. D. 1989. Multiple mutations in HIV-1 reverse transcriptase confer high-level resistance to zidovudine (AZT). *Science* 246:1155–58

70. Larder, B. A., Purifoy, D. J. M., Powell, K. L., Darby, G. 1987. Site-specific mutagenesis of AIDS virus reverse transcriptase. *Nature* 327:716–17

71. Larsson, E., Kato, N., Cohen, M. 1989. Human endogenous proviruses. *Curr. Top. Microbiol. Immunol.* 148:115–32

72. Lasky, L. A., Nakamura, G., Smith, D. H., Fennie, C., Shimasaki, C., et al.

1987. Delineation of a region of the human immunodeficiency virus type 1 gp 120 glycoprotein critical for interaction with the CD4 receptor. *Cell* 50:975–85

73. Leider, J. M., Palese, P., Smith, F. I. 1988. Determination of the mutation rate of a retrovirus. *J. Virol.* 62:3084–91

74. Leis, J., Baltimore, D., Bishop, J. M., Coffin, J., Fleissner, E., et al. 1988. Standardized and simplified nomenclature for proteins common to all retroviruses. *J. Virol.* 62:1808–9

75. Linial, M., Blair, D. 1982. Genetics of Retroviruses. See Ref. 155, pp. 649–783

76. Linial, M., Miller, A. D. 1989. Retroviral RNA packaging: Sequence requirements and implications. *Curr. Top. Microbiol. Immunol.* 157:125–51

77. Loeb, L. A., Kunkel, T. A. 1982. Fidelity of DNA synthesis. *Annu. Rev. Biochem.* 52:429–57

78. Loh, T. P., Sievert, L. A., Scott, R. W. 1988. Negative regulation of retrovirus expression in embryonal carcinoma cells mediated by an intragenic domain. *J. Virol.* 62:4086–95

79. Luo, G., Sharmeen, L., Taylor, J. 1990. Specificities involved in the initiation of retroviral plus-strand DNA. *J. Virol.* 64:592–97

80. Martinelli, S. C., Goff, S. P. 1990. Rapid reversion of a deletion mutation in Moloney murine leukemia virus by recombination with a closely related endogenous provirus. *Virology* 174:135–44

81. McClure, M. A., Johnson, M. S., Doolittle, R. F. 1987. Relocation of a protease-like gene segment between two retroviruses. *Proc. Natl. Acad. Sci. USA* 84:2693–97

82. McClure, M. A., Johnson, M. S., Feng, D.-F., Doolittle, R. F. 1988. Sequence comparisons of retroviral proteins: Relative rates of change and general phylogeny. *Proc. Natl. Acad. Sci. USA* 85:2469–73

83. McKeating, J. A., Willey, R. L. 1989. Structure and function of the HIV envelope. *AIDS* 3(Suppl. 1):S35–S41

84. Meyerhans, A., Cheynier, R., Albert, J., Seth, M., Kwok, S., et al. 1989. Temporal fluctuations in HIV quasispecies in vivo are not reflected by sequential HIV isolations. *Cell* 58:901–10

85. Miller, A. D., Buttimore, C. 1986. Redesign of retrovirus packaging cell lines to avoid recombination leading to helper virus production. *Mol. Cell. Biol.* 6:2895–902

86. Miller, C. K., Embretson, J. E., Temin,

H. M. 1988. Transforming viruses spontaneously arise from nontransforming reticuloendotheliosis virus strain T-derived viruses as a result of increased accumulation of spliced viral RNA. *J. Virol.* 62:1219–26

87. Narayan, O., Clements, J. E. 1989. Biology and pathogenesis of lentiviruses. *J. Gen. Virol.* 70:1617–39

88. Olsen, J. C., Swanstrom, R. 1985. A new pathway in the generation of defective retrovirus DNA. *J. Virol.* 56:779–89

89. O'Rear, J. J., Temin, H. M. 1982. Spontaneous changes in nucleotide sequence in proviruses of spleen necrosis virus, an avian retrovirus. *Proc. Natl. Acad. Sci. USA* 79:1230–34

90. Ott, R., Friedrich, R., Rein, A. 1990. Sequence analysis of amphotropic and 10A1 murine leukemia viruses: close relationship to mink cell focus-inducing viruses. *J. Virol.* 64:757–66

91. Özel, M., Pauli, G., Gelderblom, H. R. 1988. The organization of the envelope projections on the surface of HIV. *Arch. Virol.* 100:255–66

92. Panganiban, A. T., Fiore, D. 1988. Ordered interstrand and intrastrand DNA transfer during reverse transcription. *Science* 241:1064–69

93. Panganiban, A. T., Temin, H. M. 1984. Circles with two tandem LTRs are precursors to integrated retrovirus DNA. *Cell* 36:673–79

94. Perrino, F. W., Preston, B. D., Sandell, L. L., Loeb, L. A. 1989. Extension of mismatched 3'termini of DNA is a major determinant of the infidelity of human immunodeficiency virus type 1 reverse transcriptase. *Proc. Natl. Acad. Sci. USA* 86:8343–47

95. Peterlin, B. M., Luciw, P. A. 1988. Molecular biology of HIV. *AIDS* 2(Suppl. 1):S29–S40

96. Preston, B. D., Poiesz, B. J., Loeb, L. A. 1988. Fidelity of HIV-1 reverse transcriptase. *Science* 242:1168–71

97. Psallidopoulos, M. C., Schnittman, S. M., Thompson, L. M. III, Baseler, M., Fauci, A. S., et al. 1989. Integrated proviral human immunodeficiency virus type 1 is present in CD4+ peripheral blood lymphocytes in healthy seropositive individuals. *J. Virol.* 63:4626–31

98. Raines, M. A., Maihle, N. J., Moscovici, C., Crittenden, L., Kung, H.-J. 1988. Mechanism of c-*erb*B transduction: Newly released transducing viruses retain poly(A) tracts of *erb*B transcripts and encode C-terminally intact *erb*B proteins. *J. Virol.* 62:2437–43

99. Reanney, D. C. 1982. The evolution of RNA viruses. *Annu. Rev. Microbiol.* 36:47–73

100. Rein, A. 1982. Interference grouping of murine leukemia viruses: A distinct receptor for the MCF-recombinant viruses in mouse cells. *Virology* 120:251–57

101. Reitz, M. S. Jr., Wilson, C., Naugle, C., Gallo, R. C., Robert-Guroff, M. 1988. Generation of a neutralization-resistant variant of HIV-1 is due to selection for a point mutation in the envelope gene. *Cell* 54:57–63

102. Risser, R., Horowitz, J. M., McCubrey, J. 1983. Endogenous mouse leukemia viruses. *Annu. Rev. Genet.* 17:85–121

103. Robert-Guroff, M., Brown, M., Gallo, R. C. 1985. HTLV-III-neutralizing antibodies in patients with AIDS and AIDS-related complex. *Nature* 316:72–74

104. Roberts, J. D., Bebenek, K., Kunkel, T. A. 1988. The accuracy of reverse transcriptase from HIV-1. *Science* 242: 1171–73

105. Roberts, J. D., Preston, B. D., Johnston, L. A., Soni, A., Loeb, L. A., et al. 1989. Fidelity of two retroviral reverse transcriptases during DNA-dependent DNA synthesis in vitro. *Mol. Cell. Biol.* 9:469–76

106. Robinson, H. L., Jensen, L., Coffin, J. M. 1985. Sequences outside of the long terminal repeat determine the lymphomogenic potential of Rous-associated virus type 1. *J. Virol.* 55: 752–59

107. Roth, M., Schwartzberg, P., Goff, S. 1989. Structure of the termini of DNA intermediates in the integration of retroviral DNA: Dependence on IN function and terminal DNA sequence. *Cell* 58: 47–54

108. Rovigatti, U., Astrin, S. 1983. Avian endogenous viral genes. *Curr. Top. Microbiol. Immunol.* 103:1–22

109. Sakai, K., Dewhurst, S., Ma, X., Volsky, D. J. 1988. Differences in cytopathogenicity and host cell range among infectious molecular clones of human immunodeficiency virus type 1 simultaneously isolated from an individual. *J. Virol.* 62:4078–85

110. Salinovich, O., Payne, S. L., Montelaro, R. C., Hussain, K. A., Issel, C. J., et al. 1986. Rapid emergence of novel antigenic and genetic variants of equine infectious anemia virus during persistent infection. *J. Virol.* 57:71–80

111. Sattentau, Q. J., Weiss, R. A. 1988. The CD4 antigen: Physiological ligand and HIV receptor. *Cell* 52:631–33

112. Schwartzberg, P., Colicelli, J., Goff, S. P. 1984. Construction and analysis of deletion mutation in the *pol* gene of Moloney murine leukemia virus: A new viral function required for productive infection. *Cell* 37:1043–52

113. Scolnick, E. M., Vass, W. C., Howk, R. S., Duesberg, P. H. 1979. Defective retrovirus-like 30S RNA species of rat and mouse cells are infectious if packaged by type C helper virus. *J. Virol.* 29:964–72

114. Shippen-Lentz, D., Blackburn, E. H. 1990. Functional evidence for an RNA template in telomerase. *Science* 247: 546–52

115. Skalka, A. M. 1988. Integrative recombination of retroviral DNA. In Genetic Recombination, ed. R. Kucherlapati, R. Smith, pp. 701–24.Washington, DC: Am. Soc. Microbiol.

116. Skalka, A. M. 1989. Retroviral proteases: First glimpses at the anatomy of a processing machine. *Cell* 56:911–13

117. Skinner, M. A., Langlois, A. J., McDanal, C. B., McDougal, S., Bolognesi, D. P., et al. 1988. Neutralizing antibodies to an immunodominant envelope sequence do not prevent gp120 binding to CD4. *J. Virol.* 62:4195–200

118. Smith, D. B., Inglis, S. C. 1987. The mutation rate and variability of eukaryotic viruses: An analytical review. *J. Gen. Virol.* 68:2729–40

119. Smith, F. I., Palese, P. 1988. Influenza viruses: high rate of mutation and evolution. See Ref. 26, pp. 123–35

120. Stanley, J., Bhaduri, L. M., Narayan, O., Clements, J. E. 1987. Topographical rearrangements of visna virus envelope glycoprotein during antigenic drift. *J. Virol.* 61:1019–28

121. Starcich, B. R., Hahn, B. H., Shaw, G. M., McNeely, P. D., Modrow, S., et al. 1986. Identification and characterization of conserved and variable regions in the envelope gene of HTLV-III/LAV, the retrovirus of AIDS. *Cell* 45:637–48

122. Steinhauer, D. A., Holland, J. J. 1987. Rapid evolution of RNA viruses. *Annu. Rev. Microbiol.* 41:409–33

123. Stoye, J. P., Coffin, J. M. 1985. Endogenous retroviruses. See Ref. 156, pp. 357–404

124. Stoye, J. P., Coffin, J. M. 1988. Polymorphism of murine endogenous proviruses revealed by using virus class-specific oligonucleotide probes. *J. Virol.* 62:168–75

125. Stoye, J. P., Fenner, S., Greenoak, G. E., Moran, C., Coffin, J. M. 1988. Role of endogenous retroviruses as mutagens:

The hairless mutation of mice. *Cell* 54:383–91

126. Swanstrom, R., Parker, R. C., Varmus, H. E., Bishop, J. M. 1983. Transduction of a cellular oncogene: The genesis of Rous sarcoma virus. *Proc. Natl. Acad. Sci. USA* 80:2519–23

127. Takeuchi, Y., Nagumo, T., Hoshino, H. 1988. Low fidelity of cell-free DNA synthesis by reverse transcriptase of human immunodeficiency virus. *J. Virol.* 62:3900–2

128. Teich, N. 1982. Taxonomy of retroviruses. See Ref. 155, pp. 25–207

129. Teich, N. 1985. Taxonomy of retroviruses. See Ref. 156, pp. 1–16

130. Teich, N., Wyke, J., Kaplan, P. 1985. Pathogenesis of retroviral-induced disease. See Ref. 156, pp. 187–248

131. Teich, N., Wyke, J., Mak, T., Bernstein, A., Hardy, W. 1982. Pathogenesis of retrovirus-induced disease. See Ref. 155, pp. 785–998

132. Temin, H. M. 1974. The cellular and molecular biology of RNA tumor viruses, especially avian leukosis-sarcoma viruses, and their relatives. *Adv. Cancer Res.* 19:47–104

133. Temin, H. M. 1982. Viruses, protoviruses, development, and evolution. *J. Cell. Biochem.* 19:105–18

134. Temin, H. M. 1980. Origin of retroviruses from cellular moveable genetic elements. *Cell* 21:599–600

135. Temin, H. M. 1985. Reverse transcription in the eukaryotic genome: Retro viruses, pararetroviruses, retrotransposons, and retrotranscripts. *Mol. Biol. Evol.* 2:455–68

136. Temin, H. M. 1988. Evolution of cancer genes as a mutation-driven process. *Cancer Res.* 48:1697–701

137. Temin, H. M. 1989. Retrovirus variation and evolution. *Genome* 31:17–22

138. Temin, H. M. 1989. Is HIV unique or merely different? *J. Acquired Immune Defic. Syndr.* 2:1–9

139. Temin, H. M. 1989. Retrons in bacteria. *Nature* 339:254–55

140. Tersmette, M., de Goede, R. E. Y., Al, B. J. M., Winkel, I. N., Gruters, R. A., et al. 1988. Differential syncytium-inducing capacity of human immunodeficiency virus isolates: Frequent detection of syncytium-inducing isolates in patients with acquired immunodeficiency syndrome (AIDS) and AIDS-related complex. *J. Virol.* 62:2026–32

141. Tsichlis, P. N., Coffin, J. M. 1979. Recombination between the defective component of an acute leukemia virus and Rous associated virus-0, an endogenous virus of chickens. *Proc. Natl. Acad. Sci. USA* 76:3001–5

142. Tsichlis, P. N., Donehower, L., Hager, G., Zeller, N., Malavarca, R., et al. 1982. Sequence comparison in the cross-over region of an oncogenic avian retrovirus recombinant and its nononcogenic parent: Genetic regions that control growth rate and oncogenic potential. *Mol. Cell. Biol.* 2:1331–38

143. Varmus, H. E. 1989. Reverse transcription in bacteria. *Cell* 56:721–24

144. Varmus, H. E., Brown, P. 1989. Retroviruses. In *Mobile DNA*, ed. D. Berg, M. Howe, pp. 53–108. Washington, DC: Am. Soc. Microbiol.

145. Varmus, H. E., Swanstrom, R. 1982. Replication of retroviruses. See Ref. 155, pp. 369–512

146. Varmus, H. E., Swanstrom, R. 1985. Replication of retrovirus. See Ref. 156, pp. 75–134

147. Voynow, S. L., Coffin, J. M. 1985. Evolutionary variants of Rous sarcoma virus: Large deletion mutants do not result from homologous recombination. *J. Virol.* 55:67–78

148. Wain-Hobson, S. 1989. HIV genome variability in vivo. *AIDS* 3(Suppl. 1): S13–S18

149. Weiher, H., Barklis, E., Ostertag, W., Jaenisch, R. 1987. Two distinct sequence elements mediate retroviral gene expression in embryonal carcinoma cells. *J. Virol.* 61:2742–46

150. Weinberg, R. A. 1980. Origins and roles of endogenous retroviruses. *Cell* 22:643–44

151. Weiner, A. M., Deininger, P. L., Efstratiadis, A. 1986. Nonviral retroposons: Genes, pseudogenes, and transposable elements generated by the reverse flow of genetic information. *Annu. Rev. Biochem.* 55:631–61

152. Weiss, R. A. 1982. Experimental biology and assay of RNA tumor viruses. See Ref. 155, pp. 209–260

153. Weiss, R. A., Clapham, P. R., Weber, J. N., Dalgleish, A. G., Lasky, L. A., Berman, P. W. 1986. Variable and conserved neutralization antigens of human immunodeficiency virus. *Nature* 324: 572–75

154. Weiss, R. A., Mason, W. S., Vogt, P. K. 1973. Genetic recombinants and heterozygotes derived from endogenous and exogenous avian RNA tumor viruses. *Virology* 52:535–52

155. Weiss, R., Teich, N., Varmus, H., Coffin, J., eds. 1982. *RNA Tumor Viruses,*

Vol. 1. Cold Spring Harbor, NY: Cold Spring Harbor Lab.

156. Weiss, R., Teich, N., Varmus, H., Coffin, J., eds. 1985. *RNA Tumor Viruses,* Vol. 2. Cold Spring Harbor, NY: Cold Spring Harbor Lab.

157. White, J. M., Littman, D. R. 1989. Viral receptors of the immunoglobulin superfamily. *Cell* 56:725–28

158. Willey, R. L., Rutledge, R. A., Dias, S., Folks, T., Theodore, T., et al. 1986.

Identification of conserved and divergent domains within the envelope gene of the acquired immunodeficiency syndrome retrovirus. *Proc. Natl. Acad. Sci. USA* 83:5038–42

159. Wong-Staal, F. 1988. Variation of the HIV genome: Implications for the pathogenesis and prevention of AIDS. See Ref. 26, pp. 147–57

160. Zimmern, D. 1988. Evolution of RNA viruses. See Ref. 26, 2:211–40

Annu. Rev. Genet. 1990. 24:447–63

GENE-FOR-GENE COMPLEMENTARITY IN PLANT-PATHOGEN INTERACTIONS

N. T. Keen

Department of Plant Pathology, University of California, Riverside, California 92521

KEY WORDS: avirulence genes, disease resistance genes, elicitors, recognition, disease defense

CONTENTS

INTRODUCTION

The mechanics of disease defense in higher plants are poorly understood relative to the dramatic recent advances that have occurred in vertebrate

447

immunology. However, there are clear similarities between the defense strategies employed by plants and vertebrates. Active defense mechanisms occur in both groups of organisms that are invoked only following pathogen infection. In vertebrates this generally entails the various immune systems and in higher plants the hypersensitive response (HR). The recognitional systems modulating these active defense systems also have common features. Thus, the recognition of specific antigens in vertebrates is paralleled in plants by the recognition of analogous molecules of pathogen origin, called elicitors. In both cases the initial recognition events set off complex cascades of defense responses that executionally account for restriction of further pathogen development. These biochemical responses are very different in plants and animals, but the triggering mechanisms have much in common. In this review I consider recent progress in understanding plant recognition of pathogens leading to disease resistance, particularly where resistance is controlled by single complementary genes in the plant and pathogen. A mechanistic understanding of this gene-for-gene complementarity promises considerable insight into the basis of disease defense in plants, as well as the probability of improved disease control in practical agriculture.

ACTIVE DISEASE DEFENSE AND RECOGNITION IN PLANTS

The hypersensitive response (HR) is a widely occurring active defense system that occurs in most if not all higher plants in response to all known groups of plant pathogens—viruses, bacteria, nematodes, and fungi (for recent reviews see 11, 17, 22, 30, 41). Unlike circulating antibodies in vertebrates, active defense in plants is generally localized, involving the relatively rapid necrosis of host cells in the vicinity of an invading pathogen. These necrotic plant cells do not appear responsible per se for the restriction of pathogen development. They may function in the defense response, however, by acting as reservoirs for the accumulation of metabolites that are toxic to the pathogen (called phytoalexins) and by providing "secondary elicitors" that may stimulate defense in surrounding plant cells. These executional mechanisms associated with hypersensitive defense in plants are discussed in detail in the reviews cited above.

Before proceeding, several terms need be clarified. In plant-pathogen interactions, a resistant plant reaction denotes an incompatible plant-pathogen interaction, whereas a susceptible plant reaction results from a compatible plant-pathogen interaction. Avirulence genes of pathogens result in specific recognition by certain plant species or cultivars. Biotypes of pathogens that vary in their pattern of compatible or incompatible reactions on a set of host plant cultivars are referred to as races. Functionally, races are therefore defined by the collection of avirulence genes that they contain. Avirulence genes lead directly or indirectly to agents, called 'elicitors', that actually initiate the HR in incompatible plants. It is important to note that the term avirulence gene connotes a negative from the point of view of pathogen success on a resistant plant. We generally do not understand the function of avirulence genes in pathogens that harbor them, but their role in recognition by plants is likely gratuitous. Although the term 'avirulence gene' is historically too engrained to discard, it is confusing because the recessive alleles are generally called 'virulence genes'. However, virulence is a well-established term in microbiology denoting the quantitative aggressivity of a pathogen.

Our understanding of the chemical signals involved in plant surveillance of pathogens is limited, but recent progress has occurred (for a more detailed review, see 30). In classic work beginning in the 1930s, Flor defined the basic elements of gene-for-gene complementarity wherein single plant disease resistance genes and single complementary or matching avirulence genes in the pathogen account for plant recognition resulting in invocation of the HR (12, 15). As shown in Figure 1, plant recognition of a pathogen biotype as incompatible requires that the infected plant contain a disease resistance gene and that the complementary avirulence gene occur in the invading pathogen race. The functional alleles are generally inherited as dominant characters. If either partner lacks a functional allele, recognition and resistance do not occur and the plant becomes diseased. However, a single plant may have dozens of different resistance genes directed to a particular pathogen. Thus, for a pathogen biotype to escape surveillance, it must possess recessive alleles for all of the relevant avirulence genes.

Plant Disease Resistance Genes

Little is known about the structure and regulation of plant disease resistance genes or how they interact biochemically with pathogen avirulence genes. However, a large body of classical genetic evidence offers clues to their function. One important observation is that different resistance genes are frequently clustered and alternative functional alleles occur at some resistance gene loci. Coupled with the additional observation that significant recombination may occur between tightly linked resistance genes to generate new recognitional specificities (2, 23, 48, 51), these phenomena are suggestive of the rearrangement mechanisms that produce different recognitional specificities in vertebrate immunoglobulin genes (1, 58).

Another important feature of plant disease resistance genes, again shared with certain genes of the vertebrate immunoglobulin superfamily, such as Fc receptors and T cell receptors (33, 50), is that they encode two distinct functions—specific recognition, presumably of pathogen elicitors, and the actual triggering of the HR. Mutations in barley, called *mlo,* in maize for high

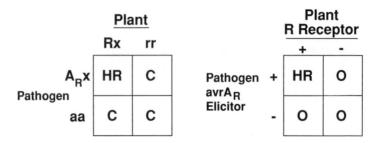

Figure 1 Left: alternative alleles of a plant disease resistance gene (R or r) and the complementary pathogen avirulence gene (A_R or a); only the occurrence of both dominant alleles results in the hypersensitive reaction (HR). Right: The elicitor-receptor model for biochemical recognition, which mirrors the genetic complementarity of avirulence gene alleles and resistance gene alleles shown on the left.

necrosis and the *Les* phenotype, and in tomato for 'autogenous necrosis' exemplify cases where the recognitional specificities of resistance genes appear to have been altered but not their ability to initiate the HR (8, 59). The simplest interpretation of these results is that resistance genes encode specific receptors with both recognitional and signal generation capabilities.

The timing of resistance responses may vary greatly in the same plant depending on the particular resistance gene/avirulence gene pairs involved (29, 30, 46). Some but not all disease resistance genes are also sensitive to temperature. The general case is for resistance to fail at higher temperatures, but examples of the converse are also known (18). While it is difficult to interpret all these observations, they are consistent with the idea that plant disease resistance genes encode receptors that detect elicitors resulting from activity of the complementary pathogen avirulence alleles. Unfortunately, these suggestions of receptor function have not been adequately tested experimentally.

Avirulence Genes

Avirulence genes in pathogens were first discovered in genetic crosses between different races of fungal pathogens that elicited or failed to elicit the HR in a plant carrying a defined resistance gene (15). Despite the accumulation of extensive data indicating that avirulence genes are widely distributed in plant pathogens, they have thus far only been cloned and characterized from bacterial and viral pathogens. The first microbial avirulence gene was cloned from race 6 of *Pseudomonas syringae* pv. *glycinea* by Staskawicz et al (54). This gene, called *avrA*, was sequenced (47) and found to encode a single protein product of 100 kd. Several other microbial avirulence genes were subsequently cloned and characterized, but neither their functions in the pathogen nor the mechanisms by which they elicit the plant HR are generally known (see 30). Indeed, the coat protein gene of tobacco mosaic virus is the only recognized avirulence gene with a known function in the pathogen (6).

Models to Explain Gene-for-Gene Specificity

Several models have been proposed to explain the molecular basis of recognition in gene-for-gene systems, but none has yet been proven by experimental evidence. Since this area is discussed in detail by Gabriel & Rolfe (17), only necessary information is presented here. One model is the elicitor-suppressor scheme (3, 10), in which pathogens are thought to produce general elicitors that initiate host defense unless a specific suppressor substance is also elaborated by a particular pathogen race. Thus, avirulence gene alleles are envisioned to dictate the production of specific suppressor substances. While the elicitor-suppressor model has found only limited experimental support in gene-for-gene systems (e.g. see (10)), it may operate in those cases where virulence rather than avirulence is dominant in the pathogen or when inhibitors of avirulence gene activity occur (e.g. see (24, 42)).

The elicitor-receptor model (4, 17, 28) is most compatible with the available genetic and biochemical data and predicts that either primary avirulence gene protein products or metabolites resulting from their catalytic activity (elicitors) are recognized by specific plant receptors encoded by disease resistance genes. Substantial evidence now exists for the production of specific elicitors by pathogens and some evidence suggests the occurrence of complementary plant receptors (see 30). However, it is not known whether the resulting elicitor-receptor complexes are translocated to the nucleus or whether second messengers lead to the expressive events of the plant HR (see 41).

AVIRULENCE GENES IN HIGHER LEVEL SPECIFICITIES

An important development has been the finding that avirulence genes cloned from certain strains of *Pseudomonas syringae* and *Xanthomonas campestris* function in related pathogens of other plants, called pathovars (pvs). Kobayashi et al (37, 38) cloned avirulence genes from *P. s.* pv. *tomato* that functioned in *P. s.* pv. *glycinea* to elicit the HR in some but not all soybean cultivars. Surprisingly, genes cloned from *P. s. tomato* behaved as race-specific avirulence genes on the nonhost plant soybean, just as the previously characterized avirulence genes isolated from *P. s. glycinea*. Indeed, one avirulence gene from *P. s. tomato* proved identical to *avrA*, originally cloned from *P.s. glycinea* race 6 (38). Whalen et al (60) performed similar experiments with *X. campestris* pathovars. They cloned an avirulence gene from

X. c. pv. *vesicatoria* that, upon introduction into *X. c. phaseoli,* caused the latter bacterium to elicit a HR on its normal host plant, bean, rather than the usual compatible reaction. In addition, this avirulence gene inhibited disease production by several other *X. campestris* pathovars on their normal hosts: pv. *glycines* on soybean, pv. *vignicola* on cowpea, pv. *alfalfae* on alfalfa, pv. *holcicola* on corn and pv. *malvacearum* on cotton. These results therefore suggest that avirulence genes may account for some cases of pathovar specificity in addition to their proven role in determining race specificity.

The demonstrations that *avr* genes from one pathovar can function in different pathovars also have important implications for our perception of plant disease resistance genes and their practical importance. The gene-for-gene interaction (Figure 1) predicts that host plants which react hypersensitively to a pathogen carrying a single, defined avirulence gene must share functionally identical disease resistance genes. This in turn implies that the same resistance gene may function in different plant species, as also suggested by some traditional crossing experiments (45). The appealing prospect emerges that resistance genes cloned from one plant may indeed function if transformed into an unrelated plant. Unfortunately, such transfers of resistance genes cannot usually be accomplished by conventional plant crossing due to sexual incompatibility barriers. Resistance genes must therefore be molecularly cloned and transformed into foreign plants.

HOMOLOGIES BETWEEN DIFFERENT AVIRULENCE GENES

With one exception, DNA sequencing data have not identified microbial avirulence genes with high homology to those in sequence data banks. However, certain avirulence genes share significant homology with other avirulence genes giving different specificities. For instance, Ronald & Staskawicz (49) sequenced the *avrBsl* gene from *X. campestris.* pv. *vesicatoria* and showed that it encoded a 50 kd protein with extensive homology to the carboxyl end of the 100 kd *avrA* protein of *P.s. glycinea,* discussed above. It is not yet known, however, if the observed homology implies a functional relatedness or if a truncated *avrA* gene would reproduce the phenotype of *avrBsl.*

Another striking example of homology between avirulence genes involves *avrB* and *avrC,* both cloned from race 0 of *P. s. glycinea* (55). The race 0 phenotype requires the presence of both *avrB* and *avrC,* but the race 1 phenotype was determined by the sole presence of *avrB.* Tamaki et al (57) sequenced *avrB* and *avrC* and showed that the protein products were 42% identical. Tamaki, Kobayashi & Keen (manuscript in preparation) constructed recombinant genes between *avrB* and *avrC* and showed that the *avrB* phenotype was dependent on a ca 500 bp internal portion of the gene while the *avrC* phenotype required a ca 600 bp internal portion of that gene. Thus, the 5' and 3' ends of the two genes were interchangeable. Recombinants constructed within the central, required regions, however, showed no avirulence phenotype, and no new recognitional specificities were observed in any of the recombinant constructions. One recombinant construct made at a *Hin*dIII site present at the 5' end of the central, required region in *avrB* yielded an atypical plant HR when introduced into *P. s. glycinea*. Soybean plants responded with a visible HR after ca 24 hr, similar to that occurring in response to bacteria carrying the wild-type *avrB* gene, but disease symptoms occurred after a few days that were typical of a compatible reaction. This recombinant avirulence gene therefore conferred an impaired *avrB* specificity. While the experiment does not illuminate the basis for avirulence gene function, it does show that the central regions of *avrB* and *avrC* confer distinct specificities despite their significant homology.

The findings with *avrB* and *avrC* are similar to those of Spaink et al (52) with the *nodE* genes from *Rhizobium trifolii* and *Rhizobium leguminosarum*. These bacteria uniquely nodulate clover or pea plants, respectively, and the central 185 amino acids of the *nodE* genes determined which plant host the bacterium would infect. Furthermore, Djordjevic et al (9) showed that mutation of the *nodE* gene of *R. trifolii* extended its host range to include pea. It therefore appears that the *nodE* genes behave as avirulence genes that restrict the bacterial host range. However, it has not yet been shown that *Rhizobium* cells carrying a certain *nodE* gene elicit the plant HR, as with recognition involving known avirulence genes. *NodE* may instead resemble the *nodH* host range gene of *R. meliloti*. This gene also behaves like an avirulence gene (14), but does not appear to interact with the plant through host recognition leading to the HR. Instead, the *nodH* gene may alter the structure of a hormone-active molecule required for nodulation of certain legume plants (14).

Kelemu & Leach (32) recently cloned an avirulence gene from *Xanthomonas campestris* pv. *oryzae* race 2 that complements the *Xa-10* disease resistance gene in rice. The avirulence gene exhibited significant homology to genomic DNA of all tested races of *X. c.* pv. *oryzae* and other *X. campestris* pathovars. The *avr10* gene was recently sequenced and found to encode a single protein product that exhibited high homology (more than 80% identical amino acids) to the *phoS* gene of *Escherichia coli* (J. Leach, personal communication). In *E. coli*, the *phoS* gene product is a periplasmic phosphate-binding protein that may be important in phosphate uptake (43). While it has not yet been shown that the *Xanthomonas avr10* gene complements *phoS* mutants in *E. coli* or that the cloned *E. coli phoS* gene exhibits avirulence gene activity, the work by Leach and associates represents the first case of a bacterial avirulence gene with significant homology to a gene of known function. Unlike previously sequenced bacterial avirulence genes, the *avr10* gene also has a signal peptide sequence (J. Leach, personal communication), indicating that it may reside in the periplasmic space or be extracellularly secreted. The *avr10* protein may therefore function as a specific elicitor per se, but this possibility has not yet been tested.

HOW AVIRULENCE GENES INTERACT WITH PLANT DISEASE RESISTANCE GENES

Viral Systems

Dawson and collaborators performed important experiments indicating that the coat protein gene of tobacco mosaic virus (TMV) functions as an avirulence gene. The N' resistance gene in tobacco is not effective against the common strain of TMV, but confers hypersensitive resistance against certain other TMV strains. The interactions of these strains or the common strain of TMV with tobacco containing either the N' or n' alleles therefore constitutes a gene-for-gene interaction (Figure 1). Knorr & Dawson (35) discovered that a single base change in the wild-type virus (C to U at position 6157) resulted in a single *ser* to *phe* amino acid substitution in the coat protein gene that accounted for the altered hypersensitive response of N' plants. Culver & Dawson (6; personal communication) subsequently transformed N' plants with several TMV coat protein genes and observed that mutations which altered amino acids at the surface of coat protein aggregates generally led to the HR, whereas changes of interior amino acids did not. This has led to the idea that the N' resistance gene permits sensing of the surface topography of TMV capside protein aggregates to elicit the HR. The work of Dawson & colleagues therefore demonstrates that the capsid protein is an elicitor of the HR in tobacco and indicates that its structural gene is an avirulence gene. This work is particularly significant since it establishes, for the first time, the biochemical function of an avirulence gene, both in the pathogen and in recognition by plants.

Microbial Avirulence Genes

The cloning and characterization of microbial avirulence genes has not yet resulted in identification of the biochemical functions of their gene products. Avirulence gene protein products have also thus far failed to exhibit elicitor activity (31, 57). However, the *avrD* gene from *Pseudomonas syringae* pv. *tomato,* to be discussed later, caused *E. coli* or *P. s. glycinea* race 4 cells to produce a low molecular weight metabolite that functioned as a race-specific elicitor of the HR in soybean leaves (31). In addition, segregating F2 and F3 progeny of a cross of the soybean cultivars Flambeau and Merit (respectively containing and lacking the putative disease resistance gene complementing *avrD*) have shown 100% linkage of their reactions to the partially purified elicitor and resistance to *P. s. glycinea* carrying the *avrD* gene (29). Since these phenotypes segregated in a 3:1 manner, a classical Mendelian disease resistance gene (designated *Rpg4*) occurs in soybean that complements *avrD*. The data are therefore consistent with the elicitor-receptor model discussed earlier and indicate that the *avrD* elicitor mediates the interaction between *avrD* and *Rpg4*. The elicitor that functions in *Rpg4* soybean is therefore very unlike the TMV capsid proteins that interact with N' tobacco.

Hrp *Genes and their Apparent Role in Disease Defense*

Mutants of bacterial pathogens have been isolated that fail to multiply in plant hosts and cause neither normal disease symptoms on a compatible plant nor the HR on an incompatible host. These pathogens harbor large clusters of so-called *hrp* genes (44). While their functions are not known, recent data suggest that *hrp* genes may be required for nutrient uptake in the plant (22). Recent work with bacterial pathogens has also raised the possibility that *hrp* genes may function as a cue for initiation of the HR in certain plants. While *hrp* clusters may be interchanged among certain bacterial pathogens (44), their introduction into other bacteria results in host-defense reactions. Thus, Huang et al (21) showed that a cosmid clone with a ca 32 kb insert from *Pseudomonas syringae* pv. *syringae* directed the HR in tobacco plants when the cloned DNA was expressed in either *P.s.* pv. *tabaci, P. fluorescens,* or *E. coli* cells. Significantly, expression of the HR phenotype required essentially the entire cosmid insert, unlike the situation normally observed for avirulence genes, which can be subcloned to relatively small DNA fragments.

Beer and associates (personal communication) also recently observed that a ca 40 kb region of DNA from *Erwinia amylovora* caused *E. coli* cells to elicit the hypersensitive reaction on tobacco plants. A cosmid clone comprising a portion of the *hrp* cluster of *P. syringae* pv. *tomato* also caused the HR in soybean plants when cloned in *P. syringae* pv. *glycinea* (J. Lorang, C. Boucher, D. Dahlbeck, B. Staskawicz & N. Keen, unpublished data). Unlike the previous cases, a ca 11 kb subcloned DNA fragment was sufficient to produce the HR. Thus, in addition to encoding functions required for pathogenicity, *hrp* clusters also initiate the plant HR when harbored by other bacteria. However, the mechanism by which *hrp* clusters elicit the HR is currently unknown. Since functional *hrp* clusters appear to be required for pathogenicity, this mechanism may constitute an effective rationale for pathogen surveillance leading to active defense in higher plants.

NONFUNCTIONAL, RECESSIVE AVIRULENCE ALLELES

Shortly after the widespread use of disease resistance genes to control plant diseases in practical agriculture, it was observed that some but not all of them were 'overcome' or defeated by the emergence of new pathogen races carrying mutations in their complementary avirulence genes. Thus, mutations in avirulence genes may not impart obvious short-term decreases in pathogen virulence or biologic fitness. If avirulence genes can be lost with no drastic detrimental effect on the pathogen, why do pathogens in nature harbor them? One idea is that certain avirulence genes may indeed contribute to virulence or pathogen fitness outside the host. This speculation has been fostered by experiments in which pathogen biotypes carrying dominant or recessive alleles of a particular avirulence gene were mixed and inoculated onto host plants lacking the complementary resistance gene. Certain avirulence alleles have been observed to become predominant in the pathogen population, suggesting that they increase fitness of the pathogen (5, 19).

Kearney & Staskawicz (25) illuminated how such stabilizing selection may function when they showed that avirulence gene avrBs2 from Xanthomonas campestris pv. vesicatoria has a role in virulence. The avrBs2 gene was cloned and employed in a marker exchange experiment to mutate the wild-type avrBs2 gene in X. c. vesicatoria. Unlike other avirulence genes studied in this bacterium (see 46), mutation of avrBs2 caused a pronounced reduction in pathogen virulence on all pepper cultivars, irrespective of whether they carried the complementing Bs2 resistance gene. Thus, avrBs2 appears to have an important role in virulence. Kearney & Staskawicz also showed that avrBs2 occurred in several other X. campestris pathovars and that its mutation also reduced virulence on their normal host plants. The results therefore indicate that the pepper resistance gene, Bs2, may be valuable in practical agriculture. Since mutations in the Xanthomonas avrBs2 gene compromise virulence, plant hosts carrying the Bs2 resistance gene should not be overcome by Xanthomonas biotypes with nonfunctional avrBs2 alleles. Several other plant disease resistance genes also exhibit stabilizing selection (see 5), suggesting that their widespread use may have considerable practical benefit for controlling plant diseases. For this to occur, however, such resistance genes must be cloned and transformed into other plant species.

Absence of Homologous DNA

When avrA was first cloned from race 6 of P. s. pv. glycinea (54), it was observed that other races of the bacterium did not harbor hybridizing DNA. This indicated that avrA DNA was simply not present in these bacteria, either because it was previously deleted or possibly because avrA had been introduced into race 6 from an outside source. Support for this latter interpretation was obtained from the observation that DNA sequences flanking the avrA gene hybridized to many bands in Southern blots with DNA from several races of P.s. pv. glycinea (54). Such a pattern is often seen with multiple copies of indigenous insertion sequences, supporting the speculation that avrA may have recently entered the *P. syringae* pv. *glycinea* genome. Races of *P.s.* pv. *glycinea* that did not exhibit *avrB* or *avrC* phenotypes also did not contain DNA that hybridized to these coding regions (57). Again, both genes also possessed flanking DNA that multiply hybridized to DNA from other *P. syringae* isolates (55). Furthermore, the GC content of the *avrB* and *avrC* genes was low, in contrast to the generally high GC content of *P. syringae* DNA. All these observations indicate that avirulence genes may readily be introduced from other sources, perhaps by IS mechanisms. Thus, the simplest mechanism for absence of an avirulence phenotype is lack of avirulence gene DNA. More subtle mechanisms are also known to modulate avirulence gene function, as discussed in the next sections.

The Story of avrBs1

Dahlbeck & Stall (7) showed that avirulent strains of *Xanthomonas campestris* pv. *vesicatoria* on pepper plants with the *Bs1* resistance gene acquired virulence at very high frequencies (ca 1 in 10^4). The basis for this extraordinary mutational frequency has recently been defined in elegant work by Stall, Staskawicz & collaborators. Stall et al (53) found that bacterial isolates exhibiting the *avrBs1* phenotype contained an indigenous plasmid that also conferred copper resistance. Swanson et al (56) cloned and characterized the plasmid-borne *avrBs1* gene and showed that mutant strains contained hybridizing DNA, thus ruling out the occurrence of plasmid curing or large deletions in the mutant strains. Kearney et al (26) sequenced a mutant *avrBs1* allele and found that it contained an insertion element that accounted for loss of the avirulence phenotype. This element, designated IS*476* (27), was only present in copper-resistant strains of *X. c.* pv. *vesicatoria* but was capable of transposition into other bacterial strains. IS*476* was also not specific for *avrBs1,* since transposition occurred into other genes. Of particular interest, the insertion element also precisely excised from the *avrBs1* gene at a sufficient frequency to permit plus/minus modulation of the avirulence phenotype by the bacterium (26). Thus, the interactions of *avrBs1* and IS*476* may allow the bacterium to adapt to changes in the occurrence of the *Bs1* resistance gene in the host population.

The avrD *Gene of* P. syringae *pv.* glycinea

Kobayashi et al (37, 38) observed that cosmid clones from *Pseudomonas syringae* pv. *tomato* caused P. *syringae* pv. *glycinea* to elicit the HR on some but not all cultivars of its normal host plant, soybean. *P. s.* pv. *tomato,* however, elicits the HR on all soybean cultivars and is therefore not able to attack this plant. One of the cosmid clones contained a single gene, called *avrD,* that accounted for the hypersensitive reactions of several soybean cultivars (39). This gene encoded a 34 kd protein product of 311 amino acids that did not contain a leader peptide sequence and did not show significant homology with proteins in the sequence data bases. The *avrD* gene comprised the first of five tandem open reading frames that appeared to be organized as an operon (39). These genes occurred on an indigenous plasmid of ca 75 kb present in all isolates of *P. syringae* pv. *tomato* examined. These observations raise the possibility that the *avrD* gene may encode an enzyme in a secondary biosynthetic pathway.

Although no biotype of *P. s.* pv. *glycinea* expressed a functional *avrD* gene, all of the tested isolates contained DNA that hybridized to the *avrD* gene from *P. s.* pv. *tomato* (39). This indicated that *P. s.* pv. *glycinea* contains a homologous recessive allele of the *avrD* gene. Kobayashi et al (40) cloned and sequenced the homologous gene from race 4 of *P. s. glycinea* and showed that it encoded a protein that read colinearly with the *avrD* protein from *P. s.* pv. *tomato* and also contained exactly 311 amino acids. The two proteins contained 86% identical amino acids with mismatches occurring throughout. This was in contrast to two downstream open reading frames of the putative operon, which encoded proteins that were 98 and 99% identical in amino acid sequence. It therefore appears that the *avrD* gene in *P. s.* pv. *glycinea* has accumulated several point mutations that eliminate its avirulence gene function.

As discussed earlier, the *avrD* gene of *P. s.* pv. *tomato* causes bacterial hosts to produce a low molecular weight elicitor substance that is recognized by soybean plants carrying the complementary resistance gene, *Rpg 4*. It is therefore believed that the *avrD* protein possesses a catalytic function that converts a normal metabolite present in several Gram-negative bacteria into the *avrD* elicitor (31). Significantly, *E. coli* cells expressing a homologous *avrD* gene from *P. s.* pv. *glycinea* also directed production of the *avrD* elicitor, but at a much lower rate than the *avrD* gene from *P. s.* pv. *tomato* (31). In addition, the *P. s. glycinea avrD* homologue was either not translated as efficiently or its protein product was less stable (40). These observations suggested that the *avrD* homologue present in *P. s.* pv. *glycinea* has been modified to lose the avirulence phenotype but may at least partially retain its bacterial function.

PROSPECTS FOR CLONING PLANT DISEASE RESISTANCE GENES

It is unfortunate that plant biologists have not directed more effort to the molecular cloning of disease resistance genes. The problem is difficult since resistance gene protein products have generally not been identified and the only scoreable phenotype is plant response to inoculation with a pathogen carrying the complementary avirulence gene. The situation is further compounded by the desirability of cloning many different resistance genes against several pathogens from dozens of important crop plants. The approaches thus far investigated are discussed by Ellis et al (13) and involve chromosome walking from linked marker genes, tagging with transposable elements, isolation of the protein product by use of affinity techniques with labeled specific elicitors and, finally, complementation cloning. All these approaches suffer from several liabilities. Due to their cumbersome nature, the walking and transposon tagging methods currently employed will likely not yield large numbers of cloned resistance genes from many different plant species.

An alternative approach may be to use the complementary specific elicitors as biochemical probes. If the elicitor is recognized by the receptor encoded by a resistance gene, as the elicitor-receptor model predicts, then it should be possible to identify this plant protein using a suitably labeled elicitor. The host-selective toxin, victorin, behaves as an elicitor that interacts only with oat plants carrying the *Pc-2* resistance gene. Wolpert & Macko (61) have shown that labeled victorin interacts with a 100 kd protein when supplied to *Pc-2* but not *pc-2* oat tissues. An antibody was prepared against the 100 kd oat protein and a DNA clone has been isolated from a *Pc-2* c-DNA library (T. Wolpert, personal communication). While it has not yet been established that the 100 kd protein is the product of the *Pc-2* gene, the work of Wolpert establishes the feasibility of using elicitors as biochemical probes for the cloning of disease resistance genes.

Plant disease-resistance genes may also be cloned by functional complementation. While this approach has generally been ignored, it has been used to clone difficult genes in vertebrates (16, 20, 50). The basic rationale involves transforming clones from a suitable library into recipient cells lacking the gene(s) of interest and monitoring them for phenotypic expression. A recent technological development, the transformation of intact plant cells with DNA coated onto microprojectiles (34) permits the functional cloning of plant disease resistance genes. Genomic or c-DNA libraries prepared from a plant cultivar carrying a large number of resistance genes would be prepared and pools of these clones transformed into suitable tissue of a recipient plant lacking the resistance genes. After time for transient gene expression, the recipient plants would be inoculated with a pathogen race carrying avirulence genes complementing the various resistance genes and subsequently scored for the rare cases where a HR occurred. Individual clones would then be rescreened from clone pools yielding the HR to select those carrying resistance genes. These clones could then be integratively transformed into the recipient plant to confirm identification of the resistance gene. While success of such functional cloning depends on the development of biological systems permitting sensitive scoring for the HR, this approach has many advantages: the donor and recipient plants need not have any genetic markers and may be polyploid or otherwise refractile to classical genetic crosses; the methodology would be fast, eliminating the time-consuming test crosses and Southern blots required in most other methods; the approach will, in principle, also work with any donor and recipient plants and these need not be of the same species; finally, since the method is based on a functional assay, any selected positive clones would have already met the prima facie criterion for resistance genes cloned via any method—they govern the HR in response to inoculation with the appropriate pathogen race.

SUMMARY AND CONCLUSIONS

The cloning of avirulence genes has greatly aided our understanding of plant-pathogen specificity. It has proven that the gene-for-gene relationship first noted by Flor (15) is correct—single avirulence genes encoding single protein products indeed are the genetic elements that interact with plant disease resistance genes. Furthermore, firm genetic evidence has provided insight into how two cloned avirulence genes (the TMV coat gene and *avrD*) cause the HR. The differences in structure of pathogen elicitors also indicates that plants have evolved diverse recognitional mechanisms to detect pathogens. It is appealing to speculate, therefore, that elicitors represent the plant equivalent of antigens in vertebrates. Another consequence of these results has been the establishment of firm genetic and biochemical evidence supporting the elicitor-receptor model for recognition of incompatible pathogen races by plants. In both TMV and bacterial pathogens, we are also beginning to understand how avirulence genes are altered to confound plant recognition of the pathogen. The next few years should yield additional information on avirulence gene structure as well as the important questions of their function in the pathogen and the molecular mechanisms whereby plant recognition occurs. The marked successes in cloning avirulence genes underscore only more forcefully the pressing need to clone and characterize plant disease resistance genes. Certainly an understanding of these genes is required to further our basic knowledge of active defense in plants and to permit their manipulation for improved control of plant diseases in practical agriculture.

Literature Cited

1. Alt, F. W., Blackwell, T. K., Yanco-poulos, G. D. 1987. Development of the primary antibody repertoire. *Science* 238:1079–87
2. Bennetzen, J. L., Qin, M-M., Ingels, S., Ellingboe, A. H. 1988. Allele-specific and *mutator*-associated instability at the *Rp1* disease-resistance locus of maize. *Nature* 332:369–70
3. Bushnell, W. R., Rowell, J. B. 1981. Suppressors of defense reactions: a mod-

el for roles in specificity. *Phytopatholo-gy* 71:1012–14
4. Callow, J. A. 1977. Recognition, resistance and the role of plant lectins in host-parasite interactions. *Adv. Bot. Res.* 4:1–49
5. Crill, P. 1977. An assessment of stabilizing selection in crop variety development. *Annu. Rev. Phytopathol.* 15:185–202
6. Culver, J. N., Dawson, W. O. 1989.

Tobacco mosaic virus coat protein: an elicitor of the hypersensitive reaction but not required for the development of mosaic symptoms in *Nicotiana sylvestris*. *Virology* 173:755–58

7. Dahlbeck, D., Stall, R. E. 1979. Mutations for change of race in cultures of *Xanthomonas vesicatoria*. *Phytopathology* 69:634–36

8. Day, P. R., Barrett, J. A., Wolfe, M. S. 1983. The evolution of host-parasite interactions. In *Genetic Engineering of Plants*, ed. T. Kosuge, C. Meredith, A. Hollaender, pp. 419–30. New York: Plenum

9. Djordjevic, M. A., Schofield, P. R., Rolfe, B. G. 1985. Tn5 mutagenesis of *Rhizobium trifolii* host-specific nodulation genes result in mutants with altered host-range ability. *Mol. Gen. Genet.* 200:463–71

10. Doke, N., Garas, N. A., Kuc, J. 1979. Partial characterization and aspects of the mode of action of a hypersensitivity-inhibiting factor (HIF) isolated from *Phytophthora infestans*. *Physiol. Plant Pathol.* 15:127–40

11. Ebel, J., Grisebach, H. 1988. Defense strategies of soybean against the fungus *Phytophthora megasperma* f.sp. *glycinea*: a molecular analysis. *Trends Biochem. Sci.* 13:23–27

12. Ellingboe, A. H. 1981. Changing concepts in host-pathogen genetics. *Annu. Rev. Phytopathol.* 19:125–43

13. Ellis, J. G., Lawrence, G. J., Peacock, W. J., Pryor, A. J. 1988. Approaches to cloning plant genes conferring resistance to fungal pathogens. *Annu. Rev. Phytopathol.* 26:245–63

14. Faucher, C., Camut, S., Denarie, J., Truchet, G. 1989. The *nodH* and *nodQ* host range genes of *Rhizobium meliloti* behave as avirulence genes in *R. leguminosarum* bv. *viciae* and determine changes in the production of plant-specific extracellular signals. *Mol. Plant-Microbe Interact.* 2:291–300

15. Flor, H. H. 1942. Inheritance of pathogenicity in *Melampsora lini*. *Phytopathology* 32:653–69

16. Frech, G. C., VanDongen, A. M. J., Schuster, G., Brown, A. M., Joho, R. H. 1989. A novel potassium channel with delayed rectifier properties isolated from rat brain by expression cloning. *Nature* 340:642–45

17. Gabriel, D. W., Rolfe, B. G. 1990. Working models of specific recognition in plant-microbe interactions. *Annu. Rev. Phytopathol.* 28:365–91

18. Gousseau, H. D. M., Deverall, B. J., McIntosh, R. A. 1985. Temperature-sensitivity of the expression of resistance to *Puccinia graminis* conferred by the *Sr15*, *Sr9b* and *Sr14* genes in wheat. *Physiol. Plant Pathol.* 27:335–43

19. Grant, M. W., Archer, S. A. 1983. Calculation of selection coefficients against unnecessary genes for virulence from field data. *Phytopathology* 73:547–51

20. Harada, N., Castle, B. E., Gorman, D. M., Itoh, N., Schreurs, J., et al. 1990. Expression cloning of a cDNA encoding the murine interleukin 4 receptor based on ligand binding. *Proc. Natl. Acad. Sci. USA* 87:857–61

21. Huang, H-C., Schuurink, R., Denny, T. P., Atkinson, M. M., Baker, C. J., et al. 1988. Molecular cloning of a *Pseudomonas syringae* gene cluster that enables *Pseudomonas fluorescens* to elicit the hypersensitive response in tobacco plants. *J. Bacteriol.* 170:4748–56

22. Hutcheson, S. W., Collmer, A., Baker, C. J. 1989. Elicitation of the hypersensitive response by *Pseudomonas syringae*. *Physiol. Plant Pathol.* 76:155–63

23. Islam, M. R., Shepherd, K. W., Mayo, G. M. E. 1989. Recombination among genes at the L group in flax conferring resistance to rust. *Theor. Appl. Genet.* 77:540–46

24. Jones, D. A. 1988. Genetic properties of inhibitor genes in flax rust that alter avirulence to virulence on flax. *Phytopathology* 78:342–44

25. Kearney, B., Staskawicz, B. J. 1989. Analysis of *avrBs2* in *Xanthomonas campestris* pathovars: conservation and possible mechanism of stable disease resistance. *Abstr. Fallen Leaf Lake Conf: Mol. Biol. Bact. Plant Pathogens*, p. 39, South Lake Tahoe, CA

26. Kearney, B., Ronald, P. C., Dahlbeck, D., Staskawicz, B. J. 1988. Molecular basis for evasion of plant host defence in bacterial spot disease of pepper. *Nature* 332:541–43

27. Kearney, B., Staskawicz, B. J. 1990. Characterization of IS476 and its role in bacterial spot disease of tomato and pepper. *J. Bacteriol.* 172:143–48

28. Keen, N. T. 1982. Specific recognition in gene-for-gene host-parasite systems. *Adv. Plant Pathol.* 1:35–82

29. Keen, N. T., Buzzell, R. I. 1990. New disease resistance genes in soybean against *Pseudomonas syringae* pv. *glycinea*: evidence that one of them interacts with a bacterial elicitor. *Theor. Appl. Genet.* Submitted In press

30. Keen, N. T., Dawson, W. O. 1990. Pathogen avirulence genes and elicitors

of plant defense. In *Genes Involved in Plant Defense*, ed. W. Boller, F. Meins. New York: Springer-Verlag. In press

31. Keen, N. T., Tamaki, S., Kobayashi, D., Gerhold, D., Stayton, M., et al. 1990. Bacterial expressing avirulence gene D produce a specific elicitor of the soybean hypersensitive reaction. *Mol. Plant-Microbe Interact.* 3:112–21

32. Kelemu, S., Leach, J. E. 1990. Cloning and characterization of an avirulence gene from *Xanthomonas campestris* pv. *oryzae*. *Mol. Plant-Microbe Interact.* 3:59–65

33. Kinet, J-P. 1989. Antibody-cell interactions: Fc receptors. *Cell* 57:351–54

34. Klein, T. M., Wolf, E. D., Wu, R., Sanford, J. C. 1987. High-velocity microprojectiles for delivering nucleic acids into living cells. *Nature* 327:70–73

35. Knorr, D. A., Dawson, W. O. 1988. A point mutation in the tobacco mosaic virus capsid protein gene induces hypersensitivity in *Nicotiana sylvestris*. *Proc. Natl. Acad. Sci. USA* 85:170–74

36. Knott, D. R. 1989. *The Wheat Rusts— Breeding for Resistance*. Berlin: Springer-Verlag, 201 pp.

37. Kobayashi, D. Y., Keen, N. T. 1985. Cloning of a factor from *Pseudomonas syringae* pv. *tomato* responsible for a hypersensitive response on soybean. *Phytopathology* 75:1355 (Abstr.)

38. Kobayashi, D. Y., Tamaki, S. J., Keen, N. T. 1989. Cloned avirulence genes from the tomato pathogen *Pseudomonas syringae* pv. *tomato* confer cultivar specificity on soybean. *Proc. Natl. Acad. Sci. USA* 86:157–61

39. Kobayashi, D. Y., Tamaki, S. J., Keen, N. T. 1990. Characterization of avirulence gene D from *Pseudomonas syringae* pv. *tomato*. *Mol. Plant-Microbe Interact.* 3:94–102

40. Kobayashi, D. Y., Tamaki, S. J., Trollinger, D. J., Gold, S., Keen, N. T. 1990. A gene from *Pseudomonas syringae* pv. *glycinea* with homology to avirulence gene D from *P.s.* pv. *tomato* but devoid of the avirulence phenotype. *Mol. Plant-Microbe Interact.* 3:103–11

41. Lamb, C. J., Lawton, M. A., Dron, M., Dixon, R. A. 1989. Signals and transduction mechanisms for activation of plant defenses against microbial attack. *Cell* 56:215–24

42. Lawrence, G. J., Mayo, G. M. E., Shepherd, K. W. 1981. Interactions between genes controlling pathogenicity in the flax rust fungus. *Phytopathology* 71:12–19

43. Levitz, R., Bittan, R., Yagil, E. 1981. Complementation tests between alkaline

phosphatase-constitutive mutants (*phoS* and *phoT*) of *Escherichia coli*. *J. Bacteriol.* 145:1432–35

44. Lindgren, P. B., Panopoulos, N. J., Staskawicz, B. J., Dahlbeck, D. 1988. Genes required for pathogenicity and hypersensitivity are conserved and interchangeable among pathovars of *Pseudomonas syringae*. *Mol. Gen. Genet.* 211:499–506

45. McIntosh, R. A. 1976. Genetics of wheat and wheat rusts since Farrer. *J. Austr. Inst. Agric. Sci.* pp. 203–16

46. Minsavage, G. V., Dahlbeck, D., Whalen, M. C., Kearney, B., Bonas, U., 1990. Gene-for-gene relationships specifying disease resistance in *Xanthomonas campestris* pv. *vesciatoria*-pepper interactions. *Mol. Plant-Microbe Interact.* 3:41–47

47. Napoli, C., Staskawicz, B. 1987. Molecular characterization and nucleic acid sequence of an avirulence gene from race 6 of *Pseudomonas syringae* pv. *glycinea*. *J. Bacteriol.* 169:572–78

48. Pryor, T. 1987. The origin and structure of fungal disease resistance genes in plants. *Trends Genet.* 3:157–61

49. Ronald, P. C., Staskawicz, B. J. 1988. The avirulence gene avrBs₁ from *Xanthomonas campestris* pv. *vesicatoria* encodes a 50-kD protein. *Mol. Plant-Microbe Interact.* 1:191–98

50. Seed, B., Aruffo, A. 1987. Molecular cloning of the CD2 antigen, and T-cell erythrocyte receptor, by a rapid immunoselection procedure. *Proc. Natl. Acad. Sci., USA* 84:3365–69

51. Shepherd, K. W., Mayo, G. M. E. 1972. Genes conferring specific plant disease resistance. *Science* 175:375–80

52. Spaink, H. P., Weinman, J., Djordjevic, M. A., Wijffelman, C. A., Okker, R. J. H., Lugtenberg, B. J. J. 1989. Genetic analysis and cellular localization of the *Rhizobium* host specificity-determining NodE protein. *EMBO J.* 8:2811–18

53. Stall, R. E., Loschke, D. C., Jones, J. B. 1986. Linkage of copper resistance and avirulence loci on a self-transmissible plasmid in *Xanthomonas campestris* pv. *vesicatoria*. *Phytopathology* 76:240–43

54. Staskawicz, B. J., Dahlbeck, D., Keen, N. T. 1984. Cloned avirulence gene of *Pseudomonas syringae* pv. *glycinea* determines race-specific incompatibility on *Glycine max* (L.) Merr. *Proc. Natl. Acad. Sci., USA* 81:6024–28

55. Staskawicz, B., Dahlbeck, D., Keen, N., Napoli, C. 1987. Molecular

characterization of cloned avirulence genes from race 0 and race 1 of *Pseudomonas syringae* pv. *glycinea*. *J. Bacteriol.* 169:5789–94

56. Swanson, J., Kearney, B., Dahlbeck, D., Staskawicz, B. 1988. Cloned avirulence gene of *Xanthomonas campestris* pv. *vesicatoria* complements spontaneous race-change mutants. *Mol. Plant-Microbe Interact.* 1:5–9

57. Tamaki, S., Dahlbeck, D., Staskawicz, B., Keen, N. T. 1988. Characterization and expression of two avirulence genes cloned from *Pseudomonas syringae* pv. *glycinea*. *J. Bacteriol.* 170:4846–54

59. Tonegawa, S. 1983. Somatic generation of antibody diversity. *Nature* 302:575–81

60. Walbot, V., Hoisington, D. A., Neuffer, M. G. 1983. Disease lesion mimic mutations. See Ref. 8, pp. 431–42

60. Whalen, M. C., Stall, R. E., Staskawicz, B. J. 1988. Characterization of a gene from a tomato pathogen determining hypersensitive resistance in non-host species and genetic analysis of this resistance in bean. *Proc. Natl. Acad. Sci., USA* 85:6743–47

61. Wolpert, T. J., Macko, V. 1989. Specific binding of victorin to a 100-kDa protein from oats. *Proc. Natl. Acad. Sci. USA* 86:4092–96

Annu. Rev. Genet. 1990. 24:465–90

INTERACTIONS BETWEEN SATELLITE BACTERIOPHAGE P4 AND ITS HELPERS

Gail E. Christie

Department of Microbiology and Immunology, Virginia Commonwealth University, Richmond, Virginia 23298-0678

Richard Calendar

Department of Molecular and Cell Biology, University of California, Berkeley, California 94720

KEY WORDS: bacteriophage P2, transactivation, immunity control, polarity suppression, capsid size determination

CONTENTS

465

0066-4197/90/1215-0465$02.00

INTRODUCTION

Satellite bacteriophage P4 is a temperate coliphage that can lysogenize or maintain itself as a multicopy plasmid in the absence of a helper (20, 36, 107), but requires a P2-related helper phage for lytic growth (105, 107). The helper phage provides the proteins needed for assembly of phage particles, packaging of P4 DNA, and lysis of the host cell (102, 106). P4 displays the host range specificity of its most recent helper (86, 102, 107). When grown with a P2 helper, the P4 particle consists of a P2 tail and an isometric capsid composed of P2 proteins but with only one third of the volume of the P2 capsid (35) to accommodate the smaller P4 genome (11 kb for P4 vs 33 kb for P2). Sequence homology between the genomes of the two phages is less than 1% and primarily limited to about 55 base pairs that determine DNA maturation and packaging specificity (73, 74; R. Ziermann & R. Calendar, manuscript in preparation). The helper functions can be provided by a coinfecting phage or in some cases by a prophage helper (108). A variety of interactions between satellite phage, helper phage, and host have evolved to support this unusual lifestyle. The biology of P4 and P2-related phages has been reviewed by Bertani and Six (9), and is only summarized briefly here as required. This review focuses on recent studies elucidating the interactions between satellite phage P4 and its helpers. The role of helper phage P2 is discussed in detail, and compared to the P2-related phage 186 where appropriate.

GENOME ORGANIZATION AND EXPRESSION

P4

P4 has a linear, double-stranded genome of 11,627 base pairs, which has been sequenced in its entirety (45). Nine genes identified by mutation have been located, and the genome contains at least four additional open reading frames of undetermined function (Figure 1; see 43 for references). The P4 DNA replication gene α, which encodes a DNA primase (4, 5, 58), is the only known P4 function that is strictly essential for P4 lytic growth. Genes to the left of α lie in a region that is nonessential as defined by deletion and substitution mutations (91, 110), and may play a role in maintaining P4 as a multicopy plasmid. The remaining genes all play some role in regulating P4 gene expression and/or the interactions involved in the exploitation of helper phages by P4, and will be discussed in this review. Five promoters have been identified by mapping of transcripts (18, 22, 32), and one by cloning and sequence analysis (85), and are also indicated in Figure 1.

P2

A map of the P2 genome is shown in Figure 2 (see 39 for references). Twenty-three genes essential for lytic multiplication have been identified;

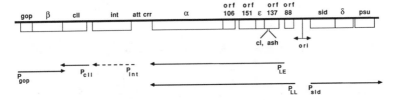

Figure 1 Map of the P4 genome, displayed as a linear DNA molecule broken at the *cos* sites. The boxes below the line indicate open reading frames. Gene names are given above the line; where no gene product has been identified the open reading frame *(orf)* is designated by the size, in amino acids, of the putative gene product. Sites necessary for the initiation of replication *(ori* and *crr)* and phage attachment *(att)* are also indicated, as are the locations of mutations within *orf*137 affecting P4 immunity *(cI)* and adaptation to a P3 helper *(ash)*. Solid arrows below the map represent transcripts for which the 5' end has been mapped; the dotted arrow represents the predicted transcript from P_{int}. (See (43) for references.)

these include twenty late genes, two early genes required for DNA replication, and the late control gene *ogr*. There are three genes involved in lysogeny, and three additional genes that play a nonessential role in the P2 life cycle. DNA sequencing of the early region (41; E. Haggård-Ljungquist, unpublished) has revealed at least six additional small open reading frames of unknown function. The P2 late genes are organized into four operons, as defined by the polar effects of amber mutations (70, 114). The four late promoters and two promoters in the early control region have been mapped (14, 15, 101), as has the promoter for the nonessential gene *old* (40). A promoter for the *ogr* gene has been identified (11, 16, 56), which is expressed at middle times during lytic infection (N. K. Birkeland, B. H. Lindqvist and G. E. Christie, manuscript in preparation).

186

The genetic arrangement of phage 186 is similar to that of P2, and viable hybrids between the two phages have been isolated (12, 50). The locations of junctions in the hybrid phages (117) suggest that the late gene regions of P2 and 186 are highly homologous and that some genes are functionally interchangeable. Heteroduplex mapping reveals no homology between the right ends of the two phage genomes (118), but regions of similarity occur between the predicted amino acid sequences of the 186 DNA replication protein (104a) and the P2 *A* gene product (E. Haggård-Ljungquist, personal communication). Like P2, 186 has a number of small open reading frames in the early region; functions have recently been assigned to several of these (93, 94). Two open reading frames of unknown function in the early region of phage 186 would encode proteins with homology to potential small proteins encoded in the P2 early region (104a; E. Haggård-Ljungquist, personal communication).

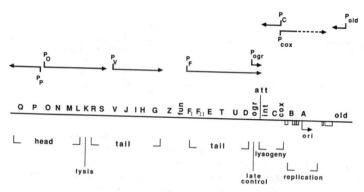

Figure 2   Genetic map of P2, oriented so that the cohesive ends are aligned with those in the P4 map. In addition to the 27 genes previously defined by mutation (for references, see 39) an additional late gene encoding the tail protein FII (66) has been identified by DNA and protein sequence analysis (L. Temple & G. E. Christie, unpublished) and an open reading frame between late genes *J* and *H*, which contains the wild-type allele of P2am78 (E. Haggård-Ljungquist, personal communication), has been designated gene *I*. Boxes below the line indicate additional open reading frames that have been located by sequence analysis (E. Haggård-Ljungquist, personal communication). Arrows show the direction and extent of transcription from known promoters; with the exception of the terminator just distal to *ogr*, none of the 3' ends has been located.

IMMUNITY CONTROL AND MUTUAL DEREPRESSION

The natural hosts for phage P4 are bacteria lysogenic for a P2-related prophage. Strains of *Escherichia coli* lysogenic for P2-like phages appear to be naturally abundant (23). P4 can activate expression of the necessary helper functions by derepressing a prophage helper. Alternatively, a resident P4 prophage can be derepressed by infection with P2 (108). This mutual derepression allows P4 to take advantage of a helper phage under varying circumstances. Before we consider the mechanism of derepression for each phage, we will summarize what is known about the normal mechanism for establishment and maintenance of lysogeny.

P2 Immunity Control

ROLE OF P2 REPRESSOR   The features of the P2 early control region are summarized in Figure 3. The lysogenic state of P2 is regulated by the P2 immunity repressor, which is the product of gene *C* (8). P2 repressor is a polypeptide of 99 amino acids that contains a putative helix-turn-helix DNA-binding domain, as deduced by comparison with well-characterized DNA-binding regulatory proteins (76). Partially purified repressor binds as a dimer to P2 DNA from the early operator region (77). The operator region has been defined by virulent mutations that render the phage insensitive to repressor. DNA sequence analysis of this region (76) has revealed two direct 8 bp repeats that lie within the region protected from DNase I digestion by repressor (101). The binding site for repressor overlaps the early promoter P_e. The P_e promoter is the initiation site for rightward transcription of the essential early operon, which includes at a minimum the genes *cox* and *B* and a small open reading frame of unknown function between them. The DNA sequence downstream of the *B* gene contains a small GC-rich inverted repeat and a stretch of T residues, suggesting that there may be a transcriptional terminator in this region (41). No experimental evidence addresses whether the early gene *A* and several additional small open reading frames located distal to *B* are part of this early operon or are transcribed separately.

In addition to the early promoter P_e and a binding site for repressor, the early control region contains a second promoter, P_c (101). The P_c promoter directs leftward transcription into the repressor gene, giving rise to a transcript that overlaps the transcript initiating at P_e by about 30 nucleotides. DNase I footprinting experiments indicate that RNA polymerase binds more strongly to P_e than to P_c, but that binding of repressor to the early operator enhances the binding of polymerase to P_c in vitro. Saha et al (101) also demonstrated activation of P_c by repressor in vivo, using fusion plasmids in which *cat* expression is controlled by P_c. P2 repressor thus positively regulates its own transcription as well as serving as a negative regulator of transcription from P_e. This dual role for P2 repressor is formally analogous to that elucidated for lambda repressor (summarized in 88). In addition, like lambda repressor, P2

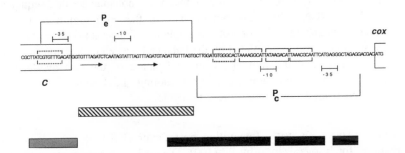

Figure 3   DNA sequence of the P2 early control region between the 5' ends of genes *C* and *cox*. Brackets show the locations of the two convergent early promoters, P_e and P_c, and the −10 and −35 regions of each promoter are indicated. The arrows designate direct repeats in the operator region; the striped box indicates the extent of repressor binding (101). Sequences with strong homology to the Cox binding consensus sequence are marked by solid rectangles, while those with weaker homology are marked by dotted rectangles. The black and grey bars below the sequence indicate regions of strong and weak Cox binding (99).

repressor further regulates its own synthesis by reducing transcription from P_c at high repressor concentrations (97). The mechanism of activation of P_c by P2 repressor appears to be quite different, however, from the mechanism of activation of transcription from p_{RM} by lambda repressor. In additional studies using CAT fusion plasmids, stimulation of transcription from P_c was also obtained in the absence of P2 repressor when P_e was inactivated by mutation (98). Furthermore, the inhibitory effect of transcription from P_e on transcription from P_c was more pronounced when the two promoters were situated in *cis* than when the P_e transcript was provided in *trans*. Transcription from P_c and transcription from P_e are thus mutually exclusive, and the activation of P_c by repressor is a consequence of preventing the convergent transcription from P_e.

ROLE OF COX   As in a number of temperate phages, the P2 lysis-lysogeny decision is regulated by two competing repressors. The first gene in the essential early operon, *cox* (for *c*ontrol *o*f e*x*cision), was originally identified by mutations deficient in prophage excision (71). The *cox* gene encodes a repressor that, like P2 immunity repressor, affects transcription from the early promoters P_e and P_c. Like lambda Cro protein, Cox inhibits transcription of the repressor gene and, at higher concentrations, also negatively regulates its own expression (98). The predicted Cox sequence of 91 amino acids (41) suggests that it too is a DNA-binding protein with a helix-turn-helix motif. DNase I protection experiments indicate that Cox binds more strongly to P_c than to P_e, which is consistent with the observed levels of repression (99). The strongly protected region, which covers the entire P_c promoter (Figure 3), contains tandem repeats of the conserved nonanucleotide sequence T(T)AAA(G)NCA. The weak binding site in the -35 region of the P_e promoter contains a single copy of this sequence. Transcription from the two early promoters is thus regulated by the balance between expression of the C and *cox* genes and the mutually exclusive nature of the two overlapping transcripts, with expression from P_c promoting the lysogenic cycle and expression from P_e promoting the lytic cycle.

INVOLVEMENT OF IHF   Integration host factor (IHF) of *E. coli* is also required for lysogenization by P2 (100). DNase I footprinting revealed IHF binding to a region 50 nucleotides upstream of the -35 region of the P_e promoter, within the C gene. In a plasmid carrying the entire early control region, including genes C and *cox*, expression is seen predominantly from P_c in a wild-type strain of *E. coli* and only from P_e in IHF-deficient mutants. This shift in promoter usage would be expected if the lytic pathway was favored in the absence of IHF. However, IHF appears to have only marginal effects on the expression of the isolated P_e or P_c promoters, as assayed in *cat* fusion plasmids. This suggests that IHF may affect binding of the phage-encoded regulatory proteins rather than directly influencing the interaction of RNA polymerase with these promoters.

COMPARISON TO 186   The arrangement of the early regulatory region of phage 186 is quite similar to that of P2. Two convergent promoters, *p*L and *p*R, generate overlapping transcripts encoding, respectively, the 186 immunity repressor *(cI)* and *apl*, whose product represses transcription from *p*L. Virulent mutations have been mapped to an inverted repeat located between the −10 and −35 regions of *p*R (62). Here, however, the similarity ends. The 186 repressor, as predicted from the DNA sequence, is nearly twice as large as and not homologous to the P2 *C* protein, and is not predicted to contain a helix-turn-helix structural motif (55). The *apl* gene product is predicted to be a helix-turn-helix DNA binding protein (94), but is not homologous to P2 Cox. There are two additional salient differences between P2 and 186 with regard to prophage derepression. First, a 186 prophage is UV-inducible whereas P2 is not (8, 116). Neither the P2 repressor nor the 186 repressor contains the characteristic sequence found in other repressor proteins cleaved by RecA-mediated proteolysis (55, 76). UV-induction of a 186 prophage is an SOS function, however, and is due to the expression of a 186 function, *tum* (*tu*rbid on *m*itomycin), which maps to the far right end of the 186 genome and is under LexA control (61). The 186 Tum system appears to have no action against P2 repressor. A second difference between P2 and 186 lies in the inability of a cell lysogenic for 186 to support the lytic growth of P4 (102). P4 is apparently unable to derepress a 186 prophage, while a P2 prophage can be derepressed following P4 infection.

Derepression of P2 by P4

P2 lysogens infected by P4 produce a burst of about 100 P4 and $\le 10^{-1}$ P2 per infected cell (107). This P4 production is due to derepression of the P2 prophage helper. That P4 infection causes derepression was first demonstrated by the loss of immunity to superinfection by P2 when a replication-deficient (α gene) mutant of P4 infects a P2-lysogenic strain under nonpermissive conditions. Derepression results in severalfold replication of the P2 prophage in situ, but does not trigger prophage excision (108). P2 prophage replication is not observed following infection with the P4 mutant am*104* (24, 29). This amber mutation defines the P4 derepression gene, ϵ. The mechanism of ϵ action has not been determined. P4 is unable to derepress a P2 prophage in a strain carrying a multicopy plasmid encoding the P2 immunity repressor (115). This suggests that the ϵ gene product may act as an anti-repressor. Cleavage of P2 repressor during derepression has not been observed (G. Dehò, personal communication), so ϵ may function by binding to repressor and inactivating it.

The P4 am*104* mutant also contains an *ash* (for "*adaptation to a second helper*") mutation. P4 *ash* mutants can grow with the help of a P3 prophage, unlike wild-type P4 (7, 68). It has been suggested (68) that the ε am*104* mutation would be lethal by itself but is suppressed by the *ash* mutation.

P4 Immunity Control

When P4 infects a bacterial host it can adopt several alternative lifestyles. If a helper phage is present, the outcome of infection is usually lytic growth. In the absence of a helper, the lytic cycle is not an option. Studies of infection by P4 alone have revealed the existence of two stable regulatory states. One is an immune state in which P4 is integrated as a prophage into the bacterial chromosome. P4 can also exist as a multicopy plasmid, with or without the presence of an integrated copy. These two states differ in expression of P4 immunity and in regulation of P4 gene expression.

THE IMMUNE-INTEGRATED STATE   Infection by wild-type P4 in the absence of a helper phage results primarily in the establishment of the immune-integrated state. P4 can also lysogenize in the presence of a helper such as prophage P2. P4 lysogenic cells are immune to P4 superinfection, regardless of the presence of a helper prophage (107, 108). P4 immunity, conferred by the *cI* locus, does not appear to involve a phage-encoded repressor. Phages carrying a mutation in *cI* were originally isolated as a class of recessive clear-plaque mutants (13). P4 immunity has been mapped to *cI* by deletion analysis of the P4 hybrid P420 (54) and by the demonstration that a cloned fragment of 827 bp from this region is sufficient to confer immunity to superinfection and to complement a P4 *cI* mutant for lysogenization (21). Mutations (*ash*) that confer upon P4 the ability to use a P3 prophage helper also map to this region, which contains an open reading frame of 137 amino acids (*orf*137; 69). However, the relationship between the *ash* and *cI* functions remains unclear. All *cI* and *ash* mutations sequenced thus far lie within a small region of *orf137*, but two *ash* mutations do not alter the amino acid sequence of the putative gene product (69), suggesting that they affect a regulatory site on the genome or alter the transcript from this region. Although indirect evidence for the existence of a *cI* amber mutant has been reported (9), Ghisotti et al (31) have cloned a 175 bp fragment that retains the ability to express P4 immunity. This fragment is internal to *orf*137, arguing strongly against the idea that this open reading frame encodes *cI*. In fact, no clear evidence indicates that the *trans*-acting P4 immunity determinant is a protein, and current efforts are directed towards investigating the possible existence of a regulatory RNA. The target for *cI* action has also not been defined clearly. The partially immunity-insensitive mutant P4 *vir1* (75) carries a point mutation that appears to increase the strength of the late leftward promoter P_{LL} and allow its expression at early times after infection (22, 69; see discussion below). An additional phenotype conferred by the *vir1* mutation is the establishment of the multicopy plasmid state with a much higher efficiency than that seen for wild-type P4.

THE MULTICOPY PLASMID STATE   P4 plasmids were originally observed in strains infected by the mutant P4 *vir1* (13, 36, 84). Although it was suggested (36) that the ability of P4 to be maintained as a plasmid was a consequence of the *vir1* mutation, wild-type P4 can also assume the multicopy plasmid state at a low frequency (20). Conversely, P4 *vir1* can form single-copy integrants at a low frequency. P4 *cl* mutants, which never form stable integrated lysogens, can also establish the multicopy plasmid state. This situation is slightly complicated by the fact that infection with P4 *cl* mutants kills most cells, due to the expression of a P4 function called *kil*, which also maps to the *cl* region and is thought to be overexpressed in *cl* mutants (F. Forti & D. Ghisotti, personal communication). Survivors of P4 *cl* infection carry P4 as a multicopy plasmid (2, 20). P4 *cl kil*⁻ mutants have been isolated; these establish the plasmid state with high efficiency (2). The copy number for P4 wild-type, *cl*, and *vir1* lysogens appears to be equivalent, 30–50 copies/cell (20), but lower copy numbers have been reported in cells carrying P4 *vir1* under various growth conditions (59). Although the plasmid state is fairly stable, P4 plasmid-carrying clones will segregate immune-integrated lysogenic cells at a low frequency. P4 lysogenic cells do not give rise to cells carrying multicopy plasmids at an appreciable frequency unless the P4 prophage is induced (20).

P4 plasmid-carrying clones frequently contain an integrated copy of P4 as well (20), but plasmid-carrying cells do not express classical P4 immunity. Several lines of evidence indicate that P4 immunity is relaxed in the plasmid state. Replication of a superinfecting phage is prevented in a P4 lysogenic strain (108); in cells carrying P4 as a plasmid, a superinfecting wild-type phage will join the pool of autonomously replicating plasmids (2). P4 in the plasmid state can also be differentiated from the immune-integrated state by the response to a superinfecting P2. Induction of a P4 prophage by a superinfecting P2 requires prophage derepression, mediated by the P2 *cox* function (see below). Induction of P4 plasmids by P2, however, is independent of *cox* (20).

The P4 multicopy plasmid state does not simply represent persistence of the initial postinfection replicative phase. The rate of P4 DNA replication appears to be higher early after infection than in the plasmid state (2). Differences in protein synthesis (2) and transcription of the P4 genome (22, 32) have been documented as well.

TRANSCRIPTION PATTERNS IN THE ALTERNATIVE INTRACELLULAR
STATES The choice between the plasmid and lysogenic pathways in the
absence of helper resembles the choice between lysis and lysogeny for other
temperate phages, and is accompanied by changes in the patterns of P4
transcription (see Figure 4). Following infection by wild-type P4, there is an
initial phase of uncommitted replication (2) in which leftward transcription
initiates from the early promoter P_{LE} (20). Functions encoded by this 4.1 kb
transcript include the immunity determinant cI and the P4 DNA primase α.
While expression of α allows replication of the P4 genome, the majority of
these infected cells become lysogens carrying an integrated copy of P4,
presumably due to cI expression. In the immune-integrated state, the 4.1 kb
transcript can no longer be detected by Northern blot analysis (21, 22).
However, there is substantial initiation of transcription from P_{LE} as shown by
S1 mapping, while a probe from the α gene is not protected. This suggests
either premature termination of transcription from P_{LE} in the immune state or
instability of the 4.1 kb transcript in the lysogen. The immunity-defective
mutant P4 $cI405$ continues to make the 4.1 kb transcript in elevated amounts
even at late times after infection. These observations led to the proposal that
cI prevents expression of the leftward early operon by causing premature
termination of transcription (22).

A second leftward transcript covering the cI-α region can be detected at low

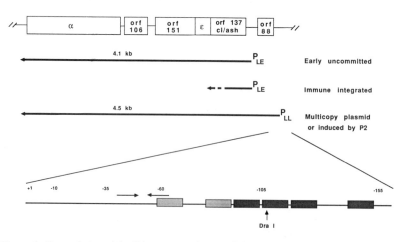

Figure 4 Transcription of the P4 α operon. A map of the region is shown at the top; arrows
below the map indicate the transcript present in each regulatory state (21, 22). The structure of the
P_{LL} promoter, which is activated by Cox or δ, is shown at the bottom. Positions upstream of the
transcription start site are given for reference. Dark rectangles indicate highly conserved copies of
the consensus Cox binding sequence; light rectangles indicate sites with weaker homology (99).
The arrows designate the inverted repeat which is the presumptive sequence required for
activation by δ. The site of DraI cleavage used to generate a truncated promoter that is still
activated by δ but no longer activated by Cox is also indicated.

abundance late after infection with wild-type P4. This 4.5 kb transcript is also present at high abundance in cells carrying P4 as a multicopy plasmid (22). Temporal appearance of this transcript correlates with the appearance of a 1.9 kb RNA corresponding to the late rightward transcript initiating from P_{sid} (18, 22). Both the 1.9 kb and 4.5 kb transcripts depend upon the P4 δ gene for their expression, as expected for late mRNAs. A low level of expression from P_{LL} was detected in P4 δ mutants at early times after infection (22); this appears to be due to δ-independent initiation of transcription 5 nucleotides upstream of P_{LL} from a weak overlapping promoter, designated P_{LL*} (21). The promoters for both the early 4.1 kb leftward transcript (P_{LE}) and the late 4.5 kb leftward transcript (P_{LL}) have been localized by S1 mapping (22). P_{LL} lies approximately 400 bp upstream of P_{LE}, giving rise to a transcript with the coding capacity for a protein of 88 amino acids (orf88) in addition to the genes encoded by the 4.1 kb transcript. The P4 vir1 mutation (69) lies in P_{LL} and leads to increased transcription from this promoter. In cells infected with P4 vir1, the 4.5 kb mRNA is present even at early times after infection and its expression no longer depends on the P4 δ gene (22). Establishment of the multicopy plasmid state thus appears to correlate with a shift from early (P_{LE}) to late (P_{LL}) leftward transcription, and early synthesis of the 4.5 kb mRNA from P_{LE} by P4 vir1 channels this mutant phage towards the plasmid state. The putative 88 amino acid protein encoded by the 4.5 kb transcript but not the 4.1 kb transcript is an attractive candidate for a gene product required for establishment or maintenance of P4 as a multicopy plasmid. Both the 4.1 and 4.5 kb transcripts include the cl determinant. Since normal P4 immunity is not expressed in the plasmid state, it is tempting to speculate that the role of the 88 amino acid protein is to interfere with expression of the immunity function.

S1 protection analysis of the 5' end of the gop-β transcript indicates that it is produced in elevated amounts from P4 plasmids, suggesting that this promoter may also be under positive control in the plasmid state (32). The products of these two genes have been proposed to play a role in P4 plasmid maintenance, by analogy to regulatory systems documented for certain other plasmids (M. Yarmolinsky, personal communication). The gop gene product kills the host when β is inactivated by mutation. If β were a functionally unstable inhibitor of gop, the loss of plasmid P4 would result in cell death because the gop protein is more stable than the β protein. Such a phenomenon would provide a selective advantage for the maintenance of P4 as a plasmid, and similar systems have been found for F and RI plasmids (30, 52).

Derepression of P4 by P2

Infection of a P4 lysogen by P2 induces the formation of P4 phages. The P2 cox gene is required for this induction (108). Replication of a P4 prophage can be induced by a plasmid containing the cox gene under lambda p_R control

(99), indicating that the *cox* function is sufficient for induction. Cox functions as an activator of transcription from the P4 late promoter P_{LL}, in contrast to its role as a repressor of transcription in P2. Using plasmids containing either the leftward early promoter P_{LE} or the leftward late promoter P_{LL} fused to the *cat* gene, Saha et al (99) have shown that expression of *cox* from a compatible plasmid results in a slight reduction in expression from P_{LE} and a marked stimulation in expression from P_{LL}. DNase I footprinting analysis demonstrates that Cox binds to both of these promoter regions. Cox protects a region upstream of the P_{LL} promoter extending from about -60 to -155 (see Figure 4); this region contains several direct repeats of the consensus sequence that is thought to be the Cox binding site. Weaker binding of Cox to sites spanning the P_{LE} promoter is also detected. A deletion derivative of the P_{LL}-*cat* fusion plasmid that removes sequences upstream of -105 is no longer activated by the Cox protein. This deletion eliminates all but one of the highly conserved copies of the Cox consensus sequence, indicating that the presence of more than one copy of this sequence is necessary for activation. This deleted promoter is still fully activated by δ (G. E. Christie, unpublished). Synthesis of the 4.5 kb transcript is thus regulated by two independent mechanisms of positive control. During the plasmid mode of replication, transcription from P_{LL} is activated by the P4 δ protein. During P4 prophage induction by an infecting P2 helper, transcription of the 4.5 kb mRNA from P_{LL} is activated by the P2 Cox protein. In both cases, the P4 replication function is expressed independently of P4 immunity control.

LATE GENE EXPRESSION AND MUTUAL TRANSACTIVATION

P2 Late Transcription

During P2 lytic infection, late gene expression requires the host RNA polymerase (73), P2 DNA replication (27a, 66, 70), and the P2 *ogr* gene product (10, 112). The *ogr* gene was defined by *trans*-dominant mutants that overcome the block to P2 late gene transcription imposed by the host *rpoA*109 mutation (112). The essential gene *B* of phage 186 encodes an analogous function that is required for 186 late gene transcription (25). The existence of viable P2-186 hybrid phages carrying P2 late genes and the *B* gene of 186 indicates that the *ogr* and *B* gene products work by the same mechanism (50). The cloned *ogr* gene also complements 186 *B* amber mutants (D. L. Anders & G. E. Christie, unpublished). Direct evidence that the P2 *ogr* gene is essential was provided by the generation in vitro of *ogr* deletion mutants, which are viable only when *ogr* is provided in *trans* (10, 44), and by the construction of an *ogr* amber mutant (G. E. Christie, unpublished).

All P2 late genes are required by satellite phage P4 when P2 is serving as the helper for P4 lytic growth (106). If P2 DNA replication is blocked by mutation in P2 genes *A* or *B* or in the host *rep* gene, preventing activation of P2 late transcription by the normal P2 pathway, P4 is able to "transactivate" the P2 late genes directly. This process requires the product of the P4 δ gene (46a, 109), which activates transcription from the same four promoters used during P2 lytic infection (14, 15). P4 can also transactivate the late genes of a coinfecting replication-defective 186 helper (102). The P4 δ gene acts independently of the P2 *ogr* or 186 *B* genes, since P4 can transactivate a P2 *ogr⁻* or 186 *Bam17* helper (44, 102).

Although transactivation is independent of helper DNA replication, an additional helper early function may be required for P4 lytic growth. Prophage derepression is clearly required for transactivation of 186 by P4. P4 cannot grow with a wild-type 186 prophage helper, but a replication-deficient 186 *CI*(ts) prophage will support P4 multiplication after derepression by temperature shift (102). Derepression appears necessary for P4 multiplication with a P2 prophage helper as well, since the derepression mutant P4 ε am*104* cannot grow in a P2-lysogenic strain (29). However, the ε am*104* phage also carries an *ash* mutation and may be affected in the expression of other P4 genes, including δ, so the relationship between derepression and transactivation of P2 by P4 has not been clearly established. P2 mutants have been isolated that impair transactivation of a P2 prophage by P4; these map to the early region (3) and may define another P2 early gene, since mutations in the P2 early genes *A,B* or *cox* do not affect transactivation by P4 (106). While it is likely these "transactivation-resistant" mutants are affected in derepression of the prophage helper, these mutants have not been characterized further. Prophage derepression cannot be required directly for P2 late promoter activation, since plasmid-encoded δ can activate late transcription from a P2 prophage (42). An early function under P2 repressor control may be involved directly or indirectly in expression of δ from a superinfecting P4.

P4 Late Transcription

The P4 late genes (*sid*, δ, and *psu*) are cotranscribed from the P4 late promoter P_{sid}; this transcription depends on the product of the P4 δ gene (18, 22) and the host RNA polymerase (57, 73), and also requires the P4 DNA replication function α (46a). Helper-independent transcription of the *cI-α* operon from P_{LL} also depends upon δ (21, 22). In the presence of a P2 helper, synthesis of P4 late proteins occurs 20 minutes earlier and in five- to tenfold greater amounts (6). This increase correlates with increased transcription from P_{sid} (18). Stimulation of transcription from P_{sid} can be obtained in the absence of a P2 helper phage if the *ogr* gene product is provided from a plasmid (18). The cloned P_{LL} promoter can also be activated in vivo by plasmid-encoded Ogr (G. E. Christie, unpublished).

The P4 δ gene was originally defined by mutants that were unable to use a replication-defective P2 helper (109). Such mutants can grow with a wild-type P2 or 186 helper, but not with a P2 ogr^- or 186 B^- helper (44, 102). P4 lytic growth thus requires either the P4 δ protein or the helper ogr (or B) gene product. P4 δ protein can also substitute for Ogr in P2 multiplication, since P2 ogr deletion phages can grow when δ is provided from a plasmid (10, 44). The P2- and P4-encoded late transcription regulatory proteins thus appear to work independently and are to some extent functionally interchangeable.

Involvement of rpoA

The $rpoA109$ mutation of E. coli substitutes a histidine for a leucine in the α subunit of RNA polymerase (26) and prevents P2 late gene transcription (14, 15, 112). This mutation also interferes with transcription from the P4 late promoter P_{sid} (18) and prevents P4 from growing with a wild-type P2 helper (102). Mutations in P2 and P4 that overcome this block have been isolated, and lie in ogr (112) and δ (46), respectively. This again indicates a similar mechanism for these two proteins, and the compensatory nature of these mutations suggests that the ogr and δ gene products modulate transcription by direct interaction with one or both α subunits of RNA polymerase. The growth of phage 186 is unaffected by the $rpoA109$ mutation (102).

Several additional $rpoA$ mutations have been described that affect transcription of specific loci in E. coli (27, 34, 78, 96). Each appears to be allele-specific; none affects the growth of P2. The only common feature of the operons affected by these different mutations in $rpoA$ is that they are all normally under positive control. Each regulatory protein may make specific contacts with the α subunit(s) of RNA polymerase to stimulate transcription.

Structure of the Transcriptional Activators

Comparison of the sequences of the P2 ogr, 186 B, and P4 δ genes confirms that these three proteins are related (Figure 5). The ogr (11, 16) and B (56) genes encode proteins of 72 amino acids that are identical at 45 positions. The predicted product of the P4 δ gene is 166 amino acids (69; corrected in 45). Homology between Ogr, B and the amino-terminal portion of δ was first described by Kalionis et al (56); a second region of homology in the C-terminal half of δ is reported by Halling et al (46). Thus, the δ gene product appears to be a covalent dimer of two 83 amino acid Ogr-like domains. Comparison of the two domains of δ to Ogr and B (42) reveals the following features: (a) most of the similarity lies in the amino-terminal portions of each domain; (b) amino acids at only seven positions are absolutely conserved among all four domains; and (c) each domain of δ is more similar to Ogr (and B) than to the other half of δ (in fact, the two domains of δ share only the seven absolutely conserved amino acid residues).

Ogr

B

δ   1 - 83

δ   84 - 166

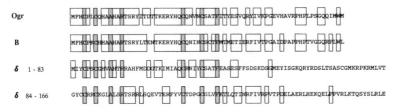

Figure 5   Amino acid sequence comparison of the P2 Ogr, 186 B and P4 δ proteins. The δ sequence is displayed as two 83 amino acid domains. The boxes indicate conserved residues at identical positions; shaded boxes designate residues that are invariant.

Ogr, B, and δ define a new class of prokaryotic transcriptional activators. They do not appear to contain a helix-turn-helix structural motif for DNA binding, nor do they resemble known σ factors. In addition, δ-dependent expression of P4 P_{sid} in vitro is inhibited by antibodies to σ^{70} and restored when σ^{70} is added back (57), demonstrating that δ does not function as an alternative sigma subunit of RNA polymerase. No other related proteins have yet been found in the GenBank or EMBL databases.

PROPOSED FUNCTIONAL DOMAINS   Based on evidence from sequence comparisons, the locations of known mutations, and biochemical studies on purified Ogr, two functional domains have been proposed (64). The first is a zinc finger domain in the highly conserved hydrophilic N-terminal region of these proteins. Four of the seven amino acid residues that are absolutely conserved among Ogr, B, and each half of δ are cysteines. These occur in repeated units of the sequence $Cys-X_{2-4}-Cys$, a motif characteristic of zinc-binding domains. Purified Ogr protein binds zinc, and the properties of Co(II)-substituted Ogr are consistent with cysteinyl coordination of a single metal ion per molecule (65). Substitution of a glycine for any of the four cysteines inactivates Ogr (K. Gebhardt & B. H. Lindqvist, manuscript in preparation).

Precedents exist for the involvement of zinc fingers in both protein:DNA and protein:protein interactions. Eukaryotic zinc finger DNA-binding proteins have been classified as Cys_2His_2 or Cys_x, based on the amino acid residues involved in coordination of the metal atom (reviewed in 53). In both classes of zinc-finger proteins, however, the size of the proposed loop is generally 12–13 amino acids, which is much smaller than the 22 amino acids found in Ogr. The UvrA protein of *E. coli* has two Cys_4 zinc fingers with loops of 19 and 20 amino acids (82); while UvrA is a DNA-binding protein, the zinc fingers have not yet been implicated directly in DNA binding. The regulatory subunit of aspartate transcarbamoylase (ATCase) contains a zinc-binding domain that, like Ogr, includes two pairs of cysteine ligands separated by 22 amino acids (81). X-ray diffraction studies indicate that this domain is the major site of interaction between the regulatory and catalytic subunits of ATCase. The zinc-binding domain in T4 gene *32* protein, originally thought to play a role in DNA binding, appears to be involved in the monomer-monomer interactions essential for the cooperative binding of gp*32* to single-stranded DNA (33). In the absence of any direct evidence for DNA binding or protein:protein interaction, the role of the zinc-binding domain in Ogr structure and function remains an open question.

A second, C-terminal domain is proposed to play a role in transcriptional activation. Transcriptional activation is generally thought to be mediated by a protein surface or domain distinct from the DNA-binding domain (89). The existence of mutations in *ogr* and δ that suppress the effects of the host *rpoA*109 mutation implies that these transcriptional activators function, at least in part, by direct interaction with RNA polymerase. Such interactions clearly have been implicated in positive regulation by the lambda (38, 49) and P22 (87) repressors and CAP protein (51, 92). All *rpoA*109-suppressing *ogr* mutations sequenced thus far are identical, and result in a Tyr to Cys change at amino acid 42 (11, 16; B. H. Lindqvist, personal communication). The four independent *rpoA*109-suppressing δ *(org)* mutations are also identical to each other, and generate a Thr to Ala change at amino acid residue 127 (46), which is equivalent to position 44 in the second Ogr-like domain of δ. This suggests that a region distal to the zinc-binding domain interacts with RNA polymerase, and may constitute the activation domain. Sequence inspection has also revealed a clustering of proline residues near the C-terminus of Ogr, B, and δ (56, 64); a proline-rich region is one motif that has been invoked in classifying existing transcriptional activators (80, 89). However, preliminary results suggest that this C-terminal proline-rich region of Ogr is not essential (K. Gebhardt & B. H. Lindqvist, personal communication).

Structure of P2 and P4 Late Promoters

The four P2 late promoters (14, 15) and the two P4 late promoters (18, 22) share features typical of positively regulated bacterial promoters recognized by RNA polymerase carrying the σ^{70} subunit (reviewed in 79, 90). None of the promoters shows good homology to the -35 consensus for σ^{70} RNA polymerase of *E. coli*. Partial homology with the -10 consensus exists to varying extents in the different promoters, with the two most highly conserved nucleotides being retained in all cases. The four P2 late promoters share some sequence similarity in the -35 region and an alternative -10 sequence (15), which are not found in the P4 late promoters.

The most striking feature of the six late promoters is homology in an upstream region centered at about -55 (Figure 6). The P4 late promoter P_{sid} contains an obvious dyad symmetry element in this sequence, and all six promoters can be aligned so that a consensus sequence, containing an imperfect dyad, can be obtained. Deletion analysis of the P2 *FETUD* operon promoter (P_F) and P_{sid} clearly implicates this upstream region in promoter function. Replacement of promoter sequences upstream of −64 eliminates P_F activity in the presence of plasmid-encoded Ogr or δ, as assayed by fusion to the *cat* gene, while an upstream deletion ending at −69 retains essentially wild-type function (37). The critical upstream boundary of P_{sid}, in a *lacZ* fusion stimulated by δ, lies between nucleotides −63 and −64, with derivatives containing vector DNA upstream of −67 or −69 retaining partial activity. A point mutation in either arm of the inverted repeat, at −62 or −48, inactivates P_{sid} (G. B. Van Bokkelen, E. C. Dale, C. Halling, R. Calendar, manuscript in preparation). A mutation at −50 in P_F, which corresponds to the same position in the inverted repeat as −48 in P_{sid}, does not impair promoter function (G. E. Christie & N. Stokes, unpublished). However, the base substitution is different (C to G in P_{sid}, C To T in P_F), which could account for the disparity. Point mutations isolated thus far that lie between the arms of the inverted repeat in P_F have no effect, while small deletions (2 or 3 bases) that remove the same nucleotides eliminate promoter activity (G. E.

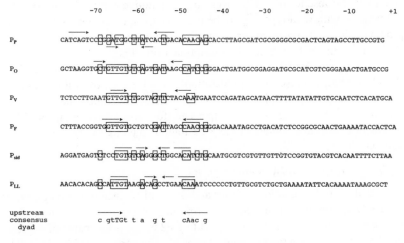

Figure 6 Comparison of the late promoters of P2 and P4, aligned at the upstream region of dyad symmetry. Each sequence ends at +1. Boxes indicate those nucleotides in or between the arms of the inverted repeat that correspond to the consensus sequence derived for this region. Capital letters in the consensus sequence indicate those nucleotides that are conserved in all six promoters; nucleotides shown in lower case are conserved in at least four of the six promoters. Arrows indicate inverted repeats present in the sequences. A better match to the downstream portion of the consensus for P_V can be obtained if the spacing between the two arms of the inverted repeat is reduced by 1 nucleotide. In addition, the upstream regions of the divergent P2 late promoters P_O and P_P overlap, so that nucleotides in the dyad symmetry element are shared.

Christie & N. Stokes, unpublished). This body of evidence is consistent with a role for the upstream inverted repeat in positive regulation of these promoters, possibly as a binding site for the regulatory proteins. However, preliminary studies with purified Ogr and RNA polymerase have failed thus far to demonstrate binding of these proteins, alone or in combination, to DNA fragments or supercoiled plasmids carrying P_F or P_{sid} (64).

In addition to the requirement for upstream sequences, evidence indicates that nucleotides in both the -10 and -35 regions are important in late promoter function, despite the generally poor homology to the consensus sequence of *E. coli*. Point mutations in either of the two highly conserved consensus nucleotides at -10 in P_{sid} impair promoter function as severely as deletion of the upstream region (G. B. Van Bokkelen et al, manuscript in preparation). Mutations in the -35 region of P_{sid} and P_F, which increase the homology to the consensus sequence of *E. coli*, also increase promoter strength. A point mutation at -34 in P_F, which results in the sequence TAGACT, increases expression about fourfold in the absence or presence of δ (G.E. Christie & N. Stokes, unpublished). When the -35 region of P_{sid} is replaced with the consensus sequence TTGACA of *E. coli*, spaced 17 nucleotides upstream of the -10 region, there is a tenfold increase in promoter activity in the absence of δ. However, this constitutive activity is still only 4% of that seen with wild-type P_{sid} in the presence of δ, even though this mutant promoter differs from a perfect consensus promoter of *E. coli* at only two positions in the -10 region. Transcription from this "-35 consensus" *sid* promoter is stimulated about 40-fold in the presence of δ (G. B. Van Bokkelen et al, manuscript in preparation). Positive control of this promoter cannot therefore be simply explained by a model in which the regulatory protein provides contacts to compensate for poor interaction of RNA polymerase with nucleotides in the -35 region, such as has been shown for lambda *cII* protein (48).

Role of Helper DNA Replication

Despite the similar functions of Ogr and δ, the difference in the requirement for P2 DNA replication in late transcription remains to be explained. One possible explanation is that a nonreplicating P2 phage is incapable of expressing the *ogr* gene, as in the late control gene C of bacteriophage Mu (111). The failure of P2 Aam129 ogr1 to support the growth of wild-type P2 in an rpoA109 host led to the suggestion that the A gene was required for expression of *ogr* (102). Transcription of the *ogr* gene begins at middle times after infection (N. K. Birkeland, B. H. Lindqvist & G. E. Christie, manuscript in preparation), as would be expected if replication were required. However, recent convincing evidence demonstrates that *ogr* expression does occur in replication-defective mutants. P2 Aam127 stimulates transcription from the P4 P_{sid} promoter as well as, or better than, P2 *vir1* (18). A coinfecting P2

Aam*127* phage complements a P2 *ogr* deletion mutant (44), and the *ogr* gene is clearly transcribed during P2 Aam*127* infection (N. K. Birkeland, B. H. Lindqvist & G. E. Christie, manuscript in preparation). The lack of complementation reported by Sauer et al (102) in the *rpoA*109 strain is probably due to interference of the *ogr*$^+$ protein produced from the replication-proficient phage with the small amount of *ogr*1 protein produced from the nonreplicating template.

The requirement for a replicating template thus appears to involve the action of Ogr rather than its synthesis. Consistent with this, the temporal appearance of P2 late mRNAs is not altered significantly by the presence of plasmid-encoded Ogr (G. E. Christie, unpublished). Coupling of transcription with DNA replication has been documented in several systems, and in a few cases the mechanism has been elucidated. Inhibition of transcription by *dam* methylation has been proposed to limit expression of the IS*10* transposase promoter to newly replicated, hemimethylated DNA (95). A similar mechanism for P2 late transcription can be ruled out by the lack of *dam* methylation sites in the late promoters. Phage T4 late transcription is stimulated by three phage-encoded replication accessory proteins that constitute part of the T4 replisome and increase the rate of open complex formation at late promoters while moving along the genome (47).

Several hypotheses have been proposed to account for the coupling of P2 replication and late transcription (42, 64): (*a*) Ogr may function much less efficiently than δ, and replication is required to generate a high number of copies of the late genes; (*b*) a topological structure generated by DNA replication is required to generate a template competent for activation by Ogr; (*c*) an additional P2 gene product expressed only by a replicating template is required for P2 late transcription; (*d*) Ogr (or a transcription complex including Ogr) interacts directly with a protein or proteins in the replisome. Existing evidence does not allow any of these possibilities to be ruled out, and they are not mutually exclusive. Plasmid-encoded Ogr is sufficient for activation of cloned P2 late promoters (37; G. E. Christie, unpublished), but about twenty-fold more activity is obtained in the presence of plasmid-encoded δ. This could reflect a difference in the efficiency with which these two proteins can act, but the levels of Ogr or δ present in these cells has not been determined. It could also indicate that the cloned P2 late promoters have not been fully uncoupled from the normal requirement for P2 DNA replication, and that maximal activation by Ogr requires an additional (and as yet unidentified) P2-encoded factor not needed by δ.

POLARITY SUPPRESSION

Certain amber mutations in the first gene of each of the four P2 late operons have strong polar effects (70). This polarity can be partially suppressed by the *suA1 (rho1)* mutation of *E. coli* (114), which implicates Rho factor in premature termination of transcription of these operons. P4 infection also suppresses the polarity of P2 polar amber mutations (113), and this effect is due to the P4 late gene *psu* (103). The *psu* gene encodes a protein of $M_r = 21,300$ (190 amino acids) that is not related to the N or Q antitermination proteins of phage lambda (18). Psu is made in large amounts (6, 103) and is found in association with P4-sized capsids but not P2-sized capsids. The *psu* gene is not essential for P4 lytic growth, but the burst size of *psu* mutants is decreased fivefold (103). P4 *psu* mutants are also able to establish the multicopy plasmid state, although the overall amount of P4 transcription as assayed by Northern analysis is reduced in P4 *psu* plasmids (22). The actual role of *psu* during P4 development remains a mystery.

P4 *psu* protein expressed constitutively from a plasmid can cause read-through of the chromosomal tryptophan operon terminator of *E. coli* (60). Plasmids carrying *psu* under the control of the lambda p_L or synthetic *tac* promoter have been constructed, and the protein has been overproduced (72). Plasmid-encoded Psu is sufficient to suppress polarity in P2 late operons, and causes readthrough of the *rho*-dependent transcription terminator in the insertion sequence IS*2* (19) cloned between the lambda p_R promoter and *lacZ* (72). Psu has no effects on initiation of transcription or translation or on mRNA stability, as assayed in vivo using a p_R-*lacZ* fusion plasmid. The effect of Psu appears to be specific for Rho-dependent terminators; no Psu-mediated antitermination was found at the Rho-independent terminators *trp a* or *rrnB* T_1. Using the p_{tac}-*psu* overproducing plasmid, Psu protein has recently been purified (M. Isaksen, personal communication), so the effect of Psu on Rho-dependent terminators can now be tested in vitro.

CAPSID SIZE DETERMINATION

Cells infected by P2 contain only capsids with 540 subunits, appropriate for the size of P2 DNA (28, 67). In contrast, P4-infected, P2 lysogenic cells contain only capsids with 240 subunits, which fit the size of P4 DNA. The major protein component of P2 and P4 capsids is a cleaved version of the P2 *N* gene product, N^*. Both capsid types contain intermediates in the cleavage of N to N^*, but P4 capsids contain substantially larger quantities of these intermediates (6). Wild-type P4 capsids also contain P4 Psu protein, as discussed above, but *psu* cannot be essential for determination of small capsid size, since an amber mutation in this gene does not block the synthesis of small capsids (103). P4 capsid size determination *(sid)* mutants (104) are defective in the synthesis of a late P4 protein that is made in quantities similar to that of the P2 capsid protein (63). The Sid protein has not been found in phage capsids. P4 *sid* mutants make minute plaques consisting of large-capsid phages that contain dimers or trimers of P4 DNA. P2 mutants (*sir*) that block *sid* function map in gene *N*, suggesting an interaction between the P4 Sid protein and the P2 capsid protein precursor (E. W. Six, M. Sunshine, J. Williams, E. Haggård-Ljungquist & B. H. Lindqvist, manuscript in preparation). Sid may serve as a scaffolding protein for small-capsid formation, or affect the processing pathway of the capsid protein precursor.

Agarwal et al (1) have shown that Sid protein, provided from a plasmid, is the only P4 function necessary to direct small-capsid formation. They further propose that Sid competes with the *O* gene product of P2 in assembly of P4-sized capsids, based on the apparent complementation of P2 *O*am279 by plasmid-encoded Sid in the packaging of an artificial P4-sized cosmid into transducing particles. Such a model, however, is difficult to reconcile with the findings of Six (106) that all of the P2 late genes, including *O*, are necessary for P4 lytic growth in a P2-lysogenic strain. Coinfection with P4 and P2 *O*am279 also yields less than 0.1 P4 progeny/infected cell (E. W. Six, personal communication), confirming the previous observation that P4 requires the P$_2$ *O* gene for normal lytic development.

SUMMARY

The helper dependence of satellite phage P4 superimposes an additional set of regulatory interactions on those required for the independent maintenance of P4 or its helpers. These interactions allow P4 to exploit a helper phage under a variety of circumstances and can affect expression of the immunity functions and late genes of both phages. The phage P2 lysis/lysogeny decision involves two competing repressors regulating mutually exclusive promoters in the early control region. In the absence of a helper phage, the P4 immunity function plays a role in the choice between lysogeny or the multicopy plasmid state. No evidence exists for a P4-encoded immunity repressor; in P4-lysogenic cells, expression of the P4 DNA replication gene α appears to be prevented by premature termination of transcription. Immunity-independent expression of α in the multicopy plasmid state involves initiation of transcription at an alternative upstream promoter that is positively regulated by P4 δ protein; the same promoter is activated by P2 Cox protein during derepression of P4 by P2. The mechanism of derepression of P2 by P4 remains to be determined, and the relationship between the P4 immunity and derepression functions and the mutations that allow P4 to grow with a P3 prophage helper is an intriguing area for further exploration. Expression of P2 and P4 late genes is regulated by phage-encoded, zinc-binding transcriptional activators that appear to interact directly with the α subunit of RNA polymerase of *E. coli*. Stimulation of P2 late transcription by P2 Ogr protein depends upon phage DNA replication, whereas activation of transcription from the same promoters by the related P4 δ gene product is replication-independent. Elucidation of the mechanisms underlying these interactions promises to provide new insights into strategies for control of gene expression.

ACKNOWLEDGMENTS

We are grateful to Elisabeth Haggård-Ljungquist, Nils Kåre Birkeland, Björn Lindqvist, Kirsti Gebhardt, Morten Isakson, Erich Six, Mel Sunshine, Douglas Anders, Susan Forsburg, Nora Linderoth, Rainer Ziermann, Gianni Dehò and Daniela Ghisotti for communication of unpublished results and comments on the manuscript. Research in our labs is supported by NIH grants GM34651 (to G.E.C.) and AI08722 (to R.C.)

Literature Cited

1. Agarwal, M., Arthur, M., Arbeit, R. D., Goldstein, R. 1990. Regulation of icosahedral virion capsid size by the in vivo activity of a cloned gene product. *Proc. Natl. Acad. Sci. USA* 87:2428–32
2. Alano, P., Dehò, G., Sironi, G., Zangrossi, S. 1986. Regulation of the plasmid state of the genetic element P4. *Mol. Gen. Genet.* 203:445–50
3. Barclay, S. L., Dove, W. F. 1978. Mutations of bacteriophage P2 which prevent activation of P2 late genes by satellite phage P4. *Virology* 91:321–35
4. Barrett, K. J., Blinkova, A., Arnold, G. 1983. The bacteriophage P4 α gene is the structural gene for bacteriophage P4-induced RNA polymerase. *J. Virol.* 48:157–69
5. Barrett, K. J., Gibbs, W., Calendar, R. 1972. A transcribing activity induced by satellite phage P4. *Proc. Natl. Acad. Sci. USA* 69:2986–90
6. Barrett, K. J., Marsh, M. L., Calendar, R. 1976. Interactions between a satellite phage and its helper. *J. Mol. Biol.* 106:683–707
7. Bertani, G. 1951. Studies on lysogenesis. I. The mode of phage liberation by lysogenic *Escherichia coli*. *J. Bacteriol.* 62:293–300
8. Bertani, L. E. 1968. Abortive induction of bacteriophage P2. *Virology* 36:87–103
9. Bertani, L. E., Six, E. W. 1988. The P2-like phages and their parasite, P4. In *The Bacteriophages*, ed. R. Calendar, 2:73–143. New York: Plenum. 760 pp.
10. Birkeland, N. K., Christie, G. E., Lindqvist, B. H. 1988. Directed mutagenesis of the bacteriophage P2 *ogr* gene defines an essential function. *Gene* 73:327–35
11. Birkeland, N. K., Lindqvist, B. H.

1986. Coliphage P2 late control gene *ogr*: DNA sequence and product identification. *J. Mol. Biol.* 188:487–90
12. Bradley, C., Ling, O. P., Egan, J. B. 1975. Isolation of phage P2-186 intervarietal hybrids and 186 insertion mutants. *Mol. Gen. Genet.* 140:123–35
13. Calendar, R., Ljungquist, E., Dehò, G., Usher, D. C., Goldstein, R., et al. 1981. Lysogenization by satellite phage P4. *Virology* 113:20–38
14. Christie, G. E., Calendar, R. 1983. Bacteriophage P2 late promoters: Transcription initiation sites for two late mRNAs. *J. Mol. Biol.* 167:773–90
15. Christie, G. E., Calendar, R. 1985. Bacteriophage P2 late promoters. II. Comparison of the four late promoter sequences. *J. Mol. Biol.* 181:373–82
16. Christie, G. E., Haggård-Ljungquist, E., Feiwell, R., Calendar, R. 1986. Regulation of bacteriophage P2 late gene expression: the *ogr* gene. *Proc. Natl. Acad. Sci. USA* 83:3238–42
17. Deleted in proof
18. Dale, E. C., Christie, G. E., Calendar, R. 1986. Organization and expression of the satellite bacteriophage P4 late gene cluster. *J. Mol. Biol.* 192:793–803
19. DeCrombrugghe, B., Adhya, S., Gottesman, M., Pastan, I. 1973. Effect of *rho* on transcription of bacterial operons. *Nature New Biol.* 241:260–64
20. Dehò, G., Ghisotti, D., Alano, P., Zangrossi, S., Borrello, M. G., et al. 1984. Plasmid mode of propagation of the genetic element P4. *J. Mol. Biol.* 178:191–207
21. Dehò, G., Ghisotti, D., Zangrossi, S., Alano, P., Neri, M. R., et al. 1988. Regulation of alternative intracellular states in the phage-plasmid P4. In *Gene*

Expression and Regulation. The Legacy of Luigi Gorini, ed. M. Bissell, G. Dehò, G. Sironi, A. Torriani, pp. 65–72. Amsterdam: Exerpta Medica. 386 pp.

22. Dehò, G., Zangrossi, S., Ghisotti, D., Sironi, G. 1988. Alternative promoters in the development of bacteriophage plasmid P4. *J. Virol.* 62:1697–704

23. Dhillon, I. E. S., Dhillon, T. S., Lam, Y. Y., Tsang, A. H. C. 1980. Temperate coliphages: Classification and correlation with habitats. *Appl. Environ. Microbiol.* 39:1046–53

24. Diana, C., Dehò, G., Geisselsoder, J., Tinelli, L., Goldstein, R. 1978. Viral interference at the level of capsid size determination by satellite phage P4. *J. Mol. Biol.* 126:447–56

25. Finnegan, G., Egan, J. B. 1981. In vivo transcription studies of coliphage 186. *J. Virol.* 38:987–95

26. Fujiki, H., Palm, P., Zillig, W., Calendar, R., Sunshine, M. G. 1976. Identification of a mutation within the structural gene for the α subunit of DNA-dependent RNA polymerase of *E. coli. Mol. Gen. Genet.* 145:19–22

27. Garrett, S., Silhavy, T. J. 1987. Isolation of mutations in the α operon of *Escherichia coli* that suppress the transcription defect conferred by a mutation in the porin regulatory gene *envZ. J. Bacteriol.* 169:1379–85

27a. Geisselsoder, J., Mandel, M., Calendar, R., Chattoraj, D. K. 1973. In vivo transcription patterns of temperate coliphage P2. *J. Mol. Biol.* 77:405–15

28. Geisselsoder, J., Sedivy, J., Walsh, R., Goldstein, R. 1982. Capsid structure of satellite phage P4 and its P2 helper. *J. Ultrastruct. Res.* 79:165–73

29. Geisselsoder, J., Youderian, P., Dehò, G., Chidambaram, M., Goldstein, R. N., et al. 1981. Mutants of satellite virus P4 that cannot derepress their P2 helper. *J. Mol. Biol.* 148:1–19

30. Gerdes, K., Helin, K., Christensen, L., Lobner-Olesen, A. 1988. Translational control and differential RNA decay are key elements of regulatory expression of the killer protein encoded by the *parB* locus of plasmid RI. *J. Mol. Biol.* 203:119–29

31. Ghisotti, D., Dehò, G., Chiaramonte, R., Zangrossi, S., Sironi, G. 1989. *Immunity of phage-plasmid P4: a preliminary identification of a novel mechanism.* Presented at UCLA West. Winter Workshop on Prokaryotic Promoters, Frisco, CO

32. Ghisotti, D., Finkel, S., Halling, C., Dehò, G., Sironi, G., et al. 1990.

33. Giedroc, D. P., Keating, K. M., Williams, K. R., Coleman, J. E. 1987. The function of zinc in gene 32 protein from T4. *Biochemistry* 26:5251–59

34. Giffard, P. M., Booth, I. R. 1988. The *rpoA*341 allele of *Escherichia coli* specifically impairs the transcription of a group of positively-regulated operons. *Mol. Gen. Genet.* 214:148–52

35. Goldstein, R., Lengyel, J., Pruss, G., Barrett, K., Calendar, R., Six, E. 1974. Head size determination and the morphogenesis of satellite phage P4. *Curr. Top Microbiol. Immunol.* 68:59–75

36. Goldstein, R., Sedivy, J., Ljungquist, E. 1982. Propagation of satellite phage P4 as a plasmid. *Proc. Natl. Acad. Sci. USA* 79:515–19

37. Grambow, N. J., Birkeland, N. K., Anders, D. L., Christie, G. E. 1990. Deletion analysis of a bacteriophage P2 late promoter. *Gene.* In press

38. Guarente, L., Nye, J. S., Hochschild, A., Ptashne, M. 1982. A mutant lambda repressor with a specific defect in its positive control function. *Proc. Natl. Acad. Sci. USA* 79:2236–39

39. Haggård-Ljungquist, E. 1990. Bacteriophage P2. See Ref. 83, pp. 63–69

40. Haggård-Ljungquist, E., Barreiro, V., Calendar, R., Kurnit, D. M., Cheng, H. 1989. The P2 phage *old* gene: sequence, transcription and translational control. *Gene* 85:25–33

41. Haggård-Ljungquist, E., Kockum, K., Bertani, L. E. 1987. DNA sequences of bacteriophage P2 early genes *cox* and *B* and their regulatory sites. *Mol. Gen. Genet.* 208:52–56

42. Halling, C. H. 1989. *Studies of the transactivation gene and the nonessential region of bacteriophage P4.* PhD thesis. Univ. Calif., Berkeley. 286 pp.

43. Halling, C. H., Calendar, R. 1990. Bacteriophage P4. See Ref. 83, pp. 70–73

44. Halling, C., Calendar, R. 1990. The bacteriophage P2 ogr and P4 δ genes act independently and are essential for P4 multiplication. *J. Bacteriol.* 172:3549–58

45. Halling, C., Calendar, R., Christie, G. E., Dale, E. C., Dehò, G., et al. 1990. DNA sequence of satellite bacteriophage P4. *Nucleic Acids Res.* 18:1649

46. Halling, C., Sunshine, M. G., Lane, K. B., Six, E. W., Calendar, R. 1990. A mutation of the transactivation gene of satellite bacteriophage P4 that sup-

presses the *rpoA109* mutation of *Escherichia coli. J. Bacteriol.* 172:3541–48

46a. Harris, J. D., Calendar, R. 1978. Transcription map of satellite coliphage P4. *Virology* 85:343–58

47. Herendeen, D. R., Kassavetis, G. A., Barry, J., Alberts, B. M., Geiduschek, E. P. 1989. Enhancement of bacteriophage T4 late transcription by components of the T4 DNA replication apparatus. *Science* 245:952–58

48. Ho, Y.-S., Wulff, D. L., Rosenberg, M. 1983. Bacteriophage lambda protein cII binds promoters on the opposite face of the DNA helix from RNA polymerase. *Nature* 304:703–8

49. Hochschild, A., Irwin, N., Ptashne, M. 1983. Repressor structure and the mechanism of positive control. *Cell* 32:319–25

50. Hocking, S. M., Egan, J. B. 1982. Genetic characterization of twelve P2-186 hybrid bacteriophages. *Mol. Gen. Genet.* 187:174–76

51. Irwin, N., Ptashne, M. 1987. Mutants of the catabolite activator protein of *Escherichia coli* that are specifically deficient in the gene-activation function. *Proc. Natl. Acad. Sci. USA* 84:8315–19

52. Jaffe, A., Ogawa, T., Hiraga, S. 1985. Effects of the *ccd* function of the F plasmid on bacterial growth. *J. Bacteriol.* 163:841–49

53. Johnson, P. F., McKnight, S. L. 1989. Eukaryotic transcriptional regulatory proteins. *Annu. Rev. Biochem.* 58:799–839

54. Kahn, M., Ow, D., Sauer, B., Rabinowitz, A., Calendar, R. 1980. Genetic analysis of bacteriophage P4 using P4-plasmid *Col*E1 hybrids. *Mol. Gen. Genet.* 177:399–412

55. Kalionis, B., Dodd, I. B., Egan, J. B. 1986. Control of gene expression in the P2-related temperate coliphages. III. DNA sequence of the major control region of phage 186. *J. Mol. Biol.* 191:199–209

56. Kalionis, B., Pritchard, M., Egan, J. B. 1986. Control of gene expression in the P2-related temperate coliphages. IV. Concerning the late control gene and control of its transcription. *J. Mol. Biol.* 191:211–20

57. Keener, J., Dale, E. C., Kustu, S., Calendar, R. 1988. In vitro transcription from the late promoter of bacteriophage P4. *J. Bacteriol.* 170:3543–46

58. Krevolin, M., Calendar, R. 1985. The replication of bacteriophage P4 DNA in vitro. Partial purification of the P4 α gene product. *J. Mol. Biol.* 182:509–17

59. Lagos, R., Goldstein, R. 1984. Phasmid

P4: Manipulation of plasmid copy number and induction from the integrated state. *Proc. Natl. Acad. Sci. USA* 158:208–15

60. Lagos, R., Jiang, R.-Z., Kim, S., Goldstein, R. 1986. Rho-dependent transcription termination of a bacterial operon is antagonized by an extrachromosomal gene product. *Proc. Natl. Acad. Sci. USA* 83:9561–65

61. Lamont, I., Brumby, A. M., Egan, J. B. 1989. UV induction of coliphage 186: prophage induction as an SOS function. *Proc. Natl. Acad. Sci. USA* 86:5492–96

62. Lamont, I., Kalionis, B., Egan, J. B. 1988. Control of gene expression in the P2-related temperate coliphages. V. The use of sequence analysis of 186 Vir mutants to indicate presumptive repressor binding sites. *J. Mol. Biol.* 199:379–82

63. Lee, S.-J. 1981. *Altered patterns of gene expression by satellite phage P4 mutants unable to regulate their P2 helper.* BA thesis. Harvard Univ., Cambridge. 106 pp.

64. Lee, T.-C. 1989. *Purification and functional characterization of the phage P2 Ogr protein: a prokaryotic zinc-binding transcriptional activator.* PhD thesis. Va. Commonw. Univ., Richmond. 163 pp.

65. Lee, T.-C., Christie, G. E. 1990. Purification and properties of the bacteriophage P2 *ogr* gene product: A prokaryotic zinc-binding transcriptional activator. *J. Biol. Chem.* 265:7472–77

66. Lengyel, J. A., Calendar, R. 1974. Control of bacteriophage P2 protein and DNA synthesis. *Virology* 57:305–13

67. Lengyel, J. A., Goldstein, R. N., Marsh, M., Sunshine, M., Calendar, R. 1973. Bacteriophage P2 head morphogenesis: cleavage of the major capsid protein. *Virology* 53:1–23

68. Lin, C.-S. 1983. *Genetic and molecular studies of bacteriophage P4: ash mutants and DNA sequence between genes psu and alpha.* PhD thesis. Univ. Iowa, Iowa City. 159 pp.

69. Lin, C.-S. 1984. Nucleotide sequence of the essential region of bacteriophage P4 DNA. *Nucleic Acids Res.* 12:8667–84

70. Lindahl, G. 1971. On the control of transcription in bacteriophage P2. *Virology* 46:620–33

71. Lindahl, G., Sunshine, M. 1972. Excision-deficient mutants of bacteriophage P2. *Virology* 49:180–87

72. Linderoth, N. 1990. *Studies on the transcription antitermination system of satellite phage P4.* PhD thesis. Univ. Calif., Berkeley. 284 pp.

73. Lindqvist, B. H. 1974. Expression of phage transcription in P2 lysogens infected with helper-dependent coliphage P4. *Proc. Natl. Acad. Sci. USA* 71:2752–55

74. Lindqvist, B. H. 1981. Recombination between satellite phage P4 and its helper P2. I. In vivo and in vitro construction of P4::P2 hybrid satellite phage. *Gene* 14:231–41

75. Lindqvist, B. H., Six, E. W. 1971. Replication of bacteriophage P4 DNA in a non-lysogenic host. *Virology* 43:1–7

76. Ljungquist, E., Kockum, K., Bertani, L. E. 1984. DNA sequences of the repressor gene and operator region of bacteriophage P2. *Proc. Natl. Acad. Sci. USA* 81:3988–92

77. Lundqvist, B., Bertani, G. 1984. Immunity repressor of bacteriophage P2: Identification and DNA-binding activity. *J. Mol. Biol.* 178:629–51

78. Matsuyama, S., Mizushima, S. 1987. Novel *rpoA* mutation that interferes with the function of OmpR and EnvZ, positive regulators of the *ompF* and *ompC* genes that code for outer-membrane proteins in *Escherichia coli* K12. *J. Mol. Biol.* 195:847–53

79. McClure, W. R. 1985. Mechanism and control of transcription initiation in prokaryotes. *Annu. Rev. Biochem.* 54:171–204

80. Mitchell, P. J., Tjian, R. 1989. Transcriptional regulation in mammalian cells by sequence-specific DNA binding proteins. *Science* 245:371–78

81. Monaco, H. L., Crawford, J. L., Lipscomb, W. N. 1978. Three-dimensional structures of aspartate carbamoyltransferase from *Eschericia coli* and of its complex with cytidine triphosphate. *Proc. Natl. Acad. Sci. USA* 75:5276–80

82. Navaratnam, S., Myles, G. M., Strange, R. W., Sancar, A. 1989. Evidence from extended X-ray absorption fine structure and site-specific mutagenesis for zinc fingers in UvrA protein of *Escherichia coli*. *J. Biol. Chem.* 264:16067–71

83. O'Brien, S. J., ed. 1990. *Genetic Maps 1990, Book 1*. New York: Cold Spring Harbor Lab. 181 pp.

84. Ow, D., Ausubel, F. M. 1980. Recombinant P4 bacteriophages propagate as viable lytic phages or as autonomous plasmids in *Klebsiella pneumoniae*. *Mol. Gen. Genet.* 180:165–75

85. Pierson, L. S. III, Kahn, M. L. 1987. Integration of satellite bacteriophage P4 in *Escherichia coli*. DNA sequences of the phage and host regions involved in site-specific recombination. *J. Mol. Biol.* 196:487–96

86. Poon, A. P. W., Dhillon, T. S. 1986. Assignment of two new host range types to the P2 family of temperate coliphages. *J. Gen. Virol.* 67:789–92

87. Poteete, A. R., Ptashne, M. 1982. Control of transcription by the bacteriophage P22 repressor. *J. Mol. Biol.* 157:21–48

88. Ptashne, M. 1986. *A Genetic Switch*. Palo Alto: Blackwell Sci. 128 pp.

89. Ptashne, M. 1988. How eukaryotic transcriptional activators work. *Nature* 335:683–89

90. Raibaud, O., Schwartz, M. 1984. Positive control of transcription initiation in bacteria. *Annu. Rev. Genet.* 18:173–206

91. Raimondi, A., Donghi, R., Montaguti, A., Pessina, A., Dehò, G. 1985. Analysis of spontaneous deletion mutants of satellite bacteriophage P4. *J. Virol.* 54:233–35

92. Ren, Y. L., Garges, S., Adhya, S., Krakow, J. S. 1988. Cooperative DNA binding of heterologous proteins: evidence for contact between the cyclic AMP receptor protein and RNA polymerase. *Proc. Natl. Acad. Sci. USA* 85:4138–42

93. Richardson, H., Egan, J. B. 1989. DNA replication studies with coliphage 186. II. Depression of host replication by a 186 gene. *J. Mol. Biol.* 206:59–68

94. Richardson, H., Puspurs, A., Egan, J. B. 1989. Control of gene expression in the P2-related temperate coliphage 186. VI. Sequence analysis of the early lytic region. *J. Mol. Biol.* 206:251–55

95. Roberts, D., Hoopes, B. C., McClure, W. R., Kleckner, N. 1985. IS*10* transposition is regulated by DNA adenine methylation. *Cell* 43:117–30

96. Rowland, G. C., Giffard, P. M., Booth, I. R. 1985. The *phs* locus of *Escherichia coli*, a mutation causing pleiotropic lesions in metabolism and pH homeostasis, is an *rpoA* allele. *J. Bacteriol.* 164:972–75

97. Saha, S. 1988. Lysogenic or lytic cycle of bacteriophage P2: a transcriptional switch. *Acta Univ. Ups. Compr. Summ. Uppsala Diss. Fac. Sci.* 153. Uppsala. 52 pp.

98. Saha, S., Haggård-Ljungquist, E., Nordstrom, K. 1987. The *cox* protein of bacteriophage P2 inhibits the formation of the repressor protein and autoregulates the early operon. *EMBO J.* 6:3191–99

99. Saha, S., Haggård-Ljungquist, E., Nordstrom, K. 1989. Activation of prophage P4 by the P2 Cox protein and the sites of action of the Cox protein on the

two phage genomes. *Proc. Natl. Acad. Sci. USA* 86:3973–77

100. Saha, S., Haggård-Ljungquist, E., Nordstron, K. 1990. Integration host factor is necessary for lysogenization of *Escherichia coli* by bacteriophage P2. *Mol. Microbiol.* 4:3–11

101. Saha, S., Lundqvist, B., Haggård-Ljungquist, E. 1987. Autoregulation of bacteriophage P2 repressor. *EMBO J.* 6:809–14

102. Sauer, B., Calendar, R., Ljungquist, E., Six, E., Sunshine, M. G. 1982. Interaction of satellite phage P4 with 186 phage helper. *Virology* 116:523–34

103. Sauer, B., Ow, D., Ling, L., Calendar, R. 1981. Mutants of satellite bacteriophage P4 that are defective in the suppression of transcriptional polarity. *J. Mol. Biol.* 145:29–46

104. Shore, D., Dehò, G., Tsipis, J., Goldstein, R. 1978. Determination of capsid size by satellite bacteriophage P4. *Proc. Natl. Acad. Sci. USA* 75:400–4

104a. Sivaprasad, A. V., Jarvinen, R., Puspurs, A., Egan, J. B. 1990. DNA replication studies with coliphage 186. III. A single phage gene is required for phage 186 replication. *J. Mol. Biol.* 213:449–63

105. Six, E. W. 1963. A defective phage depending on phage P2. *Bacteriol. Proc.* 138 (Abstr.)

106. Six, E. W. 1975. The helper dependence of satellite bacteriophage P4: which gene functions of bacteriophage P2 are needed by P4? *Virology* 67:249–63

107. Six, E. W., Klug, C. A. C. 1973. Bacteriophage P4: a satellite virus depending on a helper such as prophage P2. *Virology* 51:327–44

108. Six, E. W., Lindqvist, B. H. 1978. Mutual derepression in the P2-P4 bacteriophage system. *Virology* 87:217–30

109. Souza, L., Calendar, R., Six, E., Lindqvist, B. H. 1977. A transactivation mutant of satellite phage P4. *Virology* 81:81–90

110. Souza, L., Geisselsoder, J., Hopkins, A., Calendar, R. 1978. Physical mapping of the satellite phage P4 genome. *Virology* 85:335–42

111. Stoddard, S. F., Howe, M. M. 1990. Characterization of the *C* operon transcript of bacteriophage Mu. *J. Bacteriol.* 172:361–71

112. Sunshine, M. G., Sauer, B. 1975. A bacterial mutation blocking P2 late gene expression. *Proc. Natl. Acad. Sci. USA* 72:2770–74

113. Sunshine, M. G., Six, E. W., Barrett, K., Calendar, R. 1976. Relief of P2 phage amber mutant polarity by the satellite bacteriophage P4. *J. Mol. Biol.* 106:673–82

114. Sunshine, M. G., Thorn, M., Gibbs, W., Calendar, R., Kelly, B. 1971. P2 phage amber mutants: Characterization by use of a polarity suppressor. *Virology* 46:691–702

115. Westöö, A., Ljungquist, E. 1980. Cloning of the immunity repressor determinant of phage P2 in the pBR322 plasmid. *Mol. Gen. Genet.* 178:101–9

116. Woods, W. H., Egan, J. B. 1974. Prophage induction of noninducible coliphage 186. *J. Virol.* 14:1349–56

117. Younghusband, H. B., Egan, J. B., Inman, R. B. 1975. Characterization of the DNA from bacteriophage P2-186 hybrids and physical mapping of the 186 chromosome. *Mol. Gen. Genet.* 140:101–10

118. Younghusband, H. B., Inman, R. B. 1974. Base sequence homologies between bacteriophage P2 and 186 DNAs. *Virology* 62:530–38

Annu. Rev. Genet. 1990. 24:491–518

INTEGRATION SPECIFICITY OF RETROTRANSPOSONS AND RETROVIRUSES

Suzanne B. Sandmeyer, Lori J. Hansen, and Douglas L. Chalker

Department of Microbiology and Molecular Genetics, College of Medicine, University of California, Irvine, California 92717

KEY WORDS: Ty elements, retroelements, transposable elements, polymerase III, recombination

CONTENTS

INTRODUCTION

Classes of Retroelements

As we learn more about the integration patterns of retroelements, nonrandom integration has come to be regarded as a more general property of their lifestyle. This is perhaps not surprising, as successful propagation of these

491

elements is contingent on genomic cycling. The position and orientation of a given retroelement insertion can determine both its expression and retention in the genome. For some retroposons, sequence-specific integration is dependent upon the activity of the retroelement-encoded endonuclease (6, 140a). Retrovirus and retrotransposon integration specificity, the subject of this review, is, for the most part, not well defined. Integration of these elements depends upon the encoded integrase. Data from multiple systems, however, suggest that host chromosomal DNA, proteins, and nuclear functions are intrinsic to the targeting process.

Retrotransposons are retroviruslike elements found in various eukaryotic cells. Retrotransposons differ from retroviruses in that they do not have an obligatory extracellular phase. Retroviruses and retrotransposons have long terminal repeats (LTRs) of several hundred basepairs (bp) which flank an internal coding domain. The proteins encoded by the internal domains of retrotransposons and retroviruses are analogous in function and some are also homologous based on sequence comparison. Integration patterns of these elements vary greatly, from the apparent sequence specificity of the *Drosophila* gypsylike elements, to the position specificity of the yeast element Ty3, to the more degenerate patterns of other *Drosophila* elements, the yeast Ty1 element and animal retroviruses. Despite diverse patterns of integration, the integrases encoded by these elements have distinct similarities. In addition, we argue that many—and possibly all—retroviruses and retrotransposons exhibit some degree of integration site preference.

In vitro integration systems for animal retroviruses (18, 42) and for Ty1 (37) are yielding considerable information about the requirements of the integration reaction. The recent identification of elements with novel integration properties, together with in vitro systems with which to examine molecular mechanisms of integration, will facilitate future efforts to understand the target site selection process.

Retroviruses and Retrotransposons

Retroviruses (131) and retrotransposons (7, 9, 40) are similar in genomic organization and encoded proteins. These elements are typically between 5 and 10 kilobase pairs (kbp) in length. They are composed of an internal domain flanked by LTRs of several hundred bp in length. With a few exceptions among the retrotransposons, these elements begin and end with short inverted repeats. These short, usually imperfect, inverted repeats typically terminate in the dinucleotides TG/CA. Integrated retroviruses and retrotransposons are flanked by direct repeats of 4 to 6 bp.

Simple retroviruses are generally composed of three open reading frames (ORFs): *gag, pol,* and *env.* These ORFs typically encode the structural proteins of the nucleocapsid, catalytic proteins required for replication and integration, and envelope proteins of the virus particle, respectively. Major structural proteins include matrix (MA), capsid (CA), and nucleocapsid (NC) proteins. The catalytic proteins are protease (PR), reverse transcriptase (RT), RNaseH (RN), and integrase (IN). Retrotransposons typically have one or two ORFs encoding the structural and catalytic proteins. The upstream coding region is analogous to the *gag* gene of retroviruses and encodes structural proteins of the retrotransposon nucleocapsid. The downstream region is analogous to the retroviral *pol* gene and encodes catalytic functions. The order in which catalytic functions are encoded can differ from retroviruses, as in the arrangement PR, IN, RT, and RN, found in copia of *Drosophila* (86), Ta1 of *Arabidopsis* (134), Tnt1 of *Nicotianae* (51), and Ty1 (24) and Ty2 (60, 137) of *Saccharomyces cerevisiae* or can be generally similar, as in PR, RT, RN, and IN found in gypsylike elements of *Drosophila* (69, 83, 105, 125, 129), Ty3 of *S. cerevisiae* (58), del of *Lilium henryi* (117), and Tf1 of *Schizosaccharomyces pombe* (H. Levin & J. Boeke, personal communication).

Reverse transcription of the retroviral genome takes place inside the virion core after viral entry into the host cell. Retroviral replication is reviewed in detail by Varmus & Brown (131). The products of replication are full-length, linear molecules, and circularized forms of this product, containing one or two LTRs. For some retroviruses the linear form has been shown to be 2 bp longer on each end than the integrated provirus. Relatively little is known about the process by which replicated retroviral DNA arrives in the nucleus, or with what proteins, in addition to the integrase, it is likely to be associated at this stage. Similarities of IN structure, replication intermediates, and in vitro requirements suggest that retrotransposon integration will be fundamentally similar to that of retroviruses.

RETROVIRUS INTEGRATION

IN Function Defined In Vivo and In Vitro

Mutations in the downstream region of the *pol* genes of retroviruses have been shown to block viral infectivity at a point past replication (31, 32, 93, 99, 111). The pp32 protein in avian leukosis and sarcoma viruses (ALSV), p46 in MoMLV, and p34 of the human immunodeficiency virus (HIV) (131) have been identified as IN proteins. Alignment of the predicted protein sequences from the regions encoding these proteins has shown that they are homologous, and motifs conserved among these proteins are present in other retroviruses and retrotransposons (33, 71, 129, 140). Mutation analysis has shown that regions which encode proteins with similarity to the retrovirus IN, p90 (10, 37) and p61/58 (59) in Ty1 and Ty3, respectively, are also required for integration.

The time course of processing of full-length, extrachromosomal retroviral

DNA has been determined in MoMLV-infected rat cells. Over the initial ten hours after infection, the proportion of blunt-ended, linear species having two extra bp at each end decreases, compared to the increased amount of a species in which the 3' end is recessed by 2 nt (103). The intermediate with 3'-recessed ends was not detectable in an experiment performed with a MoMLV IN mutant, consistent with an in vivo role for IN in producing this species. While a circular two LTR species is also present in cells and can apparently act as an integration donor in some systems (94), the in vitro data support a linear integration precursor.

The development of in vitro integration systems has greatly facilitated experiments to identify proteins directly involved in the reaction. The original in vitro systems measured the ability of nucleocapsids produced from genetically tagged MoMLV (18) or Ty1 (37) to produce integrated DNA when incubated together with a naked λ DNA target. The genetic marker, a bacterial suppressor supF, allowed the target λ gtWES, which contained three amber mutations, to replicate in a wild-type host. In these assays, activity was associated with nucleocapsids and was proportional to the concentration of linear DNA (18). In addition, the reaction was dependent on wild-type IN activity (19, 37). Products and intermediates of the in vitro MoMLV reaction were consistent with ligation of 3'-recessed ends of linear donor DNA to 5'-extended ends of target DNA (19, 43). Insertions propagated by in vitro integration and recovered by biological selection were flanked by the characteristic short direct repeat. The reaction required divalent cations, but did not display dependence on dNTPs, ATP, or supercoiling of the target DNA.

More recently, efforts have focused on defining activity associated with IN (52). Characterized IN proteins are DNA-binding proteins (63, 79, 85, 104). Efforts to demonstrate endonucleolytic activity with the purified IN protein in vitro were first successful with the avian retrovirus IN. The purified AMV pp32 was shown to specifically protect the ends of the LTRs in vitro against DNase I degradation (79, 85). In addition, this IN protein and recombinant avian retroviral IN protein produced in E. coli were shown to have endonucleolytic activity, first on substrates containing the LTR/LTR junction present on circularized viral genomes (34, 35, 53–55, 127) and then on dimeric oligonucleotides representing the individual LTR termini (75). IN specifically removes the terminal dinucleotide representing the 3' end of the LTR from the dimeric oligonucleotide.

Demonstration that donor DNA in the MoMLV reaction could be separated from the nucleocapsid was achieved in vitro using a linearized plasmid with partial LTR sequences at its termini as an exogenous donor, rather than replicated DNA within the virion core (42). In this reaction, the exogenous DNA was integrated into a λ DNA target. Recently, partially purified, recombinant MoMLV IN protein produced in baculovirus has been shown to catalyze incorporation of dimeric oligonucleotides representing the ends of the MoMLV LTRs into naked target DNA (R. Craigie, personal communication). As in the original in vitro reactions, neither exogenous energy cofactors nor supercoiling of donor or target DNAs is essential. The recovered products of these reactions are consistent with insertion into a 4-bp staggered cut in the DNA target. These results are thus consistent with IN catalysis of both the donor and target cleavage and ligation of the 3' ends of donor molecules to the 5' ends of the target.

To what extent independence of the latter in vitro reaction of nucleocapsid and target proteins reflects requirements of the reaction in vivo has not yet been addressed. If the entire nucleocapsid is required for integration in vivo, it could significantly influence the logistics of association between viral DNA and chromatin. In the original retroviral and Ty1 in vitro systems, integration was assayed biologically, and insertions that interrupted essential genes in the target were not recovered. This restriction notwithstanding, integration appeared to be relatively random (18, 37, 42). Nevertheless, the intergenic regions are somewhat more AT-rich than are coding regions. It will be of interest to determine whether AT bias is a feature of target sites when products are not recovered via biological selection. As described below, the apparent randomness of in vitro insertions is at variance with in vivo observations.

Nonrandom Retroviral Integration

Although retroviruses can insert in many different sites (64), much evidence suggests that integration of retroviruses is not random in vivo. Different approaches have been taken, first to search for conserved insertion sequence motifs and second, to determine functional correlates of regions that are integration targets. The results of these analyses of in vivo integration patterns are consistent with a combination of restricted access and more specific targeting interactions.

Characteristic short direct repeats of target DNA flank retroviral insertions (29, 114, 115). The sequences of short direct repeats and flanking DNA have been inspected for common primary sequence features or potential common secondary structures for several insertions of spleen necrosis virus (114). This study showed that the direct repeats were 77% AT rich, compared to flanking cellular DNA at about 58%, but did not reveal other distinctive features. Extensive comparisons of junction sequences generated by in vivo integration of other retroviruses have not been published.

Several experiments performed in the mid-eighties with MoMLV infection of murine cells suggested that integration was nonrandom, and that DNA integrated by retroviral infection, compared with DNA introduced by transfection, tended to be in transcribed regions of the genome. Insertion of retroviruses into selectable genes allows rough determination of the frequencies of insertion into a given portion of the genome. King et al (76) measured the incidence of spontaneously occurring hypoxanthine phosphoribosyltransferase deficiency (hprt-) in F9 embryonal carcinoma cells, compared to the frequency in cells with 20 to 50 integrated MoMLV proviruses. Infection resulted in approximately a tenfold increase in the frequency of hprt- cells. Nevertheless, in a screen of 10^8 cells, if insertions within coding and intron regions disrupted hprt function, about 10^4 mutants would be expected, compared to the 70 actually recovered. These data establish that integration did not occur at this transcribed locus at the frequency predicted by random insertion throughout the genome, but they do not show where insertions did occur.

Despite the low frequency of hprt- cells generated by MoMLV insertion mutagenesis, results of other experiments have suggested that, overall, insertions are associated with transcribed regions of the genome. Expression of genes at foreign chromosomal locations reflects local transcriptional activity. These effects are attributed to the degree of condensation of chromatin (56). The expression of integrants can then be used to gauge whether insertion occurred into a region with potential for transcriptional activity. Hwang & Gilboa (66) compared the expression of integrated neo^R sequences introduced into NIH 3T3 cells by calcium phosphate transfection and by retroviral infection. The amount of transcripts generated from viral insertions did not show much variation among 20 strains and was 10- to 50-fold higher than the amount in transfected cells. These workers concluded that the results were consistent with integration into transcribed regions, while cautioning that negative effects of pBR vector DNA on expression of the transfected neo^R gene could not be ruled out. More recently, Scherdin et al (108) screened a recombinant integration library generated from MoMLV-infected NIH 3T3 cells and showed that about half of the clones contained sequences that hybridized to radioactive probes generated in nuclear run-on reactions. Although these results cannot be extrapolated to quantitate a preference, they are consistent with a significant proportion of insertions into transcriptionally active regions.

Efforts have been made to further define the association between transcribed regions and retroviral insertion sites. Transcribed or potentially transcribed regions have been shown to correspond to regions of heightened nuclease sensitivity. In addition, they contain nuclease hypersensitive regions, so-called because they are about two orders of magnitude more sensitive to nuclease digestion. Where studied, these regions of nuclease hypersensitivity are nucleosome-free and allow access of proteins to specific DNA sequences. These proteins include topoisomerases, RNA polymerases, and transcription factors. Both polymerase II- and polymerase III-transcribed genes have been associated with nuclease hypersensitive sites. Several studies have attempted to localize the positions of multiple insertions of ASLV or MoMLV within transcribed regions relative to DNase I hypersensitive sites. In a study by Robinson & Gagnon (100), 66 insertions of ALV upstream of the second exon of *c-myc* were isolated from chicken lymphoblastomas. Twenty-four of these were mapped in some detail and were positioned within 150 bp of one of five DNase I hypersensitive sites in a 3-kbp region 5' to the protein coding sequences of *c-myc*. Because these insertions were selected on the basis of the associated effect on *c-myc* expression, the conclusion that insertions are associated with DNase I hypersensitive sites is qualified. Nevertheless, these experiments suggested that insertions might be close to nucleosome-free regions with *cis*-acting function.

Two later studies addressed the problem of selection to some extent. In a study by Vijaya et al (132) in which seven insertion sites of MoMLV were mapped, three had been selected as tumor derivatives and four were considered unselected, although two of these were derived from MoMLV-induced rat thymomas and two from a rat fibroblast cell line. Each target site mapped within 500 bp of a hypersensitive site. Overall, these workers estimated a density of DNase I hypersensitive sites of about one every 6.5 kbp and the probability of this association occurring randomly at approximately 9×10^{-3}. In a related study by Rohdewohld et al (101), six MoMLV insertions into mouse early embryonic cells or into NIH 3T3 fibroblasts were analyzed for an association with DNase I hypersensitive sites. At a similar estimate of DNase I hypersensitive site density, the probability of all six sites occurring within the observed range of less than 0.7 kbp was estimated at less than 2×10^{-3}.

Although selected integrants represented a large proportion of the total sample analyzed, the uniform conclusion in the three studies was that a nonrandom association exists between insertion sites and DNase I hypersensitive sites. If this association represents actual bias in the insertion site distribution, as seems likely, it poses another, more intriguing question. Namely, which genomic feature(s) is common among regions containing DNase I hypersensitive sites and targets of integration? Clearly, even given the vagaries of restriction map analysis, these insertions are neither extremely close (within one nucleosome), nor at a uniform distance from DNase I hypersensitive sites. Thus, whether integration and these sites are specifically related or whether both associated with some common feature is not yet resolved. Results of integration studies in the yeast systems described below suggest that one such feature could be the 5' ends of transcriptionally active genes.

To directly screen for sequences common to multiple retroviral integration sites, Shih et al (113) constructed integration libraries and control libraries from RSV-infected turkey embryo fibroblasts. RSV marked with the bacterial *supF* gene was used to infect turkey embryo fibroblasts. Libraries of genomic

DNA were constructed in a derivative of λ gtWES. The integration library was recovered on a wild-type host and the control library on a suppressor-containing bacterial strain. Screening the integration library with probes derived from integration sites showed that 20% of the tested probes were represented in the independently constructed integration libraries and were therefore highly represented in these libraries, relative to control libraries. Sequence analysis of four of the highly used sites showed that they were not identical in sequence and that the two or three insertions recovered at each were into the same position at the respective loci. These insertion sites were estimated to be used once in 4,000 integrations, resulting in an overrepresentation in the integration libraries, compared to the control library, of about one-millionfold. The number of the sites present in the tested collection and the frequency of usage suggested that there are about 800 of these sites for an overally frequency of every 2.5×10^6 bp. They are less frequent than DNase I hypersensitive sites, which occur about every 6 kbp and they do not appear to be transcribed. These regions are therefore not equivalent to DNase I hypersensitive sites, although they could represent a subset of those sites. Preliminary results based on fractionation of genomic DNA suggest that the sites identified in this study cofractionate with regions of nuclear matrix attachment (J. Coffin, personal communication). Determination of the basis of preferential usage of these targets should offer interesting insights into the process of retroviral integration in vivo.

NONRANDOM RETROTRANSPOSON INTEGRATION

Drosophila Retrotransposons

Approximately ten well-characterized families of *Drosophila* retrotransposons are present in 30 to 50 copies in the *Drosophila* genome (7). The gypsylike family is a superfamily with retroviruslike organization of its coding domains and similarities among *pol* coding domains. It includes the elements gypsy (83), 17.6 (105, 129), 297 (69), tom (125), and possibly H. M. S. Beagle (118). These elements are most similar in the protein sequence of the polymerase domain of reverse transcriptase to caulimoviruses (33, 140), Ty3 (33), Tf1 (H. Levin & J. Boeke, personal communication), and del (117). These elements are distinct from some other retrotransposons and retroviruses in that the primer region is apparently not separated from the LTR by two bases (and therefore the extrachromosomal form, if it has extra bp, could not derive them from this region of the template), and the LTRs do not terminate in the TG/CA inverted repeat. Examination of sequences flanking a large number of endogenous gypsylike elements has led to the conclusion that these elements insert with sequence specificity. Although the DNA sequence at the insertion site, as well as across LTR-genomic junctions, has been determined in several cases, similarity between the presumed ends of the elements and target DNA has made unequivocal definition of the termini difficult. Despite this ambiguity, similarities exist between the short flanking direct repeats of gypsy and H. M. S. Beagle and among the flanking repeats of 17.6, 297, and tom. Analysis of several gypsy (41, 83) and one H. M. S. Beagle (118) insertion sites in variant and wild-type (uninterrupted) configurations is consistent with the target sequence 5'TAYATA3', where Y=pyrimidine, and the 4-bp direct repeat is TAYA. Analysis of insertions of 297 (67), 17.6 (68), and tom (125), suggest a target of TATATA and a repeat of (T)ATAT, depending on assignment of the element termini.

Despite the apparent sequence specificity, host proteins may also play a role in targeting insertions to the observed sites. A functional correlate of the consensus target TATATA is the *Drosophila* TATA box. In fact, several gypsy-group insertions were into the TATA boxes of known genes (67, 68, 118). Moreover, some of these elements appear to display additional preferences. The element tom is linked to a hypermutability phenotype that is most apparent for genes involved in optic morphology (62). The common feature of these genes or tissues contributing to hypermutability has not been identified, but a specific target for integration could exist.

YEAST RETROTRANSPOSONS

Four classes of retrotransposons have been described in *S. cerevisiae*. Ty1-3 (9, 12) are mentioned above and Ty4 has been described more recently (123). Insertions of isolated LTR sequences, as well as complete Ty elements, are found in the genomes of common laboratory strains. In strains expressing high levels of Ty1 (1, 45, 87) or Ty3 (L. Hansen, D. Chalker, & S. B. S., unpublished results) RNA, a nucleocapsid particle is formed that is largely composed of structural proteins encoded by the first ORF. In addition, these particles contain Ty genomic RNA, full-length DNA, and reverse transcriptase activity. These nucleocapsid particles are analogous to the virion core particle within which reverse transcription of the retroviral genome takes place. Genomic Ty elements are flanked by direct repeats of 5 bp that are generated upon integration. An in vitro integration system has been developed for Ty1 and displays requirements for activity similar to those displayed by the MoMLV in vitro integration system (37). Thus, many aspects of the transposition cycles of these elements resemble retroviral replication and integration. Of the yeast elements, the integration patterns of Ty1 and Ty3 have been examined most extensively and are consistent with the suggestion from retroviral integration that insertion is related in some way to transcription.

TY1   Ty1 is 5.9 kbp in length and is composed of an internal domain flanked by direct repeats of delta, which is 334–338 bp in length. Ty2 is closely related to Ty1, based on gene organization and predicted protein sequence, and also has delta LTRs. The minus strand primer binding site of these elements is adjacent to the LTR with no intervening bp. There are from 30 to 35 Ty1 elements in typical haploid strains of *S. cerevisiae* and over 100 delta elements (20). Ty2 is present in 5 to 15 copies (77).

Ty1 integration has been monitored in several different ways. There is a large collection of data representing analysis of endogenous Ty1 elements. Ty1 insertions are often detectable by effects on target gene expression. Insertions can inactivate, activate, or confer constitutive expression or mating-type regulation on neighboring transcriptional units (12). Recombinant techniques have facilitated the recovery of de novo insertions of the Ty1 element. Boeke et al (11) fused the inducible *GAL1* promoter to the Ty1 transcription start site and achieved about 100-fold increase in transposition frequency. In addition, Boeke et al (14) and Garfinkel et al (46) have used genetically marked elements to identify and clone insertions, making efficient recovery of otherwise unselectable insertions straightforward.

Mapping and cloning of a large number of endogenous Ty1 and Ty2 elements have allowed some conclusions to be drawn concerning their genomic distribution. First, they tend to be associated with AT-rich regions (92); second, several examples of clustered elements exist that could be partially explained by a high frequency of insertion of one element within another, or by the presence of some common target (135, 136); and third, Ty1 and Ty2, as well as Ty4, are frequently found in association with tRNA genes (38, 44, 61, 77, 90, 122). These observations, although interesting, are of limited quantitative value because of their anecdotal nature. The association of endogenous Ty1 elements with AT-rich regions could reflect the high AT-base composition of the yeast genome, coupled with the greater likelihood of retaining Ty elements in relatively AT-rich intergenic regions. Clusters of Ty1 elements that are in the vicinity of tRNA genes have been described on chromosomes X (112) and III (135). Ty1 and Ty2 (20, 77, 91), Ty3 (25), Ty4 (123), delta (20, 91) tau (23) and sigma (27, 107) elements were each discovered as polymorphisms in the vicinity of tRNA genes, and many representatives of each class are within 500 bp of tRNA genes (30, 38, 44, 90, 122). That the association is nonrandom can be inferred, because yeast has an estimated 360 tRNA genes and, for the most part, these are dispersed in the genome (57). Thus tRNA genes occur about every 40 kbp while Ty1 would occur roughly every 400 kbp and Ty2 elements even less frequently (see below). The extent of a general association between Ty1 elements and genes transcribed by polymerase III is not clear. Ty1 sequences have also been associated with 5S sequences (97, 133) and a delta element has been identified about 100 bp upstream of the single-copy U6 gene in yeast (D. Brow & C. Guthrie, personal communication). The association between tRNA genes and Ty3 is more defined and is discussed below. This association will likely differ qualitatively from the association observed between the other elements and tRNA genes.

Multiple spontaneous de novo integrations of Ty1 have been characterized in the vicinity of *HIS3* (11, 13), *ADH2* (139), *LYS2* (36, 89, 116), *URA3* (89), *CAN1* (138), and *SUP4* (48) on the basis of disruption or activation of the target gene. Interesting patterns of integration have emerged from these studies. Nevertheless, the profile of insertions in most of these systems is necessarily influenced by the nature of the selectable phenotype. This complication is particularly severe for insertions that increase expression, because of the requirement that Ty1 be oriented so that the promoter-proximal enhancer is close to the target gene and that Ty1 insertion does not directly or indirectly block transcription of the target gene.

Four insertions of Ty1 and one of Ty2 which conferred constitutive expression on *ADH2,* the gene for the glucose-repressible alcohol dehydrogenase, were mapped. These insertions occurred from -125 to -210 upstream of the translation initiation site. None of the five insertions examined was close to the characterized UAS, but two of the insertions were within 10 bp of the TATA box.

A *HIS3* gene in which λ DNA replaces the promoter region, carried on a centromere-bearing plasmid, can be activated by Ty1 insertion as a convenient method of scoring transposition. Yeast plasmids are found in nucleosome structures (95) and so should resemble chromosomal targets in at least some respects. A collection of 40 integration sites were mapped in a strain in which expression of Ty1 on a high-copy plasmid was under control of the *GAL1* promoter (13). In this study, a hotspot at which insertion of more than half of the elements occurred was located at position -60, relative to the *HIS3* ATG. Although this target is located on a plasmid and transcription in the vicinity of the *HIS3* gene has not been mapped, the gene is not expressed sufficiently to rescue a His- phenotype. This existence of such an integration hotspot argues that although the 5' flank of some target genes is used preferentially for insertion, specific initiation of transcription in the upstream region may not be absolutely essential for use of that gene as a target, or for observation of insertion hotspots.

Selection schemes based on loss of activity of the integration target extend the potential of the Ty1 system for detection of insertions within a gene, but conversely may limit recovery of insertions into the region upstream of transcription initiation. Mutants that have lost function of *LYS2, CAN1* and *URA3* can be directly selected in yeast. Eight of ten Ty1 or Ty2 insertions which caused α-aminoadipate resistance by disruption of *LYS2* were shown to be insertions in the 5'-noncoding region of that gene (36). In a more extensive study, using the galactose-inducible Ty1 element on a high-copy plasmid, 31 out of 78 resistant colonies could be attributed to Ty1 insertions in the LYS2 gene (89). These insertions were distinctly nonrandom. Of the ten insertions for which junction sequences were determined, half were in the 5'-transcribed, noncoding region (four at position -42, relative to the coding sequence and one further upstream) and the other half were at position $+226$. Although none of these insertions were upstream of the transcription start site at -82, out of a coding region of about 4 kbp, this insertion distribution was heavily biased toward the 5' region of the gene.

Ty1 insertions that disrupt the *URA3* gene and confer resistance to 5-fluoro-orotic acid (5FOA) have also been analyzed to examine positions of integration (89). Sequence analysis of the positions of 82 independent insertions showed that, as in the *LYS2* locus, insertion was biased toward the 5' end and insertion hotspots occurred. The upstream one-fourth of the gene contained more insertions than the downstream three-fourths. The pattern of insertion of at *URA3* differed from that at *LYS2*, however, in that very few insertions (2/82) were recovered in the noncoding region, probably reflecting the requirement for complete loss of activity to establish 5FOA resistance. The 82 insertions occurred at 42 positions in the 800-bp gene and one position, bp 341, was used 12 times. No particular bias in the orientation was apparent inside the coding region.

Wilke et al (138) characterized 55 spontaneous insertions of Ty1 or Ty2 elements that disrupted expression of the gene for arginine permease, *CAN1,* and thereby conferred resistance to the toxic analogue canavanine. In one strain, 10/21 of the events mapped in the 5'-noncoding region. Most other insertions were mapped in two *rad6* mutant strains. In these strains, no insertions were recovered in the 5'-noncoding region and insertions were distributed throughout the coding region. The distribution of insertions at this locus in the wild-type strain are consistent with observations at *LYS2* and *URA3* in that a bias toward insertions in the 5' end was observed. The absence of a bias in the *rad6* strain is potentially interesting, as the *RAD6* gene encodes an enzyme that polyubiquitinates histones H2A and H2B in vitro (70, 124). Therefore an effect of a mutation in *RAD6* on the pattern of Ty1 insertion could implicate nucleosomes in defining regions of retrotransposon insertion. It is also possible that *rad6* mutants simply have more tolerance for insertions in the 5'-noncoding flanking sequence of the *CAN1* gene. This could be resolved by determining the canavanine resistance phenotype of 5' Ty1 insertions crossed into the *rad6* background.

Consensus sequences summarized by Wilke et al (138) and Natsoulis et al (89) for Ty1 insertion at a number of loci suggest that the immediate insertion site sequence is barely constrained. Both studies suggested that the 5-bp repeat sequences are three-way degenerate on the ends and AT-rich in the center.

Targeting of Ty1 and Ty2 to tRNA genes was examined directly by Giroux et al (48). Mutants that had lost activity of a gene encoding a tRNA^{Tyr} ochre suppressor, *SUP4-o*, were selected on canavanine-containing medium in a genetic background *(can1–100)* where expression of arginine permease depends on an ochre suppressor. Among the 196 spontaneous mutations recovered were 12 insertions of Ty1 DNA into *SUP4:* one was at position +17, ten at position +37, and one at position +43. The hotspot and the insertion at position +43 were within a run of ten consecutive A/T bp. This region was accessible to DNase I in stable transcription complexes (74). Because of the requirement for loss of *SUP4-o* activity in this selection, it is not likely that Ty insertions upstream of *SUP4* would have been identified.

TY3   Ty3 is 5.4 kbp in length and is composed of a 4.7-kbp internal domain flanked by direct repeats of the 340-bp sigma element (25). From one to four copies of Ty3 (25) and about 30 copies of the sigma element (27, 107) are found in the genomes of typical haploid laboratory strains of *S. cerevisiae.*

Ty3 and sigma elements are closely associated with the 5' ends of tRNA genes. Southern blot analysis of 25 sigma elements from one strain showed that all were on the same restriction fragments as tRNA genes (16). This implies that about 10% of the estimated 360 tRNA genes are associated with Ty3 or sigma elements. Analysis of genomic DNA flanking two Ty3 and over thirty sigma elements from a combination of several strains, reviewed in Sandmeyer et al (106), showed that each is at position −17, plus or minus one or two bases, relative to the 5' end of a mature tRNA-coding sequence. A comparison of 20 of these sequences revealed only limited similarity among the insertion sites, however. In 11 of 20 genes, it could be inferred that the tRNA gene-proximal nick for Ty3 integration would have occurred on the nontranscribed strand within the dinucleotide sequence CA. This coincides with the presence of a common context for tRNA gene transcription initiation at the first A in the nontranscribed strand sequence PyCAACA (47). Insertions of Ty3 are unevenly distributed between the two possible orientations relative to the tRNA gene. Two-thirds of the sigma and Ty3 insertions are oriented so that transcription of the element is divergent from that of the tRNA gene. Such elements are said to be in the major orientation.

Despite the fact that Ty3 has a distinctive pattern of integration, its mode of integration is likely to be fundamentally similar to that used by retroviruses. Similarities of the yeast retrotransposons to animal retroviruses were mentioned previously. The Ty3 IN protein(s), p61/58 (59), has been identified by Western blot analysis and shown to be required for integration. The predicted protein sequence of Ty3 IN is more similar to MoMLV IN than to the other

Ty IN sequences. Ty3 IN has amino- and carboxyl-terminal extensions in addition to a large central region of similarity with characterized integrases.

To investigate the basis of the position-specific integration displayed by Ty3, a study was undertaken to identify genomic sites of de novo integration (D. Chalker & S. B. S., submitted). As in the Ty1 studies, transcription of the element was artificially elevated to enhance transposition. In this case, the *GAL1–10* upstream activating sequence (UAS) was fused upstream of the Ty3 TATA sequence (58). Transposition of a Ty3 element genetically marked with the yeast *HIS3* gene and the bacterial *neo*^{R} gene was monitored. The bacterial *neo*^{R} gene confers resistance to kanamycin in bacteria and to G418 in yeast. A total of 91 strains that had acquired one additional Ty3 element during induction of transposition were identified by Southern blot analysis. Seven genomic loci were the targets of two insertions each. This is close to the number of double insertions predicted by a Poisson distribution, if the actual number of targets is roughly equivalent to the estimated number of tRNA genes. The junctions of twelve of these insertions with host DNA were also characterized by sequence analysis, which showed that the seven loci corresponded to seven tRNA genes. Interestingly, in each case, the Ty3 insertion at a given locus was in one orientation only, although the minor orientation was represented at three loci. Therefore, some feature of the genomic target, rather than an inherent asymmetry in the transposition apparatus, determines the orientation of the insertion. The position of insertion was not absolutely conserved with respect to a given tRNA gene sequence, however, as two insertions upstream of a tRNA^{Met} gene occurred with one insertion at position -16 and one at position -17.

The 5S genes, which are also transcribed by polymerase III, have some features in common with tRNA genes. Yeast has about 150 5S genes, almost all of them included within a large tandem rDNA repeat on chromosome XII (96). These were examined to determine whether common features among polymerase III-transcribed genes were sufficient to constitute a genomic Ty3 target. Southern blot analysis showed that none of the 91 strains which acquired a single Ty3 element in the previous experiment had undergone an integration of Ty3 into a 5S gene.

To investigate a role for homology or tendency of Ty3 elements to cluster at preferred sites, Southern blot analysis was also performed using a sigma-specific probe. The 30 tRNA genes already associated with sigma elements were targets of only four insertions, which is slightly less than would be predicted if these were used randomly. Limited homology between the donor and target, in this case, therefore, did not confer preferential usage on the target. Thus, the most frequent targets of Ty3 genomic integration are tRNA genes, but there does not appear to be a highly preferred subset of those genes.

More recently, an assay was devised to allow recovery of Ty3 insertions into a plasmid-borne target (D. L. C. & S. B. S., unpublished results). The target assay has been used to identify and dissect the features of the target for Ty3 integration. Eight genes encode a single tRNA^{Tyr} isoacceptor species in yeast (49, 82, 91). The *SUP2* gene for tRNA^{Tyr} was chosen for the test target, because it is associated with sigma in some strains (107) and because the promoter sequences are relatively well-characterized (2,4). The major transcription initiation site for *SUP2* transcription by polymerase III is at position -10 (P. Kinsey & S. B. S., unpublished results). Ty3 insertions upstream of *SUP2* carried on the plasmid occurred in both orientations at positions -16 and -17. The staggered nicks inferred to accompany Ty3 insertion would have occurred between bp -15 and -16 and between -10 and -11 for a Ty3 insertion beginning at -16. Thus, the tRNA gene-proximal nick is in the vicinity of the transcription initiation site. The genomic integration pattern together with these data suggested that the common denominator marking Ty3 integration sites is related to target gene transcription.

The plasmid target assay was used to more directly investigate the relationship of Ty3 transposition to transcription of the tRNA gene target. The essential features of the tRNA gene internal control region (ICR) or promoter are the conserved box A, positions 8 to 19, and box B, positions 52 to 62, relative to the mature tRNA-coding sequence (reviewed in 47). Two transcription factors are required, in addition to polymerase III, for in vitro transcription of tRNA genes, TFIIIC and TFIIIB. TFIIIC contacts box B and to a lesser extent, box A. Transcription initiation is typically at a purine in the nontranscribed strand, the selection of which is influenced by the position of box A sequences (47). TFIIIC binding is required to load TFIIIB onto the DNA upstream of the tRNA gene, completing stable complex formation (74). Although no recognition sequence has been identified for TFIIIB, this factor footprints in DNase I protection assays over the region from -16 to -54 upstream of the tRNA^{Tyr} gene *SUP4* in the presence (74) and absence (73) of other transcription factors and is the only factor required for repeated rounds of polymerase III transcription (73). In vivo footprint analysis shows that this upstream region is protected in cells in logarithmic growth and in stationary phase (65).

Mutations were introduced to a *SUP2* gene to assess the target requirement for ability to form stable complex and to test for linkage between the transcription initiation site and the position of integration (D. L. C. & S. B. S., unpublished results). The first mutation in a target tRNA gene disrupted the ability of the tRNA gene to form a stable complex by changing a conserved base in box B (2, 4, 80, P. Kinsey & S. B. S., unpublished results). This mutation also made the tRNA gene unable to act as an integration target. To investigate the functional linkage between the position of Ty3 integration and the initiation site of tRNA gene transcription mutations that changed the context of initiation were also examined. Ty3 integration into these target plasmids occurred as into the wild-type target, near the transcription initiation site. Together the data from these mutant studies argue that the integration site is fixed by a protein-DNA complex, rather than by a specific DNA sequence.

The results of the tRNA gene target experiments implicated some feature of polymerase III transcription in Ty3 target formation. Two other yeast genes transcribed by polymerase III, the 5S gene (47) and the U6 gene (17, 85a), were therefore tested using this assay (D. L. C. & S. B. S., unpublished results). Integration of Ty3 occurred near the transcription initiation site of the plasmid-borne 5S gene. The 5S gene in yeast is dependent for expression on two regions, one bounded by positions -14 to $+8$ and box C, positions $+81$ to $+94$ (22). A conserved box A is also present, but is apparently dispensable in yeast for 5S expression. Transcription of the 5S gene is dependent on TFIIIA, in addition to TFIIIC and TFIIIB, the factors required for tRNA gene transcription, and on polymerase III (126). In DNase I protection studies, TFIIIB extends the protection of TFIIIC and TFIIIA upstream to bp 45 with partial protection at position -4 (15). When the protection of TFIIIB alone is examined after removal of other factors, it protects positions -8 through -43 (73). Thus for both the tRNA gene and for the 5S gene, the position of insertion corresponds closely to the downstream border of TFIIIB protection and is near the transcription initiation site.

Why the genomic 5S genes were not a target for transposition is not known, although there are several interesting possibilities. The integration complex may be simply excluded from the nucleolus, the structure of which is not clearly understood. The particular 5S gene used as a target was subcloned from rDNA repeat and is transcriptionally active (15). Alternatively, because the 5S genes are in the rDNA spacers, the flanking configurations or activities of promoter, enhancer, and terminator sequences (72) may somehow preclude integration. Topological constraints which operate to repress recombination within rDNA may also make rDNA refractory to integration (39). Whether or not Ty1 integration into rDNA occurs readily is not known. Ty1 sequences are underrepresented in the rDNA (92) and the delta and Ty1 sequences associated with rDNA that have been characterized are either associated with variant rDNA flanking regions and not located within the major repeat (97) or do not exhibit the flanking direct repeats characteristic of Ty1 insertions by retrotransposition (133). The R1 and R2 elements of insects which insert specifically into rDNA are also in contrast to Ty3 exclusion from this genomic target.

The yeast U6 gene is a member of a class of polymerase III-transcribed genes that share some promoter features with polymerase II-transcribed genes, including an upstream TATA box consensus (17). The required promoter elements for expression of animal cell 7SK (88) and U6 (26) genes are contained in the upstream sequences. In the yeast U6 gene, the upstream sequence, which contains a TATA consensus, is required for full activity in vivo and in vitro as observed for animal cell genes (D. Brow & C. Guthrie, personal communication). However, an internal box A consensus also exists and a box B sequence is present 120 bp downstream of the gene. The box B sequence has been shown to be required for full expression in vivo and in vitro with complete transcription extracts (D. Brow & C. Guthrie, personal communication). Moenne et al (85a) have shown that in vitro expression of the U6 gene depends on the TFIIIB transcription factor. Despite potential differences between promoter elements of the U6 and tRNA and 5S genes, a plasmid-borne copy of the U6 gene functioned as a target for insertions of Ty3 into the vicinity of the transcription initiation site. The results of experiments with the target plasmids are consistent with several potential target requirements, including features of the stable complex or even the open transcription complex.

Dictyostelium *Retrotransposons*

Several mobile elements have been characterized in the slime mold *Dictyostelium discoideum* (40). These include Tdd1-3 (Tdd1 = DIRS) and DRE1/2 (now DRE). The element Tdd-1 is 4.8 kbp in length and has inverted LTR sequences of 313 bp in length (102, 141). Insertions of this element are not flanked by direct repeats of the target sequence. Despite these differences with characterized retrotransposons, Tdd-1 encodes a reverse transcriptase that is likely to be involved in its replication. Analysis of a number of Tdd-1 genomic insertions shows that they are often located within Tdd-1 elements (21, 40). The pattern of these insertions would be difficult to completely reconcile with an homologous recombination model. An alternate possibility would be that Tdd-1 elements are themselves preferential targets for Tdd-1 insertions. Because the position of insertion is not conserved, either regional hotspots, acting as foci of multiple insertions or a specific sequence target within Tdd-1, but not necessarily at the integration site, would be consistent with an integration specificity model.

The elements Tdd-2 and Tdd-3 are about 4.5 kbp in length, but do not have the LTRs characteristic of retrotransposons. Insertions of Tdd-3 are flanked by 9–10 bp repeats of extremely AT-rich DNA. Analysis of a number of Tdd-3 insertions reveals that a number are located within Tdd-2 elements, close to a 5'22-bp region of homology between the two elements (98). Several other genomic Tdd-3 elements have been identified about 100 bp downstream of tRNA genes (84) or inserted into the 3' ends of other Tdd-3 elements. An intriguing feature shared by these hotspots for Tdd-3 insertion is that they contain consensus box B sequences (T. Dingermann & R. Marschalek, personal communication).

The elements DRE1 and DRE2, previously described by Marschalek et al (84) are the LTR and internal domain, respectively, of a retrotransposon of about 6.6 kbp in length, which is now named DRE. The internal domain of this element is composed of two ORFs which encode protein sequences with similarity to retroviral NC, PR, RT, and IN proteins. A survey of 50 out of about 200 DRE elements in the *Dictyostelium* genome shows that they are all associated with tRNA genes. Sequence analysis of the regions flanking 12 elements has shown that, as in the case of the original isolates, they occur consistently, about 50 bp upstream of the 5' ends of different tRNA genes (T. Dingermann & R. Marschalek, pesonal communication). These observations are particularly intriguing in light of the work from the Geiduschek laboratory showing that TFIIIB protects a region in the DNA about this far upstream of the tRNA gene. If this is also the case for an analogous factor in *Dictyostelium*, then this factor could be involved in target formation.

Polymerase III-Transcribed Genes as Genomic Integration Targets

The observations of a frequent association of Ty elements with tRNA genes and the relatively specific association of Ty3, Tdd-3, and DRE with tRNA genes would suggest that mechanisms have evolved to use polymerase III promoters as targets of integration events. although it is not known whether recombination hotspots are correlated with preferred integration targets, tRNA genes are also active in ectopic meiotic gene conversion (3). Whether the integration targeting functions have been adapted from cellular transcription and recombination functions can be better addressed when additional participants in these processes have been characterized at the molecular level. Recurring motifs in protein-protein and DNA-protein interactions suggest that integrases and transcription factors could interact directly or could have common domains for contacting other proteins or DNA. Versions of the metal finger motif, for instance, are common to the integrases and also occur in TFIIIA, the 5S transcription factor.

TRNA genes may be present in distinctive regions of chromatin, such as are postulated for retroviral insertions, or may even be distinguished topologically. Nuclease analysis of *Drosophila* chromatin showed that tRNA and 5S genes are associated with positions of nuclease hypersensitivity (28, 81). Microscopic analysis of tRNA and 5S genes associated with transcription proteins has produced results consistent with protein-induced changes in the structure of the DNA (5, 109, 121). DNA bends have been demonstrated to participate in integration complex formation (50, 119) and even to promote cleavage in some nonretroid transposition systems (110). Electrophoretic analysis of permuted fragments containing a yeast tRNA gene complexed with TFIIIC or TFIIIB showed evidence of an induced bend in the vicinity of the transcription initiation site (T. Léveillard, G. Kassavetis, & E. P. Geidus-chek, personal communication). One intriguing possibility is that such a conformational marker could play a role in targeting integration to genes transcribed by polymerase III.

SUMMARY

Analysis of in vivo integration patterns has provided no data to support the notion that retroelement integration is random. Rather, the diversity of insertion patterns of retroelements suggests numerous ways in which genomic DNA is identified for preferential targeting. These range from specific to general and include sequence content, removal of chromatin proteins, nuclear localization, distinctive topology, and association with particular *trans*-acting factors. Many are similar to mechanisms already demonstrated to affect activities of previously described recombinases. Moreover, such proposed targeting mechanisms could act directly or indirectly to influence integration site selection.

A variety of observations are consistent with the preferential use of open chromatin for retroelement insertion. The site-specific retroelements insert into transcribed regions. In vitro studies with retroviruses and Ty1 have shown that naked DNA can function, at least under some conditions, as a target. The association of integration sites of retroviruses and regions in which DNase I hypersensitive sites exist and the preferential integration of Ty1 at the 5' ends of some genes might suggest that regions which do not have phased nucleosomes are targets for integration. Is targeting to transcriptionally active regions essentially passive, because they are not densely associated with chromatin proteins, or is targeting active in the sense of being a more specific process?

Specific targets could be generated from DNA or protein motifs. Nucleo-some-free regions are associated with a variety of nonnucleosome proteins, including topoisomerases, nuclear matrix proteins, transcription factors, or replication proteins. These then are candidates for proteins which target integration directly, by associating with the transposition complex or, in-directly, by inducing changes in the DNA. Polymerase III-transcribed genes, which are relatively defined targets of integration for some retrotransposon systems, probably exemplify several of these mechanisms. Promoter se-quences may be directly involved in targeting some elements and positioning of the transcription complex may fix the integration sites of others.

The most common sequence feature of characterized in vivo insertion sites is that they are AT-rich. This may reflect specificity of the IN protein, particularly the gypsylike elements, increased nicking of DNA, which is relatively weakly base-paired, as appears to be the case for the FLP recombinase (130), or simply the AT content of accessible regions in chromatin. Some of these questions will be resolved by the characterization of in vitro integration sites that have been recovered by physical means, rather than by biological assay. The insertion patterns of a number of retroposons suggest that retroelements can insert with a high degree of sequence specificity. Furthermore, expression of recombinant endonucleases has shown that in a heterologous host transcription is not required for site-specific cleavage of intragenic sequences (140a).

A relatively unexplored issue is the role of nuclear localization in targeting. Three kinds of effects have been considered: First, the possibility that the position of DNA relative to the nuclear matrix might make it more or less accessible to the integration apparatus is quite intriguing. It would allow specific predictions to be made concerning what constitutes the integration complex and what physical constraints operate on its delivery to the integration site. Second, the genomic context of the target may play a role in determining the orientation of insertions. In the case of Ty3, genomic targets, but not 2 micron-based, plasmid targets, appear to specify insertion orientation. This suggests that position of the target relative to some chromosomal feature, rather than the target or immediate flanking DNA, is the orientation determinant. Third, position effects play a role in determining the frequency with which some targets are used. The observation that genomic integration of Ty3 upstream of 5S genes was not observed, yet integration upstream of a 5S gene by itself or in tandem with a tRNA gene on a plasmid, suggests that the use of 5S sequences as a target is restricted by positioning within the rDNA repeat or by the nucleolar localization of these sequences.

The extent to which recombinogenic regions are subject to elevated retroelement insertion is largely unexplored. Recombination has been correlated with transcriptional activity in several systems including immunoglobulin genes in animal cells (8), and *GAL1* (128), *HOT1* (120), and *MAT* (78) in yeast. Like the hotspots for integration, the 5' ends of several fungal genes have been shown to have hotspots for recombination, at least in meiotic cells. If transcription is a general mechanism by which regions are made accessible for integration, then parallels between recombination and integration frequencies should be observed. A study currently underway to map genomic integrations of a marked Ty1 into chromosome III will address whether broad correlations exist between transcribed regions or recombinogenic regions and integration activity. Preliminary data from this study are consistent with nonrandom integration (D. Moore, G. Natsoulis, M. Nacionales, & J. Boeke, personal communication).

Insertion specificity has been shown for members of the retroposon and retrotransposon classes of retroelements, making it clear that retroelements as a class can have highly specialized integration functions. Given the demonstrated potential of the RNA genome for diversity, it would be rather surprising if some forms of specialized integration were not identified in additional classes of retroelements.

PERSPECTIVES

Elucidation of the mechanism of retrotransposon and retrovirus integration and target site selection should proceed rapidly in the next few years. Questions to be addressed in vitro include: (a) identification of IN domains required for recognition and nicking of donor and target DNAs; (b) testing of potential roles for proteins in addition to IN, e.g. CA, NC, histones, and even transcription factors, in integration; and (c) identification of the energy source that drives the ligation of donor ends to host DNA.

Our increasing understanding of the integration reaction based on in vitro studies should benefit the design of experiments to more methodically explore in vivo integration. Central issues are (a) the effect of replication on integration; (b) the role of nuclear localization in target site discrimination; (c) the role of transcription and chromatin organization in activation or exclusion of potential integration targets; and (d) definition of the in vivo specificities of retroelements.

New tools are available in many systems for these in vivo studies. Yeast strains in which the availability of histones and transcription factors can be controlled will be useful in testing hypothetical roles of chromatin structure and transcription. Plasmid target assays using highly regulated promoters and marked yeast elements would be explicitly informative about the role of transcription and UAS sequences in target determination without requiring specific integration phenotypes. In yeast and animal cells, studies with marked elements to reduce experimentally imposed constraints on target site usage and application of the polymerase chain reaction to characterization of individual insertions should accelerate the collection of data. Integration libraries can be specifically screened with probes for some candidate high-frequency targets such as tRNA genes, regions of topoisomerase binding or nuclear matrix attachment, highly transcribed regions and oncogenes. Similarly, analysis of integration libraries could be informative about the unexplored possibility that integration patterns differ among retroviruses. A more refined profile of retrovirus insertion would facilitate their use as mutagens and genetic tags for cloning target genes. Results from in vitro and in vivo studies can be applied to the design of vectors with controlled patterns of integration; these would have applications in gene therapy and basic research. Improved understanding of retroelement integration will greatly enhance its use as a probe of the eukaryotic nucleus.

ACKNOWLEDGMENTS

We thank V. Bilanchone, J. Boeke, D. Brow, J. Coffin, R. Craigie, T. Dingermann, E. P. Geiduschek, G. W. Hatfield, P. Kinsey, T. Léveillard, H. Levin, and M. McClure for helpful discussions and for making results available prior to publication. We thank E. P. Geiduschek and D. Brow and C. Guthrie for gifts of clones containing the yeast 5S and SNR6 genes. Work in this laboratory is supported by a Public Health Service grant from the National Institutes of Health (GM33281) and by a contract from Enterprise Partners to S. B. S.

Literature Cited

1. Adams, S. E., Mellor, J., Gull, K., Sim, R. B., Tuite, M. F., et al. 1987. The functions and relationships of Ty-VLP proteins in yeast reflect those of mammalian retroviral proteins. *Cell* 49:111–19

2. Allison, D. S., Goh, S. H., Hall, B. D. 1983. The promoter sequence of a yeast tRNA^{Tyr} gene. *Cell* 34:655–64

3. Amstutz, H., Munz, P., Heyer, W.-D., Leupold, U., Kohli, J. 1985. Concerted evolution of tRNA genes: intergenic conversion among three unlinked serine tRNA genes in *S. pombe*. *Cell* 40:879–86

4. Baker, R. E., Gabrielsen, O., Hall, B. D. 1986. Effects of tRNA^{Tyr} point mutations on the binding of yeast RNA polymerase III transcription factor C. *J. Biol. Chem.* 261:5275–82

5. Bazett-Jones, D. P., Brown, M. L. 1989. Electron microscopy reveals that transcription factor TFIIIA bends 5S DNA. *Mol. Cell. Biol.* 9:336–41

6. Berg, D. E., Howe, M. M. 1989. *Mobile DNA*. Washington, DC: Am. Soc. Miocrobiol.

7. Bingham, P. M., Zachar, Z. 1989. Retrotransposons and the FB transposon from *Drosophila melanogaster*. See Ref. 6, pp. 485–502

8. Blackwell, T. K., Moore, M. W., Yancopoulos, G. D., Suh, H., Lutzker, S., et al. 1986. Recombination between immunoglobulin variable region gene segments is enhanced by transcription. *Nature* 324:585–89

9. Boeke, J. D. 1989. Transposable elements in *Saccharomyces cerevisiae*. See Ref. 6, pp. 335–74

10. Boeke, J. D., Eichinger, D., Castrillon, D., Fink, G. R. 1988. The *Saccharomyces cerevisiae* genome contains functional and nonfunctional copies of transposon Ty1. *Mol. Cell. Biol.* 8:1432–42

11. Boeke, J. D., Garfinkel, D. J., Styles, C. A., Fink, G. R. 1985. Ty elements transpose through an RNA intermediate. *Cell* 40:491–500

12. Boeke, J. D., Sandmeyer, S. B. 1990. Yeast transposable elements. In *The Molecular and Cellular Biology of the Yeast Saccharomyces cerevisiae,* ed. J. R. Broach, E. W. Jones, J. Pringle. Cold Spring Harbor, NY: Cold Spring Harbor Lab.

13. Boeke, J. D., Styles, C. A., Fink, G. R. 1986. *Saccharomyces cerevisiae SPT3* gene is required for transposition and transpositional recombination of chromosomal Ty elements. *Mol. Cell. Biol.* 6:3575–81

14. Boeke, J. D., Xu, H., Fink, G. R. 1988. A general method for the chromosomal amplification of genes in yeast. *Science* 239:280–82

15. Braun, B. R., Riggs, D. L., Kassavetis, G. A., Geiduschek, E. P. 1989. Multiple states of protein-DNA interaction in the assembly of transcription complexes on *Saccharomyces cerevisiae* 5S ribosomal RNA genes. *Proc. Natl. Acad. Sci. USA* 86:2530–34

16. Brodeur, G. M., Sandmeyer, S. B., Olson, M. V. 1983. Consistent association between sigma elements and tRNA genes in yeast. *Proc. Natl. Acad. Sci. USA* 80:3292–96

17. Brow, D. A., Guthrie, C. 1988. Spliceosomal RNA U6 is remarkably conserved from yeast to mammals. *Nature* 334:213–18

18. Brown, P. O., Bowerman, B., Varmus, H. E., Bishop, J. M. 1987. Correct integration of retroviral DNA in vitro. *Cell* 49:347–56

19. Brown, P. O., Bowerman, B., Varmus, H. E., Bishop, J. M. 1989. Retroviral integration: structure of the initial covalent product and its precursor, and a role

for the viral IN protein. *Proc. Natl. Acad. Sci. USA* 86:2525–29

20. Cameron, J. R., Loh, E. Y., Davis, R. W. 1979. Evidence for transposition of dispersed repetitive DNA families in yeast. *Cell* 16:739–51

21. Cappello, J., Cohen, S., Lodish, H. F. 1984. *Dictyostelium* transposable element DIRS-1 preferentially inserts into DIRS-1 sequences. *Mol. Cell. Biol.* 4:2207–13

22. Challice, J. M., Segall, J. 1989. Transcription of the 5S rRNA gene of *Saccharomyces cerevisiae* requires a promoter element at +1 and a 14-base pair internal control region. *J. Biol. Chem.* 264:20060–67

23. Chisholm, G. E., Genbauffe, F. S., Cooper, T. G. 1984. tau, a repeated DNA sequence in yeast. *Proc. Natl. Acad. Sci. USA* 81:2965–69

24. Clare, J., Farabaugh, P. 1985. Nucleotide sequence of a yeast Ty element: evidence for an unusual mechanism of gene expression. *Proc. Natl. Acad. Sci. USA* 82:2829–33

25. Clark, D. J., Bilanchone, V. W., Haywood, L. J., Dildine, S. L., Sandmeyer, S. B. 1988. A yeast sigma composite element, TY3, has properties of a retrotransposon. *J. Biol. Chem.* 263:1413–23

26. Das, G., Henning, D., Wright, D., Reddy, R. 1988. Upstream regulatory elements are necessary and sufficient for transcription of a U6 RNA gene by RNA polymerase III. *EMBO J.* 7:503–12

27. Del Rey, F. J., Donahue, T. F., Fink, G. R. 1982. Sigma, a repetitive element found adjacent to tRNA genes of yeast. *Proc. Natl. Acad. Sci. USA* 79:4138–42

28. DeLotto, R., Schedl, P. 1984. Internal promoter elements of transfer RNA genes are preferentially exposed in chromatin. *J. Biol. Chem.* 179:607–28

29. Dhar, R., McClements, W. L., Enquist, L. W., Vande Woude, G. F. 1980. Nucleotide sequences of integrated Moloney sarcoma provirus long terminal repeats and their host and viral junctions. *Proc. Natl. Acad. Sci. USA* 77:3937–41

30. Dobson, M. J., Kingsman, S. M., Kingsman, A. J. 1981. Sequence variation in the *LEU2* region of the *Saccharomyces cerevisiae* genome. *Gene* 16:133–39

31. Donehower, L. A. 1988. Analysis of mutant Moloney murine leukemia viruses containing linker insertion mutations in the 3' region of pol. *J. Virol.* 62:3958–64

32. Donehower, L. A., Varmus, H. E. 1984. A mutant murine leukemia virus with a single missense codon in *pol* is defective in a function affecting integration. *Proc. Natl. Acad. Sci. USA* 80:6461–65

33. Doolittle, R. F., Feng, D.-F., Johnson, M. S., McClure, M. A. 1989. Origins and evolutionary relationships of retroviruses. *Q. Rev. Biol.* 64:1–30

34. Duyk, G., Leis, J., Longiaru, M., Skalka, A. M. 1983. Selective cleavage in the avian retroviral long terminal repeat sequence by the endonuclease associated with the alpha-beta form of avian reverse transcriptase. *Proc. Natl. Acad. Sci. USA* 80:6745–49

35. Duyk, G., Longiaru, M., Cobrinik, D., Kowal, R., DeHaseth, P., et al. 1985. Circles with two tandem long terminal repeats are specifically cleaved by *pol* gene-associated endonuclease from avian sarcoma and leukosis viruses: nucleotide sequences required for site-specific cleavage. *J. Virol.* 56:589–99

36. Eibel, H., Philippsen, P. 1984. Preferential integration of yeast transposable element Ty into a promoter region. *Nature* 307:386–88

37. Eichinger, D. J., Boeke, J. D. 1988. The DNA intermediate in yeast Ty1 element transposition copurifies with virus-like particles: cell-free Ty1 transposition. *Cell* 54:955–66

38. Eigel, A., Feldmann, H. 1982. Ty1 and delta elements occur adjacent to several tRNA genes in yeast. *EMBO J.* 1:1245–50

39. Fink, G. R. 1989. A new twist to the topoisomerase problem. *Cell* 58:225–26

40. Firtel, R. A. 1989. Mobile genetic elements in the cellular slime mold *Dictyostelium discoideum*. See Ref. 6, pp. 557–66

41. Freund, R., Meselson, M. 1984. Long terminal repeat nucleotide sequence and specific insertion of the gypsy transposon. *Proc. Natl. Acad. Sci. USA* 81:4462–64

42. Fujiwara, T., Craigie, R. 1989. Integration of mini-retroviral DNA: a cell-free reaction for biochemical analysis of retroviral integration. *Proc. Natl. Acad. Sci. USA* 86:3065–69

43. Fujiwara, T., Mizuuchi, K. 1988. Retroviral DNA integration: structure of an integration intermediate. *Cell* 54:497–504

44. Gafner, J., De Robertis, E. M., Philippsen, P. 1983. Delta sequences in the 5' non-coding region of yeast tRNA genes. *EMBO J.* 2:583–91

45. Garfinkel, D. J., Boeke, J. D., Fink, G.

R. 1985. Ty element transposition: reverse transcriptase and virus-like particles. *Cell* 42:507–17

46. Garfinkel, D. J., Mastrangelo, M. F., Sanders, N. J., Shafer, B. K., Strathern, J. N. 1988. Transposon tagging using Ty elements in yeast. *Genetics* 120:95–108

47. Geiduschek, E. P., Tocchini- Valentini, G. P. 1988. Transcription by RNA polymerase III. *Annu. Rev. Biochem.* 57:873–914

48. Giroux, C. N., Mis, J. R. A., Pierce, M. K., Kohalmi, S. E., Kunz, B. A. 1988. DNA sequence analysis of spontaneous mutations in the *SUP4-o* gene of *Saccharomyces cerevisiae. Mol. Cell. Biol.* 8:978–81

49. Goodman, H. M., Olson, M. V., Hall, B. D. 1977. Nucleotide sequence of a mutant eukaryotic gene: the yeast tyrosine-inserting ochre suppressor *SUP4-o. Proc. Natl. Acad. Sci. USA* 74:5453–57

50. Goodman, S. D., Nash, H. A. 1989. Functional replacement of a protein-induced bend in a DNA recombination site. *Nature* 341:251–54

51. Grandbastien, M.-A., Spielmann, A., Caboche, M. 1989. Tnt1, a mobile retroviral-like transposable element of tobacco isolated by plant cell genetics. *Nature* 337:376–80

52. Grandgenett, D. P., Mumm, S. R. 1990. Unraveling retrovirus integration. *Cell* 60:3–4

53. Grandgenett, D. P., Vora, A. C. 1985. Site specific nicking at the avian retrovirus LTR circle junction by the viral pp32 endonuclease. *Nucleic Acids Res.* 13:6205–21

54. Grandgenett, D. P., Vora, A. C., Schiff, R. D. 1978. A 32,000-Dalton nucleic acid-binding protein from avian retravirus cores possesses DNA endonuclease activity. *Virology* 89:119–32

55. Grandgenett, D. P., Vora, A. C., Swanstrom, R., Olsen, J. 1986. Nuclease mechanism of the avian retrovirus pp32 endonuclease. *J. Virol.* 58:970–74

56. Gross, D. S., Garrard, W. T. 1988. Nuclease hypersensitive sites in chromatin. *Annu. Rev. Biochem.* 57:159–97

57. Guthrie, C., Abelson, J. 1982. Organization and expression of tRNA genes in *Saccharomyces cerevisiae.* In *The Molecular Biology of the Yeast* Saccharomyces, ed. J. N. Strathern, E. W. Jones, J. R., Broach, pp. 487–28. Cold Spring Harbor, NY: Cold Spring Harbor Lab.

58. Hansen, L. J., Chalker, D. L., Sand-

meyer, S. B. 1988. Ty3, a yeast retrotransposon associated with tRNA genes, has homology to animal retroviruses. *Mol. Cell. Biol.* 8:5245–56

59. Hansen, L. J., Sandmeyer, S. B. 1990. Characterization of a transpositionally active Ty3 element and identification of the Ty3 IN protein. *J. Virol.* 64:2599–607

60. Hauber, J., Nelbock-Hochstetter, P., Feldmann, H. 1985. Nucleotide sequence and characteristics of a Ty element from yeast. *Nucleic Acids Res.* 13:2745–58

61. Hauber, J., Stucka, R., Krieg, R., Feldmann, H. 1988. Analysis of yeast chromosomal regions carrying members of the glutamate tRNA gene family: various transposable elements are associated with them. *Nucleic Acids. Res.* 16:10623–34

62. Hinton, C. W. 1984. Morphogenetically specific mutability in *Drosophila anassae. Genetics* 106:631–53

63. Hizi, A., Hughes, S. H. 1988. Expression of the Moloney murine leukemia virus and human immunodeficiency virus integration proteins in *Escherichia coli. Virology* 167:634–38

64. Hughes, S. H., Shank, P. R., Spector, D. H., Kung, H.-J., Bishop, J. M., et al. 1978. Proviruses of avian sarcoma virus are terminally redundant, coextensive with unintegrated linear DNA and integrated at many sites. *Cell* 15:1397–410

65. Huibregtse, J. R., Engelke, D. R. 1989. Genomic footprinting of a yeast tRNA gene reveals stable complexes over the 5'-flanking region. *Mol. Cell. Biol.* 9:3244–52

66. Hwang, S. L.-H., Gilboa, E. 1984. Expression of genes introduced into cells by retroviral infection is more efficient than that of genes introduced into cells by DNA transfection. *J. Virol.* 50:417–24

67. Ikenaga, H., Saigo, K. 1982. Insertion of a movable genetic element, 297, into the T-A-T-A box for the H3 histone gene in *Drosophilia melanogaster. Proc. Natl. Acad. Sci. USA* 79:4143–47

68. Inouye, S., Yuki, S., Saigo, K. 1984. Sequence-specific insertion of the *Drosophila* transposable genetic element 17.6. *Nature* 310:332–33

69. Inouye, S., Yuki, S., Saigo, K. 1986. Complete nucleotide sequence and genome organization of a *Drosophila* transposable genetic element, 297. *Eur. J. Biochem.* 154:417–25

70. Jentsch, S., McGrath, J. P., Var-

shavsky, A. 1987. The yeast DNA repair gene *RAD6* encodes a ubiquitin-conjugating enzyme. *Nature* 329:131–34

71. Johnson, M. S., McClure, M. A., Feng, D.-F., Gray, J., Doolittle, R. F. 1986. Computer analysis of retroviral *pol* genes: assignment of enzymatic functions to specific sequences and homologies with non-viral enzymes. *Proc. Natl. Acad. Sci. USA* 83:7648–52

72. Johnson, S. P., Warner, J. R. 1989. Unusual enhancer function in yeast rRNA transcription. *Mol. Cell. Biol.* 9:4986–93

73. Kassavetis, G. A., Braun, B. R., Nguyen, L. H., Geiduschek, E. P. 1990. *S. cerevisiae* TFIIIB is the transcription initiation factor proper of RNA polymerase III, while TFIIIA and TFIIIC are assembly factors. *Cell* 60:235–45

74. Kassavetis, G. A., Riggs, D. L., Negri, R., Nguyen, L. H., Geiduschek, E. P. 1989. Transcription factor IIIB generates extended DNA interactions in RNA polymerase III transcription complexes on tRNA genes. *Mol. Cell. Biol.* 9:2551–66

75. Katzman, M., Katz, R. A., Skalka, A. M., Leis, J. 1989. The avian retroviral integration protein cleaves the terminal sequences of linear viral DNA at the in vivo sites of integration. *J. Virol.* 63:5319–27

76. King, W., Patel, M. D., Lobel, L. I., Goff, S. P., Nguyen-Huu, M. C. 1985. Insertion mutagenesis of embryonal carcinoma cells by retroviruses. *Science* 228:554–58

77. Kingsman, A. J., Gimlich, R. L., Clarke, L., Chinault, A. C., Carbon, J. 1981. Sequence variation in dispersed repetitive sequences in *Saccharomyces cerevisiae*. *J. Mol. Biol.* 145:619–32

78. Klar, A. J. S., Strathern, J. N., Abraham, J. A. 1984. Involvement of double-strand chromosomal breaks for mating-type switching in *Saccharomyces cerevisiae*. *Cold Spring Harbor Symp. Quant. Biol.* 49:77–88

79. Knaus, R. J., Hippenmeyer, P. J., Misra, T. K., Grandgenett, D. P., Müller, U. R., Fitch, W. M. 1984. Avian retrovirus pp32 DNA binding protein. Preferential binding to the promoter region of long terminal repeat DNA. *Biochemistry* 23:350–59

80. Kurjan, J., Hall, B. D., Gillam, S., Smith, M. 1980. Mutations at the yeast *SUP4* locus: DNA sequence changes in mutants lacking suppressor activity. *Cell* 20:701–6

81. Louis, C., Schedl, P., Samal, B., Worcel, A. 1980. Chromatin structure of the 5S RNA genes of *D. melanogaster*. *Cell* 22:387–92

82. Madison, J. T., Kung, H. K. 1967. Large oligonucleotides isolated from yeast tyrosine transfer ribonucleic acid after partial digestion with ribonuclease. T1. *J. Biol. Chem.* 242:1324–30

83. Marlor, R. L., Parkhurst, S. M., Corces, V. G. 1986. The *Drosophila melanogaster* gypsy transposable element encodes putative gene products homologous to retroviral proteins. *Mol. Cell. Biol.* 6:1129–34

84. Marschalek, R., Brechner, T., Amon-Bohm, E., Dingermann, T. 1989. Transfer RNA genes: landmarks for integration of mobile genetic elements in *Dictyostelium discoideum*. *Science* 244:1493–96

85. Misra, T. K., Grandgenett, D. P., Parson, J. T. 1982. Avian retrovirus pp32 DNA-binding protein I. Recognition of specific sequences on retrovirus DNA terminal repeats. *J. Virol.* 44:330–43

85a. Moenne, A., Camier, S., Anderson, G., Margottin, F., Beggs, J., Sentenac, A. 1990. The U6 gene of *Saccharomyces cerevisiae* is transcribed by RNA polymerase C (III) in vivo and in vitro. *EMBO J.* 9:271–77

86. Mount, S. M., Rubin, G. M. 1985. Complete nucleotide sequence of the *Drosophila* transposable element copia: homology between copia and retroviral proteins. *Mol. Cell. Biol.* 5:1630–38

87. Müller, F., Brühl, K. H., Friedel, K., Kowallik, K. V., Ciriacy, M. 1987. Processing of TY1 proteins and formation of Ty1 virus-like particles in *Saccharomyces cerevisiae*. *Mol. Gen. Genet.* 207:421–29

88. Murphy, S., Di Liegro, C., Melli, M. 1987. The in vitro transcription of the 7SK RNA gene by RNA polymerase III is dependent only on the presence of an upstream promoter. *Cell* 51:81–87

89. Natsoulis, G., Thomas, W., Roghmann, M-C., Winston, F., Boeke, J. D. 1989. Ty1 transposition in *Saccharomyces cerevisiae* is nonrandom. *Genetics* 123:269–79

90. Nelböck, P., Stucka, R., Feldmann, H. 1985. Different patterns of transposable elements in the vicinity of tRNA genes in yeast: a possible clue to transcriptional modulation. *Biol. Chem. Hoppe-Seyler* 366:1041–51

91. Olson, M. V., Loughney, K., Hall, B. 1979. Identification of the yeast DNA sequences that correspond to specific

tyrosine-inserting nonsense suppressor loci. *J. Mol. Biol.* 132:387–410

92. Oyen, T. B., Gabrielson, O. S. 1983. Non-random distribution of the TY1 elements within nuclear DNA of *Saccharomyces cerevisiae*. *FEBS Lett.* 161:201–6

93. Panganiban, A. T., Temin, H. M. 1984. The retrovirus *pol* gene encodes a product required for DNA integration: Identification of a retrovirus *int* locus. *Proc. Natl. Acad. Sci. USA* 81:7885–89

94. Panganiban, A. T., Temin, H. M. 1984. Circles with two tandem LTRs are precursors to integrated retrovirus DNA. *Cell* 36:673–79

95. Pérez-Ortín, J. E., Matallana, E., Franco, L. 1989. Chromatin structure of yeast genes. *Yeast* 5:219–38

96. Petes, T. D. 1979. Yeast ribosomal DNA genes are located on chromosome XII. *Proc. Natl. Acad. Sci. USA* 76:410–14

97. Piper, P. W., Lockheart, A., Patel, N. 1984. A minor class of 5S rRNA genes in *Saccharomyces cerevisiae* X2180-1B, one member of which lies adjacent to a Ty transposable element. *Nucleic Acids Res.* 12:4083–96

98. Poole, S. J., Firtel, R. A. 1984. Genomic instability and mobile genetic elements in regions surrounding two discoidin I genes of *Dictyostelium discoideum*. *Mol. Cell. Biol.* 4:671–80

99. Quinn, T. P., Grandgenett, D. P. 1988. Genetic evidence that the avian retrovirus DNA endonuclease domain of pol is necessary for viral integration. *J. Virol.* 62:2307–12

100. Robinson, H. L., Gagnon, G. C. 1986. Patterns of proviral insertion and deletion in avian leukosis virus-induced lymphomas. *J. Virol.* 57:28–36

101. Rohdewohld, H., Weiher, H., Reik, W., Jaenisch, R., Breindl, M. 1987. Retrovirus integration and chromatin structure: Moloney murine leukemia proviral integration sites map near DNase I-hypersensitive sites. *J. Virol.* 61:336–43

102. Rosen, E., Sivertsen, A., Firtel, R. A. 1983. An unusual transposon encoding heat shock inducible and developmentally regulated transcripts in *Dictyostelium*. *Cell* 35:243–51

103. Roth, M. J., Schwartzberg, P. L., Goff, S. P. 1989. Structure of the termini of DNA intermediates in the integration of retroviral DNA: dependence on IN function and terminal DNA sequence. *Cell* 58:47–54

104. Roth, M. J., Tanese, N., Goff, S. P. 1988. Gene product of Moloney murine

leukemia virus required for proviral integration is a DNA-binding protein. *J. Mol. Biol.* 203:131–39

105. Saigo, K., Kugimiya, W., Matsuo, Y., Inouye, S., Yoshioka, K., Yuki, S. 1984. Identificaion of the coding sequence for a reverse transcriptase-like enzyme in a transposable genetic element in *Drosophila melanogaster*. *Nature* 312:659–61

106. Sandmeyer, S. B., Bilanchone, V. W., Clark, D. J., Morcos, P., Carle, G. F., Brodeur, G. M. 1988. Sigma elements are position-specific for many different yeast tRNA genes. *Nucleic Acids Res.* 16:1499–515

107. Sandmeyer, S. B., Olson, M. V. 1982. Insertion of a repetitive element at the same position in the 5'-flanking regions of two dissimilar yeast tRNA genes. *Proc. Natl. Acad. Sci. USA* 79:7674–78

108. Scherdin, U., Rhodes, K., Breindl, M. 1990. Transcriptionally active genome regions are preferred targets for retrovirus integration. *J. Virol.* 64:907–12

109. Schultz, P., Marzouki, N., Marck, C., Ruet, A., Oudet, P., Sentenac, A. 1989. The two DNA-binding domains of yeast transcription factor tau as observed by scanning transmission electron microscopy. *EMBO J.* 8:3815–24

110. Schwartz, C. J., Sadowski, P. D. 1989. FLP recombinase of the 2 micron circle plasmid of *Saccharomyces cerevisiae* bends its DNA target. Isolation of FLP mutants defective in DNA bending. *J. Mol. Biol.* 205:647–58

111. Schwartzberg, P., Colicelli, J., Goff, S. P. 1984. Construction and analysis of deletion mutations in the *pol* gene of murine leukemia virus: a new viral function of required for productive infection. *Cell* 37:1043–52

112. Shalit, P., Loughney, K., Olson, M. V., Hall, B. D. 1981. Physical analysis of the *CYC1-SUP4* interval in *Saccharomyces cerevisiae*. *Mol. Cell Biol.* 1:228–36

113. Shih, C.-C., Stoye, J. P., Coffin, J. M. 1988. Highly preferred targets for retrovirus integration. *Cell* 53:531–37

114. Shimotohno, K., Temin, H. 1980. No apparent nucleotide sequence specificity in cellular DNA juxtaposed to retrovirus proviruses. *Proc. Natl. Acad. Sci. USA* 77:7357–61

115. Shoemaker, C., Goff, S., Gilboa, E., Paskind, M., Mitra, S. W., Baltimore, D. 1980. Structure of a cloned circular Moloney murine leukemia virus DNA molecule containing an inverted seg-

ment: Implications for retrovirus integration. *Proc. Natl. Acad. Sci. USA* 77:3932–36

116. Simchen, G., Winston, F., Styles, C. A., Fink, G. R. 1984. Ty-mediated gene expression of the *LYS2* and *HIS4* genes of *Saccharomyces cerevisiae* is controlled by the same *SPT* genes. *Proc. Natl. Acad. Sci. USA* 81:2431–34

117. Smyth, D. R., Kalitsis, P., Joseph, J. L., Sentry, J. W. 1989. Plant retrotransposon from *Lilium henryi* is related to Ty3 of yeast and the gypsy group of *Drosophila. Proc. Natl. Acad. Sci. USA* 86:5015–19

118. Snyder, M. P., Kimbrell, D., Hunkapiller, M., Hill, R., Fristrom, J., Davidson, N. 1982. A transposable element that splits the promoter region inactivates a *Drosophila* cuticle gene. *Proc. Natl. Acad. Sci. USA* 79:7430–34

119. Snyder, U. K., Thompson, J. F., Landy, A. 1989. Phasing of protein-induced DNA bends in a recombination complex. *Nature* 341:255–57

120. Stewart, S. E., Roeder, G. S. 1989. Transcription by RNA polymerase I stimulates mitotic recombination in *Saccharomyces cerevisiae. Mol. Cell. Biol.* 9:3464–72

121. Stillman, D. J., Better, M., Geiduschek, E. P. 1985. Electron-microscopic examination of the binding of a large RNA polymerase III transcription factor to a tRNA gene. *J. Mol. Biol.* 185:451–55

122. Stucka, R., Hauber, J., Feldmann, H. 1987. One member of the tRNA(Glu) gene family in yeast codes for a minor GAGtRNA(Glu) species and is associated with several short transposable elements. *Curr. Genet.* 12:323–28

123. Stucka, R., Lochmuller, H., Feldmann, H. 1989. Ty4, a novel low-copy number element in *Saccharomyces cerevisiae:* one copy is located in a cluster of Ty elements and tRNA genes. *Nucleic Acids Res.* 17:4993–5001

124. Sung, P., Prakash, S., Prakash, L. 1988. *RAD6* protein of *Saccharomyces cerevisiae* polyubiquitinates histones and its acidic domain mediates this activity. *Genes Dev.* 2:1476–85

125. Tanda, S., Shrimpton, A. E., Ling-Ling, C., Itayama, H., Matsubayashi, H., Saigo, K., Tobari, Y. N., Langley, C. H. 1988. Retrovirus-like features and site specific insertions of a transposable element, *tom* in *Drosophila ananassae. Mol. Gen. Genet.* 214:405–11

126. Taylor, M. J., Segall, J. 1985.

Characterization of factors and DNA sequences required for accurate transcription of the *Saccharomyces cerevisiae* 5S RNA gene. *J. Biol. Chem.* 260:4531–40

127. Terry, R., Soltis, D. A., Katzman, M., Cobrinik, D., Leis, J., Skalka, A. M. 1988. Properties of avian sarcomaleukosis virus pp32-related polendonucleases produced in *Escherichia coli. J. Virol.* 62:2358–65

128. Thomas, B. J., Rothstein, R. 1989. Elevated recombination rates in transcriptionally active DNA. *Cell* 56:619–30

129. Toh, H., Kikuno, R., Hayashida, H., Miyata, T., Kugimiya, W., et al. 1985. Close structural resemblance between putative polymerase of a *Drosophila* transposable genetic element 17.6 and *pol* gene product of Moloney murine leukemia virus. *EMBO J.* 4:1267–72

130. Umlauf, S. W., Cox, M. M. 1988. The functional significance of DNA sequence structure in a site-specific genetic recombination reaction. *EMBO J.* 7:1845–52

131. Varmus, H., Brown, P. 1989. Retroviruses. See Ref. 6, pp. 53–108

132. Vijaya, S., Steffen, D. L., Robinson, H. L. 1986. Acceptor sites for retroviral integrations map near DNase I-hypersensitive sites in chromatin. *J. Virol.* 60:683–92

133. Vincent, A., Petes, T. D. 1986. Isolation and characterization of a Ty element inserted into the ribosomal DNA of the yeast *Saccharomyces cerevisiae. Nucleic Acids Res.* 14:2939–49

134. Voytas, D. F., Ausubel, F. M. 1988. A copia-like transposable element family in *Arabidopsis thaliana. Nature* 336:242–44

135. Warmington, J. R., Anwar, R., Newlon, C. S., Waring, R. B., Davies, R. W., et al. 1986. A 'hot-spot' for Ty transposition on the left arm of yeast chromosome III. *Nucleic Acids Res.* 14:3475–85

136. Warmington, J. R., Green, R. P., Newlon, C. S., Oliver, S. G. 1987. Polymorphisms on the right arm of yeast chromosome III associated with Ty transposition and recombination events. *Nucleic Acids Res.* 15:8963–82

137. Warmington, J. R., Waring, R. B., Newlon, C. S., Indge, K. J., Oliver, S. G. 1985. Nucleotide sequence characterization of Ty 1–17, a class II transposon from yeast. *Nucleic Acids Res.* 13:6679–93

138. Wilke, C. M., Heidler, S. H., Brown, N., Liebman, S. W. 1989. Analysis of

yeast retrotransposon Ty insertions at the *CAN1* locus. *Genetics* 123:655–65

139. Williamson, V. M., Cox, D., Young, E. T., Russell, D. W., Smith, M. 1983. Characterization of transposable element-associated mutations that alter yeast alcohol dehydrogenase II expression. *Mol. Cell. Biol.* 3:20–31

140. Xiong, Y., Eickbush, T. H. 1988. Similarity of reverse transcriptase-like sequences of viruses, transposable elements, and mitochondrial introns. *Mol. Biol. Evol.* 5:675–90

140a. Xiong, Y., Eickbush, T. H. 1988. Functional expression of a sequence-specific endonuclease encoded by the retrotransposon R2Bm. *Cell* 55:235–46

141. Zuker, C., Lodish, H. F. 1981. Repetitive DNA sequences cotranscribed with developmentally regulated *Dictyostelium discoideum* mRNAs. *Proc. Natl. Acad. Sci. USA* 78:5386–90

Annu. Rev. Genet. 1990. 24:519–41

REGULATED mRNA STABILITY

Jonathan A. Atwater, Ron Wisdom, and Inder M. Verma

Molecular Biology and Virology Laboratory, The Salk Institute, P.O. Box 85800, San Diego, CA 92138

KEY WORDS: mRNA stability, gene regulation

CONTENTS

"Discovery consists of seeing what everybody has seen and thinking what nobody has thought." Albert Szent-Gyorgi

INTRODUCTION

Eukaryotic gene expression is controlled at many levels, including the abundance of mRNAs for each protein. Theoretically, the steady-state level of a particular mRNA can be controlled at any step in its biogenesis, including initiation or elongation of transcription, post-transcriptional maturation events (addition of the 5' cap, addition of 3' poly(A) residues, and splicing),

519

0066-4197/90/1215-0519$02.00

transport from the nucleus to the cytoplasm, and cytoplasmic degradation of the mature mRNA. Examples abound for the control of steady-state mRNA levels by each of these mechanisms, and our understanding of the complexity of regulatory mechanisms continues to increase. This review addresses the contribution of the mRNA degradation process to the regulation of gene expression in vertebrate eukaryotes. It is not meant to be comprehensive in scope, and instead focuses on the features learned from the detailed investigation of a few genes regulated at the level of mRNA turnover.

For cells to respond rapidly to changes in the extracellular environment, the levels of mRNAs encoding proteins that mediate these changes must be able to change rapidly. These mRNAs, in turn must be capable of rapid turnover, so that different rates of transcription are rapidly converted into changes in steady-state mRNA levels. On the other hand, it is important that the mRNAs for many structural genes be long-lived, so that constitutive, high-level transcription of these genes is not required to supply the basic components of cellular machinery. Therefore, an important first level of regulation of mRNA stability is the differential stability of different mRNAs. For example, *c-fos* mRNA, which is involved in the response of cells to changes in the external environment, is degraded rapidly in the cytoplasm, with a half-life of 8–30 min, whereas β-globin mRNA has a half-life of greater than 24 hr in erythroid cells (63, 78, 103, 116). One important control of mRNA stability is the ability to differentially regulate the turnover of mRNAs, so that some are rapidly degraded while others are stable.

A second level of regulation of gene expression by controlling mRNA turnover involves the differential stability of the same mRNA under different conditions. Examples of this are now abundant, and include, but are not limited to, altered rates of degradation of the mRNAs derived from the vitellogenin, tubulin, histone, transferrin receptor, and early response genes. The conditions that govern the rate of turnover of the particular mRNAs are extremely variable. For some genes, detailed molecular genetic data are now available, and permit some generalizations about the mechanisms that control the rate of turnover of specific mRNAs. These examples are discussed in some detail.

METHODS OF MEASURING mRNA TURNOVER: LIMITATIONS

A principal reason for our limited understanding of the regulation of mRNA turnover is the lack of good experimental techniques for measuring mRNA stability under a variety of conditions. The optimal method would be the pulse-chase technique that has been used to analyze the turnover of many other components of the cell, especially proteins. This technique has been used successfully to study the turnover of some mRNAs, notably that for vitellogenin (19, 20), but is difficult to apply widely for two reasons: it is difficult to rapidly equilibrate intracellular UTP pools; and second, the low sensitivity of this method makes it difficult to apply to mRNAs of low abundance. Therefore, most investigators have used other methods.

All other methods of measuring mRNA turnover in cells involve monitoring the decrease in mRNA following an intervention to decrease the transcription rate. As such, current methods serve to define upper limits for the turnover of a particular mRNA, since the synthesis of new mRNA molecules will result in an apparent increase in the half-life that is being measured. Furthermore, most involve major perturbations to the cell that make it difficult to extrapolate the results to a more general situation with certainty. One approach is to add a hormone or chemical that decreases the transcription of a particular species of mRNA. For example, the addition of forskolin to primary Schwann cells markedly decreases transcription of the *c-jun* gene, producing a twentyfold decrease in the steady-state level of *c-jun* mRNA with 30 min (W. Lamph & I. M. Verma, unpublished observation). One can conclude from this change that *c-jun* mRNA must turn over rapidly; if the kinetics are truly first order, the half-life is approximately 5–7 min. However, one cannot conclude from this experiment that *c-jun* mRNA is labile under other conditions. Therefore, this method is limited in the conditions to which it can be applied.

An alternative method is to use promotors whose activity can be controlled experimentally. The decay in mRNA with time is measured following withdrawal of the inductive stimulus. The relatively high level of promotor activity from many inducible promotors in the uninduced state has, in practice, restricted this method mainly to use of the Hsp-70 heat-shock promotor and the *c-fos* promotor (see 58, 73, 110, 120). As in the previous example, this method can only be applied under limited conditions (e.g. heat shock or serum induction), and one cannot generalize with certainty to other conditions.

Most commonly mRNA turnover is estimated by the use of drugs that block transcription, Actinomycin D being the most frequently used. Typically, Actinomycin D is added to cells and the amount of a particular mRNA remaining at various times after treatment is used to estimate decay vs time. However, this method results in a nonphysiologic perturbation to the cell. This seemingly hypothetical concern is indeed serious as intensive investigation of the turnover of some mRNAs as shown. For example, if iron is added to iron-starved cells transferrin receptor mRNA decays with a half-life of about six hr (79). However, if Actinomycin D is added to the cells at the same time as iron, the decay is slowed more than tenfold, implying that the Actinomycin D has an important effect other than blocking the transcription of the transferrin receptor gene (79). Such examples, wherein a particular mRNA decays slower in the presence of a transcription-blocking drug than in its absence, are quite frequent (79, 95, 110). It is not clear whether this effect is the result of the rapid decay of a labile mRNA that participates in mRNA turnover, or some other effect. Discrepencies between studies using transcription inhibitors and those using other methods are frequent enough for caution in interpreting the data.

One final method of assessing changes in mRNA stability should be mentioned. When the changes in the steady-state level of a particular mRNA under two different conditions are not matched by a comparable change in de novo transcription (as measured by nuclear run-on assays), changes in mRNA stability are often presumed. There are two important limitations of this approach: first, there is no numerical estimation of half-life, only a relative comparison; and second, no account is made for changes in mRNA maturation events or nuclear transport that could influence steady-state mRNA levels. Although this method does not permit quantitative estimates of mRNA stability, it does allow the easy analysis of mutant genes, and therefore may be useful where a stability mechanism is known to be important (46, 79, 120, 121).

Duplication of the mRNA degradation reaction in vitro should improve our understanding of the biochemical mechanisms of mRNA turnover. Ross and colleagues have used a biochemical system based on the input of polysome-bound mRNA to study the in vitro decay of histone, c-myc, and globin mRNAs, while Calame and colleagues have used a cytoplasmic extract to carry out the reaction (17, 88, 100). The fidelity of these systems has been analyzed according to several criteria. Ross and co-workers' system degraded different mRNAs (c-myc, histone, and β and δ globin) in the same rank order in vitro as they are intracellularly in the presence of Actinomycin D. The first change observed in such a degradation reaction carried out in vitro is shortening of the poly(A) tail, a result also observed in cultured cells treated with Actinomycin D, suggesting that at least some specificity is retained (77, 115, 116). At longer incubation times, RNAs are generated containing 3' truncations that extend into the body of the mRNA (16); it is not clear whether these products are the result of exonucleolytic or endonucleolytic cleavage. Since these products are not observed in cells, their significance remains obscure. Finally, the mRNA disappears without generating further truncated products that might correspond to intermediates. The specific points learned from such studies are addressed later, but the results derived from in vitro studies demand caution. The lack of extensive characterization of mutant mRNAs, both in cells and in vitro, regarding mRNA turnover makes it difficult to correlate the biochemical and molecular genetic data. In particular, much of the data derived from cells has been generated by performing chase studies with Actinomycin D, a method that may not give accurate data for some of the most extensively investigated RNAs. Secondly, translation inhibitors slow the turnover of many labile mRNAs in vivo, but not in the in vitro systems developed to date. As some evidence indicates that the effect of translation inhibitors on RNA turnover may be direct and not due to the decay of a labile protein involved in mRNA turnover, the failure to slow turnover in vitro may be important.

GENERAL DETERMINANTS OF mRNA STABILITY

Poly(A), Poly(A) Binding Protein, and mRNA Stability

Although the mechanism of addition of poly(A) to the 3' end of most newly synthesized mRNAs has been well characterized, the function(s) of poly(A) is still not completely understood. Evidence suggests, however, that polyadenylation may be involved in regulating several processes: nuclear processing and transport; modulation of mRNA translational efficiencies; and mRNA turnover (14). Although most of the evidence is based on correlation, some data suggest that poly(A) protects many mRNAs from rapid degradation. The half-lives of globin mRNAs with poly(A) tracts of varying size have been measured in Xenopus oocytes (83). Molecules with 32 or more A residues had equivalent turnover rates, whereas those with less than 30 A residues were tenfold less stable. Conversely, histone mRNA, which is normally not polyadenylated and has a very short half-life, was ten times more stable in oocytes when artificially polyadenylated (56, 90). Following the induction of metallothionein transcription, the newly synthesized mRNA has 150–200 A residues, which gradually shorten to 30 residues, but the steady-state level of metallothionein mRNA remains constant. Over the next several hours, the steady-state level declines; however, degradation intermediates shorter than those with 30 A residues are not observed, suggesting that mRNA with poly(A) tracts greater than 30 are stabilized (76). The extremely labile *fos* and *myc* oncogene mRNAs display very rapid poly(A) shortening prior to complete degradation (115, 120). Poly(A) shortening proceeds *myc* mRNA degradation in vitro as well (17). In contrast to the unstable *myc* mRNA, the poly(A) tail of β-globin mRNA remains stable during incubation in the same in vitro extract. Human growth hormone and vasopressin mRNAs exhibit elongation of their poly(A) tails in response to glucocorticoids and osmotic stress, respectively. The poly(A) elongation correlates with stabilization of these mRNAs suggesting that mechanisms exist to regulate poly(A) removal and addition, and, consequently, the rate of mRNA turnover.

Cytoplasmic mRNA exists as a ribonucleoprotein complex. Among the proteins bound to mRNA are poly(A) binding proteins (PABP) (11). PABP is very specific for binding poly(A), and digestion of poly(A)-PABP complexes with RNase results in a ladder of poly(A) fragments in multiples of 27 residues. Reconstruction experiments with poly(A) and PABP have shown that the minimal length of poly(A) necessary for PABP binding is 27–30 residues (4). This correlates well with the observation that 30 A residues at the end of mRNA are the minimum length for stabilization. In vitro experiments support the hypothesis that poly(A) in conjunction with PABP protects mRNA from rapid nucleolytic attack (9). The turnover rate of β-globin mRNA is increased sevenfold in in vitro decay extracts that have been depleted of functional PABP, either by addition of excess poly(A) competitor or by passing the extracts over a poly(A)-sepharose column. When purified PABP is added back, the mRNA is stabilized. The ability of PABP to migrate from one poly(A) tract to another may play a role in its effect on mRNA stability. It has been suggested that some mRNAs may contain internal sequences which disrupt the ability of PABP to interact with the poly(A) tail, resulting in rapid deadenylation and mRNA degradation (120).

Effect of Translation on mRNA Stability

Translation plays an important role in controlling mRNA degradation. Many of the mRNAs that have been studied in detail are stabilized in the presence of inhibitors of protein synthesis. Two possible explanations for this translation dependence have been postulated: (*a*) nucleases or other factors involved in degradation are ribosome-associated, or (*b*) a component of the degradation machinery is labile, requiring continued synthesis. For histone, tubulin, transferrin receptor and the early response genes *fos* and *myc,* ongoing translation of the mRNAs is necessary for their turnover (46, 79, 86). The evidence for this is discussed in detail for each of these genes in the following section of this review. In contrast, the mRNAs from some genes are destabilized by mutations that disrupt their normal translation process. Nonsense mutations in the β-globin genes of patients with β-thalassemia cause a 40-fold reduction in the half-life of β-globin mRNA in erythroid cells (71). The introduction of nonsense mutations in other mRNAs also reduces their stability (5, 21, 25, 33, 46, 55, 86). In some cases, polar effects are observed, for instance, immunoglobulin mRNAs with nonsense mutations in the 5' end of their coding region are as much as 100-fold less stable than mRNAs with nonsense mutations near the authentic termination codon (5). One hypothesis is that interruption of translation leaves "unprotected" regions of the mRNA that are susceptible to nucleolytic attack. A third translation-linked modulation of mRNA stability has been described during B-cell differentiation. When a B cell that displays surface IgM differentiates into a plasma cell that secretes IgM, mRNAs for both heavy and light chains increase in abundance (92). Transfection experiments with μ heavy chain genes under the control of an inducible promoter produce transcripts with half-lives of approximately 20 hr in plasmacytoma cells and approximately 3 hr in B-cell lymphomas. When a μ heavy chain gene with a mutated signal sequence is expressed in plasmacytoma cells, the half-life of the mRNA produced drops to approximately 3 hr (73). One possible explanation is that partitioning the μ heavy chain mRNA from free polysomes to membrane-bound polysomes modulates mRNA turnover.

Nucleolytic Degradation of mRNA

Very little is known about the nucleases responsible for degradation of mRNAs that exhibit regulated turnover. Two possible scenarios can be envisioned: (a) regulated, substrate-specific nucleases may exist for specific mRNAs or groups of mRNAs; or (b) regulation may occur at an earlier step in the pathway by controlling the accessability of a specific mRNA to a nonspecific mRNA nuclease. Until in vitro systems are available that will allow the characterization of purified mRNA nucleases in reconstitution experiments, it will be difficult to distinguish between these possibilities. Ross and co-workers have used an in vitro system to partially characterize an activity that degrades histone mRNA in a manner consistent with the in vivo data (90, 91, 100–102). Although the authors show substrate specificity in that histone mRNA is preferentially degraded over globin mRNA, they cannot exclude the possibility that this nucleolytic activity is a general nuclease for nonpolyadenylated mRNA. To date, there are no examples of specific nuclease recognition sites or specific endonucleolytic cleavage events to substantiate the existence of mRNA-specific nucleases.

SPECIFIC EXAMPLES OF REGULATED mRNA TURNOVER

Vitellogenin mRNA Stabilization by Estrogen

One of the earliest described examples of regulated mRNA stability is the stabilization of vitellogenin mRNA by estrogen in Xenopus (19, 20). The physiologic relevance and the use of methods that result in minimal perturbation of the system make this a clear example of regulation of gene expression by control of mRNA stability. Vitellogenin is synthesized in the liver; its presence is required for the proper development of the yolk sac. In response to estrogen, the already abundant vitellogenin mRNA increases dramatically in liver cells, such that 50,000 copies per cell of vitellogenin mRNA are present in estrogen-treated liver cells, representing an approximately 50-fold increase. The abundance of this mRNA has permitted analysis of its turnover by performing pulse-chase experiments on cultured slices of Xenopus liver in the presence and absence of estrogen. Several points have emerged from this analysis: first, although transcription of the gene is induced, the large increase observed in steady-state mRNA levels requires stabilization of the mRNA as well; second, vitellogenin mRNA is stabilized from a half-life of about 16 hr in the absence of estrogen to a half-life of about 500 hr in the presence of estrogen; third, this effect is specific for vitellogenin mRNA, as other mRNAs, including albumin, are not stabilized; fourth, this effect is reversible by withdrawal of the hormone; and, finally, this stabilization occurs in the presence of protein-synthesis inhibitors, suggesting that the synthesis of new proteins is not required (19, 20). The great advantage of this system, namely the ability to analyze an abundant mRNA under physiologically relevant conditions, has also proven to be its drawback—the analysis of mutant mRNAs is not possible. Nevertheless, this represents one of the best-characterized systems of regulated control of mRNA stability.

Autoregulation of Tubulin Synthesis

The tubulin system is one of the best-characterized examples of exquisite gene regulation at the level of mRNA stability. Tubulin exists in the cytoplasm in a dynamic flux between tubulin heterodimers (comprising one α-tubulin and one β-tubulin polypeptide) and polymerized chains of tubulin. These polymerized chains of tubulin, or microtubules, are major structural components of the cytoskeleton and mitotic spindle apparatus (for review, see 27, 31). The rate of de novo tubulin synthesis is regulated to a large extent by the intracellular concentration of tubulin heterodimers. Initial evidence for autoregulatory control of tubulin synthesis came from experiments using drugs that promote polymerization or depolymerization of cellular tubulin (8, 23, 28–30, 64, 93). Cell cultures treated with colchicine or nocodazole show a twofold increase in the intracellular concentration of tubulin heterodimers and a concurrent five- to tenfold specific repression of new tubulin synthesis. Similarly, treatment of cells with vinblastine or taxol, which reduce the intracellular concentration of tubulin heterodimers, results in a three- to fourfold increase in new tubulin synthesis (8, 86). These changes in the rate of tubulin synthesis can be accounted for by corresponding changes in tubulin mRNA levels (24, 28, 29, 64, 86). Transcription run-off experiments and experiments with enucleated cells suggest that autoregulation of tubulin mRNA abundance results from modulation of tubulin mRNA stability, rather than from changes in tubulin gene transcription (24, 28, 93).

DETERMINATION OF THE REGULATORY DOMAIN IN TUBULIN mRNA   Cleveland and co-workers have studied tubulin autoregulation by transfecting cells with various tubulin cDNA constructs under control of a heterologous promoter (44). The steady-state level of mRNA from the transfected genes is then analyzed in untreated or colchicine-treated cells. These experiments revealed the following requirements for appropriate tubulin autoregulation: (a) The element responsible for autoregulated destabilization is contained in the first 13 translated nucleotides (121); (b) The encoded tetrapeptide sequence, Met-Arg-Glu-Ile (MREI), is the important recognition signal, rather than the nucleic acid sequence (122); and (c) Translation must procede through at least 41 codons for the MREI to confer proper regulation (121).

DEGRADATION OF TUBULIN mRNA IS TRANSLATION-DEPENDENT
Treatment of cells with inhibitors of translation initiation (puromycin and pactamycin, which dissociate mRNA from polysomes), or inhibitors of translation elongation (cycloheximide, anisomycin, and emetine), results in steady-state levels of tubulin mRNA that are independent of the intracellular concentration of free tubulin heterodimers (43, 86). The necessity for ongoing translation of the tubulin mRNA is consistent with the finding that the recognition signal is the N-terminal tetrapeptide of the nascent tubulin polypeptide, and that the four amino acids only confer proper regulation if translation proceeds through at least 41 codons. Presumably, the N-terminal sequence must emerge from the ribosome before it can function as a recognition signal to activate the degradation machinery.

MODELS FOR THE AUTOREGULATED DEGRADATION OF TUBULIN mRNA   Two models for the autoregulated degradation of tubulin mRNA have been postulated by Cleveland and co-workers (32, 43, 86, 121), both of which imply a direct interaction between tubulin heterodimers and the N-terminus of the newly translated nascent tubulin peptide. The first model proposes that the interaction between the free tubulin heterodimers and the nascent tubulin peptide causes ribosome stalling, resulting in unprotected gaps on the mRNA that would be targets for endoribonucleases. The second model proposes that the tubulin heterodimer/MREI complex results in the activation of a specific, ribosome-bound nuclease.

The "ribosome-stalling model" would predict endonucleolytic cleavage, as opposed to progressive 3'–5' exonucleolytic cleavage. It should be possible to differentiate between these possibilities. Thus far, a direct interaction between tubulin heterodimers and the MREI peptide has not been shown, and none of the nucleases has yet been characterized.

Cell Cycle-Controlled Histone mRNA Stability

Histone gene expression is regulated at the level of transcription, nuclear RNA processing, and mRNA stability (2, 7, 37, 45, 50, 111–113). The abundance of histone mRNAs varies 30- to 50-fold during the cell cycle (72). Histone mRNA levels increase rapidly at the onset of DNA synthesis, and decline equally rapidly at the end of DNA replication. Precise temporal and quantitative expression of the histone genes is essential (13, 54, 72, 84, 107, 108); in yeast, inappropriate histone expression can be lethal. Under- or overexpression can perturb chromatin structure and result in abnormal transcription patterns and unequal chromosome segregation (26, 49, 74). Two components are responsible for the increased steady-state level of histone mRNA during S phase; a three- to fivefold induction of histone gene transcription at the G1/S boundary, and an approximate four- to six-fold stabilization of histone mRNA during S phase (53, 54). Histone mRNA is a relatively unstable RNA even when it is stabilized during S phase, with a half-life between 15–60 min (2, 50, 71, 107, 108). The half-life of histone mRNA is too short for reliable estimates to be made when cells are out of S phase. Evidence suggests that the inherent short half-life of histone mRNA is at least in part due to the fact that histone mRNAs are not polyadenylated (67, 70, 87).

IDENTIFICATION OF SEQUENCE ELEMENTS INVOLVED IN CELL CYCLE REGULATION OF HISTONE mRNA STABILITY   Normal cell cycle destabilization of histone mRNA can be mimicked by treatment of cells with DNA synthesis inhibitors (6, 7, 13, 16, 37, 41, 42, 50). In the presence of DNA synthesis inhibitors, histone mRNA is degraded more rapidly. The sequences responsible for destabilization have been identified by transfection experiments with hybrid genes (21, 46, 67, 70, 87). The steady-state level of RNA from transfected constructs is measured in cells that are either untreated or treated with hydroxyurea, an inhibitor of DNA synthesis. A mutation that knocks out the "S phase-regulation element" results in an mRNA with a "stable" phenotype relative to the endogenous histone mRNA. These experiments revealed that a short, conserved, stem-loop structure at the extreme 3'-terminus of histone mRNA is necessary to confer DNA synthesis-dependent RNA stabilization. The predicted stem-loop consists of a 6-bp stem and a 4-bp loop, which has been widely conserved during evolution. Although the stem-loop structure is the recognition element for cell cycle regulation, certain positional and translational requirements must be met for it to function. The stem-loop must be at the 3' end of the mRNA to be a substrate for regulated degradation. If the 3' untranslated region is elongated and polyadenylated, the stem-loop no longer functions, and the mRNA is stabilized (67, 70, 87). Also, if the distance between the translation termination codon and the stem-loop is increased to more than 300 nucleotides, regulation is lost (46). Furthermore, if the normal translation termination signal is mutated so that translation continues into the 3' untranslated region, the mRNA is no longer a substrate for cell cycle regulation (21). Treatment of cells with translation inhibitors (initiation as well as elongation inhibitors) results in stabilization of histone mRNA (46, 113, 114). Unlike tubulin, however, no specific coding sequences appear to be required to regulate histone mRNA stability.

IN VITRO STUDIES ON HISTONE mRNA DEGRADATION   A cell-free system for studying mRNA decay has been developed by Ross and co-workers, and has been used to investigate histone mRNA degradation (100). The in vitro decay rates of several mRNAs with a range of in vivo half-lives have been analyzed in these polysome-containing extracts. That the relative turnover of these mRNAs is similar in vitro and in vivo suggests that this may be a useful system for studying mRNA decay and, ultimately, for identifying and characterizing the components involved in the degradation process. Histone mRNA is degraded in a 3' to 5' direction both in vitro and in vivo. The in vitro system has been useful for detecting intermediates in the decay process. After short incubation times, an RNA is detectable that is identical to mature histone mRNA, but is lacking 5 nucleotides from the 3' end (100). A second intermediate lacking 12 nucleotides is observed after continued incubation. The first decay product maps to the base of the stem, and the second to the loop region of the predicted stem-loop structure. These intermediates have also been detected in vivo (102). An exonuclease activity with substrate specificity for histone mRNA has been partially characterized in these extracts (90, 100). The exonuclease activity requires divalent cations ($Mg++$) but does not require ATP or GTP, and functions at low monovalent cation ($K+$) concentrations. Blocking the 3' hydroxyl on the histone RNA substrate by phosphorylation results in a ten- to twentyfold reduction in the decay rate. The ribonuclease activity is bound to polysomes, and can be recovered in a high salt ribosomal wash. The ribonuclease appears to show specificity for histone RNA. While unmodified histone RNA is degraded rapidly, polyadenylated histone RNA is degraded at least ten times slower, and polyadenylated globin RNA is quite stable (99). It is still unclear whether this is the only nuclease activity necessary for complete degradation of this histone message.

The in vitro decay system has also provided data supporting the hypothesis that histones are involved in autoregulating their own synthesis. Histone mRNA is degraded four- to sixfold faster in cell-free decay reactions that are supplemented with the core histones and a postribosomal S-130 fraction (91). Neither histones nor S-130 fractions added individually augment the degradation of histone mRNA. The effect of adding histones and the S-130 to the decay reactions appears to be specific, since the degradation rate of other mRNAs is not affected.

Iron-Dependent Destabilization of Transferrin Receptor mRNA

The transferrin receptor, which mediates the cellular uptake of iron by binding to its ligand transferrin, the major iron transport protein found in serum, is regulated by modulation of mRNA stability. The intracellular level of iron acts as a negative feedback regulator of transferrin receptor expression. Treatment of cells with desferrioxamine, an iron chelator, increases transferrin receptor expression, while receptor expression is reduced in cells treated with hemin or iron salts (18, 98). These iron-dependent changes in transferrin receptor expression are reflected in the levels of transferrin receptor mRNA. Transfection experiments with the transferring receptor cDNA under control of either its own promoter or the SV40 promoter have demonstrated that iron-dependent regulation is independent of transcriptional regulation. Further studies ruled out the possibility of regulation at the level of transcriptional elongation, nuclear processing, or nuclear transport, and showed that de-sferrioxamine treatment results in a twentyfold stabilization of transferrin mRNA in the cytoplasm (79, 85). Extensive deletion analysis defined two *cis*-acting domains in the 3' untranslated region of the transferrin receptor mRNA responsible for iron-dependent regulation. The two indispensable domains are separated by about 300 nucleotides of dispensable sequence. This regulatory domain contains a predicted stem-loop of about 60 nucleotides and 5 palindromic repeats. Iron-dependent regulation is completely abolished when more than one of the palindromes is deleted or if point mutations are introduced into the stem-loop. Interestingly, homologous elements identified in the 5' untranslated region of the ferritin mRNA (which codes for the heme-storage protein ferritin) regulate the translational efficiency of ferritin mRNA in response to iron (3, 51). These elements have been termed iron response elements (IREs), and are functionally interchangeable between the 3' untranslated region of the transferrin receptor and the 5' untranslated region of ferritin. Gel-shift and UV-crosslinking experiments have identified a cytoplasmic RNA-binding protein with specificity for the palindromic IREs (52, 62, 66, 80, 104). Aside from binding to IREs, this protein is also regulated by iron and has been termed iron response factor (IRF). Following desferrioxamine treatment, there is a 25-fold increase in the pool of free IRF, and the kinetics of accumulation of IRF parallel the accumulation of transfer-rin receptor mRNA. These correlations strongly suggest that IRF stabilizes transferrin receptor mRNA by binding to its IREs. When desferrioxamine is removed, IRF is inactivated and no longer binds the IRE. This inactivation of IRF is not inhibited by either cycloheximide or Actinomycin D. In contrast, iron-dependent degradation of transferrin receptor mRNA is prevented by cycloheximide or Actinomycin D. Thus, a second component regulated dif-ferently from IRF may be involved. Since IRF does not interact with the large stem-loop structure that is indispensable for iron-dependent degradation of transferrin receptor mRNA, this putative second factor may interact with the stem-loop.

Role of mRNA Turnover in Regulation of c-fos and c-myc Expression

Among the genes most studied with regard to mRNA stability are the proto-oncogenes involved in the early response to growth factors, especially *c-fos* and c-*myc*. These genes are part of a larger group that contains a structural motif common to many unstable mRNAs, an AU-rich 3' untranslated region. The investigations to date have addressed several questions: (*a*) what are the important structural features that render these mRNAs unstable; (*b*) do the rearrangements and mutations that convert these genes to oncogenes have an important effect on mRNA stability, and; (*c*) is the increase in the steady-state level of mRNA seen after treatment of quiescent cells due, in whole or in part, to transient stabilization of the mRNA?

c-*fos* is the cellular cognate of v-*fos*, the transforming gene of FBJ- and FBR-MuSV (for review, see 117). c-*fos* mRNA is induced 20–50-fold within thirty min of the addition of serum or other mitogens to quiescent fibroblasts (63, 78). This induction occurs by the transient transcriptional activation of the c-*fos* promotor, and within two hr the message is no longer detectable (48, 63, 78). Monitoring this rapid decay has permitted the direct measurement of c-*fos* mRNA turnover under conditions of serum stimulation. Analysis of c-*fos* deletion mutants, v-*fos* genes, and c-*fos*/β-globin chimeric genes has yielded some interesting results, although some contradictions in the data make interpretation difficult. First, protein synthesis inhibitors clearly block the decay of c-*fos* mRNA, independent of any transcriptional effect (47, 77, 120). The addition of protein synthesis inhibitors when the c-*fos* promotor is no longer transcriptionally active prevents the rapid decay in the steady-state level of c-*fos* mRNA (120), a result suggesting that the turnover of c-*fos* mRNA occurs by a mechanism coupled to translation.

A series of investigations have attempted to define the structural features of c-*fos* mRNA that specify rapid turnover. Experiments to address this issue have employed mutant c-*fos* genes (both chimeric genes and deletion mutants) under the transcriptional control of either the c-*fos* promotor, in which case, decay following transient transcriptional induction with serum is monitored, or a constitutive promotor, in which case, decay is measured following the addition of transcription-blocking drugs like Actinomycin D. The results of experiments using both techniques support the hypothesis that an AU-rich sequence contained in the 3' untranslated region of c-*fos* mRNA can function as a dominant element to confer instability to an otherwise stable mRNA like β-globin (40, 59, 65, 97, 99, 110, 120). The sequence identified in these experiments is similar to sequences present in the 3' untranslated region of many unstable mRNAs, including a large number of mRNAs derived from either cytokine or early response genes that are transiently expressed at specific times (1, 22, 109). The unstable mRNAs encoded by these genes all contain AU-rich 3' untranslated regions with one to several copies of the sequence AUUUA (109). These experiments have focused attention on the 3' untranslated region of many RNAs as one that may contain information specifying the rate of mRNA turnover, with the hypothesis being advanced that a 3' untranslated region with high AU content can selectively target mRNAs for rapid turnover. However, such a hypothesis may be too simple; first, *GM-CSF* mRNA, which contains eight copies of the canonical AUUUA motif in its 3' untranslated region, is stable under some conditions and not under others (109). Second, the AU-rich sequences believed to specify rapid turnover from three different mRNAs, c-*fos*, c-*myc*, and *GM-CSF*, have widely differing effects on the turnover of heterologous reporter genes in the same cell (106). Third, several stable mRNAs, including rabbit β-globin, contain this sequence in their 3' untranslated regions. Finally, deletion of the AT-rich 3' untranslated region of c-*fos* gene does not result in a stable mRNA (58, 110; J. Atwater & I. M. Verma, unpublished observations).

In addition to increasing the turnover of certain mRNAs, the AU-rich 3' untranslated sequence accelerates the removal of A residues from the poly(A) tail. The first change observed in the metabolism of c-*fos* mRNA following serum stimulation is the removal of A residues from the 3' end of the message (77, 120). Mutant c-*fos* genes with deletions of this AU-rich sequence show a decreased rate of removal of A residues from the poly(A) tail, without the appearance of specific cleavage products upstream of the poly(A) sequence (120). This suggests that the AU-rich destabilizing element functions by promoting the removal of A residues from the 3' end of the mRNA, an event known to effect the stability of some mRNAs. It is not clear if c-*fos* mRNAs with short poly(A) tails are better substrates for the degradation machinery, or if this is merely a correlation. Two possible mechanisms for increased de-adenylation of mRNAs containing AU-rich 3' untranslated regions have been advanced: the AU-rich sequences may act as a sink for PABP and expose the normally protected poly(A) tail (17), a hypothesis inconsistent with the known binding-specificity of PABP; alternatively, the U residues base-pair with A residues from the poly(A) tract to create regions of partially double-stranded RNA that are sensitive to double-stranded RNases (120).

The results discussed above imply the existence of other structural elements capable of specifying rapid turnover of mRNA. Deletion of the 3' un-translated region of c-*fos* and replacement with a heterologous polyadenyla-tion signal derived from a stable mRNA (β-globin) does not significantly slow the decay of chimeric mRNAs following serum induction (58, 110, 120). Thus, c-*fos* mRNA contains at least two dominantly acting sequences that specify mRNA instability. Preliminary experiments suggest that the second of these sequences is contained within the c-*fos* protein coding sequence, and requires translation, but it is not yet clear whether the recognition element for this process is contained in the mRNA or the nascent c-*fos* peptide (110; J. Atwater & I. M. Verma, unpublished data). Perturbation of this pathway by protein synthesis inhibitors could be involved in the increased stability of c-*fos* mRNA in the presence of protein synthesis inhibitors, but still does not explain why the AU-rich element in the 3' untranslated region fails to permit rapid turnover when protein synthesis inhibitors are present.

c-*myc* is the cellular homolog of the transforming genes of several avian retroviruses (including MC29 and OK10), and is rearranged in a wide variety of neoplasms of both humans and rodents. c-*myc* mRNA is induced 10–30-fold when serum or growth factors are added to quiescent fibroblasts (and many other cell types) (60), but there are important differences between this induction and the similar increase observed in c-*fos* mRNA. First, the peak levels of c-*myc* mRNA expression occur 2–3 hr after the addition of serum, when c-*fos* mRNA is no longer detectable, and second, many observers find that the change in the rate of c-*myc* transcription is not sufficient to account for the change in the observed steady-state level of mRNA (10, 38, 48, 61, 68, 81). Thus, a regulated change in c-*myc* mRNA stability following serum stimulation may account for a large part of the observed change in steady-state level. Experimentally, an important difference is that the lack of a large transcriptional induction precludes using the c-*myc* promotor to study stability of the mRNA.

The instability of c-*myc* mRNA has been shown in two ways: the addition of transcription-blocking drugs such as Actinomycin D results in a rapid decay of c-*myc* mRNA in a variety of cells, with a half-life of 10–30 min (12, 34, 35, 57); second, the addition of chemicals to certain cell lines results in their differentiation, and the rapid decrease in the resulting steady-state levels of c-*myc* mRNA can be used to infer a rapid turnover. For example, the addition of hexamethylene bisacetamide (HMBA) to mouse erythroleukemia (MEL) cells results in a twentyfold decrease in steady-state levels of c-*myc* mRNA within 1 hr, a decrease that would be observed only if c-*myc* mRNA were unstable (82).

Like c-*fos*, the c-*myc* gene contains a long 3' untranslated region that is AU-rich. When c-*myc* mutant genes are analyzed for mRNA stability by performing chase experiments with Actinomycin D, deletion of the 3' untranslated region results in transcripts that are degraded more slowly, and this region can confer instability to otherwise stable chimeric mRNAs (57). Thus, the c-*myc* 3' untranslated region is capable of acting in a dominant manner to render heterologous mRNAs unstable. The analysis of poly(A)+ and poly(A)− fractions of RNA after the addition of Actinomycin D to cells shows that the first observable change in c-*myc* mRNA is the removal of A residues from the 3' end of the message (115), a result similar to that seen with c-*fos* mRNA (77, 120). Whether this is related to the turnover of c-*myc* mRNA is not certain, but possible evidence is provided by the observation that a precursor-product relationship seems to exist between poly(A)+ c-*myc* mRNA and poly(A)− c-*myc* mRNA (115). As with c-*fos* mRNA, it is not clear that this decrease in length of the poly(A) tail is a necessary precursor of degradation. These experiments suggest that the removal of A residues from the poly(A) tail is the initial event in the degradation of c-*myc* mRNA (17). In vitro experiments support the involvement of a cytosolic nucleic acid component in this reaction (16), but the relevance to intracellular mRNA turnover is not yet clear.

Like c-*fos* mRNA, determinants of mRNA instability are probably present in sequences outside the 3' untranslated region of c-*myc* mRNA. Some Burkitt's lymphoma cell lines contain c-*myc* mRNAs truncated at their 5' untranslated region (frequently with substitutions of intron 1 sequences), and some data suggest that these truncated mRNAs are degraded more slowly than the mRNA derived from the wild-type gene (39, 96). It is not clear if this effect depends on changes in translational efficiency that may result from the 5' truncation, and this effect is not reproducible upon gene transfer into all cell types (57). Finally, when the c-*myc* protein coding sequence is placed under the transcriptional control of the c-*fos* promotor, the truncated c-*myc* mRNAs undergo rapid induction and degradation following serum stimulation (R. Wisdom, unpublished data), suggesting that, at least under some circumstances, highly truncated c-*myc* mRNAs are unstable. Thus, several separate elements in the c-*myc* mRNA appear to render it unstable, and this complicates the dissection of important structural motifs. Although the hypothesis that discrete elements defined in terms of the primary and secondary structure may not exist (58) is possible, the intensive investigation of other genes makes it equally possible that such elements exist but have not yet been found.

Attempts have been made to correlate the structural rearrangements of the c-*myc* gene observed in Burkitt's lymphoma with changes in the stability of c-*myc* mRNA. It seems likely that these mutations influence promotor usage and transcriptional elongation more than mRNA turnover, although Actinomycin D experiments on some Burkitt's lymphoma cell lines suggest that 5' truncated mRNAs from the translocated c-*myc* gene are more stable than those from nonrearranged gene (39, 112a).

Both c-*fos* and c-*myc* mRNA are stabilized in the presence of inhibitors of translation, an observation that can be extended to include virtually all known early response genes (1, 60, 63, 69, 78). Two factors support this being a direct result of inhibiting translation rather than the result of the decay of a labile factor. First, the stabilization of c-*fos* mRNA is virtually instantaneous with the addition of inhibitors of translation, and second, the change in the stability of c-*myc* mRNA is not seen when the translation initiation codon is altered (120; R. Wisdom, unpublished data). The existence of mRNA-destabilizing sequences within the protein coding region of both genes and the stabilization of these mRNAs in the presence of translation inhibitors suggests that the turnover of these mRNAs may be coupled to translation, as with β-tubulin mRNA.

At this juncture, it is difficult to reconcile the data regarding c-*fos* and c-*myc* mRNA turnover into a satisfying model. One important goal is to carefully define the sequence(s) contained in the protein coding sequence that can specify rapid turnover of c-*fos* and c-*myc* mRNA. This is part of the larger issue of how cells discriminate between mRNAs stabilized by translation (e.g. β-globin), and mRNAs destabilized by translation (e.g. c-*fos*). An additional goal is to clearly define the biochemical mechanisms by which this element can function both in the presence and the absence of the 3' AU-rich sequence. The mechanism by which the AU-rich 3' untranslated region functions to destabilize the mRNA remains unclear. Answers to these questions will undoubtedly be required before we can address the importance of mRNA stability in the complex regulation of c-*fos* and c-*myc* mRNA. The significance of this question is underscored by the evidence that disruption of the 3' untranslated region of the c-*fos* gene is required for its conversion to an oncogene (75). Disruption of the 3' untranslated region presumably leads to increased mRNA stability and sustained synthesis of the protein. The consequences of over-expression of the c-*fos* protein are manifested as cellular transformation in tissue culture, bone deformities and hyperplasia of the thymus in transgenic mice, and bone tumors in mice infected with retroviruses containing v-*fos* (105, 118). A hallmark of the early response genes like c-*fos*, c-*myc*, c-*jun*, c-*rel*, etc. is the short half-life of both their mRNAs and protein products. It appears that the function of these oncoproteins is required for a short window during growth, differentiation, or development, and that sustained expression has the potential to wreak havoc on the cell.

In this review, we have elected to discuss those examples of regulated mRNA stability where there is some purchase on the molecular mechanism. The biological relevance of the control of mRNA degradation as a mechanism of control of gene expression is, however, widespread. Viral infection can influence the cytoplasmic stability of both host and viral RNAs. For example, the turnover of adenoviral transcripts changes during the course of infection (119). It remains uncertain if this is linked to changes in translational efficiency because of the presence of the tripartite leader sequence. Many viruses influence stability of mRNA by inducing interferons that, in turn, lead to the induction of specific ribonucleases. Some viruses, including influenza viruses, contain specific nucleases that cleave cell mRNAs to generate oligonucleotides containing capped 5' ends that serves as primers for mRNA synthesis (94). Presumably the decapitated cellular mRNAs are destined for rapid disposal. mRNA turnover also has an important role during differentiation and development (59). The mechanism of the extreme stability of "masked" mRNA in oocytes and very rapid degradation following fertilization remains undetermined. The regulation of mRNA stability also plays an important role during the biogenesis of specialized cells, as exemplified by the stabilization of mRNAs in eye lens cells that synthesize α- and β-crystallin protein at the expense of other mRNAs. Similarly, the predominance of globin mRNAs in erythrocytes is in part due to its slow turnover. Clearly, control of mRNA metabolism has been cleverly exploited by the cell during evolution to maintain and excel in specialized functions.

DIRECTIONS FOR FUTURE WORK

As improvements in technology have made the measurement of gene expression more accurate, the critical role of control of mRNA metabolism as a regulatory point has come into clearer focus. The methods required for careful dissection of regulatory elements are now available, and the successful application of these methods to define the structural features important for the control of degradation of some mRNAs has been presented here. The nature of the *cis*-acting elements varies from the ribosome-bound nascent peptide encoded by the targeted mRNA (β-tubulin) to special sequences and secondary structures in the mRNA itself. As the number of clearly defined elements remains small, a general paradigm has not yet emerged. It seems likely that the same methods can be successfully applied to examine the turnover of other mRNAs. Careful molecular genetic dissection of important control elements will facilitate the development of biochemical systems that mimic the important elements of in vivo mRNA turnover. By analogy with the successful approach in the analysis of eukaryotic transcription factors, it should be possible to isolate important regulatory factors based on specific binding activity.

ACKNOWLEDGMENTS

This work was supported by training fellowships from the National Institutes of Health (J. A. A. and R. W.) and grants from the National Institutes of Health and American Cancer Society (I. M. V.).

Literature Cited

1. Almendral, J. M., Sommer, D., MacDonald-Bravo, H., Burckhardt, J., Perera, J., Bravo, R. 1988. Complexity of the early genetic response to growth factors. *Mol. Cell. Biol.* 8:2140–48
2. Alterman, R-B. M., Ganguly, S., Schulze, D. H., Marzluff, W. F., Schildkraut, C. L., Skoultchi, A. I. 1984. Cell cycle regulation of mouse H3 histone mRNA metabolism. *Mol. Cell. Biol.* 4:123–32
3. Aziz, N., Munro, H. N. 1986. Both subunits of rat liver ferritin are regulated at a translation level by iron induction. *Nucleic Acids Res.* 14:915–27
4. Baer, B. W., Kornberg, R. D. 1983. The protein responsible for the repeating structure of cytoplasmic poly(A) ribonucleoprotein. *J. Cell Biol.* 96:717–21
5. Baumann, B., Potash, M. J., Kohler, G. 1985. Consequences of frameshift mutations at the immunoglobulin heavy chain locus of the mouse. *EMBO J.* 4:351–59
6. Baumbach, L. L., Marashi, F., Plumb, M., Stein, G., Stein, J. 1984. Inhibition of DNA replication coordinately reduces

cellular levels of core and H1 histone mRNAs: Requirement for protein synthesis. *Biochemistry* 23:1618–25

7. Baumbach, L. L., Stein, G. S., Stein, J. 1987. Regulation of human histone gene expression: transcriptional and post-transcriptional control in the coupling of histone messenger RNA stability with DNA replication. *Biochemistry* 26:6178–87

8. Ben-Ze'ev, A., Farmer, S. R., Penman, S. 1979. Mechanisms of regulating tubulin synthesis in cultured mammalian cells. *Cell* 17:319–25

9. Bernstein, P., Peltz, S. W., Ross, J. 1989. The poly(A)-poly(A) binding protein complex is a major determinant of mRNA stability in vitro. *Mol. Cell. Biol.* 9:659–70

10. Blanchard, J-M., Piechaczyk, M., Dani, C., Chambard, J-C., Franchi, A., et al. 1985. c-myc gene is transcribed at high rate in G₀-arrested fibroblasts and is post-transcriptionally regulated in response to growth factors. *Nature* 317:443–45

11. Blobel, G. 1973. A protein of molecular weight 78,000 bound to the polyadenylate region of eukaryotic messenger RNAs. *Proc. Natl. Acad. Sci. USA* 70:924–28

12. Bonnieu, A., Piechacyk, M., Marty, L., Cuny, M., Blanchard, J., et al. 1988. Sequence determinants of c-myc mRNA turn-over: influence of 3' and 5' noncoding regions. *Oncog. Res.* 3:155–66

13. Borun, T. W., Gabrielli, F., Ajiro, I., Zweidler, A., Baglioni, C. 1975. Further evidence of transcriptional and translational control of histone messenger RNA during the HeLa S3 cycle. *Cell* 4:59–67

14. Brawerman, G. 1981. The role of poly(A) sequence in mammalian messenger RNA. *CRC Crit. Rev. Biochem.* 10:1–38

15. Breindl, M., Gallwitz, D. 1974. On the translational control of histone synthesis. Quantitation of biologically active histone mRNA from synchronized HeLa cells and its translation in different cell-free systems. *Eur. J. Biochem.* 45:91–97

16. Brewer, G., Ross, J. 1988. Poly(A) shortening and degradation of the 3' AU-rich sequences of human c-myc mRNA in a cell-free system. *Mol. Cell. Biol.* 8:1697–708

17. Brewer, G., Ross, J. 1989. Regulation of c-myc mRNA stability *in vitro* by a labile destabilizer with an essential nucleic acid component. *Mol. Cell. Biol.* 9:1996–2006

18. Bridges, K. R., Cudkowicz, A. 1989. Effect of iron chelators on the transferrin receptor in K562 cells. *J. Biol. Chem.* 259:12970–77

19. Brock, M. L., Shapiro, D. J. 1983. Estrogen stabilizes vitellogenin mRNA against cytoplasmic degradation. *Cell* 34:207–14

20. Brock, M. L., Shapiro, D. J. 1983. Estrogen regulates the absolute rate of transcription of the *Xenopus laevis* vitellogenin genes. *J. Biol. Chem.* 258:5449–55

21. Capasso, O., Bleecker, G. C., Heintz, N. 1987. Sequences controlling histone H4 mRNA abundance. *EMBO J.* 5:1825–31

22. Caput, D., Beutler, B., Hartog, K., Thayer, R., Brown-Shimer, S., Cerami, A. 1986. Identification of a common nucleotide sequence in the 3'-untranslated region of mRNA molecules specifying inflammatory mediators. *Proc. Natl. Acad. Sci. USA* 83:1670–74

23. Caron, J., Jones, A. L., Kirschner, M. W. 1985. Autoregulation of tubulin synthesis in hepatocytes and fibroblasts. *J. Cell Biol.* 101:1763–72

24. Caron, J. M., Jones, A. L., Rall, L. B., Kirschner, M. W. 1985. Autoregulation of tubulin synthesis in enucleated cells. *Nature* 317:648–51

25. Chen, C. Y., Hitzeman, R. A. 1987. Possible translational control mechanisms associated with gene expression in yeast. In *Biological Research on Industrial Yeasts III*, ed., G. G. Stewart, I. Russell, R. D. Klein, R. R. Hiebsch. Boca Raton, FL: CRC Press

26. Clark-Adams, C. D., Norris, D., Osley, M. A., Fassler, J. S., Winston, F. 1988. Changes in histone gene dosage alter transcription in yeast. *Genes Dev.* 2:150–59

27. Cleveland, D. W. 1987. The multitubulin hypothesis revisted: what have we learned? *J. Cell. Biol.* 104:381–83

28. Cleveland, D. W., Havercroft, J. C. 1983. Is apparent autoregulatory control of tubulin synthesis nontranscriptionally controlled? *J. Cell. Biol.* 97:919–24

29. Cleveland, D. W., Lopata, M. A., Sherline, P., Kirschner, M. W. 1981. Unpolymerized tubulin modulates the level of tubulin mRNAs. *Cell* 25:537–46

30. Cleveland, D. W., Pittenger, M. F., Feramisco, L. R. 1983. Elevation of tubulin leves by microinjection suppresses new tubulin synthesis. *Nature* 305:738–40

31. Cleveland, D. W., Sullivan, D. F. 1985. Molecular biology and genetics of tubulin. *Annu. Rev. Biochem.* 54:331–65

32. Cleveland, D. W., Yen, T. J. 1989. Multiple determinants of eucaryotic mRNA stability. *New Biol.* 1:121–26

33. Daar, I. O., Maquat, L. E. 1988. Premature translation termination mediates triosephosphate isomerase mRNA degradation. *Mol. Cell. Biol.* 8:802–12

34. Dani, C., Blanchard, J. M., Piechaczyk, M., El Sabouty, S., Marty, L., Jeanteur, P. 1984. Extreme instability of myc mRNA in normal and transformed human cells. *Proc. Natl. Acad. Sci. USA* 81:7046–50

35. Dani, C., Mechti, N., Piechaczyk, M., Lebleu, B., Jeanteur, P., Blanchard, J. M. 1985. Increased rate of degradation of c-myc mRNA in interferon-treated Daudi cells. *Proc. Natl. Acad. Sci. USA* 82:4896–99

36. Deleted in proof

37. DeLisle, A. J., Graves, R. A., Marzluff, W. F., Johnson, L. F. 1983. Regulation of histone mRNA production and stability in serum-stimulated mouse 3T6 fibroblasts. *Mol. Cell. Biol.* 3:1920–29

38. Dony, C., Kessel, M., Gruss, P. 1985. Post-transcriptional control of myc and p53 expression during differentiation of the embryonal carcinoma cell line F9. *Nature* 317:636–39

39. Eick, D., Piechaczyk, M., Henglein, B., Blanchard, J. M., Traub, B., et al. 1985. Aberrant c-myc RNAs of Burkitt's lymphoma cells have longer half-lives. *EMBO J.* 4:3717–25

40. Fort, P., Rech, J., Vie, A., Piechaczyk, M., Bonnieu, A., et al. 1987. Regulation of c-fos gene expression in hamster fibroblasts: initiation and elongation of transcription and mRNA degradation. *Nucleic Acids Res.* 15:5657–67

41. Gallwitz, D. 1975. Kinetics of inactivation of histone mRNA in the cytoplasm after inhibition of DNA replication in synchronized HeLa cells. *Nature* 257:247–48

42. Gallwitz, D., Mueller, G. C. 1969. Histone synthesis in vitro by cytoplasmic microsomes from HeLa cells. *Science* 163:1351–53

43. Gay, D. A., Sisodia, S. S., Cleveland, D. W. 1989. Autoregulatory control of β-tubulin mRNA stability is linked to translation elongation. *Proc. Natl. Acad. Sci. USA* 86:5763–67

44. Gay, D. A., Yen, T. J., Lau, J. T. Y., Cleveland, D. W. 1987. Sequences that confer β-tubulin autoregulation through modulated mRNA stability reside within exon 1 of a β-tubulin mRNA. *Cell* 50:671–79

45. Graves, R. A., Marzluff, W. F. 1984. Rapid reversible changes in the rate of

histone gene transcription and histone mRNA levels in mouse myeloma cells. *Mol. Cell. Biol.* 4:351–57

46. Graves, R. A., Pandey, N. B., Chodchoy, N., Marzluff, W. F. 1987. Translation is required for regulation of histone mRNA degradation. *Cell* 48:615–26

47. Greenberg, M. E., Hermanowski, A. L., Ziff, E. B. 1986. Effect of protein synthesis inhibitors on growth factor activation of c-fos, c-myc, and actin gene transcription. *Mol. Cell. Biol.* 6:1050–57

48. Greenberg, M. E., Ziff, E. B. 1984. Stimulation of mouse 3T3 cells induces transcription of the c-fos oncogene. *Nature* 311:438–42

49. Han, M., Chang, M., Kim, U.-J., Grunstein, M. 1987. Histone H2B repression causes cell cycle-specific arrest in yeast: effects on chromosomal segregation, replication, and transcription. *Cell* 4:589–97

50. Heintz, N., Sive, H. L., Roeder, R. G. 1983. Regulation of human histone gene expression: Kinetics of accumulation and changes in the rate of synthesis and in the half-lives on individual histone mRNAs during the HeLa cell cycle. *Mol. Cell. Biol.* 3:539–50

51. Hentze, M. W., Caughman, S. W., Rouault, T. A., Barriocanal, J. G., Dancis, A., et al. 1987. Identification of the iron-responsive element for the translational regulation of human ferritin mRNA. *Science* 238:1570–73

52. Hentze, M. W., Rouault, T. A. Harford, J. B., Klausner, R. D. 1989. Oxidation-reduction and the molecular mechanism of a regulatory RNA-protein interaction. *Science* 244:357–59

53. Hereford, L., Bromley, S., Osley, M. A. 1982. Periodic transcription of yeast histone genes. *Cell* 30:305–10

54. Hereford, L. M., Osley, M. A., Ludwig, J. R., McLaughlin, C. S. 1981. Cell-cycle regulation of yeast histone mRNA. *Cell* 24:367–75

55. Hoekema, A., Kastelien, R. A., Vasser, M., DeBoer, H. A. 1987. Codon replacement in the PGK1 gene of *Saccharomyces cerevisiae:* experimental approach to study the role of biased codon usage in gene expression. *Mol. Cell. Biol.* 7:2914–24

56. Huez, G., Bruck, C., Cleuter, Y. 1981. Translational stability of native and deadenylated rabbit globin mRNA injected into HeLa cells. *Proc. Natl. Acad. Sci. USA* 78:908–11

57. Jones, T. R., Cole, M. D. 1987. Rapid cytoplasmic turnover of c-myc mRNA:

requirement of the 3' untranslated sequences. *Mol. Cell. Biol.* 7:4513–21

58. Kabnick, K. S., Housman, D. E. 1988. Determinants that contribute to cytoplasmic stability of human c-fos and beta-globin mRNAs are located at several sites in each mRNA. *Mol. Cell. Biol.* 8:3244–50

59. Kafatos, F. 1972. mRNA stability and cellular differentiation. *Acta Endocrinol.* 168:319–45

60. Kelly, K., Cochran, B. A., Stiles, C. D., Leder, P. 1983. Cell specific regulation of the c-myc gene by lymphocyte mitogens and platelet-derived growth factor. *Cell* 35:603–10

61. Knight, E., Anton, E. D., Fahey, D., Friedland, B. K., Jonak, G. K. 1985. Interferon regulated c-myc gene expression in Daudi cells at the posttranscriptional level. *Proc. Natl. Acad. Sci. USA* 82:1151–54

62. Koeller, D. M., Casey, J. L., Hentze, M. W., Gerhardt, E. M., Chan, L.-N. L., et al. 1989. A cytosolic protein binds to structural elements within the iron regulatory region of the transferrin receptor mRNA. *Proc. Natl. Acad. Sci. USA* 86:3574–78

63. Kruijer, W., Cooper, J. A., Hunter, T., Verma, I. M. 1984. Platelet-derived growth factor induces rapid but transient expression of the c-fos gene and protein. *Nature* 312:711–16

64. Lau, J. T. Y., Pittenger, M. F., Cleveland, D. W. 1985. Reconstruction of appropriate tubulin and actin gene regulation following transient transfection of cloned β-tubulin and β-actin genes. *Mol. Cell. Biol.* 5:1611–20

65. Lee, W. M. F., Lin, C., Curran, T. 1988. Activation of transforming potential of the human fos proto-oncogene requires message stabilization and results in increased amounts of partially modified fos protein. *Mol. Cell. Biol.* 8:5521–27

66. Leibold, E. A., Munro, H. N. 1988. Cytoplasmic protein binds in vitro to a highly conserved sequence in the 5' untranslated region of ferritin heavy- and light-subunit mRNAs. *Proc. Natl. Acad. Sci. USA* 85:2171–75

67. Levine, B. J., Chodchoy, N., Marzluff, W. F., Skoultchi, A. I. 1987. Coupling of replication type histone mRNA levels to DNA synthesis requires the stem-loop seqeunce at the 3' end of the mRNA. *Proc. Natl. Acad. Sci. USA* 84:6189–93

68. Levine, R. A., McCormack, J. E., Buckler, A., Sonenshein, G. E. 1986. Transcriptional and posttranscriptional control of c-myc gene expression in WEHI 231 cells. *Mol. Cell. Biol.* 6:4112–16

69. Linial, M., Gunderson, N., Groudine, M. 1985. Enhanced transcription of c-myc in bursal lymphoma cells requires continuous protein synthesis. *Science* 230:1126–30

70. Luscher, B., Stauber, C., Schindler, R., Schumperli, D. 1985. Faithful cell cycle regulation of a recombinant mouse histone H4 gene is controlled by sequences in the 3'-terminal part of the gene. *Proc. Natl. Acad. Sci. USA* 82:4389–93

71. Maquat, L. E., Kinniburgh, A. J., Rachmilewitz, E. A., Ross, J. 1981. Unstable β-globin mRNA in mRNA-deficient β⁰ thalassemia. *Cell* 27:543–53

72. Marzluff, W. F., Pandey, N. B. 1988. Multiple regulatory steps control histone mRNA concentrations. *Trends Biochem. Sci.* 13:49–52

73. Mason, J. O., Williams, G. T., Neuberger, M. S. 1988. The half-life of immunoglobulin mRNA increases during B-cell differentiation: a possible role for targeting to membrane-bound polysomes. *Genes Dev.* 2:1003–11

74. Meeks-Wagner, D., Hartwell, L. H. 1986. Normal stoichiometry of histone dimer sets is necessary for high fidelity of mitotic chromosome transmission. *Cell* 44:43–52

75. Meijlink, F., Curran, T., Miller, A. D., Verma, I. M. 1985. Removal of a 67 base-pair sequence in the non-coding region of protooncogene fos converts it to a transforming protein. *Proc. Natl. Acad. Sci. USA* 82:4987–91

76. Mercer, J. F. B., Wake, S. A. 1985. An analysis of the rate of metallothionein poly(A)-shortening using RNA blot hybridization. *Nucleic Acids Res.* 13:7929–43

77. Mitchell, R. L., Henning-Chubb, C., Huberman, E., Verma, I. M. 1986. c-fos Expression is neither sufficient nor obligatory for differentiation of monomyelocytes to macrophages. *Cell* 45:497–504

78. Muller, R., Bravo, R., Burckhardt, J., Curran, T. 1984. Induction of c-fos gene and protein by growth factors precedes activation of c-myc. *Nature* 312:716–20

79. Mullner, E. W., Kuhn, L. C. 1988. A stem-loop in the 3' untranslated region mediates iron-dependent regulation of transferrin receptor mRNA stability in the cytoplasm. *Cell* 53:815–25

80. Mullner, E. W., Neupert, B., Kuhn, L. C. 1989. A specific mRNA-binding factor regulates the iron-dependent stability of cytoplasmic transferrin receptor mRNA. *Cell* 58:373–82

81. Nepveu, A., Levine, R. A., Campisi, J., Greenberg, M. E., Ziff, E. B., Marcu, K. B. 1987. Alternative modes of c-myc regulation in growth factor-stimulated and differentiating cells. *Oncogene* 1:243–50

82. Nepveu, A., Marcu, K. B., Skoultchi, A. I., Lachman, H. M. 1987. Contributions of transcriptional and post-transcriptional mechanisms to the regulation of c-myc expression in mouse erythroleukemia cells. *Genes Dev.* 1:938–45

83. Nudel, U., Soreq, H., Littauer, U. Z., Marbaix, G., Huez, G., et al. 1976. Globin mRNA species containing poly(A) segments of different lengths. Their functional stability in *Xenopus* oocytes. *Eur. J. Biochem.* 64:115–21

84. Old, R. W., Woodland, H. R. 1984. Histone genes: Not so simple after all. *Cell* 38:624–26

85. Owen, D., Kuhn, L. C. 1987. Noncoding 3' sequences of the transferrin receptor gene are required for mRNA regulation by iron. *EMBO J.* 6:1287–93

86. Pachter, J. S., Yen, T. J., Cleveland, D. W. 1987. Autoregulation of tubulin expression is achieved through specific degradation of polysomal tubulin mRNAs. *Cell* 51:283–92

87. Pandey, N. B., Marzluff, W. F. 1987. The stem-loop structure at the 3' end of histone mRNA is necessary and sufficient for regulation of histone mRNA stability. *Mol. Cell. Biol.* 7:4557–59

88. Pei, R., Calame, K. 1988. Differential stability of c-myc mRNAs in a cell-free system. *Mol. Cell. Biol.* 8:2860–68

89. Piechaczyk, M., Yang, J-Q., Blanchard, J. M., Jeanteur, P., Marcu, K. B. 1985. Posttranscriptional mechanisms are responsible for accumulation of truncated c-myc RNAs in murine plasma cell tumors. *Cell* 42:588–97

90. Peltz, S. W., Brewer, G., Kobs, G., Ross, J. 1987. Substrate specificity of the exonuclease activity that degrades H4 histone mRNA. *J. Biol. Chem.* 262:9382–88

91. Peltz, S. W., Ross, J. 1987. Autogenous regulation of histone mRNA decay by histone proteins in a cell-free system. *Mol. Cell. Biol.* 7:4345–56

92. Perry, R. P., Kelley, D. C. 1979. Immunoglobulin messenger RNAs in murine cell lines that have characteristics of immature B lymphocytes. *Cell* 18:1333–39

93. Pittenger, M. F., Cleveland, D. W. 1985. Retention of autoregulatory control of tubulin synthesis in cytoplasts: Demonstration of a cytoplasmic mechanism that regulates the level of tubulin expression. *J. Cell Biol.* 101:1941–52

94. Plotch, S. J., Bouloy, M., Ulmanen, I., Krug, R. M. 1981. A unique cap (m7GpppXm)-dependent influenza virion endonuclease cleaves capped RNAs to generate the primers that initiate viral RNA transcription. *Cell* 23:847–58

95. Pontecorvi, A., Tata, J. R., Phyillaier, M., Robbins, J. 1988. Selective degradation of mRNA: the role of short-lived proteins in differential destabilization of insulin-induced creatine phosphokinase and myosin heavy chain mRNAs during rat skeletal muscle L6 cell differentiation. *EMBO J.* 7:1489–95

96. Rabbitts, P. H., Forster, A., Stinson, M. A., Rabbitts, T. H. 1985. Truncation of exon 1 from the c-myc gene results in prolonged c-myc mRNA stability. *EMBO J.* 4:3727–33

97. Rahmsdorf, H. J., Schonthal, A., Angel, P., Litfin, M., Ruther, U., Herrlich, P. 1987. Posttranscriptional regulation of c-fos mRNA expression. *Nucleic Acids Res.* 15:1643–59

98. Rao, K., Harford, J. B., Rouault, T., McClelland, A., Ruddle, F. H., Klausner, R. D. 1986. Transcriptional regulation by iron of the gene for the transferrin receptor. *Mol. Cell. Biol.* 6:236–240

99. Raymond, V., Atwater, J. A., Verma, I. M. 1989. Removal of an mRNA destabilizing element correlates with the increased oncogenicity of proto-oncogene fos. *Oncog. Res.* 5:1–12

100. Ross, J., Kobs, G. 1986. H4 histone messenger RNA decay in cell-free extracts initiates at or near the 3' terminus and proceeds 3' to 5'. *J. Mol. Biol.* 188:579–93

101. Ross, J., Kobs, G., Brewer, G., Peltz, S. W. 1987. Properties of the exonuclease activity that degrades H4 histone mRNA. *J. Biol. Chem.* 262:9374–81

102. Ross, J., Peltz, S. W., Kobs, G., Brewer, G. 1986. Histone mRNA degradation in vivo: The first detectable step occurs at or near the 3' terminus. *Mol. Cell. Biol.* 6:4362–71

103. Ross, J., Pizarro, A. 1983. Human beta and delta globin messenger RNAs turn over at different rates. *J. Mol. Biol.* 167:607–17

104. Rouault, T. A., Hentze, M. W., Caughman, S. W., Harford, J. B., Klausner, R. D. 1988. Binding of a cytosolic protein to the iron-responsive element of human ferritin mRNA. *Science* 241:1207–10

105. Ruther, U., Wagner, E. F., Muller, R. 1987. Deregulated c-fos expression in-

terferes with normal bone development in transgenic mice. *Nature* 325:412–16

106. Schuler, G. D., Cole, M. D. 1988. GM-CSF and oncogene mRNA stabilities are independently regulated in *trans* in a mouse monocytic tumor. *Cell* 55:1115–22

107. Schumperli, D. 1986. Cell-cycle regulation of histone gene expression. *Cell* 45:471–72

108. Schumperli, D. 1988. Multilevel regulation of replication-dependent histone genes. *Trends Gen.* 4:187–91

109. Shaw, G., Kamen, R. 1986. A conserved AU sequence from the 3' untranslated region of GM-CSF mRNA mediates selective mRNA degradation. *Cell* 46:659–67

110. Shyu, A-B., Greenberg, M. E., Belasco, J. G. 1989. The c-fos transcript is targeted for rapid decay by two distinct mRNA degradation pathways. *Genes Dev.* 3:60–72

111. Sittman, D. B., Graves, R. A., Marzluff, W. F. 1983. Histone mRNA concentrations are regulated at the level of transcription and mRNA degradation. *Proc. Natl. Acad. Sci. USA* 80:1849–53

112. Sive, H. L., Heintz, N., Roeder, R. G. 1984. Regulation of human histone gene expression during the HeLa cell cycle requires protein synthesis. *Mol. Cell. Biol.* 4:2723–34

112a. Spencer, C., LeStrange, R. C., Novak, U., Hayward, W. S., Groudine, M. 1990. The block to transcription elongation is promoter-dependent in normal and Burkitt's lymphoma *c-myc* alleles. *Genes Dev.* 4:75–88

113. Stimac, E., Groppi, V. E., Coffino, P. 1983. Increased histone mRNA levels during inhibition of protein synthesis. *Biochem. Biophys. Res. Commun.* 114:131–37

114. Stimac, E., Groppi, V. E., Coffino, P.

1984. Inhibition of protein synthesis stabilizes histone mRNA. *Mol. Cell. Biol.* 4:2082–90

115. Swartout, S. G., Kinniburgh, A. J. 1989. c-myc mRNA degradation in growing and differentiating cells: possible alternate pathways. *Mol. Cell. Biol.* 9:288–85

116. Treisman, R. 1985. Transient accumulation of c-fos RNA following serum stimulation requires a conserved 5' element and c-fos 3' sequences. *Cell* 42:889–902

117. Verma, I. M., Graham, W. R. 1987. The fos oncogene. *Adv. Cancer Res.* 49:29–52

118. Wagner, E. F., Williams, R. L., Ruther, U. 1989. c-fos and polyoma middle T oncogene expression in transgenic mice and embryonal stem cell chimeras. In *Cell to Cell Signals in Mammalian Development, NATO ASI Ser.* ed. S. W. deLatt, J. G. Bluemink, C. L. Mummery. H26:301–10. Berlin: Springer-Verlag

119. Wilson, M. C., Darnell, J. E. 1981. Control of messenger RNA concentration by differential cytoplasmic half-life. *J. Mol. Biol.* 148:231–51

120. Wilson, T., Treisman, R. 1988. Removal of poly(A) and consequent degradation of c-fos mRNA facilitated by AU-rich sequences. *Nature* 336:396–99

121. Yen, T. J., Gay, D. A., Pachter, J. S., Cleveland, D. W. 1988. Autoregulated changes in stability of polyribosome-bound β-tubulin mRNAs are specified by the first 13 translated nucleotides. *Mol. Cell. Biol.* 8:1224–35

122. Yen, T. J., Machlin, P. S., Cleveland, D. W. 1988. Autoregulated instability of β-tubulin mRNAs by recognition of the nascent amino terminus of β-tubulin. *Nature* 334:580–85

Annu. Rev. Genet. 1990. 24:543–78

MOLECULAR MECHANISMS REGULATING *DROSOPHILA* P ELEMENT TRANSPOSITION

Donald C. Rio

Whitehead Institute for Biomedical Research, Nine Cambridge Center, Cambridge, Massachusetts 02142 and Department of Biology, Massachusetts Institute of Technology, Cambridge, Massachusetts 02139

KEY WORDS: mobile DNA, RNA splicing, gene regulation, DNA rearrangements, negative control

CONTENTS

543

0066-4197/90/1215-0543$02.00

PERSPECTIVES AND SUMMARY

Transposable elements are ubiquitous among both eukaryotic and prokaryotic organisms (4). Genetic methods have allowed the detection, identification, and analysis of these mobile genetic elements in several eukaryotic organisms such as maize, worms, yeast, and *Drosophila melanogaster*. Because a large fraction of the *Drosophila* genome is composed of dispersed middle repetitive DNA sequences and because a high proportion of spontaneous mutations are caused by transposable element insertions (81), *Drosophila* has provided a wealth of information regarding the diversity of transposable elements in metazoan organisms. Various *Drosophila* transposable elements have been described that can be grouped into several different structural classes (71), of which the most distinct perhaps is the P element family. In contrast to most other types of transposons, P element transposition is regulated genetically and occurs in a tissue-specific manner (17, 20).

P transposable elements were discovered in *Drosophila* as the causative agents of a genetic syndrome called hybrid dysgenesis (9, 17, 19). Molecular analysis of P elements led to the development of a method of germline transformation in *Drosophila* (73). Today, P elements are widely used as genetic tools and have revolutionized *Drosophila* molecular genetics. Consequently, this family of transposable elements is one of the best studied eukaryotic metazoan nonretroviral transposons. Although much descriptive information has been compiled about P elements, the detailed molecular mechanisms that control their transposition and determine their tissue specificity remain largely unknown. This review summarizes recent biochemical and genetic experiments aimed at understanding the mechanism of P element transposition, its genetic control, and its tissue specificity. A more detailed discussion of the classical genetic observations concerning P elements and hybrid dysgenesis can be found in previous reviews (9, 17, 19).

P ELEMENTS AND THE GENETIC DISEASE OF HYBRID DYSGENESIS

In the early 1960s population geneticists first observed deleterious effects during interstrain crosses between wild and laboratory cultures (28). Subsequently, a diverse array of genetic abnormalities, including high rates of mutation, recombination in males, sterility resulting from aberrant gonadal development, and a variety of chromosomal rearrangements were observed when males isolated from wild populations (P strains) were mated to females from laboratory populations (M strains) (9, 17). These abnormalities were restricted to the germlines of the progeny from this so-called dysgenic cross and these phenotypes were collectively termed "hybrid dysgenesis" (36). The mutations caused by dysgenic crosses were often unstable, suggesting that they were due to transposable element insertions (9, 17).

The idea that hybrid dysgenesis resulted from mobilization of transposable elements was tested directly by transposon tagging (7) to induce mutations in the *white (w)* locus following a dysgenic cross (6, 72). These experiments led to the molecular isolation of P transposable elements and the finding that they were the causative agents responsible for hybrid dysgenesis. These studies also showed that reversion of dysgenic mutations was accompanied by loss of P element DNA insertions. Multiple P element insertions (perhaps 30 to 50) were found in P strains but were absent in M strains. In addition, some strains carried P element DNA sequences yet lacked the ability to induce hybrid dysgenesis: these strains were termed pseudo-M or M' (19, 72, 77). A subset of M' strains, called Q strains, were discovered that could suppress the gonadal sterility associated with hybrid dysgenesis, but not other phenotypes of dysgenesis (19). Some Q strains also exhibit low levels of transposase activity.

The isolation of P element DNA sequences led to the molecular characterization of the P element family of transposable elements. This analysis revealed that P elements are of two structurally distinct types: large, so-called complete elements 2907 bp in length, and smaller nonautonomous elements that are heterogeneous in size, ranging from about 500–2500 bp (57) (Figure 1A). In a typical P strain, there are 10–15 complete P elements and about 30–40 smaller elements, apparently derived from the complete elements by internal deletion. The complete 2.9 kb P elements contain four protein-coding open reading frames, designated ORF0, ORF1, ORF2, and ORF3 (57) which are transcribed to yield an mRNA species of approximately 2.5 kb (32) (Figure 2). In somatic cells, removal of two intervening sequences (IVS or introns) allows the synthesis of a 66 kd protein (Figure 2). However, this 66 kd protein-encoding mRNA may not be limited to the soma (48). In the germline, three introns are removed and an 87 kd protein is synthesized (40, 65) (Figure 2). All P elements analyzed carry perfect inverted 31 bp terminal repeats, create an 8 bp target-site duplication upon insertion, and contain an additional internal 11 bp inverted repeat near each end (57) (Figure 1B). Both the terminal 31 bp and internal 11 bp inverted repeat sequence elements play an important role in P element transposition (53).

The identification of complete 2.9 kb P elements allowed a direct test of the hypothesis that P elements were the causative agents of hybrid dysgenesis because they encoded an activity that could destabilize mutations induced during dysgenic crosses. This test involved microinjection of plasmid DNA carrying a 2.9 kb P element into embryos that had the *singed-weak (sn^w)* allele of the *singed* bristle *(sn)* locus (82). The *sn^w* allele had been isolated from a dysgenic cross and was known to be hypermutable when crossed to a P strain

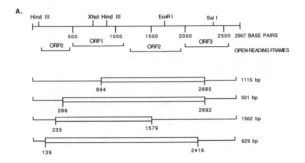

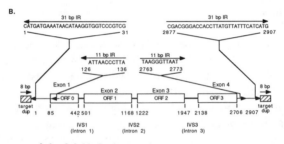

Figure 1 Structure of the 2.9 kb P element. A. 2.9 kb complete P element and several nonautonomous deletion derivatives (redrawn from data in Ref. 57). Deletions are shown in boxed areas with endpoint nucleotides indicated.   B. Location of ORFs or exons, introns (IVS), and terminal 31 bp and internal 11 bp inverted repeats (redrawn from data in Ref. 19).

(16). The sn^w mutation, caused by the tandem insertion of two small P elements at the *singed* locus (70), mutates during a dysgenic cross to either a wild-type (sn$^+$) phenotype or a more extreme (sn^e) bristle phenotype, depending upon which one of the two P elements at sn^w excises (18, 70). Mutability of the sn^w allele is a sensitive assay for P element transposase function (18), since sn^w is completely stable in an M strain background, yet the allele mutates at high frequency when crossed to strains that produce transposase activity (16, 18). Microinjection of 2.9 kb P element DNA into sn^w (M) embryos led to sn^w mutability, indicating that the 2.9 kb P elements encoded a function (transposase) capable of mobilizing the small nonautonomous P element insertions at sn^w, just as when sn^w (M) was crossed to a P strain during a dysgenic cross (82). This activity was termed transposase by analogy to the gene products of bacterial transposable elements that are required to catalyze transposition.

The finding that microinjection of 2.9 kb P element DNA into pre-blastoderm embryos induced sn^w mutability led to the development of P

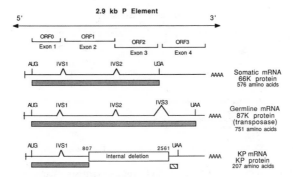

Figure 2   Structure of the P element mRNA and proteins. Four open reading frames (ORFs) or exons are shown with three introns (IVS) joining them. Translation initiation and termination codons are shown, as is the KP element internal deletion.

element-mediated germline transformation (73). This procedure involves injection of pre-blastoderm embryos in the posterior pole with two plasmid DNAs: one encoding transposase and the other carrying a phenotypic genetic marker within a small, internally deleted P element. Such a small nonautonomous P element contains the terminal *cis*-acting DNA sequences required for transposition, but does not encode transposase. Transposase, encoded by the first plasmid, catalyzes transposition of the marked P element from the injected plasmid DNA into the germline chromosomes. Following mating of the flies derived from the injected embryos, transformation is scored by the presence of genetic markers within the P element. This procedure results in a typical transformation frequency of 10–30% (80) and is now widely used in many *Drosophila* laboratories.

GENETIC CONTROL OF P ELEMENT TRANSPOSITION

The syndrome of hybrid dysgenesis is only observed when P strain males are mated to M strain females. No dysgenic traits are observed if M strain males are mated to P strain females, or in M cross M or P cross P matings, indicating that P element transposition is genetically controlled (Figure 3). These observations have led to the definition of two regulatory states: M cytotype and P cytotype (17, 19). P cytotype refers to the ability of eggs derived from P strain females to repress transposition of P elements brought into the egg with P strain sperm (Figure 3). Likewise, M cytotype refers to the permissive environment in eggs derived from M strain mothers that allows P element transposition to occur when these eggs are fertilized by P strain sperm (Figure 3). P cytotype can be thought of as exhibiting a maternal effect because the phenotype (e.g. sterility) of progeny derived from dysgenic or

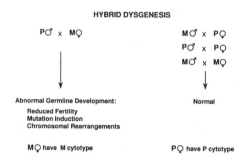

HYBRID DYSGENESIS

P♂ × M♀	M♂ × P♀
	P♂ × P♀
	M♂ × M♀
↓	↓
Abnormal Germline Development: Reduced Fertility Mutation Induction Chromosomal Rearrangements	Normal
M♀ have M cytotype	P♀ have P cytotype

Figure 3 Genetic control and maternal effect of hybrid dysgenesis. P strain males mated to M strain females result in symptoms of hybrid dysgenesis, whereas reciprocal M strain males mated to P strain females or PxP or MxM progeny are normal. P strain females have P cytotype, the ability to repress transposition of P elements brought into the oocyte by P strain sperm, whereas M strain females possess M cytotype and fail to repress transposition of P elements from P strain sperm.

nondysgenic crosses can be influenced by the female genotype (P or M). However, cytotype can also be inherited for several generations through the female germline (17, 19), in some ways resembling cytoplasmic inheritance; for convenience this aspect of cytotype will be referred to as maternal inheritance of cytotype. Both the maternal effect and maternal inheritance of P cytotype correlate with the presence of P element-containing chromosomes (17, 19) and P element DNA (84), suggesting that P cytotype is brought about by P elements themselves (17, 19).

These observations led to the proposal that, in addition to transposase, P elements encode a second function that negatively regulates transposition in P strains (57). One candidate for such a negative regulator is the 66 kd protein normally made in somatic cells from the complete 2.9 kb P elements (65) (Figure 2). Consistent with this hypothesis, it is known that P cytotype is also capable of repressing transposase activity in the soma (20, 68, 69). Recently, P element-mediated transformation has been used to show that the 66 kd protein acts as a negative regulator of transposase activity (48) (see below). However, none of the 66 kd protein-producing transformants mimic the maternal effect observed with true P strains (see below). Thus, some of the 30–40 small internally deleted P elements present in P strains may also modulate transposase activity (8, 78). The small KP element (see below) is one likely candidate for such a repressing element (8, 29).

The simple idea that M strains contain neither transposase nor repressor activity, while P strains contain both transposase activity and the repressing activity of P cytotype, was complicated by the discovery of so-called pseudo-M (or M') strains (17, 19, 72, 77). M' strains are known to carry P element DNA sequences yet usually behave as M strains during dysgenic crosses.

Table 1   Genetic assays for P cytotype repression

Assays	Tissues	References
Singed weak (sn^w) test	germline	18, 32, 82
Gonadal dysgenic sterility	germline	9, 35
Singed female sterility (cytotype-dependent alleles)	germline	68
Singed-weak (sn^w) bristle mosaics	soma	40, 68
P[w$^+$] white gene excision eye color mosaics	soma	40
P[w$^+$] white gene transposition eye color mosaics	soma	40
Modified P[w$^+$] white gene expression	soma	11
Suppression of $\Delta 2-3 \times$ Birm2 lethality	soma	20, 68
Singed bristle phenotype (cytotype-dependent alleles)	soma	68
Vestigial wing phenotype (cytotype-dependent alleles)	soma	68, 89

However, some M' strains can inhibit certain phenotypes characteristic of hybrid dysgenesis, e.g. gonadal dysgenic (GD) sterility (8, 29, 77, 78). These observations suggest that different dysgenic phenotypes might require distinct thresholds of transposase activity for the production of a phenotypic effect. For instance, sterility might require a certain level of transposase activity in a relatively large number of cells, and therefore its suppression might be easily detected, whereas other phenotypes (e.g. mutations) that occur at the cellular level but are measured in a population might be difficult to suppress, or at least their suppression might be difficult to detect. One problem with the interpretation of genetic studies of P element regulation is that a variety of different assays have been used to detect repression of transposase activity and variability has been observed among different strains depending upon the particular assay used (Table 1).

Many of the initial genetic observations on M' strains were not easy to understand. The recent molecular analysis of the internally deleted P elements present in these strains has provided some insight into the underlying basis of the partial P strain repression properties observed. For instance, some examples of M' strain transpositional repression do not exhibit the directionality typical of P strains, i.e. repression can be inherited from either males or females (8). Certain strains, such as the KP M' strain, appear to repress P element activity when mated to P strain males, yet do not exhibit the maternal inheritance of repression in subsequent generations characteristic of P cytotype (8, 29), suggesting that M' strains can affect P element transposition in a manner different from the maternally inherited P cytotype. However, recent studies examining several inbred lines derived from the Sexi M' strain indicate that other M' strains may show maternal inheritance of repression typical of P strains (78). Significant variation in repression effects was observed among the different inbred Sexi lines tested; in fact, some of these inbred lines actually exhibited low levels of transposase activity (78). The variability observed among these lines might be due to different chromosomal positions of deleted "repressor-producing" elements and of the full-length P elements in the Sexi strain. Supporting the idea that smaller internally deleted elements are producing a negative regulator of transposition is the observation that both the KP and the Sexi strains have numerous copies of one particular small P element. Although some Sexi sublines showed no repression, this observation might be explained if the genomic position of P elements differs in the lines and affects the temporal and/or spatial expression pattern of repressor proteins.

Molecular characterization of P elements in the KP M' strain indicated that a substantial fraction were of a single type termed KP elements (8). DNA sequence analysis indicated that KP elements are 1.1 kb in size and encode a 207 amino acid protein, containing the first 199 amino acids of the transposase amino terminus and an additional unrelated 8 amino acids at the carboxyl terminus due to a frameshift caused by the internal deletion (Figure 2). Interestingly, analysis of the KP protein sequence indicated a heptad leucine repeat at amino acids 101–122 (57, 66), suggesting a possible mechanism for the action of the KP protein in repression of transposition (Figure 4A and 4B). The so-called "leucine zipper" motif has been found in other proteins, including a number of DNA binding proteins that associate through alpha-helical regions as coiled coils (38, 59). In many DNA binding proteins this motif has been shown to function in protein homo- or heterodimerization rather than in the direct interaction with DNA. Thus, the presence of the leucine zipper motif suggests that the KP protein might associate with transposase to form an inactive hetero-dimer or oligomer complex (38, 59), thereby producing a dominant negative phenotype (27) perhaps by poisoning transposase protein-protein interactions with itself or with other cellular proteins involved in transposition (see section below on negative regulation of transposition; Figures 7B and 7C). The KP protein could function as a repressor without maternal inheritance of cytotype if sufficient zygotic expression of the KP protein following fertilization and the onset of zygotic transcription allowed it to associate with transposase to reduce transposition frequency. It should also be noted that the KP strain suppresses gonadal dysgenic (GD) sterility, a phenotype that first manifests itself at the larval stage. Thus, expression of a repressor protein during oogenesis might not be sufficient to suppress transpositional events in the larva that cause GD sterility, for example if the maternal repressor protein is degraded during embryogenesis. However, genetic experiments suggesting that KP elements might be responsible for the repression of GD sterility observed in M' strains have thus far not been definitive because a single KP element has not been shown to mediate repression (8, 29). The model in which a leucine zipper motif mediates protein-protein interactions would hold for proteins encoded

```
     1   MKYCKFCCKAVTGVKLIHVPKCAIKRKLWEQSLGCS    36
    37   LGENSQICDTHFNDSQWKAAPAKGQTFKRRRLNADA    72
    73   VPSKVIEPEPEKIKEGYTSGSTQTESCSᴸFNENKSᴸ   108
   109   REKIRTᴸEYEMRRᴸEQQLRESQQLEESLRKIFTDTQ   144
   145   IRILKNGGQRATFNSDDISTAICLHTAGPRAYNHLY   180
   181   KKGFPLPSRTTLYRWLSDVDIKRGCLDVVIDLMDSD   216
   217   GVDDADKLCVLAFDEMKVAAAFEYDSSADIVYEPSD   252
   253   YVQLAIVRGLKKSWKQPVFFDFNTRMDPDTᴸNNILR   288
   289   KᴸHRKGYLⱽVAIVSDᴸGTGNQKᴸWTELGISESKTWF   324
   325   SHPADDHLKIFVFSDTPHLIKLVRNHYVDSGLTING   360
   361   KKLTKKTIQEALHLCNKSDLSILFKINENHINVRSL   396
   397   AKQKVKLATQLFSNTTASSIRRCYSLGYDIENATET   432
   433   ADFFKLMNDWFDIFNSKLSTSNCIECSQPYGKQLDI   468
   469   QPDILNRMSEIMRTGILDKPKRLPFQKGᴱIVNNASᴸ   504
   505   DGLYKYᴸQENFSM@YILTSRᴱNQDIVEHFFGSMRSR   540
   541   GGQFDHPTPLQFKYRLRKYIIARNTEMLRNSGNIEE   576
                                     GMTNLKECVNKNVIP
   577   DNSESWLNLDFSSKENENKSKDDEPVDDEPVDEMLS   612
   613   NIDFTEMDELTEDAMEYIAGYVIKKLRISDKVKENL   648
   649   TFTYVDEVSHGGLIKPSEKFQEKLKELECIFLHYTN   684
   685   NNNFEITNNVKEKLILAARNVDVDKQVKSFYFKIRI   720
   721   YFRIKYFNKKIEIKNQKQKLIGNSKLLKIKL        751
```

A

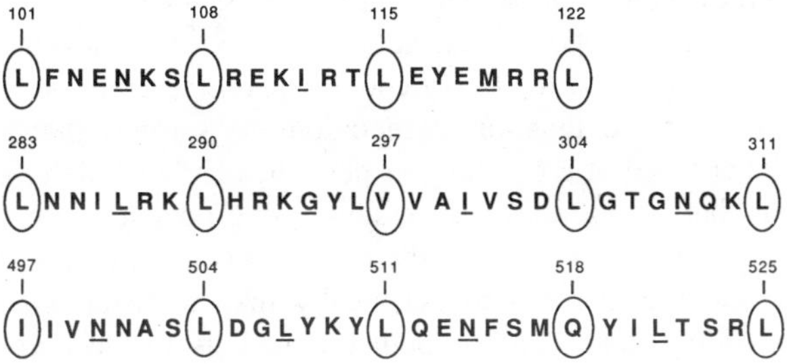

B

Figure 4 Sequence of the P element proteins. A. Sequence of transposase (751 amino acids) and 66 kd (576 amino acids) with heptad hydrophobic repeats highlighted. The 15 C-terminal 66 kd unique amino acids 562–576 (GMTNLKECVNKNVIP) are shown below the corresponding transposase sequences. B. Sequence of the regions of heptad repeats with hydrophobic amino acids circled.

by other internally deleted P elements that may lack the transposase DNA binding domain. In fact, two other leucine zipper-like motifs are found at amino acids 283–311 and 497–525 in ORF1 and ORF2 (57, 65) (Figures 4A and 4B). These three leucine zipper motifs are also shared between the 66 kd protein and transposase, and might serve to mediate other as yet undefined protein-protein interactions (Figure 2).

M' strains also lower sn^w mutability during dysgenic crosses. This effect has been hypothesized to be due to titration of transposase by the small internally deleted P elements in M' strains (77) rather than by putative negative regulatory products encoded by these elements. This alternative explanation of the M' strain genetic data is feasible, since transposase is a site-specific DNA binding protein that recognizes internal P element DNA sequences (see section on P element transposase, below). Depending upon the concentration of transposase in these strains, titration may result from nonproductive or productive binding of transposase to the termini of P elements other than those at sn^w. Although these small P elements would presumably be mobilized in the presence of transposase, they are not genetically marked; thus their mobility cannot be measured, except through the induction of insertion mutations with a recognizable phenotype or by in situ hybridization to polytene chromosomes. Therefore it is possible that the absolute transposition frequency is not reduced in M' strains, but transposase is simply shifted away from the two P elements at sn^w to other unmarked elements. The additional P element termini in M' strains, by interacting with transposase, could also reduce the effective concentration of the protein at any two termini of the same element.

Although the P element-encoded proteins are clearly important in the repression of transposition, maternal inheritance of P cytotype is still poorly understood. One proposed hypothesis is that maternal inheritance results from the insertion of multiple P elements into genomic locations that allow maternal expression during oogenesis (48, 68; Figure 9). High levels of expression of a repressor protein in the ovary could explain the ability of P cytotype to persist through more than one generation. M' strains that do not show maternal inheritance of repression, as described earlier, might have elements in positions that do not allow expression during oogenesis. In addition, low concentrations of transposase in these strains could explain the intrastrain variability and the inability to genetically map these effects to particular chromosomal positions. This hypothesis predicts that certain strains may show both P cytotype as well as KP-type repression, and that alteration of P element genomic location may change the regulatory pattern of a given strain.

Several observations have implicated larger P elements present in P strains in the repression of transposase activity and in the maternal effect and inheritance of P cytotype. One example is the identification of P cytotype-dependent alleles of the *singed (sn)* and *vestigal (vg)* loci (68, 88, 89). The expression of these cytotype-dependent alleles can be altered either positively or negatively in a P cytotype genetic background (Table 1). For example, the vg^{21-3} allele of the *vestigial* locus shows a mutant phenotype in M cytotype but is suppressed (shows a wild-type wing phenotype) in P cytotype, whereas alleles of the *singed* locus show P cytotype-dependent sterility and aberrant oocyte morphology (68). Truncated forms of transposase, including the somatically expressed 66 kd protein, can mimic the P cytotype effect on the *sn* and *vg* alleles. In addition, these proteins can substantially reduce transposase activity in both the germline and soma (48, 68) (Table 1; see below). These repressors may function differently from the KP protein, perhaps by recognizing specific DNA sequences (Figure 7A) either similar to or distinct from those bound by transposase (48, 66), or they might act similarly via protein-

protein interactions (Figure 7B). Truncated transposase proteins lacking ORF3 sequences may directly bind P element DNA sequences and in this way alter transcription of the nearby genes either positively or negatively to bring about cytotype-dependent enhancement or suppression. Consistent with the idea that these proteins directly interact with DNA (48, 68) is the fact that they can mimic this effect in the absence of transposase protein. Alternatively, the presence of three heptad hydrophobic amino acid repeats in these larger proteins (57, 66) (Figures 4A and 4B) may allow formation of inactive transposase protein complexes, reducing transposase activity or modulating expression of the cytotype-dependent alleles by affecting interactions among transcriptional regulatory proteins. Both of these ideas seem plausible since the transposase binding site near the 5' P element end overlaps the promoter TATA box-TFIID binding site (Figure 6B), and transposase itself can function as a transcriptional repressor in vitro (34; see section below on P element transposase). Given the availability of cloned DNA probes to the *singed* (70) and *vestigial* (88) loci, these different hypotheses can be tested directly by examining transcription of these genes in the P and M cytotype (89). Localization of the DNA binding domain in the 87 kd transposase protein will also help to clarify this issue (see below).

At this point, the many different genetic observations imply that at least three possible mechanisms could be operating separately or in combination to explain the negative regulatory effects of defective P elements on the various phenotypes that collectively comprise hybrid dysgenesis. First, P cytotype control could be brought about by large P element proteins including the 66 kd protein (see below) that may possess a DNA binding activity similar or identical to that of transposase (48, 66). In this model, competition for similar or nearby DNA sites might reduce transposition frequency by steric occlusion of transposase from its site on P element DNA (Figure 7A). In addition, these proteins might block transcription of the transposase mRNA in the germline, resulting in reduced rates of transposase synthesis (Figure 7A). DNA binding activity would also explain the effects on expression of cytotype-dependent alleles caused by P element insertions. The maternal effect and inheritance of P cytotype control may result from specific insertion sites in the genome that allow expression of P elements during oogenesis. Second, P cytotype control could be due to the presence of many small, internally deleted P elements that actually serve to dilute transposase by titration (79). In this model, the effective concentration of transposase at two ends of one element necessary to allow transposition falls below a critical level or threshold, thereby reducing transposition frequencies. Third, small, internally deleted P element derivatives or larger P elements may encode polypeptides capable of acting as dominant negative regulators of transposase activity by interfering with normal transposase protein-protein interactions, perhaps through specific "leucine-zipper" motifs (Figures 4A and 4B, and Figures 7B and 7C). Smaller peptides may interfere either with interactions of transposase with itself or with other proteins involved in transposition. Again, this type of repression, characteristic of M' strains, might or might not exhibit a maternal effect or inheritance as a consequence of P element genomic position.

CIS-ACTING DNA SEQUENCES INVOLVED IN P ELEMENT TRANSPOSITION

The P element termini carry perfect inverted 31 bp repeats and an 8 bp duplication of target-site DNA adjacent to these terminal repeats that is created upon insertion. These structures are characteristic of many transposable elements. However, other internal P element DNA sequences, in addition to the 31 bp terminal repeats, are required for transposition. DNA sequence analysis of internally deleted, nonautonomous P elements as well as analysis of P element transformation vectors have shown that over 100 bp are necessary in *cis* for transposition (32, 57, 74). In contrast, most bacterial transposable elements typically require only the terminal repeats in *cis* for transposition (4).

A detailed study of the *cis*-acting DNA sequence requirements for P element transposition recently used P element-mediated transformation as an assay for P element function (52). This assay utilized duplicated 3' P element ends flanked by two genetic markers. The inner 3' end was subjected to in vitro mutagenesis, while the outer end was wild-type. Transformants carried either one or both genetic markers, depending on which 3' end was used in transposition. This assay allowed a direct comparison between wild-type and mutant ends on the same molecule. Deletion analysis indicated that at the 3' P element end 163 bp of DNA were required in *cis* for efficient function of the 3' terminus (53). This deletion analysis further implicated an internal 11 bp repeat, present at both P element ends, in transposition. However, deletion of the 11 bp inverted repeat could be partially compensated for by insertion of heterologous spacer DNA segments. Related experiments showed that the 5' and 3' P element ends were not equivalent because a 5' P element end could not substitute for a 3' end (53). Furthermore, the sequence of the target site duplication played no role in transposition (53).

A more detailed analysis using clustered-point mutations and internal deletions revealed a complex sequence array in which a variety of internal P element DNA sequences had effects on transposition frequency when mutated. These mutagenesis experiments defined five regions (A–E) within 160 nt of the P element 3' terminus (Figure 5). Region A, a 55 bp region, lies proximal to the 31 bp inverted repeat and mutations within the 31 bp inverted repeat or directly adjacent to the inverted repeat in region A resulted in complete loss of transposition function. The sequence in region A that is required for transposition is the only internal sequence for which there is an absolute requirement. This sequence has been shown to interact with P element transposase (33). Taken together, these experiments provide the best evidence that interaction of transposase with these sequences is required for transposition in vivo (see below). Three of the four mutations in the transposase binding site that severely reduced transposition frequency in vivo also resulted in reduced transposase binding tested in vitro (33). Other mutations in region A resulted in only a modest reduction in transposition frequencies. Adjacent to region A is a 30 bp region (region B) that appears to play no role in transposition because mutations in this region had little effect on transposition. Furthermore, the spatial relationship between A and other internal regions was not critical for transposition, since deletions within region B had only small effects on transposition. Adjacent to region B, a 10 bp region (region C) had a modest effect on transposition frequency. Upstream of region C, a small 10 bp region (region D) was defined that, like region B, played little role in transposition. Next to region D lies the 11 bp inverted repeat, mutations in which resulted in a significant reduction in transposition frequency. A 20 bp region (region E) adjacent to the 11 bp inverted repeat showed a reduction in transposition frequency when deleted; however, when point mutations were made in region E transposition was unaffected. Taken together, these results indicate that the sequences in regions B through E might function as a type of transpositional enhancer, perhaps similar to those found in bacteriophage Mu or the *hin* inversion system (30, 31, 43, 49). The involvement of a complex array of *cis*-acting DNA sequences in P element transposition may reflect not only the necessity for interaction with P element transposase and other *Drosophila* proteins but also a requirement for a unique DNA structure built up through a variety of protein-protein or protein-nucleic acid interactions, as has been observed for bacteriophage lambda integration (39) or *Escherichia coli* and lambda DNA replication (15, 37).

P ELEMENT TRANSPOSASE—GENETICS AND BIOCHEMISTRY

The structure of P transposable elements suggested that the large 2.9 kb P elements might encode the genetically defined function, referred to as trans-

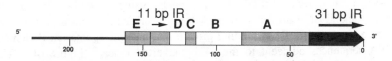

Figure 5   Diagram of 3' P element end. Regions defined by in vitro mutagenesis and transposition in vivo, A-E, and the 31 bp and 11 bp inverted repeats are shown. (Data taken from Ref. 53.)

posase, required to catalyze P element transposition. The existence of both complete 2.9 kb and smaller internally deleted P elements is similar to the maize Ac-Ds and Spm transposable elements first described by McClintock (22, 46). Genetic experiments in *Drosophilia* had suggested that P strain-derived chromosomes were responsible for the hybrid dysgenesis syndrome (9, 17, 19) and that cloned 2.9 kb P element DNA injected into *sn^w* embryos could induce *sn^w* mutability, demonstrating that 2.9 kb P elements encode transposase (82). This observation led to the development of P element-mediated germline transformation (73). Development of the wings-clipped helper plasmid (32), which allows transposase synthesis following embryo injection without integration of the helper DNA, provided a means of producing stable transformants that would not undergo secondary transposition events such as occurred in the first description of the transformation method (73). These methods facilitated a genetic analysis of the transposase gene through the construction of a genetically marked complete P element called Pc[ry] (32). Pc[ry] was then subjected to in vitro mutagenesis, introduced back into *Drosophila* using the wings-clipped helper DNA, and the various mutant derivatives assayed for their ability to produce transposase activity (32). Mutations in any one of the four protein-coding open reading frames ORF0, ORF1, ORF2, or ORF3 in the 2.9 kb P element abolished the ability of the modified P elements to produce transposase activity as assayed genetically. Furthermore, pairwise combinations of each of the ORF mutants in *trans* failed to complement for transposase activity, suggesting that each of the ORFs were joined to yield a single polypeptide. Analysis of P element cDNA clones and in vitro-modified elements expressed in *Drosophila* tissue culture cells led to the identification of an 87 kd polypeptide using P element-specific antibodies (66). Introduction of genetically marked P element-containing plasmids into cell lines expressing the 87 kd P element protein showed that the marker elements underwent excision events characteristic of those produced by transposase in vivo, verifying that the 87 kd protein possessed transposase activity (66). Previous genetic experiments had shown that in addition to catalyzing forward transposition, P element transposase can catalyze both precise and imprecise excision events (14, 75, 85, 87).

The expression of biologically active transposase in *Drosophila* tissue culture cells and the availability of P element-specific antibodies allowed the biochemical fractionation and purification of transposase (33). The transposase protein was purified to homogeneity using a combination of ion exchange and DNA affinity chromatography. Purified transposase was shown to be a site-specific DNA binding protein that recognizes internal P element DNA sequences adjacent to the 31 bp inverted repeats (Figure 6B). The protein apparently recognizes a 10 bp consensus sequence present in inverted orientation near each P element end (Figure 6A). Binding of the protein to these sites is blocked in mutants that alter the 10 bp consensus sequence. It

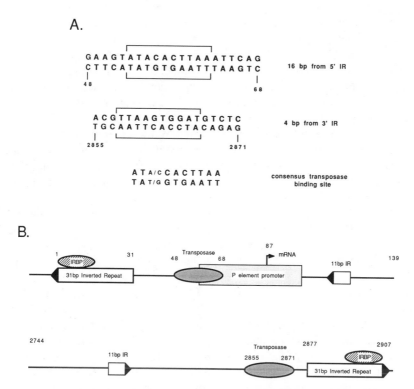

Figure 6 P element transposase binding sites. A. DNA sequences protected from DNAse I digestion at the 5' end and 3' ends. Consensus 10 bp transposase recognition sequence derived from the 5' and 3' sites. (Data taken from Ref. 33.) B. Internal location of the transposase binding sites near the P element ends. Note overlap at 5' end with the P element promoter and that transposase binds internal sites but not the inverted 31 bp terminal repeats. (Data taken from Ref. 33.)

was also found that the purified transposase possesses a very high nonspecific affinity for DNA (K_D nonspecific $\sim 10^{-8}M$) compared to site-specific DNA binding (K_D specific $\sim 10^{-9}M$). This difference in specific versus nonspecific DNA binding affinities is rather unusual, since a typical DNA binding protein such as the lac repressor has about a 10^3 to 10^6-fold difference between the affinity of the protein for specific versus nonspecific recognition sites. The nonspecific DNA binding activity of P element transposase may play a role in insertion site selection during transposition, and suggests that other *Drosophila*-encoded proteins may be required to enhance the specificity of the transposition reaction or to alter the specificity or affinity of transposase binding in vivo. This idea is supported by the fact that a mutation in the transposase binding site that did not block transposition in vivo was defective for transposase binding in vitro (33, 53).

The oligomeric form of transposase remains to be elucidated. The transposase protein might associate with itself either in the absence or presence of DNA and these associations might play a role in transposition. Such interactions could also be a target for negative control of transposition (see section above on genetic control). Biochemical data concerning the DNA binding domain of the transposase protein is not yet available; however, a region shared with the smaller 66 kd protein bears weak sequence homology to the bacterial helix-turn-helix family of DNA binding proteins (57, 66). This region also carries a "leucine zipper" heptad repeat motif adjacent to some basic amino acids (66, 86), a structural motif shown to be present and responsible for mediating dimerization and DNA binding in a number of proteins (38, 59, 86). Further biochemical studies are required to determine the structural and functional features of P element transposase.

The binding site for transposase at the 5' P element end overlaps an AT-rich sequence 20–30 bp upstream from the transcriptional initiation site at nt 87 (32) that is part of the P element promoter defined by in vitro transcription (33; Figure 6B). This AT-rich sequence appears to correspond to the TATA box sequence present in the promoters of most eukaryotic genes transcribed by RNA polymerase II. The TATA box interacts with a transcription factor, TFIID, defined in mammalian cells (54), and similar factors have been identified in *Drosophila* (61) and yeast (10). This juxtaposition of sequences suggests that transposase or perhaps the smaller derivatives such as the 66 kd protein may bind to the transposase site and function as transcriptional repressors of P element mRNA synthesis by blocking the TFIID-TATA box interaction. In fact, transposase can function as a transcriptional repressor in vitro; this effect is mediated through a binding interaction of transposase with its consensus sequence (34). The biochemical activities of the 66 kd protein are not yet known but genetic data suggest that it may also be a DNA binding protein.

THE 66 kd P ELEMENT PROTEIN AND OTHER TRUNCATED FORMS OF TRANSPOSASE ARE NEGATIVE REGULATORS OF TRANSPOSITION

The P element messenger RNA produced from complete 2.9 kb P elements in somatic cells retains the third intron (IVS3) and encodes a smaller 66 kd protein (40, 66). Genetic studies of hybrid dysgenesis had indicated that P cytotype was probably encoded by P elements themselves, suggesting that the 66 kd protein might serve as a negative regulator of transposition (17, 19, 84). This hypothesis was tested directly using P element-mediated transformation. Modified P element derivatives capable of encoding only the 66 kd protein were introduced into the *Drosophila* germline by transformation (48). Genetic experiments using marked P elements examined the effect of truncated P elements encoding the 66 kd protein on transposase activity either in the germline using the *singed-weak (sn^w)* test (18, 48) or in the soma using an eye color mosaic test with *Drosophila white (w)* gene P element transposons (P[w^+]) (40, 48). These experiments clearly indicated that the 66 kd protein-encoding P elements could cause a marked reduction of transposase activity in both the germline and soma. RNA and protein analysis confirmed that these strains were producing the 66 kd protein, although the quantitative effects of the 66 kd P elements in reducing transposase activity varied with different insertion sites, presumably reflecting differences in temporal or spatial expression patterns due to chromosomal position effects. Furthermore, no 66 kd protein-expressing lines tested showed a maternal effect or inheritance of repression, which is a distinctive characteristic of P cytotype (48). However, protein immunoblot analysis showed that P strain female ovaries and unfertilized eggs contained a significantly higher level of the 66 kd protein than those of the single copy 66 kd-encoding P element transformants used in these experiments (48). This observation led to the hypothesis that the evolution of P cytotype might involve fortuitous insertion of 66 kd-encoding elements into genomic locations that allow elevated expression levels during oogenesis, leading to a maternal effect of cytotype. This female germline expression pattern during oogenesis would probably be selected for in a population of dysgenic hybrids, since a reduction of transposition in the germline would allow increased fertility.

Another genetic study examined the ability of modified P elements with mutations in IVS3 or ORF3 to suppress the temperature-dependent pupal lethality brought about by expression of transposase in somatic cells (20, 68). Effects on cytotype-dependent alleles of the *singed* locus and *singed-weak (sn^w)* mutability were also tested (68). These genetic studies indicated that truncated transposase proteins similar or identical to the 66 kd protein could mimic some aspects of P cytotype, since the modified elements also suppressed *sn^w* activity in the germline. However, none of the modified elements showed the maternal effect typical of P cytotype regulation (68).

A different genetic approach was taken to identify repressor-producing P elements naturally found in strains with P cytotype. An M' strain capable of producing repressing activity was outcrossed to an M strain, and at each generation the strains were analyzed to determine which elements correlated with the repression of gonadal dysgenic (GD) sterility (56). Molecular analysis of M and P cytotype strains derived from outbreeding suggested a correlation between P cytotype and a single P element containing ORF0, ORF1 and ORF2 with a deletion of ORF3, resulting in synthesis of a protein similar to the 66 kd protein (56). However, other P elements were present in the strains tested, and the particular element characterized was not reintroduced by transformation into an M strain background and shown to cause

P cytotype. Therefore, these data do not definitively prove that this element was responsible for conferring P cytotype.

These molecular and genetic studies taken together demonstrate that truncated forms of transposase, the 66 kd protein being one example, can act as negative regulators of transposase activity that may be responsible for conferring P cytotype. One possible mechanism for the action of these proteins involves DNA binding (Figure 7A). It is plausible that the 66 kd protein and other relatively large truncated transposase derivatives possess the transposase DNA binding domain and block transposase action by competing for the same or nearby sites on P element DNA (Figure 7A). This idea is supported by experiments with cytotype-dependent alleles of the *singed* and *vestigial* loci whose expression is modified in the P cytotype (68, 89), even in the absence of transposase molecules. Furthermore, expression of a modified P[w$^+$] transposon can be repressed in P cytotype (11). These findings could most easily be explained by a direct interaction of the 66 kd protein and other large transposase derivatives with P element DNA sequences. For example, binding of the proteins to P element DNA sequences inserted at the cytotype-dependent loci might activate or repress transcription of the adjacent genes. In this regard, it is interesting that the transposase binding site overlaps the P element promoter (33) and that binding of transposase to this site represses P element transcription in vitro (34). Thus, the 66 kd protein could regulate transcription from the P element promoter, although it might not be able to repress expression from all P element insertions, particularly if the elements were also transcribed under the control of adjacent genomic promoters or enhancers (48). The idea of transcriptional control had been suggested earlier by genetic transformation and mutagenesis experiments with elements under the control of the hsp70 promoter (83), although no difference in P element mRNA levels was observed in somatic RNA samples from dysgenic crosses (32). It is also possible that these proteins act to modulate transposase activity by protein-protein interactions with transposase (Figure 7B).

The hypothesis that the maternal inheritance of P cytotype is caused by expression of truncated transposase proteins during oogenesis, as a consequence of insertion at specific genomic locations that trigger transcriptional activation in the ovary, seems plausible in light of recent data with P element promoter-*LacZ* fusion gene P elements (21, 25, 58). Many insertions of these P-*LacZ* elements at different genomic sites show expression of beta-galactosidase during oogenesis (see penultimate section below). Such insertions of repressor-encoding P elements would afford a selective advantage in a population by preventing hybrid dysgenesis in the germline. Immunoblotting experiments demonstrated that P strain female ovaries contain a high level of 66 kd protein but an undetectable amount of the 87 kd transposase protein (48; Figure 9), although IVS3 splicing occurs during the earliest stages of oogenesis in the germarium (41). These experiments suggest that perhaps both the 87 kd protein and the 66 kd protein are produced by full-length elements during oogenesis in P strain females, although only the 66 kd protein accumulates appreciably in the mature oocyte (48). If so, it should be possible for a few complete P elements to generate maternally inherited P cytotype by insertion into fortuitous positions in the genome. During outcrossing to M strains, the presence of these repressor-producing P strain-derived P elements expressed during oogenesis may explain the persistence of P cytotype for several generations.

Early genetic experiments on the control of hybrid dysgenesis led to the hypothesis that P elements must encode two functions: transposase and repressor. It now seems clear that at least one identified molecule, the 66 kd protein, is capable of repressing transposition (48). One model to explain cytotype also proposed that the repressor should feed back to increase its own synthesis (57), while perhaps repressing synthesis of the transposase molecule. If this were true, the nonreciprocal cross effect, that is, genotypically identical females from P male X M female (dysgenic) or M male X P female (nondysgenic) crosses transmit M or P cytotype, may be explained if maternal 66 kd protein derived from P strain females causes activation of zygotic 66 kd transcription, alterations in 66 kd protein synthesis, or enhanced 66 kd protein stability during a nondysgenic cross. The molecular basis for suppression of the vestigial vg^{21-3} phenotype in P cytotype may bear on the ability of the 66 kd protein to activate gene expression. The vg^{21-3} mutation is caused by a P element insertion and the mutant phenotype is suppressed in P cytotype implying perhaps that the 66 kd protein might act to increase vestigial gene transcription in P cytotype (89). Thus, it is conceivable that the 66 kd protein might activate gene expression of certain P element insertions, perhaps 66 kd protein-encoding elements, providing a positive feedback mechanism for 66 kd protein synthesis (57). Alternatively, the simple lack of maternal 66 kd protein in M strain females may allow high rates of P element excision (loss) and transposition (to new sites) in the dysgenic hybrids. This burst of transposition and excision would reduce the number of effective repressor-producing elements present in the fertile F1 dysgenic progeny or move these elements from positions that cause expression during oogenesis. This mobilization could cause F1 female eggs to exhibit M cytotype due to reduced 66 kd protein expression. This hypothesis is supported by the fact that single-copy 66 kd transformants show no maternal inheritance of repression (48). Thus, the position-dependent expression of these elements in P strains may explain not only the maternal effect and inheritance of P cytotype, but why eggs derived from M female dysgenic hybrids can show M cytotype in subsequent generations.

One alternative possibility regarding the mechanism of action of the 66 kd

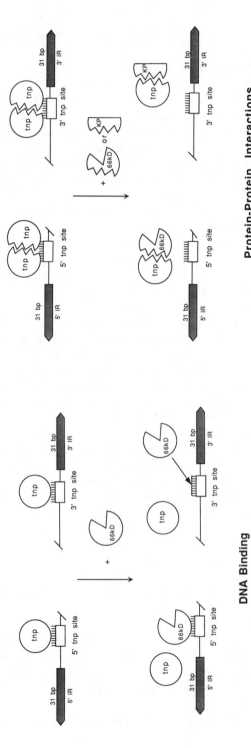

DNA Binding

Protein-Protein Interactions

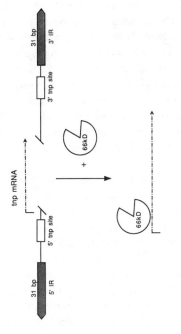

**Translational or RNA processing block
(RNA-Protein Interaction)**

Figure 7 Models for the negative control of transposition by truncated transposase proteins. A. DNA binding. Larger derivatives, such as the 66 kd protein, may bind similar, identical, or adjacent to transposase binding sites on P element DNA, thereby blocking transposase action (indicated by arrow). B. Protein-protein interactions. Transposase is depicted binding to DNA as a dimer or higher oligomer. Transposase multimer is inactivated for DNA binding by interaction with 66 kd or KP protein. C. Protein-protein-DNA interactions. Two DNA-bound transposase molecules interact. This interaction is disrupted by the 66 kd or KP proteins. D. Translational or RNA processing block. The 66 kd protein or other transposase derivatives interact with P element mRNA to block translation or processing of the transposase RNA. Alternatively, differential RNA stability of IVS3-containing 66 kd protein mRNA versus spliced IVS3-lacking transposase mRNA could account for differential 66 kd transposase synthesis.

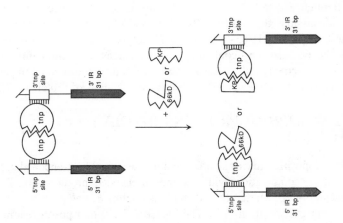

Protein-Protein-DNA Interactions

protein involves regulation at the level of RNA splicing. For example, if the 66 kd protein inhibited splicing of IVS3 in the germline, transposase synthesis would be reduced and 66 kd synthesis increased. However, experiments with IVS3-*LacZ* constructs used to study *cis*-acting sequences necessary for germline-specific splicing of IVS3 (see below) indicated that IVS3 splicing is not altered in P cytotype (41). Furthermore, genetic experiments using the Δ2–3 P element, lacking IVS3, which expresses transposase somatically, showed that in P cytotype the somatic transposase activity was reduced (48, 68). Thus, P cytotype control cannot operate solely at the level of IVS3 splicing because in these experiments the third intron had already been removed in the P[Δ2–3] construct.

Another possibility is that P cytotype control is mediated through RNA-protein interactions, perhaps to control translation, processing, stability, or nuclear transport of the P element mRNA (Figure 7D). For instance, bacteriophage Mu transposase has been proposed to be a site-specific RNA-binding protein, which binds to its own mRNA in addition to binding DNA sequences (62) required for Mu transposition. This RNA binding activity was implicated in the control of Mu transposase mRNA translation. The 66 kd protein might conceivably be an RNA-binding protein. This is also interesting in view of the maternal effect of P cytotype and the fact that one of the posterior group maternal effect genes, *vasa,* is thought to encode an RNA-binding helicase (26, 42). If the 66 kd protein controls translation of its own or the transposase mRNA, this could explain why no difference was seen in P element mRNA levels in dysgenic versus nondysgenic crosses (32). Alternatively, the 66 kd protein might function in the nucleus to facilitate the cytoplasmic export of unspliced IVS3-containing 66 kd-encoding mRNA, thereby reducing transposase synthesis and facilitating 66 kd production (Figure 7D). These issues, as well as the possible DNA-binding activity of the 66 kd protein, are being addressed through a detailed biochemical analysis of the purified protein. Finally, it is possible that when ORF2-ORF3 (IVS3) splicing occurs to produce the transposase (IVS3-lacking) mRNA, this spliced product is less stable than the 66 kd (IVS3-containing) mRNA. This differential mRNA stability would effectively reduce transposase levels and could account in part for differential steady-state levels of the transposase and 66 kd proteins (57).

OTHER *DROSOPHILA* PROTEINS INVOLVED IN TRANSPOSITION

Prokaryotic transposition reactions often require additional host-encoded proteins that act in combination with transposon-encoded proteins (4). For instance, bacteriophage Mu requires a small histonelike *E. coli* protein called HU for efficient transposition (13). The bacterial Tn10 transposon requires the *E. coli* integration host factor (IHF) or the HU protein for transposition in addition to the element-encoded transposase (51). The *E. coli* DNA replication initiation protein, dnaA, has been implicated in transposition of the bacterial Tn5 transposon (63, 91). Thus, we might expect that P element transposition would require additional *Drosophila* proteins. Genetic data have suggested the involvement of *Drosophila* genes affecting postreplication DNA repair in P element transposition, since the mutations *mei-41* (a meiotic abnormality mutant) and *mus-101* (a mutagen hypersensitive mutant) both increased the severity of embryo lethality and segregation ratio distortion during hybrid dysgenesis (79). However, no effect was observed on male recombination or on insertional mutagenesis. Thus, it is not clear whether the products of these genes are directly involved in P element transposition.

The biochemical analysis of P element transposase indicated that transposase interacts with internal P element DNA sequences adjacent to the terminal 31 bp inverted repeats (see above) (Figure 6B). However, in vitro mutagenesis experiments indicated that the terminal 31 bp repeats as well as internal sequences (other than the transposase binding sites) were required for efficient transposition (53, 74). Presumably, at least a portion of these additional DNA sequences are sites for the interaction of additional *Drosophila*-encoded proteins involved in transposition. In fact, a *Drosophila*-encoded protein present in an M strain-derived cell line was identified by its interaction with a portion of the 31 bp inverted repeats. This protein, termed inverted repeat binding protein or IRBP, interacts with the terminal 16 bp of the 31 bp repeats directly adjacent to the 8 bp duplication of target site DNA (67). The IRBP was purified to near homogeneity from *Drosophila* tissue culture cells and the binding activity was shown to reside in a 66 kd polypeptide (not encoded by the P element). The involvement of this protein in P element transposition is inferred from its interaction with the terminal repeats and by the fact that the 31 bp inverted repeats are absolutely required for P element transposition. However, the function of this gene product in transposition and its normal role in *Drosophila,* perhaps in DNA replication, repair or recombination, must await isolation of the corresponding gene and subsequent genetic and biochemical characterization. Furthermore, it is likely that other *Drosophila* proteins are involved in transposition and that they will be identified through the use of genetic screens for mutations affecting transposition frequency, as well as by the development of biochemical assays for P element transposition or excision in soluble cell-free systems.

TISSUE SPECIFICITY OF P ELEMENT TRANSPOSITION

The symptoms of hybrid dysgenesis, such as sterility and the induction of mutations, are confined to the germlines of the progeny from a dysgenic cross. Extensive genetic experiments indicated that there was an absence of somatic dysgenesis effects in P-M hybrids (17, 19, 47). This tissue-specificity might have resulted from transcriptional control in which the P element promoter was active only in germline tissues. However, this possibility was ruled out by experiments in which the P promoter was replaced by the hsp70 heat shock promoter, which is known to be active in somatic tissues. The hsp70-P element fusion gene was transcribed somatically yet yielded transposase activity only in the germline (40, 83), indicating that the block to somatic transposase expression must occur posttranscriptionally. Indeed, the P element promoter functions in somatic tissues, since P element transcripts were found in P strain adult head RNA (40). However, RNA structural analysis of somatic P element mRNA suggested that this mRNA does not encode transposase because only three of the four open reading frames were joined by removal of two small intervening sequences (IVS), while the fourth open reading frame was not joined translationally to the other reading frames (40; Figure 2). These observations first suggested the existence of an additional intron because previous genetic experiments indicated that transposase was encoded by all four open reading frames and that these reading frames were joined to encode a single polypeptide (32) (see above).

These observations led to the hypothesis that there was an additional third intron (IVS3) joining the third (ORF2) and fourth (ORF3) open reading frames (Figures 1 and 2) that was removed only in the germline. This idea was tested by identifying 5' and 3' splice site consensus sequences in the third and fourth open reading frames, which would join the open reading frames translationally when the putative intron was removed (40). Modified P element derivatives in which these splice sites had been inactivated by mutagenesis were introduced into *Drosophila* by transformation and these strains were tested for their ability to synthesize transposase activity in the germline (40). Both the 5' and 3' putative IVS3 splice site mutants failed to synthesize transposase, suggesting that these point mutations abolished IVS3 splicing. Furthermore, introduction of other modified P elements in which the entire putative IVS3 intron had been removed led to the synthesis of transposase in both the germline and soma (40). These experiments defined genetically that the tissue-specificity is brought about by the germline-specific splicing of IVS3 to join ORF2 to ORF3, allowing synthesis of the transposase mRNA and protein only in the germline. Therefore, it appeared that removal of this third intron was the only block to somatic transposition and that no germline-derived proteins or cofactors were required for transposition. Simply expressing the transposase protein in somatic cells was sufficient to allow somatic transposition. These experiments clearly showed that the germline-specificity of P element transposition and hybrid dysgenesis results from a tissue-specific RNA processing event.

GENETIC ANALYSIS OF P ELEMENT ORF2-ORF3 (IVS3) pre-mRNA SPLICING

These molecular genetic experiments led to the conclusion that the tissue-specificity of P element transposition was regulated posttranscriptionally at the level of pre-mRNA splicing (40). However, these experiments did not address whether the removal if IVS3 is activated in the germline or repressed in somatic cells. P element-mediated transformation has allowed an analysis of this tissue-specific splicing event by providing a means to introduce in vitro-modified P elements carrying mutations in IVS3 back into *Drosophila* and assay these mutant introns for splicing in the germline or soma (40, 41).

Genetic analysis of mutations in the third intron relied on the ability of modified, genetically marked, 2.9 kb complete P elements to synthesize transposase either in the germline or soma. Removal of IVS3 was assayed by the ability of the modified P elements to produce transposase activity. Germline transposase activity was monitored using the *singed-weak (sn^w)* test and somatic transposase activity was detected by scoring eye color mosaics resulting from excision of a P element derivative carrying the cell autonomous *white (w$^+$)* gene (Table 1). These tests afford sensitive but indirect genetic assays to detect transposase activity resulting from IVS3 splicing. The wild-type construct with IVS3 showed the usual tissue-specificity, normally high transposase activity only in the germline. By contrast, the Δ2–3 mutant in which IVS3 was precisely removed yielded high levels of transposase activity in both tests. In addition, a series of internal IVS3 deletion mutants were tested for their ability to express transposase in the germline and soma (41). Three internal deletions removing different but overlapping regions of IVS3 produced transposase predominantly in the germline, but unlike the wild-type IVS3, also yielded a low level of transposase activity in the soma. These results were interpreted as "leaky" somatic splicing (41) and may reflect removal of redundant negative regulatory elements located within IVS3, or possibly a structural alteration of the intron that allows its removal at a low level by the somatic splicing machinery. Two other deletion mutations abolished splicing in both germline and soma. These two mutations either removed the conserved branchpoint sequence found in all introns, or reduced the length of the intron below the minimum size to allow splicing. In addition to internal intron deletions, point mutations were introduced into both the IVS3 5' and 3' splice sites to make these sites better match the consensus 5' and 3' splice site sequences. Although changes in the polypyrimidine tract near the 3' splice site did not alter the tissue-specificity of IVS3 splicing, changes in the 5' splice site resulted in a low level of IVS3 splicing in the soma. The improved match to the 5' splice site consensus sequence might have resulted in a more stable interaction with U1 snRNP (a small nuclear ribonucleoprotein particle required for pre-mRNA splicing) (45) by increasing the complementarity to the 5' end of U1 snRNA. Indeed, the 5' splice site-U1 snRNP interaction depends at least in part on base pairing between the 5' splice site sequence and the 5' end of U1 RNA (24, 60). Furthermore, since the multiple U1 snRNA genes in *Drosophila* (52) may be expressed tissue-specifically, changes in the 5' splice site may allow recognition by a U1 snRNP present in somatic tissue, whereas normally only a germline-restricted U1 snRNP recognizes the IVS3 5' splice site.

Numerous experiments in mammalian cells have implicated exon sequences as playing a role in alternate splicing (55, 64). However, the assay described above could not be used to analyze exon mutations near IVS3 due to the constraint in the assay of maintaining the transposase open reading frame. Therefore, a second assay was devised in which a 240 bp P element DNA fragment carrying IVS3 was inserted into a hybrid gene containing the hsp70 promoter and the amino terminus of the hsp70 protein fused to the *E. coli* beta-galactosidase gene *(LacZ)*. The expression of the *LacZ* open reading frame was dependent upon removal of IVS3. This construct could be assayed for tissue-specificity in vivo using a histochemical staining procedure to detect the bacterial *LacZ* gene product (41). The wild-type construct showed tissue-specific *LacZ* activity in the ovaries of transformant females during the earliest stages of oogenesis and in transformant males at the tip of the testes (41). Unlike the assay for transposase activity, wild-type constructs in the *LacZ* assay displayed low levels of beta-galactosidase activity in the soma, presumably as a result of leaky IVS3 splicing or internal translational initiation. The analysis of mRNA from these strains should clarify whether accurate IVS3 splicing is occurring in somatic cells carrying the hsp70–IVS3-*LacZ* P element derivatives. *LacZ* activity was also observed when the wild-type hsp70–IVS3-*LacZ* construct was introduced into flies of P cytotype, suggesting that P cytotype control does not result from inhibition of IVS3 splicing (see above). These results indicate that only 240 bp of P element DNA is sufficient for tissue-specific splicing; therefore, probably no large RNA secondary or tertiary structural interactions are involved. This assay allowed the analysis of different deletions within the intron and, more importantly, of mutations in the flanking exons. Deletions within the 5' exon reduced germline *LacZ* activity without decreasing the observed low basal somatic *LacZ* activity, indicating that exon sequences are involved in regulating IVS3 splicing. Whether this regulation involves positive or negative control is unclear, although negative control is suggested from the intron deletion and 5' splice site mutations. However, taken together these experiments implicate 5' exon sequences, some internal intron sequences, and/or 3' exon sequences in controlling the tissue-specific splicing of IVS3 (41).

BIOCHEMICAL ANALYSIS OF P ELEMENT ORF2-ORF3 (IVS3) pre-mRNA SPLICING

Classical and molecular genetic analyses in *Drosophila* have identified genes whose function depends upon alternative pre-mRNA splicing. In some cases, these genetic experiments have identified *cis*-acting DNA sequence elements required for differential splicing. These analyses have also identified *trans*-acting regulatory loci, particularly genes involved in somatic sexual differentiation (1), whose protein products appear to act directly in pre-mRNA splicing by use of a conserved protein motif implicated in RNA binding (the RNP consensus sequence) (1, 2). However, elucidation of the mechanisms underlying how these splicing factors function and how germline-specific P element IVS3 splicing occurs will require the development of functional in vitro biochemical assays.

The biochemistry of pre-mRNA splicing has been studied extensively in mammalian cells and yeast because cell-free extracts are available that accurately splice mRNA precursors. These studies have indicated that pre-mRNA splicing occurs via a two-step transesterification reaction that generates branched circular lariat RNA species during the first steps of the splicing reaction (24, 60). It is also known that pre-mRNA splicing requires the participation of small ribonucleoprotein particles that assemble on the pre-mRNA in an ordered pathway to form a large, active splicing complex termed a spliceosome (24, 60, 64). Nuclear extracts derived from *Drosophila* embryos or somatic tissue culture cells are known to accurately and efficiently splice exogenous *Drosophila* synthetic pre-mRNA via the same reaction mechanism used in yeast and mammalian cells (65). The development of soluble in vitro splicing systems from *Drosophila* has already begun to shed some light on the mechanisms involved in the regulation of P element pre-mRNA splicing and should facilitate biochemical analysis of other regulated splicing events in *Drosophila*.

Attempts to detect IVS3 splicing in somatic cell extracts from either *Drosophila* tissue culture cells or embryos were unsuccessful (65, 76). This result as well as the fact that IVS3 was spliced in heterologous systems including microinjected Xenopus oocytes and human HeLa cell extracts, taken together with mutagenesis experiments of IVS3 in vivo (41), suggested that at least one aspect of IVS3 splicing regulation involved negative control of IVS3 splicing in the soma. This idea was tested directly using a biochemical complementation assay in which the effects of adding *Drosophila* somatic cell extract to the IVS3 HeLa splicing reaction were examined (76). These experiments revealed that *Drosophila* somatic components specifically inhibited IVS3 splicing in vitro, but had little or no effect on splicing of several other introns. RNA-binding studies detected a 97 kd protein that preferentially binds to IVS3 RNA within the 5' exon sequences that had been implicated genetically in the control of IVS3 splicing in vivo (41, 76).

The 5' exon sequences that specifically bind the 97 kd protein also contain a 5' splice sitelike sequence near the authentic IVS3 5' splice site that appeared to interact with a 65 kd protein, as did other RNAs with 5' splice site sequences (76) (Figure 7A). This 65 kd protein may be involved in 5' splice site recognition and may act in conjunction with U1 snRNP, which is essential for pre-mRNA splicing (24, 45, 60) and acts at an early step in the pathway of assembly of the spliceosome, by recognizing 5' splice sites (24, 45, 60). The 97 kd protein may act to either sterically occlude the 65 kd protein, 5' splice site recognition factors or U1 snRNP from recognizing the accurate 5' splice site. Alternatively, the 97 kd may act to channel the 65 kd protein, 5' splice site factors, or U1 snRNP to other sites. Inspection of IVS3 sequences revealed the presence of six additional 5' splice sitelike sequences (referred to as pseudo-5' splice sites) within and around IVS3 (76) (Figures 7A and B). This observation suggested a model in which these pseudo-5' splice sites may compete with the accurate IVS3 5' splice site for U1 snRNP or other protein factors required for accurate splice site recognition, thereby reducing splicing of IVS3 (76). This model is in agreement with the observation that IVS3 deletions removing one or more of the internal IVS3 pseudo-5' splice sites exhibited low levels of somatic IVS3 splicing in vivo (41). More evidence for the involvement of U1 snRNP comes from mutations in the accurate IVS3 5' splice site that provided a better match to the 5' splice site consensus and activated somatic splicing (41) (Figure 8B). This model is also supported by experiments in the mammalian globin system where juxtaposition of consensus 5' splice sites near an authentic site can inhibit use of both sites (55). These results suggest that at least a part of IVS3 splicing control involves somatic inhibition resulting from specific RNA-protein interactions and possibly from competition between 5' splice sites. The inhibition of IVS3 splicing may also occur during oogenesis in P strains to allow 66 kd synthesis and accumulation in the oocyte, which may explain, at least in part, the maternal effect of P cytotype (48) (Figure 9). The possibility still exists that IVS3 splicing may be activated in the germline. Biochemical tests of this idea will require a source of germline-derived cells.

FURTHER DEVELOPMENT OF P ELEMENTS AS GENETIC TOOLS

The molecular analysis of P elements led to the development of P element-mediated germline transformation (73). This in turn allowed detailed molecular genetic studies of P element function using modified elements created through in vitro mutagenesis. However, the ability to modify P elements and

A.

TTGTGGAGCATTTTTTTGGCAGCATGCGATCGAGAGGTGGACA

ATTCGACCATCCCACTCCACTGCAGTTTAAGTATAGGTTAAG

→5'SS

AAAATATATAATAGGTATGACAAATTTAAAAGAATGCGTAA

ACAAAAATGTAATTCCATGATTTATAATTGTTTAATGTTTAG

CTATATGTTTCAGGAAAGTTTCAGTTGAGAATGTAGGTAGTT

ATGTGCTGTCTATTGTGTTTTGTCTTTTATCTGTTTCTTTTCA

↓BP   3'SS◄

TTTTATTATTTAATCATTATCCTTTTGCTTATCCAGCCAGGA

ATACAGAAATGTTAAGAAATTCGGGAAATATCGAAGAGGAC

B.

SEQUENCE OF 5' SPLICE SITES

Splice Site		Sequence	Match	Location
consensus		^C_AAG/GU$_G$AGU		
P IVS 3		UAG/GUAUGA	(6/9)	nt 1945 -1953
fushi tarazu		CAG/GUAGGC	(7/9)	
P IVS3 pseudo 5' splice sites				
upstream	1	UAG/GUUAAG	(5/9)	nt 1925 - 1933
	2	GAG/GUGGAC	(5/9)	nt 1882 - 1890
downstream	1	AAU/GUAAUU	(7/9)	nt 1980 - 1988
	2	AAU/GUAGGU	(7/9)	nt 2046 - 2054
	3	UAG/GUAGUU	(6/9)	nt 2050 - 2058
	4	AUU/GUGUUU	(6/9)	nt 2070 - 2078

Figure 8 Structure of P element IVS3. A. Sequence of IVS3. 5' splice site and 3' splice site are indicated by arrows. The branchpoint (BP) is indicated. The sequences of 5' splice sitelike (pseudo-5' splice sites) sequences are underlined. (Data taken from Ref. 76.) B. Comparison of sequences of 5' splice sites. The consensus, P IVS3 and fushi tarazu 5' splice site sequences are shown at the top and the six pseudo-5' splice sites are shown below, as well as their location in the P element sequence. (Data taken from Ref. 76.)

reintroduce them into the germline has led recently to important advances in the use of P elements as genetic tools in *Drosophila*.

The introduction of modified transposase-producing P elements into the genome allowed the isolation of several so-called genomic "wings-clipped" elements which provide a stable source of transposase. The two most popular

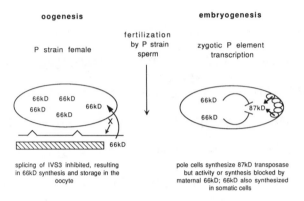

Figure 9 Model for the role of maternal 66 kd protein expression in the repression of P element transposition in P strains (P cytotype). The 66 kd protein is made during oogenesis by suppression of the ORF2-ORF3 splice (X) in the nurse cells or somatic follicle cells. This maternal store of 66 kd protein then serves to effectively repress activity or synthesis of the 87 kd transposase made in pole cells and their germline progeny.

of these are P[ry$^+$;Δ2–3](99B) (68) and Jumpstarter (12). These modified P elements can be used in several important ways. For example, the Jumpstarter element has been used to carry out single P element transposon mutagenesis (12). This element has also been used to mobilize existing P element insertions to new genomic locations, which is important since genomic position can have a profound effect on the expression of genes carried inside a P element vector (80). This genetic approach of remobilization is much simpler than generating independent transformants or mobilizing existing insertions by injection of "wings-clipped" helper DNA into transformant embryos (44). The Δ2–3(99B) element has been used to mobilize other P elements in transformant strains and to do transposon tagging-mutagenesis by crossing to an "ammunition" chromosome, usually the second chromosome of the M' Birmingham strain, which carries seventeen small, internally deleted, nonautonomous P elements (68). However, using Δ2–3 (99B) and Birm2 for mutagenesis suffers from the disadvantage of temperature-sensitive somatic lethality associated with expression of transposase in the soma by Δ2–3 in the presence of the P elements on the M' Birmingham chromosome (68). P[ry$^+$;Δ2–3] (99B) embryos can also be used as injection hosts for transformation, resulting in some cases in increased transformation frequencies in the absence of helper plasmids (20, 68). Furthermore, injection of purified transposase protein allows high rates of transformation (34). Obviously, stable genomic transposase-producing P elements have become valuable genetic tools for *Drosophila* molecular genetics.

The advantages of genetically marked transposons were appreciated very early in bacterial genetics. Phenotypes could easily be scored on petri plates by simple colorimetric assays. Transposon insertions often resulted in gene fusions in which the genetic marker on the transposable element could be used to detect gene activity of nearby genes. This classic approach has now been applied successfully in *Drosophila* using the *E. coli LacZ* gene that encodes the enzyme beta-galactosidase. Early P element transformation experiments used *LacZ* as a marker gene to which foreign transcriptional promoter sequences were fused. Tissue-specific transcriptional control elements could then be assayed by using a histochemical stain for the *E. coli* β-galactosidase enzyme. A major breakthrough, however, came when O'Kane & Gehring showed that the *LacZ* gene fused to the weak P element promoter could be used in an "enhancer trap" experiment (58). The P-*LacZ* transposon was mobilized to many new genomic locations and novel patterns of expression were detected by histochemical *LacZ* staining (58). These expression patterns were conferred by nearby transcriptional enhancer sequences that acted upon the weak P element promoter (58). These P-*LacZ* enhancer trap experiments have now been widely used to examine patterns of gene expression and to identify genes involved in oogenesis (21, 25), embryogenesis (3, 5, 58, 90) and nervous system development (23), as well as other developmental processes (50). In some cases, the P element insertion causes a mutant phenotype that is associated with the insertion (90). In most cases, the adjacent flanking genomic DNA can be isolated directly using a plasmid rescue procedure, due to the presence of a bacterial plasmid origin of replication and drug resistance marker in the P-*LacZ* transposon (90). This is a very powerful genetic tool for identifying potentially important genes by their temporal and spatial expression patterns that may not have been previously identified in classical genetic screens (50). This method should prove useful for years to come and provide new insights into *Drosophila* development and differentiation.

FUTURE DIRECTIONS

The analysis of P elements using the tools of molecular biology has allowed a general understanding of various genetic observations on hybrid dysgenesis. In addition, the development of these transposons as vectors for germline transformation and as genetic tools for *Drosophila* molecular genetics has truly made *Drosophila* the most genetically manipulable metazoan. The recent application of biochemical techniques to the problem of P element function and the regulation of transposition has led to some exciting and unexpected findings. For example, the P element transposase recognizes internal sites on P element DNA, suggesting that other *Drosophila* proteins may recognize the terminal inverted repeats and participate in transposition. Biochemical approaches should complement genetic methods in identifying and characterizing these host factors. Development of an in vitro assay for P element excision or transposition will be central in investigating the molecular mechanism of transposition, as well as defining the catalytic role of transposase in the reaction. In addition, an understanding of how the transposition reaction occurs might allow the use of P elements as genetic tools in other organisms. The genetic control of transposition seems to be complex and probably involves several types of mechanisms, including transcriptional repression, DNA binding of truncated transposase-related molecules (such as the 66 kd protein), and protein-protein interactions such as those hypothesized for the small KP protein, perhaps involving associations mediated through the "leucine zipper" motifs. In addition, although the maternal effect of P cytotype is still a mystery, one explanation could be the juxtaposition of repressor-encoding P elements near genes normally expressed during oogenesis. These ideas are both interesting and testable. The tissue-specificity of transposition, which involves control of IVS3 splicing between ORF2-ORF3, seems to involve, at least in part, somatic repression of IVS3 splicing. Identification and characterization of the regulatory molecules involved in the control of IVS3 splicing may lead to a better understanding of germline development and differentiation, as this mechanism probably operates normally on other *Drosophila* genes. Thus, the continued application of both genetic and biochemical techniques to the problems of understanding P element transposition and its regulation in *Drosophila* promises to uncover new surprises involving the molecular mechanisms governing gene expression in eukaryotes.

ACKNOWLEDGEMENTS

I thank my students, Paul Kaufman, Sima Misra, and Chris Siebel for their contributions to the work described here, for many stimulating discussions about P elements and their control, and for critical reading of the manuscript. I also thank my colleagues who contributed unpublished and inpress data, and D. Black, A. Frankel, M. Mullins, and members of my laboratory for critical reading of the manuscript and helpful suggestions, S. Misra for her thoughtful suggestions and insights, and A. Kron for preparation of the manuscript and figures. Work in my laboratory has been supported by the NIH (5 R29 HD22857-03), NSF (DMB 8857176), the Lucille P. Markey Charitable Trust, and the DuPont Center for Molecular Genetics at Whitehead Institute. I have been supported by a Lucille P. Markey Scholar Award.

Literature Cited

1. Baker, B. S. 1989. Sex in flies: the splice of life. *Nature* 340:521–24
2. Bandziulius, R. J., Swanson, M. S., Dreyfuss, G. 1989. RNA-binding proteins as developmental regulators. *Genes Dev.* 3:431–37
3. Bellen, H., O'Kane, C. J., Wilson, C., Grossniklaus, U., Kurth-Pearson, R., Gehring, W. J. 1989. P-element mediated enhancer detection: a versatile method to study development in *Drosophila*. *Genes Dev.* 3:1288–300
4. Berg, D. E., Howe, M. M., eds. 1989. *Mobile DNA*. Washington, DC: Am. Soc. Microbiol.
5. Bier, E., Vaessin, H., Shepard, S., Lee, K., McCall, K., et al. 1989. Searching for pattern and mutations in the *Drosophila* genome with a P-*LacZ* vector. *Genes Dev.* 3:1273–87
6. Bingham, P. M., Kidwell, M. G., Rubin, G. M. 1982. The molecular basis of P-M hybrid dysgenesis: the role of the P element, a P strain-specific transposon family. *Cell* 29:995–1004
7. Bingham, P. M., Levis, R., Rubin, G. M. 1981. The cloning of the DNA sequences from the *white* locus of *Drosophila melanogaster* using a novel and general method. *Cell* 25:693–704
8. Black, D. M., Jackson, M. S., Kidwell, M. G., Dover, G. A. 1987. KP elements repress P-induced hybrid dysgenesis in *D. melanogaster*. *EMBO J.* 6:4125–35
9. Bregliano, J. C., Kidwell, M. G. 1983. Hybrid dysgenesis determinants. In *Mobile Genetic Elements*, ed. J. A. Shapiro, pp. 363–410. London: Academic
10. Buratowski, S., Hahn, S., Sharp, P. A., Guarente, L. 1988. Function of a yeast TATA element binding protein in a mammalian transcription system. *Nature* 334:37–42
11. Coen, D. 1990. P element regulatory products enhance Zeste[1] repression on a P[white[duplicated]] transgene in *Drosophila melanogaster*. *Genetics*. In press
12. Cooley, L., Kelley, R., Spradling, A. 1988. Insertional mutagenesis of the *Drosophila* genome with single P elements. *Science* 239:1121–28
13. Craigie, R., Arndt-Jovin, D. J., Mizuuchi, K. 1985. A defined system for the DNA strand transfer reaction at the initiation site of bacteriophage MU transposition: protein and DNA substrate requirements. *Proc. Natl. Acad. Sci. USA* 82:7570–74
14. Daniels, S. B., McCarron, M., Love, C., Clark, S. H., Chovnick, A. 1985.

Dysgenesis induced instability of *rosy* locus transformation in *Drosophila melanogaster:* analysis of excision events and the selective recovery of control element deletions. *Genetics* 109:95–117
15. Echols, H. 1986. Multiple DNA-protein interactions governing high precision DNA transactions. *Science* 233:1050–56
16. Engels, W. R. 1979. Extrachromosomal control of mutability in *Drosophila melanogaster*. *Proc. Natl. Acad. Sci. USA* 76:4011–15
17. Engels, W. R. 1983. The P family of transposable elements in *Drosophila*. *Annu. Rev. Genet.* 17:315–44
18. Engels, W. R. 1984. A trans-acting product needed for P factor transposition in *Drosophila*. *Science* 226:1194–96
19. Engels, W. R. 1989. P elements in *Drosophila melanogaster*. See Ref. 4, pp. 437–84
20. Engels, W. R., Benz, W. K., Preston, C. R., Graham, P. L., Phillis, R. W., Robertson, H. M. 1987. Somatic effects of P element activity in *Drosophila melanogaster:* pupal lethality. *Genetics* 117:745–57
21. Fasano, L., Kerridge, S. 1988. Monitoring positional information during oogenesis in adult *Drosophila*. *Development* 104:245–53
22. Federoff, N. 1989. Maize transposable elements. See Ref. 4, pp. 375–411
23. Ghysen, A., O'Kane, C. J. 1989. Neural enhancer-like elements as specific cell markers in *Drosophila*. *Development* 105:35–52
24. Green, M. R. 1986. Pre-mRNA splicing. *Annu. Rev. Genet.* 20:671–708
25. Grossniklaus, U., Bellen, H. J., Wilson, C., Gehring, W. J. 1989. P-element-mediated enhancer detection applied to the study of oogenesis in *Drosophila*. *Development* 107:189–200
26. Hay, B., Jan, L. Y., Jan, Y. N. 1988. A protein component of *Drosophila* polar granules is encoded by *vasa* and has extensive sequence similarity to ATP-dependent helicases. *Cell* 55:577–87
27. Herskowitz, I. 1987. Functional inactivation of genes by dominant negative mutations. *Nature* 329:219–22
28. Hiraizumi, Y., Crow, J. F. 1960. Heterozygous effects on viability, fertility, rate of development and longevity of *Drosophila* chromosomes that are lethal when homozygous. *Genetics* 45:1071–83
29. Jackson, M. S., Black, D. M., Dover,

G. A. 1988. Amplification of KP elements associated with the repression of hybrid dysgenesis in *Drosophila melanogaster*. *Genetics* 120:1003–13

30. Johnson, R. C., Simon, M. I. 1985. Hin-mediated site-specific recombination requires two 26bp recombination sites and a 60bp recombination enhancer. *Cell* 41:781–91

31. Kahmann, R., Rudt, F., Koch, C., Mertens, G. 1985. G inversion in bacteriophage Mu is stimulated by a site within the invertase gene and a host factor. *Cell* 41:771–80

32. Karess, R. E., Rubin, G. M. 1984. Analysis of P transposable element functions in *Drosophila*. *Cell* 38:135–46

33. Kaufman, P. K., Doll, R. F., Rio, D. C. 1989. *Drosophila* P element transposase recognizes internal P element DNA sequences. *Cell* 59:359–71

34. Kaufman, P. K., Rio, D. C. 1990. *Drosophila* P element transposase acts as a transcriptional repressor in vitro. Submitted

35. Kidwell, M. G. 1985. Hybrid dysgenesis in *Drosophila melanogaster:* nature and inheritance of P element regulation. *Genetics* 111:337–50

36. Kidwell, M. G., Kidwell, J. F., Sved, J. A. 1977. Hybrid dysgenesis in *Drosophila melanogaster:* a syndrome of aberrant traits including mutation, sterility and male recombination. *Genetics* 86:813–33

37. Kornberg, A. 1988. DNA replication. *J. Biol. Chem.* 263:1–4

38. Landshultz, W. H., Johnson, P. F., McKnight, S. C. 1988. The leucine zipper: a hypothetical structure common to a new class of DNA binding proteins. *Science* 240:1759–64

39. Landy, A. 1989. Dynamic, structural, and regulatory aspects of λ site-specific recombination. *Annu. Rev. Biochem.* 58:913–49

40. Laski, F. A., Rio, D. C., Rubin, G. M. 1986. Tissue specificity of *Drosophila* P element transposition is regulated at the level of mRNA splicing. *Cell* 44:7–19

41. Laski, F. A., Rubin, G. M. 1989. Analysis of cis-acting requirements for germline-specific splicing of the P element ORF2-ORF3 intron. *Genes Dev.* 3:720–28

42. Lasko, P. F., Ashburner, M. 1988. The product of the *Drosophila* gene *vasa* is very similar to eukaryotic initiation factor-4A. *Nature* 335:611–17

43. Leung, P. C., Teplow, D. B., Harsley, R. M. 1989. Interaction of distinct domains in Mu transposase within Mu

DNA ends and an internal transpositional enhancer. *Nature* 338:656–58

44. Levis, R., Hazelrigg, T., Rubin, G. M. 1985. Effects of genomic position on the expression of transduced copies of the *white* gene of *Drosophila*. *Science* 229:558–61

45. Maniatis, T., Reed, R. 1987. The role of small ribonucleoprotein particles in pre-RNA splicing. *Nature* 325:673–78

46. McClintock, B. 1954. Mutations in maize and chromosome aberrations in Neurospora. *Carnegie Inst. Washington Yearb.* 53:254–60

47. McElwain, M. C. 1986. The absence of somatic effects of P-M hybrid dysgenesis in *Drosophila melanogaster*. *Genetics* 113:897–918

48. Misra, S., Rio, D. C. 1990. Cytotype control of *Drosophila* P. element transposition: the 66kD protein is a repressor of transposase activity. *Cell* 62:269–84

49. Mizuuchi, M., Mizuuchi, K. 1989. Efficient Mu transposition requires interaction of transposase with a DNA sequence at the mu operator: implications for regulation. *Cell* 58:399–408

50. Mlodzik, M., Hiromi, H., Weber, U., Goodman, C. S., Rubin, G. M. 1990. The *Drosophila seven-up* gene, a member of the steroid receptor gene superfamily, controls photoreceptor cell fates. *Cell* 60:211–24

51. Morisato, D., Kleckner, N. 1987. Tn10 transposition and circle formation in vitro. *Cell* 51:101–11

52. Mount, S. M., Steitz, J. A. 1981. Sequence of U1RNA from *Drosophila melanogaster:* implications for U1 secondary structure and possible involvement in splicing. *Nucleic Acids Res.* 9:6351–68

53. Mullins, M. C., Rio, D. C., Rubin, G. M. 1989. Cis-acting DNA sequence requirements for P-element transposition. *Genes Dev.* 3:729–38

54. Nakajima, N., Horikoshi, M., Roeder, R. G. 1989. Factors involved in specific transcription by mammalian RNA polymerase II: purification, genetic specificity and TATA box-promoter interactions of TFIID. *Mol. Cell. Biol.* 8:4028–40

55. Nelson, K. K., Green, M. R. 1988. Splice site selection and ribonucleoprotein complex assembly during in vitro pre-mRNA splicing. *Genes Dev.* 2:319–29

56. Nitasaka, E., Mukai, T., Yamazaki, T. 1987. Repressor of P elements in *Drosophila melanogaster:* cytotype de-

termination by a defective P element with only open reading frames 0 through 2. *Proc. Natl. Acad. Sci. USA* 84:7605–8

57. O'Hare, K., Rubin, G. M. 1983. Structure of P transposable elements and their sites of insertion and excision in the *Drosophila melanogaster* genome. *Cell* 34:25–35

58. O'Kane, C. J., Gehring, W. J. 1987. Detection in situ of genomic regulatory elements in *Drosophila*. *Proc. Natl. Acad. Sci. USA* 84:9123–27

59. O'Shea, E. R., Rutkowski, R., Kim, P. S. 1989. Evidence that the leucine zipper is a coiled coil. *Science* 243:538–42

60. Padgett, R. A., Grabowski, P. J., Konarska, M. M., Seiler, S., Sharp, P. A. 1986. Splicing of messenger RNA precursors. *Annu. Rev. Biochem.* 55:1119–50

61. Parker, C. S., Topol, J. 1984. A *Drosophila* RNA polymerase II transcription factor contains a promoter-region-specific DNA binding activity. *Cell* 36:357–69

62. Parsons, R. L., Harshey, R. 1988. Autoregulation of phage μ transposase at the level of translation. *Nucleic Acids Res.* 16:11285–301

63. Phadnis, S. H., Berg, D. E. 1987. Identification of base pairs in the outside end of insertion sequence IS50 that are needed for IS50 and Tn5 transposition. *Proc. Natl. Acad. Sci. USA* 84:9118–22

64. Reed, R., Maniatis, T. 1986. A role for exon sequences and splice-site proximity in splice-site selection. *Cell* 46:681–90

65. Rio, D. C. 1988. Accurate and efficient pre-mRNA splicing in *Drosophila* cell-free extracts. *Proc. Natl. Acad. Sci. USA* 85:2904–8

66. Rio, D. C., Laski, F. A., Rubin, G. M. 1986. Identification and immunochemical analysis of biologically active *Drosophila* P element transposase. *Cell* 44:21–32

67. Rio, D. C., Rubin, G. M. 1988. Identification and purification of a *Drosophila* protein that binds to the terminal 31-base-pair inverted repeats of the P transposable element. *Proc. Natl. Acad. Sci. USA* 85:8929–33

68. Robertson, H. M., Engels, W. R. 1989. Modified P elements that mimic the P cytotype in *Drosophila melanogaster*. *Genetics* 123:815–24

69. Robertson, H. M., Preston, C. R., Phillis, R. W., Johnson-Schlitz, D., Benz, W. K., Engels, W. R. 1988. A stable genomic source of P element transposase

in *Drosophila melanogaster*. *Genetics* 118:461–70

70. Roiha, H., Rubin, G. M., O'Hare, K. 1988. P element insertions and rearrangements at the *singed* locus of *Drosophila melanogaster*. *Genetics* 119:75–83

71. Rubin, G. M. 1983. Dispersed repetitive DNAs in *Drosophila*. In *Mobile Genetic Elements*, ed. J. A. Shapiro, pp. 329–61. London: Academic

72. Rubin, G. M., Kidwell, M. G., Bingham, P. M. 1982. The molecular basis of P-M hybrid dysgenesis: the nature of induced mutations. *Cell* 29:987–94

73. Rubin, G. M., Spradling, A. C. 1982. Genetic transformation of *Drosophila* with transposable element vectors. *Science* 218:348–53

74. Rubin, G. M., Spradling, A. C. 1983. Vectors of P element-mediated gene transfer in *Drosophila*. *Nucleic Acids Res.* 11:6341–51

75. Searles, L. L., Greenleaf, A. L., Kemp, W. E., Voelker, R. A. 1986. Sites of P element insertion and structures of P element deletions in the 5' region of *Drosophila melanogaster* RpII215. *Mol. Cell. Biol.* 6:3312–19

76. Siebel, C. W., Rio, D. C. 1990. Regulated splicing of the *Drosophila* P transposable element third intron in vitro: somatic repression. *Science.* 248:1200–8

77. Simmons, M. J., Bucholz, L. M. 1985. Transposase titration in *Drosophila melanogaster:* a model of cytotype in the P-M system of hybrid dysgenesis. *Proc. Natl. Acad. Sci. USA* 82:8119–23

78. Simmons, M. J., Raymond, J. D., Rasmusson, K. E., Miller, L. M., McLarnon, C. F., Zunt, J. R. 1990. Repression of P element-mediated hybrid dysgenesis in *Drosophila melanogaster*. *Genetics* 124:663–76

79. Slatko, B. E., Mason, J. M., Woodruff, R. C. 1984. The DNA transposition system of hybrid dysgenesis in *Drosophila melanogaster* can function despite defects in host DNA repair. *Genet. Res.* 43:159–71

80. Spradling, A. C. 1986. P element-mediated transformation. In *Drosophila: a Practical Approach*, ed. D. B. Roberts, pp. 175–97. Oxford: IRL Press

81. Spradling, A. C., Rubin, G. M. 1981. *Drosophila* genome organization: conserved and dynamic aspects. *Annu. Rev. Genet.* 15:219–64

82. Spradling, A. C., Rubin, G. M. 1982. Transposition of cloned P elements into

Drosophila germ line chromosomes. *Science* 218:341–47

83. Steller, H., Pirrotta, V. 1986. P transposons controlled by the heat shock promoter. *Mol. Cell. Biol.* 6:1640–49

84. Sved, J. A. 1987. Hybrid dysgenesis in *Drosophila melanogaster:* evidence from sterility and Southern hybridization that P cytotype is not maintained in the absence of chromosomal P factors. *Genetics* 115:121–27

85. Tsubota, S., Schedl, P. 1986. Hybrid dysgenesis-induced revertants of insertions at that 5' end of the rudimentary gene in *Drosophila melanogaster:* transposon-induced control mutations. *Genetics* 114:165–82

86. Vinson, C. R., Sigler, P. B., McKnight, S. L. 1989. Scissors—grip model for DNA recognition by a family of leucine zipper proteins. *Science* 246:911–16

87. Voelker, R. A., Greenleaf, A. L., Gyurkovics, H., Wisely, G. B., Huang, S., Searles, L. L. 1984. Frequent imprecise excision among reversions of a P element-caused lethal mutation in *Drosophila. Genetics* 107:279–94

88. Williams, J. A., Pappu, S. S., Bell, J. B. 1988. Molecular analysis of hybrid dysgenesis-induced derivatives of a P element allele at the *vg* locus. *Mol. Cell. Biol.* 8:1489–97

89. Williams, J. A., Pappu, S. S., Bell, J. B. 1988. Suppressible P element alleles of the *vestigial* locus in *Drosophila melanogaster. Mol. Gen. Genet.* 212:370–74

90. Wilson, C., Kurth-Pearson, R., Bellen, H. J., O'Kane, C. J., Grossniklaus, U., Gehring, W. J. 1989. P-element mediated enhancer detection: an efficient method for isolating and characterizing developmentally regulated genes in *Drosophila. Genes Dev.* 3:1301–13

91. Yin, J. C. P., Reznikoff, W. S. 1987. *dnaA*, an essential host gene, and Tn5 transposition. *J. Bacteriol.* 169:4637–45

Annu. Rev. Genet. 1990. 24:579–613

PREMEIOTIC INSTABILITY OF REPEATED SEQUENCES IN *NEUROSPORA CRASSA*[1]

Eric U. Selker

Department of Biology and Institute of Molecular Biology, University of Oregon, Eugene, Oregon 97403

KEY WORDS: repeat-induced point mutation (RIP), DNA methylation, recombination, genome evolution, rDNA

CONTENTS

[1] I dedicate this review to David D. Perkins, who by his enthusiasm, altruism, and example has been largely responsible for making Neurospora research enjoyable and rewarding.

579

INTRODUCTION

What is responsible for the size and organization of a eukaryotic genome? Genomes vary greatly in the amount of repetitive DNA they contain. Lower eukaryotes, such as fungi, tend to have small genomes with few repeated sequences. The genome of *Neurospora crassa,* for example, consists of only $\approx 4 \times 10^7$bp of DNA, of which <10% is repetitive (20, 41, 57). In contrast, higher plants and animals have genomes that are typically two orders of magnitude larger, and commonly consist of $\approx 50\%$ repetitive sequences. It is tempting to assume that the observed variability reflects different needs of the organisms. It remains possible, however, that the variability simply reflects different degrees of success in resisting the onslaught of "selfish DNA"— sequences that direct their own amplification, but serve no obvious purpose for the organism (17, 58). To assess the relative importance of these possibilities will require information of several sorts.

We need to establish the costs and benefits to an organism of carrying large amounts of repetitive DNA. Presumably, the costs will depend strongly on the biology of the organism (see 14). The success of a filamentous fungus such as Neurospora may depend on its ability to spread extremely rapidly over a food source and gather the resources. If the genome contained a large amount of unnecessary DNA, the increased metabolic cost to the organism could slow its growth. It is easy to imagine that larger nuclei, which would presumably result from a larger genome, would also be detrimental to an organism such as Neurospora that relies on cytoplasmic streaming to move nuclei through incomplete septa between the cells. Other costs of carrying repetitive DNA might be common to all organisms. For instance, any organism should be sensitive to chromosomal translocations involving dispersed repeated sequences. Mutagenesis by insertion of transposable elements might represent another example, although the immediate consequences of mutations in a haploid organism are of course qualitatively different in a diploid.

A full understanding of genome structure will require information on DNA turnover: how repeated sequences arise and become amplified; and what processes oppose their spread. In this review, which is concerned with the latter part, I summarize and discuss what is known about two processes that act on repeated DNA in the genome of *N. crassa*. The first, premeiotic recombination, eliminates tandem duplications. The second, repeat-induced point mutation, RIP, mutates both linked and unlinked duplicated sequences and often leaves the products methylated. Neither process is unique to Neurospora. Identical or closely related processes have been observed in *Ascobolus immersus* (24, 29) and other filamentous ascomycetes.

Before reviewing premeiotic recombination and RIP, it is helpful to summarize (*a*) key features of the *N. crassa* life cycle and genome, (*b*) characteristics of DNA-mediated transformation in *N. crassa,* and (*c*) early hints of instability of repeated sequences in Neurospora.

Neurospora Life Cycle and Genome

The sexual phase of Neurospora and other filamentous ascomycetes differs from that of yeasts in having a prolonged heterokaryotic phase between fertilization and karyogamy (nuclear fusion). The *N. crassa* life cycle is illustrated in Figure 1, and its main features are noted in the legend.

The Neurospora genome consists of seven chromosomes with a total genetic map length of roughly 1000 map units (60). A tandem array of ≈170 copies of a 9.3 kb DNA sequence encoding the three large RNA molecules of the ribosome, located near one tip of linkage group V, accounts for most of the repetitive DNA of the organism (7, 41). Unlike most eukaryotic organisms, the ≈100 5S rRNA genes of Neurospora are dispersed in the genome (28, 47, 82). The genes share only ≈150 bp of sequence similarity (50, 82). Available information suggests that, at least in general, protein coding genes are present as unique sequences in the *N. crassa* genome. Unidentified repeated sequences are occasionally encountered in collections of sequences from the Neurospora genome, however. Some of these possibly represent tRNA genes and relics of transposable elements or retroviruses (37, 71; P. Garrett, J. Kinsey, E. Selker, unpublished results).

DNA-Mediated Transformation

Discovery of premeiotic recombination and RIP grew from the development of DNA-mediated transformation in Neurospora. Transformation of Neurospora was first reported by Tatum and associates (48, 49, 83), who used unfractionated genomic DNA. Case et al (13) pioneered the development of an efficient procedure using cloned DNA. Zalkin and other researchers improved the procedure, and DNA-mediated transformation soon became widely applied in Neurospora research (1, 6, 8, 30, 35, 39, 43, 56, 73, 74, 79, 89). Genetic and molecular characterization of the original transformants

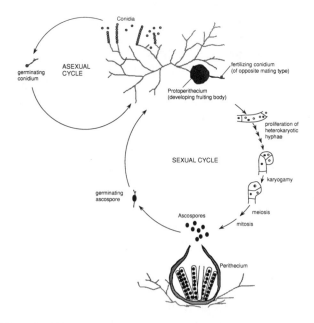

Figure 1 Life cycle of *Neurospora crassa*. The organism grows strictly as a haploid of one of two mating types (*A* or *a*). The vegetative phase is initiated when either a sexual spore (ascospore) or an asexual spore (conidium) germinates, giving rise to multinucleate cells that form branched filaments (hyphae). The hyphal system spreads out rapidly (>5 mm per hour at 37°C) to form a mycelium. Aerial hyphae grow up, and culminate in the production of the abundant orange conidia that are characteristic of the organism. The conidia, which contain one to several nuclei each, can establish new vegetative cultures, or fertilize crosses. If nutrients (principally nitrogen sources) are limiting, Neurospora enters its sexual phase by producing nascent fruiting bodies (protoperithecia), each of which originates from 100–300 vegetative nuclei (33). When a specialized hypha projecting from the protoperithecium contacts tissue of the opposite mating type, a nucleus is picked up and transported back to the protoperithecium. The heterokaryotic cells resulting from fertilization proliferate in the enlarging perithecium. Most of the proliferation involves cells having on the order of five or six nuclei (65; E. Barry, personal communication). In the final divisions, however, the cells are binucleate, containing one nucleus of each mating type. These cells bend to form hook-shaped cells (croziers) and a final conjugate mitosis occurs to produce four nuclei. Septa are laid down to produce one binucleate cell at the crook of the crozier. This cell gives rise to one ascus (65). Genetic analyses have indicated that, in general, the 100 or more asci of a perithecium are derived from a single maternal nucleus and a single paternal nucleus (33, 52). While it is generally assumed that nuclei of the initial fertilized cell undergo 7–10 divisions (22), this may be a gross underestimate (N. Raju, personal communication). Premeiotic DNA synthesis in filamentous ascomycetes occurs prior to nuclear fusion (see 65). When karyogamy finally occurs, the resulting diploid nucleus immediately enters into meiosis. Thus the diploid phase of the life cycle is limited to a single cell. The meiotic products undergo one mitotic division before being packaged as ascospores.

obtained using cloned DNA first demonstrated that Neurospora transformants fall into three classes[2]: unlinked insertions, linked insertions, and replacements (13). Most transformants of N. crassa arise by nonhomologous recombination between transforming DNA and chromosomal sequences, whether circular or linear transforming DNA is used (2, 59). Transformants frequently contain multiple copies of the transforming DNA, and two or more copies are commonly integrated at one site (6, 16, 30, 78). Frequently, transformation results in integration of essentially intact transforming DNA (2, 79; J. Kinsey, personal communication).

The least common class of transformants in Neurospora, linked insertions, arise by a single crossover between homologous sequences of a circular transforming DNA molecule and the host chromosome (12, 13, 79). In transformation experiments in which homologous recombination between transforming and host sequences can restore function of a mutated gene that is being selected, replacement[3] of the mutation is observed much more frequently than insertion of the entire plasmid (13, 16).

Hints of Instability of Repeated Sequences in Neurospora

Several classical observations are consistent with the idea that the low repetitive DNA content of the N. crassa genome is not simply an immediate result of natural selection. Indeed, the organism probably employs several mechanisms to maintain a streamlined genome. As yet, no mechanism to resist redundant DNA has been identified in the vegetative phase of the life cycle but there are reasons to suspect that one or more exist. For one thing, despite numerous attempts, N. crassa has never been successfully propagated as a diploid (see 60). This distinguishes Neurospora from some other fungi, including the filamentous ascomycete Aspergillus nidulans. Disomic ascospores form occasionally in N. crassa, but disomy breaks down rapidly to produce heterokaryons with haploid nuclei (see 60). Partial diploids, constructed by crossing translocation strains, are relatively stable during vege-

---

[2]Although some evidence of unstable autonomous replication of transforming sequences has been observed, stable autonomous replication has not been achieved in Neurospora (16, 59, 67, 85).

[3]Although replacement events formally look like products of double crossovers, in which both strands of DNA are replaced, they could also be products of heteroduplexes resulting from invasion of single strands of the transforming DNA. As expected, the relative frequency of integration of transforming DNA by homologous versus nonhomologous recombination depends on the transforming DNA. The best information available comes from a study by Asch & Kinsey (2) using the am^+ gene. They found 22 homologous recombination events among 70 transformants when 9 kb of homology was provided, but only one among 89 when 2–5 kb segments of the 9 kb region was used. It is not clear whether this difference in frequency was due to the length difference of the homology or to a sequence feature in the additional DNA.

tative growth, although they typically break down[4] eventually, generally by loss of the translocated segment (see 53). Duplications of chromosome segments are much less stable in the sexual phase of the life cycle, however. Segmental aneuploids rarely survive a cross. The duplications lead to developmental arrest at or just before karyogamy (66). This phenomenon is discussed more fully below.

Tandem duplications resulting from integration of transforming DNA by homologous recombination are lost at low but detectable frequencies during vegetative growth (30, 77). In general, however, Neurospora transformants are stable when propagated vegetatively. In striking contrast, as first noted by Tatum and associates for transformants obtained by the use of total genomic DNA (48, 49, 86, 87), and later by other researchers for transformants obtained using cloned DNA (12, 16, 30, 77), a large fraction of transformants of N. crassa are unstable in the sexual phase of the life cycle. Primary transformants of Neurospora are generally heterokaryotic, containing both transformed and untransformed nuclei. Because of this it was common practice in many laboratories to cross primary transformants to obtain homokaryotic derivatives prior to their characterization (12, 13, 16, 34, 59). In some cases this practice probably complicated the interpretations as a result of discrimination against particular types of transformants. Nevertheless, it became clear that even homokaryotic derivatives of transformants show poor transmission of transformation markers.

PREMEIOTIC RECOMBINATION

Deletion of Transforming DNA

A set of transformants generated to investigate DNA methylation in N. crassa were particularly useful for studying instability of transforming DNA in this organism. These were single-copy transformants generated with the plasmid pES174 (79). This plasmid consists of four parts: (a) the zeta-eta (ζ–η) region, a diverged (14%) direct tandem duplication of a 0.8 kb segment that is extensively methylated in N. crassa (79, 80); (b) $\approx$6 kb of uncharacterized sequences adjacent to ζ–η, that are normally unmethylated in the genome ("flank"); (c) the am^+ gene of N. crassa; and (d) sequences of the bacterial plasmid pUC8 (Figure 2). The transformation host had a deletion of the entire am gene (am_{132}; 38) and, in place of the ζ–η region, had one

---

[4]Curiously, certain mutations and growth conditions increase the rate of breakdown (54, 55). In 1977, Newmeyer & Galeazzi (53) proposed a mechanism to account for breakdown of duplications that result from quasiterminal translocations or inversions. The basic idea is that both the original rearrangement and breakdown events occur by recombination between repeated sequences located at the breakpoints. Although the model has not yet been properly tested, the finding of a repeated sequence immediately interior to a telomere (71) is consistent with it.

copy of a 0.8 kb segment that is ≈83% identical to this region (31, 81). Thus flank represented the only significant tract of homology between the transforming DNA and the host genome. One transformant analyzed, T-ES174-1, resulted from integration of the plasmid by homologous recombination, generating a local duplication of flank, as illustrated in Figure 2. Deletion of the am^+ gene occasionally occurred during vegetative growth: two of six vegetative reisolates from a culture that was initially homokaryotic revealed loss of the gene in a noticeable fraction of the cells (77). Results of Southern hybridizations suggested that deletion occurred by homologous recombination between the two copies of flank.

Greater instability of the transformation marker was observed in crosses of T-ES174-1. Only ≈16% of the progeny from a cross of an apparently homokaryotic isolate of the transformant and an am_{132} strain were Am$^+$, suggesting loss or inactivation of the gene in two thirds of the chromosomes initially carrying the transforming DNA (77). Normal segregation of the mating type genes, which are linked to the site of integration in T-ES174-1, was seen among progeny of this transformant, suggesting that loss of the marker did not involve lethality. This conclusion was confirmed and extended by analysis of progeny from individual asci. Thirty-five percent of the asci (16/55) showed Mendelian segregation of Am (4+ :4−), and sixty-five percent exhibited complete loss of the marker (9). Southern hybridization analyses on DNA of progeny from completely Am$^-$ asci revealed that the Am, ζ–η, and pUC8 sequences, and one copy of the flank sequences, had been eliminated, apparently by recombination between the two copies of flank. Restriction site polymorphisms between the host and transforming DNA sequences in the flank region revealed that the copy of flank that remains was sometimes predominantly from the host and sometimes predominantly from the transforming DNA. The transforming DNA was not lost by unequal sister chromatid exchange, since no progeny were found with two copies of the

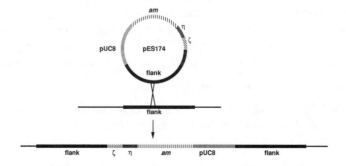

Figure 2   Generation of transformant T-ES174-1 by homologous recombination (79).

transforming sequences, which would be the reciprocal product of this event. Thus the transforming DNA was most likely deleted by intrachromatid recombination[5], a reversal of the integration reaction (Figure 2). The consistent loss of the transforming sequences from both meiotic products of the affected tetrads suggested that the recombination event had occurred prior to premeiotic DNA synthesis, and thus prior to karyogamy (see Figure 3).

Several nuclear divisions separate fertilization from karyogamy in *N. crassa*. To determine whether the high frequency of intrachromosomal recombination occurred before or after fertilization, Selker et al (77) took advantage of the fact that, in general, all the asci of a perithecium represent the products of a single fertilization event (see Figure 1 legend). If the recombination event occurred prior to fertilization in a cross between T-ES174-1 and an *am* strain, all asci from the resulting perithecium would be 0:8 (+:−) for Am. In contrast, if it occurred after fertilization, both 4:4 and 0:8 asci could be generated. Of eight perithecia in which two or more asci were analyzed, three showed just 4:4 asci, two showed just 0:8 asci, and three showed both 4:4 and 0:8 asci. Thus intrachromosomal recombination occurred after fertilization, at least in some of the perithecia. Relatively few asci were scored from most of the perithecia, and deletion of the transforming DNA may have occurred in *all* of the perithecia. The fraction of 0:8 asci fluctuated widely from perithecium to perithecium, however, suggesting that recombination could occur relatively early or late in the period between fertilization and karyogamy (E. Cambareri & E. Selker, unpublished observations).

The high frequency of this premeiotic intrachromosomal recombination raises the possibility that unlinked repeated sequences might also undergo a high level of recombination prior to meiosis. Recombination between sequences on different chromosomes would of course produce translocations, and recombination between inverted repeats on the same chromosome would give inversions. No evidence of translocations was observed among 16 progeny of strains that had ectopic copies of pES174, indicating that recombination between the unlinked copies of the flank sequences does not occur at frequencies comparable to that detected between the closely linked copies of these sequences (77). A thorough search for premeiotic interchromosomal recombination has not yet been conducted, however.

Change in rDNA Copy Number

The tandemly arranged rRNA genes are the only known natural example of linked repeated sequences in Neurospora. In several organisms, including

---

[5]Although it is also conceivable that a nonreciprocal process, possibly associated with replication (e.g. involving polymerase "jumping"), is responsible for the deletion, I refer to the deletion process as intrachromatid (or intrachromosomal) recombination.

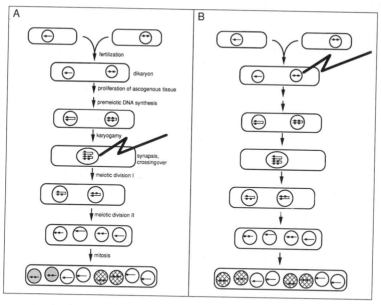

Figure 3   Manifestations of meiotic (A) and premeiotic (B) genetic instability of a duplicated sequence. The sexual cycle of *N. crassa* is diagrammed, illustrating expected outcomes of alterations (lightning bolts) of a linked duplication (duplicate squares). Different fill patterns represent different alterations in ascospores. Identical alterations of two meiotic products (four ascospores) implies that the changes occurred prior to premeiotic DNA synthesis.

Neurospora, rDNA is unusual in that recombination between homologous chromosomes is suppressed (see 68). The rRNA genes are not recombinationally inert, however. For example, in the yeast *Saccharomyces cerevisiae,* sister chromatid exchanges occur frequently during meiosis (63).

Butler & Metzenberg (7) have shown in *N. crassa* that tandem rDNA repeats of the nucleolus organizer region (NOR) are subject to copy-number changes at high frequency in the sexual phase of the life cycle. Most progeny of every cross exhibited differences from their parents in rDNA copy number. No sign of interchromosomal recombination was observed. Analyses of progeny from individual asci and from individual perithecia indicated that most alterations occur in the period between fertilization and karyogamy, the stage in which the artificial duplications of the flank region of pES174 transformants exhibited instability. All four progeny having the NOR from a given parent generally showed the same rDNA copy number. Likely evidence of changes both during and after meiosis was also seen, however. In two of 14 asci, the NOR size segregated in a 4:2:2 pattern, suggesting a copy-number change during or after premeiotic DNA synthesis. In one of 14 asci, the NOR size segregated in a 4:3:1 pattern, probably reflecting a change after meiosis.

A small cluster of rDNA repeats ($\approx$9), which had been moved to a different chromosome as a result of two sequential chromosomal translocations (61), were no less unstable, indicating that the effect was not dependent on the chromosomal location of the sequences. In 13 of 14 asci, the copy number of this ectopic rDNA was the same for all four progeny that contained it, although the copy number varied from one ascus to another (7). The ectopic rDNA in the exceptional ascus was of one size in two progeny, and one unit larger in two others. A difference of one unit could not occur by a single sister chromatid exchange. Thus this result suggested that intrachromatid recombination was responsible for the copy-number change, at least in this case. At both the normal and the ectopic NOR, most changes in rDNA copy number were reductions, consistent with the idea that many of the changes resulted from intrachromatid recombination, as in transformant T-ES174-1. Magnification of rDNA arrays may occur primarily during vegetative growth, as observed in *N. crassa* by Russell & Rodland (69). In one postmeiotic change in rDNA copy number observed by Butler & Metzenberg, chromosomes exhibiting complementary changes were identified, suggesting that the event occurred by sister chromatid exchange (7).

PREMEIOTIC MUTATION: DISCOVERY AND MAIN FEATURES OF RIP

Investigation of the fate of transforming sequences in crosses of pES174 transformants also led to the discovery of RIP. The first clue came from examining progeny derived from crosses of T-ES174-1, the transformant with the local duplication of the flank region (Figure 2). As mentioned above, normal segregation of the transformation marker, Am$^+$, was observed in those asci that did not show intrachromatid recombination. Southern hybridizations revealed, however, that 100% of Am$^+$ progeny had suffered sequence alterations in both copies of the flank region (77). Although the overall length of the sequences appeared at least roughly unchanged, the arrangement of restriction sites in the DNA showed numerous alterations. No changes were detected in the transforming DNA between the copies of flank, i.e. the pUC8, *am$^+$*, and ζ–η sequences. Use of the isoschizomers *Sau*3A, *Mbo*I, and *Dpn*I, and other restriction enzymes, produced evidence for both changes in the primary structure of the DNA and extensive de novo methylation of cytosines. A survey of random Am$^+$ progeny showed great variability in the alterations (see Figure 4).

Even unlinked copies of flank, resulting from nonhomologous integration of pES174 in the am_{132} strain, were altered in crosses. The changes occurred at a lower frequency, however, and generally appeared less radical (77). Approximately 50% of progeny from any of the four single-copy ectopic transformants tested exhibited alterations in flank. The transformation marker, am^+ showed normal segregation. As in progeny of T-ES174-1, both copies of the duplicated sequences, but none of the other tranforming sequences, appeared affected.

Timing of RIP

Biological features of Neurospora and inherent properties of RIP allowed us to pin-point when in the life cycle the process occurs (77). The diversity of the alterations in the flank region of random Am^+ progeny of the pES174 transformants allowed products of a common event to be distinguished from products of independent events. In the vast majority of asci yielding progeny with altered flank DNA, the products of sister chromatids displayed common alterations. This implied that the changes were determined prior to premeiotic DNA synthesis, and therefore prior to karyogamy and meiosis, as illustrated in Figure 3B. The phenomenon was first dubbed RIP for "rearrangement induced premeiotically" to emphasize this feature (77). In a minority of the asci, restriction pattern differences between pairs of Am^+ progeny were observed, which could have resulted from alterations during or after premeiotic DNA synthesis. Such results could also reflect heterologies in the final DNA duplex before premeiotic DNA synthesis, which were later resolved. No indication of meiotic or postmeiotic sequence alterations has yet been observed.

To determine whether RIP occurred before or after fertilization, Selker et al (77) again took advantage of the fact that the 100 or more asci of a perithecium are usually derived from a single maternal nucleus and a single paternal nucleus (33, 52). Thus, different alterations among progeny from different asci of the same perithecium would indicate that RIP occurs after fertilization. In general, different alterations were observed among progeny of individual perithecia, as shown in Figure 4. These results indicate that RIP occurs after fertilization. Identical alterations of flank sequences were observed among the three completely Am^- asci of one perithecium (number 4 of Figure 4). Results of Southern hybridizations indicated that the pUC8, ζ-η, and am sequences, and one copy of the flank sequences had been lost, presumably by intrachromosomal recombination following RIP. Finding the same pattern of alterations in multiple isolates from a given perithecium indicated that RIP can occur early during proliferation of the heterokaryotic tissue formed by fertilization.

Proof that RIP is Triggered by Duplications

These observations led Selker et al (77) to suggest that *N. crassa* has a mechanism that detects and alters sequence duplications in the haploid nuclei of the heterokaryotic tissue resulting from fertilization. Selker & Garrett (78) carried out a set of experiments to test this directly and to check whether the

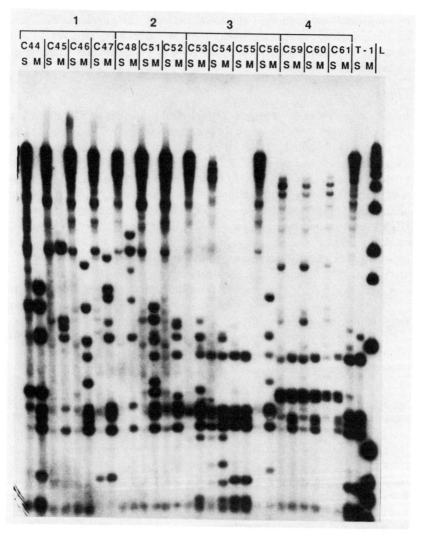

Figure 4 RIP occurs after fertilization. Progeny altered by RIP from several asci of each of four perithecia (1–4) from a cross T-ES174-1 (T-1) were analyzed by probing *Sau*3A (S) and *Mbo*I (M) digests for the flank region. *Sau*3A and *Mbo*I both recognize and cleave the sequence GATC when it is unmethylated, but *Sau*3A will not cut at this sequence if the cytosine is methylated. The enzyme *Mbo*I is insensitive to cytosine methylation, the only methylation known in *N. crassa,* but is sensitive to adenine methylation. Results are shown for one isolate per ascus. All the progeny illustrated from perithecia 1–3 except C55 were Am$^+$, and all those from perithecium 4 were Am$^-$. Perithecium 4 represents one of two cases interpreted as RIP followed by deletion observed among six perithecia in which completely Am$^-$ asci were found. RIP patterns vary greatly within perithecia 1–3 but not within perithecium 4. Radiolabeled molecular weight markers were loaded in lane L. (Reprinted with permission from Ref. 9.)

observed instability was in any way dependent on introduction into the cell of foreign DNA (pUC8), or DNA prone to methylation (ζ–η), or on disruption of the chromosome by integration of the transforming DNA. The plan, illustrated in Figure 5, took advantage of the am_{132} deletion (38). In the first phase (Figure 5A), a strain with this deletion was transformed with a 2.6 kb DNA fragment that contained the am gene but no bacterial sequences nor sequences already represented in the host. The idea was to test whether two or more copies of this unmethylated sequence, integrated in the genome without any foreign or methylated sequences, would trigger its inactivation by RIP. RIP occurred, indicating that neither methylated sequences nor non-Neurospora sequences are required for the process. Single-copy transformants, in contrast, were stable through crosses, consistent with the hypothesis that RIP is a response to sequence duplications, and showing that simple integration into the genome does not trigger the process.

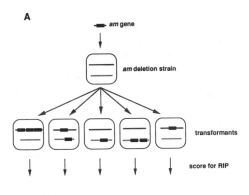

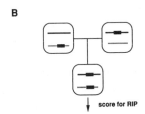

Figure 5 Scheme to test if RIP is triggered by sequence duplications.   (A) One or more copies of a Neurospora DNA fragment containing the am^+ gene were introduced into a strain with a deletion covering the entire fragment. The transformants were then crossed and scored for RIP phenotypically and at the DNA level.   (B) To test if the observed lack of RIP in some crosses was due to the single-copy status of the transformants, two of them were crossed to build a duplicate-am strain. Evidence of RIP was seen in crosses of this strain, indicating that RIP is indeed a response to sequence duplications (78).

The second phase of the study (Figure 5B) tested the possibility that stability of single-copy transformants was due to something other than their single-copy status (e.g. chromosomal position). Two such strains having single copies of am^+ at unlinked chromosomal locations were crossed to build a strain having two copies of the gene. Crosses of this duplicate-am^+ strain showed all the hallmarks of RIP, proving that RIP is triggered by duplications. Approximately 50% of the asci showed inactivation of am, and inactivation always involved both copies of the gene. This provided the first clear evidence that recognition of duplications by the cell involves interaction of the like-sequences (discussed further below). Southern hybridizations revealed sequence alterations, DNA methylation, or both, in many of the inactivated alleles (78). Experiments with strains having the am^+ gene at its normal chromosomal position plus one copy of am^+ contributed by transforming DNA gave equivalent results (K. R. Haack & E. Selker, unpublished observations). Similar experiments from several laboratories soon confirmed the suggestion that RIP operates on repeated sequences generally, and verified that the process could be used as a tool for directed, in vivo mutagenesis of specific chromosomal regions (2, 26, 27, 76).

Nature and Specificity of Alterations by RIP

Results from Southern hybridizations indicated that RIP does not, at least in general, result in large changes in the overall length of the affected sequences (77). It was impossible to tell from the Southern hybridizations, however, whether the restriction site rearrangements were due to classical rearrangements (e.g. numerous small insertions, deletions, inversions, or substitutions), point mutations, or a combination of these. Nevertheless, the repeated observation of new fragments of certain sizes among independent progeny suggested that RIP results in point mutations (78). Characterization of two ≈6 kb segments of DNA, chosen to represent mild and severe alterations by RIP, first revealed the nature of the alterations by the process (10). The mild example was from an unlinked duplication of flank that had been passed once through a cross, and, based on Southern hybridization results, seemed relatively lightly altered by RIP. The severe example was from the linked duplication of flank that had been passed twice through a cross. Its flank DNA exhibited many changed restriction fragments and seemed resistant to RIP in further crosses (see below). Both of the chosen sequences were methylated as a result of RIP.

The altered chromosomal regions were isolated and compared to their native counterparts by heteroduplex analysis to look for gross rearrangements, and by sequence analysis to detect local alterations. No evidence of classical rearrangements was observed by electron microscopy in the ≈ 6 kb heteroduplexes. In the sequences from the linked duplication, however, heteroduplexes formed under various stringency conditions exhibited evidence of substantial sequence divergence. Small bubbles were observed under conditions expected to allow pairing of molecules having up to 40% mismatch, and roughly half of the heteroduplex was opened under conditions expected to allow pairing of molecules having up to 20% mismatch (10). No bubbles were observed in heteroduplexes formed under very low stringency conditions.

DNA sequence comparisons of segments altered by RIP and their native counterparts revealed one type of mutation: G:C pairs were converted to A:T pairs[6]. In a ≈900 bp segment of the unlinked duplication, ≈10% of the G:C pairs had mutated, and in a ≈600 bp segment from one end of the linked duplication, 31% of the G:C pairs had mutated. Comparison of the melting properties of linearized plasmids having ≈6 kb of either the altered or the native sequence from the linked duplication suggested that ≈50% of the G:C pairs were lost overall (10).

Not all G:C pairs are equally mutable by RIP. The C to T mutations occur primarily at cytosines immediately 5' of adenines, and rarely at cytosines 5' of other cytosines. In the segment of the linked duplication that was sequenced, ≈64%, ≈18%, ≈13%, and ≈5% of CpA, CpT, CpG, and CpC dinucleotides, respectively, changed. The site specificity of RIP seemed equivalent in the segment from the unlinked duplication. It may not be coincidental that the observed low sequence-specificity of RIP, which is unusual for nucleic acid-enzyme interactions, is characteristic of eukaryotic DNA-methyltransferases. For example, the rat methyltransferase modifies ≈99%, ≈14%, ≈6%, and ≈0% of CpG, CpA, CpT, and CpC dinucleotides, respectively, in vitro (84). It is not yet known whether DNA methylation in *N. crassa* has specificity that parallels the site specificity of RIP.

The Rosetta Stone of RIP: The ζ–η Region

The ζ–η (zeta-eta) region, which is a diverged tandem duplication of a 794 bp segment including a 5S rRNA gene or pseudogene, was discovered in a study of 5S rRNA genes of *N. crassa* (80). The ζ and η 5S rRNA regions represent the only known exception to the rule that 5S rRNA genes are dispersed in the Neurospora genome (47, 82). A survey of laboratory strains of Neurospora indicated that not all *N. crassa* strains contain the duplication, suggesting that it arose relatively recently (81). Analysis of strains of known pedigree demonstrated, however, that the duplication of the ζ–η region from strain 74A-OR23-1VA had gone through at least six generations. Sequence comparisons between the duplicated elements of the ζ–η region, and between each of these and an unduplicated allele revealed that all of the differences were due to

---

[6]This finding led us to suggest that the acronym RIP be regarded as an abbreviation for "repeat-induced point mutation" (10).

transition mutations (31, 80). All but one of 268 inferred mutations in the region were G:C to A:T changes, suggesting that the region represented a product of RIP.

The distribution of the $\zeta-\eta$ mutations supported the idea that they resulted from RIP. The C to T changes occurred in 74%, 30%, 11%, and 2% of the CpA, CpT, CpG, and CpC dinucleotides, respectively. The data also revealed that the mutations do not occur completely without regard to the identity of the bases immediately 5', or two nucleotides 3', of the changed cytosines: C to T mutations occurred most frequently downstream of adenines, and CpApT sites were favored over all other trinucleotides (31). In the 123 nucleotides upstream of the duplicated region, the duplicated and unduplicated alleles were identical. A single G:C to A:T mutation was found about 20 bp downstream of the duplicated region.

Analysis of the $\zeta-\eta$ region also provided new information about RIP: the process is active on relatively short direct tandem duplications; and duplicate elements that trigger RIP suffer equivalent, but nonidentical, damage. Statistical analysis of the distribution of mutations in the ζ and η halves of the duplication suggested that all the CpA dinucleotides were more or less equally mutable. Visual inspection of the distribution of the mutations in the $\zeta-\eta$ region revealed some higher-order biases, however. For example, the seven CpA dinucleotides that were not altered in either the ζ or η halves of the duplication were all relatively close to the ends of the repeated segment. The ends of the repeats immediately adjacent to the unique DNA were the least mutated regions.

DNA Methylation Resulting from RIP

The bulk of the Neurospora genome is unmethylated. It is interesting, therefore, that methylation is a common feature of sequences altered by RIP (77, 78). It is important to note that the methylation resulting from RIP is detected in vegetative cells, cells that are not active for RIP. The same sequences may be methylated at the time RIP occurs, but this possibility has not yet been examined. Methylation, where observed, has usually been very heavy; most or all of the sites examined in the duplicated sequence appear methylated in a large fraction of the molecules. Aside from the tandemly repeated rDNA, which displays some methylation (61), the few methylated patches that have been encountered in wild-type strains show evidence of RIP—polarized transition mutations associated with a repeated sequence. The $\zeta-\eta$ region represents the best-characterized example. Results of restriction analyses suggest that most of the cytosines in the duplication are methylated. No evidence of methylation was found in the adjacent sequences, nor in the unduplicated allele (80, 81; E. Selker, unpublished observations).

Although sequence duplications initially trigger RIP, they are not required for the methylation of sequences altered by the process. This conclusion comes in part from the observation of methylation of unique sequences resulting from meiotic segregation. Moreover, a DNA sequence altered by RIP can direct its methylation, de novo. When the $\zeta-\eta$ region was stripped of its methylation by propagation in *Escherichia coli,* and then reintroduced into Neurospora, it became specifically and reproducibly remethylated in vegetative cells (79). A comparable experiment using a product of RIP generated in the laboratory produced equivalent results (11). We therefore conclude that the G:C to A:T mutations somehow render the sequences substrates for DNA methylation. A model to account for this has been proposed (75).

Limits of RIP

The fact that the Neurospora genome is not completely devoid of repeated sequences raised the possibility that different sequences might have different sensitivities to RIP. At one extreme, one could imagine that the process depends on special sites, analogous to promoters or recombinators, that are not present in every DNA segment. Another possibility is that some sequences can confer "immunity" to RIP on neighboring DNA. Finally, it is conceivable that certain regions of the genome might be protected from RIP. Although these possibilities have not yet been thoroughly explored, some relevant information is available.

As far as I know, every sequence of appreciable length (i.e. > 1 kb) that has been duplicated in *N. crassa* and tested, has shown signs of RIP. All of the approximately 15 Neurospora genes that have been tested in various laboratories exhibited sensitivity to the process (unpublished observations of E. Cambareri, M. Case, C. Engele, J. Fincham, L. Glass, K. Haack, H. Inoue, B. Jensen, J. Kinsey, G. Marzluf, R. Metzenberg, E. Selker, C. Staben, D. Stadler, C. Yanofsky). Foreign sequences introduced into Neurospora in multiple copies were also sensitive to RIP (27).

NATIVE-REPEATED SEQUENCES   *rDNA*   Even repeated sequences that are native to the Neurospora genome can be susceptible to RIP. That the $\zeta-\eta$ region includes relics of 5S rRNA genes suggested that these small genes are sensitive to RIP under some circumstances. The small size (<150 bp) and dispersed arrangement of the ≈100 5S rRNA genes of Neurospora probably protect them from the process (P. Garrett & E. Selker, unpublished observations). The tandemly repeated genes encoding the three large rRNA molecules are probably protected because of their particular chromosomal position rather than because of something special about their sequence. A single rDNA repeat unit introduced into an ectopic site by transformation was subject to RIP (K. Haack, B. Jensen, & E. Selker, unpublished observations). This suggests that some repeats in the NOR are available for premeiotic pairing and therefore raises the possibility that some fraction of the rDNA units of the NOR are actually susceptible to RIP. Perkins et al (61) noted that a small cluster of rDNA repeats that had been moved from their normal home in the NOR by two sequential translocations, exhibit abnormally heavy methylation. It seems possible that the methylation, which was detected in the ectopic rDNA after a cross, reflects the operation of RIP. Loss of rDNA repeats by premeiotic intrachromatid recombination, in conjunction with re-amplification and natural selection, could effectively rid the NOR of defective sequences.

TRANSPOSABLE ELEMENTS   The primary role of RIP in *N. crassa* may be to protect the organism from transposons, viruses, and selfish DNA generally. As would be expected if this were so, the Neurospora genome appears relatively clean of such sequences. Putative selfish DNA has been found in *N. crassa,* however. Schechtman identified a ≈1.6 kb sequence, "pogo", adjacent to the linkage group VR telomere that is also present at several other sites in the genome and shows some features of a transposon (71, 72). Three wild-type strains examined showed different restriction fragments hybridizing with pogo, and their segregation suggested that at least some of the copies were at different locations in the strains. Analysis of the original copy of pogo revealed 318 bp imperfect direct repeats separated by an ≈1 kb segment (72). Thirty-three of the 34 nucleotide differences between the terminal repeats are transitions and methylation is associated with the element (M. Schechtman & E. Selker, unpublished observations). Thus, the ancestor of pogo was presumably subjected to RIP.

Kinsey & Helber (37) identified an active transposon, "Tad", in a wild-type strain of *N. crassa* collected from Adiopodoumé, Ivory Coast. Tad, which is ≈7 kb in length, lacks long terminal repeats, and does not appear related to pogo. The Adiopodoumé strain was chosen for a transposon search primarily because it appeared to contain one or more factors that decreased the stability of segmental aneuploids and increased the frequency of translocations (37, 53). Ten or more copies of Tad are present in the Adiopodoumé strain, but it has not been found in any of more than 400 other wild and laboratory strains of *Neurospora* screened (36). When crossed into laboratory strains of *N. crassa,* Tad usually becomes amplified, demonstrating that at least one copy of the transposon was active. Approximately 1/3 of progeny show some evidence of alterations of the transposon, however, and sometimes all copies of the transposon appear inactivated (P. Garrett, J. Kinsey, & E. Selker, unpublished observations). The finding that Tad does not appear immune to RIP suggests that either its original host, Adiopodoumé, is defective for RIP, or that Tad represents a recent intruder into the Neurospora genome.

RECURRENCE OF RIP   Information from analysis of progeny from individual perithecia suggested that RIP can occur in any of the roughly ten cell generations between fertilization and karyogamy. This raised the possibility that duplicated sequences might be sensitive to multiple cycles of RIP. Consistent with this idea, unlinked duplications, which in contrast to linked duplications escape RIP at an appreciable frequency, exhibit less severe alterations than do linked duplications. Thus the severe alterations of linked duplications may reflect the sum of many cycles of RIP.

Cambareri et al (11) showed that sequences altered by RIP can remain susceptible to the process. Duplicated DNA of several single-copy pES174 transformants were followed through seven generations. After a few generations both linked and unlinked duplicated sequences became resistant to RIP. In one series of crosses starting with T-ES174-1, the transformant having a linked duplication of flank, five out of six progeny showed alterations in the second generation, but no further alterations were observed until five generations later, when one of six progeny exhibited evidence of RIP.

Results of an analysis of the stability of hybrids between sequences altered by RIP and their native counterpart, suggest that linked duplications diverge further than do unlinked duplications before becoming resistant to more change. DNA methylation resulting from RIP does not seem to confer resistance to the process. Rather, resistance is primarily due to sequence divergence of the homologous regions. Depletion of preferred substrates for RIP (i.e. C:G pairs in favorable sequence contexts) also plays a role. This conclusion comes from experiments in which one or more copies of a RIP-resistant derivative of flank were introduced into a strain with just the native flank sequence. The resulting transformants were then crossed and scored for RIP (11). When multiple copies of the mutated sequence happened to integrate in a cluster, they were very sensitive to RIP. In contrast, a single-copy transformant showed no evidence of RIP. The altered copy of the flank sequences was apparently unable to trigger RIP of the native copy. When an identical, second copy of the resistant DNA was crossed in from the strain that the resistant sequence was isolated from, RIP occurred, albeit at low frequency (presumably because the copies were unlinked).

Foss et al (27) obtained similar findings using the $\zeta-\eta$ region, the natural 1.6 kb tandem duplication that must have experienced numerous cycles of RIP (see above). Not surprisingly, this diverged duplication is stable in crosses of strains having single copies of the region (77). In contrast, crosses of strains having two copies of the $\zeta-\eta$ region, created by crossing single-copy transformants, exhibit evidence of RIP. A relatively low incidence of instability was observed by Foss et al presumably because of depletion of mutable sites, and because the duplicate sequences were unlinked.

THOUGHTS ON THE MECHANISM OF RIP

Detection of Duplications Probably Entails Pairing

How can the cell efficiently detect small sequence duplications, especially when the duplicate copies are at unlinked chromsomal positions? A duplication of a 2 kb segment would correspond to only $\approx 0.01\%$ of the Neurospora genome. Although detailed information is not yet available, we can start to paint a general picture of the recognition process. Recognition of sequence duplications by RIP is analogous to recognition of homologous sequences by recombination processes, and presumably RIP and recombination are mechanistically related. The temporal correlation of RIP and high frequency intrachromosomal recombination is consistent with this idea. Furthermore, observations on the pattern of inactivation of genes imply that RIP relies on pairing of duplicate sequences, as in recombination. In nuclei containing two copies of a sequence, RIP mutates either both sequences or neither sequence, but if three or more copies are present, the process frequently results in a combination of altered and unaltered copies (26, 78). All available data are consistent with the idea that RIP invariably operates on repeated sequences in a pair-wise manner.

Studies on recombination, especially in the yeasts *Schizosaccharomyces pombe* and *S. cerevisiae,* provide ample precedent for pairing of unlinked homologous sequences (see 64). In both mitotic and meiotic cells, reciprocal exchange and nonreciprocal information transfer occur nearly as frequently between gene-sized segments of homologous sequences embedded in nonhomologous chromosomes as between homologous sequences at allelic positions (32, 44). Although the products of RIP and intrachromosomal recombination appear among sexual progeny in Neurospora at frequencies at least an order of magnitude greater than the frequency of recombination between sequences of comparable size in normal meiotic yeast cells, the mechanism of pairing in all these processes may be identical. The vastly reduced amount of substrate for pairing in premeiotic (haploid) nuclei relative to meiotic (diploid) nuclei may effectively focus the pairing machinery on repeated sequences. Support for this idea comes from work of Wagstaff et al (90) using haploid yeast cells in which meiotic processes had been artificially activated. The frequency of intrachromosomal recombination observed in these haploid cells ($\approx 30\%$) was approximately tenfold higher than that characteristic of standard (diploid) meiotic cells. When considering the high frequency of RIP and premeiotic recombination in *N. crassa* it is notable that the pairing process in this fungus is probably extended over a much longer period than in yeasts; efficient pairing may occur throughout the entire period between fertilization and karyogamy.

A collection of mutants defective in RIP would help to firmly establish the relationship between premeiotic sequence recognition and meiotic processes. Identification and characterization of such mutants may be complicated, however, because they may be sterile in homozygous crosses and because RIP only occurs in tissue that has nuclei from both parents. No mutants defective in RIP have yet been identified. The mutation *mei-2*, which abolishes chromosome pairing and meiotic recombination in homozygous crosses (but produces enough ascospores for analysis), is not impaired in RIP (H. Foss, N. Raju, A. Schroeder, & E. Selker, unpublished observations). Apparently efficient pairing of homologous sequences in preparation for RIP and chromosome pairing in preparation for crossing over are not synonymous.

RIP Probably Involves a Processive DNA-cytidine Deaminase

The two chains of a segment of DNA affected by RIP sometimes show significant differences in the number of G to A (or C to T) changes. In the sequenced segment from the unlinked duplication of flank sequences, 34 G to A changes, but only 13 C to T changes, were found on one strand. A statistical evaluation that took the composition of the original sequence into consideration indicated that this difference is nonrandom. A more striking example was found in a segment of the *am* gene from an unlinked duplication that passed once through a cross: 28 C to T changes occurred in a 350 nucleotide segment of one strand, and none occurred on the other (M. Singer, R. Eyre, & E. Selker, unpublished observations). That a given strand showed just C to T or G to A changes, but not both, in spite of the fact that the original composition of the two chains were comparable, implies that RIP results from a single type of mutation and that RIP operates on one chain of DNA at a time. Analysis of a unique sequence immediately adjacent to a duplicated region that was altered by RIP showed that the process sometimes extends a short distance beyond the duplicated region (27). Taken together, these observations suggest that the RIP machinery acts in a processive manner.

In principle, G:C to A:T mutations could occur by (*a*) directed misincorporation during DNA replication, (*b*) misincorporation following an excision process (e.g. involving a glycosylase), or (*c*) direct chemical conversion of the bases. Although we do not know whether the G:C to A:T mutations of RIP result from G to A or C to T changes, the latter seems more plausible from a mechanistic standpoint. Cytosine and 5-methylcytosine (5mC) are prone to spontaneous deamination yielding uracil and thymine, respectively (see 21, 46). Under physiological conditions, spontaneous deamination of C and 5mC in double-stranded DNA probably occurs at rates considerably below 1×10^{-10} per second, and organisms have repair processes to fix G·U (see 19, 45) and G·T (see 91) mismatches before mutations become established. The high frequency of G:C to A:T changes by RIP thus suggests that the process involves enzymatic deamination of C or 5mC

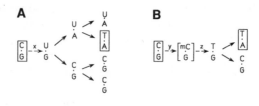

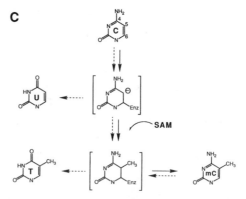

Figure 6 Possible mechanisms of G:C to A:T mutations involving enzymatic deamination. Hypothetical enzymatic reactions are shown by dashed arrows. (A) The mutations of RIP could result from enzymatic deamination of cytosine to form uracil (x) followed by DNA replication if excision of uracil by DNA-uracil glycosylase were switched off or were incomplete. (B) Alternatively, the mutations could result by cytosine methylation (y) followed by deamination (z), avoiding the uracil intermediate. (C) Enzymatic methylation of cytosine (solid arrows) involves a nucleophilic attack at the 6 position of cytosine to form a reactive 5,6-dihydrocytosine intermediate (bracketed). The resulting negative charge at the 5 position activates this previously inert carbon to accept a methyl group from S-adenosylmethionine (SAM). The dihydrocytosine intermediate is also very sensitive to displacement of the 4-amino group. Thus C to T changes could occur directly without involving a 5-methylcytosine (mC) intermediate. On mechanistic grounds, deamination would more likely follow methylation of the dihydrocytosine intermediate (as illustrated) than vice versa (D. Santi, B. Branchaud, personal communications).

residues (10). The normal repair processes might be turned off in ascogenous tissue, or simply overwhelmed by RIP (Figure 6).

Although enzymatic deamination of cytosine nucleosides and nucleotides are known, and evidence exists for RNA-cytosine deaminases (see 4, 18), no example of the postulated DNA-cytosine deaminase has been described. Curiously, the enzymatic mechanism of cytosine methylation is thought to go through an intermediate (5,6-dihydrocytosine) that is prone to spontaneous deamination. The dihydrocytosine intermediate is at least 10,000-fold more prone to deamination than cytidine (see 23, 70, 93). Thus, the C to T changes of RIP may be catalyzed by one enzyme without releasing a 5-mC intermediate, as illustrated in panel C of Figure 8. Furthermore, a common enzyme might be responsible for deamination in ascogenous tissue and for methylation in vegetative cells. Conceivably, the different outcomes might reflect different substrate concentrations, physical conditions, and/or cofactors in the respective tissues. For example, if ascogenous tissue should have an unusually low level of S-adenosylmethionine, this could prolong the lifetime of the dihydrocytosine intermediate, and thus result in deamination.

Detection and Mutation of Duplications May Be Temporally Separated

The search for homologous sequences in RIP may be relatively time consuming, while the mutagenesis may be rapid. Since all sequences are present in duplicate in the G2 phase of the cell cycle, recognition of duplications probably just occurs in G1. Although it is conceivable that the pairing machinery carries with it the enzyme(s) responsible for the mutagenesis, it seems more likely that the mutagenesis machinery would be associated with the replication machinery. The apparent processivity of RIP could simply reflect the inherent processive nature of DNA replication. Homologous sequences sharing a minimum length (e.g. 1–2 kb) would form stable pairs, which would accumulate during G1. The replication machinery, with its accessory RIP-apparatus, would then modify and dissolve the paired sequences.

EARLY SIGHTINGS OF PREMEIOTIC INSTABILITY

Instability of Transformation Markers

As mentioned above, genetic instability was noticed in early transformation studies. In 1973, Mishra et al (48, 49) reported that a large fraction of Inl^+ strains, obtained by treating an inl (inositol-requiring) strain with massive quantities of DNA from a wild-type strain, showed non-Mendelian transmission of inl^+ in crosses. Approximately one half of the Inl^+ strains transmitted the marker to 2% or less of their progeny, and about one third of the strains transmitted the marker to ≈20–30% of their progeny. The remaining Inl^+ strains showed normal transmission of inl^+. (These might have resulted from replacement or true reversion of the inl mutation in the recipient.) All Inl^+ strains obtained in experiments without DNA treatment, or with DNA treatment using DNA from the inositol-requiring strain, exhibited normal transmission of the marker. Poor transmission did not depend on which strain was used as the female. Furthermore, non-Mendelian transmission was not limited to the first cross, but was seen again when Inl^+ F_1 progeny were crossed, indicating that poor transmission could not be attributed to heterokaryosis. Based on these findings, an "exosome" model of transformation was proposed in which the introduced DNA did not integrate into the genome, and was therefore readily lost during meiosis (48). Analysis of progeny from individual asci revealed that, invariably, the introduced marker was transmitted to either zero or four of the eight isolates. This finding was taken as evidence that the mechanism removing exosomes was either not active or was fully active in meiosis.

The observed instability most likely resulted from RIP. The strains that transmitted inl^+ to roughly 25% of their progeny behaved like single-copy transformants having the introduced DNA at sites unlinked to the inl gene of the host. The strains that transmitted the marker at lower frequencies behaved as if they had linked copies of the transforming gene. Based on what is known about the fate of transforming DNA in Neurospora, it seems unlikely that a sizable fraction of the transformants had linked duplications, however. The extremely poor transmission of inl^+ may have been due to lethality resulting from RIP acting on critical sequences that happened to be taken up by the transformants.

Working with a similar set of transformants, Szabó & Schablik (86) obtained more or less equivalent results. In crosses of the Inl^+ transformants with an Inl^- strain, most of the transformants were barren on medium lacking inositol but fertile on medium containing inositol, consistent with the idea that the inl^+ gene was inactivated premeiotically. Crosses between a wild-type strain and the Inl^+ transformants were fertile on either medium, presumably because of complementation of the inactivated alleles by the inl^+ gene of the wild-type.

In 1984, Grant et al (30) demonstrated instability of a transformation marker, am^+, in strains obtained by transformation of an am deletion strain using the cloned gene. None of 33 Am^+ strains obtained transmitted the gene to more than 2% of their progeny in crosses. Results of Southern hybridizations demonstrated that the strains had multiple copies of the transforming DNA, arranged mostly in clusters of tandem arrays. Two possible explanations for the results were suggested by the authors—either the am^+ sequences were chromosomally integrated but then eliminated during meiosis, or they were present on autonomously replicating plasmids that were lost in the cross. Although restriction analysis revealed changes that appeared partially attributable to loss of autonomous plasmids, in retrospect, it seems likely that the transforming DNA was all integrated in the genome. Loss of the transforming marker probably resulted from a combination of RIP and intrachromosomal recombination. Occasional transformants that showed Mendelian segregation of am^+ were obtained in transformations of a strain with a point mutation in the am gene. Such transformants probably arose by clean replacement of the mutation.

Soon thereafter, Dhawale & Marzluf (16) and Case (12) reported sexual instability of the $qa\text{-}2^+$ gene introduced into a qa (quinic acid requiring) strain by transformation. Approximately 74% of the transformants in one study showed poor transmission of the marker (16). In an extensive survey of Qa-2$^+$ transformants that arose in a variety of ways, Case (12) observed loss of gene expression in pairs of meiotic products, as reported for the Inl$^+$ transformants obtained using total Neurospora DNA (48), as well as for transformants obtained using pooled DNA from a cosmid collection (25). Results of Southern hybridizations showed that loss of qa-2$^+$ function was due to inactivation rather than loss of the transforming DNA (12). Subsequent analysis of DNA from representative strains confirmed that inactivation reulted from RIP (E. Foss, M. Case, & E. Selker, unpublished observations).

In retrospect, possible manifestations of RIP and premeiotic intrachromosomal recombination are evident in the first report of Neurospora transformation using cloned DNA (13). The transformants were obtained by treating a qa-2 strain with plasmid DNA including the qa-2$^+$ gene and the adjacent gene, qa-4$^+$ (13, 74). Putative transformants were crossed before analysis to obtain homokaryotic isolates. Two of 14 Qa-2$^+$ strains obtained in this way integrated at or very near the native qa-2 locus. Both showed inactivation of qa-4$^+$. It is likely that the qa-4$^+$ gene was inactivated by RIP. Interestingly, two out of three sexual isolates of these two "linked-insertion" type transformants lacked all bacterial sequences. Presumably the sequences were lost by premeiotic intrachromosomal recombination.

Recurrent Mutation of Cya-8

There is no reason to think that RIP is limited to duplications created by transformation. As discussed above, the ζ–η region apparently represents a case of RIP of a natural tandem duplication. D. Perkins & V. Pollard (personal communication) have collected a set of observations that may reflect the operation of RIP on an unlinked duplication involving two structural genes. First, they noted that ≈20% of progeny from crosses in which one or both partners carry the eas mutation (easily wettable conidia) grew very poorly. Characterization of representative progeny demonstrated that they all suffered from mutations at a single locus unlinked to eas, and Bertrand showed that they were deficient in cytochrome aa_1. Mutations in the new locus, cya-8, occurred only in the nucleus that contained the eas mutation, and occurred in the period between fertilization and karyogamy. These observations are consistent with the idea that the eas mutation resulted from an insertion that included cya-8 sequences. Standard laboratory strains having this mutation would thus have two copies of cya-8 sequences, and would therefore be subject to RIP.

Barren Phenotype of Segmental Aneuploids

Strains containing duplications of segments of chromosomes can be constructed readily in Neurospora by crossing appropriate translocation strains (see 60). As first noted by St. Lawrence in 1953 (see 60), partial diploids display great instability in the sexual cycle. In general, they produce "barren" perithecia, i.e. perithecia that produce no or few ascospores[7] (60). Infertility of segmental aneuploids in Neurospora contrasts with the fertility of duplication strains in the homothallic fungus A. *nidulans* (3), and with the fertility exhibited by the vast majority of Neurospora strains having balanced rearrangements, i.e. those not having duplications (60).

The barren phenotype of segmental aneuploids has been thoroughly studied by Perkins and associates (see 60, 66, 88). Raju & Perkins (66) noted that the degree of barrenness appears to be independent of the genic content and size of the duplication and that barren perithecia are usually more or less normal in size and abundance. A careful cytological study of crosses involving nine different segmental aneuploids revealed that perithecial development is initiated but sexual development is arrested before meiosis (66). All strains seemed blocked during crozier formation or about the time of karyogamy. Each strain showed some variability in the stage of arrest, however, and some variability between strains was also observed. This study clearly indicated that duplications, per se, lead to developmental abnormalities prior to meiosis. The authors noted: "The question remains unanswered, what type of genetic element could be responsible for barrenness, that is distributed so ubiquitously through the chromosomes as to be present even in short duplicated segments, and whose dosage effect is so potent that it interferes with karyogamy or meiosis."

Most likely the barren phenotype of segmental aneuploids is an indirect result of RIP. A typical duplication covers 10–50% of a chromosome and probably includes a number of critical genes. Presumably, the RIP machinery efficiently detects these large duplications, and then wreaks havoc throughout them. Duplications of chromsome segments are occasionally transmitted through a cross without apparent alterations (88). It would be interesting to know if the nontranslocation chromosome of the duplication is generally (or always) from the nonduplication parent, as one would expect if RIP caused serious damage in the duplicated sequences. Limited analysis of survivors from crosses of duplication strains has not revealed mutations in genes covered by the duplication at frequencies typical of RIP (D. Perkins, M. Singer, E. Cambareri, & E. Selker, unpublished observations). Nevertheless, using a chromosomal duplication covering just two known genes, Perkins (personal communication) has found evidence of RIP among the rare progeny from such a cross. The relatively low frequency of mutations recovered could indicate tht RIP operates inefficiently on large duplications. Alternatively, it

---

[7]The barren phenotype itself is to varying degrees unstable, as might be expected considering that duplications are somewhat unstable during vegetative growth. Loss of the barren phenotype is invariably associated with breakdown of the duplication.

could reflect frequent lethality associated with the operation of RIP on such duplications.

It is interesting to consider how RIP could result in the barren phenotype. The general expectation is that so long as only one partner in a cross has a duplication, the mutations resulting from RIP would be complemented by the unaffected genes of the other nucleus. This appears to be generally the case for small duplications created by transformation. It is not uncommon, however, to observe poor fertility in crosses involving a strain with a duplication of just 3–10 kb. Furthermore, one characteristic of barren perithecia, namely the poor development of beaks (the structure surrounding the opening through which the ascospores are shot), is also frequently seen in crosses of strains harboring small duplications (E. Cambareri, K. Haack, P. Garrett, & E. Selker, unpublished observations). RIP might block sexual development in several ways. One possibility is that genes having nucleus-limited functions required during sexual development are widely distributed in the genome. In some cases production of defective products by genes mutated by RIP could poison the cell. Insufficient dosage of gene products may be serious in some cases as well. Finally, it seems possible that general deleterious effects might result from RIP. For example, changes in chromatin structure and/or new DNA methylation in the duplicated regions could affect DNA replication.

EVOLUTIONARY IMPLICATIONS OF RIP AND PREMEIOTIC DELETION

All observations to date suggest that RIP inactivates both copies of duplicated genes. Nevertheless, because of the temporal position of RIP in the Neurospora life cycle, duplication of an essential gene need not lead to death of the organism. Since RIP occurs in cells having nuclei from each parent, except where both parents harbor the same duplication, complementation may prevent cell death. Moreover, in the case of unlinked duplications, a sizable fraction of duplications escape recognition. Linked duplications are recognized very efficiently, but premeiotic deletion will rescue one member of a direct tandem duplication in many of the progeny. A substantial number of ascospores from a cross in which one parent has a duplication of an essential gene will be dead, of course, but this should not represent a large sacrifice on the part of the species.

Generation of Diversity

GENE FAMILIES   RIP may be responsible for the single-copy status of Neurospora genes that are represented as small families of identical or closely related genes in other organisms. The histone genes represent one example (92). Since gene duplications are generally regarded as a starting point for the evolution of new genes, it might seem that RIP should limit diversity. RIP may actually promote diversity, however. In general, 25% of progeny from a cross in which one parent harbors a small unlinked duplication should contain one native and one mutated copy of the sequence (Figure 7). Unlinked duplicated sequences appear to become resistant to subsequence cycles of RIP after fairly limited divergence (i.e. 10–20%). Sequences altered by RIP might therefore provide useful raw material for the evolution of new genes. Thus, while RIP may suppress the generation of gene families composed of very similar members, the process may spur the development of gene families with divergent members.

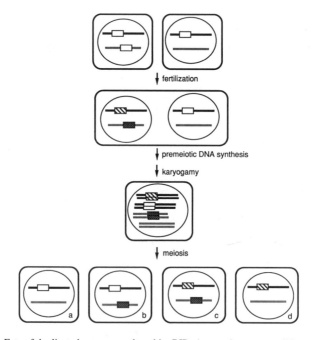

Figure 7   Fate of duplicated sequences altered by RIP. A cross between a wild-type strain (top right) and a strain harboring an unlinked duplication of a chromosomal segment is illustrated. Single and duplicate copies of the segment are indicated by boxes. In the period between fertilization and karyogamy, both copies of the duplicated sequence are altered by RIP (hatching) in the nucleus harboring the duplication. The homologous sequence in the other nucleus (empty box) is unaltered and, in general, should complement mutations resulting from RIP. The four possible combinations of chromosomes harboring the altered or unaltered homologues are shown. If the duplicated segment included essential genes, two of the meiotic products (c and d) should be inviable. One of the two viable products (b) would receive one altered and one unaltered sequence.

"SPONTANEOUS" MUTATION   The existence of processes such as somatic hypermutation of antibody genes (40) and RIP suggests that we should reexamine the meaning of "spontaneous" (in the sense of accidental) mutation. Cells can *cause* mutations to occur. Thus, some mutations thought to result from environmental insults to DNA, chemical instability of the nucleotides, or mistakes in DNA replication or repair, may actually result from enzyme-catalyzed conversions. The postulated DNA-cytosine deaminase of RIP may be more widespread than RIP itself. Considerable evidence in both prokaryotes and eukaryotes indicates that cytosines that are subject to methylation are unusually mutable (see references in 80). While it is assumed that this is due to failure to repair G·T mismatches resulting from spontaneous deamination of 5-mC residues, such mutations could be due to enzymatic deamination. A DNA-cytosine deaminase might normally work in conjunction with a G·T to G·C correction system, such as that recently discovered in animal cells (91), to demethylate DNA (mC:G $\rightarrow$ T·G $\rightarrow$ C:G). Any mismatches left uncorrected would lead to mutations. Furthermore, since the unstable 5,6-dihydrocytidine is thought to be an intermediate in enzymatic methylation (70, 93), the methylase itself should act as a deaminase and may directly catalyze C to T changes.

Mutation and Elimination of Repeated Sequences

Presumably both premeiotic recombination and RIP control the spread of repeated sequences. Deletion by premeiotic recombination should minimize the number of tandemly repeated sequences in the genome, and RIP is ideally suited to counter all types of selfish DNA. However, RIP does not directly eliminate selfish genes; it simply inactivates them, and thus generates pseudogenes. Although sequences altered by RIP could be preferentially deleted from the genome, no evidence for this idea is available. Sequences altered by RIP might confer a selective disadvantage on an individual beyond that expected from loss of function. If so, RIP would both inactivate selfish DNA and given natural selection a tool to eliminate it from a population.

Control of Genome Structure

In addition to inactivating genes, mutations resulting from RIP suppress subsequent pairing of the affected sequences. Thus RIP might thwart synapsis of nonhomologous chromosomes in meiosis. RIP should also prevent recombination between dispersed repeated sequences, which would result in translocations and other rearrangements. The observation that linked sequences that have been altered by RIP are poor substrates for intrachromosomal recombination supports this idea (11).

Chromosomal rearrangements that occur in spite of RIP may create duplications, either directly, or after recombination involving parental and rearranged chromosomes (see 60). As discussed above, the resulting segmental duplications are nearly sterile, probably because of RIP. Thus, RIP should indirectly rid a population of certain chromosomal rearrangements. This should help to preserve the gross organization of the genome (78), and may be responsible for the rarity of chromosomal rearrangements among Neurospora strains collected from nature (62). Although no evidence is available suggesting that RIP operates outside of the perithecium, the intolerance of diploidy and disomy exhibited by Neurospora might be due to a low frequency of a RIP-like process operating in vegetative tissue.

PREMEIOTIC INSTABILITY IN OTHER FUNGI

RIP may be limited to a subset of the filamentous ascomycetes. The process has not been detected in basidiomycetes (5, 51), and duplicate sequences are not inactivated in the well-studied yeast *S. cerevisiae*, or in the yeast *S. pombe* (E. Selker, unpublished observations), neither of which have a heterokaryotic phase in their life cycle. In contrast, duplicate sequences are inactivated in the heterothallic filamentous ascomycete *Ascobolus immersus*, which, like *N. crassa,* has an extended stage in which haploid nuclei of different strains share a common cytoplasm (24, 29). As in *N. crassa,* inactivation in *A. immersus* occurs in the interval between fertilization and karyogamy, appears to involve pairing, acts on both linked and unlinked sequences, and results in heavy methylation. *A. immersus* also shows high frequency premeiotic intrachromosomal recombination. The two organisms exhibit some differences, however. In *A. immersus,* a linked duplication sometimes survives a cross unaltered. Furthermore, in some cases, genes inactivated in Ascobolus appear to revert, especially if the organism is treated with the drug 5-azacytidine, which interferes with DNA methylation. The RIP process may result in considerably fewer mutations in *A. immersus* than in *N. crassa*. Results of Southern hybridizations are consistent with this idea (29), but no sequence information is yet available. The inactivation process of *A. immersus* could even involve only methylation, although this would imply that this fungus, unlike Neurospora, has an efficient mechanism to maintain methylation patterns during vegetative propagation (see 75).

Preliminary evidence suggests that RIP also operates in the plant pathogens, *Gibberella pulicaris* (Y. Salch & M. Beremand, personal communication) and *Gibberella fujikuroi* (M. Dickman & J. Leslie, personal communication), both of which are also heterothallic filamentous ascomycetes with a dikaryotic phase preceding karyogamy. In contrast, RIP does not occur, at least at high frequency, in two other heterothallic filamentous ascomycetes, *Magnaporthe grisea* (F. Chumley & B. Valent, unpublished observations) and *Cochliobolus heterostrophus* (O. Yoder, personal communication), both of which are also plant pathogens.

RIP has not been found in the two homothallic filamentous ascomycetes that have been examined, *Sordaria macrospora* (42) and *Aspergillus nidulans* (C. Scazzochio, personal communication), nor in the functionally homothallic ascomycete *Podospora anserina* (15). Nevertheless, all three organisms show high frequency deletion of tandem repeats. Considering that RIP inactivates both members of a duplication, it seems reasonable from an evolutionary standpoint that RIP would be limited to heterokaryotic cells in outbreeding organisms (77). All the nuclei in the premeiotic tissue of a homothallic fungus usually come from a single parent. If that parent happened to have a duplication of an essential gene, RIP might result in cell death (see Figure 7). The situation would be comparable to the operation of RIP in vegetative cells. Premeiotic intrachromosomal recombination, which eliminates just one member of a tandem duplication, is more conservative than RIP, and should not cause problems for a homothallic organism.

SUMMARY

Maintenance of a steamlined genome is probably important to a free-living fungus. The period between fertilization and karyogamy in the life cycle of Neurospora and related fungi provides an ideal time for "genome-cleaning". Premeiotic intrachromosomal recombination deletes tandem repeats at high frequency in both homothallic and heterothallic filamentous ascomycetes. This eliminates excess copies of tandemly repeated genes and at the same time favors their homogenization. Heterothallic fungi such as Neurospora also take the bolder steps of mutating and modifying both copies of duplicated sequences, linked or unlinked, by RIP. Because these organisms are outbreeders, and because RIP operates immediately prior to meiosis in cells having nuclei from both parents, the process does not cause much lethality or loss of genetic information. RIP should effectively counter selfish and redundant DNA, and at the same time generate raw material for evolution. In addition, RIP should both prevent chromosomal rearrangements by causing divergence of dispersed repeated sequences and rid a population of duplication-generating rearrangements. Thus, this form of genetic instability potentially stabilizes the gross organization of the genome.

ACKNOWLEDGEMENTS

I thank Ed Barry, Bruce Branchaud, Ira Herskowitz, Jack Kinsey, Mike Lynch, Bob Metzenberg, Dot Newmeyer, David Perkins, Daniel Santi, Frank Stahl, Charles Yanofsky, and present and past members of my laboratory for stimulating discussions. I am grateful to Ed Cambareri, Jette Foss, Jeff Irelan, Jack Kinsey, Bob Metzenberg, Vivian Miao, Rik Meyers, Dot Newmeyer, David Perkins, and Jeannie Selker for comments and useful suggestions on the developing manuscript. I am also grateful to all my colleagues who kindly communicated their unpublished results. The work from my laboratory was supported by National Science Foundation Grant DCB 8718163 and Public Health Services grant GM-35690 from the National Institutes of Health. Part of this work was done during the tenure of an Established Investigatorship of the American Heart Association.

Literature Cited

1. Akins, R. A., Lambowitz, A. M. 1985. General method for cloning *Neurospora crassa* nuclear genes by complementation of mutants. *Mol. Cell. Biol.* 5:2272–78
2. Asch, D. K., Kinsey, J. A. 1990. Relationship of vector insert size to homologous integration during transformation of *Neurospora crassa* with the cloned *am* (GDH) gene. *Mol. Gen. Genet.* 22:37–43
3. Bainbridge, B. W., Roper, J. A. 1966. Observations on the effects of a duplication in *Aspergillus nidulans*. *J. Gen. Microbiol.* 42:417–24
4. Bass, B. L., Weintraub, H. 1988. An unwinding activity that covalently modifies its double-stranded RNA substrate. *Cell* 55:1089–98
5. Binninger, D. M., Skrzynia, C., Pukkila, P. J., Casselton, L. A. 1987. DNA-mediated transformation of the basidiomycete *Coprinus cinereus*. *EMBO J.* 6:835–40
6. Bull, J., Wootton, J. C. 1984. Heavily methylated amplified DNA in transformants of *Neurospora crassa*. *Nature* 310:701–4
7. Butler, D. K., Metzenberg, R. L. 1989. Premeiotic change of nucleolus organizer size in Neurospora. *Genetics* 122:783–91
8. Buxton, F. P., Radford, A. 1984. The transformation of mycelial spheroplasts of *Neurospora crassa* and the attempted isolation of an autonomous replicator. *Mol. Gen. Genet.* 196:339–44
9. Cambareri, E. B. 1987. *A study of a novel genetic process in* Neurospora crassa: *RIP (Rearrangement induced premeiotically) by tetrad and Southern analyses.* MS thesis. Univ. Oregon, Eugene. 36 pp.
10. Cambareri, E. B., Jensen, B. C., Schabtach, E., Selker, E. U. 1989. Repeat-induced G-C to A-T mutations in Neurospora. *Science* 244:1571–75
11. Cambareri, E. B., Singer, M. J., Selker, E. U. Recurrence of repeat-induced point mutation (RIP) in *Neurospora crassa*. Genetics. In press
12. Case, M. E. 1986. Genetical and molecular analyses of *qa-2* transformants in *Neurospora crassa*. *Genetics* 113:569–87
13. Case, M. E., Schweizer, M., Kushner, S. R., Giles, N. H. 1979. Efficient transformation of *Neurospora crassa* utilizing hybrid plasmid DNA. *Proc. Natl. Acad. Sci. USA* 76:5259–63
14. Cavalier-Smith, T. 1978. Nuclear volume control by nucleoskeletal DNA, selection for cell volume and cell growth rate, and the solution of the DNA C-value paradox. *J. Cell Sci.* 34:247–78
15. Coppin-Raynal, E., Picard, M., Arnaise, S. 1989. Transformation by integration in *Podospora anserina* III. Replacement of a chromosome segment by a two-step process. *Mol. Gen. Genet.* 219:270–76
16. Dhawale, S. S., Marzluf, G. A. 1985. Transformation of *Neurospora crassa* with circular and linear DNA and analysis of the fate of the transforming DNA. *Curr. Genet.* 10:205–12
17. Doolittle, W. F., Sapienza, C. 1980. Selfish genes, the phenotype paradigm and genome evolution. *Nature* 284:601–3
18. Driscoll, D. M., Wynne, J. K., Wallis, S. C., Scott, J. 1989. An in vitro system for the editing of apolipoprotein B mRNA. *Cell* 58:519–24
19. Duncan, B. K., Weiss, B. 1982. Specific mutator effects of *ung* (uracil-DNA) glycosylase) mutations in *Escherichia coli*. *J. Bacteriol.* 151:750–55
20. Durán, R., Gray, P. M. 1989. Nuclear DNA, an adjunct to morphology in fungal taxonomy. *Mycotaxon* 36:205–19
21. Ehrlich, M., Norris, K. F., Wang, R. Y.-H., Kuo, K. C., Gehrke, C. W. 1986. DNA cytosine methylation and heat-induced deamination. *Biosci. Rep.* 6:387–93
22. Emerson, S. 1966. Mechanisms of inheritance. 1. Mendelian. In *The Fungi: An Advanced Treatise,* ed. G. C. Ainsworth, A. S. Sussman, 2:513–66. New York: Academic
23. Evans, B. E., Mitchell, G. N., Wolfen-

den, R. 1975. The action of bacterial cytidine deaminase on 5,6-dihydrocytidine. *Biochemistry* 14:621–24
24. Faugeron, G., Rhounim, L., Rossignol, J.-L. 1990. How does the cell count the number of ectopic copies of a gene in the premeiotic inactivation process acting in *Ascobolus immersus?* *Genetics* 124: 585–91
25. Fhehér, Z., Schablik, M., Kiss, A., Zsindely, A., Szabó, G. 1986. Characterization of *inl*$^+$ transformants of *Neurospora crassa* obtained with a recombinant cosmid-pool. *Curr. Genet.* 11:131–37
26. Fincham, J. R. S., Connerton, I. F., Notarianni, E., Harrington, K. 1989. Premeiotic disruption of duplicated and triplicated copies of the *Neurospora crassa am* (glutamate dehydrogenase) gene. *Curr. Genet.* 15:327–34
27. Foss, E. J., Kinsey, J. A., Garrett, P. W., Selker, E. U. Specificity of repeat-induced point mutation (RIP) in Neurospora: Sensitivity of unique DNA adjacent to a duplicated region, non-Neurospora sequences, and a natural diverged tandem duplication. *Genetics.* In press
28. Free, S. J., Rice, P. W., Metzenberg, R. L. 1979. Arrangement of the genes coding for ribosomal ribonucleic acids in *Neurospora crassa*. *J. Bacteriol.* 137: 1219–26
29. Goyon, C., Faugeron, G. 1989. Targeted transformation of *Ascobolus immersus* and de novo methylation of the resulting duplicated DNA sequences. *Mol. Cell. Biol.* 9:2818–27
30. Grant, D. M., Lambowitz, A. M., Rambosek, J. A., Kinsey, J. A. 1984. Transformation of *Neurospora crassa* with recombinant plasmids containing the cloned glutamate dehydrogenase *(am)* gene: Evidence for autonomous replication of the transforming plasmid. *Mol. Cell. Biol.* 4:2041–51
31. Grayburn, W. S., Selker, E. U. 1989. A natural case of RIP: Degeneration of DNA sequence in an ancestral tandem duplication. *Mol. Cell. Biol.* 9:4416–21
32. Jinks-Robertson, S., Petes, T. D. 1986. Chromosomal translocations generated by high-frequency meiotic recombination between repeated yeast genes. *Genetics* 114:731–52
33. Johnson, T. E. 1976. Analysis of pattern formation in Neurospora perithecial development using genetic mosaics. *Dev. Biol.* 54:23–36
34. Kim, S. Y., Marzluf, G. A. 1988. Transformation of *Neurospora crassa* with *trp-1* gene and the effect of host

strain upon the fate of the transforming DNA. *Curr. Genet.* 13:65–70
35. Kinnaird, J. H., Keighren, M. A., Kinsey, J. A., Eaton, M., Fincham, J. R. S. 1982. Cloning of the am (glutamate dehydrogenase) gene of *Neurospora crassa* through the use of a synthetic DNA probe. *Gene* 20:387–96
36. Kinsey, J. A. 1989. Restricted distribution of the Tad transposon of Neurospora. *Curr. Genet.* 15:271–75
37. Kinsey, J. A., Helber, J. 1989. Isolation of a transposable element from *Neurospora crassa*. *Proc. Natl. Acad. Sci. USA* 86:1929–33
38. Kinsey, J. A., Hung, B. T. 1981. Mutation at the *am* locus of *Neurospora crassa*. *Genetics* 99:405–14
39. Kinsey, J. A., Rambosek, J. A. 1984. Transformation of *Neurospora crassa* with the cloned *am* (glutamate dehydrogenase) gene. *Mol. Cell. Biol.* 4:117–22
40. Kocks, C., Rajewsky, K. 1989. Stable expression and somatic hypermutation of antibody V regions in B-cell developmental pathways. *Annu. Rev. Immunol.* 7:537–59
41. Krumlauf, R., Marzluf, G. A. 1980. Genome organization and characterization of the repetitive and inverted DNA sequences in *Neurospora crassa*. *J. Biol. Chem.* 255:1138–45
42. Le Chevanton, L., Leblon, G., Lebilcot, S. 1989. Duplications created by transformation in *Sordaria macrospora* are not inactivated during meiosis. *Mol. Gen. Genet.* 218:390–96
43. Legerton, T. L., Yanofsky, C. 1985. Cloning and characterization of the multifunctional *his-3* gene of *Neurospora crassa*. *Gene* 39:129–40
44. Lichten, M., Borts, R. H., Haber, J. E. 1987. Meiotic gene conversion and crossing over between dispersed homologous sequences occurs frequently in *Saccharomyces cerevisiae*. *Genetics* 115:233–46
45. Lindahl, T. 1976. New class of enzymes acting on damaged DNA. *Nature* 259:64–66
46. Lindahl, T., Nyberg, B. 1974. Heat-induced deamination of cytosine residues in deoxyribonucleic acid. *Biochemistry* 13:3405–10
47. Metzenberg, R. L., Stevens, J. N., Selker, E. U., Morzycka-Wroblewska, E. 1985. Identificaion and chromosomal distribution of 5S rRNA genes in *Neurospora crassa*. *Proc. Natl. Acad. Sci. USA* 82:2067–71
48. Mishra, N. C., Tatum, E. L. 1973. Non-Mendelian inheritance of DNA-induced

inositol independence in Neurospora. *Proc. Natl. Acad. Sci. USA* 70:3875–79

49. Mishra, N. C., Szabó, G., Tatum, E. L. 1973. Nucleic acid-induced genetic changes in Neurospora. In *The Role of RNA in Reproduction and Development*, ed. M. C. Niu, S. J. Segal, pp. 259–68. Amsterdam: Elsevier, North-Holland

50. Morzycka-Wroblewska, E., Selker, E. U., Stevens, J. N., Metzenberg, R. L. 1985. Concerted evolution of dispersed *Neurospora crassa* 5S RNA genes: pattern of sequence conservation between allelic and nonallelic genes. *Mol. Cell. Biol.* 5:46–51

51. Munoz-Rivas, A., Specht, C. A., Drummond, B. J., Froeliger, E., Novotny, C. P. 1986. Transformation of the basidiomycete, *Schizophyllum commune*. *Mol. Gen. Genet.* 205:103–6

52. Nakamura, K., Egashira, T. 1961. Genetically mixed perithecia in Neurospora. *Nature* 190:1129–30

53. Newmeyer, D., Galeazzi, D. R. 1977. The instability of Neurospora duplication *Dp(IL→IR)H4250*, and its genetic control. *Genetics* 85:461–87

54. Newmeyer, D., Galeazzi, D. R. 1978. A meiotic UV-sensitive mutant that causes deletion of duplications in Neurospora. *Genetics* 89:245–69

55. Newmeyer, D., Schroeder, A. L., Galeazzi, D. R. 1978. An apparent connection between histidine, recombination, and repair in Neurospora. *Genetics* 89:271–79

56. Orbach, M. J., Porro, E. B., Yanofsky, C. 1986. Cloning and characterization of the gene for β-tubulin from a benomyl-resistant mutant of *Neurospora crassa* and its use as a dominant selectable marker. *Mol. Cell. Biol.* 6:2452–61

57. Orbach, M. J., Vollrath, D., Davis, R. W., Yanofsky, C. 1988. An electrophoretic karyotype of *Neurospora crassa*. *Mol. Cell. Biol.* 8:1469–73

58. Orgel, L. E., Crick, F. H. C. 1980. Selfish DNA: the ultimate parasite. *Nature* 284:604–7

59. Paietta, J., Marzluf, G. A. 1985. Plasmid recovery from transformants and the isolation of chromosomal DNA segments improving plasmid replication in *Neurospora crassa*. *Curr. Genet.* 9:383–88

60. Perkins, D. D., Barry, E. G. 1977. The cytogenetics of Neurospora. *Adv. Genet.* 19:133–285

61. Perkins, D. D., Metzenberg, R. L., Raju, N. B., Selker, E. U., Barry, E. G. 1986. Reversal of a Neurospora translocation by crossing over involving displaced rDNA, and methylation of the rDNA segments that result from recombination. *Genetics* 114:791–817

62. Perkins, D. D., Turner, B. C. 1988. *Neurospora* from natural populations: Toward the population biology of a haploid eukaryote. *Exp. Mycol.* 12:91–131

63. Petes, T. D. 1980. Unequal meiotic recombination within tandem arrays of yeast ribosomal DNA genes. *Cell* 19:765–74

64. Petes, T. D., Hill, C. W. 1988. Recombination between repeated genes in microorganisms. *Annu. Rev. Genet.* 22:147–68

65. Raju, N. B. 1980. Meiosis and ascospore genesis in Neurospora. *Eur. J. Cell Biol.* 23:208–23

66. Raju, N. B., Perkins, D. D. 1978. Barren perithecia in *Neurospora crassa*. *Can. J. Genet. Cytol.* 20:41–59

67. Rossier, C., Pugin, A., Turian, G. 1985. Genetic anaysis of transformation in a microconidiating strain of *Neurospora crassa*. *Curr. Genet.* 10:313–20

68. Russell, P. J., Petersen, R. C., Wagner, S. 1988. Ribosomal DNA inheritance and recombination in *Neurospora crassa*. *Mol. Gen. Genet.* 211:541–44

69. Russell, P. J., Rodland, K. D. 1986. Magnification of rRNA gene number in a *Neurospora crassa* strain with a partial deletion of the nucleolus organizer. *Chromosoma* 93:337–40

70. Santi, D. V., Wataya, Y., Matsuda, A. 1978. Approaches to the design of mechanism-based inhibitors of pyrimidine metabolism. In *Enzyme-Activated Irreversible Inhibitors*, ed. N. Seiler, M. J. Jung, J. Koch-Weser, pp. 291–303. Amsterdam/New York/Oxford: Elsevier/ North-Holland Biomedical

71. Schechtman, M. G. 1987. Isolation of telomere DNA from *Neurospora crassa*. *Mol. Cell. Biol.* 7:3168–77

72. Schechtman, M. G. 1990. Characterization of telomere DNA from *Neurospora crassa*. *Gene* 88:159–65

73. Schechtman, M. G., Yanofsky, C. 1983. Structure of the trifunctional *trp-1* gene from *Neurospora crassa* and its aberrant expression in *Escherichia coli*. *J. Mol. Appl. Gen.* 2:83–99

74. Schweizer, M., Case, M. E., Dykstra, C. C., Giles, N. H., Kushner, S. R. 1981. Identification and characterization of recombinant plasmids carrying the complete *qa* gene cluster from *Neurospora crassa* including the *qa-1*⁺ regulatory gene. *Proc. Natl. Acad. Sci. USA* 78:5086–90

75. Selker, E. U. 1990. DNA methylation and chromatin structure: a view from below. *Trends Biochem. Sci.* 15:103–7

76. Selker, E. U., Cambareri, E. B., Garrett, P. W., Jensen, B. C., Haack, K. R., et al. 1989. Use of RIP to inactivate genes in *Neurospora crassa*. *Fungal Genet. Newsl.* 36:76–77
77. Selker, E. U., Cambareri, E. B., Jensen, B. C., Haack, K. R. 1987. Rearrangement of duplicated DNA in specialized cells of Neurospora. *Cell* 51:741–52
78. Selker, E. U., Garrett, P. W. 1988. DNA sequence duplications trigger gene inactivation in *Neurospora crassa*. *Proc. Natl. Acad. Sci. USA* 85:6870–74
79. Selker, E. U., Jensen, B. C., Richardson, G. A. 1987. A portable signal causing faithful DNA methylation *de novo* in *Neurospora crassa*. *Science* 238:48–53
80. Selker, E. U., Stevens, J. N. 1985. DNA methylation at asymmetric sites is associated with numerous transition mutations. *Proc. Natl. Acad. Sci. USA* 82:8114–18
81. Selker, E. U., Stevens, J. N. 1987. Signal for DNA methylation associated with tandem duplication in *Neurospora crassa*. *Mol. Cell. Biol.* 7:1032–38
82. Selker, E. U., Yanofsky, C., Driftmier, K., Metzenberg, R. L., Alzner-DeWeerd, B., RajBhandary, U. L. 1981. Dispersed 5S RNA genes in *N. crassa:* Structure, expression and evolution. *Cell* 24:819–28
83. Shockley, T. E., Tatum, E. L. 1962. A search for genetic transformation in *Neurospora crassa*. *Biochem. Biophys. Acta* 61:567–72
84. Simon, D., Stuhlmann, H., Jahner, D., Wagner, H., Werner, E., Jaenisch, R. 1983. Retrovirus genomes methylated by mammalian but not bacterial methylase are non-infectious. *Nature* 304:275–77
85. Stohl, L. L., Lambowtiz, A. M. 1983. Construction of a shuttle vector for the filamentous fungus *Neurospora crassa*. *Proc. Natl. Acad. Sci. USA* 80:1058–62
86. Szabó G., Schablik, M. 1982. Behaviour of DNA-induced inositol-independent transformants of *Neurospora crassa* in sexual crosses. *Theor. Appl. Genet.* 61:171–75
87. Szabó, G., Schablik, M., Fekete, Z., Zsindely, A. 1978. A comparative study of DNA-induced transformants and spontaneous revertants of inositolless *Neurospora crassa*. *Acta Biol. Acad. Sci. Hung.* 29:375–84
88. Turner, B. C., Taylor, C. W., Perkins, D. D., Newmeyer, D. 1969. New duplication-generating inversions in *Neurospora*. *Can. J. Genet. Cytol.* 11:622–38
89. Vollmer, S. J., Yanofsky, C. 1986. Efficient cloning of genes of *Neurospora crassa*. *Proc. Natl. Acad. Sci. USA* 83:4869–73
90. Wagstaff, J. E., Klapholz, S., Waddell, C. S., Jensen, L., Esposito, R. E. 1985. Meiotic exchange within and between chromosomes requires a common rec function in *Saccharomyces cerevisiae*. *Mol. Cell. Biol.* 5:3532–44
91. Wiebauer, K., Jiricny, J. 1989. In vitro correction of G·T mispairs to G·C pairs in nuclear extracts from human cells. *nature* 339:234–36
92. Woudt, L. P., Pastink, A., Kempers-Veenstra, A. E., Jansen, A. E. M., Mager, W. H., Planta, R. J. 1983. The genes coding for histone H3 and H4 in *Neurospora crassa* are unique and contain intervening sequences. *Nucleic Acids Res.* 11:5347–60
93. Wu, J. C., Santi, D. V. 1987. Kinetic and catalytic mechanism of *Hha*I methyltransferase. *J. Biol. Chem.* 262:4778–86

Annu. Rev. Genet. 1990. 24:615–57

HUMAN TUMOR SUPPRESSOR GENES

Eric J. Stanbridge

Department of Microbiology and Molecular Genetics, California College of Medicine, University of California at Irvine, Irvine, California 92717

KEY WORDS: somatic cell hybrid, microcell fusion, tumor suppression, p53 gene, retinoblastoma gene

CONTENTS

0066-4197/90/1215-0615$02.00

INTRODUCTION

Approximately a decade ago, the molecular genetic era of human cancer was ushered in with the discovery of "dominantly acting" activated cellular oncogenes (112, 187). The first activated oncogenes were isolated by transfection of DNA from human cancer cells into mouse NIH 3T3 cells, a process that resulted in neoplastic transformation. The activated oncogenes were quickly found to be homologs of retroviral transforming genes (42, 159). This finding, which was predicted on the basis of the seminal studies that showed that the avian retroviral src oncogene had evolved from the capture of a cellular protooncogene (206), led to the further identification of numerous candidate cellular oncogenes.

The discovery that activated oncogenes could be found in 10 to 30 percent of human cancers led to theories that activation of single or multiple cooperating cellular oncogenes was in itself sufficient to create a cancerous cell. These theories were all the more attractive when it was found that the expanding list of oncogene functions included growth factors, growth factor receptors, signal transducers, protein kinases, and transcriptional activators—all of which, when behaving aberrantly, might lead to uncontrolled cell proliferation.

A tacit assumption in many interpretations of these studies was the dominant nature of activated oncogenes. However, earlier studies with somatic cell hybrids had clearly shown that when malignant cells were fused with normal cells, the resulting hybrid cells were nontumorigenic. This phenomenon of tumor suppression indicated that a gene (or genes) from a normal cell might replace a defective function in the cancer cell and render it responsive to normal regulators of cell growth. Such genetic elements have been termed tumor suppressor genes (198).

In this review I present the evidence for tumor suppressor genes and discuss their role in the control of neoplasia.

SOMATIC CELL HYBRID STUDIES

The earliest experiments using cell hybrids to undertake a genetic analysis of malignancy predate the discovery of human oncogenes by some twenty years. Early investigators in this field, most notably Barski, Ephrussi and colleagues (6, 180), fused mouse cells of high and low malignant potential and noted that the resulting hybrid cells were as malignant as the highly malignant parent, thereby leading to the interpretation that malignancy behaves as a dominant trait. This finding would certainly agree with the concept of "dominantly acting" oncogenes. However, Harris, Klein and colleagues (23, 87), in a more extensive and critical series of experiments, demonstrated that when malignant mouse cells were fused with nonmalignant mouse cells, the resulting hybrid cells were unable to form tumors, i.e. the tumorigenic phenotype was suppressed—a finding contrary to that of Barski & Ephrussi as well as to the concept of "dominantly acting" oncogenes. These apparently conflicting experimental findings, and the subsequent differences in interpretation of data, were due to the chromosomal instability of the hybrid cells; chromosomes were lost rapidly from the hybrid cells and tumorigenic segregants rapidly arose, presumably as a consequence of the loss of chromosome(s) containing "tumor-suppressor" genes. Other investigators encountered similar chromosome instability in both intraspecies rodent cell hybrids and interspecies human x rodent cell hybrids (reviewed in 113, 140).

Furthermore, this rapid chromosome loss makes it extremely hard to identify specific chromosomes that possibly control the expression of the tumorigenic phenotype.

The problem of chromosomal instability was eventually overcome by using intraspecies human cell hybrids (196, 202). Extensive analyses of human carcinoma (HeLa) x human diploid fibroblast hybrid cells confirmed stable retention of both sets of parental chromosomes. At the same time, tumor formation was completely and stably suppressed, with tumorigenic segregants arising only rarely. Suppression of tumorigenicity has now been demonstrated in various human malignant x human normal cell crosses, suggesting that tumor suppression is a fairly general result of introducing genetic information from normal cells into cancer cells. This allows the further conclusion that many types of tumors arise through loss of critical growth-regulating genes from their genomes.

Although the phenomenon of tumor suppression has been demonstrated in a wide array of malignant x normal cell crosses (173, 199), it was nevertheless felt that hybridomas (e.g. normal lymphocyte x malignant plasmacytoma hybrids) might represent a significant exception to this general finding. Despite there being no a priori reason why the malignant phenotype should invariably be suppressed by introducing genetic information from normal cells, recent publications now show that both mouse and human lymphoma x lymphocyte hybrids are suppressed (114, 174). The large number of mouse hybridomas that form ascites tumors may reflect chromosomal instability. Human hybridomas form ascites tumors more rarely than mouse hybridomas in immune deficient animals, possibly because chromosomes carrying tumor suppressor genes are more stably retained.

Although tumor-forming ability is suppressed in malignant x normal hybrid cells, many cells continue to display certain phenotypes of transformation in vitro (reviewed in 202). The normal genes operating in these hybrid cells can apparently normalize some but not all traits of the malignant parent. This is in consonance with a model of tumor development that involves multiple genetic lesions in the tumor cell genome, only some of which are corrected by the normal genes expressed in the hybrids. Since these mutations are all presumably essential for tumorigenesis, the correction of any one following cell hybridization may cause the hybrid cell to lose its tumor-forming ability. In studies of certain malignant x normal cell fusions, all in vitro transformation phenotypes are suppressed, resulting in hybrid cells that behave as fully normal cells in culture and eventually senesce and die (162).

Investigation of the growth of hybrid cells in the host immune deficient animal revealed why such cells express so many phenotypic traits associated with the transformed state in vitro and yet fail to form tumors in vivo. Whereas the nontumorigenic cells express relatively few traits of differentiation in culture, they are induced to terminally differentiate in vivo (161, 201). The program of differentiation expressed is that of the normal parental cell irrespective of the nature of the malignant parental cell (161). For example, nontumorigenic HeLa x fibroblast hybrids differentiate into connective tissue (85, 201) and HeLa x keratinocyte hybrids differentiate into keratinizing tissue (161). In both instances, tumorigenic segregants form undifferentiated carcinomas indistinguishable from the parent HeLa. Thus, the biological mechanism of tumor suppression in somatic cell hybrids appears to be restoration of an inducible program of terminal differentiation.

Specific Chromosomes are Associated with Control of Tumorigenic Expression

Specific chromosomes from the normal parent that may be involved in the suppression of tumorigenic phenotype in somatic cell hybrids have been difficult to identify, particularly for intraspecies cell hybrids. Several investigators have used interspecies cell hybrids, particularly those of rodent x human hybrid cells, where human chromosomes are lost preferentially and can be readily distinguished from those of the rodent complement. However, the human chromosomes can be lost so rapidly that suppression of tumorigenicity is sometimes not seen or may be only transient.

Harris, Klein and colleagues, in early studies, were unable to establish the association of any particular chromosome with suppression of the tumorigenic phenotype. Although they did suggest a role for chromosome 4, the authors felt that the evidence was not adequate (99). Subsequently, Harris and colleagues (54) reexamined the role of the normal chromosome 4 in the suppression of malignancy using natural polymorphisms of the centromeric heterochromatin to identify the parental origin of the chromosomes 4 in the hybrid cells. They found that the normal chromosome 4 was involved in the suppression of malignancy in all tumors examined, including carcinoma, melanoma, sarcoma, and lymphoma. In all crosses between malignant and normal cells a selective pressure operated in vivo against chromosome 4 derived from the normal cell and in favor of the chromosome 4 derived from the malignant cell. Thus, chromosomes 4 in all tumors isolated apparently function differently from the chromosomes 4 of the normal parent. It was suggested that the reappearance in hybrids of malignancy that was initially suppressed may result from a reduction in the number of copies of normal chromosome 4 and an increase in the number of copies of malignant chromosome 4. Thus, the gene or genes on the normal chromosome 4 responsible for the suppression of malignancy may act in a dose-dependent manner.

In studies with hamster x human hybrid cells, several chromosomes have been implicated as carrying tumor suppressor genes (103, 104). Bouck and her colleagues ascertained that human chromosome 1 serves both to suppress anchorage independence (208) and activate an inhibitor of angiogenesis (165) when present in transformed baby hamster kidney (BHK) cells. The chromosome 1 seems to activate the expression of a gp140 hamster protein that has very strong homology to human thrombospondin, a known inhibitor of angiogenesis. The human thrombospondin gene maps to chromosome 15; therefore, the presence of human chromosome 1 presumably serves to activate the endogenous hamster thrombospondin-equivalent gene. The association of expression of an inhibitor of angiogenesis and tumor suppression is attractive, based upon Folkman's seminal studies that show that lack of neovascularization inhibits tumor growth (61). The BHK/chromosome 1 cells also fail to induce neovascularization of the cornea when implanted into the rat eye and they fail to form tumors. However, inoculation of the same cells subcutaneously does not suppress tumor formation (N. Bouck, personal communication), possibly because their chromosomal instability leads to significant loss of human chromosome 1. Formal proof of tumor suppression should come from introduction and expression of the thrombospondin gene in transformed BHK cells.

Investigators studying intraspecies hamster x hamster hybrid cells have noted the association between loss of chromosome 3 and chromosome 15 with reexpression of tumorigenicity in Chinese hamster and Syrian hamster hybrid cells, respectively (156, 173).

The strongest evidence that a specific chromosome is implicated in tumor suppression came from intraspecies studies with HeLa x human diploid fibroblast hybrids. Studies combining cytogenetic and molecular restriction fragment length polymorphism (RFLP) showed that the fibroblast chromosome 11 was implicated in tumor suppression. When both copies of fibroblast chromosome 11 were present in the HeLa x fibroblast hybrid, the cells were completely nontumorigenic. Loss of one or other copy of this chromosome was associated with reexpression of tumorigenicity (101, 103, 195, 204). As discussed below, microcell transfer of a normal chromosome 11 into tumorigenic HeLa x fibroblast hybrids restored suppression (179), thereby providing strong evidence for the presence of a tumor suppressor gene on this chromosome.

In related studies, Benedict and colleagues (11) analyzed paired combinations of nontumorigenic and tumorigenic segregant hybrids derived from the fusion of HT1080 human fibrosarcoma cells with human diploid fibroblasts. In this case, the loss of chromosome 1, and possibly chromosome 4, was associated with the restoration of tumorigenic potential in these hybrids.

In these and other studies, reexpression of tumorigenicity in nontumorigenic hybrid cells is associated with the discrete loss of specific chromosomes from the normal parent. Therefore, tumor suppression is correlated with the presence of normal chromosomes that presumably carry genetic information responsible for this effect, i.e. tumor suppressor genes. Furthermore, the data also indicate that at least one genetic alteration responsible for the multistep progression of a normal cell to a malignant cell is the loss of function of a critical controlling gene (23, 202). Such gene deletion or inactivation could be corrected (complemented) by introduction of an intact gene derived from the normal partner via cell fusion.

The multiplicity of chromosomes involved in tumor suppression of the various malignant cells described above also indicated that multiple tumor suppressor genes exist with distinct specific controlling functions for specific malignancies. If different malignant cells suffer from genetic defects in different tumor suppressor genes then one could expect complementation when two different malignant cells are fused together, resulting in a nontumorigenic hybrid cell. This is precisely what occurs with human malignant x malignant cell fusions (68, 229). Certain combinations remain tumorigenic, indicating a common genetic defect, whereas others are no longer able to form tumors. HeLa x HT1080 fibrosarcoma hybrids—where chromosome 11 and chromosome 1, respectively, are thought to carry the inactivated or deleted tumor suppressor genes—are completely nontumorigenic, which indicates complementation of the defective genes. Conversely, with malignant x malignant hybrid crosses between mouse cells, where loss of genetic information on chromosome 4 seemed common, most hybrids remained tumorigenic (235). Obviously, another interpretation of these latter results, other than failure to complement, would be chromosomal instability.

In summary, the early somatic cell hybrid experiments strongly indicated the existence of genetic elements capable of negatively regulating malignant growth and have provided suggestive evidence that loss of function of such elements is required for progression to the malignant state. Although these experiments provided strong evidence for such genetic loss of function, they were laboratory artifacts. The next step was to seek evidence for loss of gene function in naturally occurring human cancers.

CYTOGENETIC AND MOLECULAR GENETIC EVIDENCE FOR GENE LOSS IN HUMAN CANCER

Chromosomal rearrangements, including deletions, are often found in human cancer cells. An inkling that chromosomal rearrangement may also underlie the genetic losses occurring during formation of human tumors came from the studies of Boveri early in this century. His own observations of the development of sea urchin eggs and the studies of others who had found frequent nuclear and mitotic irregularities in carcinoma cells led Boveri (21) to posit that chromosomal abnormalities were the key cellular changes in converting a normal cell into a malignant derivative. Data bearing on this hypothesis were acquired during subsequent decades through cytologic determination of chromosomal numbers and structures in many types of malignant cells. Such studies have uncovered a bewildering array of deviations from genetic normality. Nonetheless, careful analyses of chromosomal structure, including the microstructure revealed by chromosome banding patterns, have revealed specific chromosomal abnormalities associated at high frequency with particular malignant conditions. Examples of such abnormalities include chromosomal translocations, amplifications, and deletions.

There is an extensive body of evidence that has received excellent and detailed review elsewhere (121, 143). The very enormity of this descriptive literature covering the cytogenetic analyses of dozens of malignancies left only the most dedicated cytogeneticists convinced of the significance of specific chromosomal abnormalities associated with specific malignancies. Heim & Mitelman (88), in their invaluable catalog of the seemingly random reports of cancer-related chromosomal abnormalities into specific cancer-type categories, identified specific chromosome deletions and rearrangements. With the advent of DNA probes able to detect RFLP markers scattered throughout the genome (19), molecular geneticists initially focused their attention upon malignancies with defined chromosomal abnormalities.

The first malignancy to be so investigated was retinoblastoma. This cancer is of considerable scientific interest because, among other notable features, there are both sporadic and familial forms. The tumor arises from cells of the embryonal neural retina and occurs only in young children. In most cases, retinoblastoma arises sporadically with a worldwide incidence of approximately 1:20,000, but in approximately one third of the cases, the tumor is heritable, with the inherited predisposition behaving as a highly penetrant autosomal dominant trait. These properties led Knudson to postulate his now classical "two-hit" theory (105) in which he proposed that all types of retinoblastoma involve two separate mutations carried by all retinoblastoma tumor cells. In sporadic retinoblastoma, he argued, both mutations occur somatically in the same retinal precursor cell, whereas in heritable retinoblastoma one of the mutations is germinal and the second somatic.

Cytogenetic observations that a proportion of retinoblastoma tumors had deletions of chromosome 13q14 (245) pointed to the possible deletion or inactivation of a gene. The biochemical genetics further identified the importance of this region in retinoblastoma: the gene encoding the polymorphic enzyme, esterase D, mapped very closely to the 13q14 region (193, 194). The esterase D gene encodes two allelic forms distinguishable by electrophoresis. By following the genetic transmission of these esterase D alleles, the close link between the disease-predisposing retinoblastoma allele, transmitted in familial cases of retinoblastoma, and esterase D became clear (9, 77, 194). Thus, the gene conferring familial predisposition was located at 13q14. However, the second target postulated by Knudson remained unknown. His original hypothesis did not identify the mutation of both alleles of the same gene as the critical events. Other investigators, however, hypothesized the recessive nature of the retinoblastoma gene and suggested that both alleles of this gene would need to be inactivated for retinoblastomas to arise (9, 10). Confirmation of this hypothesis came from RFLP studies.

Cavenee and colleagues used a series of chromosome 13-specific DNA probes that mapped to the long arm of chromosome 13 and were informative, i.e. heterozygous, with respect to RFLP bands on a Southern blot. Whereas several DNA markers mapping in the 13q14 region were present in a heterozygous state in the normal cells of a patient predisposed to familial retinoblastoma, the same markers were often found in a homozygous state in tumor tissue derived from the same individual (27). This loss of heterozygosity (LOH), or reduction to homozygosity or hemizygosity, involved loss of sequences from the chromosome 13 not carrying the affected retinoblastoma (Rb-1) locus, or loss of the entire chromosome 13 in question, with duplication (in the reduction to homozygosity) of the copy containing the defective Rb-1 allele. Thus, in this particular tumor, the second hit postulated by Knudson involved the other Rb-1 allele. The sporadic retinoblastomas also arise frequently through the generation of homozygosity at the Rb-1 locus, the difference being the involvement of two somatic events compared to one germinal and one somatic event in familial cases.

The combined evidence indicated that retinoblastomas arise through loss of both functional Rb-1 alleles and therefore the gene for retinoblastoma can be considered recessive. The chromosome alterations by which Rb-1 alleles are reduced to homozygosity or the wild-type allele is otherwise eliminated are illustrated in Figure 1. They include: mitotic recombination between the chromosomal homologs with a breakpoint between the tumor locus and the centromere; mitotic nondisjunction with loss of the chromosome containing the wild-type Rb-1 allele, either with or without duplication of the mutant chromosome; mitotic or regional second events such as gene conversion, deletion, or point mutation (28, 148). The loss of both wild-type alleles is seen in both familial and sporadic forms of retinoblastoma. In the familial

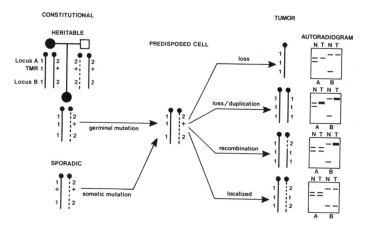

Figure 1 Chromosomal mechanisms that could effect homozygous losses of genetic information. (Top left) Inheritance of a chromosome 13 that carries a recessive defect at the tumor (TMR) locus designated "t" results in a child who is genotypically t/+ in all her cells. A retinoblastoma might develop by eliminating the dominant wild-type alleles at the TMR locus by one of the indicated mechanisms. (Top right) A recessive mutation occurring in a single retinal cell could also be unmasked by one of the same mechanisms. (Far right) The chromosomal mechanism involved in tumor formation can be discerned by comparing genotypes at loci nearer the centromere and the telomere of chromosome 13 in normal (N) and tumor (T) tissues; experimental predictions are shown at the far right. The use of marker loci in this way does not allow the detection or discrimination of the other regional mechanisms shown at bottom right. (Adapted from reference 200a, with permission)

form, one germinal mutation is inherited, followed by a second somatic event, whereas in the sporadic form both events are somatic.

Individuals with heritable retinoblastoma have a high probability of developing second malignancies, particularly soft-tissue sarcomas and osteogenic sarcomas. Loss or inactivation of wild-type alleles of the Rb-1 gene also occur in these tumors (82).

The finding of loss of genetic information at the Rb-1 locus associated with retinoblastoma formation spurred investigators to examine other malignancies for LOH at Rb-1 and other chromosomal locations. Once again, the choice of RFLP DNA probes mapping to specific chromosomal regions was generally dictated by evidence of deletions at discrete chromosomal sites identified by cytogenetic analyses. Numerous malignancies show LOH, listed in Table 1, and the number continues to grow. A few of these cancers will now be discussed in some detail.

Wilms' Tumor

Wilms' tumor is a pediatric nephroblastoma with an incidence of approximately 1:10,000; it accounts for 85% of all childhood kidney cancer (136).

Table 1   Evidence of loss of genetic information in human cancers[a,b]

Tumor type	Chromosome region(s) involved	Reference
Wilms' tumor–sporadic	11p13, 11p15	109, 167
Wilms' tumor–familial	unknown	80, 94
Retinoblastoma	13q14	27, 245
Osteogenic sarcoma[c]	13q14, 17p	82, 216
Soft tissue sarcoma[c]	13q14	82
Neuroblastoma	1p, 14q, 17	24, 72
Glioblastoma multiforme (Astrocytoma)	10, 17p	52, 97
Bladder carcinoma	9q, 11p, 17p	218
Breast carcinoma	1q, 11p, 13q, 17p	1, 29, 44, 133
Colorectal carcinoma	5q, 17p, 18q	13, 192, 223
Renal cell carcinoma	3p	32, 110, 111
Multiple endocrine neoplasia type 1	11q	149, 215, 243
Multiple endocrine neoplasia type 2	1p, 10, 22	135
Tumors associated with bilateral acoustic neurofibromatosis	22q	48
Uveal melanoma	2	146
Melanoma	1, 6	45a, 217a
Myeloid leukemia	5q	120
Small cell lung cancer	3p, 13q, 17p	56, 83, 241
Non small cell lung cancer	3p, 11p, 13q, 17p	230, 241

[a] Data derived from cytogenetic and RFLP analyses.
[b] See Table 2 for further information.
[c] Second malignancies in familial retinoblastoma patients.

The Wilms' tumor originates from embryonal kidney stem cells and may consist of a classical form, composed of embryonic blastema cells, or a differentiated form that contains striated muscle, squamous epithelial, cartilagenous, and other cell types.

Wilms' tumor, like retinoblastoma, may arise sporadically or be familial and dominantly heritable. The sporadic form accounts for >95% of all Wilms' tumors and generally presents as unilateral foci while the familial form consists of unilateral or bilateral foci (137).

The association between Wilms' tumor and aniridia (lack of an iris) focused attention on a locus in chromosome 11p13. There is an approximately 50:50 ratio of sporadic to familial forms of aniridia. A significant proportion (30–50%) of individuals with the sporadic form of aniridia develop a Wilms' tumor. Children with both conditions are often mentally retarded and may suffer from other, genitourinary abnormalities (141). Individuals with this syndrome, termed WAGR syndrome, often have cytogenetically detectable deletions that include all or part of band 11p13 (64, 191). These data prompted a molecular analysis of this region using a series of RFLP markers mapping to the 11p chromosome. Several groups of investigators demon-

Table 2   Genetic alterations in the short arm of chromosome 11 in human tumors

Tumor	Chromosome region	Reference
Wilms'	11p13, 11p15	109, 167
Hepatoblastoma	11p15	108
Rhabdomyosarcoma	11p15	185
Hepatocellular carcinoma	11p	225
Bladder carcinoma	11p15	218
Beckwith-Wiedemann syndrome (Wilms', adrenocortical carcinoma, etc.)	11p15	128a
Breast carcinoma	11p15	1
Non small cell bronchogenic carcinoma	11p13-11pter	230
Testicular cancer	11p	129
Ovarian carcinoma	11p15	123

strated a reduction to homozygosity of DNA markers on 11p in Wilms' tumor tissue compared to normal tissue from the same individual (59, 109, 155). Although many of the probes used had not at that time been carefully mapped to the 11p13 region, the conclusion was that Wilms' tumor resembles retinoblastoma in having a recessive tumor suppressor gene locus at 11p13 which is deleted or inactivated in the malignant cells. Further refinements in molecular mapping of these RFLP probes continued to support the notion of LOH in the 11p13 region (76, 220). Thus, as with retinoblastoma, cytogenetic and molecular evidence converged to apparently indicate loss of gene(s) in the 11p13 region associated with Wilms' tumor.

This simple association was soon dispelled. Various research groups have now found evidence for loss of heterozygosity at 11p15 without involvement of 11p13 (134, 167). Reduction to homozygosity, that is LOH, involving the 11p15 region has been noted in diverse human cancers (Table 2), suggesting that a critical tumor suppressor gene mapping to this region may be involved in several malignancies, a possibility already confirmed with other tumor suppressor genes (see below). However, in Wilms' tumor, certain familial forms do not appear to map at all to chromosome 11 (80, 94).

Thus, genetic subtypes of Wilms' tumor may exist that can progress to the malignant state via inactivation of different tumor suppressor genes. Alternatively, some type of cooperative effect may occur between genes located at 11p13 and 11p15 that can be abrogated by loss of function of either one or both loci (89). Functional analyses are needed to resolve these and other possibilities, but the genetics of Wilms' tumor is in all likelihood more complicated than the simple two-hit inactivation of Rb-1 alleles in retinoblastoma.

Colorectal Cancer

In both retinoblastoma and Wilms' tumor there were early indications that only a single tumor suppressor gene locus was involved. Investigations of other, more common, malignancies, e.g. lung, breast, and colon, indicated that multiple loci may be involved. One of the most extensively studied malignancies is colorectal cancer.

Colorectal tumors progress through easily recognizable clinical stages that allow study of tumor progression. In their extensive investigations, Vogelstein and colleagues have provided evidence for certain genetic alterations associated with tumor progression outlined in Figure 2. Tumorigenesis is preceded by widespread cellular hyperproliferation from which foci of benign adenomas arise. The adenomas then progress in size and dysplasia, and often change in morphology from tubular to villous. If an adenoma cell acquires the ability to proliferate and invade through the basement membrane, it becomes by definition a carcinoma. Further progression involves the capacity to metastasize to distant sites. This progression may take decades and the final malignant tumor is monoclonal (58, 223, 224).

The genetic alterations that accompany this progression include oncogene activation, DNA hypomethylation and loss or mutation of tumor suppressor genes. *Ras* gene (predominantly Ki-ras) mutations occur in approximately 50% of colorectal carcinomas, as well as in intermediate and late-stage adenomas, but in only about 10% of small adenomas (18, 62). This suggests that *ras* gene activation is perhaps a necessary genetic alteration but is clearly not sufficient for progression to the carcinomatous state. The other genetic alterations involve loss of specific chromosomal regions and appear to follow a temporal order of loss (223).

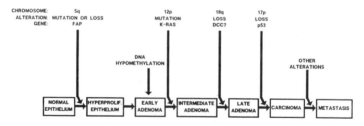

Figure 2 A model for colorectal tumorigenesis. Tumorigenesis proceeds through a series of genetic alterations involving tumor suppressor genes (particularly those on chromosomes 5, 17, and 18) and oncogenes *(ras)*. Although these alterations occur during characteristic phases of tumor development, the total accumulation of alterations, rather than their order with respect to one another, seems most important for determining tumor behavior. (Figure provided by Dr. B. Vogelstein.)

The earliest genetic loss seen in sporadic colorectal cancers involves allelic deletion of sequences on the long arm of chromosome 5 (192, 223). This region of 5q previously had been implicated in familial colorectal cancer. Familial adenomatous polyposis (FAP) is an autosomal dominant syndrome in which literally hundreds of colorectal adenomatous polyps develop in affected persons. Mapping of the locus segregating with this disease (the *fap* locus) was derived from a combination of cytogenetics, where a 5q interstitial deletion was noted in an individual with FAP (90), and RFLP analysis using DNA probes mapping to this region (13, 127). The same allele loss was seen in approximately 40% of sporadic colorectal carcinomas. Careful examination of adenoma versus carcinoma tissue showed that 5q allelic loss was seen very early in progression in sporadic colorectal carcinoma, at the small adenoma stage, whereas in FAP loss was not observed until carcinomas were formed (13, 223). The significance of the timing of 5q allelic loss in FAP versus sporadic colorectal carcinoma is not known.

Similarly, cytogenetic and RFLP allelic deletions of 17p12-p13 and 18q21-qter were found in sporadic and familial colorectal carcinomas (15, 144, 147, 223). In both, LOH was seen late in the progression to the malignant condition (223). It was soon found that the 17p region corresponded to the p53 gene that had previously been cloned and described as an oncogene (3)! This gene is discussed in more detail below.

Thus, loss of function of multiple candidate tumor suppressor genes seems to be involved in colorectal carcinomas. Although these genetic alterations follow a distinct temporal order exceptions to this sequence clearly exist. Hence, the accumulation of changes rather than their sequence (223, 224) probably is more important. Furthermore, as the frequencies of such alterations imply, not all colorectal carcinomas may need to accrue all these genetic alterations to arrive at the malignant state. A caveat here, however, is that most data for allelic loss are based upon RFLP analysis, which is only a crude measure of genetic change.

Lung Carcinomas

Bronchogenic carcinomas are classified into four major categories: squamous cell carcinoma, adenocarcinoma, undifferentiated small cell carcinoma (SCLC), and large cell carcinoma. Although squamous cell carcinomas are the most common, the overwhelming majority of lung tumors that have been extensively karyotyped are of the SCLC type. Although numerous structural and numerical cytogenetic abnormalities have been detected, the most prominent is a deletion of chromosome region 3p14-p23 (56, 231). Chromosome 3p changes have been found in nearly all SCLC and in a large fraction of non-SCLC. RFLP analysis confirmed allelic loss of 3p sequences in virtually 100% of SCLC and frequent loss in non-SCLC (22, 145, 150, 230, 241). In light of the allelic loss of Rb and p53 loci seen in other tumors, allelic deletions of these loci were also examined. Abnormalities involving the 13q14 chromosome region (containing the Rb-1 locus) have been found in both SCLC and non-SCLC (83, 230, 241). Loss of heterozygosity at 17p13 was detected in >50% of cases of SCLC (230, 241). More heterogeneity was found in non-SCLC; for example, consistent LOH at 17p13 was seen in squamous cell carcinomas but not in adenocarcinomas of the lung (230). Interestingly, LOH of both the 11p13 and 11p15 chromosome regions were found in non-SCLC (230). Thus, loss of function of multiple candidate tumor suppressor genes is also implicated in lung cancer and, also, evidence is accumulating that certain tumor suppressor genes, e.g. p53 and Rb-1, may play critical roles in multiple malignancies. The molecular evidence discussed below also supports this notion.

Renal Cell Carcinoma

Renal cell carcinoma (RCC) is the most common adult form of malignancy in the human kidney. The disease is usually sporadic but may occur heritably. The hereditary form is characterized by an earlier age of onset, the frequent appearance of multiple primary tumors, and bilateral kidney involvement. Cytogenetic studies identified the short arm of chromosome 3 as the most frequent site of nonrandom aberrations (111, 227, 242). Cohen and colleagues (32) described a large family with a constitutional 3:8 translocation that had a high incidence of RCC. The translocation breakpoint was localized to 3p14.2 (226). Thus, both sporadic and familial forms of RCC may involve inactivation of a tumor suppressor gene on chromosome 3p.

RFLP analyses also confirmed LOH at 3p in a majority of RCC cases (110, 246). It will be interesting to determine if multiple tumor suppressor genes map to this 3p region or if the same gene is involved in RCC and SCLC.

Neuroblastoma

Neuroblastoma is a tumor of the postganglionic sympathetic neurons and is the most common solid tumor in children (244). A subset of patients exhibit a predisposition to develop neuroblastoma, and this predisposition follows an autosomal dominant pattern of inheritance. Various chromosome changes have been noted in neuroblastomas but the most characteristic cytogenetic abnormality is deletion of the short arm of chromosome 1 (24, 71). Alterations of chromosome 17 have also been noted, but not with the frequency seen with 1p deletions (72).

In addition to chromosome deletions, double minutes and homogeneously staining regions (HSR) on chromosomes have been noted (4) and associated with amplification of the N-myc oncogene, which maps to chromosome 2 (183, 184). Amplification and overexpression of N-myc is associated with poor prognosis in this disease (186). Thus, once again, we see indications of a combination of oncogene expression combined with possible inactivation of a tumor suppressor gene(s) involved in tumor progression.

Carcinoma of the Breast

Many structural and numerical chromosome changes have also been reported for breast cancers. This multiplicity of cytogenetic alterations is also reflected in RFLP analysis. Significant frequencies of LOH of alleles were noted for the chromosome regions 1q23-q32 (29), 3p14-p21 (44), 11p15 (1), 13q14 (44, 131) and 17p13 (133). In one extensive analysis, concurrent LOH at 2 to 4 loci (excluding 1q) was noted in approximately 50% of tumors showing LOH (44). Allele losses did not correlate with the degree of malignancy of the breast tumors. At present, no correlation has been made between these DNA markers and possible linkage for high-risk breast cancer families.

FUNCTIONAL EVIDENCE FOR TUMOR SUPPRESSOR GENES

Although the high frequency of specific chromosome abnormalities and LOH provided persuasive evidence that inactivation of candidate tumor suppressor genes was necessary for a cell to become cancerous, this is really only "guilt by association". In only a few instances was LOH of a specific chromosome locus supported by family linkage analysis (e.g. the *fap* locus in Gardner syndrome). In many cases LOH was observed only in sporadic malignancies.

Thus, functional evidence was needed for the putative tumor suppressor genes. The ultimate proof comes from introducing an intact cloned tumor suppressor gene into a cancer cell that lacks a functional copy. This has been accomplished for two such genes but only with enormous difficulties.

The somatic cell hybrid data described earlier constitute strong circumstantial evidence for tumor suppressor genes. However, this represents the introduction of a complete genome from a normal cell and is, therefore, only marginally helpful in identifying specific effects. Before candidate tumor suppressor genes were cloned, a method was developed to transfer single chromosomes from normal to cancer cells to more specifically identify tumor suppressor gene activity.

Monochromosome Transfer Studies

Single chromosome transfer was made possible by the development of the technique of microcell transfer (63). The key feature of this technique was that the transferred chromosome was retained as a complete structural unit in succeeding generations of recipient cells, unlike the techniques of metaphase chromosome transfer (138), where the transferred chromosome is rapidly degraded. The drawback was that chromosome transfer is essentially random. To allow for transfer and selective retention of single specific chromosomes, dominant selectable markers, e.g. the bacterial gpt or neo genes, were integrated into individual human chromosomes via plasmid DNA transfection or retroviral infection (178, 219).

Chromosomes tagged with dominant selectable markers could then be transferred from normal cells into cancer cells previously shown to have deletions or LOH at specific chromosome regions. The first such studies showed that a normal copy of chromosome 11 derived from human diploid fibroblasts suppressed the tumor-forming ability of HeLa cells (179). This was quickly followed by the demonstration that the same chromosome 11 when transferred into Wilms' tumor cells suppressed their tumorigenic phenotype (228). In both cases, chromosome 11 was predicted to carry the candidate tumor suppressor gene, although evidence indicates that different genes are involved in suppression of HeLa versus Wilms' tumor (142). The specificity of tumor suppressor gene activity was shown by transferring "irrelevant" chromosomes, e.g. chromosomes 13 and X, which had no effect on the tumorigenic potential of HeLa or Wilms' tumor cells. In both cases, tumor suppression occurred although the microcell hybrids continued to behave as transformed cells in culture, a result reminiscent of the somatic cell hybrid results described earlier (202).

In one instance, the transfer of chromosome 13, containing an intact Rb-1 gene, into retinoblastoma cells containing a defective Rb-1 gene resulted in strong growth suppression in culture in addition to tumor suppression (8, 200). The growth suppression was associated with expression of wild-type Rb protein and cells that escaped this suppression and resumed proliferation no longer expressed the Rb protein (8).

Various single chromosome transfers have now been reported, as outlined in Table 3. In most cases, the predicted chromosome (based upon chromosome deletion and RFLP studies) has indeed carried the tumor suppressor genetic information. However, occasionally, transfer of a chromosome considered to be "irrelevant" has a dramatic tumor-suppressing effect, e.g. chromosome 17 into neuroblastoma cells carrying a chromosome 1p⁻ deletion (S. Bader & E. J. Stanbridge, in preparation), thereby revealing the presence of an unsuspected tumor suppressor gene.

In most cases, transfer of a single copy of the chromosome carrying a tumor

Table 3 Tumor suppression associated with the transfer of single human chromosomes via microcell fusion

| Tumor cell line | Tumor-suppressing chromosome | | Reference |
	Expected*	Observed	
HeLa	11	11	179
Wilms' tumor	11	11	157, 228
Retinoblastoma	13	13	9, 200
SiHa cervical carcinoma	—	11	107
Endometrial carcinoma	—	1, 6, 9	237
Renal cell carcinoma	3	3	188
Neuroblastoma	1	17	Bader et al (unpublished)
Melanoma	6	6	217a

*Predicted from cytogenetic and RFLP analyses.

suppressor gene is sufficient to suppress tumor-forming ability of the malignant cells in question. This is in keeping with the recessive nature of these genes, where loss of function of both alleles seems to be required for expression of the neoplastic phenotype and, therefore, introduction of one wild-type allele may be expected to correct this defect. However, in renal cell carcinoma, a malignancy associated with a high frequency of 3p⁻ deletions, tumor suppression requires the presence of two copies of transferred chromosome 3 (188). This suggestion of a chromosome (gene) dosage effect has also been noted occasionally in somatic cell hybrid experiments (11).

Most recently, evidence for the dominant nature of at least one "senescence" gene has been provided by transferring human chromosome 1 into immortalized hamster cells, which resulted in their reversion to nonproliferating senescent cells (210).

The transfer of single chromosomes carrying tumor suppressor gene activity represents a welcome decrease from the genetic complexity of whole genome transfers effected by somatic cell fusion. However, many genes reside on a single chromosome and further regionalization of the respective tumor suppressor gene activity is desirable. Extensive fragmentation of chromosomes by high doses of ionizing radiation followed by cell fusion was originally described by Goss & Harris (78) as a method for fine order mapping of chromosomal genetic loci. This is very useful for that purpose (35, 75) but not as a primary method for ascertaining the location of tumor suppressor genes because there is no facile method of assaying for tumor suppressor function other than the in vivo tumorigenicity testing in immune deficient mice (197). The large number of cell clones that would need to be tested for the stable property of tumor suppression renders this an uneconomical approach.

A less devastating approach than that of Goss & Harris is to treat cells with low doses of ionizing radiation to generate discrete interstitial deletions. Such treatment often leads to both intra- and interchromosomal rearrangements. Although intrachromosomal rearrangements may be acceptable, the latter would clearly not be. To circumvent this problem, microcells containing single chromosomes are used as the target for ionizing radiation followed by fusion to recipient whole cells. In developing this procedure to obtain radiation-reduction chromosomes, Dowdy and colleagues (45) have obtained copies of chromosome 11 deleted in the 11p13 or 11p15 regions. These chromosomes will be helpful in clarifying the issue of the location of the Wilms' tumor suppressor gene locus/loci in 11p13 and/or 11p15.

MOLECULAR GENETIC EVIDENCE FOR TUMOR SUPPRESSION

Molecular cloning of tumor suppressor genes is far more difficult than the cloning of cellular oncogenes. In the case of activated oncogenes, a positive selection method was available in the form of mouse NIH3T3 cells. This cell line is aneuploid and immortal but behaves in culture much like normal cells; as the cells reach confluency they cease to divide and form an orderly monolayer. Transfection of the cells with DNA containing an activated oncogene results in discrete foci of transformed cells that are released from contact inhibition and form multilayered colonies (112, 187). Unfortunately, the somatic cell hybrid and monochromosome transfer data suggest that, in most cases, tumor suppressor genes do not alter the transformed phenotype of the cells. Thus, one would be looking for the rare cell that is nontumorigenic but still transformed against a background of transformed, tumorigenic cells.

Despite exceptions to this scenario, many investigators have attempted to clone tumor suppressor genes using the strategy of reverse genetics coupled with the "brute force" approach of chromosome walking and jumping.

The Rb-1 Retinoblastoma Gene

Having identified the region on chromosome 13q14 as the most likely locus for the retinoblastoma gene, efforts began to clone the Rb-1 gene by chromosome walking from flanking markers. The esterase-D gene was used as one such starting point because of its known close linkage to the candidate Rb-1 locus (193). In addition, random probes were prepared from a flow-sorted population of chromosome 13 metaphase chromosomes (115). A 1.5-kb DNA fragment, termed H3-8, selected from the chromosome 13 lambda phage library, detected deletions involving 13q14 in 3 out of 37 retinoblastomas (47). Chromosome-walking techniques were used to isolate sequences surrounding this DNA sequence. One probe was isolated that hybridized to a

4.7-kb RNA transcript in retinal cells but failed to detect a transcript in certain retinoblastoma cells. This probe was then used to isolate a cDNA that turned out to be the Rb-1 gene (65). This finding was quickly confirmed by other independent groups (67, 125). A key indication that this was indeed the Rb-1 gene was the finding of internal deletions contained within the Rb-1 locus in both retinoblastomas and osteosarcomas (66, 67, 212).

Further analysis showed that the Rb-1 gene is very large, containing 27 exons dispersed within 200 kb of genomic DNA (92, 213). Surprisingly, the 4.7-kb mRNA transcript is expressed in essentially all tissues examined. Studies of both retinoblastomas and osteosarcomas (osteosarcomas are the most common second malignancies seen in patients with the hereditary form of retinoblastoma) have found both large and small deletions of the Rb-1 gene as well as point mutations (49, 93, 238).

High-affinity antibodies generated against predicted Rb-1 peptide sequences identified a discrete protein variously identified as p105 or p110 (126), referred to here as p110. The protein was localized in the nucleus and although no obvious metal ion/DNA-binding domains were found in the protein sequence, the Rb-1 protein does appear to be a DNA-binding phosphoprotein. As yet, no specific DNA-binding sites have been identified.

The Rb-1 protein is found in both unphosphorylated and phosphorylated states, and protein gels revealed multiple forms of the protein, ranging from the unphosphorylated p110 to the highly phosphorylated pp114 form. Careful analyses of cycling cells indicated that the phosphorylation state of the Rb-1 protein changes in a cell cycle-dependent manner (25, 30, 41). The Rb-1 protein is found in the unphosphorylated form in G0 and G1 and progressively more phosphorylated forms are seen in the S and G2/M phases. If one accepts the evidence obtained from genetic studies that the Rb-1 protein acts as a negative regulator of cellular proliferation, then for cells to divide, the Rb-1 protein must be regulated posttranslationally. Phosphorylation is clearly a candidate for such regulation and it is reasonable to assume that the unphosphorylated p110 form is the active form that holds cells in G0 or G1. Phosphorylation of p110 to pp114 releases cells from this control and allows them to proceed through DNA synthesis and mitosis (Figure 3). Nothing is yet known about the kinases and phosphatases involved in control of the phosphorylated state of Rb-1 protein but they are likely to be critical determinants for cell-cycle control.

Evidence also suggests that the retinoblastoma protein may play some role in the induction of differentiation. When human leukemic cell lines were treated with retinoic acid or tumor promoting agent (TPA), the cells were induced to differentiate and this was associated with hypophosphorylation of the Rb-1 protein (30).

Given that the Rb-1 protein is expressed in virtually all normal cells and

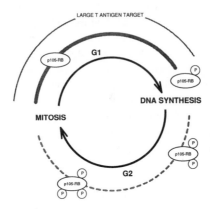

Figure 3   RB phosphorylation and the cell cycle. The 105-RB protein is synthesized throughout the cell cycle. In the G1 phase, it is not detectably phosphorylated; in S phase and G2/M phase, it is phosphorylated. The sequential addition of phosphate groups is diagrammed to represent one possible mechanism of generating multiple phosphatase-sensitive p105-RB species. SV40 T antigen binds only to the G1 phase, unphosphorylated species, whereas the adenovirus E1A proteins bind both the phosphorylated and unphosphorylated forms. (Adapted from reference 25, with permission).

appears to play a role in control of cycling cells, inactivation of the Rb-1 gene may well be involved in other malignancies. This prediction has been borne out. Deleted or mutant Rb-1 alleles have now been found in SCLC and non-SCLC lung malignancies, breast, prostate, and bladder cancers (16, 122, 221, 230, 240), but not in other malignancies such as colorectal cancer (223).

Although genetic alterations are found in both Rb-1 alleles in virtually all retinoblastomas, such alterations are found in only a fraction of the malignancies noted above. This suggests that mechanisms other than Rb-1 inactivation in these cancers may result in progression to the malignant state. However, SCLC tumors are informative in this regard. Although only a fraction of the tumors show evidence of deletions or rearrangements of the Rb-1 locus, others show no mRNA transcript, or express the mRNA but do not express Rb-1 protein (83, 240). Thus, virtually 100% of SCLCs fail to express a functional Rb-1 protein, an indication that both transcriptional and translational control of expression of Rb-1 is possible.

A fascinating property of the Rb-1 protein is its association with the transforming proteins of DNA tumor viruses. Adenovirus E1A, SV40 large T, polyoma large T, and human papillomavirus (HPV) E7 transforming proteins all complex with Rb-1 (40, 51, 232).

The E1A protein binds both unphosphorylated and phosphorylated forms of Rb-1. In addition to Rb-1, several other cellular proteins including p300, p107 and p60 complex with E1A (50, 84, 234). Intact domains 1 (amino acids

30–60) and 2 (amino acids 121–127) of E1A protein are required for the binding of the Rb-1 protein and p107. Mutations in these domains abolish both Rb-1 binding and the transforming activity of the viral oncogene (128, 233). p60 also binds to these domains of the E1A protein (A. Giordano & E. Harlow, personal communication). It has now been demonstrated that a large region located in the C terminal half of Rb-1, encompassing residues 379 to 792, is required to be relatively intact for SV40 T antigen and E1A binding to occur (101a). The large size of this domain is likely to indicate the requirement for maintenance of functional conformation since only a fraction of the amino acid residues actually make contact with the viral proteins.

The cellular p60 that binds to E1A also complexes with p34 (73a). This p34 protein recently has been identified as cdc2 kinase (46, 124), and is homologous to the yeast $cdc2^+$/CDC28 protein kinase that plays a central role in regulating the cell cycle in these lower eukaryotes. Again, the phosphorylated-dephosphorylated species of this enzyme appear to be distinct functional states. The mammalian cdc2 kinase also associates with histone H1 kinase (2). This enzyme complex has been termed the "M phase-specific" H1 kinase and is transiently activated at the G2/M transition in a wide variety of dividing cells. Thus, several cellular proteins that complex with E1A protein may play critical roles in regulating the mammalian cell cycle. One of the major goals of adenoviral early protein expression is to drive resting cells into S-phase. In so doing they abrogate normal cell cycle control and force cells into a proliferating state—one of the hallmarks of a neoplastic cell.

The SV40 large T antigen also forms a specific complex with the Rb-1 protein. Here too, the Rb-1 protein is one of several proteins, including the p107 cellular protein that binds to E1A, that bind to SV40 T antigen (55). The unphosphorylated form of the Rb-1 protein binds preferentially to SV40 T antigen, providing further evidence that this may be the active form of the protein (130). The association of the SV40 T with Rb-1 depends on the intact nature of the 105-114 amino acid sequence of the T antigen. Mutations or deletions in this domain abolish both Rb-1 binding and the transforming activity of SV40 T antigen (55), as well as SV40 T binding to a 118–120 kd cellular protein (since shown to be indistinguishable from p107 protein that binds to E1A). This domain of SV40 T is structurally similar to regions conserved in other papovaviral T antigens, to domain 2 of E1A and the E7 protein of HPV16.

These results have led to the hypothesis that the transforming genes of these DNA tumor viruses exert their transforming effect, in part, by complexing with Rb-1 protein and other cellular proteins, thereby effectively neutralizing their negative control over cell proliferation (50, 55, 234).

Formal proof that Rb-1 functions as a tumor suppressor gene came from the infection of Rb-1 deficient retinoblastoma and osteosarcoma cells with a retroviral construct containing a full-length Rb-1 cDNA (93a). Expression of this gene had a dramatic growth-suppressing effect in vitro and suppressed tumor formation in vivo, a finding similar to that seen with the transfer of a normal chromosome 13 (8, 200). More recently, the same construct introduced into Rb-1 deficient prostatic carcinoma cells showed no effect on growth behavior in vitro but did suppress tumorigenicity (16). Thus, Rb-1 has different effects on the ability of cells to proliferate in vitro but in all cases examined functioned as a tumor suppressor gene. When the Rb-1 gene was introduced into cancer cells already expressing wild-type Rb-1 protein, there was no effect on proliferation or tumor-forming ability (93a).

p53 Gene

Vogelstein and colleagues noted that a previously cloned p53-encoding oncogene mapped to the 17p13 region frequently deleted in colorectal cancer. Mapping studies showed that one copy of the p53 gene is deleted in all colorectal cancers with any detectable deletions of chromosome 17p (3). In virtually all tumors examined, the remaining p53 allele contained at least one point mutation resulting in amino acid substitution. Thus, p53 also seems to fit the category of a recessive tumor suppressor gene, even though p53 had already been identified as an oncogene!

p53, a 375 amino acid nuclear phosphoprotein, was first identified as a cellular protein that complexed with SV40 large T antigen and coimmunoprecipitated with T antigen when extracts of SV40-transformed cells were mixed with anti-SV40 T antibodies (118). Adenovirus E1B protein and HPV E6 protein also form complexes with p53 (177). The p53 protein was found in very low amounts in normal cells, but was usually more abundant in transformed cells in culture and also in human tumors (37). When first identified, overexpression of p53 was thought to contribute in some way to the neoplastic condition of the tumor cells. However, recent data indicate that point mutations at multiple sites in the p53 gene are responsible for the apparent increase in expression and reflect significant differences in protein stability. Whereas wild-type p53 protein has a half-life of about 15 minutes, various mutant p53 proteins have half-lives measured in hours (117).

The p53 gene has been conserved during evolution. After it was successfully cloned, the murine p53 gene was found to cooperate with activated ras genes in transfection studies to transform primary rodent cells (158). Furthermore, p53 alone served to immortalize normal rat cells (169). Thus, p53 behaved as an oncogene and fell into the same "cooperating" class as c-myc, another candidate nuclear oncogene (116). In these studies, plasmid constructs were used that led to high levels of expression of p53—again indicating an oncogenic role for overexpression of what was thought to be wild-type p53 protein.

This interpretation was incorrect. It turned out that the original cloned p53 gene contained a point mutation. Studies by several groups have now clearly shown that only p53 genes that carry mutations act as cooperating oncogenes and that the wild-type gene actually serves to suppress transformation. Finlay and colleagues (60) and Eliyahu et al (53) showed that plasmid constructs containing the wild-type p53 gene inhibited the ability of E1A plus *ras* or mutant p53 plus *ras* oncogenes to transform primary rat embryo fibroblasts. The rare transformed foci that arise have either not retained the wild-type gene, do not express it, or have mutated the wild-type p53.

Several interesting features of mutant p53 proteins have emerged. In addition to their extended half-life, murine mutant p53 protein (but not wild-type) complexes with a constitutively expressed member of the heat-shock family, hsc 70 (209), the mutant forms also fail to bind SV40 T as efficiently as wild type and are defective in their ability to inhibit SV40 replication (31). In rodent cells, mutation of only a single allele is necessary to abrogate normal p53 function. The p53 protein functions as a homodimer and it is thought that when a mutant:wild-type homodimer is formed, it complexes with hsc 70 and is effectively neutralized. The significantly extended stability of mutant forms would favor such an event, and this has led to the hypothesis that a mutation of the p53 gene would function as a dominant negative mutation (60), in a fashion similar to that proposed by Herskowitz (91).

The evolutionary conservation of the p53 gene extends to five domains of amino acid sequence (98, 117). The point mutations found in human cancers cluster in these conserved domain regions, as do those found in rodent tumors (117, 151). The initial identification of human p53 mutations was determined in colorectal cancers. Although admittedly small in number, the prevailing finding is that each colorectal cancer has lost an allele by deletion and has also acquired a point mutation in the other allele. Thus, the notion of dominant negative mutations has yet to be proven in human cancers. However, mutation of a single allele of p53 could convey some selective in vivo advantage on the colonic epithelial cell that will favor its further progression toward the malignant state. If so, one would predict that the initial event would be a point mutation, followed subsequently by deletion or inactivation of the remaining wild-type allele. This chronologic series has yet to be determined.

Colorectal cancer is not the only human malignancy with deleted or mutated p53 genes. The combination of cytogenetic, RFLP, and sequencing analyses has identified other malignancies that exhibit p53 aberrations, including lung, bladder, brain, neurofibrosarcomas, and breast tumors, and chronic myeloid leukemia. (17, 97, 151, 214).

An as yet unanswered question is whether mutation of the p53 gene represents solely an inactivation of an important tumor suppressor gene or a possible gain of function, with a positive transforming function, i.e. an oncogene?

Although functional evidence supports a role for murine p53 as a tumor suppressor in rodent cells (53, 60), formal proof has not yet been obtained with human p53. To date, introduction of plasmid or retroviral vectors containing the wild-type p53 gene into colorectal cancer cells has resulted in either no viable transfectants or in cells that do not express wild-type p53 protein (B. Vogelstein, personal communication; M. Goyette & E. J. Stanbridge, unpublished observations). This result is reminiscent of those with murine p53 and suggests that this gene product has a dramatic negative effect on the proliferation of certain cancer cells.

One must once again assume that the normal role of p53 is not to act as a tumor suppressor. Several examples from both rodent and human tumors indicate that the cells can maintain viability and proliferate in the total absence of p53 expression (7, 236). Currently, the most logical choice is that p53, like the Rb-1 protein, is somehow involved in cell cycle regulation. Recently, p53 has been shown to be a substrate of the cdc2 protein kinase (12a). A serine at residue 315 is phosphorylated by both p60-cdc2 and cyclin B-cdc2. The phosphorylation is cell-cycle dependent and the level of p53 protein also varies in the cell cycle, being largely absent from cells that have just completed division and accumulating during G1 phase. Thus, as with Rb-1 protein, p53 phosphorylation-dephosphorylation states may contribute to regulation of the cell cycle.

Inactivation of p53 function by antibody microinjection or antisense mRNA also prevents the normal mitotic response of untransformed cells to serum growth factors (139, 190). The conservation of DNA sequences in the early genes of various DNA tumor viruses, whose protein products bind and presumably neutralize the normal regulatory effects of p53 and Rb-1, would argue for such a role. The primary function of these viruses is not to transform host cells, but to replicate themselves. In so doing they require host DNA replication functions. By evolving mechanisms to release cells from normal cell cycle control, they also provide a biochemical means for uncontrolled proliferation.

The binding of p53 and Rb-1 to the transforming proteins of various DNA tumor viruses suggests a significant role for these interactions, especially since mutations that prevent binding also cause the virus to lose its transforming potential. Thus, one model is to suggest that the complexing of the viral transforming proteins with p53 and/or Rb-1 (and other cellular proteins) blocks their regulatory role in maintaining normal cell cycle control. Release from this control could then allow for uncontrolled proliferation, an essential step on the progression to immortalization and neoplasia.

This is a particularly attractive hypothesis since, for the first time, it provides for a direct interaction between tumor suppressor gene and oncogene products. However, it predicts that the viral transforming proteins would neutralize the activity of the tumor suppressor proteins in a dominant or dosage-dependent fashion. Although no evidence supports or refutes this hypothesis, it is important to recognize that transformation of normal human cells by these DNA tumor viruses is an extremely rare event (164). Infection of human diploid fibroblasts with SV40 results in 100% of cells expressing T antigen in the nuclei. However, the vast majority of these cells enter senescence and only the rare cell survives the senescence crisis and becomes immortal (164). From this one is led to conclude that complexing of Rb-1 and other cellular proteins with the viral transforming proteins is of itself insufficient to induce immortalization of normal diploid human cells. Also, not only do these viral transforming genes rarely immortalize human cells, they almost invariably fail to convert them to neoplastic cells (207).

The Wilms' Tumor Gene

Cloning the Rb-1 gene was deceptively simple because one of the earliest anonymous DNA probes isolated was found to map within the region of the Rb-1 locus. The reality of the arduous nature of chromosome walking and jumping came with attempts to clone the Wilms' tumor (WT) gene. Cytogenetic and RFLP analyses had, for the most part, implicated chromosome region 11p13 as the most likely locus of the WT gene.

Using a combination of genetic resources, including somatic cell hybrids containing only an intact human chromosome 11, or fragments thereof (75, 220), flow-sorted chromosome 11 libraries (115), and chromosome-mediated gene transfer (12), a large number of probes were identified that mapped to 11p13. Eventually, the technique of pulsed-field gel electrophoresis (PFGE), which provides a means of separating large DNA fragments and constructing long-range restriction maps of chromosomal regions (33, 39, 74), proved successful in isolating a candidate WT gene.

Two groups independently used PFGE to isolate a DNA fragment with the following properties: it mapped to 11p13; it was partially deleted in Wilms' tumors that had homozygous deletions in 11p13; and it recognized a 3.0-kb transcript in fetal kidney and not in other tissues (26, 70, 170). cDNA clones were obtained and sequence analysis indicates a protein with four zinc finger domains and a region rich in proline and glutamine. The amino acid sequence of the predicted polypeptide also shows significant homology to two growth-regulated mammalian polypeptides, EGR1 and EGR2. These early growth response genes have been implicated in pathways controlling cell proliferation (100). Overall, the sequence analysis suggests a protein that may be a transcription factor involved in regulation of gene expression.

There are some significant contrasts between the WT and Rb-1 or p53 gene products. The WT gene seems to have a restricted tissue expression whereas Rb-1 and p53 are expressed in virtually all tissues. No evidence currently exists to implicate the WT gene in malignancies other than Wilms' tumor, whereas Rb-1 and p53 abnormalities are found in various malignancies. Even more strikingly, in most Wilms' tumors examined by both research groups there is no obvious rearrangement or deletion of the WT gene and most tumors express an unaltered 3.0-kb transcript.

Notwithstanding the lack of obvious alterations in the WT gene, a significant proportion of Wilms' tumors may contain point mutations in this gene, or some form of posttranscriptional control of protein expression may occur, as is seen with Rb-1 protein expression in SCLC. However, alternatively a significant proportion of Wilms' tumors may involve alterations of the gene(s) in the 11p15 region. These questions will be answered by sequence analysis of the WT gene in Wilms' tumors and introduction of the wild-type gene into Wilms' tumor cells. Irrespective of the outcome of such experiments, this gene seems likely to be involved in some way in the WAGR syndrome and kidney development.

The DCC Gene

Recently, Fearon and colleagues (57) made progress in cloning the candidate colorectal carcinoma tumor suppressor gene on chromosome 18q21-qter. In a technical tour de force, Fearon et al used probes that mapped to the chromosome 18q21-qter region and found with one probe a carcinoma that had a homozygous deletion in this region. In other cases they also noted allelic loss, and even one "gain" of heterozygosity. The compilation of data was sufficient for them to attempt cloning of expressed sequences in the region. After many futile attempts with candidate exon probes, they attempted to improve the sensitivity of the expression assays using the polymerase chain reaction in a novel exon-connection strategy. This finally proved successful and allowed for the isolation of a partial cDNA clone mapping to 18q21. The gene is expressed in many tissues, including normal colonic mucosa, with the highest expression found in brain tissue. However, there is absent or drastically reduced expression in colorectal carcinomas.

The gene is very large, with an mRNA transcript size of 10 to 12 kb. Somatic mutations within the gene have been observed in many colorectal cancers; these include a homozygous deletion at the 5' end of the gene, a point mutation within one of the introns, and ten examples of DNA insertions within a fragment immediately downstream of one of the exons. The significance of these alterations is unresolved, since a full-length cDNA or genomic clone is not yet available.

Intriguingly, this gene, termed DCC (deleted in colorectal carcinomas), shows significant homology to neural cell adhesion molecules (CAMs) and other related cell surface glycoproteins. The DCC gene contains four immunoglobulinlike domains of the C2 class and a fibronectin type III-related domain similar to the domains present in N-CAM, LI, and other members of this family of CAM.

That DCC is related to genes involved in cell-surface interactions is clearly provocative. Abundant evidence indicates that disruptions of cell adhesion and cell communication are critical events in neoplastic transformation. Disruption of normal cell-cell contacts is often noted in the process of metastasis, and intercellular adhesion mediated by CAMs directly influences differentiation (171), a process often disrupted in malignancy. As yet, alterations of the DCC gene have been found only in colorectal carcinomas. However, given its high expression in brain and other tissues, defects in this gene may well be found in tumors originating in these anatomical sites.

Another interesting possibility is that the DCC gene may correspond to the HNPCC (heritable non-polyposis colorectal carcinoma) syndrome described by Lynch et al (132). This is a syndrome with autosomal dominant inheritance in which affected individuals are predisposed to carcinoma of the colon and other organs. The gene involved in this syndrome has been linked to the Kidd blood group on chromosome 18q11-q21 (14).

Other Candidate Tumor Suppressor Genes

Noda and colleagues took the novel approach of using rat cells transformed by a viral Ki-ras oncogene, since such cells are capable of reverse transformation to a state of morphology and growth behavior somewhat akin to that of normal cells (154). They then transfected them with a pooled cDNA library derived from human fibroblasts and selected for flat revertants (which were also only weakly tumorigenic) and isolated several cDNA clones from these. One, Krev-1 (102, 152), which was independently cloned by other investigators and termed rap-1 (163), shares significant homology with *ras* genes, particularly those regions shown to be important for *ras*-transforming activity (152).

The Krev-1 gene has been considered a tumor suppressor gene because of its rather modest tumor suppressing effect on Ki-ras transformed rodent cells. However, it requires substantial overexpression of the gene (20–30-fold greater than normal) to obtain tumor suppression. The effect is seen even though there is no down regulation of viral Ki-ras expression (102). Interestingly, mutation of the wild-type Krev-1 gene at codons 12 and 59 (gly→val and ala→thr, respectively) leads to a more highly potent tumor suppressing/reverse transformation effect, whereas mutation at codon 61 (thr→lys) completely abolishes the effect (153).

Current thinking is that Krev-1 may act by competing for either an upstream or downstream target of *ras*. One such candidate is the GTPase activating protein (GAP), first described by McCormick and colleagues (217).

No naturally occurring cases of involvement of Krev-1 in human cancers have been documented and it is difficult to envisage such an effect without mutationally enhanced activity.

Other candidate tumor suppressor genes have been described using similar selection strategies (181), but their functional significance is still unknown.

TUMOR SUPPRESSOR GENES ARE NOT ANTIONCOGENES

This bold statement highlights the reviewer's concern that certain tacit assumptions are being made about tumor suppressor genes without adequate scientific documentation. In response to the now clear evidence of tumor suppression, there has been a tendency to assume that tumor suppressor genes in some way directly affect oncogene action. This tendency extends to the use of the descriptive term "antioncogenes" (79, 106). However, both a conundrum and a paradox presented by this position is that both oncogenes and tumor suppressor genes appear to function in a dominant fashion (199). Data derived from somatic cell hybrids and flat revertants provide evidence both for and against any such mechanism of action. In one experimental system, rat cells were transformed with Rous sarcoma virus (RSV), and these transformed cells were then fused with normal rat cells. The resulting hybrid cells retained the integrated RSV provirus but had a normal morphology, were nontumorigenic, and expressed little or no v-*src* protein. This is suggestive of an antioncogene effect with significant down-regulation of the viral oncogene expression (79).

A less compelling example is that dealing with the human fibrosarcoma cell line HT1080. These cells contain one activated allele of the *N-ras* gene. It has been reported that flat nontumorigenic revertants of these cells express a slightly lower (up to twofold) level of expression of activated *N-ras* p21 protein compared to their tumorigenic counterparts (160).

Many more examples exist that argue against a direct effect of tumor suppressor genes on oncogene expression. Some investigators have shown that when rodent cells, transformed with viral *ras* oncogenes or activated cellular *ras* oncogenes, are fused with normal cells, the resulting hybrid cells are nontumorigenic and yet continue to express the activated p21 protein at high levels in the appropriate plasma membrane location (36, 154). In an illustrative case using human tumor cells, the EJ bladder carcinoma cell line, which contains an endogenous activated *c-Ha-ras* oncogene, was fused with normal diploid fibroblasts. The resulting hybrid cells retained their transformed phenotype in culture but were completely nontumorigenic—a finding reminiscent of HeLa x fibroblast hybrids. Rare tumorigenic segregants were isolated. All hybrid cells, irrespective of their tumorigenic potential, expressed high levels of the mutant Ha-ras p21 protein (69). Similar results have been obtained with HT1080 x fibroblast somatic cell hybrids.

Several points may be implied from these data. Oncogenes may be dominantly expressed in these situations, but are clearly not dominantly acting, in the sense that their expression is not sufficient to endow a cell with malignant properties. The tumor suppressor function does appear to be dominantly acting but does not directly affect expression of the oncogene and, therefore, in this sense does not behave as an antioncogene. Of course, the products of the tumor suppressor genes may act in a negative sense on targets of oncogene products. Until these targets are known, this remains speculation. However, most nontumorigenic hybrid cells retain their transformed properties in culture. Thus, activation of the oncogenes in question may contribute to and, in fact, be necessary for, transformation but insufficient for the progression to the malignant state.

Evidence has also been accumulating from examination of tissues representing preneoplastic stages of human cancer (18, 34), experimental skin carcinogenesis models (5) and transgenic mouse models (81), that expression of activated oncogenes is not sufficient to induce neoplastic growth and that further genetic changes are necessary to reach the malignant state, including the inactivation of tumor suppressor genes.

The information described above raises questions concerning the dominant role claimed for activated oncogenes in the causality of cancer. The notion of this dominance has been based entirely on studies where activated human cellular oncogenes, e.g. myc and ras, have been introduced into rodent cells. Here, it was initially surmised that single (187), and later two or more cooperating oncogenes were able to cause neoplastic transformation (116). However, despite intensive efforts to transform normal human diploid fibroblasts or epithelial cells with varying combinations of activated cellular oncogenes, the results have been uniformly negative (95, 172). Two instances of immortalization of normal human cells by viral oncogenes (v-ras and v-myc) have been reported, but these represent very rare events (96, 239). More recently, it has been shown that nontumorigenic, immortalized, aneuploid human cells (analogous to mouse NIH3T3 cells) become tumorigenic when transfected with activated cellular ras oncogenes (20, 168). However, only a fraction of the cells progress to this stage, although all express the activated oncogene.

DO TUMOR SUPPRESSOR GENES CONTROL EXPRESSION OF A NEW CLASS OF ONCOGENES?

Despite evidence that tumor suppressor genes do not function as antioncogenes, this statement applies to the conventional "dominantly acting" oncogenes discussed above. Tumor suppressor genes may indeed act to negatively regulate other classes of oncogenes. One such example has been documented with HeLa x fibroblast hybrids. As mentioned earlier, both nontumorigenic and tumorigenic segregant hybrids behave as transformed cells in culture, and express many phenotypic traits in common that are associated with transformation (202). A few traits differ in the two sets of hybrid cells, including the expression of a 75-kd membrane-associated glycophosphoprotein (43, 211). HeLa cells express this antigen, fibroblasts and nontumorigenic HeLa x fibroblast hybrids are negative, but 100% of tumorigenic segregants reexpress the antigen (43). Further analysis, using ionizing radiation, which produces chromosomal deletions, and retroviral insertional mutagenesis have shown a virtual 100% correlation between the expression of this antigen and tumorigenic expression (43, 166, 203). One possible explanation is that p75 is a critical oncogene that is negatively regulated by a tumor suppressor gene. Loss of chromosome 11 containing the candidate tumor suppressor gene, deletion or inactivation of the same by ionizing radiation or retroviral insertional mutagenesis, respectively, would result in reexpression of this critical oncogene. Alternatively, p75 may be only an indicator of tumorigenicity; if so, both p75 and the critical oncogene may well be negatively regulated by the tumor suppressor gene and reexpressed when this gene is inactivated. Recently, p75 has been cloned and found to be identical to intestinal alkaline phosphatase (IAP) (119). This may at first glance seem to be an odd candidate for an oncogene. However, a large corpus of literature deals with the association of ectopic expression of alkaline phosphatases (placental, intestinal, or germ cell) in human cancers (86). Irrespective of the role of IAP as an oncogene or as an indicator of tumorigenesis, the speculation remains that certain oncogenes may be negatively regulated in *trans* by tumor suppressor gene products. This would predict that certain tumor suppressor genes would negatively regulate oncogenes at the transcriptional or posttranscriptional level. Intestinal alkaline phosphatase is negatively regulated at the transcriptional level (119), and certain candidate tumor suppressor gene products are DNA-binding proteins with features of transcription factors. Further studies will test this hypothesis.

CONCLUSIONS AND FUTURE PROSPECTS

The phenomenon of tumor suppression and its genetic basis are now firmly established. Molecular identification of tumor suppressor genes is in its infancy, yet current knowledge clearly indicates that multiple such genes exist. Already, there are indications that certain tumor suppressor genes, e.g.

p53 and Rb-1, may be involved in a variety of malignancies whereas others, e.g. the DCC gene, may be restricted to a single type of cancer. Certain malignancies, e.g. retinoblastoma, exhibit loss of function of a single tumor suppressor gene in accord with the Knudson two-hit hypothesis, but others, e.g. colorectal carcinoma, seem to require inactivation of multiple candidate tumor suppressor genes to progress to the fully malignant state. An intriguing question in this latter case is whether correction of any single tumor suppressor gene defect is sufficient to control tumorigenic expression or whether some combination (including all) of the deleted genes will require complementation? The answer in the short term will probably come from a combination of gene transfer (e.g. p53) and transfer of single normal chromosomes via microcell fusion.

Most studies indicate the recessive nature of mutations of tumor suppressor genes and the apparent requirement for inactivation of both alleles as a prerequisite for neoplastic expression. Some support for this notion is obtained from single chromosome studies where introduction of a haploid copy of the wild-type gene seems to be sufficient for tumor suppression (8, 228). However, notable exceptions include somatic cell hybrids of HT1080 fibrosarcoma x fibroblasts and single chromosome transfer of chromosome 3 into renal carcinoma cells that suggest a gene-dosage effect (11, 188). Furthermore, there is the possibility of dominant negative mutations, illustrated by mutations in the p53 encoding gene in rodent cells (60). Whether the mutations are recessive or semi-dominant will probably vary according to the gene(s) involved and is likely to have important bearings on diagnosis and treatment of cancers associated with alterations of these genes.

The admittedly few functional studies of human tumor suppressor genes completed to date indicate that loss of function is the genetic event associated with progression to the malignant state. This is clearly the converse of "dominantly acting" oncogenes, where mutation of the protooncogene may lead to altered or aberrant function, or to inappropriate expression of normal gene product due to deregulation as a consequence of, e.g. chromosome translocation or overexpression of normal gene product due to amplification, etc, may provide a positive stimulus for cell proliferation (222). However, an example from the murine model of p53 suggests that one consequence of mutation in a tumor suppressor gene could be interference with function of the remaining wild-type allele (60). In a recent report, a second example of this "dominant negative" effect has been seen with the thyroid hormone receptor (THR), which is the cellular homolog of the avian erythroblastosis retrovirus erbA oncogene (175). The wild-type THR protein acts as a DNA-binding transcriptional factor that activates genes required for differentiation and cessation of proliferation. The product of the v-erb-A oncogene, a mutated form of the THR gene, binds to the same DNA sites as the wild-type THR protein but fails to activate gene expression. The v-erb-A protein competes effectively with THR protein for its DNA-binding sites, prevents activation of the genes responsible for differentiation, and therefore ablates the wild-type function of THR (38, 176). Other oncogenes should be reevaluated for their potential to function as tumor suppressor genes when present as homologous wild-type alleles.

A further prediction is that a significant proportion of tumor suppressor genes may function as transcriptional factors, acting to both positively and negatively regulate expression of genes critical in the control of carcinogenesis.

As mentioned earlier, the molecular identification of human tumor suppressor genes remains an arduous task owing to the lack of good positive selection procedures. Most investigators will continue to use the regimen of reverse genetics, guided in the first instance by evidence derived from cytogenetic and RFLP analyses. However, other approaches may prove fruitful, including subtractive cDNA hybridization, genomic DNA and cDNA transfections with selection for nontransformed variants, and retroviral insertional mutagenesis (152, 181, 182, 203, 205). These methods have been attempted with varying success and will require clever selection protocols.

To my mind, one of the more important spin-offs from the scientific inquiry into tumor suppressor genes is to highlight the need for reevaluation of the role of "dominantly acting" oncogenes in human cancer progression. Evidence from various experimental sources (but certainly predicted from somatic cell hybrid studies) clearly indicates that oncogene activation and expression is insufficient to endow a cell with tumor-forming properties. Here, I refer to "conventional" oncogenes such as myc, ras and src. The problem, in part, has been to rely too heavily on rodent cell experimental systems. Rodent cells are far more prone to spontaneous transformation and are much more sensitive to the transforming effects of carcinogens than are human cells. This extends to cellular oncogenes where normal human cells are particularly resistant to the transforming effects of activated *human* cellular oncogenes. It would be naive in the extreme to suggest that oncogene activation plays no role in human cancer—the frequency with which activated oncogenes are seen in these malignancies certainly suggests some contributory role. One pattern emerging from studies in which activated human oncogenes are introduced into *human* cells is that expression of the oncoproteins seems to provide a limited proliferative advantage to these cells, and may also be a perturbing influence on the cellular genome (20, 73), thereby leading to further genetic alterations critical for tumor progression, including inactivation of tumor suppressor genes.

I believe that to more fully understand the role of both oncogenes and tumor suppressor genes in the causality of human cancer, it will be necessary to develop more experimental model systems that use human cells rather than rodent cells (199a).

The molecular evidence over the past two decades provides the strongest proof that genetic changes are important in the genesis of cancer. A more complete knowledge of the functional roles that both tumor suppressor genes and oncogenes play is critical for our understanding of cancer progression. One should not be surprised to find yet other genetic elements and epigenetic phenomena interacting in a complex interplay to divert the normal cell to the wayward cancerous state.

Literature Cited

1. Ali, I. U., Lidereau, R., Theillet, C., Callahan, R. 1987. Reduction to homozygosity of genes on chromosome 11 in human breast neoplasia. *Science* 238:185–88

2. Arion, D., Meijer, L., Brizuela, L., Beach, D. 1988. cdc2 is a component of the M phase-specific histone H1 kinase: Evidence for identity with MPF. *Cell* 55:371–78

3. Baker, S. J., Fearon, E. R., Nigro, J. M., Hamilton, S. R., Preisinger, A. C., et al. 1989. Chromosome 17 deletions and p53 mutations in colorectal carcinomas. *Science* 244:217–21

4. Balaban-Malenbaum, G., Gilbert, F. 1977. Double minute chromosomes and the homogeneously staining regions in chromosomes of a human neuroblastoma cell line. *Science* 198:739–41

5. Balmain, A., Brown, K. 1988. Oncogene activation in chemical carcinogenesis. *Adv. Cancer Res.* 51:147–82

6. Barski, G., Cornefert, F. 1962. Characteristics of "hybrid"-type clonal cell lines obtained from mixed cultures in vitro. *J. Natl. Cancer Inst.* 28:801–21

7. Ben-David, Y., Prideaux, V. R., Chow, V., Benchimol, S., Bernstein, A. 1988. Inactivation of the p53 oncogene by internal deletion or retroviral integration in erythroleukemic cells induced by Friend leukemia virus. *Oncogene* 3:179–85

8. Benedict, W., Banerjee, A., Araujo, D., Xu, H.-J., Hu, S.-X., et al. 1990. Microcell transfer of chromosome 13: Reexpression of the retinoblastoma gene inhibits tumorigenicity. *Cell Growth Differ.* Submitted

9. Benedict, W. F., Murphree, A. L. 1983. Chromosomal and genetic basis for cloning the retinoblastoma gene within 13q14. In *Banbury Report 14: Recombinant DNA Applications to Human Disease,* pp. 135–39. Cold Spring Harbor, NY: Cold Spring Harbor Lab.

10. Benedict, W. F., Murphree, A. L., Banerjee, A., Spina, C. A., Sparkes, M. C., et al. 1983. Patient with 13 chromosome deletion: Evidence that the retinoblastoma gene is a recessive cancer gene. *Science* 219:973–75

11. Benedict, W. F., Weissman, B. E., Mark, C., Stanbridge, E. J. 1984. Tumorigenicity of human HT1080 fibrosarcoma x normal fibroblast hybrids is chromosome dosage dependent. *Cancer Res.* 44:3471–79

12. Bickmore, W. A., Maule, J. C., van Heyningen, V., Porteous, D. J. 1989. Long-range structure of H-ras 1-selected transgenomes. *Somat. Cell Mol. Genet.* 15:229–35

12a. Bischoff, J. R., Friedman, P. N., Marshak, D. R., Prives, C., Beach, D. 1990. Human p53 is phosphorylated by p60-cdc2 and cyclin B-cdc2. *Proc. Natl. Acad. Sci. USA* 87:4766–70

13. Bodmer, W. R., Bailey, C. J., Bodmer, J., Bussey, H. J. R., Ellis, A., et al. 1987. Localization of the gene for familial adenomatous polyposis on chromosome 5. *Nature* 328:614–16

14. Boman, B. M., Lynch, H. T., Kimberling, W. J., Wildrick, D. M. 1988. Reassignment of a cancer family syndrome gene to chromosome 18. *Cancer Genet. Cytogenet.* 34:153–54

15. Boman, B. M., Wildrick, D. M., Alfaro, S. R. 1988. Chromosome 18 allele loss at the D18S6 locus in human colorectal carcinomas. *Biochem. Biophys. Res. Commun.* 155:463–69

16. Bookstein, R., Shew, J.-Y., Chen, P.-L., Scully, P., Lee, W.-H. 1990. Suppression of tumorigenicity of human prostate carcinoma cells by replacing a mutated RB gene. *Science* 247:712–15

17. Borgstrom, G. H., Vuopio, P., de la Chapella, A. 1982. Abnormalities of chromosome No. 17 in myeloprolifera-

tive disorders. *Cancer Genet. Cytogenet.* 5:123–25
18. Bos, J. L., Fearon, E. R., Hamilton, S. R., Verlaan-de Vries, M., van Boom, J. H., et al. 1987. Prevalence of ras gene mutations in human colorectal cancers. *Nature* 327:293–97
19. Botstein, D., White, R. L., Skolnick, M., Davis, R. W. 1980. Construction of a genetic linkage map in man using restriction fragment length polymorphisms. *Am. J. Hum. Genet.* 32:314–31
20. Boukamp, P., Stanbridge, E. J., Foo, D. Y., Cerutti, P., Fusenig, N. 1990. Lack of correlation between activated c-Ha-ras oncogene expression and malignant transformation or altered differentiation potential of the human HaCaT keratinocyte cell line. *Cancer Res.* 50:2840–47
21. Boveri, T. 1914. *Zur frage der Erstehung Maligner Tumoren.* Jena: Gustav Fischer
22. Brauch, H., Johnson, B., Hovis, J., Yano, T., Gazdar, A., et al. 1987. Molecular analysis of the short arm of chromosome 3 in small-cell and non-small cell carcinoma of the lung. *New Engl. J. Med.* 317:1109–13
23. Bregula, U., Klein, G., Harris, H. 1971. The analysis of malignancy by cell fusion II. Hybrids between Ehrlich cells and normal diploid cells. *J. Cell Sci.* 8:673–80
24. Brodeur, G. M., Sekhon, G. S., Goldstein, M. N. 1977. Chromosomal aberrations in human neuroblastomas. *Cancer* 40:2256–63
25. Buchkovich, K., Duffy, L., Harlow, E. 1989. The retinoblastoma protein is phosphorylated during specific phases of the cell cycle. *Cell* 58:1097–105
26. Call, K. M., Glaser, T., Ito, C. Y., Buckler, A. J., Pelletier, J., et al. 1990. Isolation and characterization of a zinc finger polypeptide gene at the human chromosome 11 Wilms' tumor locus. *Cell* 60:509–20
26a. Cavenee, W., Hastie, N., Stanbridge, E., eds. 1989. *Recessive Oncogenes and Tumor Suppression. Current Communications in Molecular Biology.* Cold Spring Harbor, NY: Cold Spring Harbor Lab. 234 pp.
27. Cavenee, W. K., Dryja, T. P., Phillips, R. A., Benedict, W. F., Godbout, R., et al. 1983. Expression of recessive alleles by chromosomal mechanisms in retinoblastoma. *Nature* 305:779–84
28. Cavenee, W. K., Hansen, M. F., Nördenskjold, M., Kock, E., Maumenee, I., et al. 1985. Genetic origin of muta-

tions predisposing to retinoblastoma. *Science* 228:501–3
29. Chen, L.-C., Dollbaum, C., Smith, H. S. 1989. Loss of heterozygosity on chromosome 1q in human breast cancer. *Proc. Natl. Acad. Sci. USA* 86:7204–7
30. Chen, P.-L., Scully, P., Shew, J.-Y., Wang, J. Y. J., Lee, W.-H. 1989. Phosphorylation of the retinoblastoma gene product is modulated during the cell cycle and cellular differentiation. *Cell* 58:1193–98
31. Clarke, C. F., Cheng, K., Frey, A. B., Stein, R., Hinds, P. W., et al. 1988. Purification of complexes of nuclear oncogene p53 with rat and *Escherichia coli* heat shock proteins: In vitro dissociation of hsc70 and dnaK from murine p53 by atp. *Mol. Cell. Biol.* 8:1206–15
32. Cohen, A. J., Li, F. P., Berg, S., Marchetto, D. J., Tsai, S., et al. 1979. Hereditary renal-cell carcinoma associated with a chromosomal translocation. *New Engl. J. Med.* 301:592–95
33. Compton, D., Weil, M. M., Jones, C., Riccardi, V. M., Strong, L. C., et al. 1988. Long range physical map of the Wilms' tumor-aniridia region on human chromosome 11. *Cell* 55:827–36
34. Corominas, M., Kamino, H., Leon, J., Pellicer, A. 1989. Oncogene activation in human benign tumors of the skin (keratoacanthomas): is HRAS involved in differentiation as well as proliferation? *Proc. Natl. Acad. Sci. USA* 86:6372–76
35. Cox, D. R., Pritchard, C. A., Uglum, E., Casher, D., Kobori, J., et al. 1989. Segregation of the Huntington disease region of human chromosome 4 in a somatic cell hybrid. *Genomics* 4:397–407
36. Craig, R. W., Sager, R. 1985. Suppression of tumorigenicity in hybrids of normal and oncogene-transformed CHEF cells. *Proc. Natl. Acad. Sci. USA* 82:2062–66
37. Crawford, L. V., Pim, D. C., Lamb, P. 1984. The cellular protein p53 in human tumors. *Mol. Biol. Med.* 2:261–72
38. Damm, K., Thompson, C. C., Evans, R. M. 1989. Protein encoded by v-erb-A functions as thyroid hormone receptor antagonist. *Nature* 339:593–97
39. Davis, L. M., Byers, M. G., Fukushima, Y., Qin, S., Nowak, N. J., et al. 1988. Four new DNA markers are assigned to the WAGR region of 11p13: Isolation and regional assignment of 112 chromosome 11 anonymous segments. *Genomics* 3:264–71

40. DeCaprio, J. A., Ludlow, J. W., Figge, J., Shew, J.-P., Huang, C.-M., et al. 1988. SV40 large T antigen forms a specific complex with the product of the retinoblastoma susceptibility gene. *Cell* 54:275–83

41. DeCaprio, J. A., Ludlow, J. W., Lynch, D., Furukawa, Y., Giffin, J., et al. 1989. The product of the retinoblastoma susceptibility gene has properties of a cell cycle regulatory element. *Cell* 58:1085–95

42. Der, C. J., Krontiris, T. G., Cooper, G. M. 1982. Transforming genes of human bladder and lung carcinoma cell lines are homologous to the *ras* genes of Harvey and Kirsten sarcoma viruses. *Proc. Natl. Acad. Sci. USA* 79:3637–40

43. Der, C. J., Stanbridge, E. J. 1981. A tumor-specific membrane phosphoprotein marker in human cell hybrids. *Cell* 26:429–39

44. Devilee, P., van den Broek, M., Kuipers-Dijkshoorn, N., Kolluri, R., Meera-Khan, P., et al. 1989. At least four different chromosomal regions are involved in loss of heterozygosity in human breast carcinoma. *Genomics* 5:554–60

45. Dowdy, S., Fasching, C., Stanbridge, E. J. 1990. Discrete interstitial deletions introduced into radiation-reduced microcell hybrids: A useful method for inactivation of tumor suppressor genes. *Genes, Chromosomes, Cancer.* In press

45a. Dracopoli, N. C., Harnett, P., Bale, S. J., Stanger, B. Z., Tucker, M. A., et al. 1989. Loss of alleles from the distal short arm of chromosome 1 occurs late in melanoma tumor progression. *Proc. Natl. Acad. Sci. USA* 86:4614–18

46. Draetta, G., Beach, D. 1988. Activation of cdc2 protein kinase during mitosis in human cells: Cell cycle-dependent phosphorylation and subunit rearrangement. *Cell* 54:17–26

47. Dryja, T., Papaport, J. M., Joyce, J. M., Petersen, R. A. 1986. Molecular detection of deletions involving band 14q of chromosome 13 in retinoblastomas. *Proc. Natl. Acad. Sci. USA* 83:7391–94

48. Dumanski, J. P., Carlbom, E., Collins, V. P., Nordenskjöld, M. 1987. Deletion mapping of a locus on human chromosome 22 involved in the oncogenesis of meningioma. *Proc. Natl. Acad. Sci. USA* 84:9275–79

49. Dunn, J. M., Phillips, R. A., Becker, A. J., Gallie, B. L. 1988. Identification of germline and somatic mutations

50. Dyson, N., Buchkovich, K., Whyte, P., Harlow, E. 1989. The cellular 107K protein that binds to adenovirus E1a also associates with the large T antigens of SV40 and JC virus. *Cell* 58:249–55

51. Dyson, N., Howley, P. M., Munger, K., Harlow, E. 1989. The human papilloma virus-16 E7 oncoprotein is able to bind to the retinoblastoma gene product. *Science* 243:934–37

52. El-Azouzi, M., Chung, R. Y., Farmer, G. E., Martuza, R. L., Black, R. M., et al. 1989. Loss of distinct regions on the short arm of chromosome 17 associated with tumorigenesis of human astrocytomas. *Proc. Natl. Acad. Sci. USA* 86:7186–90

53. Eliyahu, D., Michalovitz, D., Eliyahu, S., Punhasi-Kimhi, O., Oren, M. 1989. Wild-type p53 can inhibit oncogene-mediated focus formation. *Proc. Natl. Acad. Sci. USA* 86:8763–67

54. Evans, E. P., Burtenshaw, M. D., Brown, B. B., Hennion, R., Harris, H. 1982. The analysis of malignancy by cell fusion. IX. Reexamination and clarification of the cytogenetic problem. *J. Cell Sci.* 56:113–30

55. Ewen, M. E., Ludlow, J. W., Marsilio, E., DeCaprio, J. A., Millikan, R. C., et al. 1989. An N-terminal transformation-governing sequence of SV40 large T antigen contributes to the binding of both p110^{Rb} and a second cellular protein, p120. *Cell* 58:257–67

56. Falor, W., Ward-Skinner, R., Wegryn, S. 1985. A 3p deletion in small cell lung carcinoma. *Cancer Genet. Cytogenet.* 16:175–77

57. Fearon, E. R., Cho, K. R., Nigro, J. M., Kern, S. E., Simons, J. W., et al. 1990. Identification of a chromosome 18q gene that is altered in colorectal cancers. *Science* 247:49–56

58. Fearon, E. R., Hamilton, S. R., Vogelstein, B. 1987. Clonal analysis of human colorectal tumors. *Science* 238:193–97

59. Fearon, E. R., Vogelstein, B., Feinberg, A. P. 1984. Somatic deletion and duplication of genes on chromosome 11 in Wilms' tumor. *Nature* 309:176–78

60. Finlay, C. A., Hinds, P. W., Levine, A. J. 1989. The p53 proto-oncogene can act as a suppressor of transformation. *Cell* 57:1083–93

61. Folkman, J., Klagsbrun, M. 1987. Angiogenic factors. *Science* 235:442–47

62. Forrester, K., Almoguera, C., Han, K., Grizzle, W. E., Perucho, M. 1987. Detection of high incidence of K-*ras*

affecting the retinoblastoma gene. *Science* 241:1797–1800

oncogenes during human colon tumor-igenesis. *Nature* 327:298–303

63. Fournier, R. E. K., Ruddle, F. H. 1977. Microcell-mediated transfer of murine chromosomes into mouse, Chinese hamster, and human somatic cells. *Proc. Natl. Acad. Sci. USA* 74:319–23

64. Franke, U., Holmes, L. B., Atkins, L., Riccardi, V. M. 1979. Aniridia-Wilms' tumor association: Evidence for specific deletion of 11p13. *Cytogenet. Cell Genet.* 24:185–92

65. Friend, S. H., Bernards, R., Rogelj, S., Weinberg, R. A., Rapaport, J. M., et al. 1986. A human DNA segment with properties of the gene that predisposes to retinoblastoma and osteosarcoma. *Nature* 323:643–46

66. Friend, S. H., Horowitz, J. M., Gerber, M. R., Wang, X. F., Bogenmann, E., et al. 1987. Deletions of a DNA sequence in retinoblastomas and mesenchymal tumors: Organization of the sequence and its encoded protein. *Proc. Natl. Acad. Sci. USA* 84:9059–63

67. Fung, Y.-K. T., Murphree, A. L., T'Ang, A., Qian, J., Hinrichs, S. H., et al. 1987. Structural evidence for the authenticity of the human retinoblastoma gene. *Science* 236:1657–61

68. Geiser, A. G., Anderson, M. J., Stanbridge, E. J. 1989. Suppression of tumorigenicity in human cell hybrids derived from cell lines expressing different activated *ras* oncogenes. *Cancer Res.* 49:1572–77

69. Geiser, A. G., Der, C. J., Marshall, C. J., Stanbridge, E. J. 1986. Suppression of tumorigenicity with continued expression of the c-Ha-ras oncogene in EJ bladder carcinoma-human fibroblast hybrid cells. *Proc. Natl. Acad. Sci. USA* 83:5209–13

70. Gessler, M., Poustka, A., Cavenee, W., Neve, R. L., Orkin, S. H., et al. 1990. Homozygous deletion in Wilms' tumors of a zinc-finger gene identified by chromosome jumping. *Nature* 343:774–78

71. Gilbert, F., Balaban, G., Moorhead, P., Bianchi, D., Schlesinger, H. 1982. Abnormalities of chromosome 1p in human neuroblastoma tumors and cell lines. *Cancer Genet. Cytogenet.* 7:33–42

72. Gilbert, F., Feder, M., Balaban, G., Brangman, D., Lurie, D. K., et al. 1984. Human neuroblastomas and abnormalities of chromosome 1 and 17. *Cancer Res.* 44:5444–49

73. Gilbert, P. X., Harris, H. 1988. The role of the ras oncogene in the formation of tumors. *J. Cell Sci.* 90:433–46

73a. Giordano, A., Whyte, P., Harlow, E., Franza, B. R. Jr., Beach, D., et al. 1989. A 60 kd cdc 2-associated polypeptide complexes with the E1A proteins in adenovirus-infected cells. *Cell* 58:981–90

74. Glaser, T., Driscoll, D., Antonarakis, S., Kazazian, H., Housman, D. 1989. A highly polymorphic locus cloned from the breakpoint of a chromosome 11p13 deletion associated with the WAGR syndrome. *Genomics* 5:880–93

75. Glaser, T., Housman, D., Lewis, W. H., Gerhard, D., Jones, C. 1989. A fine-structure deletion map of human chromosome 11p: Analysis of J1 series hybrids. *Somat. Cell Mol. Genet.* 15:477–501

76. Glaser, T., Lewis, W. H., Bruns, G. A. P., Watkins, P. C., Rogler, C. E., et al. 1986. The B-subunit of follicle-stimulating hormone is deleted in patients with aniridia and Wilms' tumour, allowing a further definition of the WAGR locus. *Nature* 321:882–87

77. Godbout, R., Dryja, T. P., Squire, J., Gallie, B. L., Phillips, R. A. 1983. Somatic inactivation of genes on chromosome 13 is a common event in retinoblastoma. *Nature* 304:451–53

78. Goss, S. J., Harris, H. 1975. New method for mapping genes in human chromosomes. *Nature* 255:680–84

79. Green, A. R., Wyke, J. A. 1985. Anti-oncogenes. A subset of regulatory genes involved in carcinogenesis? *Lancet* 2:475–77

80. Grundy, P., Koufos, A., Morgan, K., Li, F. P., Meadows, A. T., et al. 1988. Familial predisposition to Wilms' tumour does not map to the short arm of chromosome 11. *Nature* 336:377–78

81. Hanahan, D. 1988. Dissecting multistep tumorigenesis in transgenic mice. *Annu. Rev. Genet.* 22:479–519

82. Hansen, M. F., Koufos, A., Gallie, B. L., Phillips, R. A., Fodstad, O., et al. 1985. Osteosarcoma and retinoblastoma: A shared chromosomal mechanism revealing recessive predisposition. *Proc. Natl. Acad. Sci. USA* 82:6216–20

83. Harbour, J. W., Lai, S. L., Whang-Peng, J., Gazdar, A. F., Minna, J., et al. 1988. Abnormalities in structure and expression of the human retinoblastoma gene in SCLC. *Science* 241:353–57

84. Harlow, E., Whyte, P., Frainza, B. R. Jr., Schley, C. 1986. Association of adenovirus early-region 1A proteins with cellular polypeptides. *Mol. Cell. Biol.* 6:1579–89

85. Harris, H. 1985. Suppression of malig-

nancy in hybrid cells: the mechanism. *J. Cell Sci.* 79:83–94

86. Harris, H., 1990. The human alkaline phosphatases: what we know and what we don't know. *Clin. Chim. Acta* 186: 133–50

87. Harris, H., Miller, O. J., Klein, G., Worst, P., Tachibana, T. 1969. Suppression of malignancy by cell fusion. *Nature* 223:363–68

88. Heim, S., Mitelman, F. 1987. *Cancer Cytogenetics.* New York: Liss

89. Henry, I., Couillin, G. P., Barichard, F., Huerre-Jeanpierre, C., Glaster, T., et al. 1989. Tumor-specific loss of 11p15.5 alleles in de11p13 Wilms' tumor and in familial adrenocortical carcinoma. *Proc. Natl. Acad. Sci. USA* 86:3247–51

90. Herrera, L., Kakati, S., Gibas, L., Pietrzak, E., Sandberg, A. A. 1986. Gardner syndrome in a man with an interstitial deletion of 5q. *Am. J. Med. Genet.* 25:473–76

91. Herskowitz, I. 1987. Functional inactivation of genes by dominant negative mutations. *Nature* 329:219–22

92. Hong, F. D., Huang, H.-J. S., To, H., Young, L. J., Oro, A., et al. 1989. Structure of the human retinoblastoma gene. *Proc. Natl. Acad. Sci. USA* 86:5502–6

93. Horowitz, J. M., Yandell, D. W., Park, S.-H., Canning, S., Whyte, P., et al. 1989. Point mutational inactivation of the retinoblastoma antioncogene. *Science* 243:937–40

93a. Huang, H.-J. S., Yee, J. K., Shew, J.-Y., Chen, P.-L., Bookstein, R., et al. 1988. Suppression of the neoplastic phenotype by replacement of the RB gene in human cancer cells. *Science* 242:1563–66

94. Huff, V., Compton, D. A., Chao, L. Y., Strong, L. C., Geiser, C. F., et al. 1988. Lack of linkage of familial Wilms' tumour to chromosomal band 11p13. *Nature* 336:377–78

95. Hurlin, P. J., Fry, D. G., Moher, V. M., McCormick, J. J. 1987. Morphological transformation, focus formation, and anchorage independence induced in diploid human fibroblasts by expression of a transfected H-ras oncogene. *Cancer Res.* 47:5752–57

96. Hurlin, P. J., Maher, V. M., McCormick, J. J. 1989. Malignant transformation of human fibroblasts caused by expression of a transfected T24 HRAS oncogene. *Proc. Natl. Acad. Sci. USA* 86:187–91

97. James, C. D., Carlbom, E., Nordenskjold, M., Collins, V. P., Cavanee, W.

K. 1989. Mitotic recombination of chromosomes 17 in astrocytomas. *Proc. Natl. Acad. Sci. USA* 86:2858–62

98. Jenkins, J. R., Sturzbecher, H. 1988. The p53 oncogene. In *The Oncogene Handbook,* ed. E. P. Reddy, A. M. Skalka, T. Curran, pp. 403–23. Amsterdam: Elsevier

99. Jonasson, J., Povey, S., Harris, H. 1977. The analysis of malignancy by cell fusion VII. Cytogenetic analysis of hybrids between malignant and diploid cells and of tumours derived from them. *J. Cell Sci.* 24:217–54

100. Joseph, L. J., LeBeau, M. M., Jamieson, G. A., Archarya, S., Shows, T., et al. 1988. Molecular cloning, sequencing, and mapping of EGR2, a human early growth response gene encoding a protein with "zinc-binding finger" structure. *Proc. Natl. Acad. Sci. USA* 85:7164–68

101. Kaelbling, M., Klinger, H. P. 1986. Suppression of tumorigenicity in somatic cells. III. Cosegregation of human chromosome 11 of a normal cell and suppression of tumorigenicity in intraspecies hybrids of normal diploid x malignant cells. *Cytogenet. Cell Genet.* 41:67–70

101a. Kaelin, W. G. Jr., Ewen, M. E., Livingston, D. M. 1990. Definition of the minimal Simian virus 40 large T antigen and Adenovirus E1A-binding domain in the retinoblastoma gene product. *Mol. Cell Biol.* 10:3761–69

102. Kitayama, H., Sugimoto, Y., Matsuzaki, T., Ikawa, Y., Noda, M. 1989. A *ras*-related gene with transformation suppressor activity. *Cell* 56:77–84

103. Klinger, H. P. 1982. Suppression of tumorigenicity. *Cytogenet. Cell Genet.* 32:68–84

104. Klinger, H. P., Shows, T. B. 1983. Suppression of tumorigenicity in somatic cell hybrids. II. Human chromosomes implicated as suppressors of tumorigenicity in hybrids with Chinese hamster ovary cells. *J. Natl. Cancer Inst.* 79:559–69

105. Knudson, A. G. 1971. Mutation and cancer: Statistical study of retinoblastoma. *Proc. Natl. Acad. Sci. USA* 68:820–23

106. Knudson, A. G. Jr. 1985. Hereditary cancer, oncogenes, and antioncogenes. *Cancer Res.* 45:1437–43

107. Koi, M., Morita, H., Yamada, H., Satoh, H., Barrett, J. C., et al. 1989. Normal human chromosome 11 suppresses tumorigenicity of human cervical tumor cell line S1Ha. *Mol. Carcinogen.* 2:12–21

108. Koufos, A., Hansen, M. F., Copeland, N. G., Jenkins, N. A., Lampkin, B. C., et al. 1985. Loss of heterozygosity in three embryonal tumors suggests a common pathogenetic mechanism. *Nature* 316:330–34

109. Koufos, A., Hansen, M. F., Lampkin, B. C., Workman, M. L., Copeland, N. G., et al. 1984. Loss of alleles at loci on human chromosome 11 during genesis of Wilms' tumor. *Nature* 309:170–72

110. Kovacs, G., Erlandsson, R., Boldog, F., Ingvarsson, S., Müller-Brechlin, R., et al. 1988. Consistent chromosome 3p deletion and loss of heterozygosity in renal cell carcinoma. *Proc. Natl. Acad. Sci. USA* 85:1571–75

111. Kovacs, G., Szucs, S., de Riese, W., Baumgartel, H. 1987. Specific chromosome aberration in human renal cell carcinoma. *Int. J. Cancer* 40:171–78

112. Krontiris, T. G., Cooper, G. M. 1981. Transforming activity of human tumour DNAs. *Proc. Natl. Acad. Sci. USA* 78:1181–84

113. Kucherlapati, R., Shin, S. 1979. Genetic control of tumorigenicity in interspecific mammalian cell hybrids. *Cell* 16:639–48

114. Kubota, K., Katoh, H. 1990. Cessation of autonomous proliferation of mouse lymphoma EL4 by fusion with a T cell line. *Int. J. Cancer.* In press

115. Lalande, M., Fryja, T. P., Schreck, R. R., Shipley, J., Flint, A., et al. 1984. Isolation of human chromosome 13-specific DNA sequences cloned from flow sorted chromosomes and potentially linked to the retinoblastoma locus. *Cancer Genet. Cytogenet.* 13:283–95

116. Land, H., Parada, L. F., Weinberg, R. A. 1983. Tumorigenic conversion of primary embryo fibroblasts requires at least two cooperating oncogenes. *Nature* 304:596–602

117. Lane, D. P., Benchimol, B. 1990. p53: Oncogene or anti-oncogene? *Genes Dev.* 4:1–8

118. Lane, D. P., Crawford, L. V. 1979. T-antigen is bound to host protein in SV40-transformed cells. *Nature* 278:261–63

119. Latham, K. M., Stanbridge, E. J. 1990. Identification of the HeLa tumor-associated antigen p75/150 as intestinal alkaline phosphatase and evidence for its transcriptional regulation. *Proc. Natl. Acad. Sci. USA* 87:1263–67

120. LeBeau, M. M., Epstein, N. D., O'Brien, S. J., Neinhuis, A. W., Yang, Y.-C., et al. 1987. The interleukin 3 gene is located on human chromosome 5 and is deleted in myeloid leukemias with a deletion of 5q. *Proc. Natl. Acad. Sci. USA* 84:5913–17

121. LeBeau, M. M., Rowley, J. D. 1986. Chromosomal abnormalities in leukemia and lymphoma: Clinical and biological significance. *Adv. Hum. Genet.* 15:1–54

122. Lee, E. Y. H. P., To, H., Shew, J. Y., Bookstein, R., Scully, P., et al. 1988. Inactivation of the retinoblastoma susceptibility gene in human breast cancers. *Science* 241:218–21

123. Lee, J. H., Kavarayl, J. J., Wharton, J. T., Wildride, D. M., Blick, M. 1989. Allele loss at the C-Ha-*ras* I locus in human ovarian carcinoma. *Cancer Res.* 49:1220–22

124. Lee, M., Nurse, P. 1987. Complementation used to clone a human homologue of the fission yeast cell cycle control gene cdc2⁺. *Nature* 327:31–35

125. Lee, W.-H., Bookstein, R., Hong, F., Young, L.-J., Shew, J.-Y., et al. 1987. Human retinoblastoma susceptibility gene: Cloning, identification and sequence. *Science* 235:1394–99

126. Lee, W.-H., Shew, J.-Y., Hong, F. D., Sery, T. W., Donosco, L. A., et al. 1987. The retinoblastoma susceptibility gene encodes a nuclear phosphoprotein associated with DNA binding activity. *Nature* 329:642–45

127. Leppert, M., Dobbs, M., Scambler, P., O'Connell, P., Nakamura, Y., et al. 1987. The gene for familial polyposis coli maps to the long arm of chromosome 5. *Science* 238:1411

128. Lillie, J. W., Loewenstein, P. M., Green, M. R., Green, M. 1987. Functional domains of adenovirus type 5 E1A proteins. *Cell* 50:1091–100

128a. Little, M. H., Thompson, D. B., Hayward, N. K., Smith, P. J. 1988. Loss of alleles on the short arm of chromosome 11 in a hepatoblastoma from a child with Beckwith-Wiedemann syndrome. *Hum. Genet.* 79:186–89

129. Lothe, R. A., Fossa, S. D., Stenwig, A. E., Nakamura, Y., White, R., et al. 1989. Loss of 3p or 11p alleles is associated with testicular cancer tumors. *Genomics* 5:134–38

130. Ludlow, J. W., DeCaprio, J. A., Huang, C.-M., Lee W.-H., Paucha, E., et al. 1989. SV40 large T antigen binds preferentially to an underphosphorylated member of the retinoblastoma susceptibility gene product family. *Cell* 56:57–65

131. Lundberg, C., Skoog, L., Cavenee, W. K., Nordenskjold, M. 1987. Loss of heterozygosity in human ductal breast tumors indicates a recessive mutation on

chromosome 13. *Proc. Natl. Acad. Sci. USA* 84:2372–76

132. Lynch, H. T., Schuelke, G. S., Kimberling, W. J., Albano, W. A., Lynch, J. F., et al. 1985. Hereditary nonpolyposis colorectal cancer (Lynch Syndromes I and II): II. Biomarker studies. *Cancer* 56:939–51

133. Mackay, J., Steel, M. C., Elder, P. A., Forrest, A. P. M., Evans, H. J. 1988. Allele loss on the short arm of chromosome 17 in breast cancers. *Lancet* 1:1384–85

134. Mannens, M., Slater, R. M., Heyting, C., Bliek, J., de Kraker, J., et al. 1988. Molecular nature of genetic changes resulting in loss of heterozygosity of chromosome 11 in Wilms' tumor. *Hum. Genet.* 81:41–48

135. Mathew, C. G. P., Smith, B. A., Thorpe, K., Wong, Z., Royle, N. J., et al. 1987. Deletion of genes on chromosome 1 in endocrine neoplasia. *Nature* 328:524–26

136. Matsungaga, I. 1981. Genetics of Wilms' tumor. *Hum. Genet.* 57:231–46

137. Maurer, H. S., Pendergrass, T. W., Borges, W. 1979. The role of genetic factors in the etiology of Wilms' tumor. *Cancer* 43:205–8

138. McBride, O. W., Ozer, H. L. 1973. Transfer of genetic information by purified metaphase chromosomes. *Proc. Natl. Acad. Sci. USA* 70:1258–62

138a. Menon, A. G., Anderson, K. M., Riccardi, V. M., Chung, R. Y., Whaley, J. M., et al. 1990. Chromosome 17p deletions and p53 mutations associated with the formation of malignant neurofibrosarcomas in von Recklinghausen neurofibromatosis. *Proc. Natl. Acad. Sci. USA* 87:5435–39

139. Mercer, W. E., Avignolo, C., Baserga, R. 1984. Role of the p53 protein in cell proliferation as studied by microinjection of monoclonal antibodies. *Mol. Cell. Biol.* 4:276–81

140. Miller, D. A., Miller, O. J. 1983. Chromosomes and cancer in the mouse: Studies in tumours, established cell lines, and cell hybrids. *Adv. Cancer Res.* 39:153–82

141. Miller, R. W., Fraumeni, J. F., Manning, M. D. 1964. Association of Wilms' tumor with aniridia, hemihypertrophy, and other congenital malformations. *New Engl. J. Med.* 270:922–27

142. Misra, B. C., Srivatsan, E. S. 1989. Localization of the HeLa cell tumor-suppressor gene to the long arm of chromosome 11. *Am. J. Hum. Genet.* 45:565–77

143. Mitelman, F. 1985. *Catalogue of Chromosomal Observations in Cancer*. New York: Liss

144. Monpezat, J. P., Delattre, O., Bernard, A., Grunwald, D., Remvikos, Y., et al. 1988. Loss of alleles on chromosome 18 and on the short arm of chromosome 17 in polyploid colorectal carcinomas. *Int. J. Cancer* 41:404–8

145. Mooibroek, H., Osinga, J., Postmus, P. E., Carritt, B., Buys, C. H. C. M. 1987. Loss of heterozygosity for a chromosome 3 sequence presumably at 3p21 in small cell lung cancer. *Cancer Genet. Cytogenet.* 27:361–65

146. Mukai, S., Dryja, T. P. 1986. Loss of alleles at polymorphic loci on chromosome 2 in uveal melanoma. *Cancer Genet. Cytogenet.* 22:45–53

147. Muleris, M., Salmon, R.-J., Dutrillaux, A.-M., Vielh, P., Safrani, B., et al. 1987. Characteristic chromosomal inbalances in 18 near-diploid colorectal tumors. *Cancer Genet. Cytogenet.* 29: 289–301

148. Murphree, A. L., Benedict, W. F. 1984. Retinoblastoma: Clues to human oncogenesis. *Science* 223:1028–33

149. Nakamura, Y., Larsson, C., Julier, C., Byström, C., Skogseid, B., et al. 1989. Localization of the genetic defect of multiple endocrine neoplasia type I within a small region of chromosome 11. *Am. J. Hum. Genet.* 44:751–55

150. Naylor, S., Johnson, B., Minna, J., Sakaguchi, A. 1987. Loss of heterozygosity of chromosome 3p markers in small-cell lung cancer. *Nature* 329:451–54

151. Nigro, J. M., Baker, S. J., Preisinger, A. C., Jessup, J. M., Hostetter, R., et al. 1989. Mutations in p53 gene occur in diverse human tumor types. *Nature* 342:705–8

152. Noda, M., Kitayama, H., Matsuzaki, T., Sugimoto, Y., Okayama, H., et al. 1989. Detection of genes with a potential for suppressing the transformed phenotype associated with activated *ras* genes. *Proc. Natl. Acad. Sci. USA* 86: 162–66

153. Noda, M., Kitayama, H., Sugimoto, Y., Matsuzaki, T., Nagano, Y., et al. 1989. Biological activities of the transformation suppressor gene Krev-1 and its mutants. See Ref. 26a, pp. 159–62

154. Noda, M., Selinger, Z., Scolnick, E. M., Bassin, R. H. 1983. Flat revertants isolated from Kirsten sarcoma virus-transformed cells are resistant to the action of specific oncogenes. *Proc. Natl. Acad. Sci. USA* 80:5602–6

155. Orkin, S. H., Goldman, D. S., Sallan,

S. E. 1984. Development of homozygosity for chromosome 11p markers in Wilms' tumor. *Nature* 309:172–74

156. Oshimura, M., Gilmer, T. M., Barrett, J. C. 1985. Nonrandom loss of chromosome 15 in Syrian hamster tumors induced by v-Ha-*ras* plus v-*myc* oncogenes. *Nature* 316:636–39

157. Oshimura, M., Kugho, H., Koi, M., Shimizu, M., Yamada, H., et al. 1990. Transfer of a normal human chromosome 11 suppresses tumorigenicity of some but not all tumor cell lines. *J. Cell. Biochem.* In press

158. Parada, L. F., Land, H., Weinberg, R. A., Wolf, D., Rotter, W. 1984. Cooperation between gene encoding p53 tumour antigen and *ras* in cellular transformation. *Nature* 312:649–51

159. Parada, L. F., Tabin, C. J., Shih, C., Weinberg, R. A. 1982. Human EJ bladder carcinoma oncogene is homologue of Harvey sarcoma virus *ras* gene. *Nature* 297:474–78

160. Paterson, H., Reeves, B., Brown, R., Hall, A., Further, M., et al. 1987. Activated N-*ras* controls the transformed phenotype of HT1080 human fibrosarcoma cells. *Cell* 51:803–12

161. Peehl, D. M., Stanbridge, E. J. 1982. The role of differentiation in the control of tumorigenic expression in human cell hybrids. *Int. J. Cancer* 30:113–20

162. Pereira-Smith, O. M., Smith, J. R. 1988. Genetic analysis of indefinite division in human cells: Identification of four complementation groups. *Proc. Natl. Acad. Sci. USA* 85:6042–46

163. Pizon, V., Chardin, P., Lerosey, I., Olofsson, B., Tavitian, A. 1988. Human cDNAs *rap1* and *rap2* homologous to the Drosophila gene Dras3 encode proteins closely related to *ras* in the "effector" domain. *Oncogene* 3:201–4

164. Ponten, J., Jensen, F., Koprowski, H. 1963. Morphological and virological investigation of human tissue cultures transformed with SV40. *J. Cell. Comp. Physiol.* 61:145–53

165. Rastinejad, F., Polverini, P. J., Bouck, N. P. 1989. Regulation of the activity of a new inhibitor of angiogenesis by a cancer suppressor gene. *Cell* 56:345–55

166. Redpath, J. L., Sun, C., Colman, M., Stanbridge, E. J. 1987. Neoplastic transformation of human hybrid cells by gamma radiation: A quantitative assay. *Radiat. Res.* 110:468–72

167. Reeve, A. E., Sih, S. A., Raizis, A. M., Feinberg, A. P. 1989. Loss of heterozygosity at a second locus on chromosome 11 in sporadic Wilms'

tumor cells. *Mol. Cell. Biol.* 9:1799–803

168. Rhim, J. S. 1989. Neoplastic transformation of human epithelial cells in vitro. *Anticancer Res.* 9:1345–65

169. Robinski, B., Benchimol, S. 1988. Immortalization of rat embryo fibroblasts by the cellular p53 oncogene. *Oncogene* 2:445–52

170. Rose, E. A., Glaser, T., Jones, C., Smith, C. L., Lewis, W. H., et al. 1990. Complete physical map of the WAGR region of 11p13 localizes a candidate Wilms' tumor gene. *Cell* 60:495–508

171. Ruoslahti, E. 1988. Fibronectin and its receptors. *Annu. Rev. Biochem.* 57:375–413

172. Sager, R. 1984. Resistance of human cells to oncogenic transformation. *Cancer Cells* 2:487–93

173. Sager, R. 1985. Genetic suppression of tumor formation. *Adv. Cancer Res.* 44:43–68

174. Saltman, D., Ross, J. A., Gordon, J. M., Krajewski, A. S., Ross, A. R., et al. 1990. Suppression of malignancy in human lymphoid cell hybrids: The role of differentiation. *J. Cell Sci.* In press

175. Sap, J., Munoz, A., Damm, K., Goldberg, Y., Ghysdael, J., et al. 1986. The c-erb-A protein is a high affinity receptor for thyroid hormone. *Nature* 324:635–40

176. Sap, J., Munoz, A., Schmitt, J., Vennström, B. 1989. Repression of transcription mediated at a thyroid hormone response element by the v-erb-A oncogene product. *Nature* 340:242–44

177. Sarnow, P., Ho, Y. S., Williams, J., Levine, A. J. 1982. Adenovirus E1b-58kd tumor antigen and SV40 large tumor antigen are physically associated with the same 54Kd cellular protein in transformed cells. *Cell* 28:387–94

178. Saxon, P. J., Srivatsan, E. S., Leipzig, G. V., Sameshima, J. H., Stanbridge, E. J. 1985. Selective transfer of individual human chromosomes to recipient cells. *Mol. Cell. Biol.* 5:140–46

179. Saxon, P. J., Srivatsan, E. S., Stanbridge, E. J. 1986. Introduction of human chromosome 11 via microcell transfer controls tumorigenic expression of HeLa cells. *EMBO J.* 5:3461–66

180. Scaletta, L. J., Ephrussi, B. 1965. Hybridization of normal and neoplastic cells in vitro. *Nature* 205:1169–71

181. Schaefer, R., Iyer, J., Iten, E., Nirkko, A. C. 1988. Partial reversion of the transformed phenotype in HRAS-transfected tumorigenic cells by transfer of a human gene. *Proc. Natl. Acad. Sci. USA* 85:1590–94

182. Schneider, C., King, R. M., Philipson, L. 1988. Genes specifically expressed at growth arrest of mammalian cells. *Cell* 54:787–93

183. Schwab, M., Alitalo, K., Klempnauer, K. H., Varmus, H. E., Bishop, J. M., et al. 1983. Amplified DNA with limited homology to *myc* cellular oncogene is shared by human neuroblastoma cell lines and a neuroblastoma tumour. *Nature* 305:245–48

184. Schwab, M., Varmus, H. E., Bishop, J. M., Grzeschik, K. H., Naylor, S. L., et al. 1984. Chromosome localization in normal human cells and neuroblastomas of a gene related to c-*myc*. *Nature* 308:288–91

185. Scrable, H. J., Witte, D. P., Lampkin, B. C., Cavenee, W. K. 1987. Chromosomal localization of the human rhabdomyosarcoma locus by mitotic recombination mapping. *Nature* 329:645–47

186. Seeger, R. C., Brodeur, G. M., Sather, H., Dalton, A., Siegel, S. E., et al. 1985. Association of multiple copies of the N-*myc* oncogene with rapid progression of neuroblastomas. *New Engl. J. Med.* 313:1111–16

187. Shih, C., Padhy, L. C., Murray, M., Weinberg, R. A. 1981. Transforming genes of carcinomas and neuroblastomas introduced into mouse fibroblasts. *Nature* 290:261–64

188. Shimizu, M., Yokota, J., Mori, N., Shiun, T., Shinoda, M., et al. 1990. Introduction of normal chromosome 3p modulates the tumorigenicity of a human renal cell line YCR. *Oncogene* 5:185–94

189. Deleted in proof

190. Shobat, O., Greenberg, M., Reisman, D., Oren, M., Rotter, V. 1987. Inhibition of cell growth mediated by plasmids encoding p53 anti-sense. *Oncogene* 1:277–83

191. Slater, R. M., deKrater, J. 1982. Chromosome number 11 and Wilms' tumor. *Cancer Genet. Cytogenet.* 5: 237–45

192. Solomon, E., Voss, R., Hall, V., Bodmer, W. F., Jass, J. R., et al. 1987. Chromosome 5 allele loss in human colorectal carcinomas. *Nature* 328:616–19

193. Sparkes, R. S., Murphree, A. L., Lingua, R. W., Sparkes, M. C., Field, L. L., et al. 1983. Gene for hereditary retinoblastoma assigned to human chromosome 13 by linkage to esterase D. *Science* 219:971–79

194. Sparkes, R. S., Sparkes, M. C., Wilson, M. G., Towner, J. W., Benedict, W. F., et al. 1980. Regional assignment of genes for human esterase D and retino-

blastoma to chromosome band 13q14. *Science* 208:1042–43

195. Srivatsan, E. S., Benedict, W. F., Stanbridge, E. J. 1986. Implication of chromosome 11 in the suppression of neoplastic expression in human cell hybrids. *Cancer Res.* 46:6174–79

196. Stanbridge, E. J. 1976. Suppression of malignancy in human cells. *Nature* 260:17–20

197. Stanbridge, E. J. 1983. The use and validity of tumorigenicity assays in immune-deficient animals. In *4th Int. Workshop on Immune-Deficient Animals*, ed. B. Sordat, pp. 196–208. New York/Berlin: Springer Verlag

198. Stanbridge, E. J. 1985. A case for human tumor-suppressor genes. *Bioessays* 3:252–55

199. Stanbridge, E. J. 1987. Genetic regulation of tumorigenic expression in somatic cell hybrids. *Adv. Viral Oncol.* 6:83–101

199a. Stanbridge, E. J. 1989. An argument for using human cells in the study of the molecular genetic basis of human cancer. In *Cell Transformation and Radiation-Induced Cancer*, ed. K. Chadwick, C. Seymour, B. Barnhart, pp. 1–10. New York: Adam Hilger

200. Stanbridge, E. J. 1989. The genetic basis of tumor suppression. *CIBA Symp.* 142:149–59. New York: Wiley

200a. Stanbridge, E. J., Cavenee, W. K. 1989. Heritable cancer and tumor suppressor genes: a tentative connection. In *Oncogenes and the Molecular Origins of Cancer*, ed. R. A. Weinberg, pp. 281–306. Cold Spring Harbor, NY:Cold Spring Harbor Lab.

201. Stanbridge, E. J., Ceredig, R. L. 1981. Growth regulatory control of human cell hybrids in nude mice. *Cancer Res.* 41:573–80

202. Stanbridge, E. J., Der, C. J., Doersen, C. J., Nishimi, R. Y., Peehl, D. M., et al. 1982. Human cell hybrids: Analysis of transformation and tumorigenicity. *Science* 215:252–59

203. Stanbridge, E. J., Dowdy, S. F., Latham, K. M., Müller, M. M., Gross, M. M. 1989. Strategies for cloning human tumor suppressor genes. See Ref. 26a, pp. 189–96

204. Stanbridge, E. J., Flandermeyer, R. R., Daniels, D. W., Nelson-Rees, W. A. 1981. Specific chromosome loss associated with the expression of tumorigenicity in human cell hybrids. *Somat. Cell Mol. Genet.* 7:699–712

205. Steeg, P. S., Bevilacqua, G. B., Kopper, L., Thorgeirsson, U. P., Talmadge, J. E., et al. 1988. A novel gene is

associated with low tumor metastatic potential. *J. Natl. Cancer Inst.* 80:200–4

206. Stehelin, D., Varmus, H. E., Bishop, J. M., Vogt, P. K. 1976. DNA related to the transforming gene(s) of avian sarcoma viruses is present in normal avian DNA. *Nature* 260:170–73

207. Stiles, C. D., Desmond, W. Jr., Sato, G., Saier, M. H. Jr. 1975. Failure of human cells transformed by simian virus 40 to form tumors in athymic nude mice. *Proc. Natl. Acad. Sci. USA* 72:4971–75

208. Stoler, A., Bouck, N. 1985. Identification of a single chromosome in the normal human genome essential for suppression of hamster cell transformation. *Proc. Natl. Acad. Sci. USA* 80:570–74

209. Sturzbecher, H.-W., Addison, C., Jenkins, J. R. 1988. Characterization of mutant p53-hsp72/73 protein-protein complexes by transient expression in monkey COS cells. *Mol. Cell. Biol.* 8:3740–47

210. Sugawara, O., Oshimura, M., Koi, M., Annab, L. A., Barrett, J. C. 1990. Induction of cellular senescence in immortalized cells by human chromosome 1. *Science* 247:707–10

211. Sutherland, D. R., Bicknell, D. C., Downward, J., Greaves, M. F., Baker, M. A., et al. 1986. Structural and functional features of a cell surface phosphoglycoprotein associated with tumorigenic phenotype in human fibroblast x HeLa cell hybrids. *J. Biol. Chem.* 15:2418–24

212. T'Ang, A., Varley, J. M., Chakraborty, S., Murphree, A. L., Fung, Y.-K. T. 1988. Structural evidence for the authenticity of the human retinoblastoma gene. *Science* 235:1394–99

213. T'Ang, A., Wu, K.-J., Hashimoto, T., Liu, W.-Y., Takahashi, R., et al. 1989. Genomic organization of the human retinoblastoma gene. *Oncogene* 4:401–7

214. Takahashi, T., Nau, M. M., Chiba, I., Birrer, M. J., Rosenberg, R. K., et al. 1989. p53: A frequent target for genetic abnormalities in lung cancer. *Science* 246:491–94

215. Thakker, R. J., Bouloux, P., Wooding, C., Chotai, K., Broad, P. M., et al. 1989. Association of parathyroid tumors in multiple endocrine neoplasia type I with loss of alleles on chromosome 11. *New Engl. J. Med.* 321:218–29

216. Toguchida, J., Ishizaki, K., Nakamura, Y., Sasaki, M. S., Ikenaga, M., et al. 1989. Assignment of common allele loss in osteosarcoma to the subregion of 17p13. *Cancer Res.* 49:6247–51

217. Trakey, M., McCormick, F. 1987. A cytoplasmic protein stimulates normal N-ras p21 GTPase, but does not affect oncogenic mutants. *Science* 238:542–45

217a. Trent, J. M., Stanbridge, E. J., McBride, H. L., Meese, E. U., Casey, G., et al. 1990. Tumorigenicity in human melanoma cell lines controlled by introduction of human chromosome 6. *Science* 247:568–71

218. Tsai, Y. C., Nichols, P. W., Hiti, A. L., Williams, Z., Skinner, D. G., et al. 1990. Allelic losses of chromosomes 9, 11 and 17 in human bladder cancer. *Cancer Res.* 50:44–47

219. Tunnacliffe, A., Parker, M., Povey, S., Bengtsson, B. O., Stanley, K., et al. 1983. Integration of Eco-gpt and SV40 early-region sequences into human chromosome 17: A dominant selection system in whole-cell and microcell human-mouse hybrids. *EMBO J.* 2:1577–84

220. van Heyningen, V., Boyd, P. A., Seawright, A., Fletcher, J. M., Fantes, J. A., et al. 1985. Molecular analysis of chromosome 11 deletions in aniridia-Wilms' tumor syndrome. *Proc. Natl. Acad. Sci. USA* 82:8592–96

221. Varley, J. M., Armour, J., Swallow, J. E., Jeffreys, A. J., Ponder, B. A. J., et al. 1989. The retinoblastoma gene is frequently altered leading to loss of expression in primary breast tumours. *Oncogene* 4:725–29

222. Varmus, H. E. 1984. The molecular genetics of cellular oncogenes. *Annu. Rev. Genet.* 18:553–612

223. Vogelstein, B., Fearon, E. R., Hamilton, S. R., Kern, S. E., Preisinger, A. C., et al. 1988. Genetic alterations during colorectal tumor development. *New Engl. J. Med.* 319:525–32

224. Vogelstein, B., Fearon, E. R., Kern, S. E., Hamilton, S. R., Preisinger, A. C., et al. 1989. Allotype of colorectal carcinomas. *Science* 244:207–11

225. Wang, H. P., Rogler, C. E. 1988. Deletions in human chromosome arms 11p and 13q in primary hepatocellular carcinomas. *Cytogenet. Cell Genet.* 48:72–78

226. Wang, N., Perkins, K. L. 1984. Involvement of band 3p14 in t(3 : 8) hereditary renal carcinoma. *Cancer Genet. Cytogenet.* 11:479–81

227. Wang, N., Soldat, L., Fan, S., Figemshau, S., Clayman, R., et al. 1983. The consistent involvement of chromosome 3 and 6 aberrations in renal cell carcinoma. *Am. J. Hum. Genet.* 35:73A

228. Weissman, B. E., Saxon, P. J., Pasquale, S. R., Jones, G. R., Geiser, A. G., et al. 1987. Introduction of a normal

human chromosome 11 into a Wilms' tumor cell line controls its tumorigenic expression. *Science* 236:175–80

229. Weissman, B., Stanbridge, E. J. 1982. Complementation of the tumorigenic phenotype in human cell hybrids. *J. Natl. Cancer Inst.* 70:667–73

229a. Werness, B. A., Levine, A. J., Howley, P. M. 1990. Association of human papillomavirus types 16 and 18 E6 proteins with p53. *Science* 248:76–79

230. Weston, A., Willey, J. C., Modali, R., Sugimura, H., McDowell, E. M., et al. 1989. Differential DNA sequence deletions from chromosomes 3, 11, 13 and 17 in squamous cell carcinoma, large cell carcinoma, and adenocarcinoma of the human lung. *Proc. Natl. Acad. Sci. USA* 86:5099–103

231. Whang-Peng, J., Bunn, P. Jr., Kao-Shan, C., Lee, E., Carney, D., et al. 1982. A non-random chromosomal abnormality, del 3p (14–23) in human small cell lung cancer. *Cancer Genet. Cytogenet.* 6:119–34

232. Whyte, P., Buchkovich, K. J., Horowitz, J. M., Friend, S. H., Raybuck, M., et al. 1988. Association between an oncogene and an antioncogene: The adenovirus E1a proteins bind to the retinoblastoma gene product. *Nature* 334:124–29

233. Whyte, P., Ruley, H. E., Harlow, E. 1988. Two regions of the adenovirus early region 1A proteins are required for transformation. *J. Virol.* 62:257–65

234. Whyte, P., Williamson, N. M., Harlow, E. 1989. Cellular targets for transformation by the adenovirus E1A proteins. *Cell* 56:67–75

235. Wiener, F., Klein, G., Harris, H. 1974. The analysis of malignancy by cell fusion. VI. Hybrids between different tumour cells. *J. Cell Sci.* 16:189–98

236. Wolf, D., Admon, S., Oren, M., Rotter, V. 1984. Major deletions in the gene encoding the p53 tumor antigen cause lack of p53 expression in HL-60 cells. *Proc. Natl. Acad. Sci. USA* 86:790–94

237. Yamada, H., Wake, N., Fujimoto, S., Barrett, J. C., Oshimura, M. 1990. Multiple chromosomes carrying tumor suppressor activity for a uterine endometrial

carcinoma cell line identified by microcell mediated chromosome transfer. *Oncogene*. In press

238. Yandel, D. W., Dayton, S. H., Campbell, T. A., Dayton, S. H., Petersen, R., et al. 1989. Oncogenic point mutations in the human retinoblastoma gene: their applications to genetic counselling. *New Engl. J. Med.* 321:1689–95

239. Yoakum, G. H., Lechner, J. F., Gabrielson, M. G., Shamsuddin, A. M., Trump, B. F., et al. 1985. Transformation of human bronchial epithelial cells transfected by Harvey ras oncogene. *Science* 227:1174–79

240. Yokota, J., Akiyama, T., Fung, Y. K. T., Benedict, W. F., Namba, Y., et al. 1988. Altered expression of the retinoblastoma (RB) gene in small-cell carcinoma of the lung. *Oncogene* 3:471–75

241. Yokota, J., Wada, M., Shimosato, Y., Terada, M., Sugimura, T. 1987. Loss of heterozygosity on chromosomes 3, 13, 17 in small cell carcinoma and on chromosome 3 in adenocarcinoma of the lung. *Proc. Natl. Acad. Sci. USA* 84:9252–56

242. Yoshida, M. A., Ohyashiki, K., Ochi, H., Gibas, Z., Pontes, J. E., et al. 1986. Cytogenetic studies of tumor-tissue from patients with nonfamilial renal cell carcinoma. *Cancer Res.* 46:2139–47

243. Yoshimoto, K., Iizuka, M., Iwahana, H., Yamasaki, R., Saito, H., et al. 1989. Loss of the same alleles of HRAS1 and D11S151 in two independent pancreatic cancers from a patient with multiple endocrine neoplasia type 1. *Cancer Res.* 49:2716–21

244. Young, J. L. Jr., Ries, L. G., Silverberg, E., Horm, J. W., Miller, R. W. 1986. Cancer incidence, survival, and mortality for children younger than 15 years. *Cancer* 58:598–602

245. Yunis, J. J., Ramsay, N. 1978. Retinoblastoma and subband deletion of chromosome 13. *Am. J. Dis. Child.* 132:161–63

246. Zbar, B., Brauch, H., Talmadge, C., Linehan, M. 1987. Loss of alleles of loci on the short arm of chromosome 3 in renal cell carcinoma. *Nature* 327:721–24

Annu. Rev. Genet. 1990. 24:659–97

GENETICS OF CIRCADIAN RHYTHMS

Jeffrey C. Hall

Department of Biology, Brandeis University, Waltham Massachusetts 02254

KEY WORDS: *Drosophila, Neurospora,* rodents, clock mutants, cycling RNAs and proteins

CONTENTS

INTRODUCTION

Contemporary genetic studies of biological rhythms involve the isolation and application of mutants as well as the cloning and manipulation of DNA sequences—some but not all of which correspond to loci defined by the "clock mutations." These mutations have, in the main, resulted from screens for strains exhibiting altered circadian rhythms or their apparent absence. Indeed, several mutants with shorter-than-normal or anomalously long circadian periods, and those that are at least superficially arrhythmic, have been isolated. Other mutants were found by simply noticing something amiss in conjunction with rhythm experiments, or were found by testing the rhythmicity of certain visible and biochemical mutants.

659

0066-4197/90/1215-0659$02.00

One idea behind these genetic approaches is that they could provide one more strategy for understanding what biological clocks consist of. Maybe, then, studies of rhythm mutations, as well as the nucleic acids and encoded proteins defined by them, will increase chronobiologists' chances of elucidating something about the intriguing, but so far intractable, mechanisms by which so many different kinds of organisms keep time.

Some recent investigations in the purely genetic area have expanded the types of rhythm variants available and the kinds of organisms in which single-gene mutations affecting time-dependent phenomena have now been identified. Some of the molecular studies involving rhythms are continuing to analyze mutationally defined clock genes and trying to understand what the products of these loci are doing; other studies in this area are beginning to isolate genes known initially from the standpoint of how their encoded RNAs are expressed.

This review discusses recent findings in the genetic area, the molecular one, and where these two intersect chronobiologically. An extensive treatment of the background information concerning genes and rhythms will be avoided, as this has been covered in several recent articles (27, 31, 43–47, 63, 64, 69, 77, 114, 139, 146–148).

GENETIC VARIANTS USED TO ANALYZE RHYTHMS

Drosophila

PERIOD MUTANTS   The most salient clock mutants in these dipteran insects remain, alas, those encoded by mutations at the X-chromosomal *period (per)* locus of *Drosophila melanogaster*. The basic features of these genetic variants and the phenotypes they cause have been reviewed ad nauseam. The most remarkable aspects of these early studies are that the first set of systematically induced rhythm mutants (in any organism) included a short-period, a long-period, and an arrhythmic strain, and that these three independently isolated *per* mutations were by definition all allelic (65). A large debt is owed by all who work in the area of clock genetics to R. J. Konopka, who led this initial screen for rhythm mutants and has continued to induce and isolate them during the past 20 years.

Some of Konopka's *per* mutants have not (or have barely been) reported. Their basic properties are described here, because they are pertinent to certain fundamental properties of biological clocks: "temperature compensation," on the one hand; and "entrainment" by, or "phase-shift responses" to, environmental stimuli, on the other. (See 90 or 145 for text-book treatments of these and other details of circadian rhythms.)

One of the relatively new *per* mutants is caused by a long-period allele, *per^{L2}*. The circadian locomotor activity rhythms of adults expressing this mutation are basically like those affected by per^{L1} (cf 48, 65). Thus, per^{L1} and per^{L2} exhibit 28–30 h periodicities ("taus," or τs), at 25° C. (See 24, 46 for brief accounts or examples of per^{L2}-influenced behavior.) When such "free-running" behavioral rhythms were monitored in constant darkness (called "DD"), at lower or higher temperatures than the standard 25° C, both per^{L1} (66) and per^{L2} (33) adults exhibited shortened or lengthened τs, respectively. The degrees to which each of these long-period mutants has lost the excellent temperature compensation associated with wild-type rhythms of this poikilothermic organism are very similar (33).

A new short-period per mutant has also been found (R. J. Konopka, personal communication). This mutant's free-running period, from the same kinds of activity monitorings referred to above, is an astonishing 16 h, a value 3 h shorter than in the original per^S mutant (cf 65, 151). The latter, whose circadian pacemaker already runs 5 h faster than normal (= 24 h in *D. melanogaster*), can nevertheless be reset each day to exhibit 24.0 rhythmicity in conditions (called "LD") of cycling light (12 h) vs darkness (12 h). Thus, this altered-τ mutant entrains to the LD cycles (see below, Figure 1). Yet, per^Ss behavior in these quasi-natural conditions is not like that of wild-type: The "evening peaks" of locomotor activity are in the daytime, instead of being at "lights-off" as in wild-type (M. Hamblen-Coyle, D. A. Wheeler, M. Rosbash, J. C. Hall, in preparation). Conversely, per^{L1} and per^{L2} display evening peaks in the night, though these mutants, too, are driven into exhibiting 24.0 h periodicities in 12 : 12 LD (Hamblen-Coyle et al, in preparation).

It was not necessarily expected that these short- and long-τ mutants would be able to be reset, each day, to an LD regime whose cycle durations lie far from the endogenous (free-running) periods specified by these genotypes (also see section below on mammalian mutants). That such mutants can be reset indicates that per^S has the phase of rhythm delayed fully 5 h each day, and that the "driven" rhythmicities of per^{L1} and per^{L2} are phase-advanced by the same amount. These LD phenomena (Figure 1) reflect the underlying "phase-response" system (associated with essentially all pacemakers found in organisms), because they are only *circa*-dian and hence must be reset daily to respond to the 24 h cycles of the environment.

Curiously, the determination of a phase response curve (PRC) for *Drosophila* locomotor activity rhythms—now so widely used in studying circadian rhythms in these species of insect—had never been reported in a primary publication (though see Ref. 44). This has now been remedied (28). As is usual for any eukaryotic organism, part of the wild-type PRC for *Drosophila* defines a light-insensitive phase in DD, i.e. a 12-h segment of a free-running day during which light pulses lead to no phase shifts; if the flies had been in 12:12 LD before proceeding into DD, the insensitive portion of the PRC, called the "subjective day," corresponds to when the lights were on during entrainment. Light pulses delivered during approximately the first half of the subjective night led to phase delays, whereas pulses given during late subjective night induced phase advances (28).

per mutations are pleiotropic in their effects on several temporally related phenotypes. These include phenomena whose time-scales are far outside the circadian range, in either direction. In addition, some of the *per* mutant phenotypes reported (e.g. learning or visual-response abnormalities) do not have any obvious temporal dependence (for review, see 43). The more recently discovered, or revised, elements of *per*'s pleiotropy as it relates to time-based phenotypes warrants some further discussion:

per's action during development Developmental timing appears to be altered by per^{L1}, per^S, and *per-zero* (per^{01}) mutations (75). The first of these had the most obvious effects and indeed noticeably lengthened the durations of larval and pupal stages. per^S had the opposite effects, though they tended to be less pronounced. per^{01} erratically lengthened or shortened a given developmental stage compared to the wild-type durations, depending on the rearing conditions (DD, LD, or LL, the last of these meaning constant light).

The end of *Drosophila*'s development is its senescence. In this regard, everyone asks (and rightly so) if *per* mutations affect adult lifespans (e.g. for per^S, perhaps the mutant flies live fast, die young). The answer is apparently "no," at least in the sense that, in two sets of studies, there were no systematic lengthenings of per^{L1}'s lifespan and no shortenings effected by per^S (33, 64). At high temperature, however, per^{01} adults lived a few days fewer than did flies expressing the rhythmic genotypes—the two mutants just noted, as well as coisogenic per^+ controls (33). This mildly abnormal phenotype, caused by an "arrhythmic" *per* allele, could, however, be a rather nonspecific viability problem (see Ref. 4 for a general discussion), as opposed to a "time-of-life" deficit.

To attempt an interpretation of the positive results obtained from assaying the effects of *per* mutations on preimaginal stages, Kyriacou et al (75) note that *embryonic* expression of *per* mRNA and protein in the central nervous system (CNS)—as demonstrated by James et al (60), Liu et al (81), and Siwicki et al (127); or conceivably that which was detected in the salivary glands, as shown by Bargiello et al (7)—could influence the timing of *subsequent* developmental stages. In this regard, *per* expression during the larval period itself is minimal [by Northern blotting, and by in situ detection of the encoded protein in salivary glands only (7)], or has been undetectable (60, 81, 127); the same holds for about the first half of pupation (60, 81, 120).

Pupal expression of *per* can, with no difficulty, be hypothesized as involved in the control of periodic eclosion (for discussion see 8, 81, 120, 127).

Likewise, the gene's adult expression is readily rationalized as one of the factors underlying the fly's locomotor activity rhythm. But is the embryonic expression also important for one or more of the *per*-influenced phenotypes that occur later in the life cycle? One way to attempt answering this kind of question is to perform temperature-shift and heat-pulse experiments on a conditional mutant. None existed for *per,* whereby the gene's function is "off" at restrictive temperature and "on" at permissive. So a temperature-sensitive mutant was created by generating transformants whose transduced DNA has a "heat-shock promoter" (*hsp*) fused to *per* coding sequences (34). Subjecting the developing transformed animals, whose genetic background was per^{01}, to various hot and cold regimes indicated that the gene's action during preimaginal stages was neither necessary nor sufficient for locomotor activity rhythms to be exhibited by adults. For example, if these transformants developed at low temperature (with *per* off), and were "LD entrained" under that condition, they were nevertheless rhythmic in their adult behavior if the temperature was turned up once the flies proceeded into constant darkness (34).

These conditional-mutant experiments have been extended. For example, it is of interest to ask whether a circadian clock is running in these *hsp-per* transformants during the LD cycles that preceed heat treatment and transfer to DD conditions. Recall, as background, experiments showing wild-type flies that were put through their development in DD and then were left in this condition as adults exhibited very weak or no rhythms in their locomotor activity (25). Thus, at least one light-dark transition is necessary to start a strong circadian clock running in this organism, as well as to set its phase.

It was hypothesized (33) that a clock function, as kicked into action by exposing the flies to LD cycles, could be already operating in animals whose *per* allele is off, i.e. in the unheated *hsp-per* transformants. Thus, when the gene's activity is subsequently switched on, it would be involved in something like linking the activity of this pacemaker to "oscillator output," which is part of the pathway mediating the final phenotype (sleep/wake cycles) but is not part of the central clock. This hypothesis about *per*'s action was seriously undermined by the finding that the phase of activity rhythms exhibited by these transformants did *not* depend on that of the LD cycle to which the flies were exposed, prior to turning the gene on at the beginning of DD (33). Therefore, this gene's action seems to be necessary for pacemaker functioning itself.

To bolster the conclusion that *per* expression is necessary only for the manifestation of the adult fly's rest/activity cycles—and hence is not used at earlier developmental stages for setting up the nervous system to run the rhythms later on—Ewer et al (33) monitored the *hsp-per* transformants' behavior during LD at low temperature and then continued to follow them in DD at high temperature. The question was: Is there any "leakage" of heat-

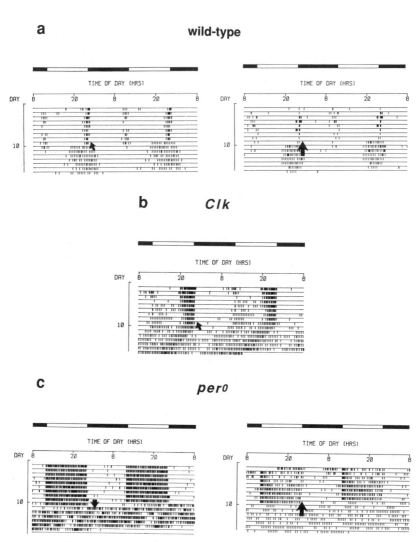

Figure 1 Locomotor activity rhythms of adult *Drosophila*. General movements of these individual *D. melanogaster* flies were monitored automatically (re infrared light-beam breakages)
as in Hamblen et al (48). Successive days of the activity events (each such event being
represented as a vertical mark, dropping below one of the horizontal lines) are plotted left-to-right
and also top-to-bottom (hence, days 1–2 on the top horizontal line, days 2–3 on the next, and so
forth). For about the first 10 days of monitoring in each case (LD conditions), the lights cycled on
(open portions of horizontal bars on top of each "actogram") and off (filled portions of bars).
"Lights-on" times were at noon, i.e. 4 h after the left-most "8" (= 8 am) in each plot. After the
arrows, the lights were turned off (DD conditions). (*a*) Wild-type actograms, after Dushay et al
(28) [left panel] and Hamblen-Coyle et al (49) [right]; each record was from monitoring a male's

shock-promoted *per* expression at low temperature, during LD? If so, then some of these flies, whose genetic background was *per*^{01}, should have been somewhat wild-typelike in their LD behavior (cf 28, 49, 96), i.e. they would have *anticipated* the times of L-to-D and D-to-L, by becoming gradually more active in advance of these environmental transitions, and would have been relatively inactive during the middle of the lights-on and lights-off periods. In contrast, *per*^{01}'s behavior in LD seems mostly to be a *response* to the environment (Figure 1c): rather little activity in the dark, then a rapid (sustained) increase at (and during) the light (see 52 for another viewpoint). The answer from these further *hsp-per* experiments was that many of the transformed flies that expressed a rhythmic phenotype following a postentrainment temperature-upshift had nevertheless behaved in a notably "*per*^{0}-like" manner (cf Figure 1c) during the prior monitoring of their activity in LD at low temperature (33).

When considering *per* expression in the developing CNS of embryos from a purely descriptive point of view, note that the in situ localizations of the gene's transcript, its protein, or stainings involving a *per*-promoted reporter protein have been examined (in published reports) only at low resolution (60, 81, 127). Curiously, the regions of such expression (ventral/medial) within a given ganglion are near to those where a "*per* relative" is also expressed. This is *single-minded (sim)*, defined originally by embryonic lethal mutations causing defective development of the CNS's ventral midline (136). A stretch of relatively N-terminal, *per*-encoded amino acids (ca 15% of the total on-paper polypeptide, which is about 1200 amino acids long) is weakly similar to a relatively N-terminal region (ca 20% of the total) within the *sim* sequence (19). Note (and recall later) that this region of similarity occurs in a region of the *per* sequence that is highly conserved among *Drosophila* species (17, 135, 144).

The *sim* protein is expressed in nuclei of developing cells of the CNS (19),

---

activity; during LD, the flies tended to become active at least an hour in advance of lights-on or lights off; during DD, the "morning peaks" of behavior largely went away, and the "evening peaks" expanded. (*b*) Actogram for a *Clk* mutant male (after 28); the fly was "forced" into 24.0 h rhythmicity during LD, in which conditions the evening peaks were earlier than those observed for wild-types; during DD, the "free-running," short-period rhythms caused by this mutation were manifested, wherein the most active portions of a given cycle began and ended at least an hour earlier on successive days. (*c*) Actograms of flies expressing a "*per*-zero" mutation—left: a male hemizygous for *per*^{01} (after 29); right: a female heterozygous for this mutation and a deletion of the *period* gene (after 49); in LD, the beginnings and ends of the dense segments of activity markings tend to coincide with the L portions of a given environmental cycle; in subsequent DD, these forced rhythms degenerated into arrhythmicity.

and *per* also has a nuclear distribution in certain tissues of the adult, such as photoreceptors, cells of the gut, and in the fly's excretory organ (81, 120, 127). In adult neurons, however, the clock gene product appears to be a cytoplasmic protein (127, 152). Recent examinations of *per*-promoted β-galactosidase activity in the "reporter transformants" (cf 81) indicate that the fusion protein (approximately the N-terminal half of *per*, with the C-terminal half replaced by the *lacZ* gene of *E. coli*) is similarly expressed throughout the cytoplasm of adult neurons; the stainings seem not to be present in the nuclei of these cell bodies in the brain (X. Liu, personal communication). *per* signals in tissue types for which the fusion protein is found in nuclei (see above) also seem to be in the cytoplasm of these cells (X. Liu, personal communication).

The same reporter transformants exhibit, at embryonic stages, cytoplasmic subcellular localization in the neurons of the developing CNS (S. T. Crews, personal communication). These observations included the demonstration of a separate, nonoverlapping set of CNS cells expressing the *per* and the *sim* proteins; moreover, the midline cells removed by the action of a *sim* mutation left *per*'s cellular expression normal (S. T. Crews, personal communication). Therefore, the *per-sim* similarity remains an uninterpretable curiosity. This includes the fact that the predicted proteins encoded within these two genes are only somewhat similar to each other, i.e. to nothing else in databases. Finally, in contrast to the case of *sim*, a *per*-null genotype allows not only for full viability (all the way to adulthood), but causes no gross morphological defects in the CNS of embryos (or any other stages).

Salivary gland rhythms   A variety of rhythmicities associated with these glands have been reported in *Drosophila* larvae, albeit with rather scanty data (104). One reasonably detailed report showed an apparent circadian rhythm of membrane potential in cells of the larval salivaries, as determined by application of a voltage-sensitive dye (143). Provocatively, this rhythmicity was said to be "damped" or absent in salivary cells of *per*^{01} larvae. These phenomena have recently been called into question: Direct recordings of membrane potentials from salivary-gland cells of wild-type larvae, over the course of a given day or even longer, revealed no rhythmic fluctuations in this physiological parameter (G. D. Block & M. W. Young, cited in 57).

Another phenotypic effect of *per* mutations on salivary gland gland physiology, which does not overtly connect to a rhythm, involves intercellular communication among electrically coupled cells within the gland (7). The product of this gene could therefore be a coupling agent. The possible rhythm connection is the following speculation: *per*-mediated communication among neurons might be the way in which the gene's action participates in building or operating a neural pacemaker (for reviews, see 147, 148). Additional findings relevant to this hypothesis involve the 5–15 h periodicities that can be extracted by spectral methods from locomotor activity records of *per*^{0} flies (e.g. 22, 24, 49). These hidden rhythmicities in the so-called arrhythmic mutants have been interpreted to mean that the pacemaker underlying these rest/activity cycles is noncircadian and that the *per* gene product serves to couple the fundamental (relatively high-frequency) oscillator (for review, see 26). It could follow from this view that *per* promotes clock function but is not per se a component of it. However, what if the hypothetical coupling function occurred within individual neurons, instead of among them? None of the results pertinent to, nor the inferences stemming from, the hidden rhythms of *per⁰* demand that the neuronal coupling agent acts intercellularly. It could just as well function within individual neurons, each one of which would be a circadian pacemaker by virtue of *per's* action. Thus, the gene would be back in the fold as a key component of the central elements of the clock.

Ovarian diapause   It has recently been asked if *per's* influence on temporally dependent phenomena extends to helping the flies discriminate short (relatively little daylight in a 24 h cycle) from long days. The latter condition leads to ovarian diapause in many forms, including *D. melanogaster,* as demonstrated in experiments performed at very low temperature (122). All the *per* variants tested for this phenotype did well: Saunders (121) found that females expressing either a long (*L2*)- or a short (*S*)-period allele were identical to wild-type in their "critical day lengths," or CDLs (whereby 14 h of daylight is "diapause-averting"). The two arrhythmic types that were tested might have been predicted not to be able to detect 14-h (or other) diurnal periods, but they did do. Yet, they were not like wild-type, in that *per⁰¹* females had CDLs 2–3 h shorter than normal (121, 123); and *per⁻* females, who are homozygously deleted of the gene (and, necessarily, of two neighboring X-chromosomal transcription units as well; 8, 103), exhibited 5 h shorter-than-normal CDLs (121). In spite of these mutant phenotypes, the investigators concluded that *per* expression is not "causally" involved in day- or night-length measurements (121, 123), which therefore would be controlled by the action of a "photoperiodic clock" whose actions are only at best "modulated" by an absence of this clock gene.

High-frequency rhythms   Certain *per* mutants have been shown, or suggested, to cause defects in various high-frequency ultradian rhythms. The fastest such rhythm is the ca 3 Hz heartbeat of 3rd instar larvae (for review, see 109). Heartbeating was said to be erratic in *per⁰¹* animals, though this has been reported in abstract form only (23, 82). A more recent study of this problem was performed by J. Ewer (personal communication). He found that the heartbeat of a given animal (which was monitored noninvasively, using a method based on light-path interruptions) could be a clean 3 Hz beating, or could consist of periods of regular beating interspersed with periods of erratic signals and interruptions. Both bouts of erratically timed beats and the interruptions were of variable, and sometimes quite long, durations. In addition, the heartbeat of a given larva could change dramatically over the duration of the recording period (usually 20 min). However, none of these types of records, or temporally dependent trends, were consistently associated with a particular *per* genotype. This included the fact that there was no more occasional irregularity in per^{O1} or per^- records than in wild-type larvae (J. Ewer, personal communication). Incidentally, no noticeable expression of *per* gene products is detectable in the larval heart—or in any larval tissues in one series of studies (81, 127); though others (7, 120) report *per* protein and transcripts in the salivary glands of larvae. In any event, *per* pleiotropy, in terms of mutational effects, does not extend to the very high frequency range of rhythmicity in the larval heart.

There are, on the other hand, many reports (most recently reviewed in 70) of *per* mutant males exhibiting defects in a rhythm of courtship song, whose ultradian periods in wild-type *D. melanogaster* are about one minute. Males expressing per^S sing with about 40 s rhythms; per^{L1} leads to ca 80 s cycles; and per^{O1} males do not exhibit any apparently regular fluctuation of the relevant song parameter (rate of tone pulse production, accompanying the courtship wing vibrations). These mutant and normal phenotypes have been extensively discussed in recent years. One reason for this is that some investigators have attempted to show (20, 35) or argue (10) that the singing rhythm is not a feature of these males' wing-vibration-generated sounds.

The validity of the song rhythms has recently been reiterated through several experimentally and analytically based arguments (74, 76). Portions of these studies showed that Crossley (20), as well as A. W. Ewing (in some pilot experiments performed by that investigator, prior to his study described in 35), have indeed recorded rhythmic songs; furthermore, the periods were in the usual (wild-type) range for this species. The two reports (74, 76, respectively) that include reanalyses of Crossley's and Ewing's data also discuss the several technical and algorithmic reasons for these two other investigators' claims that original rhythm results (71) could not be replicated. (For detailed discussion, see 10, 43, 69, 70, 124).

If we assume for the time being that the song rhythms are real, there are some new genotype-phenotype correlations. Certain of these stem from the fact that the periods of such rhythms are species-specific, in that *D. simulans* males vary their rates of song pulse production over cycle durations in the 30–40 second range (71, 73, 144); and the *D. yakuba* song rhythm period is on the order of 70–80 s (134). [These three species are all members of the *melanogaster* subgroup (80).]

With regard to *D. simulans* and *D. melanogaster,* the genetic etiology of the song rhythm difference was shown to map solely to the X chromosome, from analyzing the singing of reciprocally hybrid males (73). These studies of the hybrids' behaviors suggested that interspecific differences in the X-chromosomal *period* genes' informational content or expression could be the factor(s) underlying the presumed evolutionary divergence of this behavioral phenotype (which, incidentally, has adaptive significance in terms of reproductive isolation) (72, 73).

All this has been confirmed by some clone and sequence experiments augmented by interspecific gene transfers. Wheeler et al (144) isolated per^+ DNA from *D. simulans* (via homology to *D. melanogaster* probes) and introduced this material into *D. melanogaster,* by creating the appropriate germ-line transformants (cf 16, 150); the genetic background of these barely partially hybrid males included hemizygosity for per^{01}. The flies sang as if they were *D. simulans* (144) in that they expressed rather fast-running "song clocks," i.e. the periodicities were squarely in the 30–40 s *simulans* range—that for *melanogaster* being 50–65 s. Therefore, the source of this behavioral difference between these two species of *Drosophila* resides at only one locus.

Sequence comparisons of portions of *simulans per* to the corresponding regions of the *melanogaster* gene showed minimal divergence, except within a subset of the gene's largest exon (which corresponds to ca 570 amino acids, i.e. about 45% of the total protein). The ca 360-bp segment encodes a series of alternating threonine-glycine pairs, which is (at least descriptively) a hallmark of this gene's informational content in some, but not all, *Drosophila* species (e.g. 17, 58, 102). The Thr-Gly repeat, moreover, has been implicated as being especially important in the control of song, but not circadian, rhythms (150). Thus, further interspecific gene transfer experiments were performed to produce transformed males carrying *per* of *D. melanogaster* from which 700 bp (including the aforementioned 360) had been removed and replaced by the corresponding cassette from *D. simulans;* the reciprocal chimeric transformant type was also generated. The former males sang as if the whole *per* gene was from *D. simulans* and the latter as if they carried a per^+ DNA fragment entirely from *D. melanogaster* (144). Finally, this study showed that intra- and interspecific variations in the length of the Thr-Gly repeat were probably not responsible for the single-gene control of the song rhythm difference between *melanogaster* and *simulans;* but instead that 1–4 amino acid substitutions (just C-terminal to the repeat region) have occurred over evolutionary time and define the intragenic etiology of these species-specific courtship song rhythm differences.

CLOCK MUTATIONS AT LOCI OTHER THAN *PER*   Certain engineered *per* mutants are song rhythm variants only (cf 150). Indeed, all the transformed types noted above exhibited similar circadian rhythms of locomotor activity; the same is found when comparing the behavior of wild-type *D. melanogaster* and *D. simulans* adults (144). Reciprocal types of rhythm mutants are now known: The X-chromosomal *Clock* mutation, isolated (64) by virtue of a ca 1.5 h shortening of the circadian rhythms of locomotor activity (Figure 1b), leads to essentially normal one-minute song rhythms when *Clk* is hemizygous in males (28). Also, *disconnected* mutants (so named because of their eye-brain disconnections; 132) sing in an essentially normal manner (70); yet their locomotor activity is largely arrhythmic (29), i.e. similar to that of *per⁰* adults (Figure 1c).

These findings on the *Clk* mutant could mean that the fly's song and circadian clocks share some but not all mechanistic components (and of course these pacemaker mechanisms could not be the self-same entities, given that the relevant periodicities differ by 10^3). *disco*'s rhythm defect implies that the CNS abnormalities, which are probably caused by these mutations (see below), are more important in the brain than in more posterior portions of the adult nervous system. This inference is based on the following: (*a*) Eye-brain connectivity, or even the presence of eyes, is largely irrelevant to the fly's circadian system (reviewed in 43); it follows that this aspect of the *disco* phenotype is not responsible for the mutant's circadian-arrhythmic behavior (see discussion in 29); thus, other features of the fly's head neuroanatomy are likely to be deranged in *disco*. (*b*) The pacemaker for the courtship song—as defined by the focus of a *per*-mutation's effects on the behavior of genetic mosaics—seems to be in the male's *thoracic* nervous system (R. J. Konopka, C. P. Kyriacou, J. C. Hall, unpublished data, cited in 42), whereas the circadian pacemaker is almost certainly in the head (given the mosaic results reported in 68).

The recently determined details of rhythm defects caused by *Clk* were accompanied by a fairly fine-level localization of the mutation responsible for these phenotypes. *Clk* may not define a novel gene because it maps so very near the *per* locus (28). Whether this variant is mutated in a close neighbor of *per* (and there are some molecular candidates, cf. 8, 83, 103), or within the confines of that locus remains undetermined. If the latter is true, where and how *Clk* is mutated would be interesting, given that it could be: (*a*) associated with amino acid substitution in an informatively different location from that responsible for the other intragenically mapped short-period mutant, i.e. the *per^S*-defined serine (which became substituted by asparagine on induction of this mutation; 9, 151); or (*b*) a "hyper-expression" regulatory variant, based on the negative correlation of *per⁺* dosage with circadian period (18, 130, 152); and (*c*) instructive about the control of circadian and song rhythms—which are usually affected in parallel by *per* variants (the other exception, if "*per^Clk*" turns out to be one too, appeared in 150).

Another X-chromosomal rhythm mutation, *Andante (And)*, is like *per* in that it affects circadian rhythms (67) and probably leads to abnormal song cycling as well. Thus, *And* causes a "moderately slow" circadian clock (by definition) and slightly longer-than-normal song rhythm periodicities in males hemizygous for this mutation (70). *And*'s basic phenotype (with respect to which the mutant was isolated by Konopka et al.; 67) is that, for circadian rhythms of either eclosion or adult locomotor activity, the periods are 1–1.5 h longer than normal. The mutation is semidominant, as are essentially all circadian rhythm mutations (summarized in 27). *And* is also associated with a pigmentation mutation, long known in *D. melanogaster,* called *dusky (dy):* Not only did *And* map to the *dy* locus (by recombination and using chromosomal aberrations), but it is per se *dusky*-winged (67). Yet, a classical *dy* mutant was found to exhibit normal circadian rhythms (67). However, some newly induced *dy*s (59; J. M. Ringo, H. B. Dowse, cited in 57) do lead to 25–26 h rhythms, which are essentially the same as those observed in tests of the original *And* mutant (67). On the other hand, one new *dy* exhibits short-period circadian rhythmicity (J. M. Ringo, H. B. Dowse, personal communication), and still others (57, 59) have normal rhythms (cf. 67). Despite the genetic complexities implied by these results, the rhythm and pigment functions defined by these mutations could be the same, or at least subsets of a "complex locus" of related (primary) gene actions. Is *And-dy,* then, interesting, or could a biochemical defect (related to biogenic amines, hence body color and neurochemistry; cf. 107) indirectly affect rhythms in a manner destined not to be incisively informative? Alternatively, perhaps the relevant amine-containing substance (if any) is involved in actual pacemaker functioning.

Both possibilities should be kept in mind with regard to some recent findings involving *ebony* (*e*) body color mutants in *D. melanogaster:* Flies homozygous for a given mutant allele exhibit a variety of locomotor rhythm anomalies, both in DD and in LD (94). Interestingly, eclosion rhythms, determined for *e* cultures, were indistinguishable from wild-type (94), so that these mutants are the most specific yet in their rhythm abnormalities. Yet, *e* mutants have long been known to be pleiotropically defective in other adult behaviors, which are not apparently related to clock functions; and there are some data on *e*-induced "amine abnormalities" as well (see 94 and 107 for background literature). The classical visual-response defects exhibited by *e* flies were suggested to be involved in the rhythm defects, because blockage of all light input through external photoreceptors—effected by construction of *no receptor-potential-A* (*norpA*) + *e* double mutants—partly suppressed the aberrant locomotor activity associated with the latter (94). This result seems to argue that the relatively mild activity rhythm defects expressed by other visual system mutants (e.g. 29) might result from inadequate or scrambled pacemaker-modulatory cues, whose input would be via the external eyes. An alternative is that these visual system mutants have relatively cryptic anatomical abnormalities in their central brain, as well as more obvious disruptions of their eye and/or optic lobe morphologies.

That the *disco* visual-system mutants *are* centrally, as well as peripherally, aberrant in their neuroanatomy (see above) is indicated by the fact that they are the only visual system mutants known to be essentially arrhythmic in eclosion and adult activity (29). *disco* is also exceptional in other ways: Male song rhythms are essentially normal (see above), and the *disco*-mutant alleles are completely recessive with respect to defects in locomotor rhythmicity (29). [*e* mutations were also more or less recessive in this regard (94).] Since it is hypothesized (see above) that *disco*'s circadian clock runs so poorly (22), or apparently not at all (29), because of brain damage, its recessivity may be the exception that proves the rule: A rhythm abnormality, as effected by mutation, might usually be semidominant, if something about the ongoing, physiologically based *operation* of the clock itself is genetically aberrant (this routinely encountered semidominance is telling us something, but we do not yet know what). But a *developmental* genetic defect (documented for *disco;* 132) that damages various tissue morphologies, including perhaps pacemaker structure, could well be expected to be recessive.

disco is indeed anatomically pleiotropic, beyond its "eye-brain" defect: There are disruptions of the embryonic peripheral nervous system (PNS), and the mutant's overall viability is woeful (29, 132). The morphological abnormalities associated with *disco* include at least one region of the adult central brain: Certain "lateral neurons" (LNs), which stain prominently with an anti-*per* antibody (called anti-S, or αS; 127), are essentially undetectable in sections through the heads of *disco* adults (152). In *disco,* these LNs are either missing, badly misplaced, or cannot express the *per*-encoded antigen. This study also showed that there is a prominent staining rhythm in the LNs, as well as in other cells of the CNS and PNS of the wild-type fly (this rhythmicity is discussed in a later section).

Interestingly, the rhythm of *per*-related immunoreactivity in the photoreceptors was apparently autonomous, since it occurred in the near absence of behavioral rhythmicity (cf. 22, 29) and was readily observable (152) in *disco* eyes regardless of whether they were connected to or disconnected from the brain (cf. the incomplete penetrance for *disco*'s eye-brain anatomical phenotype; 29, 132). The same kind of quasi-normal staining cycles were observed by Zerr et al (152) in certain *per*-expressing CNS cells of *disco* brains; these appear glia-like (in the mutant or in wild-type) in their morphology and intraganglionic locations (cf. 127).

Expression of the *per*-encoded protein in the eyes is puzzling—since these external photoreceptive structures are dispensable with regard to light-mediated influences on *Drosophila*'s circadian rhythms (see above). Perhaps, however, this example of *per*'s widespread spatial expression means that an *autonomous* circadian oscillator exists in cells of the eye (cf. 131). In fact, this argument has been tentatively put forward (81, 120, 127) with regard to the presence of this gene's products in a wide range of adult organ systems (see 39 for information on an autonomous circadian pacemaker, running in a moth's testis—this being one of the tissues that expresses *per* in the fly). Yet, the first attempt at demonstrating an effect of *per* mutations on the daily rhythm of rhabdomere turnover in the adult eyes (131) has been most equivocal (15).

An intriguing aspect of this antibody-mediated protein staining is that the cycles could be elicited in all photoreceptors of mutant flies whose eyes are completely unresponsive to light (152). The relevant genetic tools were the totally blind *norpA* and the partially blind *ninaE* mutants. The former are missing a phospholipase-C (PLC) in all their external photoreceptors (11); the latter are devoid of rhodopsin in all their "outer" photoreceptors (for review, see 89), which ring each facet in the compound eyes. That all of the cell types in question express *per* and can readily have their cyclical expression of this gene's protein product turned on (see below) in either mutant implies that a novel photoreception/transduction mechanism may be operating in the circadian system; this could parallel the fact that some kind of extraocular (and hence highly specialized?) photoreceptors mediate light inputs to this system. The correlated results from light-regime "turn-on" experiments were the following: Exposing *Drosophila* to light dark cycles or even to but one light-dark transition led to robust and cyclical *per* protein expression; but rearing the flies in constant light, then monitoring their locomotor activity in this continuing condition, left them behaviorally arrhythmic (*per^0* or *disco* phenocopies) and also non-staining in the PNS or CNS of adults, in immunohistochemical experiments using αS (152).

One more element of these experiments involving visual mutants and rhythms concerns the effects or *norpA* mutations on the adult's activity rhythm (29). This could be because the PLC-encoding transcript of this gene appears not to be entirely photoreceptor-specific: The RNA was detected in situ in the optic ganglia and central brain as well as in the eye (11); this fits with the fact that eyelessness removes only 90% of PLC enzymatic activity measurable in head extracts, whereas *norpA* mutations can nearly eliminate such activity (54). Therefore, the apparent presence of this enzyme in the brain, and hence absence in a *norpA*-null mutant, could be the basis of the slightly fast circadian pacemaker running in these flies (29). Alteratively, light input to the eyes could somehow modulate the pace of the clock (see discussion above about the *ebony* mutant), which pacemaking would occur in a more internal tissue location.

The only other rhythm mutants in *Drosophila* for which any recent data have been collected are the arrhythmics found in *D. pseudoobscura*. The relevant five mutants were isolated as X-chromosomal eclosion-arrhythmic variants (reviewed in 97). Each mutation was apparently an independent isolate; the five such variants fell into two complementation groups with regard to the eclosion phenotype. These findings have largely been confirmed, in that the four extant mutant strains are all *per^0*- or *disco*-like in their adult locomotor activity (J. C. Hall, M. Hamblen-Coyle, unpublished observations). Moreover, each of these mutations segregates 1:1, again, from behavioral tests of adults; and monitoring the various doubly heterozygous female types confirmed that probably two X-chromosomal loci are involved. It would be interesting to ascertain if one of these loci coincides with the *per* gene from *D. pseudoobscura*—which has been cloned and sequenced (17). Analysis of these data revealed that the *per* gene of *pseudoobscura,* which is on the X chromosome (17), has diverged dramatically in several coding regions from that of *melanogaster*. Other such regions, however, are quite conserved (17). Could these latter regions be especially important for the basic functioning of the gene product? Such an argument could be buttressed if it could be shown that an evolutionarily conserved "domain" had suffered a protein-inactivating missense mutation in one or more of the X-linked arrhythmic mutants of *D. pseudoobscura*. In this regard, keep in mind that the first three *per^0*'s isolated (48, 65) are mere nonsense mutations, all at the same nucleotide pair (9, 49, 151). Thus, they are not very useful for mutational dissections of "structure-function" relationships in the protein. One hopes that *per^{04}* (which *is* different from *per$^{01-03}$;* 49), as well as one or more of the *D. pseudoobscura* mutants, will be more informative.

These new (49) and resurrected (cf. 97) mutants and their molecular promise are mentioned mostly because, after 20 years of mutagenizing and screening for *Drosophila* strains with aberrant rhythms, only about 15 mutants have been isolated. Furthermore, the number of genes known to be able to mutate such that the fly's rhythms go away or exhibit anomalous periodicities are few and far between (for a listing of these loci, see 27). Unlike the case of, say, visual mutants (e.g. 89) the *D. melanogaster* genome is infinitely far from being "saturated" for putative clock genes, defined by mutants or in any other way. So, any new rhythm mutation—from whatever *Drosophila* species, and especially at a locus that is approachable molecularly—is welcome.

Mammals

THE *TAU* MUTANT OF GOLDEN HAMSTER   The first single-gene rhythm mutant in a mammal was reported by Ralph & Menaker (101). The mutation, called *tau,* indeed causes τs to be altered, in the short-period direction. *tau* is semidominant (naturally): It effects ca 22 h periods of locomotor activity (specifically, "wheel-running") rhythms, when the mutation is heterozygous; and ca 20 h periods (hence, more than four h shorter than normal) when it is homozygous (101). As one might imagine, the mutant was discovered as a heterozygote: The animal's activity in LD-cycling conditions was noticed to anticipate lights-off appreciably earlier than normal. Thus, its phase is early, as for the short-period *per^S* and *Clk* mutants of *Drosophila* when they are monitored in LD (see above). Subsequently, *tau*-expressing hetero- or homozygotes were found to exhibit the shorter-than-normal free running periods noted above (101). Incidentally, all the other rhythm mutations—which routinely turned out to be semidominant—were isolated in microbial eukaryotes or invertebrates based on the hemi- or homozygous expression of the mutations (see reviews cited in the Introduction).

The rhythm phenotypes exhibited with the *tau* mutant, including the degree of semidominance observed in heterozygotes, make this new mutation formally similar to *per*^{S} in *D. melanogaster*. One difference is that some *tau*/*tau* individuals are unable to entrain to 12:12 LD cycles (101), in that their activity cycles drift into short periodicities under these conditions. This indicates that the hamster mutant's endogenous clock runs at a pace that is beyond the "limits of entrainment." In contrast, each *per*^{S} fly tested so far (Hamblen-Coyle et al, in preparation) has been able to lock in to the environmental cycling (i.e. reset its clock by 5 h, every day).

The impact of this mammalian rhythm mutant, including the fact that its circadian periodicities are so far from normal, soon became *neurobiological* as well as behavioral (unlike the current status of most studies of rhythm mutants in *Drosophila*). First, consider that the suprachiasmatic nucleus (SCN), located in the mammalian hypothalamus, contains *the*—or at least *a*—circadian pacemaker for the organism (reviewed in 88, 118). For example, SCN lesion leads to arrhythmicity, which can be "rescued" to normality by transplanting a genetically normal SCN from another individual of the same (rodent) species (e.g. 21, 79). But such experiments only demonstrate that the SCN is necessary for rhythmicity (i.e. information from the real pacemaker may need to go through this hypothalamic nucleus, which could have another role in the system).

Does the *tau* mutation affect the structure and/or function of the SCN? A positive answer was beautifully achieved by Ralph et al (100) transplanted a mutant SCN into a lesioned host and found that, once the graft "took", the activity rhythm period was characteristic of the donor tissue's genotype. The reciprocal transplant worked as well. The power of applying the *tau* mutant in these transplant experiments is that they also nailed the SCN as a pacemaker per se (also see 30).

MULTIGENIC RHYTHM VARIANTS   "Strains" of mammals, usually mice, have been selected for, or run across, whose circadian rhythms are somewhat different from what is nominally normal for that species. The most recent, and probably most systematic, study of this kind is that of Schwartz & Zimmerman (126; see this report and refs. 46, 47 for citations of previous publications in this area). 12 inbred mouse strains were used, in which the extremes for free-running τ values were about an hour different—not very impressive perhaps, irrespective of the genetic intractability (multigenicity) implicit in this study at the outset. In fact, these investigators (126) also demonstrated that the two most extremely different strains do not have a monogenic etiology for the τ difference. Nevertheless, it was mildly informative to show that some of the interstrain differences in endogenous periodicities (with regard to wheel-running activity) were associated with altered PRCs (e.g. a "short-τ" strain displayed relatively less advance phase-shifts, resulting from light pulses delivered during late subjective night).

Similar genetic connections that suggest common mechanisms underlying different components of certain circadian rhythmicities have been found before in multigenic contexts. A variety of insect strain differences, some derived from selection, have resulted in a nonselected rhythmic character changing in parallel with the selected one (for review, see 43, with more recent examples appearing in 2 and 3).

BRATTLEBORO RATS   That the Brattleboro strain of rat exhibits diabetes insipidus (reviewed in 105) is attributable to its being "vasopressin-minus", as a result of a single base-pair deletion (125) within the gene encoding this peptide hormone (whose complete name is "arginine vasopressin," or AVP). This mutant strain is mentioned because vasopressin is found in the SCN (plus other hypothalamic nuclei); and fluctuating levels of this substance in the cerebral spinal fluid (CSF) define a circadian rhythm, which obviously cannot occur (at the peptide level) in Brattleboros (105). Furthermore, the vasopressin-null mutation has been applied as a genetic control in certain molecularly based experiments, which were performed for chronobiological reasons (see below). This mutant has also been used to discount the possibility of a causal connection between the CSF rhythm of AVP levels and the rat's basic, or general, circadian rhythmicities: Various behavioral and biochemical rhythms of this kind are "relatively normal" in Brattleboros (see 105). The most recent study of this kind examined circadian sleep/arousal rhythms in Brattleboro rats; these rhythmicities were readily detectable in the relevant EEG recordings from the mutant, but certain elements had significantly reduced amplitudes (12). Therefore, AVP synthesis in the hypothalamus is not necessary for the generation of these circadian sleep rhythms, but the strength of such oscillations could be, as these authors surmise, modulated by circadian release of this peptide into the CSF (12).

Microbial Organisms

Neurospora   The bread mold *Neurospora crassa* exhibits some of the most powerfully analyzable circadian rhythms known. A key aspect of such studies is the array of rhythm mutations of *Neurospora,* which have been recovered over the same period of time during which the *Drosophila* mutants have been induced and isolated—in the same manner in both organisms (mutagenesis, followed by brute-force rhythm monitorings).

Figure 2 shows the conidiation rhythm associated with growth of *N. crassa*. It also illustrates aberrant rhythmicities caused by various mutant alleles of the most salient clock gene in this fungus, *frequency (frq)*. Little new information has been reported about these mutant, or mutant vs normal, phenotypes, though more is now known about the gene (see below). Thus, the reviews

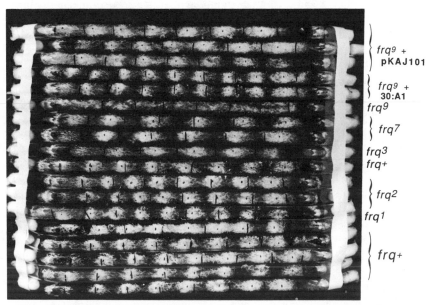

Figure 2   Conidiation rhythms in *Neurospora*. Cultures of *N. crassa* were inoculated into "race tubes," the bottom views of which (N = 18, taped together) were photographed after about 9 days of growth. (A side view of this kind of culture container, in which the agar-based medium is spread along the bottom of the tube lengthwise, is diagrammed in 27.) Growth—whose rate is more or less constant—was from right to left in these experiments. Each 24 h time span was marked by a vertical line. In the "wild-type" (but see below), the growing mycelia differentiate bands of aerial hyphae and conidia about every 21–22 h; the centers of these are marked by dots. Conidial banding is much more readily observable in cultures expressing the *band (bd)* mutation (see 36), which served as the (control) genetic background for the following: Cultures expressing various alleles of the *frequency (frq)* clock gene, or a *frq* mutation and a piece of *frq⁺*-containing DNA transduced into their genomes, had growth and bandings determined. These genotypes are indicated on the right. The following τ values (in h) are typically caused by these *frq* mutations (see 27, for further details): *frq¹*: 16; *frq²*: 19.3; *frq⁴*: 24; *frq⁷*: 29 [for the first two of these, the fast clocks are reflected by the dots being at progressively earlier times with respect to the vertical lines; for the second two, the dots are later and later]; the *frq⁹* allele leads essentially to arrhythmicity—no banding—under the usual growth conditions and within this many days of growth; the two transformant types shown (see 87 for details) had a *bd, frq⁹* genetic background and carried either a 7.7-kb DNA "insert" (**pKAJ101**) cloned from the *frq⁺* locus, or a larger *frq⁺*-containing insert cloned in a cosmid vector (**30 : 1A**). The rhythm periods exhibited by these five transformed cultures were essentially normal.

listed in the Introduction (particularly 27, 77) should be consulted for background information on *frq* and other *Neurospora* rhythm mutants. Included in these summaries are discussions of the various genetic and other manipulations that have been performed over the past 20 years and have been aimed at deducing more about the relevant gene actions than could be revealed by simple comparisons of the normal (ca 21.5 h), vs fast, vs slow, vs essentially nonexistent rhythms exemplified in the frq^+ vs *frq*-mutant rhythms shown in Figure 2.

The most substantial recent advance in this system of fungal chronobiology was made by McClung et al (87), who cloned DNA from the *frq* locus and identified the gene by chromosomal walking and by generating transformants, certain of whose incoming pieces of (cosmid-cloned) DNA "rescued" the arrhythmic frq^9 mutation (Figure 2). This, by the way, is more or less how *per* was cloned and identified in *Drosophila* (see 113, 149).

The consensus segment of VIIth chromosomal DNA that includes most (possibly all) of the *frq* locus was defined as the region of overlap among frq^9-rescuing cosmids. The segment was thus found to be 13 kb in length, of which a 7.7-kb subset was subsequently shown to be capable of restoring rhythms to this mutant (87; Figure 2). About 8.6 kb of genomic DNA (the aforementioned 7.7-kb region, plus some extra) has been sequenced and also used in Northern blotting experiments. The results and implications of these findings are taken up in a later section.

Podospora   There are a few "clock mutants," as they are termed in a review by Esser (32), in the fungus *Podospora anserina;* but they are old and neither well known nor well studied. A new, spontaneous, mutant in this category has recently surfaced: mbc^r. It was initially noticed by its resistance to a fungicide (86). In this strain as in the older such variants (32), growth is rhythmically banded, unlike what is observed for the wild-type (86). This mutant is therefore like *band (bd)* in *N. crassa* (see Figure 2 legend), a mutation that allows the conidiation rhythm readily to be revealed to the eye and measured—with *bd* being difficult to view as an actual clock mutant (see 36). Are the growth rhythms of *Podospora* controlled by an underlying pacemaker (with properties like those described elsewhere in this review), and should mutations like mbc^r be prefaced with the description "clock?" Positive answers are difficult: whereas 24 h growth rhythms are observed in LD cycles, the mutant's banding periodicities range from 33–72 h in constant darkness (86). Banding rhythmicity of *Neurospora,* in contrast, free-runs with circadian τs and otherwise has all the attributes of being run by a bona fide clock (27, 36).

Euglena   Some rhythm studies using genetic variants (but not "clock mutants" per se) as experimental tools have been recently performed with the flagellated alga *Euglena gracilis*. Edmunds and colleagues have applied a photosynthesis-deficient variant, ZC, that is completely devoid of chloroplasts (13, 78, 133). These microbial biologists asked the same question as that posed by application of various visual mutants in *Drosophila* (see above and 43). In *Euglena,* LD regimes could readily stimulate ZC subsequently to exhibit free-running rhythms of cell division; moreover, light pulses phaseshifted the rhythm in an essentially wild-type manner (13). Therefore, the light perception and processing involved in getting that environmental information to this unicellular organism's pacemaker appears to operate independently of functional chloroplasts. That this genetic experiment would reveal such dependence was not anticipated a priori: *Euglena* grown such that they become "protists" bleached of their chloroplasts are still entrainable and rhythmic (see 31). These microbial findings are reminiscent of the normal *per* protein cyclings observed in *Drosophila* photoreceptors genetically gutted of their normal sensory reception and transduction (152).

The ZC strain has also been used in some chronobiologically based enzyme studies, in which the investigators wished simply to measure (rhythmic) changes in NAD kinase and NADP phosphatase activities without interference from chloroplast enzymes (78). The various aspects of the cyclically fluctuating levels of these activities found in this study were discussed in terms of a previously presented model (41). It involves these two *Euglena* enzymes, regulation of Ca^{++} levels (see also 133), and how these pieces of chemistry could represent clock gears, in the sense of constituting a selfsustained oscillating loop that would underly the organism's circadian rhythms (31, 78). Such biochemical loops and feedbacks are an important feature of the following molecular genetic discussion.

MOLECULAR CHRONOBIOLOGY

Clone and Sequence

It is safe to say that determination of the amino-acid encoding information within the six clock genes that have been cloned and sequenced has come nowhere near solving the "clock problem." This discussion, however, attempts to put in perspective the more salient points from recent molecular studies of rhythms.

Five *per* genes from *Drosophila* have been cloned and sequenced (16, 17, 58, 135, 144). These were isolated from four *Drosophila* species other than *melanogaster,* by homology to the latter's gene; and in two instances they were confirmed as having "*per* activity" by bioassaying the pieces of DNA via germ-line transformants using *D. melanogaster per^0* hosts (96, 144).

The *per* coding sequence of *D. melanogaster* has, from the mid 1980s to the present, been similar to no other meaningful proteins—if one discounts, as one should, the case of *single-minded* discussed earlier (the *sim* sequence being homologous to nothing except *per;* see above). Certain regions of *per*'s interspecific *Drosophila* relatives (see above) are divergent from, or in some instances not in the least similar to, the corresponding region of the *melanogaster* information. Yet, data base searches using these partly non-conserved sequences have not revealed any protein similarities (17).

The other known relative of some of the *Drosophila per*s is the *frq* gene of *N. crassa*. First, note that the 7.7 kb *frq*-containing DNA fragment is complementary to two RNA species, one ca 5 kb in length, the other ca 1.5 kb (87). (Both such transcripts may be necessary for the overall clock function associated with this locus.) The longest ORF found from sequencing the relevant genomic DNA (see above) almost certainly corresponds to information encoded within the larger of these two mRNAs. Astonishingly, a portion of this ORF pulled out the *per* sequence of *D. melanogaster* (87) when GENBANK was searched (it would have so done with regard to *per* from *D. yakuba* and *D. simulans* as well; cf. 135 and 144, respectively; and see below). The conceptually translated products of both of these rhythm-influencing genes include a repeat of alternating threonine-glycine (TG) pairs, though the motif is relatively short and "modified" in the *frq* gene. In addition, some amino-acid similarities exist between regions of the two genes that flank the respective repeats (87).

Is this inter-clock-gene similarity a mere coincidence? Time will tell, but consider the following for now: (*a*) Two of the five sequenced *Drosophila per* genes, cloned from species far away from the *melanogaster* subgroup, do not have TG repeats, though they do retain a few TG pairs in the relevant region of the predicted protein (17). (*b*) The repeat is dispensable intraspecifically, as well, since circadian rhythms remained basically normal in the pertinent germ-line transformants even after the repeat was deleted in vitro (150), although the deletion did leave behind some TG pairs, and locomotor activity rhythms became somewhat temperature-sensitive [i.e. τ values were longer than normal at high temperature (33)]. Furthermore, there was a dramatic period-shortening effect on courtship song rhythms (150). Thus, the possibility remains that TG repeats or at least some TG dimers are hallmarks of certain proteins whose structures and functions mediate clock functions in pacemaker cells—for reasons that are completely unknown biochemically.

Yet, the discovery of *per*'s TG repeat has led to some biochemical experiments. These stemmed from the fact that such a repeat is similar (at the amino- and nucleic-acid levels) to serine-glycine repeats found in the core proteins of some vertebrate proteoglycans (reviewed in 117). Using anti-*per* antibodies, two separate investigations showed that the *per* protein of *D. melanogaster* may have some proteoglycan-like qualities, whereby the material was very large and anionic until the fly extracts were treated with a glycosamino-glycan (GAG) side-chain-attacking enzyme called heparitinase (7, 102).

The following difficult questions remain: If this clock gene product is a glycoprotein with (sulfated) GAG side-chains, is the *per* protein one of very few such substances known in insects? Yes (cf. 37). Therefore, guessing the function of such substances, which are poorly characterized in invertebrates at present, is a dubious undertaking. Does *per*-as-proteoglycan suggest *anything* about the biochemical function of this gene's encoded protein or how and why it helps mediate biological rhythms? Probably not (see 43, 117), but conceivably so (see 7).

Some remarks about a putative mammalian relative of *per:* van den Pol et al (141) reported the predicted amino-acid sequence for a nerve growth factor-regulated gene in rat, called VGF. An antibody (αVGF) was prepared against this fusion protein (starting with the production of a VGF-*lacZ* fusion polypeptide). It led to strong staining of the rat's SCN. These signals were colocalized to AVP-immunoreactive cells; but that peptide was neither responsible for, nor did it influence, the VGF signals because Brattleboro SCNs showed the same αVGF-mediated staining. Given that the anti-*per* antibody (αS) effected stainings of mammalian SCNs (128; see below)—the question arises as to whether VGF could be in some way similar to *per*. The real reason to think this may be so is because there was a notable 38% similarity between the *per*- and VGF-encoded amino acid sequences (141). However, the significance of this matching collapsed when, after computer-randomizing the VGF sequence, the similarity to *per* of *D. melanogaster* remained fully 30%. Hence, the "homology" here is none, and these numbers reflect only a similarity in amino-acid compositions—which, for *per,* is indeed rather odd (see discussion in 151).

Cyclical Expression of Substances Related to Rhythms

CIRCADIAN FLUCTUATIONS OF CLOCK-CONTROLLED GENE PRODUCTS Early work in this area has involved measuring circadian variations of enzyme activities and/or protein concentrations examples of which were alluded to above (31, 78). One recent case involves some protein antigens in a snail, whose quantities and/or qualities exhibit circadian fluctuations (111). These biochemically determined rhythms are mentioned, because the proteins were extracted from the eyes of *Bulla*—a most interesting tissue chronobiologically (e.g. 56, 62, 129; see below).

Many of the rhythmically varying substances, known in a wide range of animals, plants, and microorganisms, were chosen for assays at various times of day (in LD and/or DD), because one suspected them to underlie a given circadian clock's output. Such an output function or substance was often known from earlier studies, which were carried out at levels rather far removed from gene action [e.g. monitorings of circadian release of melatonin from vertebrate pineal glands (reviewed in 51) or of "glow" rhythms in a marine alga (reviewed in 92)]. The pertinent gene products, which were subsequently shown to cycle as well, would also, it seems, not be part of "the clock." Instead the relevant transcription and/or translation rates (see below) would presumably be controlled by a pacemaker function that, figuratively at least, is located more centrally.

One prediction, then, is that a mutation in a gene encoding such a "rhythmic protein"—whose expression is probably not very close to the central pacemaker—would not affect general features of the organism's rhythms. As was discussed earlier, this has been more or less realized for the AVP peptide in a mutant strain of rat (12, 105).

The converse is equally important: Can a centralized clock mutation affect the expression of other "downstream" RNAs? This has been answered by Loros et al (84), with regard to certain circadian-cycling RNAs in *Neurospora*. The normal (21.5 h) circadian rhythms of these transcripts' abundances had their periodicities increased by the effects of a long-period *frequency* mutation called *frq^7* (84; cf Figure 2). Thus, if *frq* can be reasonably assumed to directly affect the central pacemaker of *Neurospora*, expression of certain other genes is influenced by that clock.

The opposite effect may occur, at least in mammals. That is, the expression of certain genes appears to influence the clock. This was inferred from the recent experiments showing that light pulses, which reset rat or hamster rhythms in the usual manner (see above), also induce the expression of transcription factors in the SCN of these rodents (119, 153; see also 5). These findings imply that the activation of certain genes is involved in phase-shifting of circadian pacemakers. There is enough time for these intermediate events to occur, in the sense that the effects of environmental inputs on a circadian pacemaker need not be the instant responses often elicited by light stimuli. But it is too early to speculate about the kinds of genes in cells of an SCN undergoing phase-shifting that would be activated by the turned-on transcription factors (119, 153).

It is also not known what the two clock-controlled fungal genes (84) are doing or what they encode: (*a*) as yet no mutations have been found at either of these two loci; and (*b*) the sequences of the cDNAs in question have not led to any database matches (J. J. Loros & J. C. Dunlap, personal communication). But this pioneering investigation of cycling transcripts has some object lessons.

First, consider that circadian rhythms of RNA abundances, found in deliberate screens like this one (which involved "subtraction hybridization," (84)) may uncover novel functions beyond those already suspected to be rhythmically expressed (see above). The other examples of cycling trans-cripts—most of them in fact being the usual suspects, in a variety of meta-zoan animals or in higher plants—have recently been reviewed (61, 114; see also 1, 55, 95). So these cases of clock *output* (no doubt) will not be dis-cussed at length. Note, however, that a circadian-cycling protein in this cate-gory does not necessarily mean that the RNA cycles as well. The canonical cases for this caveat include (*a*) the demonstration that *translational* con-trol underlies luciferin-binding-protein (LBP) cycling in *Gonyaulax*, via cloning of the relevant nucleic acid sequences and using them to monitor a temporally unchanging transcript abundance (reviewed in 92) [LBP, whose fluctuating levels define a strong circadian rhythm, is a key factor in the bioluminescence rhythmicity of this alga]; and (*b*) inferences about this same level of control in *Acetabularia*, the stalks of which express circa-dian rhythms involving photosynthesis, even when cut away from the portion of the cell containing the nucleus or when treated with transcrip-tion inhibitors (reviewed in 31). The nuclear genome-independence of the biologically and biochemically defined rhythms in this organism tends to undermine certain grand schemes that might hope to point the way to a glo-bal understanding of clock mechanisms at the molecular level (see below, Figure 3b).

Second, a screen for cycling RNAs might pull out half the genome (see below), or some such destined-to-be intractable nightmare. But Loros et al (84) found very few putatives during their screen, which came down to the two cases noted above; and these clones were independently isolated several times. Yet, perhaps additional genes would have been so found if "differen-tial," rather than "subtractive" schemes had been applied; the former might disclose RNA abundance rhythms with lower amplitudes than those revealed subtractively. Nevertheless, these investigators (84) suggested that the two functions implied are special—such as being semidedicated to a clock-output function. A previous related study claimed that gross RNA amounts in *Gonyaulax* undergo detectable fluctuations over the course of circadian cycles (142). This observation has not been reproducible, despite extensive further experiments (P. M. Milos & J. W. Hastings, personal communication). Total *Gonyaulax* RNA, harvested from cells at 6 different circadian times (CTs), was electrophoresed, and the gels were stained with ethidium bromide. The total amounts of RNA and the stained profiles did not appreciably vary as a function of CT. These investigators (P. M. Milos & J. W. Hastings, personal communication) speculate that this discrepancy could relate to differential degradation in the RNA preparations involved in the previous experiments (142), as judged by the apparent absence of ribosomal RNA in some of the reported gels.

Third, fluctuating abundances of specifically probed transcripts could be due to temporally dependent oscillations in transcription rates. This level of control for the two *Neurospora* transcripts (cf. 84) has recently been demonstrated in nuclear run-on experiments (85).

Analogous demonstrations have been accomplished with certain plant genes (reviewed in 61; see also 108, 137) involved, for example, in light-harvesting, such as the chlorophyll$_{a,b}$ (Cab) proteins. In several species, the transcripts encoding Cabs fluctuate in abundance with circadian periodicities (61). The uncovering of these RNA rhythmicities suggested a possible 5'-flanking control of transcription-rate variations. Indeed, transformants in which a reporter RNA was fused to the promoter region of a Cab locus led to the same kind of reporter cycling as found for the primary product of the endogenous gene (93).

These results on circadian gene regulation suggest ways for using transformants, like those created in this plant, to select for new clock mutations (see 27, 84). Loros et al (84) and Dunlap (27) should be consulted for details about the idea (and credited with coming up with it). Suffice it to say here that one could end up "plating out" (or at least enriching for) the relevant fungi or flies, by selecting for or against expression of the reporter, which (*a*) would be fused to the 5' flanking material, (*b*) would when translated confer positive or negative selectability, and (*c*) could be at an anomalousy high or low level at the *wrong time* of a circadian cycle, *because its expression is being clock-controlled by a newly induced mutation* in a gene of interest. If a diploid organism is screened, the agony of rendering the mutagenized chromosomes homozygous could be averted, i.e. in the animals that also will carry the relevant fusion gene, because so many circadian rhythm mutations are semi-dominant for their effects on period (see earlier sections of this review for examples). These contemplated screens, then, might lead to the isolation of perhaps 10 new rhythms mutants per year, per organism; instead of < 1 per year, as has been the case up until now (27). Only such a substantial upswing in the rate of clock-gene identifications will presage the achievement of genome saturations regarding loci influencing circadian rhythms.

CYCLICAL EXPRESSION OF THE *PERIOD* GENE   As was previewed in the earlier discussion of this gene's expression in the arrhythmic *disco* mutant, staining of the *per* protein exhibits regular fluctuations of intensities (reviewed in 115). In the brain and the visual system of wild-type adults, the staining decreases from the middle to the end of the day, then becomes increasingly intense from the middle to the end of the night (127, 152). Cyclical stainability of the protein in the CNS is evident in brain lateral neurons ("LNs," as mentioned above) and also in what seem to be glial-like cells distributed widely in many of the fly's ganglia (127, 152). Brain expression of this protein may be relevant to its influence on the fly's rest/activity cycles, at least (42, 68). Furthermore, the LN cells could be of primary importance (see discussion in 127, 152). Yet, there are no definitive experimental data on this point, such as results from behaviorally monitoring genetic mosaics carrying an internal marker for CNS cells carrying a given *per* genotype (cf. 38). Thus, the role of *per*-expressing glia cannot be ruled out; in this regard, the relatively high density of glial cells in the mammalian SCN (e.g. 91, 140) might be kept in mind.

These *per* immunoreactivity rhythms in the CNS and the visual system had one peak—at dawn—per daily cycle; and they persisted in constant darkness [in which conditions the τ was estimated as ca 24 h, though the amplitude was less than in LD (152)]. One of the most intriguing results of these immunohistochemical studies came from running *per*^{S} and wild-type flies in parallel. The trough of the rhythm for this mutant was considerably earlier, in LD, than that for wild-type; and the free-running period of *per*^{S}'s staining cycles was ca 4 h shorter than normal (152). Thus, the nature of the protein product encoded by a given *per* allele influences its own cycling.

These findings suggest that a feedback loop could be part of the fly's circadian clock. This was shown to be so in a series of experiments that monitored abundance of *per* mRNA (50). These determinations (usually involving RNase protection experiments) revealed a dramatic circadian rhythm for RNA extracted from the adult head. Body RNA transcribed from this gene showed small peak vs trough differences (50, 115), and this partly explains why the abundance rhythm of *per* mRNAS was missed in earlier attempts to look for it (see 103, 149; see also 83).

Support for the idea of a feedback loop, whereby the *per* protein influences the abundance of the transcript giving rise to it, came from testing the effects of *per*-mutations (50): (*a*) In LD, the abundance peak for RNA extracted from *per*^{S} heads was considerably earlier than the *per*^{L1} peak; (*b*) in these same conditions, no appreciable abundance oscillations were detected for *per*^{01}; and (*c*) in DD, the τ for the *per*^{S} RNA rhythm was shorter than the wild-type cycle duration (Figure 3a).

Thus, the nature and the presence of this gene's final product exerts a dramatic influence on the changing abundance of the mRNA, which further promotes the hypothesis that the action of *per* involves direct or indirect feedback (Figure 3b). In contrast, an RNA rhythm in the SCN of rat, regarding the transcript encoding vasopressin (14, 106, 138), seems normal in the "AVP-minus" Brattleboro mutant discussed above.

At what level of gene action is RNA cycling of *per* controlled? It is not yet known whether the feedback loop is "tight" (Figure 3b), meaning that the product translated from the mRNA may directly affect the transcript's stability; or whether the *per* protein directly interacts with the gene and influences its transcription rate (note that this protein is nuclear in some cell types of the adult though is apparently cytoplasmic in others, including CNS neurons).

Alternatively, the loop might be so "wide" that a whole series of steps

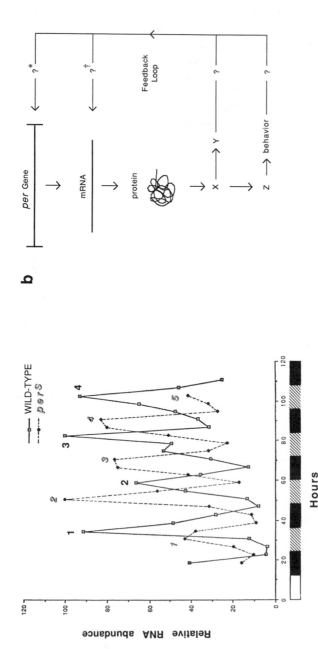

Figure 3 Examples and implications of cycling *period* gene products. (*a*) *per* mRNA abundance oscillations, after Hardin et al (50), who determined these cycling transcript levels by quantifying the results of RNase protection experiments. The RNA was extracted from the heads of genetically normal and short-period mutant adults, which had been entrained by exposure to LD cycles before proceeding into DD. The "L" phase of the last entrainment cycle is indicated by the open portion of the horizontal bar below the abscissa; filled portions of the bar indicate "subjective nights," during DD, and shaded portions "subjective days." The abundance peaks are numbered separately for the wild-type (Roman) vs *per*^s (italics) cycles. The mRNA fluctuation associated with this mutant has a periodicity considerably shorter than in the wild-type. Specifically, and from formal analysis of these data (26), the *per*^s τ is 19.3 h and the control value 22.0 h. Whereas the wild-type abundance oscillations were not really normal (i.e. 24 h), the two rhythms superimposed here were completely out of phase by the end of this experiment. (*b*) Feedback effects of the *per* protein on fluctuating levels of the mRNA (after 50). Since the mutation shown in a, and other mutant *per* alleles, affect the transcript cycling (50, 115), it follows that the protein—or effects of its actions (mediated by the hypothetical steps called "X," "Y," "Z," and "behavior")—influences the transcription of this gene ("?*") or the longevity of the mRNA ("?†") in a time-dependent manner.

intervenes between the initial "activity" of the *per* protein and its effects on RNA abundance. The clock mechanism progressing through such steps ultimately, of course, leads to rhythmic behavior, which could be necessary for the RNA cycling (Figure 3b; see further discussion in 115).

Any future hypotheses must take into account that the mRNA peak phase leads the peak of immunoreactivity by about 6 h (compare 152 to 50). Robinson et al (112) should also be consulted for consideration of other pertinent issues, in the context of the AVP mRNA rhythm, which they showed is not necessarily transcriptional in origin.

One goal of future studies in this area would be to determine if an obligatory relationship exists between the *per* mRNA and protein cyclings and one or more biologically rhythmic phenotypes—such as the adult's locomotor rhythms. In this regard, the more or less head-specific transcript cycling (50) may be correlated with a "head focus" for the influence of *per* alleles on this behavioral phenotype, as revealed by genetic mosaic analysis (42, 68). A more recent, albeit preliminary, result is that the cryptic or nonexistent activity rhythms expressed by dark-reared/dark-monitored flies (cf 25) was correlated with fairly robust expression of *per* protein staining (152); the opposite histochemical result was found in constant light (152), in which condition *Drosophila* is behaviorally arrhythmic (66, 152). Yet no evidence of cyclical staining could be discerned in the "all DD" experiment (152). Therefore, no *per* expression (as in per^0, per^-, or in LL), no activity rhythms, but good—although overly constant—*per* expression (as in DD), weak rhythmicity.

A final implication of these molecular cycles reflects back on the *per* dosage-effect experiments, mentioned earlier. Genetic manipulations that would be expected to reduce or increase the amount of the gene's products— i.e. hemizygosity for per^+ in a female or extra doses of per^+ in either sex (see below)—lead to longer and shorter activity rhythm τs, respectively (18, 130, 152). A further aspect of these experiments, involving various copy numbers of a per^+ allele, concerns dosage *compensation:* A given dose of this X-chromosomal gene's normal allele in a male (achieved via chromosome aberrations or germ-line transformants involving per^+ DNA) leads to shorter τs than the same dosage in a female (16, 130; M. K. Cooper, M. Hamblen-Coyle, X. Liu, M. Rosbash, J. C. Hall, in preparation).

The flies of these various dosage-altered *per* genotypes may have had altered peak values associated with the molecular cyclings in question. Yet there are no biochemical data on this point, with reference to the dosage alteration/rhythm monitoring studies just cited. There are, however, molecular results related to another kind of expression level manipulation. This involves the array of germ-line transformants, generated (6) with a piece of per^+ DNA, one of whose ends is within the gene corresponding to the 3' untranslated region of the mRNA. In the various genomic locations where it landed, this 7.1-kb fragment rescued the behavioral effects of a per^0 mutation (9) with a wide range of interstrain periodicities (ca 25–40 h). There was, moreover, a consistent relationship between abundance of *per* transcript and these varying τ values: The lower the mRNA concentration, the longer the rhythm period (9). These results fit with the previous findings (e.g. 130; see 18 for formal analyses of "gene expression/circadian τ" relationships). However, the results that showed correlations between the behaviors and the *per* expression levels in the "7.1" transformants could be regarded as unexpected, since the *times* when the mRNA was taken from these transformed animals (9, 50) were apparently not considered (cf. 50). Yet, these abundance oscillations are by far more prominent for head RNA than for RNA extracted from adult bodies (115). Perhaps the fact that Baylies et al (9) determined *per* transcript abundances using whole-fly extracts minimized the importance of controlling for times of extraction.

With this further reference to the cyclical expression of the *per* gene's products, we return to a consideration of the implied feedback loop: What does the *rest of it* (the left-hand portion of Figure 3b)—other than the parts of it concerning temporal regulation of mRNA levels (right-hand portion)—consist of? What are the actions of genes like the various *per*'s, or *And*, or *Clk*, or *tau*, or *frq*, in terms of their effects on cell biochemistry and physiology? This remains the mystery of clocks, which includes, in the context of this review, the fact that there are few clues about what the actions of clock genes are at the cellular level.

Here, however, is a possible way to address these vexing problems. The following findings and proposals might allow the feedback mechanisms under discussion to be filled in regarding terms of the several sequential sections of the loop (Figure 3b) that could be relevant:

INTERSPECIFIC RELATIVES OF *PER*   The biochemistry and physiology of proteins like that encoded by the period genes of *Drosophila* may be approachable by characterizing and manipulating *per* protein relatives in *other* organisms, i.e. those in which circadian pacemakers have been identified as to tissue location and are accessible in cellular terms. The relatives in question have been inferred from the immunohistochemical stainings, effected by the αS antibody already mentioned (127, 152). Thus, this reagent, generated using an immunogen (127) that is indeed from an evolutionary conserved region of the protein among *Drosophila* species (17), stains pacemaker structures in a wide range of organisms. This was first found in molluscs by Siwicki et al (129). Analogous immunohistochemical results from applying αS have more recently been obtained in the *Xenopus* (R. G. Foster, personal communication); the lizard *Anolis* (R. G. Foster, personal communication);

the fish *Gambusia* (R. G. Foster, pesonal communication); in mammals (128; K. K. Siwicki & D. Welch, personal communication; R. G. Foster, personal communication); and, less surprisingly, in a variety of arthropod species, ranging from other insects such as the moth *Manduca* (K. K. Siwicki & P. F. Copenhaver, personal communication); the beetle *Pachymorpha* (B. Frisch, J. C. Hall, M. Rosbash, G. Fleissner, and G. Fleissner, unpublished observations); to a scorpion (Frisch et al, unpublished observations). The neurons in the organisms screened with this antibody have stained with what appear to be cytoplasmic-filling, nonnuclear signals, as far as discernible by low-resolution microscopy (similar, in general, to the pattern of αS-mediated staining seen in *Drosophila* brain neurons, 127, 152). Thus far, the negative results—no staining of anything after sectioning CNS tissue and applying αS—have been obtained with bee (B. Frisch, personal communication), as well as with lamprey, salmon, chicken, quail, and mouse (R. G. Foster, personal communication).

Some of the pacemaker cells stained by this anti-*per* reagent, notably those in two species of mollusc (129) and in two species of mammal (rat: 128; hamster: R. G. Foster, personal communication), are amenable to characterizations and experimental manipulations in organ cultures (reviews: 40, 56). It has therefore been proposed (114, 128, 129) that further study of these inferred *per*-like proteins might elucidate their effects on the many biochemical and physiological parameters uncovered in the context of manipulating inputs to and monitoring the circadian rhythmic outputs from these cultured pacemakers (see 40, 56, 62, 91, 98, 99, 110, 111 for information on the relevant neurobiology and neurochemistry).

To effect changes in the quality and quantity of the *per*-like proteins, it will almost certainly be necessary to isolate the genes that encode the interspecific relatives. This cloned material could then be used (*a*) to generate antibodies against the native (putative) clock protein (e.g. as encoded by a molluscan gene), which could be a more effective clock-disrupting reagent to inject into the pacemaker neurons—most accessibly, in the snail *Bulla* (cf 56, 62, 110)—than would αS itself; and (*b*) in "anti-sense" or in vitro mutagenesis experiments based on injection of the relevant constructs into the pacemakers. Although the antisense approach is problematical, and conceptual difficulties are involved in other methods of altering phenotypes by introducing in vitro-mutagenized genes into genetically normal cells, consider these sanguine features of clock genes, their products, and the rhythms they mediate: (*a*) The levels of the pertinent products likely would not have to approach zero for readily detectable defects to be realized, given the dosage effects discussed recently (e.g., a mere twofold drop in dosage of *per* lengthens circadian periods by an hour; see above). (*b*) Rhythm mutations, in organisms, are semidominant.

In this respect, cloned material from a rodent, which would be partly similar to the *per* genes of *Drosophila,* could be used in reverse-genetic strategies aimed at making mouse mutations (reviewed in 116), in this case, rhythm variants. Such potential mutants could be studied not only at the organismal level (cf 100, 101), but also by recording from and otherwise manipulating the genetically aberrant SCN tissues (cf 40).

Such molluscan, mammalian, etc, clones will help determine how similar these relatives are among each other and to *per* in *Drosophila.* We believe that some informative similarities will be revealed by these clone and sequence efforts. An alternative view might be that the αS reagent stains random proteins, which are only by chance located in the right locations [e.g. in the basal retinal neurons of *Bulla* (129); or in cells of the SCN of rat (128)]. In fact, in most of these interspecific staining studies, the signals are not limited to the known or highly suspected pacemaker structures and cells. [Thus, there is molluscan staining in a few cerebral ganglionic cells of unknown function (129); and mammalian staining in hypothalamic nuclei other than the SCN, as well as in the hippocampus and the cerebral cortex (128)]. Given the alarmingly widespread *per* expression in several neural and nonneural locations within *Drosophila* itself (as listed in 43), what would one expect?

Does any independent evidence suggest that a protein detected in some other form by the anti-*per* material has something to do with circadian rhythms? Yes: Abundance fluctuations of the *per*-relative in *Aplysia* (129), have been found in Western blotting experiments, using eye extracts (as in *Bulla,* the sea hare's circadian clock is in its eye, 56). A single band is readily detected by αS on these blots (129), though the antibody detects nothing in *Drosophila* extracts (127), necessitating the immunohistochemical monitoring of the fly's *per* protein rhythmicity (152). The eye band of *Aplysia* was most intense at the end of the night, and there was indeed one peak plus one trough per day in LD (129); the peak time for this biochemical rhythm was very similar to that determined histochemically in *Drosophila* (152). Thus, not only do these proteins in both species define their own circadian rhythms, but they even have phases that are suspiciously similar to each other. Thus, a further parallel can be predicted: Once the *Aplysia* gene encoding this eye protein is in hand, the pertinent probes will be used to show that the RNA transcribed from that gene cycles in its abundance.

CONCLUSION

Circadian rhythms are ubiquitous. Such low-frequency oscillations are even found in certain prokaryotes (most recently, 53). Their ostensible universality is one reason why the biological clocks seem interesting and important. Genetic studies of factors underlying circadian pacemaking are beginning to indicate that clock mechanisms, defined in part via products of the genes described or inferred in this review, may also be ancient entities that have remained highly conserved in evolution. Therefore, cracking the clock in one or a very few forms may well perform this service for most of the others.

ACKNOWLEDGMENTS

Several of the studies reviewed here have been carried out by members of my laboratory, in collaboration with M. Rosbash, C. P. Kyriacou, R. J. Konopka, H. B. Dowse, J. M. Ringo, J. J. Jacklet, S. Strack, and W. J. Schwartz. I greatly appreciate the participation of those valued colleagues in these many experiments. Financial support for the work done at Brandeis University, in collaboration with M. Rosbash, was provided by NIH grant GM-33205 (J. C. H. & M. R.) and the Howard Hughes Medical Institute (M. R.). I thank S. T. Crews, H. B. Dowse, J. C. Dunlap, J. Ewer, R. G. Foster, B. Frisch, J. W. Hastings, R. J. Konopka, X. Liu, J. J. Loros, P. M. Milos, J. M. Ringo, K. K. Siwicki, and M. W. Young for communicating unpublished results and J. C. Dunlap for providing the photograph for Figure 2. I am grateful to B. S. Baker, J. C. Dunlap, J. Ewer, R. G. Foster, and M. Rosbash for reading the manuscript and suggesting several improvements.

Literature Cited

1. Albers, H. E., Stopa, E. G., Zoeller, R. T., Kauer, J. S., King, J. C., et al. 1990. Day-night variation in prepro vasoactive intestinal peptide/peptide histidine isoleucine mRNA within the rat suprachiasmatic nucleus. *Mol. Brain Res.* 7:85–89
2. Allemand, R., Biston, J., Fouillet, P. 1989. Variabilité génétique du profil circadien d'activité de la Drosophile dans une population naturelle. *C. R. Acad. Sci. Sér. III* 309:477–83
3. Allemand, R., Boulétreau-Merle, J. 1989. Correlated responses in lines of *Drosophila melanogaster* selected for different oviposition behaviors. *Experientia* 45:1147–50
4. Arking, R., Dudas, S. P. 1989. Review of genetic investigations into the aging process of Drosophila. *J. Am. Geriatr. Soc.* 37:757–73
5. Aronin, N., Sagar, S. M., Sharp, F. R., Schwartz, W. J. 1990. Light regulates expression of a Fos-related protein in rat suprachiasmatic nuclei. *Proc. Natl. Acad. Sci. USA.* 87:5959–62
6. Bargiello, T. A., Jackson, F. R., Young, M. W. 1984. Restoration of circadian behavioural rhythms by gene transfer in *Drosophila. Nature* 31:752–54
7. Bargiello, T. A., Saez, L., Baylies, M. K., Gasic, G., Young, M. W., Spray, D. C. 1987. The *Drosophila* clock gene *per* affects intercellular junctional communication. *Nature* 328:686–91
8. Bargiello, T. A., Young, M. W. 1984. Molecular genetics of a biological clock in *Drosophila. Proc. Natl. Acad. Sci. USA* 81:2142–46
9. Baylies, M. K., Bargiello, T. A., Jackson, F. R., Young, M. W. 1987. Changes in abundance and structure of the *per* gene product can alter periodicity of the *Drosophila* clock. *Nature* 326: 390–92
10. Bennet-Clark, H. C. 1990. Do the song pulses of *Drosophila* show cyclic fluctuations? *Trends Ecol. Evol.* 5:93–97
11. Bloomquist, B. T., Shortridge, R. D., Schneuwly, S., Perdew, M., Montell, C., et al. 1988. Isolation of a putative phospholipase C gene of Drosophila, *norpA*, and its role in phototransduction. *Cell* 54:723–33
12. Brown, M. H., Nunez, A. A. 1989. Vasopressin-deficient rats show a reduced amplitude of the circadian sleep rhythm. *Physiol. Behav.* 46:759–62
13. Carré, I. A., Oster, A. S., Laval-Martin, D. L., Edmunds, L. N. Jr. 1989. Entrainment and phase shifting of

the circadian rhythm of cell division by light in cultures of the achlorophyllous ZC mutant of *Euglena gracilis*. *Curr. Microbiol.* 19:223–29

14. Carter, D. A., Murphy, D. 1989. Diurnal rhythm of vasopressin mRNA species in the rat suprochiasmatic nucleus: independence of neuroendocrine modulation and maintenance in explant culture. *Mol. Brain Res.* 6:233–39

15. Chen, D.-M., Christianson, J. S., Sapp, R. J., Stark, W. S. 1990. Circadian facets of sensitivity, rhodopsin and rhabdomere in *Drosophila*. *Neurosci. Abstr.* 16:In press

16. Citri, Y., Colot, H. V., Jacquier, A. C., Yu, Q., Hall, J. C., et al. 1987. A family of unusually spliced and biologically active transcripts encoded by a *Drosophila* clock gene. *Nature* 326:42–47

17. Colot, H. V., Hall, J. C., Rosbash, M. 1988. Interspecific comparison of the *period* gene of *Drosophila* reveals large blocks of non-conserved coding DNA. *EMBO J.* 7:3929–27

18. Coté, G. G., Brody, S. 1986. Circadian rhythms in *Drosophila melanogaster:* analysis of period as a function of gene dosage at the *per* (period) locus. *J. Theor. Biol.* 121:487–503

19. Crews, S. T., Thomas, J. B., Goodman, C. S. 1988. The Drosophila *single-minded* gene encodes a nuclear protein with sequence similarity to the *per* gene product. *Cell* 52:143–52

20. Crossley, S. A. 1988. Failure to confirm rhythms in *Drosophila* courtship song. *Anim. Behav.* 36:1098–109

21. DeCoursey, P. J., Buggy, J. 1989. Circadian rhythmicity after neural transplant to hamster third ventricle: specificity of suprachiasmatic nuclei. *Brain Res.* 500:263–75

22. Dowse, H. B., Dushay, M. S., Hall, J. C., Ringo, J. M. 1989. High resolution analysis of locomotor activity rhythms in *disconnected*, a visual system mutant of *Drosophila melanogaster*. *Behav. Genet.* 19:529–42

23. Dowse, H. B., Kass, L., Ringo, J. M. 1988. Studies on a congenital heart defect in *Drosophila*. *Behav. Genet.* 18:714–15 (Abstr.)

24. Dowse, H. B., Ringo, J. M. 1987. Further evidence that the circadian clock in *Drosophila* is a population of coupled oscillators. *J. Biol. Rhythms* 2:65–76

25. Dowse, H. B., Ringo, J. M. 1989. Rearing *Drosophila* in constant darkness produces phenocopies of *period* circadian clock mutants. *Physiol. Entomol.* 62: 785–803

26. Dowse, H. B., Ringo, J. M. 1990. Is the

circadian clock a "metaoscillator?" Evidence from studies of ultradian rhythms in *Drosophila*. See Ref. 146

27. Dunlap, J. C. 1990. Closely watched clocks: molecular analysis of circadian rhythms in *Neurospora* and *Drosophila*. *Trends Genet.* 6:159–65

28. Dushay, M. S., Konopka, R. J., Orr, D., Greenacre, M. L., Kyriacou, C. P., et al. 1990. Phenotypic and genetic analysis of *Clock*, a new circadian rhythm mutant in *Drosophila melanogaster*. *Genetics* 125:557–78

29. Dushay, M. S., Rosbash, M., Hall, J. C. 1989. The *disconnected* visual system mutations in *Drosophila melanogaster* drastically disrupt circadian rhythms. *J. Biol. Rhythms* 4:1–27

30. Earnest, D. J., Sladek, C. D., Gash, D. M., Wiegand, S. J. 1989. Specificity of circadian function in transplants of the fetal suprachiasmatic nucleus. *J. Neurosci.* 9:2671–77

31. Edmunds, L. N. Jr. 1988. *Cellular and Molecular Bases of Biological Clocks*. New York: Springer-Verlag. 497 pp.

32. Esser, K. 1974. *Podospora anserina*. In *Handbook of Genetics*, ed. R. C. King, 1:531–51. New York: Plenum. 676 pp.

33. Ewer, J., Hamblen-Coyle, M., Rosbash, M., Hall, J. C. 1990. Requirement for *period* gene expression in the adult and not during development for locomotor activity rhythms of imaginal *Drosophila melanogaster*. *J. Neurogenet.* 7:31–73

34. Ewer, J., Rosbash, M., Hall, J. C. 1988. An inducible promoter fused to the *period* gene in *Drosophila* conditionally rescues *per*-mutant arrhythmicity. *Nature* 333:82–84

35. Ewing, A. W. 1988. Cycles in the courtship song of male *Drosophila melanogaster* have not been detected. *Anim. Behav.* 36:1091–97

36. Feldman, J. F., Dunlap, J. C. 1983. *Neurospora crassa*: a unique system for studying circadian rhythms. *Photochem. Photobiol. Rev.* 7:319–68

37. Fessler, J. H., Fessler, L. I. 1989. *Drosophila* extracellular matrix. *Annu. Rev. Cell Biol.* 5:309–39

38. Fischbach, K. F., Technau, G. 1984. Cell degeneration in the developing optic lobes of the sine oculis and small-optic-lobes mutants of *Drosophila melanogaster*. *Dev. Biol.* 104:219–39

39. Giebultowicz, J. M., Riemann, J. G., Raina, A. K., Ridgway, R. L. 1989. Circadian system controlling release of sperm in the insect testes. *Science* 245:1098–100

40. Gillette, M. U. 1990. SCN elec-

trophysiology *in vitro:* rhythmic activity and endogenous clock properties. In *SCN: The Mind's Clock,* ed. D. C. Klein, R. Y. Moore, S. M. Reppert. New York: Oxford Univ. Press. In press

41. Goto, K., Laval-Martin, D. L., Edmunds, L. N. Jr. 1985. Biochemical modeling of an autonomously oscillatory circadian clock in *Euglena. Science* 228:1284–88

42. Hall, J. C. 1984. Complex brain and behavior functions disrupted by mutations in Drosophila. *Dev. Genet.* 4:355–78

43. Hall, J. C., Kyriacou, C. P. 1990. Genetics of biological rhythms in *Drosophila. Adv. Insect Physiol.* 22:221–98

44. Hall, J. C., Rosbash, M. 1987. Genes and biological rhythms. *Trends Genet.* 3:185–91

45. Hall, J. C., Rosbash, M. 1987. Genetics and molecular biology of rhythms. *BioEssays* 7:108–12

46. Hall, J. C., Rosbash, M. 1987. Genetic and molecular analysis of biological rhythms. *J. Biol. Rhythms* 2:153–78

47. Hall, J. C., Rosbash, M. 1988. Mutations and molecules influencing biological rhythms. *Annu. Rev. Neurosci.* 11:373–93

48. Hamblen, M., Zehring, W. A., Kyriacou, C. P., Reddy, P., Yu, Q., et al. 1986. Germ-line transformation involving DNA from the *period* locus in *Drosophila melanogaster:* overlapping genomic fragments that restore circadian and ultradian rhythmicity to *per°* and *per⁻* mutants. *J. Neurogenet.* 3:249–91

49. Hamblen-Coyle, M., Konopka, R. J., Zwiebel, L. J., Colot, H. V., Dowse, H. B., et al. 1989. A new mutation at the *period* locus with some novel effects on circadian rhythms. *J. Neurogenet.* 5:229–56

50. Hardin, P. E., Hall, J. C., Rosbash, M. 1990. Feedback of the *Drosophila period* gene product on circadian cycling of its messenger RNA levels. *Nature* 343:536–40

51. Hastings, M. H., ed. 1989. Phylogeny and function of the pineal gland (multiauthor review). *Experientia* 45:903–1008

52. Helfrich, C., Engelmann, W. 1987. Evidences for circadian rhythmicity in the *per°* mutant of *Drosophila melanogaster. Z. Naturforsch. Teil C* 42:1335–38

53. Huang, T.-C., Tu, J., Chow, T.-J., Chen, T.-H. 1990. Circadian rhythm of the prokaryote *Synechococcus* sp. RF-1. *Plant Physiol.* 92:531–33

54. Inoue, H., Yoshioka, T., Hotta, Y. 1985. Genetic study of inositol trisphos-phate involvement in phototransduction using *Drosophila* mutants. *Biochem. Biophys. Res. Commun.* 132:513–19

55. Iovanna, J., Dusetti, Calvo, E., Cardinali, D. P. 1990. Diurnal changes in actin mRNA levels and incorporation of ³⁵S-methionine into actin in the rat hypothalamus. *Cell Mol. Neurobiol.* 10:207–16

56. Jacklet, J. W. 1989. Cellular neuronal oscillators. In *Neuronal and Cellular Oscillators,* ed. J. W. Jacklet, pp. 482–527. New York: Dekker. 553 pp.

57. Jackson, F. R. 1990. Circadian rhythm mutants of *Drosophila.* See Ref. 146

58. Jackson, F. R., Bargiello, T. A., Yun, S.-H., Young, M. W. 1986. Product of *per* locus of *Drosophila* shares homology with proteoglycans. *Nature* 320:185–88

59. Jackson, F. R., Newby, L. M., DiBartolomeis, S. M. 1989. Drosophila dusky (dy) mutations alter circadian rhythms. *Neurosci. Abstr.* 15:461

60. James, A. A., Ewer, J., Reddy, P., Hall, J. C., Rosbash, M. 1986. Embryonic expression of the *period* clock gene in the central nervous system of *Drosophila melanogaster. EMBO J.* 5:2313–20

61. Kay, S. A., Millar, A. J. 1990. Circadian regulated *Cab* gene transcription in higher plants. See Ref. 146

62. Khalsa, S. B. S., Block, G. D. 1989. Calcium in phase control of the *Bulla*-circadian pacemaker. *Brain Res.* 506:40–45

63. Konopka, R. J. 1987. Genetics of biological rhythms in *Drosophila. Annu. Rev. Genet.* 21:227–36

64. Konopka, R. J. 1987. Neurogenetics of *Drosophila* circadian rhythms. In *Evolutionary Genetics of Invertebrate Behavior,* ed. M. D. Huettel, pp. 215–21. New York: Plenum. 335 pp.

65. Konopka, R. J., Benzer, S. 1971. Clock mutants of *Drosophila melanogaster. Proc. Natl. Acad. Sci. USA* 68:2112–16

66. Konopka, R. J., Pittendrigh, C., Orr, D. 1989. Reciprocal behavior associated with altered homeostasis and photosensitivity of *Drosophila* clock mutants. *J. Neurogenet.* 6:1–10

67. Konopka, R. J., Smith, R. F., Orr, D. 1990. Characterization of *Andante,* a new *Drosophila* clock mutant, and its interactions with other clock mutants. *J. Neurogenet.* In press

68. Konopka, R., Wells, S., Lee, T. 1983. Mosaic analysis of a *Drosophila* clock mutant. *Mol. Gen. Genet.* 190:284–88

69. Kyriacou, C. P. 1990. The molecular

ethology of the *period* gene in *Drosophila*. *Behav. Genet.* 20:191–211

70. Kyriacou, C. P., Greenacre, M. L., Thackeray, J. R., Hall, J. C. 1990. Genetic and molecular analysis of courtship song rhythms in *Drosophila*. See Ref. 146

71. Kyriacou, C. P., Hall, J. C. 1980. Circadian rhythm mutations in *Drosophila melanogaster* affect short-term fluctuations in the male's courtship song. *Proc. Natl. Acad. Sci. USA* 77:6929–33

72. Kyriacou, C. P., Hall, J. C. 1982. The function of courtship song rhythms in *Drosophila*. *Anim. Behav.* 30:794–801

73. Kyriacou, C. P., Hall, J. C. 1986. Interspecific genetic control of courtship song production and reception in *Drosophila*. *Science* 232:494–97

74. Kyriacou, C. P., Hall, J. C. 1989. Spectral analysis of *Drosophila* courtship song rhythms. *Anim. Behav.* 37:850–59

75. Kyriacou, C. P., Oldroyd, M., Wood, J., Sharp, M., Hill, M. 1990. Clock mutations alter developmental timing in *Drosophila*. *Heredity* 64:395–401

76. Kyriacou, C. P., van den Berg, M. J., Hall, J. C. 1990. *Drosophila* courtship song cycles in normal and *period* mutant males revisited. *Behav. Genet.* 20:631–58

77. Lakin-Thomas, P. L., Coté, G., Brody, S. 1990. Circadian rhythms in *Neurospora crassa:* biochemistry and genetics. *Crit. Rev. Microbiol.* In press

78. Laval-Martin, D. L., Carré, I. A., Barbera, S. J., Edmunds, L. N. Jr. 1990. Rhythmic changes in the activities of NAD kinase and NADP phosphatase in the achlorophyllus ZC mutant of *Euglena gracilis* Klebs (Strain Z). *Arch. Biochem. Biophys.* 276:433–41

79. Lehman, M. N., Silver, R., Gladstone, W. R., Khan, R. M., Gibson, M., Bittman, E. L. 1987. Circadian rhythmicity restored by neural transplant. Immunocytochemical characterization of the graft and its integration with host brain. *J. Neurosci.* 7:1626–38

80. Lemeunier, F., David, J. R., Tsacas, L., Ashburner, M. 1986. The *melanogaster* species group. In *The Genetics and Biology of Drosophila,* ed. M. Ashburner, H. L. Carlson, J. N. Thompson, 3:148–256. London: Academic. 548 pp.

81. Liu, X., Lorenz, L., Yu, Q., Hall, J. C., Rosbash, M. 1988. Spatial and temporal expression of the *period* gene in *Drosophila melanogaster*. *Genes Dev.* 2:228–38

82. Livingstone, M. S. 1981. Two mutations in *Drosophila* affect the synthesis of octopamine, dopamine, and serotonin by altering the activities of two amino-acid decarboxylases. *Neurosci. Abstr.* 7:351 (Abstr.)

83. Lorenz, L. J., Hall, J. C., Rosbash, M. 1989. Expression of a *Drosophila* mRNA is under circadian clock control during pupation. *Development* 107:869–80

84. Loros, J. J., Denome, S. A., Dunlap, J. C. 1989. Molecular cloning of genes under control of the circadian clock in *Neurospora*. *Science* 243:385–88

85. Loros, J. J., Dunlap, J. C. 1990. *Neurospora* clock-controlled genes are regulated at the level of transcription. Submitted

86. Lysak, G., Hohmeyer, H., Veltkamp, C. J. 1990. A spontaneous benomyl-resistant mutant of *Podospora anserina* exhibiting a diurnal growth rhythm. *Experientia* 46:517–20

87. McClung, C. R., Fox, B. A., Dunlap, J. C. 1989. The *Neurospora* clock gene *frequency* shares a sequence element with the *Drosophila* clock gene *period*. *Nature* 339:558–62

88. Meijer, J. H., Rietveld, W. J. 1989. Neurophysiology of the suprachiasmatic circadian pacemaker in rodents. *Physiol. Rev.* 69:671–707

89. Montell, C., Mismer, D., Fortini, M. E., Zuker, C. S., Rubin, G. M. 1988. Identification and expression of *Drosophila* phototransduction genes. In *Molecular Biology of the Eye: Genes, Vision, and Ocular Disease,* ed. J. Piatigorsky, T. Shinohara, P. S. Zelenka, pp. 277–91. New York: Liss. 471 pp.

90. Moore-Ede, M. C., Sulzman, F. M., Fuller, C. A. 1982. *The Clocks that Time Us*. Cambridge, MA: Harvard Univ. Press. 448 pp.

91. Morin, L. P., Johnson, R. F., Moore, R. Y. 1989. Two brain nuclei controlling circadian rhythms are identified by GFAP immunoreactivity in hamsters and rats. *Neurosci. Lett.* 99:55–60

92. Morse, D. S., Fritz, L., Hastings, J. W. 1990. What is the clock? Translational regulation of circadian bioluminescence. *Trends Biochem. Sci.* 15:262–65

93. Nagy, F., Kay, S. A., Chua, N.-H. 1988. A circadian clock regulates transcription of the wheat *Cab-1* gene. *Genes Dev.* 2:376–82

94. Newby, L. M., Jackson, F. R. 1990. *Drosophila ebony* mutants have altered circadian activity but normal eclosion. *J. Neurogenet.* In press

95. Noshiro, M., Nishimoto, M., Okuda,

K. 1990. Rat liver cholesterol 7α-hydroxylase. Pretranslational regulation for circadian rhythm. *J. Biol. Chem.* 265:10036–41

96. Petersen, G., Hall, J. C., Rosbash, M. 1988. The *period* gene of *Drosophila* carries species-specific behavioral instructions. *EMBO J.* 7:3939–47

97. Pittendrigh, C. S. 1974. Circadian oscillations in cells and the circadian organization of multicellular systems. In *The Neurosciences Third Study Program*, ed. F. O. Schmitt, F. G. Worden, pp. 437–58. Cambridge, MA: MIT Press. 1107 pp.

98. Raju, U., Yeung, S. J., Eskin, A. 1989. Involvement of proteins in light resetting ocular circadian oscillators of *Aplysia*. *Am. J. Physiol.* 528:R256–62

99. Ralph, M. R., Block, G. D. 1990. Circadian and light-induced conductance changes in putative pacemaker cells of *Bulla gouldiana*. *J. Comp. Physiol.* 166:589–95

100. Ralph, M. R., Foster, R. G., Davis, F. C., Menaker, M. 1990. Transplanted suprachiasmatic nucleus determines circadian period. *Science* 247:975–78

101. Ralph, M. R., Menaker, M. 1988. A mutation of the circadian system in golden hamsters. *Science* 241:1225–27

102. Reddy, P., Jacquier, A. C., Abovich, N., Petersen, G., Rosbash, M. 1986. The *period* clock locus of D. melanogaster codes for a proteoglycan. *Cell* 46:53–61

103. Reddy, P., Zehring, W. A., Wheeler, D. A., Pirrotta, V., Hadfield, C., et al. 1984. Molecular analysis of the *period* locus in Drosophila melanogaster and identification of a transcript involved in biological rhythms. *Cell* 38:701–10

104. Rensing, L. 1971. Hormonal control of circadian rhythms in *Drosophila*. In *Biochronometry*, ed. M. Menaker, pp. 527–40. Washington, DC: Natl. Acad. Sci. 662 pp.

105. Reppert, S. M., Schwartz, W. J., Uhl, G. R. 1987. Arginine vasopressin: a novel peptide in cerebrospinal fluid. *Trends Neurosci.* 10:76–80

106. Reppert, S. M., Uhl, G. R. 1987. Vasopressin messenger ribonucleic acid in supraoptic and suprachiasmatic nuclei: appearance and circadian regulation during development. *Endocrinology* 120:2483–87

107. Restifo, L. L., White, K. 1990. Molecular and genetic approaches to neurotransmitter and neuromodulator systems in *Drosophila*. *Adv. Insect Physiol.* 22:115–219

108. Riesselmann, S., Piechulla, B. 1990.

Effect of dark phases and temperature on the chlorophyll a/b binding protein mRNA level oscillations in tomato seedlings. *Plant Mol. Biol.* 14:605–16

109. Rizki, T. M. 1978. The circulatory system and associated cells and tissues. In *The Genetics and Biology of Drosophila*, ed. M. Ashburner, T. R. F. Wright, 2:397–452. London: Academic. 601 pp.

110. Roberts, M. H., Bedian, V., Chen, Y. 1989. Kinase inhibition lengthens the period of the circadian pacemaker in the eye of *Bulla gouldiana*. *Brain Res.* 504:211–15

111. Roberts, M. H., Chen, Y., Bedian, V. 1989. Temporally varying antigens in the eye of *Bulla gouldiana*. *Neurosci. Abstr.* 15:462

112. Robinson, B. G., Frim, D. M., Schwartz, W. J., Majzoub, J. A. 1988. Vasopressin mRNA in the suprachiasmatic nuclei: daily regulation of polyadenylate tail length. *Science* 241:342–44

113. Rosbash, M., Hall, J. C. 1985. Biological clocks in Drosophila: finding the molecules that make them tick. *Cell* 43:3–4

114. Rosbash, M., Hall, J. C. 1989. The molecular biology of circadian rhythms. *Neuron* 3:387–98

115. Rosbash, M., Hall, J. C., Hardin, P. 1990. Circadian oscillations in protein and mRNA levels of the *period* gene of *Drosophila melanogaster*. See Ref. 146

116. Rossant, J. 1990. Manipulating the mouse genome: implications for neurobiology. *Neuron* 4:323–34

117. Ruoslahti, E. 1988. Structure and biology of proteoglycans. *Annu. Rev. Cell Biol.* 4:229–55

118. Rusak, B., Bina, K. G. 1990. Neurotransmitters in the mammalian circadian system. *Annu. Rev. Neurosci.* 13:387–401

119. Rusak, B., Robertson, H. A., Wisden, W., Hunt, S. P. 1990. Light pulses that shift rhythms induce gene expression in the suprachiasmatic nucleus. *Science* 248:1237–40

120. Saez, L., Young, M. W. 1988. In situ localization of the per clock protein during development of *Drosophila melanogaster*. *Mol. Cell. Biol.* 8:5378–85

121. Saunders, D. S. 1990. The circadian basis of ovarian diapause regulation in *Drosophila melanogaster:* Is the *period* gene causally involved in photoperiodic time measurement? *J. Biol. Rhythms.* In press

122. Saunders, D. S., Gilbert, L. I. 1990.

Regulation of avarian diapause in *Drosophila melanogaster* by photoperiod and moderately low temperature. *J. Insect Physiol.* 36:195–200

123. Saunders, D. S., Henrich, V. C., Gilbert, L. I. 1989. Induction of diapause in *Drosophila melanogaster:* photoperiodic regulation and the impact of arrhythmic clock mutations on time measurement. *Proc. Natl. Acad. Sci. USA* 86:3748–52

124. Schilcher, F. V. 1989. Have cycles in the courtship song of *Drosophila* been detected? *Trends Neurosci.* 12:311–13

125. Schmale, H., Richter, D. 1984. Single base deletion in the vasopressin gene is the cause of diabetes insipidus in Brattleboro rats. *Nature* 308:705–9

126. Schwartz, W. J., Zimmerman, P. 1990. Circadian timekeeping in BALB/c and C57BL/6 inbred mouse strains. *J. Neurosci.* In press

127. Siwicki, K. K., Eastman, C., Petersen, G., Rosbash, M., Hall, J. C. 1988. Antibodies to the *period* gene product of Drosophila reveal diverse tissue distribution and rhythmic changes in the visual system. *Neuron* 1:141–50

128. Siwicki, K. K., Hall, J. C., Rosbash, M., Schwartz, W. J. 1990. An antibody to the *Drosophila period* protein labels cell bodies in the suprachiasmatic nucleus of the rat. Submitted

129. Siwicki, K. K., Strack, S., Rosbash, M., Hall, J. C., Jacklet, J. W. 1989. An antibody to the *Drosophila period* protein recognizes circadian pacemaker neurons in *Aplysia* and *Bulla. Neuron* 3:51–58

130. Smith, R. F., Konopka, R. J. 1982. Effects of dosage alterations at the *per* locus on the period of the circadian clock of *Drosophila. Mol. Gen. Genet.* 185: 30–36

131. Stark, W. S., Sapp, R., Schilly, D. 1988. Rhabdomere turnover and rhodopsin cycle: maintenance of retinula cells in *Drosophila melanogaster. J. Neurocytol.* 17:499–509

132. Steller, H., Fischbach, K.-F., Rubin, G. M. 1987. *disconnected:* a locus required for neuronal pathway formation in the visual system of Drosophila. *Cell* 50: 1139–53

133. Tamponnet, C., Edmunds, L. N. Jr. 1990. Entrainment and phase-shifting of the circadian rhythm of cell division by calcium in synchronous cultures of the wild-type Z strain and of the ZC achlorophyllous mutant of *Euglena gracilis. Plant Physiol.* 93:425–31

134. Thackeray, J. R., Kyriacou, C. P. 1988. Clock genes and song rhythms in *Drosophila* species. *J. Int. Cycle Res.* Suppl. 19:212–13 (Abstr.)

135. Thackeray, J. R., Kyriacou, C. P. 1990. Molecular evolution in the *Drosophila yakuba period* locus. *J. Mol. Evol.* In press

136. Thomas, J. B., Crews, S. T., Goodman, C. S. 1988. Molecular genetics of the *single-minded* locus: a gene involved in the development of the Drosophila nervous system. *Cell* 52:133–41

137. Thomas, M., Cretin, C., Vidal, J., Keryer, E., Gadal, P., Monsinger, E. 1990. Light-regulation of phosphoenolpyruvate carboxylase mRNA in leaves of C₄ plants: evidence for phytochrome control of transcription during greening and for rhythmicity. *Plant Sci.* 69:65–78

138. Uhl, G. R., Reppert, S. M. 1986. Suprachiasmatic nucleus vasopressin messenger RNA: circadian variation in normal and Brattleboro rats. *Science* 232:390–93

139. Vanden Driessche, T. 1987. The molecular mechanism of circadian rhythms. *Arch. Int. Physiol. Biochim.* 97:1–11

140. van den Pol, A. N. 1980. The hypothalamic suprachiasmatic nucleus of the rat: intrinsic anatomy. *J. Comp. Neurol.* 191:661–702

141. van den Pol, A. N., Decavel, C., Paterson, B. 1989. Hypothalamic expression of a novel gene product, VGF: immunocytochemical analysis. *J. Neurosci.* 9:4122–37

142. Walz, B., Walz, A., Sweeney, B. M. 1983. A circadian rhythm in RNA in the dinoflagellate *Gonyaulax polyedra. J. Comp. Physiol.* 151:207–13

143. Weitzel, G., Rensing, L. 1981. Evidence for cellular circadian rhythms in isolated fluorescent dye-labelled salivary glands of wild type and an arryhthmic mutant of *Drosophila melanogaster. J. Comp. Physiol.* 143:229–35

144. Wheeler, D. A., Kyriacou, C. P., Greenacre, M. L., Yu, Q., Rutila, J. E., et al. 1990. Molecular transfer of a species-specific courtship behavior from *Drosophila simulans* to *D. melanogaster.* Submitted

145. Winfree, A. T. 1987. *The Timing of Biological Clocks.* New York: Scientific Am. Library, 199 pp.

146. Young, M. W., ed. 1990. *Molecular Genetics of Biological Rhythms.* New York: Dekker. In press

147. Young, M. W., Bargiello, T. A., Baylies, M. K., Saez, L., Spray, D. C. 1989. The molecular-genetic approach

to a study of biological rhythms in *Drosophila*. In *Trends in Chronobiology, Advances in the Biosciences,* ed. W. J. Rietveld, 73:43–53. London: Pergamon. 386 pp.

148. Young, M. W., Bargiello, T. A., Baylies, M. K., Saez, L., Spray, D. C. 1989. Molecular biology of the *Drosophila* clock. See Ref. 56, pp. 529–42

149. Young, M. W., Jackson, F. R., Shin, H.-S., Bargiello, T. A. 1985. A biological clock in *Drosophila*. *Cold Spring Harbor Symp. Quant. Biol.* 50:865–75

150. Yu, Q., Colot, H. V., Kyriacou, C. P., Hall, J. C., Rosbash, M. 1987. Behaviour modification by *in vitro* mutagenesis of a variable region within

the *period* gene of *Drosophila Nature* 326:765–69

151. Yu, Q., Jacquier, A. C., Citri, Y., Hamblen, M., Hall, J. C., Rosbash, M. 1987. Molecular mapping of point mutations in the period gene that stop or speed up biological clocks in *Drosophila melanogaster*. *Proc. Natl. Acad. Sci. USA* 84:784–88

152. Zerr, D. M., Hall, J. C., Rosbash, M., Siwicki, K. K. 1990. Circadian fluctuations of *period* protein immunoreactivity in the CNS and the visual system of *Drosophila*. *J. Neurosci.* 10:2749–62

153. Kornhauser, J. M., Nelson, D. E., Mayo, K. E., Takahashi, J. S. 1990. Photic and circadian regulation of *c-fos* in the hamster suprachiasmatic nucleus. *Neuron* 5:127–34

SUBJECT INDEX

A

AAF
See 2-(N-Acetoxy-N-acetylamino) fluorene

2-(N-Acetoxy-N-acetylamino) fluorene (AAF)
frameshift mutation and, 196

Acquired immunodeficiency syndrome (AIDS)
HIV-neutralizing antibodies in, 434

Actinomycetes
polyketide biosynthesis in, 39, 45
genetics of, 46-48

Actinomycin D
c-myc mRNA and, 533
iron response factor and, 530
mRNA turnover and, 521-22

Actinorhodin
biosynthesis in *Streptomyces coelicolor*
genes for, 45
structure of, 38

Adenovirus E1A
transforming proteins of
Rb-1 retinoblastoma protein and, 634-35

Aerolysin
extracellular secretion by *Aeromonas salmonicida*, 69, 81

Aeromonas salmonicida
extracellular secretion of aerolysin by, 69, 81

Aflatoxin
frameshift mutation and, 196
occurrence and biological properties of, 39

Aflatoxin B
structure of, 38

AIDS
See Acquired immunodeficiency syndrome

Albertini, R. J., 305-21

Aleurone, 288

Alfalfa
nodule metabolism in, 293

Amycolatopisis
polyketide biosynthesis in
genes for, 45, 46-48

Aneuploidy
segmental in *Neurospora crassa*
barren phenotype of, 603-5
sex-chromosome
acute radiation exposure and, 333-34

Angiogenesis
inhibition of
chromosome 1 and, 619

Angiosperms
polyketide biosynthesis in, 39
genetics of, 46

Anisomysin
tubulin mRNA and, 527

Ansamycins
occurrence and biological properties of, 39

Anthocyanin
synthesis in flowering plants, 49-50
synthesis in parsley, 44

Anthracyclines
occurrence and biological properties of, 39

Antibiotics
biosynthetic genes for, 45
macrolide
biosynthesis of, 57-58
occurrence and biological properties of, 39
polyketide
biosynthesis by Actinomycetes, 45

Antibody genes
somatic hypermutation of, 607

Antioncogenes, 642-43

Antirrhinum majus
anthocyanin/phytoalexin biosynthesis in, 49-50
defA gene expression in, 283

α-1-Antitrypsin
inborn errors of metabolism and, 148

Arabidopsis
agamous mutation in, 283
2S albumin seed-storage gene of, 288
12S-storage protein of, 285

Ascobolus immersus
premeiotic recombination in, 581
repeat-induced point mutation and, 608

Ascomycetes
premeiotic recombination in, 581
propagation as diploid, 583

ASLVs
See Avian sarcoma-leukosis viruses

Aspergillus, 5-31
conidiation in, 9-13
developmental competence and, 14-17

developmental induction and, 17-19
model for genetic regulation of, 30-31
mutational studies of, 19-21
regulation of, 14-30
developmentally important genes of
molecular cloning of, 22-23
growth and development in, 6-14
control of, 7-8
life cycle of, 6-7
molecular genetic manipulations of, 13-14
regulated genes of
cloning and characterization of, 25-30
molecular estimates of, 21-22
spore-specific genes of
clustering of, 23-25

Aspergillus nidulans
catabolic alcohol dehydrogenase gene of, 13
conidiation in, 9
environmental signals inducing, 17-19
light dependence of, 8
red light and, 18
conidiophore of, 10
DNA-mediated transformation system of, 13
hyphae of
5-azacytidine and, 17
life cycle of, 6-7
methylcytosine and, 17
propagation as diploid, 583
repeat-induced point mutation and, 609

Ataxia telangiectasia, 312

Atherosclerosis
genetics of, 171-83

Atwater, J. A., 519-36

Aurantinin
occurrence and biological properties of, 39

Avermectin
biosynthesis of, 57-58
genes for, 45
occurrence and biological properties of, 39
production by *Streptomyces avermitilis*, 43

Avian retroviruses
transforming genes of
cellular homolog of, 533

Avian sarcoma-leukosis viruses (ASLVs), 413
pp32 protein of, 493

699

PAL gene
    See Phenylalanine ammonia
        lyase (PAL) gene
Pamamycin
    aerial mycelium formation
        and, 40
PAPBs
    See Poly(A) binding proteins
Parsley
    anthocyanin synthesis in, 44
Patatin
    potato tubers and, 290-91
Pathogen avirulence alleles
    nonfunctional, recessive, 455-
        58
Pathogen avirulence genes, 448-
    51
    in higher level specificities,
        451-52
    homologies between, 452-54
    plant disease resistance genes
        and, 454-55
Pathovars, 451-52
Patulin
    occurrence and biological
        properties of, 39
P cytotype repression
    genetic assays for, 549
Pea
    bacteria nodulating, 453
    nodule metabolism in, 293
P element proteins
    sequence of, 551
    66 kd
        P element transposition
            and, 558-64
P elements
    development as genetic tools,
        570-73
    hybrid dysgenesis and, 544-47
    IVS3
        structure of, 571
    transposition of
        cis-acting DNA sequences
            and, 554-55
        Drosophila proteins and,
            564-65
        genetic control of, 547-54
        mRNA splicing and, 567-
            70
        66 kd P element protein
            and, 558-64
        tissue specificity of, 565-66
P element transposase
    binding sites of, 557
    genetics and biochemistry of,
        555-58
    trunctated forms of
        P element transposition
            and, 558-64
Penicillium patulum
    type I polyketide synthase of,
        52-53

Petunia
    chalcone synthase genes in,
        49
    glycine-rich cell proteins of
        genes encoding, 279
Phage
    See Bacteriophage
β-Phaseolin
    embryo-specific gene expres-
        sion and, 285
Phaseolus
    chalcone synthase genes in,
        49
    glycine-rich cell proteins of
        genes encoding, 279
    nodule metabolism in, 293
    phenylalanine ammonia lyase
        gene family of, 283-84
Phaseolus vulgaris
    seeds of
        lectin-storage protein in,
            286
Phenylalanine ammonia lyase
    (PAL) gene
    petal-specific expression of,
        283-84
Phenylpropanoid gene
    expression in floral organs,
        284
Phialides, 11
Phytoalexin
    biosynthesis of
        flowering plants and, 49-
            50
        resveratrol synthase and,
            58-59
    occurrence and biological
        properties of, 39
Phytohemagglutinin
    as lectin-storage protein, 286
PI-II
    See Proteinase inhibitor II
PKS
    See Polyketide synthase
Plant-pathogen interactions
    gene-for-gene complementar-
        ity in, 447-60
    pathogen avirulence genes
        and, 451-58
Plant pathogens
    repeat-induced point mutation
        and, 608
Plants
    dicotyledonous
        cell wall proteins in, 279-
            80
    disease defense and recogni-
        tion in, 448-51
    disease resistance genes of,
        449-50
        avirulence genes and, 454-
            55
        cloning of, 458-60

fertilization of
    genes controlling, 281-82
flowering
    anthocyanin/phytoalexin
        biosynthesis in, 49-50
    gene expression in, 275-94
    aleurone-specific, 288
    embryo-specific, 285-86
    endosperm-specific, 286-88
    floral-specific, 284-85
    fruit-specific, 289-90
    leaf-specific, 277-79
    nodule-specific, 292-94
    petal-specific, 283-84
    root-specific, 280-81
    stamen-/pistil-specific, 282-
        83
    stem-specific, 279-80
    tuber-specific, 290-92
    metabolites of
        structures of, 38
    reproductive organs of
        gene expression in, 281-85
    type II fatty acid synthase
        systems of, 54-55
    vegetative organs of
        gene expression in, 277-81
Plant transcription factor, 281
Podospora anserina, 6
    circadian rhythms of, 678
    repeat-induced point mutation
        and, 609
Poly(A) binding proteins
    (PABPs)
    mRNA stability and, 523-24
Polyethers
    occurrence and biological
        properties of, 39
Polyketides, 37-61
    biosynthesis of
        biochemistry of, 40-44
        genes for, 44-50
        genetics of, 46-48
        molecular genetics of, 50-
            59
        schematic of, 41
    occurrence and biological
        roles of, 39-40
Polyketide synthase (PKS)
    specificity of, 59-61
Polyoma large T antigen
    Rb-1 retinoblastoma protein
        and, 634-35
Polypropionate
    occurrence and biological
        properties of, 39
Pregnancy outcome
    acute radiation exposure and,
        330-31
Premeiotic deletion
    evolutionary implications of,
        605-8
Premeiotic instability, 601-5

CUMULATIVE INDEXES

CONTRIBUTING AUTHORS, VOLUMES 20–24

CHAPTER TITLES, VOLUMES 20–24